GREAT WHITE SHARKS
The Biology of Carcharodon carcharias

GREAT WHITE SHARKS
The Biology of Carcharodon carcharias

Edited by

A. Peter Klimley

Bodega Marine Laboratory
University of California, Davis
Bodega Bay, California

David G. Ainley

H. T. Harvey & Associates
Alviso, California

Academic Press
An Imprint of Elsevier

San Diego   London   Boston   New York   Sydney   Tokyo   Toronto

Cover Photograph by Howard Hall.

This book is printed on acid-free paper.

Academic Press
An Imprint of Elsevier
525 B Street, Suite 1900, San Diego, California 92101-4495, USA
http://www.academicpress.com

Academic Press
Harcourt Place, 32 Jamestown Road, London NW1 7BY, UK
http://www.academicpress.com

Library of Congress Cataloging-in-Publication Data

Great white sharks : the biology of Carcharodon carcharias / edited by
   A. Peter Klimley, David G. Ainley.
         p.   cm.
   Includes bibliographical references and index.
   ISBN-13: 978-0-12-415031-7 ISBN-10: 0-12-415031-4 (pb)

   1. White shark. I. Klimley, A. Peter. II. Ainley, David G.
QL638.95.L3G73 1996
597'.31—dc20   96-3068
                                          CIP

ISBN-13: 978-0-12-415031-7
ISBN-10: 0-12-415031-4 (pb)

Transferred to Digital Printing 2011

Francis G. Carey

1937–1994

Frank was a good scientist and friend, and like
us studied the white shark. We will miss him
and dedicate this book to his memory.

Contents

PART

VI

ECOLOGY AND DISTRIBUTION

PART

VII

POPULATION BIOLOGY

PART

VIII

INTERACTIONS WITH HUMANS

Contributors

Numbers in parentheses indicate the pages on which the authors' contributions begin.

David G. Ainley (3, 223, 375), H. T. Harvey & Associates, Alviso, California 95002

Jack A. Ames (309), California Department of Fish & Game, Monterey, California 93940

Scot D. Anderson (111, 175, 223, 241, 275, 281, 365, 375), Inverness, California 94937

Shelton P. Applegate (19), Instituto de Geología, Universidad Nacional Autonóma de México, Mexico D.G. 04510, Mexico

H. David Baldridge, Jr. (477), Sarasota, Florida 34239

Raymond Bandar (263), Department of Ornithology and Mammology, California Academy of Sciences, San Francisco, California 94118

George W. Barlow (257), Department of Integrative Biology and Museum of Vertebrate Zoology, University of California, Berkeley, California 94720

S. Curtis Bowman (55), Hughes-Bowman Design Group, St. Augustine, Florida

Barry D. Bruce (401), CSIRO Division of Fisheries Research, Hobart, Tasmania 7001, Australia

Elinor M. Bullen (393), Oceanographic Research Institute, Durban 4000, South Africa

George H. Burgess (457), International Shark Attack File, Florida Museum of Natural History, University of Florida, Gainesville, Florida 32611

Gregor M. Cailliet (81, 91, 415), Moss Landing Marine Laboratory, Moss Landing, California 95039

Matthew Callahan (457), International Shark Attack File, Florida Museum of Natural History, University of Florida, Gainesville, Florida 32611

Geremy Cliff (351, 393), Natal Sharks Board, Umhlanga Rocks, South Africa

Ralph S. Collier (217), Shark Research Committee, Van Nuys, California 91407

Leonard J. V. Compagno (55, 91), Shark Research Center, South African Museum, Cape Town 8000, South Africa

Daniel E. Crocker (193), Department of Biology, and Institute for Marine Sciences, University of California, Santa Cruz, California 96064

Leo S. Dempski (121), Division of Natural Sciences, New College, University of South Florida, Sarasota, Florida 34243

Sheldon F. J. Dudley (351), Natal Sharks Board, Umhlanga Rocks, South Africa

Dave A. Ebert (91), US Abalone, Davenport, California

Luis Espinosa-Arrubarrena (19), Instituto de Geología, Universidad Nacional Autonóma de México, Mexico D.G. 04510, Mexico

Ian K. Fergusson (321), European Shark Research Bureau, Herts AL7 2NN, Welwyn Garden City, England

Craig Ferreira (381), South African White Shark Research Institute, Capetown 8000, South Africa

Theo P. Ferreira (381), South African White Shark Research Institute, Capetown 8000, South Africa

Malcolm P. Francis (157), National Institute of Water and Atmospheric Research Ltd., Wellington, New Zealand

Otto B. F. Gadig (347), Departamento de Sistematica e Ecologia, Campus Universitario, Universidade Federal de Paraiba, CCEN, Joao Pessoa, Paraiba, Brazil

John J. Geibel (309), California Department of Fish and Game, Menlo Park, California 94025

Marci Glazer (481), Center for Marine Conservation, San Francisco, California 94104

Kenneth J. Goldman (111), Department of Biology, San Francisco State University, San Francisco, California 94132

Michael D. Gottfried (55), Calvert Marine Museum, Solomons, Maryland 20688

Anesh Govender (393), Oceanographic Research Institute, Durban 4000, South Africa

Krista D. Hanni (263), The Marine Mammal Center, Sausalito, California 94965

R. Philip Henderson (223, 275, 281), Point Reyes Bird Observatory, Stinson Beach, California 94970

Burr Heneman (481), Living Resources, Bolinas, California 94924

Gordon Hubbell (9), Jaws International, Key Biscayne, Florida 33149

Robert E. Jones (263, 293), Museum of Vertebrate Zoology, University of California, Berkeley, California 94720

Mark R. Jury (351), Oceanography Department, University of Cape Town, Rondebosch 7700, South Africa

A. Peter Klimley (3, 91, 111, 175, 241, 275, 281, 365), Bodega Marine Laboratory, University of California, Davis, Bodega Bay, California 94923

Robert N. Lea (419), Marine Resources Division, California Department of Fish and Game, Monterey, California 93940, and California Academy of Sciences, San Francisco, California 94118

Burney J. Le Boeuf (193), Department of Biology and Institute for Marine Sciences, University of California, Santa Cruz, California 96064

Marie Levine (435), Shark Research Institute, Princeton, New Jersey 08540

Douglas J. Long (37, 263, 293, 317), Department of Icthyology, California Academy of Sciences, Golden Gate Park, San Francisco, California 94118

Mark Marks (217), Shark Research Center, South African Museum, Capetown 8000, Republic of South Africa

Andrew P. Martin (49), Department of Biology, University of Nevada, Las Vegas, Las Vegas, Nevada 89154

John E. McCosker (111, 419), Steinhart Aquarium, California Academy of Sciences, Golden Gate Park, San Francisco, California 94118

Henry F. Mollet (81, 91), Monterey Bay Aquarium, Monterey, California 93940

Richard C. Murphy (5, 401), The Cousteau Society, Chesapeake, Virginia 23320

Donald R. Nelson (401, 471), Department of Biological Sciences, California State University, Long Beach, California 90840

R. Glenn Northcutt (121), Department of Neurosciences & Neurobiology, Unit of Scripps Institution of Oceanography, University of California, San Diego, La Jolla, California 92093

Christine A. Pattison (309), California Department of Fish and Game, Morro Bay, California 93442

Harold L. Pratt, Jr. (131), Narragansett Laboratory, Narragansett, Rhode Island 02882

Robert W. Purdy (67), Department of Paleobiology, National Museum of Natural History, Washington, DC 20560

Peter Pyle (175, 223, 241, 263, 275, 281, 375), Point Reyes Bird Observatory, Stinson Beach, California 94970

Jan Roletto (263), Gulf of the Farallones National Marine Sanctuary, San Francisco, California 94123

Ricardo S. Rosa (347), Departamento de Sistematica e Ecologia, Campus Universitario, Universidad Federal de Paraiba, CCEN, Joao Pessoa, Paraiba, Brazil

Wesley R. Strong Jr. (207, 229, 401, 471), Department of Biology, University of California, Santa Barbara, Santa Barbara, California 93106 and The Cousteau Society, Chesapeake, Virginia 23320

Kazuyuki Teshima (139), Tohoku National Fisheries Research Institute, Shinhama-cho, Shiogama, Miyagi, Japan

Antonio D. Testi (91), Italian Shark Research Project, I-20148 Milano MI, Italy

Minoru Toda (139), Okinawa Expo Aquarium, Motobu-cho, Okinawa 905-03, Japan

Senzo Uchida (139), Okinawa Expo Aquarium, Motobu-cho, Okinawa 905-03, Japan

Rudy P. Van der Elst (393), Oceanographic Research Institute, Durban 4000, South Africa

Benjamin M. Waggoner (37), Museum of Paleontology, University of California, Berkeley, Berkeley, California 94720

Ronald W. Warner (217), California Department of Fish and Game, Eureka, California 95501

Frederich E. Wendell (309), California Department of Fish and Game, Morro Bay, California 93442

John West (449), Taronga Zoo, Masman NSW, 2088 Sidney, Australia

Trail K. Witthuhn (393), Struis Bay 7285, South Africa

Kazunari Yano (139), Seikai National Fisheries Research Institute, Ishigaki Tropical Station, Fukai Ota, Ishigaki, Okinawa, Japan

Acknowledgments

Some of the chapters in this book originated from talks given at a symposium on the biology of the white shark that was held at the Bodega Marine Laboratory of the University of California, Davis, on March 4–7, 1993. For that reason, we acknowledge those persons who helped us organize that exciting meeting and the organizations that provided the financial support for it. Mr. Ed Ueber, Director, Gulf of the Farallones National Marine Sanctuary, and Ms. Jan Roletto, Scientific Coordinator, helped greatly in planning and obtained much of the financial support needed. Dr. James Clegg, Director, Bodega Marine Laboratory, also worked closely with us and made the university's facilities available. Other organizations that contributed funds include the following: Bodega Bay Sea Urchin Association, California Academy of Sciences, Cousteau Society, Marine Mammal Commission, National Audubon Society, Sea World of San Diego, American Elasmobranch Society, and the U.S. Department of the Navy (Office of Naval Research).

In bringing this book together, we are especially grateful to all the reviewers, to Ms. Diane Cosgrove, who converted all the tables into a consistent format, and to Academic Press for their assistance in planning and publishing this important volume.

In order to keep the cost of this book as affordable as possible, we canvassed the scientific community for funds to help defray production expenses. The magnitude of the response to our plea was overwhelming. The following organizations provided funds: Bodega Marine Laboratory, California Department of Boating and Waterways, David and Lucille Packard Foundation, Gulf of the Farallones National Marine Sanctuary, Monterey Bay National Marine Sanctuary, Natal Sharks Board of South Africa, National Audubon Society, Primary Industries South Australia Fisheries, Point Reyes Bird Observatory, Sea World of San Diego, Shark Research Institute of Princeton (New Jersey), the South Australian Research and Development Institute, and Discovery Channel. The insignias of these institutions are shown below. We thank the colleagues and institutions mentioned above, and finally Pat and Susan for their patience and support while we produced this volume.

A. Peter Klimley
Bodega Bay, California
David G. Ainley
Alviso, California

Introduction

1

White Shark Research in the Past: A Perspective

A. PETER KLIMLEY
Bodega Marine Laboratory
University of California, Davis
Bodega Bay, California

DAVID G. AINLEY
H. T. Harvey & Associates
Alviso, California

There are over 350 species of sharks, with members of the group as diverse as the huge plankton-feeding whale and basking sharks, tiny deep-water sharks that are bioluminescent or that take "cookie cutter" bites out of seals and whales, and the bizarre hammerhead sharks, with their greatly expanded rostrum. Yet humans often ignore this great diversity and think of members of the group as conforming to a single image, that of the white shark *Carcharodon carcharias*. The public's fascination with this species is reflected by record attendance of the movie *Jaws* and its many sequels. The reason for this obsession is a morbid one. Humans consider the white shark dangerous—it feeds on us. One of us was possessed by the strong emotional response aroused by this predator 10 years ago while witnessing a predatory attack on a seal not more than 10 m from shore at Southeast Farallon Island. The water was crimson red with blood when the shark appeared, swimming back and forth, high in the water. The experience sent a shiver down the spine and cold sweat to the forehead. Although all of us regularly experience the human feelings of thirst, hunger, anger, and sexual arousal, rarely do we have the fear of being eaten. We feel this when watching a white shark devour a seal.

Despite the human obsession with this predator, we have been slower to learn about the biology of this species. Since World War II, many studies have been conducted on the biology of sharks, primarily members of the family Carcharhinidae. This has been due to the interest and financial support of the U.S. Navy. However, it was only 10 years ago, on May 7, 1983, that a symposium was held at the California State University, Fullerton campus on the biology of the white shark. Although a few presentations at this meeting were quantitative, most sets of observations were small, and the other presentations consisted largely of anecdotal stories. Yet this meeting was a milestone. In any scientific discipline, we learn in a stepwise manner. Our first observations are anecdotal, only later to become increasingly quantitative as we gather sufficient information to begin to ask testable questions. Empirical studies then lead to experimental ones. After that first meeting, relatively little on the white shark appeared in the scientific literature until the symposium we organized at the Bodega Marine Laboratory of the University of California on March 4–7, 1993. This meeting drew scientists from six continents to talk on topics ranging from the evolution of the white shark to its behavior.

A favorite topic of the press, newspaper articles about white sharks often contain ideas rather than facts, and this has frustrated those who are willing to obtain data in the slow, methodical manner characteristic of science. It takes great restraint not to say something that seems correct but which is based on observations too few to make a firm conclusion. Disseminating such ideas can lead to trouble later; only in the process of disproving ideas does science advance closer to the truth.

The public's misconception regarding this mythical species even occurs in the name of the shark, "white shark." The first sharks that were seen were dead, lying on their back on the deck of a boat, and the species was given a popular name based on the coloration of its belly. However, from an ecological perspective, "black shark" is probably a better name. This black coloration matches the color of a rocky or vegetated bottom or the darkness over deep water. Adult sharks can swim under their favorite prey, seals and sea lions, without being detected and ambush them from below. Even the other popular name, "maneater," is called to question in this book. In Chapter 22, Klimley *et al.* hypothesize that sharks often spit out humans, birds, and sea otters after seizing them, because their bodies lack the energy-rich layers of fat possessed by pinnipeds and cetaceans.

In organizing the symposium on the biology of the white shark, we first wanted to provide an opportunity for scientists to make traditional scientific presentations. For this reason, we scheduled four sessions on different aspects of the biology of the shark. These resulted in papers that are now grouped to round out the eight sections of the book: Evolution, Anatomy, Physiology, Behavior, Ecology and Distribution, Population Biology, and Interactions with Humans.

However, we also wanted to provide a forum for evaluating several controversial ideas that had appeared repeatedly in newspapers and, for that reason, are believed by the public to be facts. One such idea was that the population of white sharks off the western coast of North America was growing because seals and sea lions were steadily becoming more abundant. In contrast, sharks were decreasing off the eastern coast as they became increasingly overfished. In fact, this was happening to white sharks there as well as elsewhere in the world. There was concern that this could occur on the West Coast. Another common speculation was that white sharks attack people because they look like seals swimming at the surface of the water. This may seem obvious, but is not a tested fact yet—much disagreement exists in this book between the conclusions of Collier *et al.* (Chapter 19), Anderson *et al.* (Chapter 20), and Strong (Chapter 21). These and other controversial ideas were evaluated during the plenary sessions on population dynamics, predatory strategies, and attack avoidance. Participants gave brief talks, a moderator solicited questions from the audience, and a rapporteur gave a critical evaluation of the discussions presented. Three chapters in the book contain the comments of rapporteurs: Chapter 23 (by Barlow), Chapter 38 (by Cailliet), and Chapter 44 (by Baldridge). Through this process, we hoped to develop a consensus among knowledgeable persons on these very controversial topics.

Academic Press has been very supportive, letting us present all of this information to you in a single volume. We hope that you enjoy reading the many articles as much as we did while preparing the volume.

2

A Plea for White Shark Conservation

RICHARD C. MURPHY

The Cousteau Society
Chesapeake, Virginia

When the Cousteaus began their expeditions in Australia, they did not plan to focus their attention on white sharks. However, after we arrived in South Australia and began investigating the situation, we discovered that there were many questions not being addressed. We had the good fortune to meet Mr. Barry Bruce of the South Australian Department of Fisheries and offered to participate in their white shark research program. Our involvement included the contribution of physical resources and some expertise, including that of Don Nelson and Rocky Strong. This work in South Australia resulted in the production of one television special, a coffee table book, and data which will be developed into a number of papers. I would like to offer some thoughts on great whites based on our experiences with them.

Throughout history, undeserved reputations of mythical proportions have beleaguered many fierce animal species. Today, infamous acclaim continues to take a serious toll on the sharks, in particular, the white shark. Just how serious a toll, however, has been difficult to determine. At the outset of our expedition in Australia, we asked the same question and found only anecdotes for answers. Because we feared that direct as well as incidental fishing pressure might be having a significant effect on the local population of white sharks, we offered to investigate this possibility further and, toward this end, carried out five expeditions between 1989 and 1991.

An important objective of our work was to tag as many white sharks as possible and to provide an opportunity for a long-term assessment of population dynamics. Using standardized baiting, mark and identification, and telemetry tracking, we found that some sharks regularly used a surprisingly small expanse of ocean, remaining in or passing by the same localities over a period of months and even years. In all, we encountered 67 different white sharks. From tagging and resighting data, we estimated the population to consist of only a few hundred sharks in the entire study region. As no historical basis for comparison exists, we can only guess as to whether this small population of sharks is depleted. What we do know, however, is that individuals of the species are relatively rare and their rate of capture by humans is alarmingly high. During the first year of our project, 3 of the 18 sharks we tagged were returned dead after 78 days at liberty. Also, the male and female segments of the population appear to be segregated, with the females frequenting areas that are generally more accessible to fishermen.

We are deeply concerned about the increasing monetary value of white shark jaws and teeth. This creates a market for the capture of white sharks. Shark populations will inevitably dwindle unless prudent controls are enacted. Fortunately, the governments of both South Africa and Australia have already taken up this cause on behalf of the white shark. Our reasons for wanting and needing to protect this species are, in fact, the same as those applied to the protection of other terrestrial top carnivores. In addition to being increasingly rare, they are majestic preeminent participants in a complicated food web which we, as yet, only partially understand. Perhaps of greater concern is that our willingness and ability to properly manage white sharks and other creatures

5

are indicators of the economic, political, and sociological health of our own species. We hope that future management will establish strong safeguards against intentional and unintentional threats to the well-being of the white shark. I am pleased to have joined the distinguished team of experts at the symposium in approaching the difficult task of establishing a rational conservation and management program and promoting an appreciation for members of this unique species, alive, in the wild, and performing their ecological functions.

Evolution

Using Tooth Structure to Determine the Evolutionary History of the White Shark

GORDON HUBBELL
Key Biscayne, Florida

Introduction

Shark teeth are the most commonly collected vertebrate fossil. They appear on beaches, prairies, mountaintops, and deserts, as well as in riverbeds. They have also been found in the Antarctic and the deepest part of the ocean. In the living shark, teeth are constantly produced and shed; a typical carcharhinid, such as the lemon shark *Negaprion brevirostris*, may produce 20,000 teeth in its first 25 years, and may live as long as 50 years. Sharks have been very common inhabitants of our oceans for about 400 million years, and shark teeth produce excellent fossils. Thus, it is easy to understand why millions of fossilized shark teeth exist in marine deposits throughout the world. However, for the paleontologist trying to trace the evolution of white sharks using only fossilized teeth—which is a virtual necessity, as other portions of shark bodies have not fossilized well—the task can be very frustrating, because one cannot make accurate deductions about the body length of the fossil specimens, placement of the fins, or the relationship between the fins and other body structures. Moreover, white shark teeth do not vary between the sexes. Finally, disassociated fossil teeth in marine deposits are difficult to assign to a geological age.

In order to completely understand tooth variability and deformity, the scientist needs to study extant white shark jaws and tooth sets as well as large numbers of fossil white shark teeth of known geological age. It is a dangerous practice to identify a fossil shark species on the basis of examining one or two teeth. It is, however, very difficult to find large concentrations of fossil white shark teeth in one stratigraphic horizon. There are numerous places that produce thousands of fossil teeth from the bull *Carcharhinus leucas*, *N. brevirostris*, and dusky shark *Carcharhinus obscurus*, but few areas have appreciable numbers of white shark teeth. Relying on specimens collected by amateur or commercial collectors is problematic, as many have little or no experience collecting data and exact geological ages are unknown. The compact shape and smooth enamel surface of fossil teeth offer little resistance to slippery hillsides or riverbanks, which allows them to settle into deposits from which they did not originate; a rushing creek could displace them far from their original site. Because of the number of teeth that a single shark can produce and shed in its lifetime, it is safe to assume that almost all of the teeth that are collected are teeth that were shed when the shark was still living. Since shark teeth do not develop in sockets or are not fused to the jaw, as they are in other vertebrates, there is nothing to hold them in place when the animal dies. The teeth become scattered and their association is quickly lost. The cartilaginous skeleton does not fossilize well and usually disintegrates rapidly, eliminating any skeletal remains to tie the teeth to a particular geological formation. Natu-

9

rally associated sets of teeth, together with vertebrae and other skeletal remains, are an extremely rare find for tertiary sharks. Yet these naturally associated sets represent the key to unlocking the mystery of the evolution of the white shark.

In this chapter, I review the structure of white shark teeth, as well as the variation therein and the factors that cause them to vary, with the purpose of establishing the basis on which white shark teeth can be used to evaluate the fossil record of this species and its ancestors (see also Chapter 4, by Applegate and Espinosa-Arrubarrena; Chapter 5, by Long and Waggoner; and Chapter 8, by Purdy).

Materials and Methods

I examined the jaws and tooth sets from 40 white sharks ranging from 122 to 594 cm in total length (TL). I purchased these from fishermen who had accidentally captured the sharks in their gill nets. In most cases, the fresh jaws were packed in salt and shipped to me by air. They came from Western Australia (21), southern California (11), Florida (4), South Australia (2), the northeastern United States (1), and eastern Canada (1) and were collected between July 1980 and February 1994, except for six specimens, for which the date of collection was not known. I removed the muscle and loose connective tissue from the jaws upon arrival. Then I bleached them in hydrogen peroxide and dried them in an air-conditioned room after they had been stretched out on a cloth-covered board. I received 11 jaws in such bad condition that all I was

able to salvage were the tooth sets. I also studied the first upper right tooth (A-2) from an additional 18 specimens.

The upper teeth of a white shark show a noticeable variation in size and shape according to their position in the jaw. I analyzed the characteristics of the tooth crown (enamel) of the first five tooth positions, starting from the front of the upper jaw in adult specimens (Table I). I refer here to tooth positions using the nomenclature proposed by Applegate and Espinosa-Arrubarrena (Chapter 4).

Six measurements were made of each tooth (Fig. 1). *Crown height:* vertical distance between straight line touching the lower extensions of the enamel adjacent to the root at the base of the crown and a parallel line touching the tip of the enamel. *Crown width:* widest part of the enamel at the base of the crown. *Medial margin:* number of serrations along edge of the enamel toward the center of the mouth. *Lateral margin:* number of serrations along the edge of the enamel toward the side of the mouth. *Degree of slant:* The angle between a line drawn perpendicular to a straight line touching the lower extensions of the enamel and beginning at the midpoint of the crown width and a line beginning at the same point and passing through the tip of the enamel. The angle was positive if the slant was toward the side of the mouth (i.e., lateral) and negative toward the center of the mouth (i.e., medial). All linear measurements were normalized by dividing the TL of the longest shark (594 cm) by that of each of the other sharks in order to derive an adjustment factor for each specimen. Each measurement was then multiplied by the adjustment factor for the corresponding

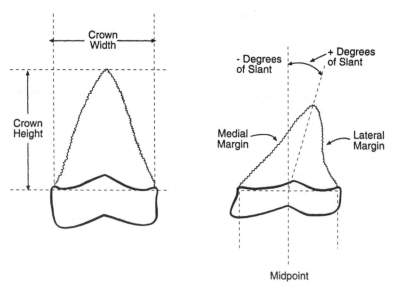

FIGURE 1 Measurements used for tooth description and comparison (viewed from labial or outer surface).

TABLE I Crown Height and Angle for the First Five Upper Tooth Positions

Tooth No.	Specimen ID	Sex	TL (m)	Second anterior (A-2)			Third anterior (A-3)			Intermediate (I)			First lateral (L-1)			Second lateral (L-2)		
				Ht (mm)	N:Ht (mm)[a]	Angle (deg)[b]	Ht (mm)	N:Ht (mm)	Angle (deg)[b]	Ht (mm)	N:Ht (mm)	Angle (deg)[b]	Ht (mm)	N:Ht (mm)	Angle (deg)	Ht (mm)	N:Ht (mm)	Angle (deg)
1	T42392	M	3.79	32.5	50.9	1.0	31.5	49.4	4.0	21.5	33.7	−3.0	27.0	42.3	13.0	27.5	43.1	6.0
2	SP22294	F	3.91	32.5	49.4	2.0	33.0	50.1	3.0	23.0	34.9	−9.0	27.0	41.0	9.0	27.0	41.0	9.5
3	H93086	M	4.17	38.5	54.8	4.0	38.0	54.1	0.5	26.5	37.7	−9.0	31.0	44.2	16.0	34.0	48.4	9.0
4	F122793	M	4.27	35.0	48.7	4.5	34.0	47.3	2.0	25.0	34.8	−2.0	28.5	39.6	16.0	29.5	41.0	13.0
5	ADCJ	U	4.27	36.0	50.1	2.5	35.5	49.4	3.0	24.0	33.4	−5.0	27.5	38.3	18.0	31.5	43.8	10.0
6	H32089	M	4.42	40.5	54.4	2.0	40.0	53.8	0.5	26.0	34.9	−8.5	29.5	39.6	10.0	34.0	45.7	7.5
7	H72589	U	4.57	37.0	48.1	3.0	37.0	48.1	1.0	27.5	35.7	−9.0	29.5	38.3	13.5	32.5	42.2	11.5
8	F9682	M	4.57	42.0	54.6	0	41.0	53.3	2.0	31.0	40.3	−9.0	34.5	44.8	16.0	37.0	48.1	10.0
9	F51089	F	4.57	35.0	45.5	−3.0	37.0	48.1	2.0	26.5	34.4	−8.0	31.5	40.9	12.5	33.5	43.5	6.5
10	SW41285	M	4.71	39.5	49.8	3.0	40.0	50.4	4.5	29.5	37.2	−14.5	35.5	44.8	14.0	37.5	47.3	8.0
11	G121382	M	4.72	38.5	48.5	7.0	38.5	48.5	0.0	28.0	35.2	−8.0	33.0	41.5	5.0	35.0	44.0	7.0
12	SP1394	M	4.74	43.5	54.5	5.0	43.0	53.9	3.0	34.0	42.6	−4.5				38.5	48.2	11.0
13	H8993	M	5.18	40.5	46.4	−3.5	38.0	43.6	7.5	30.0	34.4	−3.5	32.0	36.7	17.5	34.5	39.6	16.0
14	F22490	U	5.18	46.0	52.7	1.5	45.5	52.2	7.0	32.5	37.3	−8.0	37.0	42.4	11.0	39.5	45.3	13.0
15	F7687	F	5.18	48.5	55.6	1.0	47.5	54.5	4.0	33.5	38.4	−10.0	38.0	43.6	9.0	43.0	49.3	5.0
16	N11693	F	5.23	45.0	51.1	0	45.0	51.1	3.0	34.0	38.6	−11.0	38.0	43.2	17.0	41.5	47.1	8.0
17	R92185	F	5.36	47.5	52.6	3.5	46.0	51.0	1.5	29.5	32.7	−18.0	38.5	42.7	16.0	41.0	45.4	7.0
18	X11384B	F	5.47	45.0	48.9	7.5	44.5	48.3	0	35.5	38.6	−5.0	38.0	41.3	13.0	39.5	42.9	13.0
19	H10991	F	5.54	43.0	46.1	0.5	42.0	45.0	3.5	29.0	31.1	−4.0	36.0	38.6	10.0	37.5	40.2	5.0
20	F82380	F	5.54	49.0	52.5	7.0	46.0	49.3	1.5	35.0	37.5	−7.0	41.5	44.5	13.0	43.5	46.6	9.0
21	L11685	F	5.63	48.0	50.6	4.0	45.5	48.0	4.5	31.5	33.2	−15.0	38.5	40.6	13.0	43.0	45.4	9.5
22	SW31287	U	5.64	52.0	54.8	0	51.5	54.2	2.5	33.5	35.3	−9.0				44.5	46.9	8.0
23	M91683	F	5.76	45.5	46.9	4.5	45.0	46.4	6.0	32.0	33.0	−5.0	37.0	38.2	10.0	42.5	43.8	8.5
24	H5384	F	5.94	49.0	49.0	3.0	48.5	48.5	0	32.5	32.5	−13.0	37.5	37.5	22.0	39.0	39.0	13.5
25	H112690	M	5.94	45.0	45.0	6.0	46.0	46.0	1.5	34.0	34.0	−7.0	37.5	37.5	20.0	38.5	38.5	13.5
26	F7282	F	5.94	44.5	44.5	4.0	44.5	44.5	−2.5	31.0	31.0	−6.5	36.5	36.5	16.0	36.0	36.0	12.0
N				26			26			26			24			26		
Mean				50.2			49.6			35.5			40.8			43.9		
SD				3.4			3.2			2.8			2.7			3.5		
CV[a]				6.8			6.5			7.9			6.6			8.0		

TL, Total length; Ht, height.

[a]Normalized N = measurement × (TL of largest specimen/TL of specimen with measurement); coefficient of variation (CV) = SD(100)/x̄. x̄ refers to deviation from perpendicular without regard for direction.

[b]Negative angles are in the medial direction; positive angles, in the lateral direction.

shark to produce measurements that would exist if all the shark specimens had been 594 cm long.

Results

Dimensions of the tooth crown are shown for 26 adult specimens ranging from 379 to 594 cm TL. The first two tooth positions (A-2 and A-3) in the upper jaw were very similar, with mean crown heights of 50.2 and 49.6 mm, respectively. The first position, A-2, averaged slightly longer; however, in seven specimens, the crown height of A-3 was at least that of A-2. In both positions, the teeth were nearly perpendicular, but in some specimens, the teeth in either position slanted a few degrees medially (toward the front of the mouth) or laterally (toward the back of the mouth). The slant angles averaged 2.7° for A-2 and 2.5° for A-3. The extremes were 7.5° lateral slant and 3.5° medial slant. The tooth in the third position, which is the intermediate tooth (I), differed greatly from the two anterior teeth. This tooth was shorter, averaging only 35.5 mm in perpendicular crown height (normalized). It was always slanted medially an average of 8.1° and showed the greatest recurvature of any of the tooth positions in 9 of the 26 jaws. The fourth position represents the first lateral tooth (L-1); this tooth always slanted laterally an average of 13.8°. It was slightly longer (40.8 mm) than the intermediate tooth, but still was shorter than A-2. The

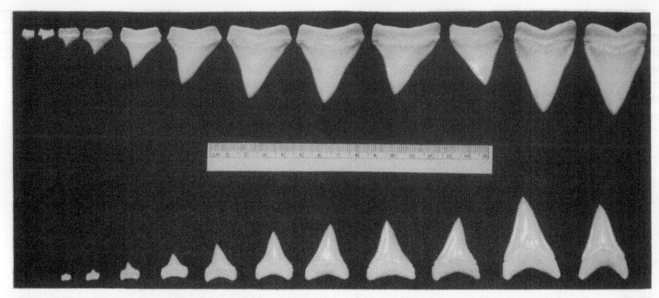

FIGURE 2 Tooth size and shape according to position in the jaw, labial view, front row, right side, from a 440 cm TL male white shark.

second lateral tooth (L-2) was the largest of the four lateral teeth, with a mean normalized height of 43.9 mm. This tooth also always slanted laterally an average of 9.6°. The remaining lateral and posterior teeth become progressively smaller toward the back of the mouth. This variation in size and structure of the first five tooth positions in the upper jaw is typical of the sharks of the order Lamniformes, but is noticeably greater than one would expect in other large sharks, such as those in the order Carcharhiniformes. Therefore, before a scientist can make any assumptions on the placement of individual teeth, he must have a thorough knowledge of the dental pattern (Fig. 2).

A thorough comprehension of the dental pattern is complicated by the fact that the teeth change shape as the white shark grows. In a juvenile (<150 cm TL), the tooth crowns are usually narrow, with lateral cusplets and sometimes lacking serrations (Fig. 3). The serrations that are present tend to be coarse. An examination of A-2 in five specimens of <150 cm TL (Table II) revealed 18–22 serrations on the medial side of the tooth and 17–22 on the lateral side (not includ-

TABLE II **Measurements of the Crown of the First Upper Right Tooth [Second Anterior (A-2)] in White Sharks <1.55 m Total Length**

| Specimen | | | TL (m) | Height (mm) | N:Height (mm)[a] | Width (mm) | N:Width (mm) | Ht/Wd ratio | Serrations | | Angle (deg) | Serrations per cusplet | |
No.	ID	Sex							Medial	Lateral		Medial	Lateral
1	BHB	F	1.22	12.0	14.4	8.0	9.6	1.50	20	19	9.0	1	1
2	RC81487	M	1.24	11.0	13.0	8.5	10.0	1.29	18	17	2.0	1	1
3	BH71087	M	1.35	11.5	12.4	8.5	9.2	1.35	20	17	2.0	1	1
4	BH71590	M	1.42	13.5	13.9	10.5	10.8	1.29	19	22	5.0	2	2
5	CA8564	M	1.46	14.0	14.0	11.0	11.0	1.27	22	20	6.0	1	1
N					5		5	5				5	
Minimum					12.4		9.2	1.27				1.0	
Maximum					14.4		11.0	1.50				2.0	
Mean					13.5		10.1	1.34				1.2	
SD					0.8		0.8	0.09				0.4	
CV[a]					5.9		7.9	6.71				33.3	

TL, Total length; Ht, height; Wd, width.
[a]Normalized N = measurement × (TL of largest specimen/TL of specimen with measurement); coefficient of variation (CV = SD(100)/$\bar{x}$.

ing the lateral denticle, which could add as many as 2 more per side). The overall appearance of the teeth in these small sharks is similar to the late Paleocene white shark *Carcharodon orientalis*. The standard deviations (SDs) and the coefficients of variation (CVs) for the height, width, and height/width ratio demonstrated little variation in the A-2 tooth in specimens of <150 cm TL.

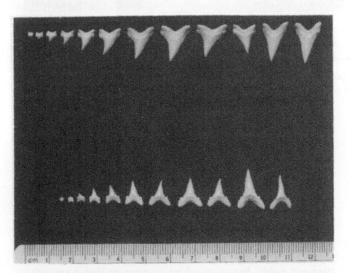

FIGURE 3 Front row teeth, labial view, from a 142 cm TL male white shark, showing lateral cusplets, narrow teeth, and, in the first lower anterior, lack of serrations.

Among white sharks that have grown to 300–400 cm TL, the crown has become slightly wider and the lateral cusplets have disappeared. Serrations are apparent, although finer, on all of the teeth, with the possible exception of the smallest posteriors, which may totally lack serrations (Fig. 4). An examination of A-2 in nine specimens of 229–379 cm TL (Table III) revealed 30–39 serrations on the medial side and 29–41 on the lateral side. Through this stage of growth, tooth size and shape were predictable in pattern.

Among sexually mature white sharks (≥380 cm TL in the male and ≥450 cm TL in the female), tooth structure varies much more. Some individuals have relatively long, narrow teeth (Fig. 5), while others have short wide teeth (Fig. 6). In a study of the upper right A-2 in 44 white sharks of >380 cm TL (Table IV), the crown height/width ratio varied from 1.03 to 1.39. It should be noted that these two extremes were also found in specimens that each measured approximately 520 cm TL (specimens H3884 and F7687). As the white shark grows in length, A-2 widens, as evidenced by the gradual diminishing of the height/width ratio, that is, from 1.34 in sharks of <150 cm TL to 1.27 in those of 150–380 cm TL, and to 1.17 in those of >380 cm TL. This ontogenetic change is subtle, as shown by the fact that the SD and the CV remain fairly constant during the process. A subtle increase in variability is also apparent, as shown by the fact that only 1 of 14 specimens with a TL of <380 cm had

TABLE III Measurements of the Crown of the First Upper Right Tooth [Second Anterior (A-2)] in White Sharks ≥1.5 and <3.8 m Total Length

Specimen			TL (m)	Height (mm)	N:Height (mm)[a]	Width (mm)	N:Width (mm)	Ht/Wd ratio	Serrations		Angle (deg)[b]
No.	ID	Sex							Medial	Lateral	
1	F122793-2	M	2.29	21.5	35.6	16.5	27.3	1.30	38	34	5.0
2	F102281	F	2.44	22.0	34.2	18.5	28.7	1.19	34	34	7.0
3	F61583A	M	2.59	23.5	34.4	17.5	25.6	1.34	35	33	2.0
4	J9281	M	2.72	22.0	30.7	19.0	26.5	1.16	30	29	3.0
5	C41185	F	2.82	23.0	30.9	19.5	26.2	1.18	33	30	1.0
6	F6785	M	3.35	28.0	31.7	21.0	23.8	1.33	36	39	0.0
7	F61583B	F	3.35	28.5	32.2	21.0	23.8	1.36	35	37	−5.0
8	F6785A	M	3.35	30.0	33.9	23.0	26.0	1.30	39	38	0.5
9	T42392	M	3.79	32.5	32.5	25.0	25.0	1.30	37	41	1.0
N					9		9	9			
Minimum					30.7		23.8	1.16			
Maximum					35.6		28.7	1.36			
Mean					32.9		25.9	1.27			
SD					1.7		1.6	0.08			
CV[a]					5.2		6.2	6.30			

TL, Total length; Ht, height; Wd, width.

[a]Normalized N = measurement × (TL of largest specimen/TL of specimen with measurement); coefficient of variation (CV) = SD(100)/x̄. x̄ refers to deviation from perpendicular without regard for direction).

[b]Negative angles are in the medial direction; positive angles, in the lateral direction.

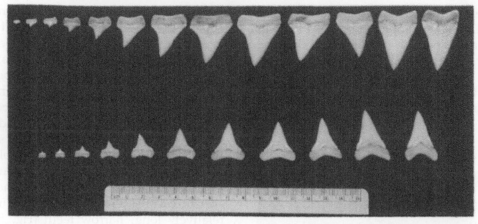

FIGURE 4 Front row teeth, labial view, from a 366 cm TL white shark, showing an absence of lateral cusplets, wider contour, and uniformity of serrations.

a height/width ratio that deviated more than 10% from the mean. However, 10 of 44 specimens >380 cm in TL deviated more than 10% from the mean. In 8 (18%) of the white sharks of >380 cm TL, the lateral margins of the crown of A-2 were concave. In 3 (7%), both margins were convex. In 8 (18%), both were straight; the remaining 25 (57%) had a mixture of concave, convex, and straight margins. The crown of the tooth was recurved labially or was flat, and the angle ranged from an 8° lateral slant to a 3.5° medial slant. The number of serrations on A-2 ranged from 39 to 72 on the medial side and were 37–68 on the lateral side. Although the height, width, and height/width ratio for most specimens >380 cm in TL were fairly constant, as shown by the SDs and CVs, some specimens demonstrated enough variation in normal tooth

structure that, were it the only criterion for identifying species, more than one species of extant white shark likely would have been named.

Up to this point, we have not considered tooth deformities. Of 123 modern white shark jaws I examined over the past 15 years, 10 possessed deformed teeth. The deformities consisted of tips that were hooked lingually (Fig. 7A), a tooth separated into two distinct teeth because of an injury to the gum tissue (Fig. 8), a tooth having a hole through the center of the crown, caused by a stingray barb that had penetrated the soft developing tooth bud (Fig. 9), a tooth that had grown together with the tooth behind it (Fig. 7B), and other teeth that were twisted, wrinkled (Fig. 7C), or notched (Fig. 7D). Deformed teeth represented 0.25% of the fully developed teeth in these 123

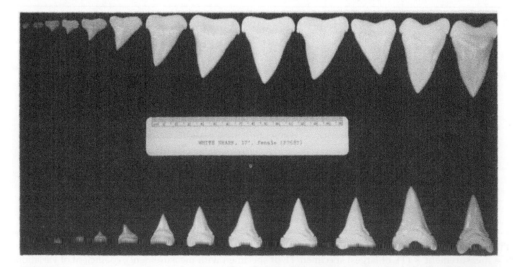

FIGURE 5 Front row teeth, labial view, right side, of a 518 cm TL female white shark, showing an example of long narrow teeth.

TABLE IV **Measurements of the Crown of the First Upper Right Tooth
[Second Anterior (A-2)] in White Sharks ≥3.8 m Total Length**

Specimen No.	ID	Sex	TL (m)	Height (mm)	N:Height (mm)[a]	Width (mm)	N:Width (mm)	Ht/Wd ratio	Serrations Medial	Serrations Lateral	Margin Medial	Margin Lateral	Angle (deg)[b]
1	SP22294	F	3.91	32.5	49.4	28	42.5	1.16	41	37	S	CC	2.0
2	F6785B	M	3.96	33.0	49.5	25	37.5	1.32	45	45	CX	SCC	−1.0
3	F83083	M	3.96	33.0	49.5	30	45.0	1.10	45	42	CC	CC	2.0
4	F2680B	M	3.96	35.5	53.3	31	46.5	1.15	47	45	S	SCC	2.0
5	F12484B	F	4.11	33.5	48.4	25	36.1	1.35	44	43	S	CC	5.5
6	H31683	U	4.11	34.0	49.1	28	39.7	1.24	48	43	CC	CC	4.0
7	H93086	M	4.17	38.5	54.8	35	49.9	1.10	48	44	CC	CC	4.0
8	F4683	M	4.27	33.0	45.9	30	41.7	1.10	40	38	CC	CC	1.0
9	F9886A	U	4.27	34.0	47.3	28	39.0	1.21	49	44	S	CC	4.0
10	X11384C	U	4.27	34.5	48.0	30	41.7	1.15	41	43	CC	S	0.0
11	F81287	U	4.27	35.0	48.7	27	37.6	1.30	39	44	CX	S	−1.0
12	ADJC	U	4.27	36.0	50.1	29	40.3	1.24	44	47	CX	S	2.5
13	H32089	M	4.42	40.5	54.4	33	44.3	1.21	46	44	CX	S	2.0
14	F51089	F	4.57	35.0	45.5	28	36.4	1.25	41	45	CX	CX	−3.0
15	H72689	U	4.57	37.0	48.1	34	43.5	1.10	43	42	S	CX	3.0
16	F9682	M	4.57	42.0	54.6	34	43.5	1.25	60	58	S	S	0.0
17	SW41285	M	4.71	39.5	49.8	39	48.6	1.03	51	50	CC	CC	3.0
18	G121382	M	4.72	38.5	48.5	33	41.5	1.17	44	42	S	CC	7.0
19	SP1394	M	4.74	43.5	54.5	40	49.5	1.10	53	48	CX	S	5.0
20	F111882	M	4.88	38.0	46.3	35	42.0	1.10	55	49	S	S	2.0
21	F10882	M	4.88	39.5	48.1	34	40.8	1.18	50	47	S	S	1.0
22	H2888	M	4.88	39.5	48.1	36	43.8	1.10	48	48	CC	CC	2.0
23	F92982	M	4.88	42.5	51.7	37	45.0	1.15	51	45	CX	CC/D	3.5
24	H10886	M	5.03	43.5	51.4	35	41.3	1.24	53	47	S	S	1.5
25	X10892	M	5.16	44.5	51.2	38	43.2	1.19	55	52	CC/CX	CC/CX	8.0
26	H3884	M	5.18	40.0	45.9	39	44.7	1.03	56	48	CC	CC	5.0
27	H8993	M	5.18	40.5	46.4	37	42.4	1.09	48	52	CC	CX/D	−3.5
28	F82187	M	5.18	44.5	51.0	39	44.1	1.16	58	52	S	S	2.5
29	F22490	U	5.18	46.0	52.7	39	44.1	1.19	51	57	CC/CX	CC/CX	1.5
30	F7687	F	5.18	48.5	55.6	35	40.1	1.39	67	66	CX	CX	1.0
31	N11693	F	5.23	45.0	51.1	43	48.8	1.05	52	53	CC/CX	CC/CX	0.0
32	R92185	F	5.36	47.5	52.6	42	46.5	1.13	60	56	CC	CC	3.5
33	F52683	M	5.38	39.5	43.6	38	41.4	1.05	54	50	S	S	2.0
34	X11384B	F	5.47	45.0	48.9	37	40.2	1.22	58	58	CC	CC	7.5
35	X11384A	F	5.54	41.0	44.0	37	39.1	1.12	57	55	S	CC	5.0
36	H10991	F	5.54	43.0	46.1	36	38.6	1.19	50	52	CX	CX	1.5
37	F82380	F	5.54	49.0	52.5	41	44.0	1.20	62	55	CX	S	7.0
38	L11685	F	5.63	48.0	50.6	43	45.4	1.12	60	58	S	CC	4.0
39	F42787	F	5.64	48.0	50.6	36	37.9	1.33	58	56	CX	S	3.5
40	SW31287	U	5.64	52.0	54.8	46	47.9	1.14	59	66	S	S	0.0
41	M91683	F	5.76	45.5	46.9	42	43.3	1.08	57	55	S	SCC	4.5
42	F7282	F	5.94	44.5	44.5	39	39.0	1.14	60	55	CX	S	4.0
43	H112690	M	5.94	45.0	45.0	40	40.0	1.13	64	62	CX	CC	6.0
44	H5384	F	5.94	49.0	49.0	36	36.0	1.36	72	68	S	S	3.0
N		44	44	44	44	44	44	44	44	44			44.0
Mean			4.93	40.8	49.3	35.2	42.4	1.17	51.9	50.1			2.7
SD			0.64	5.3	3.4	5.6	3.6	0.09	7.7	7.4			2.6
CV[a]			13.04	13.0	6.8	14.7	8.5	7.69	14.9	14.8			94.6

TL, Total length; S, straight; CX, convex; CC, concave; SCC, slightly concave; D, deformed.

[a]Normalized N = measurement × (TL of largest specimen/TL of specimen with measurement); coefficient of variation (CV) = SD(100)/x̄ (x̄ refers to deviation from perpendicular without regard for direction).

[b]Negative values slant in the medial direction; positive values, in the lateral direction.

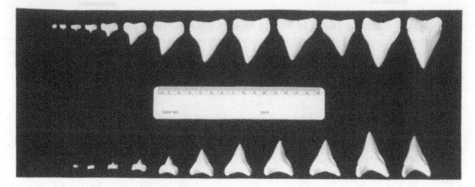

FIGURE 6 Front row teeth, labial view, right side, of a 488 cm TL male white shark, showing an example of short wide teeth.

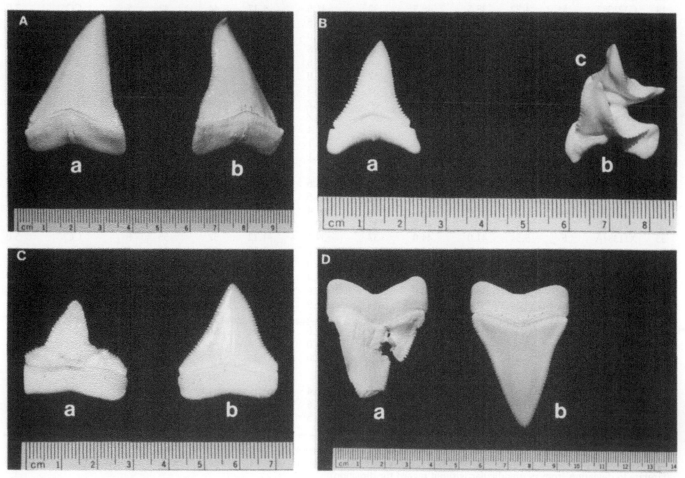

FIGURE 7 Deformed teeth compared to normal teeth from the same position on the opposite side of the jaw of a white shark. (A) Second upper laterals from a 594 cm TL female [the upper right tooth (b) has a lingually hooked tip; the upper left tooth (a) is normal]. (B) Third lower anteriors from a 427 cm TL female [the lower left tooth (b) is fused to the tooth behind it (c); the lower right tooth (a) is normal]. (C) Upper intermediates from a 488 cm TL male [the upper right tooth (a) is noticeably deformed; the upper left tooth (b) is normal]. (D) First upper anteriors from a 500 cm TL male [the upper right tooth (a) is notched and twisted; the upper left tooth (b) is normal].

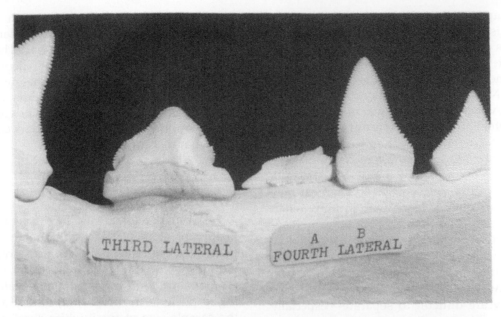

FIGURE 8 Upper right mandible of a 536 cm TL female white shark, showing a deformed third lateral tooth and a fourth lateral split into two partial teeth (the result of an old jaw injury).

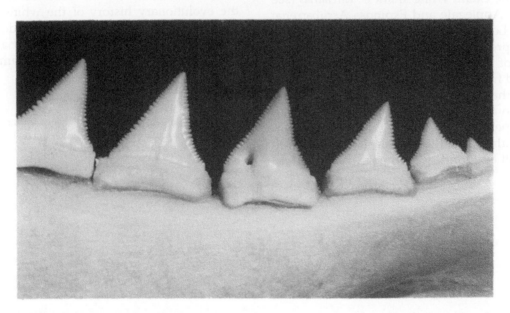

FIGURE 9 Upper right mandible of a 244 cm TL female white shark, showing the third lateral tooth with a hole through the enamel (caused by a stingray barb when the tooth was developing).

jaws. Interestingly, of the 13,151 fossil *Carcharodon carcharias* teeth collected from a 1-acre site in Sacaco, Peru, 0.25 were deformed. Deformities are also present in the megatooth white shark *Carcharodon megalodon*. I presently have 559 deformed *C. megalodon* and *Carcharodon auriculatus* teeth in my collection and have seen many more in other collections.

Discussion

Some species of extinct white sharks are rare in fossil collections. For example, the late Paleocene white shark, known variously as *Paleocarcharodon landanensis*, *Paleocarcharodon orientalis*, or simply *C. ori-*

entalis (see Chapter 4, by Applegate and Espinosa-Arrubarrena), is found in only four or five widely separated areas of the world. My research collection, containing just 92 specimens from Charles County, Maryland, is the world's largest collection of the seldom-collected North American form.

The rate of evolutionary change in many sharks is slow (see Chapter 6, by Martin). It is not unusual for a species to have existed for more than 10 million years, some for possibly 100 million years. Tracing white shark evolution is like watching a movie in slow motion. The change from one species to the next may have taken several million years, with primitive characteristics such as lateral denticles periodically reappearing later in more modern forms. This gradual change from one species to the next brings fewer steps in the white shark evolutionary scheme, and differences between succeeding species may be so subtle that they are difficult to separate—especially when all that remains are the teeth.

Different theories have been proposed about the evolution of the extant white shark *C. carcharias* (see Chapter 4, by Applegate and Espinosa-Arrubarrena; Chapter 5, by Long and Waggoner; and Chapter 6, by Martin). One proposes that it evolved from the mega-toothed line of white sharks, and another suggests that it evolved from the Miocene mako *Isurus hastalis*. At this time, there is insufficient reliable fossil material to come to agreement on the lineage. However, through a cooperative effort among scientists and amateur shark enthusiasts, the answer to this evolutionary puzzle will be found. It is imperative that scientifically significant specimens—both modern and fossil—be properly collected and preserved, and that the information gained is freely shared by researchers. By enlisting the aid of amateur paleontologists and carefully training them in the proper methods of fossil collecting, there could be a substantial increase in scientific reference material. This will give us a better understanding of how the white shark came into being and what its future may hold.

Summary

Fossil shark teeth are used to trace the evolutionary history of sharks. The technique works well with many shark species, but becomes a complex process with respect to the white shark *C. carcharias*, because (1) tooth structure varies ontogenetically and according to position in the jaw; (2) at any geological site, the sample size of fossil white shark teeth is small; and (3) there is an almost total lack of fossil white shark remains other than teeth. On the other hand, the evolutionary history of the white shark can be successfully constructed after consideration of the normal dental pattern, ontogenetic changes in tooth shape, and variation in tooth shape among adults.

Acknowledgments

I thank the many people who have contributed to my work on shark teeth over the years, the anonymous reviewers of the manuscript, and the editors who patiently worked with me to its completion.

4

The Fossil History of Carcharodon and Its Possible Ancestor, Cretolamna: A Study in Tooth Identification

SHELTON P. APPLEGATE *and* LUIS ESPINOSA-ARRUBARRENA

Instituto de Geología
Universidad Nacional Autonóma de México
Mexico City, Mexico

Introduction

It may come as a surprise to some that the lineage of the white shark *Carcharodon carcharias* (Linnaeus, 1758) has an excellent fossil record, consisting almost entirely of teeth. Other structures, such as vertebrae, jaws, and endocrania, are found but rarely reported. Much that could give us vital information is lost in poor collecting methods, and what is saved often remains in museums or in private hands, unstudied. Therefore, even though to some paleontologists (as is our case), the fossil record of the *Carcharodon* lineage is complete enough to warrant detailed paleobiological studies and taxonomic analyses, for others, the record is far from clear and construction of a phylogeny seems hopeless. Even among the proponents of the first view, some see a single lineage and others see a mixture of several taxa. It is important, therefore, to wrest the maximum information from the teeth available.

The only known natural fossil set of *Carcharodon* teeth is in the private collection of Gordon Hubbell, and it belongs to *C. carcharias* (see Chapter 3, by Hubbell). The set is from the Pliocene of Peru. In regard to

Carcharodon megalodon (Agassiz, 1843), only two associated sets are known. One was described by Uyeno *et al.* (1989), and the other was recently collected at Lee Creek Mine, North Carolina, and is illustrated in the work of Purdy *et al.* (1996). Storms (1901) described a partial tooth set of *Carcharodon auriculatus* (Blainville, 1818) and mentioned the existence of associated vertebrae. Leriche (1926) figured two partially associated sets, one corresponding to *Carcharodon angustidens* (Agassiz, 1843) and the other to *Carcharodon turgidus* (Agassiz, 1843).

Although the above materials seem insufficient for a study such as the one undertaken here, an extensive literature exists on the subject, as well as fossil teeth present in numerous collections. Therefore, aside from the natural and associated sets mentioned, we constructed new sets based on figured materials complemented with the fossil teeth from museums and private collections of the United States, Great Britain, and Mexico.

This chapter was motivated by three concerns: (1) to discover whether the *Carcharodon* lineage corresponds to a natural genus comprising more than just the Recent species, (2) to propose a possible Cretaceous ancestor of this genus and, (3) to determine how closely related *Carcharodon* is to other taxa.

Materials and Methods

Recent Material Examined

Over 100 *C. carcharias* jaws were examined. These included embryonic, newborn, subadult, and adult stages of both sexes.

The Recent jaw samples studied are housed in both scientific and private collections, that is, in the California Academy of Sciences, the Natural History Museum of Los Angeles County, the Geological Institute of the University of Mexico, and the private collection of Gordon Hubbell of Key Biscayne, Florida.

When possible, the teeth of these jaws were removed for inspection and photographed on the lingual side; particular attention was paid to every aspect pertaining to tooth variation (e.g., serrations, dental bands, crown, and tooth shape) in relation to growth stages and to the position of the teeth in the jaws (see Chapter 3, by Hubbell).

The photographs, drawings, and descriptions reported in the literature included those of Bigelow and Schroeder (1948), Bass *et al.* (1975), and Compagno (1984a), among others. As this information proved to be consistent with the observations obtained from the jaws examined, we were able to establish the parameters of tooth variation in the living species and apply them to the fossil forms of *Carcharodon*.

Fossil Materials Examined

The type and referred material for each species discussed in the text was examined with reference to the collections in the following places: the U.S. National Museum of Natural History, the Natural History Museum of Los Angeles County, the British Museum (of Natural History), the Geological Institute of the University of Mexico, and the private fossil shark teeth collection of Gordon Hubbell.

Also, as with the Recent materials, a literature search was performed to compare our observations on the actual fossil teeth with what had been reported by such authors as Agassiz (1833–1844), Leriche (1910, 1926), Arambourg and Signeux (1952), Casier (1950, 1960a), Cappetta (1987), and others.

Once the fossil teeth had been compared with those of the living forms, it was possible, in most cases, to place them in the proper position within the jaw. With all of the tooth types adequately identified and arranged, a reconstruction of the dental sets was possible and the original characterization of the species (based on single teeth) was enhanced.

Natural and Artificial Tooth Sets

One of the important rules we followed in our work was to give little weight to either single-tooth taxonomy or single-character classifications. As is done in other taxa that have teeth (e.g., mammals), before a tooth is considered, it must be referred to at least its possible location in the jaw, realizing, of course, that the exact position of some dental types may cause difficulties. We illustrate the teeth from the inner side (the lingual view) and within the context of tooth sets (Applegate, 1965). These sets may be (1) natural, with all teeth in place in the jaws; (2) associated, that is, when all the teeth come from a single individual but were not found in place; or (3) artificial, when all of the teeth are from a single locality or several localities but can be referred with some confidence to a single species.

The Dental Formula

Lamniform sharks possess a number of different kinds of teeth, depending on the position in the jaws and tooth morphology (Applegate, 1965; Hubbell, Chapter 3). In brief, these teeth are symphyseals, denoted by the letter *S* (although, important for this chapter, this position is not present in *Carcharodon*); anteriors, *A*; intermediates, *Int*; laterals, *L*; and posteriors, *P*. The symphyseals are small teeth that border the midline. The anteriors are large teeth positioned in the front of the jaw; the crowns of the upper teeth tend to be flattened, whereas the lower teeth are more rounded in cross section. In *Carcharodon*, the single intermediate is "reversed," which means that it points to the symphysis instead of toward the gape of the jaw; the upper laterals may number 4–7; and these teeth are curved, pointing to the corners of the jaws.

The lower laterals also vary in number and are spaced farther apart. They are erect and narrowly pointed. The posteriors are small and highly variable; only root shape and the straightness of the crown serve to distinguish the upper posteriors from the lowers (Applegate, 1965; Hubbell, Chapter 3). The general dental formula for *Carcharodon* is

Symphysis				
	A2	Int1	L4-7	P3-4
	A3		L4-7	P3-4

To illustrate the dental positions in the jaw, Fig. 1a shows the labial view of an upper and lower jaw of *C. carcharias.* Figure 1b diagrams a single tooth of *Carcharodon* (*C. angustidens*), in lingual view, to illustrate the most frequent dental terminology used throughout the chapter.

The Anterior Dental Formula

Among the Lamniforms, the anterior teeth that show dignathic heterodonty can be separated from each other not only by position but also by their characteristic morphology and symmetry (see also the discussions of tooth morphology in Chapter 3, by Hubbell, and Chapter 5, by Long and Waggoner). Applegate (1965: Fig. 3) illustrated the dental elements of the anterior teeth of *Carcharias taurus* (Rafinesque, 1810), placing them in order from the midline of the jaw: first an upper, then a lower, designated by the letters *A, B, C, D, E,* and *F.* However, since uppercase letters have been used to designate the larger categories of tooth position, for the sake of clarity for

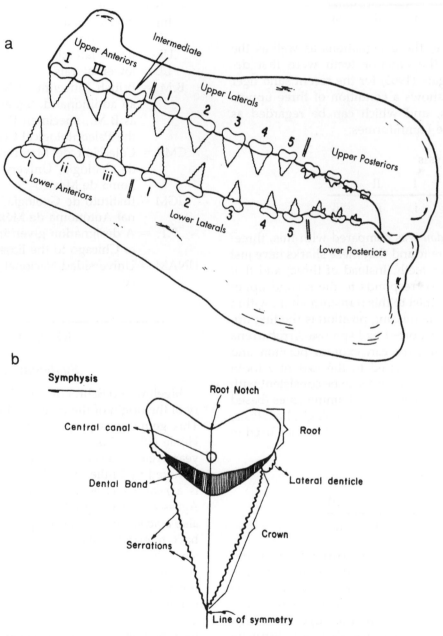

FIGURE 1 (a) The jaws of *Carcharodon carcharias,* left side, showing the nomenclature used for tooth positions. (b) The parts of a typical *Carcharodon* tooth, illustrated by a first upper anterior, I, of the Oligocene species *C. angustidens.*

this chapter, Applegate's (1965) terminology will be modified, and instead of using letters for both the general (A, Int, L, P, etc.) and the more specific anterior formulas (A, B, C, D, E, and F), the latter will be indicated with upper- and lower-case Roman numerals. The first anterior will be designated as position I, the first lower anterior as i, the second upper as II, the second lower as ii, and so on. A graphic representation of this condition could be shown as

Symphysis

	I	II	III
	i	ii	iii

As stated above, the designations as well as the characteristics of the anterior teeth were first described by Applegate (1965) for the recent sand tiger *C. taurus*, which shows a condition of three uppers and three lowers, and which can be regarded as "primitive" for the Lamniformes:

Symphysis

	I	II	III
	i	ii	iii

When *Carcharodon* was compared with this, three/three condition, we found that these sharks have just two upper anterior teeth, instead of three, and that the missing tooth corresponds to the second upper position (II). In contrast to this transformation, within the genus *Isurus*, the missing position is the first upper anterior (I), as reported by Espinosa-Arrubarrena (1987). Dental symmetry can express position and evolutionary patterns related to the loss of a tooth type (Fig. 2). Therefore, in order to be consistent with the general standard within the Lamniformes (based on *Carcharias*) and with our morphological analysis, the modified anterior formula for *Carcharodon* used in this chapter is

Symphysis

	I		III
	i	ii	iii

The third upper anterior tooth (III), which has not been mentioned, is present in all Lamniforms that show dignathic heterodonty. This similarity in the shape of this tooth position is characterized by a  straight, or nearly straight, anterior edge and a high degree of asymmetry, features that are distinctly shared by *Carcharias*, *Isurus*, and *Carcharodon* (Fig. 2).

Summary of Abbreviations

The following is a list of abbreviations used in this discussion of shark teeth.

    I = first upper anterior (as in *C. taurus*)

    II = second upper anterior (corresponds to the first in *Isurus oxyrinchus*)

    III = third upper anterior (as in all Lamniformes with dignathic heterodonty)

    i = first lower anterior

    ii = second lower anterior

    iii = third lower anterior

    S = symphyseals

    Int = intermediate

    P = posteriors

    UL = upper laterals

    LL = lower laterals

    B.M. = British Museum of Natural History, with an additional designation used in the text as B.M., specimen P., the *P.* standing for the Paleontological Collection

    CMR = Colección de Material Reciente, Instituto de Geología, Universidad Nacional Autónoma de México

    IGM = Instituto de Geología, Universidad Nacional Autónoma de México

    P.F. = A designation given in the Field Museum of Chicago to the Fossil Fish Collection

    UNAM = Universidad Nacional Autónoma de México

RESULTS

Systematics

Bigelow and Schroeder (1948) gave a brief summary of the origin of the name of the genus *Carcharodon*. This genus was proposed by Smith in Müller and Henle (1838), but with no type species. In the same year, Agassiz designated the type *Carcharodon smithii*, as noted by Müller and Henle (1838). This, however, is a *nomen nudem*, not used by Müller and Henle. Agassiz later designated the type *Carcharodon verus*, also a *nomen nudem*. Subsequently, many researchers have referred to Agassiz as the author of the genus, since, in 1835, he had illustrated a tooth set undoubtedly from *Squalus carcharias* Linnaeus. Today, Smith, not Agassiz, is considered to be the author of the genus (Compagno, 1984). With such a confusing beginning, it is not surprising that the fossil species have had an even more checkered history.

In 1923, Jordan and Hannibal erected a new genus,

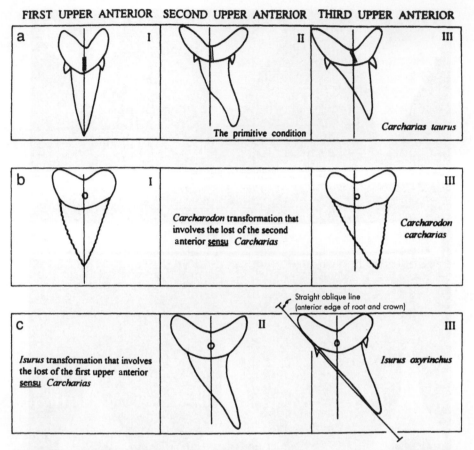

FIRST UPPER ANTERIOR   SECOND UPPER ANTERIOR   THIRD UPPER ANTERIOR

FIGURE 2 The upper anterior teeth of *Carcharias, Carcharodon,* and *Isurus,* right side (lingual view), demonstrating the ideas on tooth morphology and evolutionary patterns discussed in the text.

Carcharocles, the orthotype being *C. auriculatus* (Blainville, 1818). This was said to be similar to *Carcharodon,* but with a strong denticle on each side of the crown base. The teeth were narrower and more erect than in *C. carcharias,* but also finely serrated. In this group, Jordan and Hannibal placed *C. auriculatus, C. angustidens,* and *C. megalodon.* We must add here that *C. carcharias* has lateral denticles in its immature stage (Figs. 3 and 4), and *C. auriculatus* and other denticulated forms can have serrations as large as those of *C. carcharias.* Only *C. megalodon* has consistently small serrations.

Casier (1960a) again split the genus *Carcharodon* into two other genera, aside from *C. carcharias: Palaeocarcharodon* and *Procarcharodon. Palaeocarcharodon* teeth were characterized as being very compressed in a labial–lingual direction, with irregular serrations on the sides and the presence of lateral denticles. The roots were said to be poorly developed. The type species of *Palaeocarcharodon* was *Carcharodon landanesis,* a synonym of the older name *Carcharodon orientalis* (Sin-

zow, 1899). Casier and others who followed his work, however, were evidently not aware of how similar *C. orientalis* teeth are to the immature teeth in the living *C. carcharias* (cf. Figs. 3–5).

The other genus of Casier, *Procarcharodon,* basically included the same forms that Jordan and Hannibal (1923) had named *Carcharocles,* that is, *C. auriculatus* and a number of its synonyms, plus *C. angustidens* along with *C. megalodon.* Casier (1960a) believed that lateral denticles were present in Eocene and Oligocene forms, but not in the more recent members of his genus, that is, *C. carcharias* and *C. megalodon.* However, this is not quite true: all immature *C. carcharias* have lateral denticles, and occasional teeth of *C. megalodon* also show them. Aside from this, another idea we rejected is Casier's (1960a) belief that modern *Carcharodon* was derived from *Isurus hastalis* (Agassiz, 1843), through *Isurus escheri* (Agassiz, 1844), which shows delicately serrated sides to the crowns of its teeth.

Later, based on the above ideas, Gluckmann (1964)

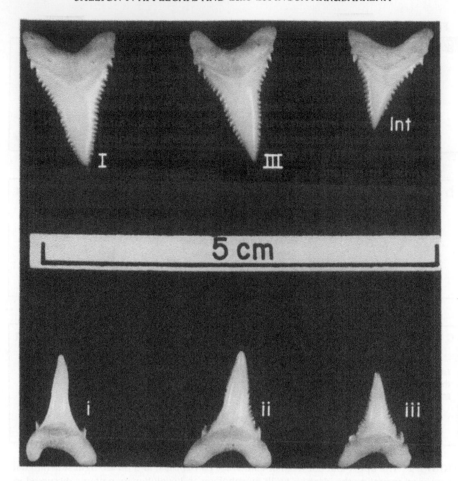

FIGURE 3 The anteriors of CMR-2, an immature *Carcharodon carcharias*, 100 cm in total length.

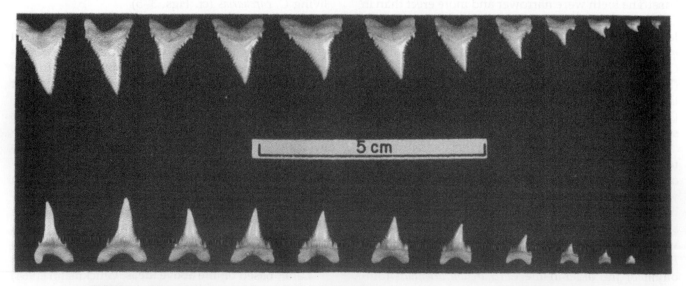

FIGURE 4 The complete right side of CMR-2, an immature *Carcharodon carcharias*, 100 cm in total length.

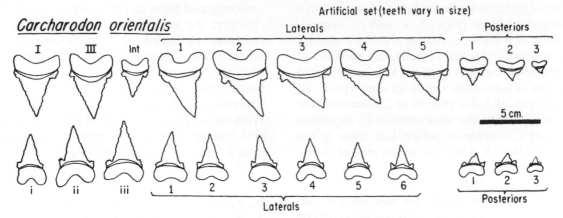

Carcharodon orientalis

FIGURE 5 An artificial tooth set of *Carcharodon orientalis*.

erected two new genera related to this problem. The first, *Cosmopolitodus*, was created to account for the ancestor of *C. carcharias*. His type species, however, corresponded to *I. hastalis*, a valid species properly (we believe) placed within *Isurus*. Consistent with this line of thought, Cappetta (1987) noted that worn *C. carcharias* teeth look very much like the teeth of *I. hastalis*. We would agree with this statement, but "to look like" does not denote a careful comparison such as the one that we require.

The second genus of Gluckmann, *Megaselachus*, used for *C. megalodon*, can also be dismissed from serious consideration, because it was based only on large size, the lack of lateral denticles, and small serrations. Size is not even a good species character. We have seen, in various collections, several good examples of *C. megalodon* teeth having perfectly good lateral denticles. The matter of serrations is correct only in the sense that *C. megalodon* lacks large serrations. Other species, such as *C. auriculatus*, which Gluckmann (1964) included in *Otodus*, at times can show equally small serrations.

The Family Carcharodontidae

For our purposes, the genus *Carcharodon* belongs in its own family, the Carcharodontidae (Applegate, 1991a). The fundamental reasons for this rest on both biological and paleontological data that, when contemplated in the context of the order Lamniformes, point out a need for a complete revision of this higher taxonomic group. This is beyond the scope of this chapter, however.

Within the traditional concept of the Lamnidae (*sensu* Compagno 1984 and others), which includes *Lamna*, *Isurus*, and *Carcharodon*, the latter can be shown to differ in the following aspects.

1. The paleontological record of the *Carcharodon* lineage, including its possible ancestor, *Cretolamna appendiculata* (Agassiz, 1838), ranges from the Albian (Early Cretaceous) to present times. With this unique geological record, it is possible to exclude *Isurus* and *Lamna* from this taxon, simply based on the large amount of time (>110 million years) they have been separated, and because neither is derived from *C. appendiculata*.

2. The anatomical evidence of the living species of *Isurus*, *Lamna*, and *Carcharodon* involves at least differences in the vertebral structures, the overall shapes of the endocrania (including the rostrum), and the positions of the fins.

3. Among the three genera, the morphology of the teeth involves differences in overall tooth shape (including the symmetrical pattern of the first upper anterior teeth), as well as in the presence or absence of serrations and lateral denticles.

Regarding the paleontological record, the validity of our interpretation is discussed in the remainder of this chapter. The biological evidence related to the anatomical differences among *Isurus*, *Lamna*, and *Carcharodon* is extensive enough that it should be treated in a separate chapter involving a detailed revision of the Lamnidae. Nevertheless, as far as tooth morphology is concerned, in order to set up a framework for further discussions, we review here the features that separate *Carcharodon* from *Isurus* and *Lamna*.

In the recent *I. oxyrinchus* (Rafinesque, 1810) and *Isurus paucus* (Guitart Manday, 1966), the teeth overall are spike-like. This shape in *Isurus* differs from the living species of *Lamna*, in which the teeth tend to be short and narrowly triangular (see the work of Bigelow and Schroeder, 1948; Bass *et al.*, 1975; Compagno, 1984). When these genera are compared with *C. carcharias*, the difference is striking, due to the great width of the crowns in the latter, expressed in a broad triangular shape (Figs. 3, 4, and 6). Second, serrated teeth is a feature, among the three genera, that is unique to *Carcharodon*. Finally, the presence of lateral denticles is found in all of the known fossil and recent species of *Lamna* (including all stages of development). In turn, all fossil species of *Carcharodon* bear lateral denticles, with the exception of *C. megalodon* and the living *C. carcharias*, which lack them in the adult stage (cf. Figs. 3, 4, and 6). In the case of *Isurus*, the lack of denticles is evident and readily separates this genus from the other two.

With these ideas in mind, we discuss the *Carcharodon* lineage, starting with what we believe to be the ancestor, and continue with the different species within the genus, as they appear in the fossil record. This analysis also includes several problematic taxa that some researchers have referred to this taxonomic group.

Cretolamna appendiculata (Agassiz, 1843)

No natural set of teeth of *C. appendiculata* has ever been described. Two associated sets were figured by Woodward (1911: Figs. 63 and 64; B.M. No. 45 and B.M. No. 39053). Both are from the *Holaster subglobosus* zone (Late Cretaceous), the first from Dover and the second from Maidstone, England. Ar-

ambourg and Signeux (1952) figure teeth of this species from the Montrian of Louis Gentil and the Maestrichtian of Ouled Abdoun, Morocco. *Cretolamna appendiculata* is supposed to have a geological range from the Albian to the lower part of the Eocene (Cappetta, 1987). Some evidence indicates that the Early Cretaceous forms might belong to a different species. With the works of Woodward (1911) and Arambourg and Signeux (1952), it is possible to reconstruct at least a tentative artificial set (Fig. 7).

That *C. appendiculata* possessed symphyseals is evident. There are three above and three below, or S_1, S_3, S_5/S_2, S_4, and S_6. Some of the upper symphyseals, S_1, S_2, and S_3, are shown in the work of Woodward (1911: Fig. 63). In them, the root interspace varies from almost absent to a very large open U shape; root height is almost equal to the crown height. The outermost symphyseals (S_1 and S_2) seem to lack lateral denticles and have a very open root interspace. The two inner upper symphyseals (S_4 and S_6) have narrow roots, similar to those of *C. taurus*, with a high interspace that is almost a closed V shape.

The upper anterior dental formula is I, III. The II position is missing, and the lower anteriors show the typical i, ii, and iii positions. The first upper anterior, I, is the most symmetrical tooth. This is true for all of the I's in *Carcharodon*, except for the occasional abnor-

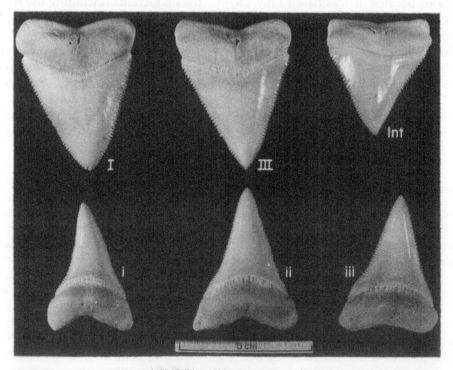

FIGURE 6 The anteriors of CMR-54, a 630-cm specimen of *Carcharodon carcharias* from near Cabo San Lucas, Baja California, Mexico.

mal teeth. The second upper anterior, III, has a low posterior root, whereas the anterior is higher, more elongate, and triangular.

In the lower jaw, the first anterior, i, has a symphyseal limb of the root that is narrow, similar to the condition found in *C. taurus;* the posterior root is bulbous with a rounded end. In ii, the anterior root is curved and a bit elongate, though not as much as in *C. taurus.* This feature is shared with most of the *Carcharodon* species, but is rare in other lamniform genera. The third lower anterior, iii, has a relatively thin root, similar to that of ii, but the interspace is much more open and the anterior root is very elongate, as in ii. The anterior root tip of iii is pointed, whereas the posterior is rounded, in contrast to the laterals, which show a rounded anterior root point and a posterior end that tends to be square. The size of iii is similar to that of ii, not smaller, as in *Carcharias* and *Isurus.*

All teeth of *C. appendiculata,* except the first symphyseals, S_1 and S_2, have low dental bands and lateral denticles on each side of the crown. The lateral denticles are short and triangular, instead of being needle-like as in *Carcharias* (Applegate, 1965).

There exists a single small tooth, an intermediate that shows the kind of root associated with III and the first lateral, yet the root sides are more perpendicular. The first upper lateral is of equal size or longer and is much straighter than the second lateral. This condition differs from that in *Carcharias* and *Isurus,* in which the second, third, or fourth lateral is usually the largest (Applegate, 1965; Espinosa-Arrubarrena, 1987).

In general, the morphology of the teeth of *C. appendiculata* is very close to that of *Carcharodon orientalis,* the earliest (Paleocene–Eocene) known form of the genus (cf. Figs. 5 and 7, as well as the other *Carcharodon* species recognized in this chapter). *Cretolamna* differs from *Carcharodon* in (1) the presence of symphyseals, (2) the absence of serrations (a feature common to all species of *Carcharodon*), and (3) the teeth of *Cretolamna* being somewhat thicker than in most early species of *Carcharodon.* In spite of these differences, *C. appendiculata* shows its close relationship with *Carcharodon orientalis* in having similar tooth shapes, particularly those of the crowns (both upper and lower). Lateral denticles are present in both species, and the overall shape of the roots is also similar, except that *Cretolamna* has deeper notches in these structures.

Carcharodon orientalis (Sinzow, 1899)

Carcharodon orientalis, known from the middle Paleocene and the lower Eocene, is the oldest known species of *Carcharodon.* It seems to be related to *Cretolamna,* although the teeth are not as thick as in *Cretolamna.* Otherwise, it resembles this genus, aside from the lack of symphyseal teeth and the presence of serrations. This species was named by Sinzow (1899), although the taxon has also been called *C. landanesis* Leriche, 1910 (Cappetta, 1987). The artificial set shown in Fig. 3 is constructed mainly from teeth studied in the Hubbell collection. The serrations in *C. orientalis* are irregular and there are prominent lateral denticles, as in the immature *C. carcharias* (cf. Figs. 3–5). Both species demonstrate the basic *Carcharodon* tooth morphology. To judge from the size of the teeth (Fig. 5), *C. orientalis* likely attained a total length of 5–6 m, comparable with adult specimens of extant *C. carcharias.*

Otodus obliquus (Agassiz, 1843)

Another species that has been thought to be related to *Carcharodon* is *Otodus obliquus.* No natural tooth sets of *Otodus* are known, even though these teeth are quite common in Paleogene sediments. Our

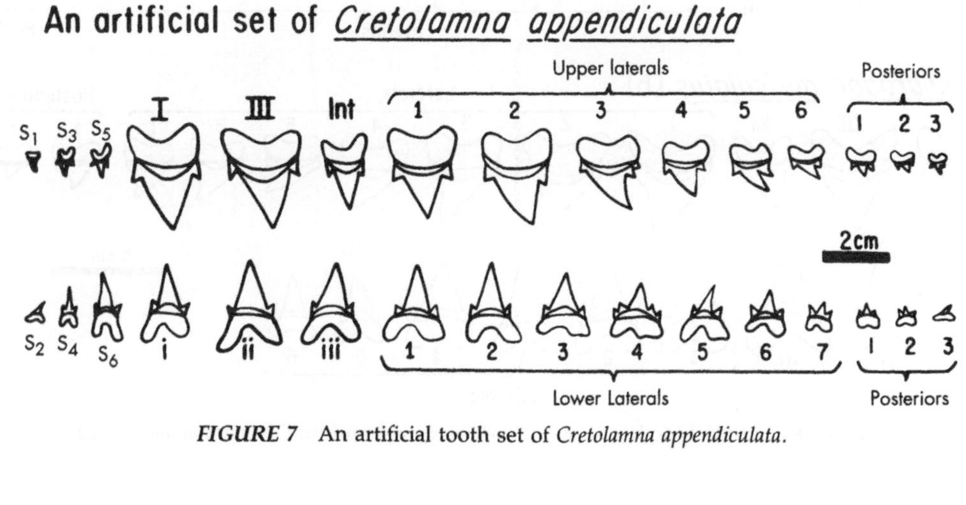

An artificial set of *Cretolamna appendiculata*

FIGURE 7   An artificial tooth set of *Cretolamna appendiculata.*

artificial set (Fig. 8a) was based on actual specimens and figures from the literature. Therefore, it is subject to revision.

Casier (1950) illustrated teeth that he then placed in the genus *Lamna*, but today, beyond any question, it is accepted that they belong to *Otodus* (Cappetta, 1987). The fact that the specimens showed serrations made Casier believe that these forms could be a transition between *O. obliquus* and *C. auriculatus*. When the teeth in question are examined, however, it is evident that the presence of fine serrations and dental bands (Fig. 1b) are the only characters in common with *Carcharodon*. In fact, the characters of *Otodus* that differ markedly from those of *Carcharodon* are: (1) the roots of the teeth are thicker and swollen with rounded ends, whereas in *Carcharodon* these structures are flattened and have square-shaped ends; (2) the root interspace is deeper, particularly in the lowers, whereas in *Carcharodon* the spaces are shallow; (3) the crowns are proportionally longer, contrary to those of *Carcharodon*, which are shorter; and (4) there are elongated nonserrate denticles in *Otodus*, in contrast to the short and serrated denticles of *Carcharodon*.

Although not without some striking similarities to *Carcharodon*, we believe that *Otodus* is a lamniform derivative, closer to *Carcharias* than to the white shark lineage. The reason for this interpretation rests on the presence of three upper anterior teeth (I, II, and III), such as in *C. taurus* (the "primitive" condition), as opposed to the ("derived") condition of only two upper anteriors (I and III) found in *Cretolamna* and *Carcharodon*. Supplementary to this condition, there is also some evidence of a lower symphyseal in the genus *Otodus* (based on B.M. specimen P 27651, from Khoumbga, Morocco), a characteristic known in *Carcharias*, yet completely unknown in *Carcharodon*.

Therefore, we interpret *Otodus* as a taxon originating in the Cretaceous (closely related to *Carcharias*); although successfully paralleling the evolution of *Carcharodon*, the genus became extinct by the middle Eocene.

Carcharodon auriculatus (Blainville, 1818)

Carcharodon auriculatus, as recognized in this chapter, is known from the middle Eocene. Its taxonomic history is characterized by numerous synonyms, on the bases of teeth from different positions in the jaws; among these are *Carcharodon disauris* (Agassiz, 1843), *Carcharodon debrayi* (Leriche, 1906), and recently *Carcharodon nodai*, from Japan (Yabumoto, 1987). Blainville's (1818) illustration of the type tooth of

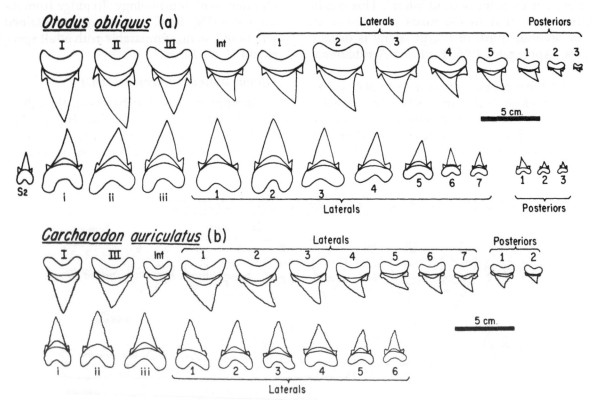

FIGURE 8 Artificial tooth sets of (a) *Otodus obliquus*, and (b) *Carcharodon auriculatus*.

C. auriculatus shows the same extreme curvature of a posterior upper lateral, as figured by Yabumoto (1987) for *C. nodai*. These tooth types are identical in overall shape to the fourth or fifth laterals in our artificial set (Fig. 8b). We are inclined to keep Blainville's (1818) name, *C. auriculatus*, until a natural or associated set is discovered. We recommend our artificial set as a standard for practical purposes of identification. This middle Eocene shark, if we interpret it correctly, is perhaps the smallest species of *Carcharodon*, with an adult maximum length of about 4 m. This size interpretation is based on a comparison between the largest first upper anterior (I) of *C. auriculatus* and a tooth of the same size and position in *C. carcharias*. In spite of its differences in size and tooth shape, this species has long been confused with the Oligocene *C. angustidens* and another, even larger, Eocene species, which we refer to in this chapter as *Carcharodon* sp. (Fig. 9a), discussed in the next section. The artificial set of *C. auriculatus* (Fig. 8b) was assembled from material housed in the U.S. National Museum of Natural History and the British Museum of Natural History.

Carcharodon sp.

Much confusion exists in differentiating specimens of *C. auriculatus* and *C. angustidens*. Part of this problem is caused by the presence of a large *Carcharodon* (with teeth as large as those of the extant *C. carcharias*) during the middle and late Eocene. Evidently, this new species has not been adequately characterized due to the tendency among specialists to refer this unknown form either to *C. auriculatus* or to *C. angustidens*, without realizing their differences in size and tooth shapes (see Figs. 8b and 9a and what follows for *C. angustidens*). The crowns of all the teeth tend to be more elongated than the ones of *C. auriculatus* (Fig. 9a). The lower teeth are large and massive with thick rounded roots, each having deep and wide interspaces. The uppers have rectangular roots with shallow root notches.

Carcharodon sokolowi (Jaekel, 1895)

Jaekel (1895) named *Carcharodon sokolowi* from specimens collected in southern Russia. Later, the species was reported from northern Germany (Leriche, 1910). Leriche felt that *C. sokolowi* was an intermediate taxon between *C. auriculatus*, the middle Eocene form, and *C. angustidens*, from the middle Oligocene. Recently, *C. sokolowi* has been found in the lower Oligocene deposits of Baja California Sur (our unpublished data).

In general, the teeth of this species show narrow but thick crowns and moderately expanded and ro-

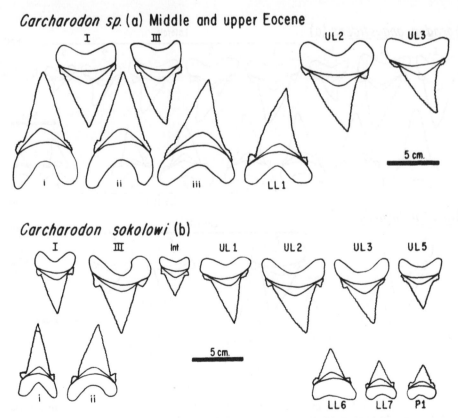

FIGURE 9 Artificial tooth sets of (a) *Carcharodon* species, and (b) *Carcharodon sokolowi*.

bust roots (Fig. 9b). The first upper anterior, I, has an open V-shaped root interspace. The lateral denticles flare out to the sides and bear strong though irregular serrations. The serrations of the crowns are fine, although not as fine as in *C. megalodon*. The second upper anterior, III, has very large lateral denticles; its uniquely asymmetrical root shape, ending in a conspicuous bulb, represents the most characteristic feature for the species. The first upper lateral has an almost perpendicular crown. The second lateral is inclined toward the gape of the jaws, and a deep interspace exists between the roots of this lateral. The third lateral has a very shallow smoothly curved root interspace. The first lower anterior, i, has roots that are almost equal in size; the interspace is deep and V-shaped. The second lower anterior, ii, has an elongate anterior root (the posterior branch is rounded) and its crown is straight. When the teeth of *C. sokolowi* are examined in relation to *C. carcharias*, it is evident that this wide-ranging species must have reached at least 6 m in length.

Carcharodon angustidens (Agassiz, 1843)

On the basis of the description by Leriche (1910) and from work we performed in museum collections, an almost complete, partially associated set of *C. an-*

gustidens teeth has been assembled (Fig. 10a). The teeth are characterized by narrow lance-like crowns and large, thick, and short roots. The sides of both the upper and lower anteriors, as well as those of the first lateral, are straight. The second and succeeding laterals are strongly curved in a posterior direction. All teeth have strong prominent serrations. *Carcharodon angustidens* was a moderately large shark, reaching almost 6 m in length; it existed during the middle Oligocene.

Carcharodon turgidus (Agassiz, 1843)

Carcharodon turgidus is known from the Oligocene of Russia and the middle Oligocene of Belgium. It is a form having broad teeth and very flattened crowns that are finely serrated (Fig. 10b). The lateral denticles are very small, and the roots are short and stocky. With the above characteristics, this is a logical ancestor for *C. megalodon*, and may be the same as *Carcharodon praemegalodon* (Weiler, 1928) from the Oligocene of Germany. The latter type is difficult to place, as it is known only by a single lower first anterior. The great width and straight sides of the teeth, along with the small lateral denticles, also make *C. turgidus* a possible ancestor for *C. carcharias*.

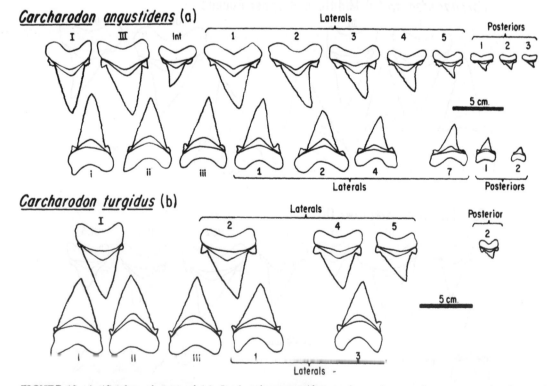

FIGURE 10 Artificial tooth sets of (a) *Carcharodon angustidens*, redrawn in part from an associated set of Leriche (1910); and (b) *Carcharodon turgidus*, redrawn from Leriche (1910).

Carcharodon subauriculatus (Agassiz, 1843)

Carcharodon subauriculatus is found in the early Miocene beds at Lee Creek Mine, North Carolina. Some teeth have strongly serrated lateral denticles, yet on the other side of the crown, these denticles may be missing or some teeth simply lack them. The species is not illustrated here, but an artificial set in the U.S. National Museum of Natural History has been figured by Purdy *et al.* (1996). Other unnamed forms of the same age that may correspond to this species are those found at Pecten Point and Pyramid Hill, Kern Country, California. Also, *Carcharodon mexicanus* and *C. productus* may ultimately have to be regarded as synonyms of *C. subauriculatus*. As was pointed out for *C. turgidus*, *C. subauriculatus* can also be considered as a precursor of *C. megalodon* and/or *C. carcharias*.

Carcharodon megalodon (Agassiz, 1843)

Carcharodon megalodon, because of the beauty and large size of its teeth, has received much attention in the scientific and popular literature. However, the estimated maximum total length for this shark has been exaggerated to reach 39 m; about 17 m is more realistic (see Chapter 7 by Gottfried *et al.*). Considered by some to have existed from the Eocene into the Pleistocene, well-documented fossils place this species not below the middle Miocene nor above the upper Pliocene. Although *C. megalodon* may be related to *C. carcharias*, we do not see one being the direct ancestor of the other. Nevertheless, some workers, such as Jordan and Hannibal (1923), Casier (1960a), Gluckmann (1964), and Cappetta (1987), have placed *C. megalodon* in three other genera, denying any close relationship to *C. carcharias*.

Using the first upper anterior, I, in the Field Museum of Chicago specimen PF 1168 (Fig. 11, No. 1), we were able to estimate a length of 12 m. This was done comparing I's of known lengths of the Recent *C. carcharias* with this specimen. The estimate compares favorably with the work of Gottfried *et al.* (Chapter 7). We have constructed tooth sets, based proportionally

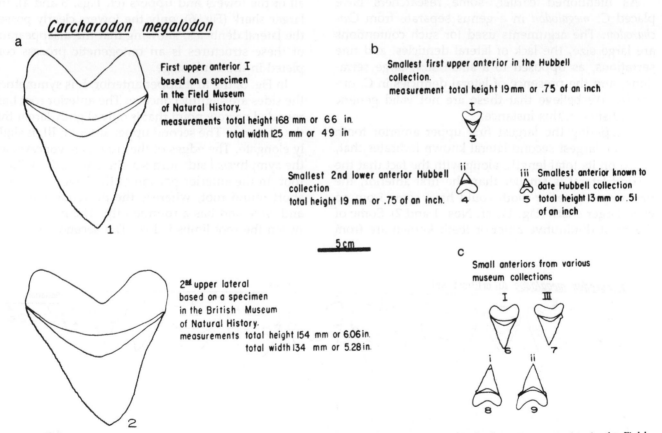

Carcharodon megalodon

First upper anterior I based on a specimen in the Field Museum of Natural History. measurements total height 168 mm or 6.6 in. total width 125 mm or 4.9 in

Smallest first upper anterior in the Hubbell collection. measurement total height 19 mm or .75 of an inch

Smallest 2nd lower anterior Hubbell collection total height 19 mm or .75 of an inch.

iii Smallest anterior known to date Hubbell collection 5 total height 13 mm or .51 of an inch

5 cm

2ⁿᵈ upper lateral based on a specimen in the British Museum of Natural History. measurements total height 154 mm or 6.06 in. total width 134 mm or 5.28 in.

Small anteriors from various museum collections

FIGURE 11 (a) Large tooth specimens of *Carcharodon megalodon*: 1, the largest known tooth, first upper anterior, in the Field Museum of Natural History, specimen PF 1168; and 2, the largest known second upper lateral from the British Museum of Natural History, specimen No. P10725. (b) The smallest *C. megalodon* teeth in the Hubbell Collection: 3, a 19 mm total height, heart-shaped I; 4, a 19 mm total height ii; and 5, a 13 mm total height iii. (c) The anteriors of a very small *C. megalodon* specimen: 6, a first upper anterior, I; 7, a second upper anterior, III; 8, a first lower anterior, i; and 9, a second lower anterior, ii.

on this tooth, that fit nicely in the jaw of a shark of about 12 m.

Keyes (1972) reported *C. megalodon* from the Eocene of New Zealand. The fossil teeth used were old Geologic Survey materials that lack reliable stratigraphic data and should be disregarded until they can be validated. Therefore, the record of *C. megalodon* in New Zealand, tens of millions of years before it appears elsewhere, is highly unlikely.

In Baja California Sur, the oldest *C. megalodon* that we know is from the lower part of the middle Miocene, at La Cocina (IGM Loc. 283). It seems that in older beds, *C. megalodon* is anteceded by *C. subauriculatus* or other species. We do not believe that this species occurred in the Eocene, Oligocene, or lower Miocene. The youngest *C. megalodon* that we know is from the late Pliocene, locality Rancho Algodones (IGM Loc. 92), near San José del Cabo, Baja California Sur. The so called "Pleistocene" records that exist in the literature are unproven and represent either reworked materials or teeth dredged from the ocean floor, where this could easily be Pliocene, not Pleistocene.

As mentioned earlier, some researchers have placed *C. megalodon* in a genus separate from *Carcharodon*. The arguments used for such contentions are large size, the lack of lateral denticles, and fine serrations, as opposed to a small size, large serrations, and the presence of lateral denticles in *C. carcharias*. We believe that these are not valid generic charactistics in this instance.

Comparing the largest first upper anterior tooth with the largest second lateral known indicates that, based on its total length, along with the fact that the second lateral is smaller than the first anterior, the first upper anterior tooth could have come from an even larger shark (Fig. 11, cf. Nos. 1 and 2). Some of the most diminutive anterior teeth known are from

the Hubbell collection (Fig. 11, Nos. 3–5). These teeth clearly demonstrate that the tooth morphology is retained even at these sizes, and no lateral denticles are present. Number 3 is a heart-shaped I (total length 19 mm); number 4 is a second lower anterior, ii, with the same total length, and number 5, the smallest tooth, corresponds to a third lower anterior, position iii (total length 13 mm). In the same figure, slightly larger teeth are shown (Nos. 6–9). Number 6 is a I; 7, a III; 8, a first lower i; and 9, a second lower ii. Shown in Fig. 12 is an artificial set of teeth from *C. megalodon*.

The irregular serrations, lateral denticles, and narrow teeth characteristic of juvenile *C. carcharias* do not exist in juvenile *C. megalodon*. This, as suggested by David Ward (personal communication), may be an example of heterochrony, in which the immature stage may have been cut out by an instantaneous change during the history of this species.

Carcharodon carcharias (Linnaeus, 1758)

Carcharodon carcharias is known from the Upper Miocene of Baja California, Mexico, and California. Very small lateral denticles and serrations are present on all of the lowers and uppers (cf. Figs. 3 and 4). In a larger shark (Fig. 6), only the lowers clearly possess the lateral denticles, showing that the disappearance of these structures is an ontogenetic process completed in the adult stage.

In Fig. 3, the first upper anterior, I, is symmetrical; the sides are slightly concave. The anterior root has a rounded narrow point that is only slightly larger than the posterior. The second upper anterior, III, is slightly elongate. The edges of the crown are concave, and the symphyseal side forms a straight border with the root. In the anterior portion of the lower jaw, i has a short round root, whereas the posterior is elongate and wide and has a rounded tip. The interspace between the root limbs is low. The second lower ante-

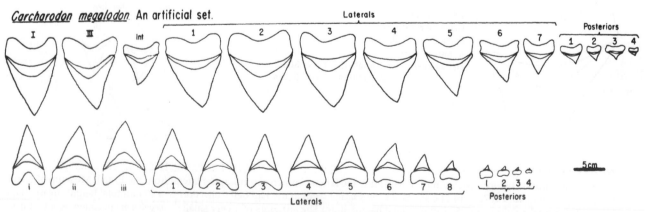

FIGURE 12 An artificial tooth set of *Carcharodon megalodon*.

rior, ii, has an elongate symphyseal branch, and both roots have broadly rounded points. The third lower anterior, iii, has a slightly elongate anterior root, and the short posterior limb has a straight posterior edge.

In Fig. 4, the intermediate tooth, Int, is reversed. Its anterior root edge is almost straight and the posterior is longer. The anterior edge of the crown is almost perpendicular, and the posterior side is angled toward the front of the jaw. The first upper lateral bears an almost straight crown and has wide roots that form an open V-shaped interspace. The second upper lateral is the largest lateral and the widest tooth in the jaw. The anterior edge of its crown is concave, and the tip of the posterior edge is convex. The third lateral is straighter than the second, but it angles slightly toward the gape of the jaw. The total length of this tooth is greater than the total width. The total height in the fourth lateral is only slightly greater than the width of the roots. The fifth lateral is much narrower with fine serrations. The four posteriors have a root width that is greater than the tooth height, and their crowns are strongly curved toward the gape. The lower laterals are erect with straight sides. Their anterior roots are elongate, with rounded tips. The posterior roots are short with their posterior edges almost vertical. The root interspace between all the lower laterals is low and nearly symmetrical. As is the case with other species of *Carcharodon*, posteriors have straight root sides and the interspace is symmetrical.

Figure 6 shows the anterior teeth and intermediate of a 4.85-m female specimen collected near Cabo San Lucas, Baja California Sur, Mexico. The roots are massive compared with the immature specimen. A narrow dental band is present in all of these teeth. The first anterior, I, is symmetrical, as is the case with all of the teeth in this position within the genus *Carcharodon*. The sides of the crowns are convex, and the root's width is almost equal to the crown's base. In the second upper anterior, III, the symphyseal edge of the root and crown describes a straight line. The anterior limb of the root is longer than the posterior. The intermediate, Int, has roots of almost equal size, and the tip of the anterior edge is convex. The crown in this "reversed" tooth is somewhat distorted, being curved instead of flat. The first lower, i, has a deep root interspace. The anterior root is wide and approximately the same length as the posterior limb. The second lower anterior, ii, has a shallow interspace. The anterior root is only slightly elongated and pointed, while the posterior root is shorter and rounded at its tip. Of all lower anteriors in *C. carcharias*, iii has the shallowest interspace; the anterior limb of the root does not project as far downward as

in ii, and the posterior root does not go past the base of its crown (cf. *C. carcharias* and *I. hastalis, C. carcharias* supposed ancestral species, in Fig. 13).

Isurus hastalis (Agassiz, 1843)

As with many other fossil sharks species, *I. hastalis* has its own taxonomic problems, and it may be that the correct name for this species is *Isurus xiphodon* (R. Purdy, personal communication). Nevertheless, until this ambiguity is resolved, we prefer to keep the widely used name *I. hastalis*. Here, the species is discussed because of the proposal that *C. carcharias* was derived from *I. hastalis*, not from another species within the genus *Carcharodon*. This concept, held by some workers (e.g., E. Casier and H. Cappeta), is based on the labiolingual flattening of the teeth and apparent similarities in tooth shape between *I. hastalis* and *C. carcharias*. We consider these "resemblances" to be only superficial and, as a consequence of convergence in feeding habits, similar to what has been discussed as part of the *Otodus* problem.

When the dentitions of *C. carcharias* and *I. hastalis* are studied and compared, the first element to consider is that in *I. hastalis* (as in *I. oxyrinchus* as well), the first upper anterior is a II and not a I, as exists in *Carcharodon* (see Fig. 2). This is true for all species of *Isurus*. The first upper anterior II in *I. hastalis* is characterized by marked asymmetry at all stages of development (Fig. 13b, bottom). Leriche (1926: Plate XXXI) demonstrated the same asymmetry. The second upper anterior, III, is directed toward the gape of the jaw. The tooth is triangular and its roots are equal and square-sided. In all of the uppers, the root interspace is so shallow as to be almost nonexistent, in marked contrast to the equivalent tooth positions in *Carcharodon*, in which the interspace is deep. The intermediate, Int, is strongly curved toward the jaw gape. The first lateral is triangular and smaller than the second lateral. The third lateral is the largest lateral tooth, a condition similar to that in *I. oxyrinchus*. The posteriors have low and strongly curved crowns. No dental band is present.

In the lower teeth, the roots are elongate and narrow, with a deep V-shaped interspace (Fig. 13b). Only the last lower laterals and posteriors have shallow interspaces. The lower anterior and lateral crowns are much more spike-like than in *Carcharodon*. In the first lower anterior, i, the roots are elongate and pointed. The posterior root, however, is short and bears a rounded tip. The second lower, ii, has a very elongate, narrow, and acutely pointed anterior root. The third anterior, iii, has an anterior root that is much shorter than that of ii.

The lowers are unique. Although no natural sets

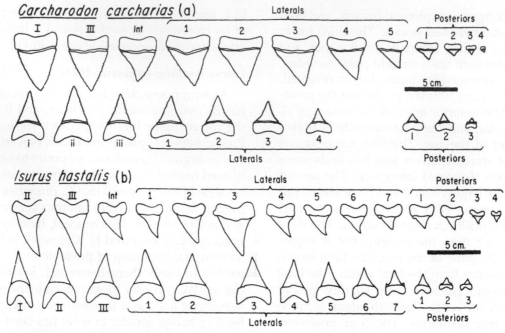

FIGURE 13   For comparison: (a) a natural tooth set of *Carcharodon carcharias*, specimen CMR-52; and (b) an artificial set of *Isurus hastalis*.

have been figured in the literature, a privately held specimen from the Miocene Valmonte Diatomite of California shows the anterior part of such a set (our unpublished observation). Thus, there is no question that Leriche (1926) was correct in the placement of the lowers with the uppers in *I. hastalis*.

Leriche (1910) illustrated teeth of two Agassizian species, *I. hastalis* and *I. escheri*, the latter of which he renamed *I. hastalis* var. *escheri*. These two forms have very similar teeth; nonetheless, *I. escheri* shows weak serrations on its crowns. Thus, we regard it as a valid species, although distant from the *C. carcharias* lineage. The presence in the fossil record of an isurid with even weak serrations has led scientists to believe that the transition between *I. hastalis* and *C. carcharias* was (the serrated form) *I. escheri* acting as the intermediate stage. Although an appealing idea, it is invalidated by a detailed morphological analysis of the teeth. Besides the presence of a II instead of a I in *I. hastalis* and *I. escheri*, the tooth morphology of the lower anteriors eliminates these two species from being a direct ancestor of *C. carcharias*. Hence, the progenitor of the white shark, *C. carcharias*, must be sought within the genus *Carcharodon*, as we have portrayed here.

A Possible Carcharodon Phylogeny

We present a schematic fossil record of what we consider to be all of the valid representatives within the family Carcharodontidae (Fig. 14). We do not claim

that every form of *Carcharodon* is included, although the species considered are those that existed, as confirmed by natural, associated, and/or artificial tooth sets (Applegate, 1965; Espinosa-Arrubarrena, 1987).

The ages in Fig. 14 are given in millions of years and represent the beginning of each epoch. The stratigraphic ranges of the species, except for *C. appendiculata*, are the most conservative and have definite "logical" evidence recorded in the literature. Regarding the range of *C. carcharias*, known to exist since the late Miocene, there are specimens that, upon eventual analysis, may extend this species back into the middle Miocene.

Some of the lines in the evolutionary patterns of *Carcharodon* require additional critical material. For these lineages, we used question marks in Fig. 14 to note the several possible pathways of descent. We feel that, for all intended purposes, *C. appendiculata* is the best possible ancestral form within the family Cretoxyrhinidae, due to its remarkable similarity to the Paleocene *C. orientalis*, the oldest taxon recorded in the genus (cf. Figs. 5 and 7).

The ideas and paleontologic information presented in Fig. 14 reflect not only the present state of paleontological research, as we see it, but also the Cenozoic fossil record rich in "*Carcharodon*" teeth. With work such as ours, it should be obvious that when any evolutionary study on this taxon is undertaken, these fossil shark teeth should be given much more serious consideration than has been done in the past.

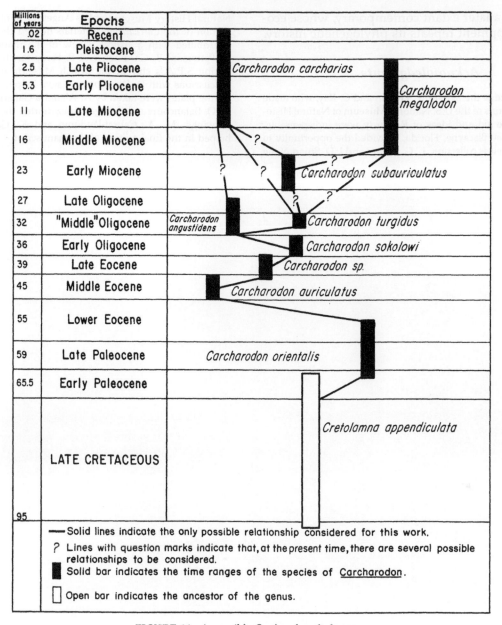

Millions of years	Epochs			
.02	Recent			
1.6	Pleistocene			
2.5	Late Pliocene		*Carcharodon carcharias*	
5.3	Early Pliocene			*Carcharodon megalodon*
11	Late Miocene			
16	Middle Miocene			
23	Early Miocene		*Carcharodon subauriculatus*	
27	Late Oligocene			
32	"Middle" Oligocene	*Carcharodon angustidens*	*Carcharodon turgidus*	
36	Early Oligocene		*Carcharodon sokolowi*	
39	Late Eocene		*Carcharodon sp.*	
45	Middle Eocene		*Carcharodon auriculatus*	
55	Lower Eocene			
59	Late Paleocene		*Carcharodon orientalis*	
65.5	Early Paleocene			
	LATE CRETACEOUS		*Cretolamna appendiculata*	
95				

— Solid lines indicate the only possible relationship considered for this work.

? Lines with question marks indicate that, at the present time, there are several possible relationships to be considered.

■ Solid bar indicates the time ranges of the species of <u>Carcharodon</u>.

□ Open bar indicates the ancestor of the genus.

FIGURE 14 A possible *Carcharodon* phylogeny.

There is no doubt that, as we progress in understanding the lines of evolution within the Lamniformes (based on new fossil discoveries and more information derived from the modern representatives of this taxonomic group), the ideas expressed here will be subject to change. Nevertheless, this attempt should serve as a foundation from which other hypotheses of relationship can be tested.

Summary

The genus *Carcharodon* has been subdivided into a number of genera: *Palaeocarcharodon, Megaselachus,* and *Carcharocles,* none of which we consider to be valid. The fossil record of the genus *Carcharodon* begins in the Paleocene and runs to the Recent. Neither *Isurus* nor the fossil genus *Otodus* is closely related to *Carcharodon.* The best candidate for an ancestor lies within the family Cretoxyrhinidae, particularly in the genus *Cretolamna.* Using data derived from teeth and other physical features, we place *Carcharodon* as a monotypic genus within its own family, the Carcharodontidae. Tooth structure and size indicate that *Carcharodon* began primarily as a fish feeder and evolved to feed on marine mammals. This taxon reached its highest specialization with *C. megalodon,* in the lower–middle Miocene, and was survived by

C. *carcharias*, a later extant contemporary, whose ecological development repeats its phylogenetic history.

Acknowledgments

We acknowledge the help of N. Hotton and C. Ray, who made the fossil collections of the U.S. National Museum of Natural History available. The aid and ideas of R. Purdy were indispensable. G. Hubbell of Key Biscayne, Florida, provided the opportunity to examine his priceless collection. L. G. Barnes and J. D. Stuart of the Natural History Museum of Los Angeles County permitted examination of the *Carcharodon* materials housed there. L. Barnes also contributed with a critical review of the manuscript. We thank C. Patterson, B. Gardiner, and P. Forey for their kind assistance during S.P.A.'s stay at the British Museum of Natural History. Discussions with D. Ward proved most helpful. The photographs for the plates were taken by A. Altamira Gallardo and G. Chavez. A. O. Betancourt and F. Vega-López drafted the final copies of the drawings. M. A. Cabral-Perdomo and K. González-Rodríguez helped in the final editing of the manuscript.

5

Evolutionary Relationships of the White Shark: A Phylogeny of Lamniform Sharks Based on Dental Morphology

DOUGLAS J. LONG
Department of Ichthyology
California Academy of Sciences
Golden Gate Park
San Francisco, California

BENJAMIN M. WAGGONER
Museum of Paleontology
University of California
Berkeley, California

Introduction

Evolutionary relationships within and among different groups of sharks have long been discussed by taxonomists, but phylogenies (evolutionary relationships) were originally drawn up in an interpretive manner emphasizing few characters. Several earlier writers (e.g., Garman, 1913; Daniel, 1934; White, 1936, 1937) provided more detailed anatomical descriptions of several sharks that led to better interpretations about elasmobranch relationships and greatly assisted later workers. Only recently have elasmobranch evolution and interpretation of phylogenetic relationships been studied cladistically. Studies by Compagno (1970, 1973, 1977, 1988, 1990b) clarified many elasmobranch relationships and provided a more accurate phylogenetic framework to better understand shark evolution.

Interrelationships within some groups, however, are not entirely agreed on, primarily because of different interpretation of characters. Early studies of the relationships of lamniform sharks began primarily with Jordan (1898), and lamniform sharks were later organized by Bigelow and Schroeder (1948). Phy-logenetic studies did not begin until Compagno's (1973, 1977) phylogenetic overviews of shark relationships. Maisey (1985) used characters of the chondrocranium and jaw suspension to develop a cladogram (a phylogenetic tree) showing relationships of a few species of lamniform sharks in an attempt to assess the relationships of *Megachasma* and *Cetorhinus*. Compagno (1988, 1990), using a greater number of cranial, jaw, and anatomical characters than did Maisey, developed a cladogram to explain relationships of all species of lamniform sharks (Fig. 1).

Teeth have rarely been used as the sole characters in shark phylogenies, but Compagno (1990b) listed a few dental characters in his character matrix. Unfortunately, he did not define or explain these characters well. Within lamniform sharks as a group, different genera show a wide variety of morphological variation, but Casier (1947a–c) demonstrated that morphological characters of shark teeth are often taxon specific. Even though some teeth may be derived or highly modified, teeth and dental patterns (Fig. 2) exhibit particular synapomorphies (characters or character states shared by two or more taxa, indicating evolutionary relationship) that align a species

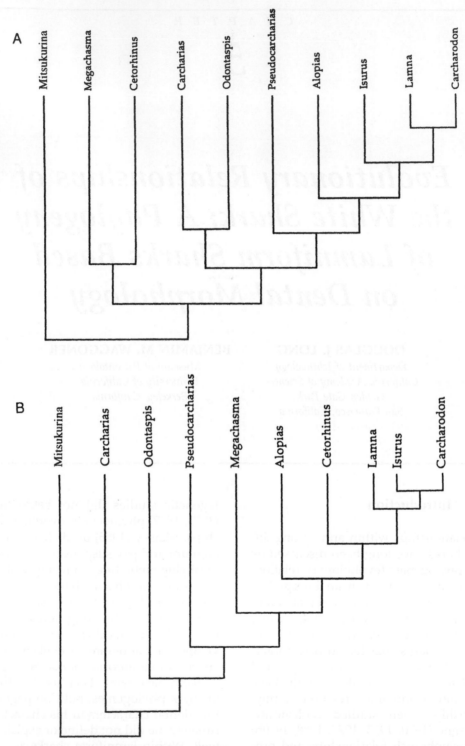

FIGURE 1 (A) Cladogram showing the phylogeny of the 10 lamniform shark genera, based on 23 dental characters (Table I). (B) Cladogram showing the phylogeny of the 10 lamnoid genera, based on different aspects of anatomy and reproduction. (Simplified from Compagno, 1990b.)

within a particular clade. Therefore, dental characters may be variable within a species or group of species (see Chapter 3, by Hubbell), but still may be reliable indicators of phylogenetic relationship.

Cladistic analysis is a quantitative method used to assess the phylogenetic relationships between organisms and is used to infer the evolutionary history of these organisms (see Brooks and McLennan, 1991;

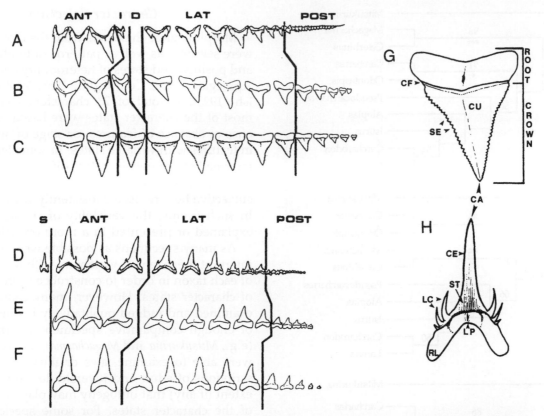

FIGURE 2 Dental terminology of lamniform sharks. Upper (top series) and lower (bottom series) dentitions of (A and D) *Carcharias taurus*, (B and E) *Isurus oxyrinchus*, and (C and F) *Carcharodon carcharias*. ANT, Anterior tooth position; I, intermediate tooth position; D, diastema; LAT, lateral tooth position; POST, posterior tooth position; S, symphyseal tooth position. (Right) Also shown is the tooth terminology for (G) *C. carcharias* upper lateral tooth and (H) *Odontaspis ferox* lower anterolateral tooth. CA, Crown apex; CE, cutting edge; CF, crown foot; CU, cusp; LC, lateral cusplets; LP, lingual protuberance; RL, root lobe; SE, serrations; ST, striations.

Maddison and Maddison, 1992). Cladograms based on dental characters found in lamnoid sharks were constructed (Figs. 1A, 3, and 4) to provide a basis for understanding the phyletic relationships of extant lamniform sharks in general and to determine the phylogenetic relationships of *Carcharodon* within the Lamnidae. This chapter provides a discussion of the dental features in lamniform sharks and how they were used to construct an evolutionary scenario of this group in a cladistic framework (Figs. 1, 3, and 4).

Materials and Methods

Comparative Specimens

We examined the dentition of species within each genus of lamniform shark (as defined by Compagno, 1990b) from dried jaws and tooth sets in research and museum collections. We looked at the dentition of

Mitsukurina owstoni (Mitsukurinidae; n = 1), *Carcharias taurus* (Odontaspididae; n = 8), *Odontaspis ferox* (Odontaspididae; n = 6), *Pseudocarcharias kamohari* (Pseudocarchariidae; n = 3), *Megachasma pelagios* (Megachasmidae; n = 1), *Alopias pelagicus* (Alopiidae; n = 2), *Alopias superciliosus* (Alopiidae; n = 8), *Alopias vulpinus* (Alopiidae; n = 17), *Cetorhinus maximus* (Cetorhinidae; n = 6), *Isurus oxyrinchus* (Lamnidae; n = 28), *Isurus paucus* (Lamnidae; n = 2), *Lamna ditropis* (Lamnidae; n = 7), *Lamna nasus* (Lamnidae; n = 2), and *Carcharodon carcharias* (Lamnidae; n = 27).

Published descriptions and dental illustrations provided additional information on the above taxa, including those by Antunes (1970), Applegate (1965, 1966, 1977), Bass *et al.* (1975), Berry and Hutchins (1990), Bigelow and Schroeder (1948), Branstetter and McEachran (1986), Cadenat and Blanche (1981), Cigala-Fulgosi (1983b, 1992), Compagno (1984a,b), D'Aubrey (1964), Daugherty (1964), De Beaumont

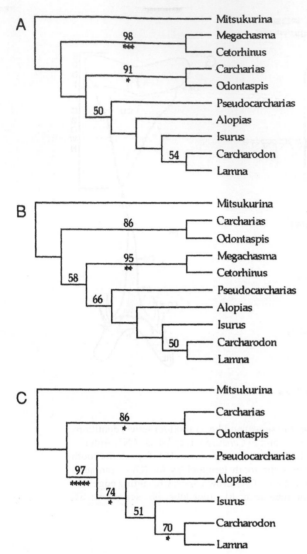

FIGURE 3 Cladograms of lamniform shark phylogeny, with numbers of asterisks indicating the degree of bootstrap support (≥50% for a clade). (A) The most parsimonious tree from the original data matrix of 23 characters, also equivalent to one of the two most parsimonious trees from the recoded matrix of 20 characters. (B) One of two most parsimonious trees from the recoded matrix of 20 characters (the other most parsimonious tree is identical to that shown in A). (C) The most parsimonious tree from the recoded data matrix, with *Megachasma* and *Cetorhinus* deleted.

(1959), Desbrosses (1930), Dodrill and Gilmore (1979), Garrick (1967, 1974), Gruber and Compagno (1982), Gubanov (1974, 1985), Herman (1979), Hussakoff (1909), Jordan (1898), Matthews and Parker (1951), Maul (1955), Moreno and Moron (1992a), Nakaya (1971), Parker and Boseman (1954), Sadowsky (1970), Seigel and Compagno (1986), Springer and Gilbert (1976), Stevens and Paxton (1985), Taniuchi (1970), Taylor *et al.* (1983), Uyeno and Matsushima (1979), and Whitley (1950).

Character Selection

Morphological characters from the crown and root were selected, as well as patterns of tooth placement and position and states of heterodonty (as defined by Applegate, 1965; Compagno, 1988; Welton and Farish, 1993). A total of 23 characters were selected; most of the character states were based on the presence or absence of features, or stage of development in features with variable states. In some features with intermediate character states, different stages of development for some characters may be somewhat subjective but are used consistently with each taxon. In such events, the variability of character states is explained or presented in a more objective context.

As many specimens as possible were examined to interpret the extent of variability within the dentition of each taxon in order to construct a more precise list of character states. However, many characters were examined and judged qualitatively because of a lack of enough comparative specimens of some species (e.g., *Mitsukurina* and *Megachasma*). In most cases, we were able to examine jaws or illustrations and descriptions of adult and juvenile dentition to see the extent (if any) that ontogeny may play in the validity of the character states. For some species, such as some *Alopias* species and *Megachasma*, juvenile dentition is unknown or poorly described. However, character selection was based on dentition of mature individuals. Definitions and dental terminologies are discussed below (see Fig. 2).

Selection of Outgroup, Characters, and Character States

Lamniform sharks are regarded as a monophyletic group, as defined by Compagno (1973, 1977, 1990b). *Mitsukurina* is selected as the outgroup for the Lamniformes, and Compagno (1990b) listed and discussed the morphological characters that make it the most likely outgroup to the Lamniformes. Dental synapomorphies of *Mitsukurina* and other Lamniformes include elongate cusps, lateral cusplets, elongate root lobes, a lingual protuberance, monognathic heterodonty, dignathic heterodonty not developed, a diastema in the upper jaw, and symphyseal tooth positions in the lower jaw (Tables I and II). Characters and character states are defined as follows.

1. *Monognathic heterodonty:* The division of teeth into positions or disjunct sets within a single jaw, estimated on the basis of morphological differentiation and size disparity between adjacent sets; weak heterodonty shows little differentiation, but strong heterodonty shows an abrupt change in size between

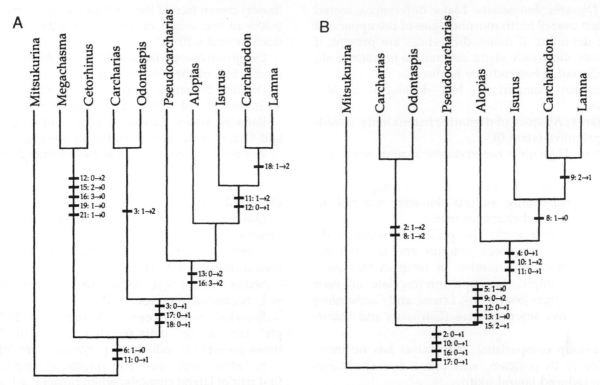

FIGURE 4 The most parsimonious trees from (A) the original data matrix of 23 characters and (B) the recoded data matrix of 20 characters, with *Megachasma* and *Cetorhinus* deleted. Unambiguous synapomorphies are indicated; homoplasies and autapomorphies are not shown.

tooth sets (i.e., a >20% reduction in cusp length between anteriors and laterals).

Outgroup comparison: Moderate monognathic heterodonty in *Mitsukurina*, with anteriors slightly longer than laterals and posteriors having low crowns—taxa that show pronounced heterodonty considered to be developed; those that show abrupt changes between tooth sets considered to be strong,

those that show little or no obvious pattern considered to be weak.

Polarity: Weak monognathic heterodonty considered primitive (state 0).

States: Weak = 0, moderate = 1, developed = 2, strong = 3.

Recoded states: Weak to moderate = 1, developed = 2, strong = 3.

TABLE I Character Matrix Based on 23 Dental Characters

| Common name | Species | Character number and state[a] | | | | | | | | | | | | | | | | | | | | | | |
		1	2	3	4	5	6	7	8	9	10	11	12	13	14	15	16	17	18	19	20	21	22	23
Goblin shark	*Mutsukurina owstoni*	1	0	0	1	0	2	1	1	1	0	0	2	0	1	2	3	0	0	1	0	1	0	0
Sand tiger shark	*Carcharias taurus*	2	0	2	1	0	1	1	2	2	0	1	2	0	1	1	3	1	1	1	0	1	0	0
Smalltooth sand tiger	*Odontaspis ferox*	2	0	2	0	0	1	2	2	2	0	1	2	0	1	2	3	1	1	1	0	1	0	0
Crocodile shark	*Pseudocarcharias kamohari*	3	0	1	1	0	0	1	1	1	2	1	2	2	0	2	2	1	1	1	1	1	0	0
Megamouth shark	*Megachasma pelagios*	1	0	0	0	2	0	3	1	2	1	1	0	0	0	0	0	0	0	0	0	0	0	0
Thresher sharks	*Alopias* spp.	1	0	3	0	1	0	3	1	1	2	2	1	1	2	2	1	1	1	2	2	1	0	0
Basking shark	*Cetorhinus maximus*	0	0	0	0	2	0	0	0	0	0	1	0	0	0	0	0	0	0	0	0	0	0	0
White shark	*Carcharodon carcharias*	2	1	1	0	2	0	3	1	0	1	3	1	2	0	2	2	1	2	1	0	2	1	1
Mako sharks	*Isurus* spp.	3	0	1	1	2	0	3	1	0	2	2	1	2	0	2	2	1	1	1	1	1	0	0
Mackerel sharks	*Lamna* spp.	2	0	1	1	2	0	1	2	0	1	2	1	2	0	2	2	1	2	1	0	1	0	0

[a]Characters, character states, and coding criteria are discussed in the text. Autapomorphic characters were eliminated.

2. *Dignathic heterodonty:* Major differences, scored between overall tooth morphologies of the upper and lower dentition, if major differences are present; if the teeth show only slight differences or none at all, then dignathic heterodonty is absent.

Outgroup comparison: Not developed in *Mitsukurina*.

Polarity: Absence of dignathic heterodonty considered primitive (state 0).

States: Dignathic heterodonty absent = 0, dignathic heterodonty present = 1.

Recoded states: This character is an autapomorphy for *Carcharodon* only, so this character was not included in recoded character matrix.

3. *Intermediate tooth:* The position showing a diminutive tooth between anterior and lateral teeth varies, as does the number of teeth in that position, for example, just one intermediate on each side of the upper jaw (*Isurus, Lamna,* and *Carcharodon*) to one to five intermediates (*Carcharias* and *Odontaspis*).

Outgroup comparison: *Mitsukurina* has no intermediate tooth position, although a few specimens have a reduced lateral tooth.

Polarity: Absence of an intermediate tooth position considered primitive (state 0).

States: Absent = 0, one only = 1, one or more = 2, occasionally present = 3.

Recoded states: Absent = 0, one only = 1, one or more = 2; "occasionally present" state recoded as a polymorphism (01).

4. *Diastema:* The presence of spacing between the intermediate and adjacent first lateral teeth varies between some groups; if adjacent root lobes are touching or are separated by a very small gap, the diastema is not present.

Outgroup comparison: Present in *Mitsukurina*.

Polarity: Presence considered to be primitive (state 0).

States: Absent =0, present = 1.

Recoded states: No change in character coding.

5. *Lateral root lobes:* The relative length of the root lobes extending laterally or basolaterally from the crown foot (25–30% of the length of the crown height); short root lobes do not extend far laterally.

Outgroup comparison: Lateral teeth of *Mitsukurina* show long root lobes that may be >30% of the total cusp height.

Polarity: Long root lobes considered primitive (state 0).

States: Long = 0, moderate = 1, short = 2.

Recoded states: Long = 0, short = 1; "moderate" character recoded as long.

6. *Striations:* Fine vertical grooves on the lingual (inner) crown face of the teeth; long striations extend ≥30% of the length of the cusp height, short striations extend ≤20%.

Outgroup comparison: Striations in *Mitsukurina* extend ≥50% of the total cusp height.

Polarity: Striations considered primitive (state 1).

States: Absent = 0, long = 1, short = 2.

Recoded states: Absent = 0, present = 1; "short" and "long" states both recoded as present.

7. *Lateral cusplets:* The number of lateral cusplets listed either as absent or by number of pairs; where present during development or in aberrant individuals, listed as occasionally present.

Outgroup comparison: A single pair in *Mitsukurina*.

Polarity: Presence of single lateral denticles considered primitive (state 1).

States: Absent = 0, single pair = 1, multiple pairs = 2, occasionally present = 3.

Recoded states: Absent = 0, present = 1; "multiple" and "single" pairs recoded as present, "sometimes present" recoded as a polymorphism (01).

8. *Lateral cusplet size:* The apicobasal length of the first pair of lateral cusplets, when present, adjacent to the crown relative to the size of the primary cusp in the lateral teeth; cusplets >10% of the primary cusp height considered to be moderate, but if shorter, considered to be minute.

Outgroup comparison: When present, minute in *Mitsukurina*.

Polarity: Minute lateral denticles considered primitive (state 1).

States: Absent = 0, minute = 1, moderate = 2.

Recoded states: No change in character coding.

9. *Lingual protuberance:* Lingual root bulge in the upper lateral teeth scored as variable, either well developed (a prominent bump) or almost flat.

Outgroup comparison: Moderately developed in the root of *Mitsukurina* (character state 1).

Polarity: Moderate protuberance considered primitive (state 1).

States: Weak = 0, moderate = 1, strong = 2.

Recoded states: No change in character coding.

10. *Distal inclination of lateral teeth:* Degree of vertical inclination of the central cusp in relation to the jaw; erect teeth, or ones with an inclination of <10°, listed as having weak inclination, 10–20° as moderate, and >20° as strong.

Outgroup comparison: Lateral teeth in *Mitsukurina* show little or no distal inclination.

Polarity: Erect cusps considered primitive (state 0).

States: Absent or weak = 0, moderate = 1, strong = 2.

Recoded states: No change in character coding.

11. *Lateral cusp width:* The ratio of the total height to the basal width of the primary cusp in upper anterolateral teeth; width of <25% of the total cusp height considered low, 25–50% moderate, 50–90% strong, and >95%, or if the cusp is wider than it is high, considered very wide.

Outgroup comparison: <25% in *Mitsukurina*.

Polarity: Low tooth width considered primitive (state 0).

States: Low = 0, moderate = 1, strong = 2, very strong = 3.

Recoded states: Low = 0, moderate = 1, strong = 2; "very strong" recoded as strong.

12. *Anterolateral cusp height:* In upper anterolateral teeth, the ratio of the total cusp height to the total tooth height.

Outgroup comparison: Cusps extremely long in *Mitsukurina*.

Polarity: Long cusps considered primitive (state 0).

States: Short = 0, moderate = 1, long = 2.

Recoded states: No change in character coding.

13. *Reduction of posterior teeth:* The total number of posterior teeth—9 to 15 indicates no reduction, 6–9 some decrease, ≤5 much decrease.

Outgroup comparison: About 10 posterior teeth in *Mitsukurina*.

Polarity: Multiple posterior teeth considered primitive (state 0).

States: Little or no reduction = 0, moderate = 1, much reduction = 2.

Recoded states: Little or no reduction = 0, reduction = 1; "moderate" and "strong" reductions both recoded as reduction.

14. *Symphyseal teeth:* Small teeth at the connection between the left and right sides of the lower jaw, but presence varies among taxa.

Outgroup comparison: A single symphyseal tooth in *Mitsukurina*.

Polarity: Presence considered primitive (state 1).

States: Absent = 0, present = 1, occasionally present = 2.

Recoded states: Absent = 0, present = 1; "sometimes present" recoded as a polymorphism (01).

15. *Upper anterior tooth number:* The number of teeth in the upper anterior position.

Outgroup comparison: Two in the upper jaw of *Mitsukurina* (state 2).

Polarity: Two considered primitive.

States: None = 0, three = 1, two = 2.

Recoded states: None = 0, present = 1; "two" and "three" recoded as present.

16. *Lower anterior tooth number:* The number of teeth in the lower anterior position.

Outgroup comparison: Three in *Mitsukurina* (state 3).

Polarity: Three considered primitive (state 3).

States: None = 0, one = 1, two = 2, three = 3.

Recoded states: None = 0, one or two = 1, three = 2.

17. *Upper anterior tooth size:* First anterior tooth longer or shorter than the adjacent second tooth, including consistency within groups.

Outgroup comparison: Longer than the second tooth in *Mitsukurina*.

Polarity: Longer first upper anterior considered primitive (state 0).

States: First anterior longer = 0, shorter = 1.

Recoded states: Longer = 0, shorter = 1; "variable" recoded as a polymorphism (01).

18. *Lower anterior tooth size:* Relative to the adjacent second tooth, consistency of size within groups.

Outgroup comparison: Longer than the second tooth in *Mitsukurina*.

Polarity: Longer considered primitive (state 0).

States: Longer = 0, shorter = 1, variable = 2.

Recoded states: Longer = 0, shorter = 1; "variable" recoded as a polymorphism (01).

19. *Upper anterolateral teeth:* Total apicobasal size of the first lateral tooth relative to the adjacent second lateral; if consistently smaller within the normal process of tooth replacement, listed as such, but if sometimes smaller or larger, listed as occasionally smaller.

Outgroup comparison: First lateral tooth smaller in *Mitsukurina*.

Polarity: Smaller considered primitive (state 1).

States: No size discrepancy = 0, smaller = 1, occasionally smaller = 2.

Recoded states: No size discrepancy = 0, smaller = 1; "occasionally smaller" recoded as a polymorphism (01).

20. *Distal blade:* Basal distal cutting edge of the lateral teeth sometimes expanded into a blade.

Outgroup comparison: No distal blades in *Mitsukurina*.

Polarity: No distal blade considered primitive (state 0).

States: Absent = 0, slight = 1, occasionally well developed = 2.

Recoded states: Absent = 0, present = 1; "weakly developed" or "occasionally well-developed" blades both scored as present.

21. *Enlarged intermediate tooth:* An intermediate tooth ≥75% of the cusp length of the adjacent first lateral; relative size does not change with growth.

Outgroup comparison: In *Mitsukurina*, no intermediate tooth position to compare.

Polarity: Reduced intermediate considered primitive (state 0).

States: Absent = 0, reduced = 1, enlarged = 2.

Recoded states: Absent = 0, present = 1; "reduced" and "enlarged" teeth both recoded as present.

22. *Reversed intermediate tooth:* Change in inclination distally to medially, and not aberrant, called a stable reversal.

Outgroup comparison: No intermediate tooth to compare in *Mitsukurina*, but all teeth erect or slightly inclined distally.

Polarity: Distally inclined (nonreversed) considered primitive (state 0).

States: Distally inclined = 0, reversed = 1.

Recoded states: An autapomorphy for *Carcharodon* only; thus, not included in the recoded character matrix.

23. *Serrations:* Presence on the lateral cutting edges of all tooth positions.

Outgroup comparison: Smooth in *Mitsukurina*.

Polarity: Smooth cutting edge considered primitive.

States: Smooth = 0, serrated = 1.

Recoded states: An autapomorphy for *Carcharodon* only; thus, not included in the recoded character matrix.

Procedure

We analyzed original data sets of the 23 characters (Table I). Sets containing autapomorphies (characters and character states found in only one taxon, namely, *Carcharodon*), provided no information about phylogeny. We therefore produced a second character data matrix in which autapomorphies were deleted, leaving a total of 20 characters. We also recoded characters that varied within taxa as polymorphisms

(multistate characters), instead of coding them as separate character states (Table II). With good reason, we suspected convergence between *Cetorhinus* and *Megachasma* (see Discussion), and therefore we analyzed the recoded data set both with and without these two taxa. Both of these data sets were analyzed, using PAUP 3.1.1 (Swofford, 1993) and the branch-and-bound search algorithms, to find the most parsimonious trees. All characters were unordered. Characters were optimized using the accelerated transformation (ACCTRANS) option of PAUP.

To determine tree robustness, we performed bootstrap analyses with 500 replicate heuristic searches on both data sets. We also performed decay analyses, for which support for a clade ($\geq$50%) is measured by whether or not it is present in all trees one step longer, two steps longer, etc., than the most parsimonious tree. Once the most parsimonious trees had been found, the characters were reweighted by the rescaled consistency index (RCI) of each character, defined as consistency index (CI) $\times$ retention index (RI) (see Farris, 1989). The data set was then analyzed three times by the branch-and-bound algorithm. The minimum value of the RCI was used when more than one tree was present.

The phylogenetic information content of a data set was assessed by the degree of skew in the distribution of all tree lengths, measured by the g_1 statistic (Huelsenbeck, 1991; Hillis and Huelsenbeck, 1992). The more negative the value of g_1, the more left-skewed the distribution, and the better the quality of the phylogenetic data. This method of analysis was intended for use with molecular data (see Chapter 6, by Martin), but it has also been applied to morpholog-

---

TABLE II   Character Matrix Based on 20 Recoded Characters

Common name	Species	1	2	3	4	5	6	7	8	9	10	11	12	13	14	15	16	17	18	19	20	21	22	23
Goblin shark	*Mutsukurina owstoni*	1	*[a]	0	1	0	1	1	1	1	0	0	2	0	1	1	2	0	0	1	0	1	*	*
Sand tiger shark	*Carcharias taurus*	2	*	2	1	0	1	1	2	2	0	1	2	0	1	1	2	1	1	1	0	1	*	*
Smalltooth sand tiger	*Odontaspis ferox*	2	*	2	0	0	1	1	2	2	0	1	2	0	1	1	2	1	1	1	0	1	*	*
Crocodile shark	*Pseudocarcharias kamohari*	3	*	1	1	0	0	1	1	1	2	1	2	1	0	1	1	1	1	1	1	1	*	*
Megamouth shark	*Megachasma pelagios*	1	*	0	0	1	0	01[b]	1	2	1	1	0	0	0	0	0	0	0	0	0	0	*	*
Thresher sharks	*Alopias* spp.	1	*	01	0	1	0	01	1	1	2	2	1	1	01	1	1	1	1	01	2	1	*	*
Basking shark	*Cetorhinus maximus*	1	*	0	0	1	0	0	0	0	0	1	0	0	0	0	0	0	0	0	0	0	*	*
White shark	*Carcharodon carcharias*	2	*	1	0	1	0	01	1	0	1	2	1	1	0	1	1	1	01	1	0	1	*	*
Mako sharks	*Isurus* spp.	3	*	1	1	1	0	01	1	0	2	2	1	1	0	1	1	1	1	1	1	1	*	*
Mackerel sharks	*Lamna* spp.	2	*	1	1	1	0	1	2	0	1	2	1	1	0	1	1	1	01	1	0	1	*	*

[a]Asterisks denote characters not used in the recoded character matrix.
[b]Many variable character states were recoded as polymorphisms.

ical data (e.g., that of Jamieson, 1994). For all data sets, we generated 10,000 random trees and calculated the g_1 value. We repeated this five times, calculated the mean g_1 value, and compared it with the table of critical values in the article by Hillis and Huelsenbeck (1992).

Results

Branch-and-bound searching on the original data set yielded a single most parsimonious tree with a length of 60 (Figs. 1A and 3); the CI was 0.767 (0.754 excluding uninformative characters), and the RI was 0.720. The bootstrap consensus tree collapsed some branches to polytomies but did not contradict the most parsimonious tree in any way. Decay analysis showed support for only two clades: *Carcharias* plus *Odontaspis* appeared in all trees of length 61, while *Megachasma* plus *Cetorhinus* was supported by all trees up to length 64. Three successive reweightings by the RCI yielded single trees identical to the most parsimonious tree.

Analysis of the recoded data set yielded two most parsimonious trees, one of which was identical to the single tree derived from the original data set (Fig. 3A). On the other tree (Fig. 3B), the clade *Carcharias* plus *Odontaspis* was closer to the root than was *Cetorhinus* plus *Megachasma*. Both trees were 50 steps long, with a CI of 0.740 and an RI of 0.745. Decay analysis showed strong support for only one clade, *Cetorhinus* plus *Megachasma*, which was supported by all trees of lengths 51 and 52. In the bootstrap consensus tree, *Carcharias* plus *Odontaspis* was closer to the root than was *Cetorhinus* plus *Megachasma*, but this was not strongly supported (58%). Three successive reweightings yielded single trees identical to the most parsimonious tree of the original data set.

When *Cetorhinus* and *Megachasma* were deleted from the data set, analysis yielded a single most parsimonious tree (Figs. 3C and 4B) compatible with the results of the other two analyses. The length was 35 steps; the CI was 0.829 (0.667 excluding autapomorphies), and the RI was 0.806. Successive reweightings did not alter the tree topology. The bootstrap consensus tree was fully resolved and identical to the most parsimonious tree. Decay analysis showed that most clades were supported by the two trees one step longer than the most parsimonious; relationships within the Lamnidae were supported by all trees up to five steps longer. We depict this tree (Fig. 4B), noting unambiguous synapomorphies; autapomorphies and homoplasies (morphological convergence) are not shown.

For the 10 lamniform genera, the critical value of g_1 is -0.59 for 10 binary characters and -0.39 for 50 binary characters; g_1 lower than this means that the data set is significantly ($p = 0.01$) more informative than a random data set (Hillis and Huelsenbeck, 1992). Our data set is not exclusively binary, but the use of g_1 values for binary character sets gives a conservative estimate of the probability of significant structure in the data set. The mean g_1 was -0.670796 for the original set, -0.764281 for the recoded data set, and -0.851596 for the recoded data set with *Cetorhinus* and *Megachasma* removed. We conclude that both sets contain phylogenetic information.

Sanderson and Donoghue (1989) showed that CIs very directly with the number of taxa. They empirically derived the formula $CI_{expected} = 0.90 - 0.022n + 0.000213n^2$ from a large sample of published cladograms. For 10 genera, $CI_{expected} = 0.701$. Analyses of both the original and recoded data sets yielded a CI greater than the $CI_{expected}$, well within the scatter of Sanderson and Donoghue's data.

Discussion

Based strictly on dental characters, the cladogram generated from this study showed many overall similarities to that of Compagno (1990b) (Fig. 1B), but with several major exceptions. First, Compagno (1990b) asserted that, in contradiction to Maisey's (1985) conclusions, *M. pelagios* and *C. maximus* were not closely related. He showed *Cetorhinus* to be a sister taxon to the Lamnidae and *Megachasma* to be a sister taxon to *Alopias* plus *Cetorhinus* plus Lamnidae (Fig. 1B). In this study, however, these two species fell out together, away from most of the other lamnoid sharks, in a relationship strongly supported by all measures of clade robustness (Figs. 3A and 4A).

Both of these species, however, have become planktivores, and their small teeth are not used in feeding (Matthews and Parker, 1951; Van Denise and Adriani, 1953; Parker and Boseman, 1954; Hallacher, 1977; Taylor *et al.*, 1983; Lavenberg, 1991). Not surprisingly, the teeth of these two genera probably have become vestigial, and many dental characters have been so reduced, modified, or lost that they contribute little to phylogenetic understanding. In fact, earlier forms of these taxa show characters that are later reduced, such as larger root lobes and a wider crown face on fossil *Cetorhinus* (Herman, 1979) and longer root lobes and larger lateral cusplets in fossil *Megachasma* (Stewart, in Lavenberg, 1991). We feel that *Megachasma* and *Cetorhinus* are not a valid primitive sister group to other lamnid sharks, and thus that Compagno's (1990b) cladogram is probably correct.

Figure 4B is a revision of our original cladogram, excluding the problematical taxa *Megachasma* and *Cetorhinus* and showing placement of characters at the nodes.

The relationships of *Carcharodon*, *Lamna*, and *Isurus* within the Lamnidae are also disputed, and two phyletic arrangements have been proposed. Compagno (1990b) suggested that *Carcharodon* and *Isurus* were closely related, with *Lamna* as a sister taxon (Fig. 5). Synapomorphies of the *Isurus–Carcharodon* group include two dental features (enlarged anterior teeth and loss of lateral cusplets in adults), intestinal valve counts, and vertebral counts; autapomorphies of *Lamna* include the development of secondary caudal keels and several features of the chondrocranium.

Our cladogram (Fig. 1A) shows more discrepancies compared to a partial cladogram of lamniform sharks based on mDNA (Martin *et al.*, 1992). The latter authors show *Carcharias* and *Alopias* to be a clade, separate from the Lamnidae; within the Lamnidae, they show *Lamna* to be a sister taxon to *Isurus* and *Carcharodon* (Fig. 5A). However, unlike the cladograms generated by Compagno (1990b) (Fig. 1B) and by us, that of Martin *et al.* does not present data from all of the lamniform genera. The contrary phylogenetic relationships seen in the work of Martin *et al.* (1992) may be the result of a phylogenetically incomplete data set (lacking *Mitsukurina*, *Odontaspis*, *Megachasma*, *Pseudocarcharias*, and *Cetorhinus*). Chapter 6 (by Martin) also shows a phylogeny of the Lamnidae (with fewer taxa), with associations as in the article by Martin *et al.* (1992). However, their additional cladistic analysis indicates that an alternate phylogenetic scenario, with *Isurus* as a sister taxon to *Lamna* and *Carcharodon*, cannot be refuted. Here, too, data from other lamniform sharks would complete their cladistic analysis.

The dental analysis we have presented provides another interpretation of the phyletic relationships of the Lamnidae: *Lamna* and *Carcharodon* are more closely related, with *Isurus* being the sister taxon (Fig. 5B). Two major synapomorphies separate the first two taxa from *Isurus*. Contrary to the work of Compagno (1990b), we found the anterior teeth of *Lamna* to be comparatively smaller apicobasally than the anteriors seen in *Isurus*; additionally, the distal inclination of the teeth in *Lamna* and *Carcharodon* are reduced, making the crown angle more erect. At least five major autapomorphies separate *Carcharodon* from *Lamna*, including mesiodistally widened upper teeth (causing dignathic heterodonty), increased size of the intermediate tooth (causing a loss of the diastema), reversal of the intermediate tooth, serrations on the cutting edges of the teeth, and loss of the lateral cusplets in the adult (Fig. 2C and F). These autapomorphies probably relate to trophic adaptations. The larger, stronger, serrated teeth of *Carcharodon* allow the fast, high-impact piercing, slicing, cutting, and fracturing needed when preying on large marine vertebrates (see Preuschoft *et al.*, 1974; Frazzetta, 1988; see also numerous chapter in this volume). Both *Isurus* and *Lamna* have long narrow teeth with smooth cutting edges. These are adaptations for feeding on smaller fast-moving fishes and cephalopods (Frazzetta, 1988) that are the bulk of their diet (Compagno, 1984a).

As illustrated here, the use of dental characters in a cladistic analysis helps to create a phylogeny that supports or contradicts other phylogenies based on other characters. When this phylogeny is compared to Compagno's (1990b) cladogram of lamnoid sharks, despite the different sister taxon identified for *Carcharodon*, the overall similarity between cladograms strengthens the validity of dental characters as phylogenetic tools (see Chapter 3, by Hubbell; Chapter 4, by Applegate and Espinosa-Arrubarrena). Problematic are species having a highly reduced or vestigial dentition, in which many key characters have been lost or were so modified that they cannot be easily recognized. Characters from *Megachasma* and *Cetorhinus* lead to an erroneous assumption of relatedness, and a grouping distant from that of other lamnoid sharks.

More importantly, a dental-based phylogeny may be used for extinct forms of lamnoid sharks in which only the teeth are available (see Chapter 3, by Hubbell; Chapter 4, by Applegate and Espinosa-Arrubarrena). Many enigmatic or problematic fossil lamnoids have been shuffled around among different groups (see

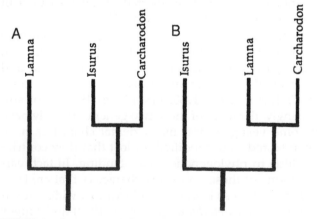

FIGURE 5 Competing cladograms of generic relationships in the Lamnidae. (A) *Lamna* as the sister taxon to *Carcharodon* and *Isurus* [simplified from that of Compagno (1990), based on selected anatomical characters]. (B) *Isurus* as the sister taxon to *Carcharodon* and *Lamna*, based on a matrix of dental characters analyzed by PAUP (from this study).

Cappetta, 1987; Ward, 1988; Compagno, 1990b; Long, 1992b, 1994; Welton and Farish, 1993; Siverson, 1992, 1995), and their phylogenetic relationships are debated. Additionally, relationships of *C. carcharias* to extinct forms are under debate (DeMuizon and DeVries, 1985; Long, 1992b; see also several chapters in this volume). If these taxa are viewed within the same cladistic framework, however, a project we currently have under way, then their relationships can be more adequately interpreted to better understand the long evolutionary history of lamniform sharks.

Summary

Cladistic analysis of the 10 shark genera in the order Lamniformes, on the basis of 23 dental characters, was carried out using a computer program (PAUP 3.1.1) to infer phylogenetic relationships within the group. *Mitsukurina* was selected as the outgroup, and a single most parsimonious cladogram was generated. This cladogram, supported by four different statistical methods for assessing tree robustness, showed the Odontaspididae (*Carcharias* plus *Odontaspis*) to be a primitive clade. The Lamnidae proved to be the most derived clade, with *Carcharodon* plus *Lamna* as a clade and *Isurus* as the sister taxon. *Carcharodon* showed several autapomorphic dental characters related to trophic adaptations. The placement of *Megachasma* plus *Cetorhinus* as a clade within this cladogram is considered dubious, because both taxa may have independently reduced their teeth as an adaptation for filter feeding, and hence lost many dental characters necessary for accurate phylogenetic assessment. The cladogram generated from this study is compared with several recent analyses of extant lamniform shark relationships.

Acknowledgments

We thank the following people who provided comparative material for this study: T. Iwamoto and D. Catania (California Academy of Sciences), J. Seigel and C. Swift (Los Angeles County Museum of Natural History), H. J. Walker, Jr. (Scripps Institution of Oceanography), D. Buth (University of California, Los Angeles), M. Roeder (San Diego Natural History Museum), and the fishermen of the Newport Beach dory fleet and those from the fishing cooperative of Ensenada, Baja California, Mexico. We also thank those who reviewed portions of this study and the manuscript: D. Polly, and D. Lindberg (Museum of Paleontology, University of California), anonymous reviewers, and the editors.

Systematics of the Lamnidae and the Origination Time of Carcharodon carcharias Inferred from the Comparative Analysis of Mitochondrial DNA Sequences

ANDREW P. MARTIN

Department of Biology
University of Nevada–Las Vegas
Las Vegas, Nevada

Introduction

The evolutionary history of the white shark *Carcharodon carcharias* has been controversial (Cappetta, 1987). Recent work on fossil shark teeth has cleared up many ambiguities, however, and indicates that the genus is relatively ancient (Siverson, 1992; Applegate and Espinosa-Arrubarrena, Chapter 4; Purdy, Chapter 8). Although it is possible to identify the origination of *Carcharodon* from fossilized remains, paleontologists have not yet established the phylogeny of the Lamnidae, and alternative hypotheses of the relationships among living species have been reported. On the basis of a survey of morphological characteristics, Compagno (1990b) proposed that *Isurus* (mako sharks) and *Carcharodon* are sister taxa, and that *Lamna* (porbeagle and salmon sharks) is the most ancestral genus of the family (Fig. 1). Alternatively, analysis of tooth characters led Long and Waggoner (Chapter 5) to propose that *Carcharodon* and *Lamna* are sister taxa.

Ideally, inference of relationships among extant taxa can be related to first appearance times (FATs) in the fossil record in order to reconstruct the phylogeny of the living species. Molecular data (i.e., DNA or amino acid sequences) can provide very powerful information about relationships (Felsenstein, 1988). Moreover, because mutations accumulate in a clock-like manner over time, divergence times between lineages can be dated with the aid of a well-calibrated molecular clock (Martin *et al.*, 1992). Agreement between molecular and fossil estimates of origination and divergence times provides compelling evidence that can be used to refute alternative hypotheses in favor of the one best supported by the data. Moreover, concordance between phylogenetic inferences and the stratigraphic record justifies calibration of molecular clocks using fossils. By contrast, disparity between molecular and fossil depictions of history indicates either that the molecules do not contain information about evolutionary history or that the fossil record is incomplete or misinterpreted.

Phylogenetic analysis using parsimony indicates that the hypothesis proposed by Compagno (1990b),

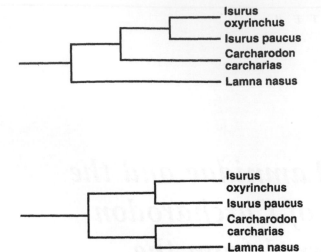

FIGURE 1 Alternative hypotheses of relationships among genera of the Lamnidae from (top) Compagno (1990b) and (bottom) Long and Waggoner (Chapter 5).

namely, that *Carcharodon* and *Isurus* are sister taxa, is best supported by the data, but that the alternative (*Carcharodon* and *Lamna* are sister taxa) cannot be confidently refuted. I present here a molecular systematic analysis of four species of the family Lamnidae: *C. carcharias*, *Isurus oxyrinchus*, *Isurus paucus*, and *Lamna nasus*.

Materials and Methods

The data consist of complete gene sequences (1146 base pairs) of the cytochrome *b* gene for *C. carcharias*, two species of *Isurus* (*I. oxyrinchus* and *I. paucus*), *L. nasus*, and two outgroups, *Heterodontus francisci* (Heterodontiformes) and *Galeocerdo cuvier* (Carcharhiniformes; GenBank accession numbers L08031 and L08036–L08038). In addition, cytochrome *b* gene sequences for members of the genera *Carcharhinus*, *Negaprion*, *Sphyrna*, and *Galeocerdo* were also used to calibrate a molecular clock (see Martin and Palumbi, 1993). The cytochrome *b* gene is encoded by the mitochondrial genome. Lack of recombination of the mitochondrial genome provides biologists with a superb marker of phylogenetic relationships (Avise, 1994). In addition, a great deal is known about the rate and pattern of mitochondrial DNA evolution in general (Avise, 1994) and cytochrome *b* in particular (Irwin *et al.*, 1991; Martin and Palumbi, 1993; Martin, 1995). These features probably account for the great utility of the cytochrome *b* gene for inferring phylogenetic relationships across considerable time spans (see,

e.g., Meyer and Wilson, 1990; Irwin *et al.*, 1991; Milinkovitch *et al.*, 1994).

Phylogenetic Analysis

Phylogenetic analysis of the data was accomplished using PAUP (Swofford, 1990), a computer program that constructs topologies of relationships among taxa and determines the number and types of changes for each tree. Because only six taxa were included in the analyses, all possible trees to relate the species were constructed (for six taxa, there are 105 bifurcating trees). Various permutations of the data were investigated. First, all changes were weighted equally. Weighting is a means of assigning importance to character state changes, based on observed or expected probabilities of change (Hillis *et al.*, 1994). Second, different types of mutations were assigned different weights. In mDNA, overall rates of nucleotide substitution are relatively fast (Avise, 1994; Martin *et al.*, 1992), and rates of transition interchanges (G–A and T–C) can be 10- to 20-fold greater than rates of transversion interchanges (all other substitutions) (Moritz *et al.*, 1987; Irwin *et al.*, 1991; Martin, 1995). When comparing highly divergent sequences (e.g., between different genera or families), multiple transition substitutions at a site may obscure the signal of evolutionary history (Kimura, 1980). By contrast, the signal of evolutionary history is often retained in the shared derived transversion changes (Hillis *et al.*, 1994; Martin, 1995). Thus, omission of transition changes (i.e., assigning a weight of zero to transitions) can often improve the accuracy of phylogenetic inference. Finally, the DNA sequence data were translated into amino acids for phylogenetic analysis. Bootstrap resampling of the data was done to evaluate the strength of support in the data for nodes of the tree (Felsenstein, 1985). Although there is disagreement about the meaning of bootstrap *p* values (Hillis and Bull, 1993; Felsenstein and Kishino, 1993), the emerging consensus is that *p* values are conservative estimates of inference accuracy (Hillis and Bull, 1993).

For all phylogenetic analyses, parsimony was adopted as the criterion for evaluating alternative topologies; the minimum-length tree was defined as the "best" phylogenetic hypothesis. However, it is important to recognize that alternative phylogenetic hypotheses may not be significantly less parsimonious than the "best" tree. Previous hypotheses of the relationships among genera and species of lamnid sharks have been proposed (Compagno, 1990b; Long and Waggoner, Chapter 5). A conservative nonparametric Wilcoxon sign-rank test (Templeton, 1983) was used to test whether alternative hypotheses were significantly different.

Estimation of the Origin of Carcharodon

Previous studies have demonstrated that mutations accumulate in vertebrates in a clocklike manner (Martin *et al.*, 1992). The existence of molecular clocks provides a means to estimate divergence and origination times for lineages. It is necessary, however, to calibrate the molecular clock for sharks. There is a good published record of FATs for extant lineages of Carcharhiniform sharks. In addition, geological evidence indicates that some species that inhabit the tropical eastern Pacific Ocean and western Atlantic Ocean (Compagno, 1984a) were separated by the rise of the Isthmus of Panama approximately 3–7 million years ago (Ma) (Knowlton *et al.*, 1993). Cytochrome *b* gene sequences were determined for *Galeocerdo* (FAT ≈ 56 Ma), *Negaprion* (FAT ≈ 42 Ma), two species of *Carcharhinus* (FAT ≈ 42 Ma), *Sphyrna lewini*, and two subspecies of *Sphyrna* separated by the Isthmus of Panama: *S. t. tiburo* and *S. t. verspertina* (FAT ≈ 5 Ma) (Martin *et al.*, 1992; Martin and Palumbi, 1993). The number of transversion substitutions at the third positions of codons, corrected for multiple substitutions at a site (Kimura, 1980), was estimated for each pairwise comparison. [The number of transversion substitutions per site measures the sequence difference between two sequences; transversion substitutions are mutations involving a change from purine to pyrimidine or from pyrimidine to purine. They tend to accumulate linearly over time (Martin *et al.*, 1992; Martin, 1995).] Only third positions of codons were included, because these sites have been shown not to be subject to selection; therefore, the accumulation of sequence differences at these sites between species is a function of mutation rate and the time since divergence from a common ancestor.

Among the pairwise divergence estimates, the average pairwise difference for each divergence time was calculated. The slope of the regression of divergence time on the average number of substitutions per site defined the rate at which the molecular clock ticked; 95% confidence intervals, calculated for all pairwise comparisons, were used to estimate the range of divergence times between lineages. Because the FATs of lineages were minimum origination dates, the calibrated molecular clock estimated the maximum rate of nucleotide substitution; therefore, estimates of age using the molecular clock represented minimum origination (or divergence) times. Molecular estimates of origination times were compared to fossils to assess whether inferences of evolutionary history match the record preserved in stone. In particular, I questioned whether evidence for a relatively ancient origination time for *Carcharodon* (Applegate and Espinosa-Arrubarrena, Chapter 4; Purdy, Chapter 8) is also evident in the molecules.

Results and Discussion

Phylogenetic Systematics of Lamnidae

When all types of changes were assigned equal weights, there were 221 phylogenetically informative sites (of the 1146 nucleotide positions surveyed). Pronounced left skew of the tree-length (TL) distribution of all possible topologies indicated significant phylogenetic information ($g_1 = -1.21$, $p < 0.01$) (Hillis and Huelsenbeck, 1992). A single most parsimonious tree was found with a TL of 437, identical to the phylogenetic hypothesis of Compagno (1990b) (Fig. 1). Two trees were seven mutations longer than the minimum TL (444), one of which united *Carcharodon* with *Lamna* (Fig. 1). For the transversions, there were 63 phylogenetically informative sites, significant left skew in the TL distribution ($g_1 = -1.71$, $p < 0.01$), and a single minimum TL (88), identical in topology with the minimum TL obtained from analysis of all changes (transitions and transversions). Analysis of the amino acids produced similar results.

Bootstrap analyses offered very strong support ($p = 100$) for the monophyly of the Lamnidae (Fig. 2) [*monophyly* refers to a group(s) of species that are descendant from a common ancestor]. Given that bootstrap values are conservative estimates of accuracy (Hillis and Bull, 1993), there is reasonably strong support for the monophyly of the two species of *Isurus*,

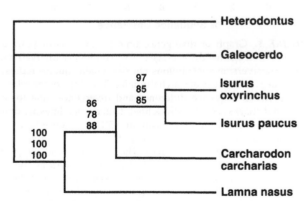

FIGURE 2 Minimum-length topology derived from exhaustive searches of all possible trees for the six taxa. Bootstrap values are shown for analysis of (from top to bottom) all substitution types, transversion changes only, and amino acid changes.

and the monophyly *Isurus* plus *Carcharodon*. Thus, the analysis strongly indicated that *Isurus* and *Carcharodon* are sister taxa, with *Lamna* as the most ancestral extant genus of the family (as proposed by Compagno, 1990b). However, lack of significant Wilcoxon sign-rank test results did not permit confident refutation of the hypothesis that *Lamna* and *Carcharodon* are sister taxa ($z = -1.3$, $p = 0.19$ and $z = -1.134$, $p = 0.25$ for all sites and transversions-only analyses, respectively).

Origination Time of Carcharodon

A molecular clock for sharks was calibrated based on the number of transversion differences between two sequences (Fig. 3). The molecular clock calibration indicates that a 1% sequence difference due to transversion mutations corresponds to approximately 6 million years. Corrected divergence estimates between the species of Lamnidae can be converted to divergence time, providing a minimum estimate of origination time for extant lineages of lamnid sharks. Thus, using this calibration, *Carcharodon* diverged

from its common ancestor with *Isurus* at least 43 Ma. Molecular clocks calibrated from origination times provide estimates of the minimum divergence time between lineages; therefore, the origination of *Carcharodon* (from the common ancestor with *Isurus*) probably occurred sometime earlier, perhaps in the Paleocene or early Eocene (Fig. 4).

Evolutionary History of the Lamnidae

Phylogenetic analysis of complete cytochrome *b* gene sequences provided support for the hypothesis

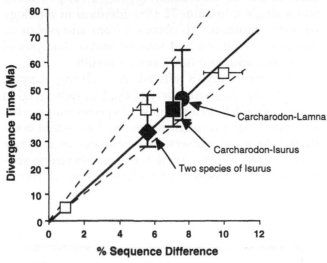

FIGURE 3 Graph of divergence time [in millions of years ago (Ma)] plotted against the average (±95% confidence interval) number of transversion substitutions per site. Open squares represent the values for lineages of Carcharhinidae. The slope of the relationship between sequence difference and divergence time for carcharhinid sharks is shown (thick line). Thin dashed lines are drawn through the minimum and maximum 95% confidence limits and represent an envelope for defining the range of probable divergence times between lineages. The solid diamond designates the divergence estimate between the two species of *Isurus*; the solid square, the average divergence between *Carcharodon* and the two species of *Isurus*; the solid circle, the average divergence between *Lamna* and the other three species. Linear regression demonstrated a significant association between time and DNA evolution ($n = 3$, $r^2 = 0.998$, $p = 0.027$; genetic distance data log-transformed), supporting claims for the existence of a molecular clock in sharks.

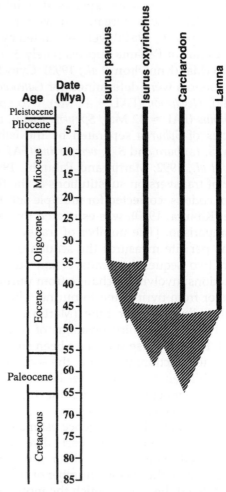

FIGURE 4 Phylogenetic hypothesis of the Lamnidae in relation to geological time scale [in millions of years ago (Mya)]. Shaded areas indicate uncertainty about dates of divergence. The molecular data indicate that the two species of *Isurus* diverged from a common ancestor 34–48 Mya, that *Isurus* and *Carcharodon* split 43–60 Mya, and that *Lamna* and the common ancestor of *Isurus* and *Carcharodon* diverged 46–65 Mya. These interpretations of the molecular data are consistent with (1) a pre-Oligocene divergence between the two extant species of *Isurus* (Espinosa-Arrubarrena, 1987) and (2) a Paleocene or early Eocene emergence of the extant genera of lamnid sharks (Cappetta, 1987; Siverson, 1992; Applegate and Espinosa-Arrubarrena, Chapter 4; Purdy, Chapter 8).

of relationships among genera proposed by Compagno (1990b), namely, that *Carcharodon* is the sister taxon of *Isurus*. However, an alternative hypothesis that places *Lamna* and *Carcharodon* as sister taxa (Long and Waggoner, Chapter 5) was not refuted. One reason for the inability to confidently refute Long and Waggoner's hypothesis in favor of Compagno's hypothesis may be that the three extant genera diverged from each other in a relatively brief period a long time ago. The latter hypothesis is supported by the fact that the estimates of sequence divergence, due to transversion substitutions among the three genera, are very similar (*Isurus–Carcharodon* = 7.1%, *Isurus–Lamna* = 8.4%, and *Carcharodon–Lamna* = 6.2%). In addition, using a molecular clock calibrated for sharks (based on divergence times and sequence differences between carcharhinid sharks), the estimated divergence times among the three genera were nearly identical and relatively ancient (Paleocene or early Eocene) (Fig. 4).

The molecular data indicated that the origin of *Carcharodon*, *Isurus* and *Lamna* may have occurred sometime in the Paleocene or early Eocene. Fossils indicate that lamnid sharks were abundant and diverse in the late Cretaceous and Paleocene (Cappetta, 1987; Siverson, 1992). During the late Cretaceous, two different groups of lamnid-like sharks existed: one group consisting of the genera *Cretolamna* and *Cretoxyrhina*, and the other of *Archeolamna* (Siverson, 1992). All three genera were well differentiated. Cappetta (1987, p. 97) noted that the origin of *Lamna* " . . . is probably to be found among the numerous species of *Cretolamna*" (see also Chapter 4, by Applegate and Espinosa-Arrubarrena). Because the extant genera diverged from a common ancestor in the Paleocene or early Eocene, they must be descendants of only one of these genera; therefore, Cappetta's statement on the origin of *Lamna* may be appropriate for characterizing the origin of the family (not just *Lamna*). Siverson

(1992, p. 526) noted, however, that the " . . . extant lamnids *Carcharodon*, *Isurus*, and *Lamna* . . . are probably derived from an *Archeolamna*-like ancestor. . . ." Clearly, the solution to the controversy regarding the affinity of extant and extinct lineages remains within the paleontologist's arena. The molecules can, however, focus our attention on the most likely time to search for the common ancestor of lamnid sharks.

Summary

Phylogenetic analysis of complete cytochrome *b* gene sequences for four species of the family Lamnidae (*C. carcharias*, *I. oxyrinchus*, *I. paucus*, and *L. nasus*) using two outgroups (*G. cuvier* and *H. francisci*) indicates that the genera *Isurus* and *Carcharodon* are sister taxa, *Lamna* being the most ancestral lineage of the family. Application of a molecular clock indicates that the divergence times between extant lineages dates back to the early Eocene or Paleocene, supporting recent claims for a relatively ancient origin of the genus *Carcharodon*.

Acknowledgments

I am grateful to the editors for the invitation to contribute a molecular perspective on the origin and evolution of white sharks, and also for their fair and thoughtful review of an early draft. Two anonymous reviewers and E. Heist provided critical comments. Many thanks are extended to G. Naylor for continued assistance and enthusiasm in all stages of molecular systematic research on sharks. I also want to acknowledge G. Naylor, G. Cliff, J. Stevens, A. Amorim, M. Miya, J. Caira, C. Chen, K. Goldman, E. Heist, and K. Yano for participation in the collection and preservation of shark tissue samples worldwide. The eventual harvest of international collaboration will be a description of the global population structure for the white shark. Individuals interested in participating in, or contributing to, the global white shark project should contact me.

Size and Skeletal Anatomy of the Giant "Megatooth" Shark Carcharodon megalodon

MICHAEL D. GOTTFRIED
Calvert Marine Museum
Solomons, Maryland

LEONARD J. V. COMPAGNO
South African Museum
Cape Town, South Africa

S. CURTIS BOWMAN
Hughes-Bowman Design Group
St. Augustine, Florida

Introduction

The extinct lamnid shark *Carcharodon megalodon* (Agassiz, 1835), variously referred to as the "megatooth shark," "great-tooth shark," or "megalodon," is the largest macropredatory shark to ever live and is among the largest known fishes. *Carcharodon megalodon* is arguably the most spectacular marine predator of the Cenozoic and has accordingly received much attention from the public and popular scientific media. Large teeth of *C. megalodon* are prized by fossil collectors, amateur paleontologists, and commercial fossil dealers. Based on tooth size, the megatooth reached a maximum estimated total length (TL) of at least 1500 cm (Randall, 1973; Purdy *et al.*, in preparation; Applegate and Espinosa-Arrubarrena, Chapter 4), more than twice the commonly reported maximum TL of 640 cm (Bigelow and Schroeder, 1948; Randall, 1973, 1987; Compagno, 1984a) for the extant white shark *Carcharodon carcharias* (Linnaeus, 1758). Fossils of *C. megalodon* range in age from the Miocene into the early Pliocene and have been found in North and South America, Europe, Africa, Japan, and Australasia. The wide occurrence of megatooth sharks in subtropical to temperate latitudes of the world's oceans, and their apparent preference for coastal habitats, are broadly similar to the distribution pattern of

C. carcharias (as summarized by Compagno, 1984a; Applegate and Espinosa-Arrubarrena, Chapter 4), keeping in mind that absence of *C. megalodon* in more tropical depositional environments may be an artifact of preservational and/or collection bias.

The robust, triangular, evenly serrated teeth of *C. megalodon* (Fig. 1A) reach a height of 168 mm, although unsubstantiated reports of even larger teeth persist. Associated megatooth dentitions have been recovered from the Miocene Pungo River and Pliocene Yorktown formations at Lee Creek, North Carolina (Purdy *et al.*, in preparation) and from the Miocene of Saitama Prefecture, Japan (Uyeno *et al.*, 1989). A relatively complete Lee Creek tooth set was used by the U.S. National Museum of Natural History (USNM) as the primary basis for a reconstruction of the entire dentition and jaws. Copies are exhibited at USNM and several other museums (Fig. 1B). In addition to teeth, vertebral centra have occasionally been recovered. The most complete specimen (discussed later) is an associated column of approximately 150 individual centra, ranging from nearly complete to fragmentary, from the Miocene of the Antwerp Basin, Belgium (Leriche, 1926); isolated megatooth centra have also been found in eastern North America (Fig. 1C) and Japan. Assignment of isolated vertebral centra to *C. megalodon* is based on

Carcharodon megalodon

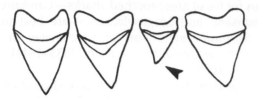

Carcharodon carcharias

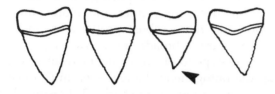

Isurus hastalis (= I. xiphodon)

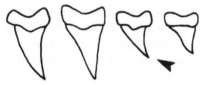

FIGURE 2 First four upper teeth in Carcharodon megalodon, Carcharodon carcharias, and Isurus hastalis (=Isurus xiphodon). Note the similar shape and reversed orientation of the intermediate teeth (arrows) in the two Carcharodon species versus Isurus, the symmetrical upper anterior teeth (first tooth in the series) in Carcharodon spp. versus the asymmetrical tooth in Isurus, and the more rounded root lobes in Carcharodon spp. versus the angular lobes in Isurus. (Drawn after Applegate and Espinosa-Arrubarrena, Chapter 4.)

roots of upper and lower teeth of makos (Isurus, including I. xiphodon) (Fig. 2) have angular lobes, unlike the more rounded lobes of white and megatooth sharks (Fig. 2) and mackerel sharks Lamna. Angular lobes are especially noticeable in the lower anterior and lateral tooth rows and may be a unique derived state for makos.

The best-supported hypothesis is the traditional one: megatooth is properly assigned to Carcharodon. It is incumbent on proponents of assigning the megatooth to a separate genus to summarize and present their arguments in favor of revising this classification.

Materials and Methods

Carcharodon megalodon is represented in the fossil record by teeth (Fig. 1A and B) and by vertebral cen-

tra (Fig. 1C). Isolated pieces of prismatic cartilage that are occasionally found in late Cenozoic marine deposits may also pertain to this shark, but they are not diagnostic enough to permit assignment. The lack of complete or even partial fossil skeletons forces us to rely on the morphology of the megatooth's closest living analog as the initial basis for interpreting missing parts of the skeleton. Because even partially complete fossil skeletons of C. megalodon are lacking, the first step in attempting to reconstruct the skeleton was to conduct a close examination of the skeletal anatomy of the extant species, and then make the appropriate modifications for the larger and more robust fossil form. To this end, we dissected and examined skeletonized specimens of C. carcharias in the SAM and from the research collection of the second author (L.J.V.C.). The primary specimen studied was a 294 cm TL male (LJVC 911219B), caught in 1991 off the KwaZulu–Natal coast of South Africa by the Natal Sharks Board (their reference number NSB ZIN 91-048). The complete skeleton of this specimen was dissected from its surrounding muscle and connective tissue following hot maceration in a large commercial cooker; it was then placed in a 50% solution of isopropyl alcohol. A complete set of measurements was taken, and drawings, black-and-white photographs, and color slides were made of the skeleton. In addition to this primary specimen, a second individual (LJVC 911219A), a female of 283 cm TL also caught by the NSB, was partially dissected to examine the structure of the hyoid arch and the branchial arches, the most complex region of the skeleton. Skeletonized material of several other specimens from South Africa, California, and Baja California, Mexico, supplemented these two individuals. Along with the drawings and photographs, video footage was taken of the various stages of the dissection and the skeletal parts, and X-rays were taken in order to examine the jaws, hyoid elements, and branchial arches in articulation.

After documenting the skeleton, the drawings were modified to reflect our interpretation of differences in the skeletal anatomy of C. megalodon relative to that of C. carcharias (see the following section). These inferences are based largely on observing how the skeleton of C. carcharias changes ontogenetically as the sharks grow larger, and then extrapolating the observations to an adult-sized megatooth shark with an even more robust dentition. The inferences, therefore, depend on the assumption that the megatooth shark was quite similar in its general skeletal morphology to the living species, which is the most parsimonious assumption, given the inadequacy of the fossil record. The similar morphology of the teeth and vertebral centra in fossil and living species of *Carcharodon* lends credibility to this assumption.

The next step toward producing the full-sized reconstruction involved using the modified drawings, supplemented by enlargements of the photographs and the X-rays, to construct a three-dimensional one-eighth scale model of the *C. megalodon* skeleton. Individual skeletal parts were sculpted of modeling clay, around which silicon rubber molds were poured; the molds were then used to cast the finished skeleton out of sculptable polyester resin. This component of the project was carried out at the exhibits studio and modelmaking shop at CMM.

The size of the full-scale reconstruction is predicated on measurements of the teeth and reconstructed megatooth jaws, on exhibit at CMM (Fig. 1B), which were cast from the original Smithsonian molds. These jaws will be modified somewhat and incorporated into the finished skeleton. Based on scaling factors derived from measurements of *C. carcharias* jaws, and the size of the largest tooth in the megatooth jaws (the upper second anterior, which is 127 mm high) and of the jaws themselves, the existing jaw reconstruction corresponds to a shark of approximately 1130 cm TL (37.2 feet). A shark of this size corresponds to a medium-sized adult male, according to the sizing assumptions described in the following section.

Results

Scaling Assumptions for Carcharodon megalodon

Living white sharks have large teeth relative to body size when compared to most other sharks, including large-toothed macropredatory carcharhinids such as the similar-sized tiger shark *Galeocerdo cuvier* and the smaller but very large-toothed bull shark *Carcharhinus leucas*. The megatooth dentition is apparently similar in morphology and number of tooth rows to that of *C. carcharias* (Uyeno *et al.*, 1989; Applegate and Espinosa-Arrubarrena, Chapter 4; L. J. V. Compagno, unpublished data). It is unlikely that the megatooth had teeth much larger in proportion to its body size than the white shark, based on the extreme relative size of the white shark's teeth and the size requirements for jaws of sufficient size to operate such gigantic teeth. The jaws would have to be at least as large and likely somewhat larger than the white shark's jaws, in order to serve as adequate mounts for the large functional teeth and to accommodate the tooth rows. We argue that the megatooth

jaws functionally complemented the teeth in being somewhat more robust, larger, and thicker, and with correspondingly more massive muscles to operate them, relative to those of the white shark. In addition, adequate depth is necessary on the lingual surface of each jaw for the tooth bays necessary to accommodate the thick replacement teeth and the dental (odontogenic) membrane. Huge and very thick teeth and the massive jaws and musculature to operate them, in turn, require a massive body in order for the megatooth shark to function effectively. The megatooth would likely have had a streamlined, fusiform shape similar to, but more robust than, the white shark and other lamnids, with more bulging jaws and a broader, blunter, and relatively more massive head. These inferred proportional differences between sharks in the same genus are similar to those observed in the large-toothed big-jawed bull shark in comparison to weaker-jawed *Carcharhinus* species with similarly shaped but smaller teeth (e.g., *Carcharhinus obscurus*, *Carcharhinus plumbeus*, and *Carcharhinus altimus*). Massive jaws and teeth imply a correspondingly large branchial region for respiration, a relatively big abdomen with a large stomach and intestine for processing large prey, a big liver to provide energy storage and flotation, large strong fins for propulsion and maneuvering, and massive segmental musculature to propel the shark in slow-cruise and fast-dash propulsion modes.

Direct Sizing of Megatooth Sharks from White Shark Teeth

We used measurements on the distinctive second upper anterior teeth (UA2, equivalent to the third upper anterior in *Carcharias taurus*) (see Applegate, 1965; Applegate and Espinosa-Arrubarrena, Chapter 4) in sizing the megatooth shark. All tooth height measurements given are the maximum height of the tooth, measured as a vertical line from the tip of the crown to the bottom of the lobes of the root, parallel to the long axis of the tooth. We assume that the ratio of UA2 size to TL in megatooth sharks is similar to that in white sharks, and that tooth size increased as TL increased in a similar fashion in the two species (see Mollet and Cailliet, Chapter 9). UA2 teeth in living white sharks range in size from 14.2 mm maximum height and 11.3 mm maximum width for a newborn or young of the year (129 cm TL) up to 56–63 mm height (n = 10, mean 59.2 mm) and 40–50 mm width (n = 5) for subadult and adult females 490–600 cm TL (L. J. V. Compagno, G. Hubbell, and S. A. Gruber, unpublished data). The largest UA2 teeth of megatooth sharks are at least 162–168 mm high.

A simple method of obtaining a relative TL ratio of

megatooth to white shark is by determining the tooth height ratio of the largest white and megatooth shark teeth. Using the average UA2 height (59.2 mm) and average TL (550 cm) from the 10 large female white sharks for which we had tooth height data, and the greatest UA2 height (168 mm) from a megatooth shark, the following equation gives an average TL estimation for large female megatooth sharks (tooth ratio 2.84:1): 5.5 m × 168/59.2 mm = 15.6 m TL.

Using the tooth and body size of a 600-cm white shark from Gans Bay, South Africa (the largest white shark with a reasonably trustworthy TL measurement available to us), the estimate for a corresponding maximum TL for an adult female megatooth shark (tooth ratio 2.80:1) is 6.0 m × 168/60 mm = 16.8 m TL.

On the basis of two larger female white sharks for which tooth size is available—a 640-cm female from Cojimar, Cuba, and a 720-cm female from Malta—estimates of maximum TL for exceptionally large adult female megatooth sharks (tooth ratio 2.85:1) are 6.4 m × 168/59 mm = 18.2 m TL and 7.1 m × 168/59 mm = 20.2 m TL, respectively.

Extrapolations of Megatooth Total Length from Regressions of White Shark Tooth Height Versus Body Size

Using UA2 height (UA2H) measurements on 73 white sharks ranging from 129 to 600 cm TL (ques-

tionable 640-cm and 720-cm sharks omitted), a linear least-squares regression (Fig. 3) was calculated using Quattro Pro 4.0, and the formula

$$TL\ (m) = a + b[UA2H\ (mm)]$$

where a and b, as calculated, were −0.22 and 0.096, respectively ($r^2 = 0.96$). Using 168 mm as a maximum UA2H for the megatooth shark, −0.22 + 0.096(168) = 15.9 m TL.

We consider 1590 cm (52.2 feet) a conservative maximum TL estimate for *C. megalodon*.

Extrapolation of Megatooth Total Length from the Vertebral Width of White Sharks

We assume that the width of the megatooth vertebral column has the same size relationship to TL as it does in the white shark. Using maximum vertebral widths (MVWs) from the vertebral columns of 16 white sharks ranging from 190 to 370 cm TL, a linear regression was calculated with the formula

$$TL\ (m) = a + b(MVW)$$

where a and b, as calculated, were 0.22 and 0.058, respectively ($r^2 = 0.97$). Using the MVW (155 mm) of the largest centrum from the most complete associated megatooth vertebral column [Institut Royal d'Histoire Naturelle de Belgique in Brussels (IRSNB) 3121, centrum 4; see Discussion], we calculated TL as follows: 0.22 + 0.058(155) = 9.21 m TL.

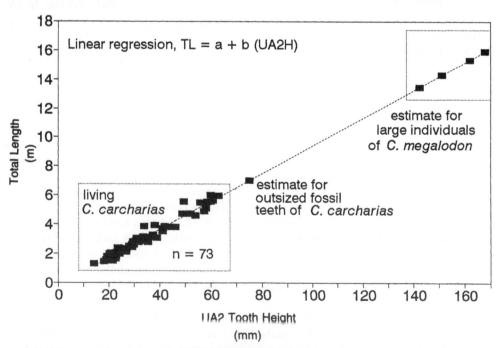

FIGURE 3 Linear relationship between the maximum height of the upper second anterior tooth (UA2H) and the total length (TL) in *Carcharodon carcharias*, with extrapolated values for *Carcharodon megalodon*.

Provided that megatooth vertebrae have the same width-to-TL ratio as white shark vertebrae, these results indicate that vertebral centra of megatooth sharks could attain a much larger width than those of the Belgian specimen. Using the TL estimate of 1590 cm from the tooth height regression, a maximum size for megatooth centra can be reverse-estimated: 155 mm $\times$ 15.9/9.21 = 268 mm.

Unfortunately, IRSNB 3121 could not be used to estimate megatooth size using precaudal column length or maximum vertebral length due to crushed, incomplete, and possibly missing centra as well as uncertainties as to the number of precaudal centra included in the specimen.

Mass of the Megatooth Shark

Using a sample of 175 white sharks at various growth stages with length–mass data (Mollet *et al.*, Chapter 10; L. J. V. Compagno, unpublished data) (Table I), an exponential regression curve (Fig. 4), according to the equation weight (kg) = $a[TL (m)^b]$, was fitted by the method of least squares, where a = $3.29E^{-06}$ and b = 3.174. Extrapolation of this white shark curve to the TL estimated above for megatooth sharks yields the estimated mass for *C. megalodon* at

various stages in its life cycle (Table I). Mass estimates are also given in Table I for the CMM/SAM megatooth restoration and for the conservative maximum megatooth TL estimate of 1590 cm generated by the tooth height–total length regression.

Discussion

Skeletal Anatomy of Carcharodon megalodon

We discuss here conservative inferences concerning skeletal features of *C. megalodon* as compared to *C. carcharias*.

Chondrocranium (Fig. 5A and B)

In comparison to *C. carcharias*, we reconstruct the chondrocranium of *C. megalodon* as broader across the orbits, more domed along its dorsal surface, and with a relatively shorter, broader, and more bluntly rounded rostrum, changes based on extrapolating from the allometric growth of the chondrocranium in *C. carcharias* (L. J. V. Compagno *et al.*, unpublished data). In addition, the orbital sinus, as reconstructed, is somewhat reduced and less elevated, with relatively small-

TABLE I **Body Length and Mass of the White and Megatooth Sharks at Various Life Stages**

| | White shark | | | | Megatooth shark | | | | |
| | TL | | BM | | TL[a] | | BM[b] | | |
Stage	m	ft	kg	lb	m	ft	kg	lb	tons
Largest fetus[c]	1.5	5.0			4.2	13.8	745	1644	0.8
Smallest neonate[c]	1.3	4.2	23	51	3.6	11.8	430	949	0.5
Largest immature male	3.2	10.5	320	704	9.1	29.8	8079	17,835	8.9
Smallest mature male	3.7	12.1	380	838	10.5	34.3	12,590	27,793	13.9
Largest mature male	5.0	16.5			14.3	46.8	33,947	74,939	37.5
Largest immmature female	4.9	16.1			13.9	45.6	31,240	68,963	34.5
Smallest mature female	4.7	15.4			13.3	43.8	27,370	60,418	30.2
Largest mature female	6.0	19.7	1241	2734	17.0	55.9	59,413	131,154	65.6
Largest mature female?	7.1	23.4	2300	5066	20.3	66.5	103,197	227,808	114.0
Mature 11.3 TL male[d]					11.3	37.2	16,313	35,889	18.0
Large female[e]					15.9	52.2	47,690	104,918	52.6

[a]Total lengths (TL) are based on the linear regression in Fig. 3.

[b]Body masses (BM) are based on the exponential regression in Fig. 4.

[c]The estimated lengths and mass for megatooth fetuses and neonates given here are speculative and may be excessive as scaled up from the white shark fetuses. Because smaller live-bearing sharks tend to have fewer and proportionately larger young than do larger species, it seems reasonable to assume that the megatooth shark had absolutely larger young than the white shark, but that megatooth young were smaller relative to the mother than is the case in the white shark, perhaps 200–300 cm TL and <250 kg mass.

[d]Calvert Marine Museum/South African Museum restoration.

[e]Based on 168-mm height of the first upper tooth (the second anterior, A-2).

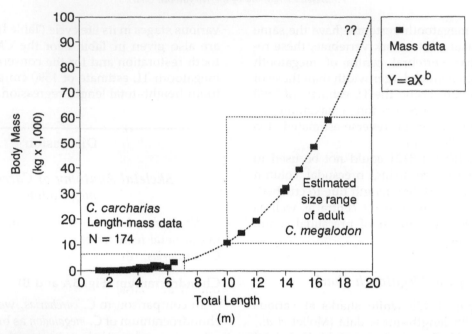

FIGURE 4 Exponential regression curve showing the length–mass relationship of *Carcharodon carcharias*, with extrapolated values for *Carcharodon megalodon* generated from linear regression of tooth size (see Fig. 2 and text).

er sclerotic capsules and eyes. In overall aspect, the megatooth chondrocranium has a blockier, more robust appearance than that of the white shark, reflecting the more massive jaws and dentition of *C. megalodon*.

Jaws (Fig. 5A and B)

As noted above, the very large and robust teeth of *C. megalodon* indicate that the megatooth jaws, in or-

der to functionally support such a massive dentition, would have been even more strongly developed than those of the white shark, which possess a formidable but somewhat more gracile dentition. In addition, *C. carcharias* jaws become more robust ontogenetically as the sharks grow larger and their teeth approach the proportions of those in *C. megalodon*. We believe that the existing reconstruction of the megatooth jaws (Fig. 1B), developed originally for the USNM, is not

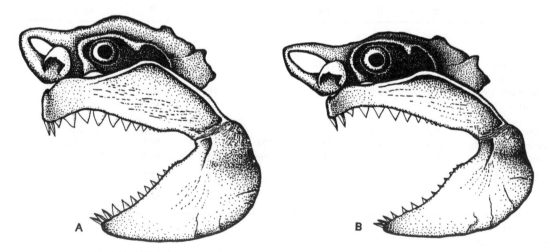

FIGURE 5 (A) Reconstructed chondrocranium and jaws of *Carcharodon megalodon* compared (not to scale) with (B) *Carcharodon carcharias* (based on LJVC 911219B). Note for *C. megalodon* the greater robustness and overall blockier appearance of the chondrocranium, the deeper jaws, and the relatively larger and more robust dentition.

deep enough through the lower jaw (Meckels cartilage), probably because this earlier reconstruction was based primarily on dried, and therefore somewhat distorted and misleading, *C. carcharias* jaws. Freshly dissected and alcohol-preserved wet specimens of *C. carcharias* have significantly deeper jaws than those in the USNM reconstruction, and thus we suggest that the megatooth jaws were more massive still. The CMM/SAM reconstruction incorporates noticeably deeper and more robust jaws. Along with the proportional inconsistencies, the USNM reconstructed jaws are opened to an exaggerated gape angle of approximately 100°. This probably resulted from modeling these jaws from dried *C. carcharias* specimens that were opened unnaturally wide for dramatic effect. This gape angle is well in excess of the maximum gape observed on photographs and films of *C. carcharias* feeding sequences. The gape angle of the modified jaws in our megatooth reconstruction will be approximately 75°, more in keeping with observations of living white sharks. In addition, close examination of the tooth arrangement on the original USNM reconstruction, and on CMM's copy, reveals problems with the arrangement of the dentition, including the incorrect placement of upper anterior and/or lateral teeth, and an intermediate tooth, on the lower jaw. The dentition on the USNM jaw was reconstructed prior to the discovery of important natural tooth sets at Lee Creek Mine, North Carolina, and in Florida, that help to clarify the arrangement of the teeth in *C. megalodon*. These problematic areas will be modified in the CMM/SAM reconstruction.

Axial Skeleton

Centra of *C. megalodon* are relatively uncommon but have been found in Miocene deposits along the east coast of North America (Fig. 1C), in Belgium (Leriche, 1926), and in Japan (Uyeno and Sakamoto, 1984). The most complete megatooth vertebral specimen is an associated column consisting of approximately 150 nearly complete to fragmentary centra (Fig. 6), collected from the Miocene of Belgium. These centra are mentioned but not described in detail by Leriche (1926), who also figures one centrum. They were on display for many years at the IRSNB in Brussels (IRSNB 3121) (Fig. 6). Leriche (1926, p. 427) wrote that the centra were recovered from Miocene sediments during construction of fortifications around Antwerp in the 1860s (van Beneden, 1882), but gave no specifics on the precise locality or the collector(s). IRSNB 3121 was recently examined by one of us (M.D.G.); it was clear, from the identical color, surface texture, degree of preservation, and regular size gradation, that the centra derived from a single individual, as was also suggested by Leriche (1926). The Miocene age, large size, and morphological similarity of the Belgian centra to those of *C. carcharias* support the assignment of this specimen to *C. megalodon*. In addition, teeth of *C. megalodon* are regularly found in the same Antwerp Basin deposits. The other Miocene lamnoid shark that could possess centra of comparable size is the basking shark *Cetorhinus* sp., but *Cetorhinus* centra are far fewer in number (109–116 in *Cetorhinus maximus* versus 170–187 in *C. carcharias*), are more elongate in proportion, and have thin and flaky centra walls, more prominent annular rings, and a fairly prominent perforation in the center of each centrum. These features readily distinguish basking shark centra from those of *Carcharodon* (Kozuch and Fitzgerald, 1989; Gottfried, 1995; L. J. V. Compagno, unpublished data). Whale shark *Rhiniodon* vertebrae are also very large, but are quite different in morphology from those of any large lamnoid, including *Carcharodon* and *Cetorhinus* (White, 1930).

The largest of the Antwerp centra, IRSNB 3121, centrum 4, measures 155 mm in diameter, corresponding to a shark of 920 cm TL (see above). This is well below the TL estimate of nearly 1600 cm derived

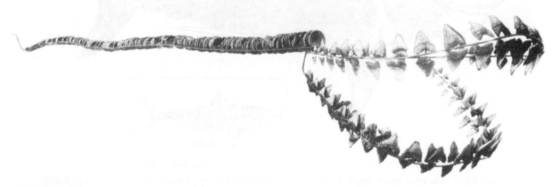

FIGURE 6 Associated vertebral column of *Carcharodon megalodon* (IRSNB 3121), as displayed at the Musée Royal d'Histoire Naturelle in Brussels; the "jaws" are a composite of nonassociated individual teeth.

from the largest teeth of *C. megalodon*, indicating that considerably larger megatooth centra may occur. One difficulty in interpreting these centra is that they are numbered 1–150, with the largest centra positioned anteriormost at the time they were mounted for exhibit. These numbers almost certainly do not reflect the order in which the centra occurred naturally in life. The largest centrum in the fossil series (centrum 4) is among the first as numbered, whereas in 15 white shark vertebral columns examined at the California Academy of Sciences (CAS), the widest centra are monospondylous precaudal centra numbered 37–66, posterior to the neurocranium. It is possible that part of the anterior portion of the column is missing in IRSNB 3121, but it is more likely that the centra have been arranged and numbered out of order. Furthermore, the smallest centrum of IRSNB 3121 (centrum 150), which has a diameter of 55 mm, is proportionately large enough that it could not have come from near the posterior tip of the vertebral column, although it might have come from near the caudal fin base. The width ratio of the last precaudal diplospondylous centrum to the widest precaudal centrum in the 15 white sharks at CAS (expressed as a percentage) is 60–69% (mean 63.9%), whereas the width ratio of IRSNB 3121, centrum 150, to the widest centrum (number 4) is 36%. Using the mean ratio of last precaudal centrum width to widest precaudal centrum (63.9%, SD = 2.5, range 60–69%), we can estimate the width of the last precaudal centrum in IRSNB 3121 from the widest precaudal centrum (63.9% of 155 mm) as 99 mm (93–107 mm). Given the variation in

this size ratio in living white sharks, IRSNB 3121, centrum 142, might be approximately equivalent to the last precaudal centrum, with the remaining smaller centra being caudals. The caudal fin in white sharks incorporates 66–84 progressively smaller centra, which become quite tiny at the posterior tip of the fin. Whatever the precise position was of the smallest centra in IRSNB 3121, these observations indicate that *C. megalodon* had a significantly higher vertebral count than the 170–187 total and 101–108 precaudal centra present in *C. carcharias* (Compagno *et al.*, 1996). *Carcharodon megalodon* probably had well over 130 precaudal centra and 200+ total centra if the Belgian specimen is, in fact, missing a part of the anterior column as well as many of the caudal centra.

Scaling and Proportions of the Megatooth

Our skeletal reconstruction of *C. megalodon* (Fig. 7) conforms to a very robust shark that is more massively proportioned, particularly in the head region, than the living white shark. In comparison to living white sharks, the head of the megatooth is reconstructed as being blunter and wider (signifying correspondingly stouter jaws and greater adductor musculature mass to power them), and with a somewhat pig-eyed appearance (based on the allometric decrease in the size of the orbit as living white sharks grow larger). The fins are interpreted as being quite similar, although proportionately slightly larger than those of living white sharks [fin growth in *C. carcharias* is approximately isometric (L. J. V. Compagno,

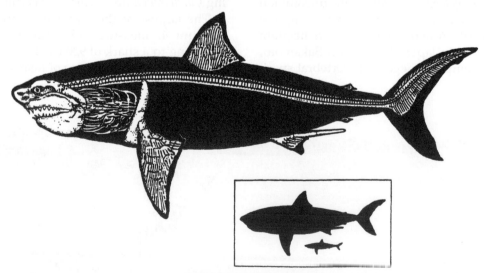

FIGURE 7   Skeletal reconstruction of *Carcharodon megalodon* (approximately 1100 cm TL). (Inset) Body outline of an 1100 cm *C. megalodon* (drawn to scale) with a 300 cm TL *Carcharodon carcharias* to show proportional differences.

unpublished data)]. We have interpreted the megatooth fins as being slightly larger and thicker in proportion because the megatooth shark is so massive that it is reasonable to assume relatively larger fins for propulsion and control of its movements.

Paleobiological Considerations

The megatooth reconstruction and size estimates discussed here naturally lead to speculation on the paleobiology of these enormous sharks (see Chapter 4, by Applegate and Espinosa-Arrubarrena). The nearly panoceanic occurrence of *C. megalodon* in nearshore deposits at subtropical to moderately high-temperate latitudes indicates that it occurred in environmental conditions broadly similar to those favored by the living species, *C. carcharias*: biologically rich nearshore and coastal shelf areas where large prey are relatively abundant. Megatooth sharks may also have occurred in other environments, including more tropical habitats, that have gone unrecognized due to preservational and/or collecting biases.

The presence of bite marks made by large serrated teeth on fossil cetacean remains (e.g., Demèrè and Cerutti, 1982; Cigala-Fulgosi, 1990) indicates that whales and dolphins were regular prey and/or scavenging targets for fossil *Carcharodon*, as they are today for *C. carcharias* (Compagno, 1984a; Applegate and Espinosa-Arrubarrena, Chapter 4; Long and Jones, Chapter 27). A survey of the diet of white sharks along the Natal coast of South Africa (Cliff *et al.*, 1989) shows that fish and other sharks are actually more common prey items than marine mammals for smaller individuals, and that white sharks shift more to larger marine mammals as the sharks grow larger, while not confining their efforts to such prey [a pattern consistent with field observations from other parts of the white shark's range (e.g., Klimley, 1985b). Klimley *et al.* (Chapter 16) suggest that white sharks select prey on the basis of high fat content, a pattern that accounts for their preference for blubber-rich marine mammals. One possible implication of these observations is that the giant adult megatooth sharks may have relied on large marine mammals for an even greater percentage of their diet than do living white sharks.

A seeming anomaly in the above speculation, which implies that the distribution of megatooth sharks was closely linked to the presence of cetaceans, pinnipeds, and possibly other large marine vertebrates as food resources, is the observation that megatooth sharks apparently became extinct before the end of the Pliocene. By the Pliocene, the largest cetaceans—the balaenopterid and balaenid mysticetes—were well established, along with diverse odontocetes, pinni-

peds, and other large marine animals that would have been suitable megatooth prey. The extinction of *C. megalodon* near the end of the Cenozoic was, therefore, likely due to another factor, or combination of factors, such as climatic change, competition from large odontocetes, or a distributional shift of large marine mammals to colder, high-latitude conditions for which the megatooth was unsuited.

Summary

The giant "megatooth" shark *C. megalodon* (Lamnidae) is represented in the late Cenozoic fossil record by large distinctive teeth and occasional vertebral centra. Dental characters support assignment of this form to *Carcharodon*, not *Carcharocles*, as some authors have maintained. We made inferences on the skeletal anatomy of *C. megalodon* on the basis of comparisons with ontogenetic trends in the Recent species, *C. carcharias*, for the purpose of exhibiting a skeletal reconstruction of the megatooth. We inferred that, in comparison to *C. carcharias*, *C. megalodon* had a broader chondrocranium with a more domed cranial roof and a shorter rostrum, less elevated orbits, more massive jaws coupled with a more robust dentition, a higher vertebral count (170–187 versus 200+, respectively), and proportionately larger fins. Based on the largest available teeth, we conservatively estimate a maximum TL of 1590 cm for adult female *C. megalodon*.

Acknowledgments

The staff of the SAM Department of Marine Biology, particularly M. van der Merwe and R. Alexander, provided much help during the course of this project. J. F. Langley (CMM) assisted in producing the reconstruction and very skillfully crafted the scale model of the skeleton. R. Purdy (USNM) generously shared information he has gathered, discussed many aspects of fossil and living sharks, and graciously allowed us to mention observations from his research. The Natal Sharks Board (Umhlanga Rocks, Natal Province, South Africa) cooperated in providing specimens for dissection or measurement. Allowing us to examine specimens in their care were D. Nolf (IRSNB, Brussels); A. E. Sanders (Charleston Museum, Charleston, South Carolina); the late W. I. Follett, the late L. Dempster, W. D. Eschmeyer, P. Sonoda, and D. Catania (CAS); and T. Uyeno (National Museum of Science, Tokyo). We are grateful to A. P. Klimley (Bodega Marine Laboratory) and D. G. Ainley (PRBO International Biological Research) for organizing and inviting us to participate in the 1993 Biology of the White Shark Symposium at Bodega Marine Laboratory and in the proceedings volume. In addition to them, helpful comments and input were provided by S. P. Applegate (Instituto de Geología, Mexico City), G. Cliff (Natal Sharks Board), H. Fink (CMM), G. Hubbell (Key Biscayne, Florida), and P. Rasmussen (USNM), as well as by two anonymous reviewers.

Financial support has been provided by a partnership grant to the CMM and the SAM from the American Association of Museums' International Partnership Among Museums program and by an exhibit planning grant to the CMM from the Maryland Historical and Cultural Museums Assistance Program. Additional support from the Calvert Marine Museum Society, the Department of Marine Biology at the SAM (including the Shark Research Center), and the South African Foundation for Research Development is gratefully acknowledged.

Paleoecology of Fossil White Sharks

ROBERT W. PURDY

Department of Paleobiology
National Museum of Natural History
Washington, D.C.

Introduction

Does the Tertiary fossil record provide adequate evidence about the evolution of species in the genus *Carcharodon*? I believe not. Until now, the interpretations of the fossil history of *Carcharodon* were based on assumptions that the fossil record of these sharks is represented by remains of adults, that those species not present in the sediments of the different Tertiary epochs were extinct or had not yet evolved, and that these sharks were restricted to warm waters. In the past 30 years, studies of the ecology and behavior of the extant *Carcharodon carcharias* and new discoveries in the fossil record suggest that these assumptions are wrong.

Elton (1927) recognized the importance of ecology in interpreting the life history of an animal:

In solving ecological problems we are concerned with what animals do in their capacity as whole, living animals, not as dead animals or as a series of parts of animals. We have next to study the circumstances under which they do these things, and most important of all, the limiting factors which prevent them from doing certain other things. By solving these questions it is possible to discover the reasons for the distribution and number of animals in nature.

Elton's observations about the importance of ecology, as Long (1992a) also noted, apply as well to the study of the fossil record. Until now, the life history of fossil *Carcharodon* has remained unknown, because our basis for interpreting it, the life history of the extant species, has only recently become known. The paleontological evidence, reviewed in light of the extant

species' ecology, achieves new significance and begins to explain the distribution and faunal association of the fossil species.

What inferences can we make about the fossil species from the behavior and distribution of the extant species *C. carcharias*? In the extant species, which Compagno (1984a) stated is coastal and mostly amphitemperate, I looked for ecological patterns and behavior that might also be apparent in the fossil record. Among the patterns and behaviors that we know, the following questions may be answered on the basis of the fossil record or may have important bearing on the distribution of the fossil species.

1. *Carcharodon carcharias* >3 m in total length (TL) prefers to prey on pinnipeds and whales rather than on fish, so they inhabit areas where these prey are abundant (Arnold, 1972; Klimley, 1985a; Pratt *et al.*, 1982). Is the distribution of the fossil species related to that of fossil whales and pinnipeds?

2. Off the California coast, white sharks move south of Point Conception to bear their young; as the young reach adulthood, they move northward (Klimley, 1985a). Is there fossil evidence for nursery areas?

3. The modified circulatory system, which permits *C. carcharias* to retain body heat (Carey *et al.*, 1982), enables it to frequent cold waters, such as the Bering Sea (Cook and Brzycki, 1986, personal communication; Goldman *et al.*, Chapter 11). When did *Carcharodon* adapt to cold water? What influence would this have on its fossil record?

4. When white sharks 4–5 m TL are feeding on a

whale carcass, they abandon it when a larger individual approaches (Pratt *et al.*, 1982). Would small-toothed species of *Carcharodon* avoid areas frequented by larger giant-toothed species?

Materials and Methods

This chapter is based on a review of the literature of the occurrences of fossil marine mammals and of fossil *Carcharodon* spp., and of the collections of Gordon Hubbell (Key Biscayne, Florida), the National Museum of Natural History (USNM), the Charleston Museum, and the Florida State Museum (Fig. 1). The localities listed in Fig. 1 were limited to those (1) which were heavily collected over a long period, (2) in which the relative abundance of *Carcharodon* and marine mammals could be estimated, and (3) in which the stratigraphic origin of the specimens was documented.

Certain limitations hamper the interpretation of these data. First, most of the sediments bearing these fossil remains have not been bulk-sampled, and there-

+ - MANY FOUND
□ - SOME FOUND
R - RARE
O - NONE FOUND

Epoch	Locality	TROPICAL	SUBTROP	WARM TEMP	COOL TEMP	UPWELLING	Small JUV	Small ADULT	Giant JUV	Giant ADULT	XIPHODON	ARCHAEO	ODONTO	MYSTI	PINNIP	PALEONTOLOGIC & PHYSICAL EVENTS
PLEISTO	29 Champlain Sea				X		O	O	O	O	O		+	+	O	START OF GLACIATION OF N. AMERICA & EUROPE
PLEISTO	28 Waccamaw Fm, NC				X		O	□	O	O	O		R	O	R	
PLEISTO	27 Ridgeville, SC			X			O	O	O	O	+		+	+	O	
PLIOC	26 Raysor Fm, SC			X			O	R	O	O	O		R	R	O	CLOSURE OF MIDDLE AMERICAN SEAWAY
PLIOC	25 Goose Creek Fm, SC			X		X	O	R	O	O	O		+	O	O	
PLIOC	24 Yorktown Fm, NC			X		X	O	R	R	+	+		+	+	+	
PLIOC	23 Belgium Scaldisian				X	X	O	+	O	O	R?		+	+	+	
MIOCENE	22 Belgium Diestian		X				O	R	O	+	+		+	+	R	FIRST APPEARANCE OF MODERN MYSTICETI
MIOCENE	21 Bone Valley Fm, FL		X				O	O	+	R	+		+	□	□	
MIOCENE	20 Eastover Fm, VA		X			X	O	O	O	+	+		+	+	R	
MIOCENE	19 Choptank-St. Mary's Fm, MD		X				O	O	O	□	□		R	□	R	
MIOCENE	18 Belgium Anversian		X				O	R	□	+	+		+	+	□	
MIOCENE	17 Calvert Fm, MD		X				O	R	+	□	+		+	+	□	
MIOCENE	16 Gatun Fm, Panama		X				O	R	O	R	O		?	?	?	
MIOCENE	15 Pungo River Fm, NC		X			X	O	O	+	O	R		+	O	R	
MIOCENE	14 Calvert Fm, DE		X				O	O	□	O	□		□	O	R	
MIOCENE	13 Belgrade Fm, NC		X				O	O	□	O	O		□	O	O	
OLIG	12 Chandler Bridge Fm, SC		X				O	R	+	□	O		+	+	R	FIRST APPEARANCE OF PINNIPEDS & ODONTOCETES
OLIG	11 Belgium Rupelian		X				O	O	?	+						
EOCENE	10 Twiggs Clay, GA	X					O	O	□	O		□				START OF CIRCUMPOLAR ANTARCTIC CURRENT & STRATIFICATION OF OCEAN
EOCENE	9 Castle Hayne Fm, NC	X					O	□	□	□		R				
EOCENE	8 Santee Fm, SC	X					O	O	□	□		R				
EOCENE	7 Jackson Fm, AL	X					O	O	?	+		+				FIRST APPEARANCE OF MYSTICETES
EOCENE	6 Fayum, Egypt	X					O	O	?	+		+				
EOCENE	5 Barton Clays, England	X					?	+	?	+		□				FIRST APPEARANCE OF ARCHAEOCETES
EOCENE	4 Belgium Bruxellian	X					O	O	?	+		O				
PALEOC	3 Morocco	X					+	+								
PALEOC	2 Aquia Fm, MD	X					□	□								
PALEOC	1 St. Stephens, SC	X					□	□								

FIGURE 1 Localities at which the relative abundance of fossil marine mammals and *Carcharodon* are known [paleontological and physical events taken from Whitmore (1994)]. Localities are identified only by formation or by stratigraphic age. WATER TEMP., Water temperature (otherwise, TEMP refers to temperate); SUBTROP, subtropical; JUV, juvenile; MAR MAM, marine mammals; ARCHAEO, Archaeoceti; ODONTO, Odontoceti; MYSTI, Mysticeti; PINNIP, Pinnipedia; PLEISTO, Pleistocene; Fm, Formation; PLIOC, Pliocene; OLIG, Oligocene; PALEOC, Paleocene.

fore the relative abundance of these taxa is based on their representation in the above collections. Second, because bulk-sampling techniques have not been applied, rare taxa, such as the small-toothed species of *Carcharodon*, are underrepresented in museum collections. Finally, in my search for bite marks on fossil marine mammal bones, *Carcharodon* did not leave many telltale signs of predation; bones with serrated tooth marks are not common.

The shark teeth were measured as follows: *height* represents the distance from the apex of the crown to a line tangent to the basal lobes of the root. The tooth terminology used here is that of Applegate (1965) and Compagno (1988). Distributions of sharks were taken in part from the work of Cappetta (1987).

Taxonomy

Before considering the paleoecology of these sharks, I should remark about their taxonomy and history. For reasons listed below, I include in the genus *Carcharodon* fossil species that were recently assigned to other genera by some paleontologists; a more thorough discussion about the taxonomy and taxonomic history of these sharks will be contained in a future paper (see also Chapter 4, by Applegate and Espinosa-Arrubarrena). First, the first tooth in the upper jaw, the second anterior (Compagno, 1990a) (the first anterior tooth of Applegate and Espinosa-Arrubarrena, Chapter 4), normally asymmetrical in other lamnoids, is usually symmetrical in both the fossil and extant species (see also Chapter 3, by Hubbell). Second, this same tooth, rather than the second lower anterior tooth, as in all other lamnoids except *Cetorhinus* and the alopiids, is the largest tooth in the dentition. Both characters are unique to *Carcharodon*.

The taxonomic identity of several of the fossil species is not yet resolved. Thus, for ease of discussion, I refer here to the large species of *Carcharodon*, including *C. "auriculatus"* (=*C. sokolowi*), *C. angustidens*, *C. subauriculatus*, and *C. megalodon*, as the giant-toothed species (or giant-toothed line), and those species with compressed small teeth (as in the extant species) I refer to as the small-toothed species (or small-toothed line).

Evolutionary History

About 60 million years ago, when the seas were warm even at high latitudes (Berggren and Hollister, 1974), the ancestor of the extant white shark and the extinct giant-toothed great white sharks (Fig. 2) first appeared in the Paleocene seas of southern Russia, Morocco, Angola, and the United States (Arambourg,

1927; Arambourg and Signaux, 1952; Casier, 1960a; Sintsov, 1899). This first white shark, *Carcharodon orientalis*, had teeth within the size range of the extant species, and except for the presence of lateral cusplets in the adult teeth, the teeth of these two species are very similar.

During the early Eocene, in addition to the small-toothed species, a giant-toothed species evolved and had a worldwide tropical distribution; the latter species is usually called *C. auriculatus*. Applegate and Espinosa-Arrubarrena (Chapter 4) and Case and Cappetta (1990) determined that this species belongs to the small-toothed line; these authors assign the giant-toothed species form to *C. sokolowi*, and the former species was recently named *Carcharodon nodai* (=*C. auriculatus*) (Yabumoto, 1989). Adults of both species lines possess lateral cusplets; in both, cusplets were not lost until the middle Miocene.

The teeth of Eocene and Oligocene giant-toothed species are usually found in the same beds as the remains of cetaceans. None of these teeth exceed 13 cm in height. In the Miocene, about 15 million years ago, the large-toothed line culminated in its largest form, *C. megalodon*. It was not until the Pliocene, however, that teeth of this species became very common in the deposits of the Atlantic coastal plain of North America. In the marine Pliocene outcrops of Europe, this species is rare or absent. The purported earliest specimens of this species, about 10,000–40,000 years ago, were dredged from late Pleistocene sediments off Madagascar (Roux and Geistodoerfer, 1988).

The small-toothed line is rare in the fossil record from the Eocene to the Miocene. In the North American Atlantic coastal plain, teeth of these species were recently discovered in the early Eocene Nanjemoy Formation (Fig. 3A), the middle Eocene Castle Hayne Formation (Fig. 3B), the late Oligocene Chandler Bridge Formation (Fig. 3C), and the middle Miocene Calvert Formation (Fig. 3D); in Baja California, they occur in the middle Miocene Rosarito Beach Formation (Vazquez and Zubillaga, 1989) and in Switzerland, in the middle Miocene Molasse (Leriche, 1927). Gillette (1984) identified *C. carcharias* as occurring in the middle Miocene Gatun Formation of Panama, but Vokes (1970, 1992), on the basis of gastropods, determined the age of this formation to be early Pliocene. (Since Vokes' reassignment of the Gatun Formation was not discovered until after completion of this study, the Gatun Formation is still listed in the Miocene portion of Fig. 1.) In 1861, Michelotti also described a new species (*Carcharodon gibbesi*) of this line from the early Miocene of Italy (Fig. 3E). In the early Pliocene, about 5 million years ago, except in the west-

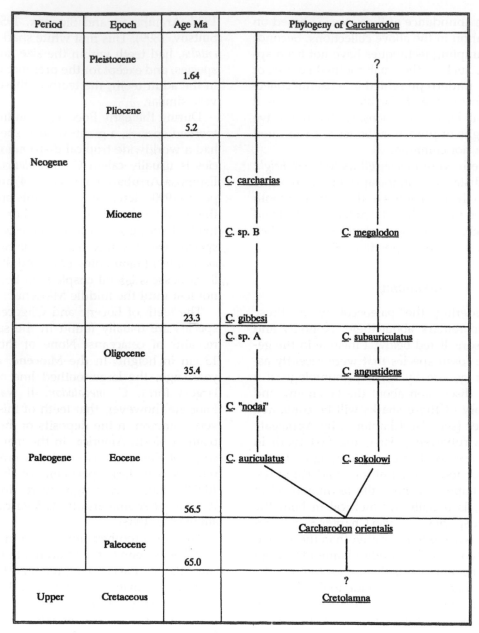

Period	Epoch	Age Ma	Phylogeny of <u>Carcharodon</u>
Neogene	Pleistocene	1.64	?
	Pliocene	5.2	
	Miocene	23.3	<u>C. carcharias</u> <u>C.</u> sp. B <u>C. megalodon</u> <u>C. gibbesi</u>
Paleogene	Oligocene	35.4	<u>C.</u> sp. A <u>C. subauriculatus</u> <u>C. angustidens</u>
	Eocene	56.5	<u>C. "nodai"</u> <u>C. auriculatus</u> <u>C. sokolowi</u>
	Paleocene	65.0	<u>Carcharodon orientalis</u>
Upper	Cretaceous		? <u>Cretolamna</u>

FIGURE 2   Cenozoic time chart with a phylogenetic scenario for *Carcharodon*.

ern North Atlantic, the small-toothed species suddenly became more abundant (Leriche, 1926, 1936).

Results

Prey Relationships

Was the distribution of the fossil white sharks related to the availability of prey? Since shark teeth are so abundant in the fossil record, paleontologists as-

sumed that the fossil record for Tertiary sharks was fairly complete. In the past 20 years, studies of the behavior and distribution of the extant white shark, however, suggest that this assumption is erroneous, that these sharks have patchy distributions in the seas [for fossil species, see Cione and Reguero (1994), Long (1992a), and R. W. Purdy *et al.* (unpublished data)], and that their distribution may be governed by the availability of prey.

In the fossil record, confirming this relationship in distribution between predator and prey is difficult,

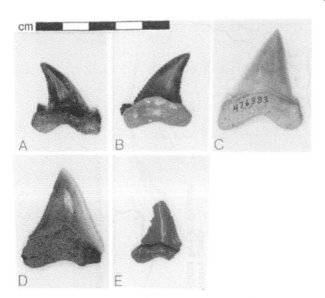

FIGURE 3 Carcharodon teeth of the small-toothed species line. (A) Carcharodon auriculatus: lingual view of a tooth from the Nanjemoy Formation, Popes Creek, Maryland (USNM 25197). (B) Carcharodon auriculatus: lingual view of a tooth from the Castle Hayne Formation, Rose Hill, North Carolina (USNM 392102). (C) Carcharodon sp. A: lingual view of a tooth from the Chandler Bridge Formation, Summerville, South Carolina (USNM 476833). (D) Carcharodon sp. B: lingual view of a tooth from the Calvert Formation, Kaufman Camp, Maryland (USNM 336204). (E) Carcharodon gibbesi, cast of type: lingual view of Michelotti's tooth from the lower Miocene of Italy (USNM 476379). All ×1.

for two reasons. First, the record of these sharks begins in the Paleocene, and at this time, before whales had evolved, their distribution was extensive in the Atlantic basin. Second, in Eocene and Oligocene deposits, fossil marine mammals, which are not common fossils in these sediments, occur in areas already inhabited by white sharks. Despite these weaknesses in the fossil record, fossil evidence indicates a possible relationship between the occurrence of marine mammals and white sharks.

Giant-Toothed Carcharodon

In the Eocene, a correlation seems to exist between the abundance of archaeocete remains and those of the giant-toothed Carcharodon teeth (see Figs. 1 and 4). The remains of these sharks and large archaeocetes are most abundant in Alabama and Mississippi as well as in the Fayum of Egypt (Case and Cappetta, 1990; Gingerich, 1992). At this time, both of these areas were covered by warm epicontinental seas, the former being an arm of the Gulf of Mexico and the latter a part of the Tethys Sea. Elsewhere in sediments of equivalent age and water temperature, the remains of both are less common. These occurrences suggest

a possible predator–prey relationship between the giant-toothed Carcharodon and large archaeocetes. The Belgian Bruxellian Carcharodon apparently existed before archaeocetes appeared.

In the Oligocene, with the exception of the Chandler Bridge Formation of South Carolina and the Astoria Formation of Oregon, the fossil record for marine mammals and giant-toothed Carcharodon is scant. In the Oligocene, the inundation of the continental margins was not as great as during either the Eocene or the Miocene. Of the two exceptions named, only the Chandler Bridge faunas have been studied (A. Sanders and L. G. Barnes, unpublished data; R. W. Purdy, unpublished data); in this fauna, toothed whales, small mysticete whales, and juvenile giant-toothed Carcharodon are abundant, but adults of these sharks are rare and large mysticete whales are absent. The seas may have cooled, as the presence of phosphate in these sediments (Ward et al., 1979) indicates upwelling (Riggs, 1984). Teeth of these sharks appear to be common in the early Oligocene (Rupelian) sediments of Belgium (Leriche, 1910), but fossil whales from these sediments have yet to be reported.

The Miocene record, particularly in the Belgian Anversian (middle Miocene) and Diestian (latest Miocene) (Leriche, 1926; Whitmore, 1994), the Calvert Formation (lower to early middle Miocene) of Maryland and the Eastover Formation (late Miocene) of Virginia, and the Yorktown Formation (early Pliocene) at Lee Creek (see Figs. 1, 5, and 6) indicate that where large mysticetes (as well as large sperm whales in the Anversian and Yorktown) are abundant, giant-toothed Carcharodon are also abundant. It is from the middle Miocene through the early Pliocene (Fig. 6) that the fossil record indicates the greatest diversification of marine mammals. It is at this time also that the cooling of the North Atlantic commenced (Whitmore, 1994), a trend consistent with increased productivity of the ocean. This increased productivity may account for the abundance of the remains of large whales and giant-toothed Carcharodon in these sediments.

The converse also seems to be true. In the Bone Valley Formation (late Miocene) of Florida, the Belgrade and Pungo River formations (both early Miocene) of North Carolina, and the Choptank and St. Mary's formations, which were deposited in subtropical seas (both middle Miocene) of Maryland, mysticetes and other large whales and adult giant-toothed Carcharodon are rare or absent. In the Bone Valley and Pungo River formations, however, which were subtropical areas of upwelling (Riggs, 1984), juvenile giant-toothed Carcharodon and odontocetes are common. It may be, however, that these upwelled

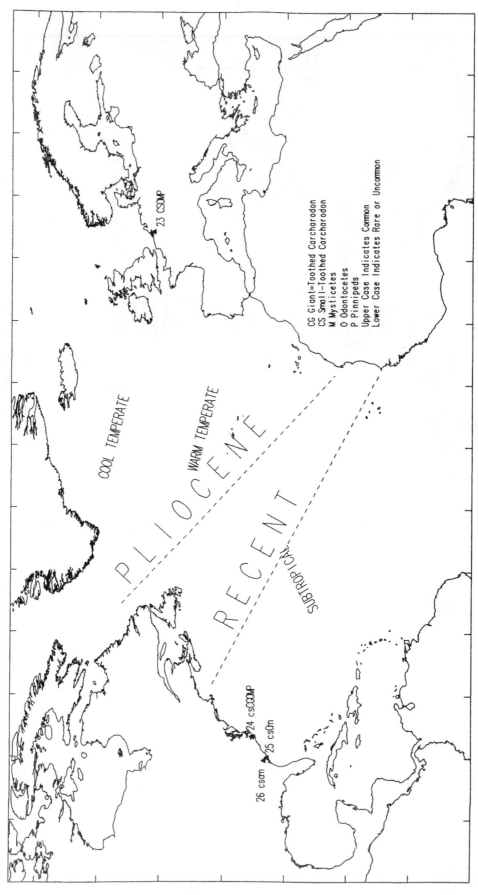

FIGURE 6 The occurrence of Pliocene *Carcharodon*, marine mammals, and the Pliocene and Recent (summer) boundaries between subtropical and temperate waters (see Fig. 1 for identification of locality numbers).

waters did not reach the surface, where large mysticetes fed (Whitmore, 1994). In the Belgrade, Choptank, and St. Mary's formations, there is no evidence of upwelling. This suggests that areas of upwelling were important feeding areas for juvenile giant-toothed *Carcharodon*, and for juvenile *Carcharodon*, an important prey may have been small whales.

At Lee Creek in the Pliocene Yorktown Formation, giant-toothed *Carcharodon*, along with mysticete whales and sperm whales, occurs more abundantly than in any of the Miocene formations mentioned above. This appears to be the first record of a high-use area in the fossil record of marine vertebrates; such areas occur today along the edge of the continental shelf between Cape Hatteras and the Gulf of Maine (Kenney and Winn, 1986; Payne *et al.*, 1984). Although the Lee Creek locality is not at the edge of the continental shelf, a submarine embayment extended from the edge of the continental shelf to the mouth of the Pamlico River; it is near the mouth of the Pamlico, near the head of this submarine embayment, that the Lee Creek Mine is located. This submarine embayment permitted an eddy of the Gulf Stream to drag cold upwelled water farther onto the continental shelf (Popenoe, 1985; Riggs, 1984). This nutrient-rich water supported one of the largest marine fossil vertebrate faunas ever found (R. W. Purdy *et al.*, unpublished data). In addition to warm-water forms following the Gulf Stream, such as blue marlin *Makaira nigricans*, tiger *Galeocerdo* sp., and mako sharks *Isurus* spp., abundant also were cold-water fish, such as hake *Merluccius bilinearis* and tilefish *Lopholatilus* sp. Tunas, sea birds, monk seals *Monachus*, toothed whales, including oceanic sperm whales, one species of cetothere, and several modern baleen whales were abundant as well as large *C. megalodon*. Anterior teeth of the latter were often >13 cm in height.

The stratigraphic distribution of the fossil marine vertebrates and our collection experience for the past 23 years at Lee Creek indicate a correlation between the giant-toothed *Carcharodon* and high-use areas of marine vertebrates. First, this fauna is restricted to the basal lithological unit of the Yorktown Formation, the only Yorktown unit with evidence of upwelling—the presence of phosphate (Riggs, 1984). In the beds immediately overlying this basal unit, seals, toothed whales, makos, and carcharhiniform sharks still appear, but mysticete whales, sperm whales, and *C. megalodon* are rare or absent. The eddy of the Gulf Stream was, evidently, no longer strong enough to bring upwelled water into the surface waters in the Lee Creek area of the Yorktown sea.

Second, in the basal unit of the Yorktown Formation, the lateral distribution of large fossil whale re-

mains and *C. megalodon* has changed. As phosphate mining has progressed southward, away from the river and the Pliocene submarine embayment, the remains of these fossil vertebrates became less abundant. Today, with mining about 1 mile from the river, the remains of these fossil vertebrates are rare.

Associated with this rich fauna is more concrete evidence for a predator–prey relationship between marine mammals and giant-toothed *Carcharodon*; this evidence is large whale bones, principally tail vertebrae and flipper bones, bearing bite marks made by very large serrated teeth that fit the size of *C. megalodon* (Fig. 7). These occurrences and shark-bitten whale bones suggest that adult *C. megalodon* preferred cetacean high-use areas and fed on large whales.

I should note here that in the Scaldisian (early Pliocene) of Belgium, although the remains of mysticetes are abundant, the teeth of giant-toothed *Carcharodon* are rare or absent. This may have been due to the cooler temperatures of the northern seas (see below).

The Lee Creek evidence and the data in Fig. 1 indicate that juvenile giant-toothed *Carcharodon* preferred areas where small whales were abundant and that the adult giant-toothed *Carcharodon* line preferred areas with abundant large whales;, these data also suggest that these preferences may have developed as early as the middle Eocene.

Although mysticetes and sperm whales persist to the present time, before the late Pliocene, the giant-toothed *Carcharodon* disappeared from the North Atlantic.

Small-Toothed Carcharodon

Although *C. carcharias* does not feed solely on pinnipeds, Ainley *et al.* (1981, 1985; Long *et al.* and Chapter 24) showed that they are important prey for white sharks living off the California coast and that their abundance is directly related to that of pinnipeds at certain localities). Additional evidence is contained in the work of Compagno (1984a), Klimley (1994), and Klimley *et al.* (1992).

Although tropical monk seals were abundant in the Lee Creek fauna, the remains of *C. carcharias* are very rare. Another lamnid shark, a large wide-toothed mako *Isurus xiphodon* (R. W. Purdy *et al.*, unpublished data) having teeth convergent with those of the white shark may have been the principal predator on these seals. Two seal specimens from Lee Creek exhibit bite marks, including the tip of a lower anterior tooth, made by this mako. In the early Pliocene of Europe, however, *C. carcharias*, *I. xiphodon*, and monachine and phocine seal remains are abun-

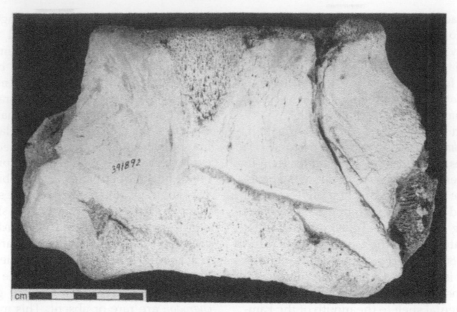

FIGURE 7 Baleen whale caudal vertebra bitten by *Carcharodon megalodon* (USNM 391892) from the Pliocene of North Carolina.

dant (Figs. 1 and 6) (Leriche, 1927; Ray, 1976). The greater abundance of phocine seals, important prey for the extant species of white shark, may account for the abundance of *C. carcharias*; likewise, in the early Pliocene at Lee Creek, the rarity of phocine seals may account for the rarity of *C. carcharias*.

In view of the feeding habits of the extant *Carcharodon*, these few examples from the fossil record indicate a positive relationship between fossil marine mammals and the small-toothed line of *Carcharodon*. From the middle Eocene, small whales and in the late Oligocene pinnipeds (C. E. Ray, personal communication) were available as prey for the small-toothed *Carcharodon*. It is possible, then, that marine mammals became important prey for these sharks as early as the Eocene. If this is true, the availability of fossil marine mammals must have influenced the distribution of *Carcharodon*, and this raises some questions. In Paleogene sediments, is the scarcity of small-toothed *Carcharodon* related to the scarcity of small whales? If marine mammal high-use areas are rarely represented (or accessible) in the fossil record, how well do we know the distribution of the giant-toothed line? Further systematic collecting of the available Tertiary marine sediments may shed some light on these questions, but they may not be answered until collecting on the continental shelf becomes possible.

Nursery Areas

Is there evidence of nursery areas for fossil *Carcharodon*? In the fossil record, the first evidence for a

nursery area is found in the late Oligocene Chandler Bridge Formation of Summerville, South Carolina; here, Albert Sanders (Sanders, 1980) of the Charleston Museum excavated a bone bed that contained about a dozen primitive odontocete and small mysticete skulls and about 100 *Carcharodon* teeth of juvenile giant-toothed species. The anterior teeth measured 4.4–6.2 cm in height (mean 5.2, n = 8); these, as well as the laterals, possess lateral cusplets (Fig. 8). Only a few teeth of adults were found. These were without lateral cusplets, and one of these is equal in size to the teeth from some of the largest individuals of *C. megalodon* from the early Pliocene of North Carolina. This

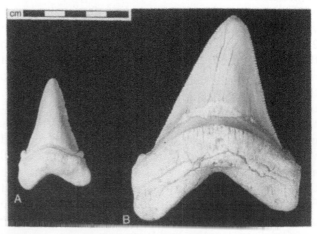

FIGURE 8 Teeth of (A) a juvenile and (B) an adult of the giant-toothed *Carcharodon* from an Oligocene nursery area near Charleston, South Carolina.

occurrence suggests that South Carolina was a nursery area for the giant-toothed white shark.

Casey and Pratt's (1985) observations on the distribution of *C. carcharias* in the western North Atlantic indicate a contrast between the nursery areas used by the giant-toothed and small-toothed lines. In the Middle Atlantic Bight, they noted that more small and intermediate-sized *C. carcharias* sightings occurred between Cape Hatteras and Cape Cod (35°00′ N, 43°00′ N) than in any other portion of the region; this portion is also a high-use area for cetaceans. Casey and Pratt suggested, "Apparently, young sharks have a lower tolerance for cooler waters that limits their distribution north of Cape Cod, and they may have an intolerance to higher temperatures that limits their distribution off the southeastern United States. The occurrence of small and intermediate size white sharks in continental shelf waters of the Middle Atlantic Bight suggests this area served as a nursery area for juveniles." During the early Pliocene, comparable nursery waters for this species would have been north of Labrador. This contrasts with the case of *C. megalodon*, whose nursery areas, at least in the western North Atlantic, were in warm waters south of Cape Hatteras.

Distribution

What effect did the cooling of the oceans have on the distribution of *Carcharodon*? At the end of the Eocene, Antarctic circumpolar circulation and the northward migration of cold water in the southern oceans began (Savin *et al.*, 1975). More cold waters increased the productivity of the oceans, increasing the number of whales and fish. If, as in the extant species, the fossil species possessed a circulatory system adapted for conserving body heat, it would have permitted both species lines of *Carcharodon* to follow their prey, archaeocete whales and probably basking sharks, into cooler waters at higher latitudes. If at this time the small-toothed species line developed its preference for feeding in cooler waters, this would account for the rarity of its teeth in the Tertiary fossil record, which represents predominantly warm-water environments. It is not until the late Miocene or early Pliocene, about 24 million years later, that cooler waters are represented in the fossil record, resulting in the greater abundance of the small-toothed line.

Judging from the available fossil evidence, in the late Eocene, the giant-toothed species, along with archaeocete whales (Mitchell, 1989), inhabited the temperate waters of Antarctica (Welton and Zinsmeister, 1980; Long, 1992), but whether or not these sharks continued to live in these waters as they became more boreal is not preserved in the fossil record. The available data indicate that they preferred warm seas (Fig. 1).

The abundance of *C. carcharias* in the Pliocene may be related to the spread of cold waters toward the Equator (Fig. 5). In the Miocene Atlantic Ocean, subtropical waters reached to the Denmark Strait, Iceland, and the Norwegian Sea; by the early Pliocene, this boundary had retreated to a line from Spanish Sahara to Labrador (Cifelli, 1979). Leriche (1927) noted that in the early Pliocene of Europe, a cool-water environment, *C. carcharias* seems to replace *C. megalodon*. The Miocene occurrences of the small-toothed *Carcharodon* are also in areas where cold upwelled waters and marine mammals were present. If these occurrences are not coincidental, the small-toothed *Carcharodon* preferred cooler marine environments than are generally preserved in the fossil record.

Segregation by Size

Segregation by size occurs in the extant species and may also have affected the distribution of the small species line. McCosker (1985) and Pratt *et al.* (1982) observed white sharks feeding on large floating whale carcasses (see also Chapter 23, by Barlow). They noted that no more than two white sharks were feeding together, and these disappeared suddenly if a larger individual approached. The smaller white sharks did not return to feed until the large one left. McCosker suggested, " . . . feeding adults exclude other sharks and smaller conspecifics."

It is not unreasonable, then, to infer that *C. megalodon* decreased the presence of *C. carcharias*; by feeding in an area at the same time as *C. megalodon*, *C. carcharias* would jeopardize its own life. This would account for their either-or distribution in the late Neogene, as well as the rarity of the small-toothed species in the Paleogene and earlier Neogene fossil record. In the late Pliocene of Baja California, where the two species do occur together (S. P. Applegate, personal communication), the resolution of the stratigraphic record is not fine enough to determine the degree of zoogeographic overlap. A possible scenario is that they were following different migrating prey, which visited the area at different times.

Summary

The fossil record of the genus *Carcharodon* cannot be accurately interpreted on the basis of fossils alone. Only in light of the ecology of extant species does the fossil evidence achieve new significance and explain the distribution and faunal association of the fossil species. On the basis of the feeding and behavioral patterns of the extant white shark *C. carcharias* as well as fossil evidence, the following inferences can be made about the preferred prey, nursery areas, probable time of acclimation to cold water, and distribution of the extinct species, for both the small-toothed and giant-toothed lines. (1) Early in the evolution of *Carcharodon*, marine mammals became important prey for these sharks; in the fossil record, the abundance of these sharks appears to correlate positively with the abundance of marine mammals. (2) In the western North Atlantic basin, the giant-toothed *Carcharodon* used warm-water areas for nurseries, while the small-toothed species line used cooler waters. (3) Giant-toothed *Carcharodon* inhabited the temperate

waters of Antarctica in the late Eocene, and by the early Pliocene, the small-toothed *Carcharodon* was more common in cool temperate waters than in warm waters. (4) The small-toothed species preferred cool temperate waters and avoided areas preferred by the giant-toothed species.

Acknowledgments

I thank C. E. Ray and A. P. Klimley for encouragement and helpful comments that improved this chapter. I also thank D. G. Ainley, D. J. Bohaska, W. DiMichele, F. C. Whitmore, and two anonymous reviewers for their helpful comments. The oral presentation of this paper at the white shark symposium at the Bodega Marine Laboratory was supported by the Remington and Marguerite Kellogg Fund.

Anatomy

Using Allometry to Predict Body Mass from Linear Measurements of the White Shark

HENRY F. MOLLET
Monterey Bay Aquarium
Monterey, California

GREGOR M. CAILLIET
Moss Landing Marine Laboratory
Moss Landing, California

Introduction

Morphometric proportions have been used by many authors in studies of morphology, taxonomy, and growth. Garrick (1960) reviewed changes in the proportional dimensions in sharks and reported qualitative features that various shark species have in common, including the observation that sharks showed accelerated growth (positive allometry) in the trunk compared to the head and tail. Clark and von Schmidt (1965) observed positive allometry for the trunk in a number of species belonging to disparate families. Bass (1973) and Bass *et al.* (1975) gave extensive quantitative summaries of proportional dimensions for sharks from the east coast of southern Africa, including complete morphometrics for white sharks *Carcharodon carcharias* of total length (TL) ranging from 1.70 to 3.91 m.

The linear size-on-size regression $y = a + bx$ has been used successfully as an aid to shark taxonomy (Steffens and D'Aubrey, 1967). Casey and Pratt (1985) and Cliff *et al.* (1989) used it to relate precaudal length (PCL, also called standard length), fork length (FL), and TL in the white shark.

Parr (1956) proposed that the ratio-on-size regressions $y/x = a + bx$ and $y/x = a + bx + bx^2$ were very convenient for graphical presentations and for practical purposes of taxonomy. Using trial-and-error fit-

ting with different ratio-on-size polynomial equations, Bass (1973) found the equation $y/x = a + bx + c/x$ to be the most suitable for describing a large number of relative growth patterns, ranging from isometric to (simple) allometric to curvilinear, with and without minima or maxima.

Surprisingly, the allometric size-on-size equation $y = ax^b$ (power function) has rarely been used to report elasmobranch morphometry [except for reporting mass (M)–TL relationships]. However, von Bertalanffy (1960), Gould (1966), and Ricker (1979) all proposed it to be the most suitable equation for characterizing allometry.

For large sharks, such as the white shark, there has been much controversy about record sizes, especially when conversions were required from a smaller or different kind of measurement, such as PCL or M. Initially, we simply wanted to be able to convert reported M–TL equations to M–PCL equations, which requires an equation of the same form. Subsequently, we hoped to demonstrate the advantages of the allometric equation over linear size-on-size and linear ratio-on-size equations from both a theoretical and practical point of view.

Similarly, we theorized that a prolate spheroid model, obtained by rotating an ellipse around its long axis, might be useful for predicting M from length measurements (PCL) for the white shark. This model, which assumes isometry of all dimensions, produced excellent agreement between calculated and observed M (Van Dykhuizen and Mollet, 1992) for captive sevengill sharks *Notorynchus cepedianus*. In this species, the exponent b in the M–length power equation changed from 3.33 (allometry) to 3.00 (isometry) when TL was replaced with PCL as the length variable. The white shark, which has a general body shape similar to that of the sevengill, appeared to be a suitable candidate on which to test this model.

Casey and Pratt (1985) correctly indicated that M estimates for white sharks based solely on length were questionable because of considerable variations in girth (G). They reported equations for M based on TL and G for three TL ranges. We attempt to show that a single equation, instead of three equations, could be used to predict the M of a white shark.

Materials and Methods

Data Sources

Raw morphometric data on 327 white sharks from four sources were used (Table I). We obtained data for western North Atlantic white sharks from the National Marine Fisheries Service Narragansett Laboratory. Worldwide data, including those from Australia, were obtained from two game fishing associations (J. Casey and H. L. Pratt, personal communication). Data from

the west coast of North America were obtained from three additional sources. Finally, worldwide data published by Tricas and McCosker (1984) were compared with the above data sources, and only non-overlapping data were used. Not all measurements were available from each shark; thus, the totals for some measurements are <327.

Data Analysis

We used five general equations to relate all possible combinations of morphometric measurements, each equation providing criteria for isometry and (positive) allometry (Table II). First, we used standard linear or polynomial regression to fit equations 1–4 and 5b to the data, including residual analyses (Sokal and Rohlf, 1981; Neter *et al.*, 1983). These calculations were carried out with the MGLH module of SYSTAT (Wilkinson, 1988a). This also applied to allometric equation 1, a nonlinear power function that was linearized by a logarithmic transformation: $\ln y = \ln a + b \ln y$.

Second, the geometrical mean (GM) regression was calculated for the size-on-size (and ln size-on-ln size) data (Ricker, 1973; McArdle, 1988). The ratio-on-size data do not require the GM regression, because the relative error of the ratio y/x is much larger than that of x (Ricker, 1973; McArdle, 1988). Thus, the ordinary (y-on-x) regression yielded the final result.

TABLE I Numbers of Morphometric Measurements for the White Shark from Various Sources

No.	Region	PCL	FL	TL	G	M	PCL range (m)	Reference
1	Eastern coast of North America	59	75	71	47	61	0.96–4.47	H. L. Pratt, Jr. (personal communication), Casey and Pratt (1985), J. G. Casey (personal communication)
2	Western coast of North America			2		2		Roedel and Ripley (1950)
3	Western coast of North America		1			1		Kenyon (1959)
4	Western coast of North America			67		67	1.02–4.65	Klimley (1985a)
5	Australia		78		78	78	2.58–4.49	Casey and Pratt (1985), J. G. Casey (personal communication)
6	Worldwide			102		102	1.00–4.65	Tricas and McCosker (1984)
7	Worldwide[a]		16	3	16	16	1.17–4.31	Casey and Pratt (1985), J. G. Casey (personal communication)
	Totals[a]	327	327	327	141[b]	327	0.96–4.65	

PCL, Precaudal length; FL, fork length; TL, total length; G, girth; M, body mass.

[a]Missing PCL, FL, and TL estimates calculated using geometrical mean power regression based on measurements of western North Atlantic white sharks.

[b]G and M were available for 140 sharks.

TABLE II   Form and Properties of Regression Equations

No.	Equation	Isometry	Positive allometry	"Good" properties	Reference
1a	$y = ax^b$	$b = 1^*$	$b > 1^*$	$y = a$ when $x = 1$[†];	Gould (1966)
1b	$\ln y = \ln a + b \ln x$			log transformation of 1a	
2	$y = a + bx$	$a = 0$	$a < 0$ (negative)	$y/x = b$ when $x = $ infinity	Steffens and D'Aubrey (1967)
3a	$y/x = a + bx$	$b = 0$	$b > 0$	$y/x = a$ when $x = 0$ and $y = 0$;	Bass (1973)
3b	$y = ax + bx^2$			equivalent polynomial equation[‡]	
4a	$y/x = a + bx + cx^2$	$b = c = 0$	$c = 0, b > 0$	General case is curvilinear;	Parr (1956)
4b	$y = ax + bx^2 + cx^3$			equivalent polynomial equation[‡]	
5a	$y/x = a + bx + c/x$	$b = c = 0$	$c = 0, b > 0$	General case is curvilinear;	Bass (1973)
5b	$y = c + ax + bx^2$			equivalent polynomial equation[‡]	

[*] $b = 3$ and $b > 3$ mass versus length regression.
[†] Suggests meters as the most convenient length unit.
[‡] should r^2 be used to evaluate the quality of the fit, then use equations 3b, 4b, and 5b, not 3a, 4a, and 5a.

Equation 5a was fit to the data with the help of the NONLIN module of SYSTAT (Wilkinson, 1988a) through specification of the model equation.

The GM regression parameters were calculated with the NONLIN module of SYSTAT through specification of the loss function (Fleury, 1991). They can also be obtained from the ordinary regression results, using $b(GM) = b/r$ and $a(GM) = $ mean $y - b(GM)$ mean x. However, the NONLIN module does not produce the necessary data for a residual analysis. Therefore, residuals had to be analyzed using the results of the y-on-x regression. Similarly, the SYGRAPH module (Wilkinson, 1988b) plots the regression line with 95% confidence bands (CBs) for the y-on-x regression, but not for the GM regression. Therefore, the results of the GM regression were added to the figures. The 95% CB for the line in SYGRAPH is calculated according to the method of Neter *et al.* (1983) and is based on W statistics ($W^2 = 2F(1 - \alpha, 2, n - 2)$, rather than on t statistics. It is approximately 20–30% wider than the CB based on t statistics (Wallis and Roberts, 1956).

The quality of the fits using different fitting equations for the same data set was evaluated with the help of the t statistic of intercept and slope, standard error of estimate (SEE), residual analysis (studentized residual and leverage), and last coefficient of determination r^2 (Wilkinson, 1988a). However, the coefficient r^2 is relatively ineffective for expressing the closeness of fit and reliability of estimation of data (Whittaker and Woodwell, 1968; Bass, 1973).

The GM regression results were also used for predictions, following the recommendations and specific examples for fishery data in the work of Ricker (1973). Approximate 95% CBs for these predictions were calculated following the method of Whittaker and Woodwell (1968).

Prolate Spheroid Approximation

Assuming isometry, maximum G at 0.5 PCL, and a weightless caudal fin, the M of a shark was approximated by assuming a prolate spheroid shape (Voellmy, 1955) and a density of 1.0 gcm^{-3}. The long axis was assumed to be the PCL of the shark. The short axis was assumed to be the diameter, $d = G/\pi$, of the shark. The available white shark data suggested that the G/PCL ratio was roughly constant (0.621 ± 0.006 SE, N = 141), which allowed elimination of G from the M formula and yielded M = 20.5 $(kgm^{-3})PCL^3$.

Results

Regression Results

The three two-parameter methods used to analyze the variable pairs (PCL–FL, FL–TL, PCL–TL, and G–PCL) produced equally reasonable fits (Table III). This included the ratio-on-size regressions (equation 3a), as indicated by the $r^2 = 1.0$ values of the equivalent polynomial regressions (equation 3b). Because the results were similar for all four variable pairs, we concentrated on PCL–TL. The graphical comparison of the three methods for PCL–TL indicates that the ratio-on-size regression produced the worst fit (Fig. 1). However, the ratio-on-size plot magnifies the residuals. It provided the best graphical presentation of the raw data and identified two outlying values,

TABLE III Results of Five Regression Analyses
Performed on White Shark Morphometric Measurements

No.	Equation	Description	Pair	ln a	a	b	CB	N	r^2	SEE
1	$\ln y = \ln a + \ln x$	Allometric size-on-size GM power regression	ln PCL–ln FL	−0.1332	0.8753	1.0169	1.003–1.030	57	0.997	0.0213
			ln FL–ln TL	−0.1048	0.9005	1.0197	1.007–1.032	69	0.997	0.0206
			ln PCL–ln TL	−0.2427	0.7845	1.0382	1.019–1.057	58	0.995	0.0291
			ln G–ln PCL	−0.5585	0.5721	1.0798	1.038–1.121	141	0.948	0.110
			ln M–ln PCL	2.789	16.26	2.985	2.932–3.038	327	0.974	0.216
			ln M–ln TL	2.068	7.914	3.096	3.040–3.151	327	0.973	0.218
2	$y = a + bx$	Linear size-on-size GM regression; CB applies to intercept a, which determines allometry	PCL–FL		−0.038	0.9061	−0.063–(−0.013)	57	0.998	0.0424
			FL–TL		−0.057	0.9426	−0.082–(−0.032)	69	0.988	0.0465
			PCL–TL		−0.095	0.8550	−0.130–(−0.061)	58	0.996	0.0581
			G–PCL		−0.169	0.6860	−0.272–(−0.066)	141	0.914	0.204
3a	$y/x = a + bx$	Linear ratio-on-size ordinary regression	PCL/FL–FL		0.8734	0.0055	0.0008–0.010	57	0.089	0.0187
			FL/TL–TL		0.8986	0.0066	0.0022–0.011	69	0.120	0.0186
			PCL/TL–TL		0.7815	0.0112	0.0058–0.017	58	0.235	0.0229
			G/PCL–PCL		0.5817	0.0140	0.0026–0.025	141	0.041	0.0686
			M/PCL^3–PCL		17.35	−0.336	−0.73–0.06 (NS)	327	0.009	3.699
			M/TL^3–TL		8.387	0.193	0.008–0.378	327	0.013	2.064
3b	$y = ax + bx^2$	Equivalent polynomial regression of equation 3a	PCL–FL		0.8734	0.0055	0.0036–0.009	57	1.000	0.0421
			FL–TL		0.8933	0.0083	0.0052–0.011	69	1.000	0.0439
			PCL–TL		0.7750	0.0133	0.0093–0.017	58	0.999	0.0534
			G–PCL		0.5654	0.0189	0.0018–0.036	141	0.989	0.203
4a	$y/x = a + bx + cx^2$		PCL/TL–TL	0.8080	−0.010	0.0035	b (NS), c (NS)	58	0.255	0.0228
4b	$y = ax + bx^2 + cx^3$	Equivalent polynomial regression	PCL–TL	0.8265	−0.024	0.0056	b (NS)	58	0.999	0.0513
5a	$y/x = a + bx + c/x$		PCL/TL–TL	0.7580	0.015	0.028	b (NS), c (NS)	58	0.239	0.0230
5b	$y = c + ax + bx^2$	Equivalent polynomial regression	PCL–TL	0.6990	0.025	0.098	c (NS)	58	0.997	0.0523

a, y intercept of regression; b, slope; CB, 95% confidence band; N, sample size; r^2, coefficient of determiniation; SEE, standard error of estimate; GM, geometrical mean; PCL, precaudal length; FL, fork length; TL, total length; G, girth; M, body mass; NS, not significant.

which had almost identical and low PCL/TL = 0.73 without need of an analytical regression analysis. The ratio-on-size plot in Fig. 1C compared the regression lines of all three methods and indicates that they were very similar within the data range. A probability plot of the residuals and a residuals-versus-TL plot confirmed that the three methods yielded similar normal residual distributions and homoscedastic residual variances, respectively.

The GM power regression using the allometric equation gave small but significant positive allometry for PCL–FL, FL–TL, PCL–TL, and G–PCL, because the exponents, b, were all larger than 1.0: PCL = 0.8753 FL$^{1.0169}$, FL = 0.9005 TL$^{1.0197}$, PCL = 0.7845 TL$^{1.0382}$, and G = 0.5721 PCL$^{1.0798}$. The coefficient of determination r^2 indicated a very good fit, but the exponent, b, was barely significantly different from 1.0.

Likewise, the linear size-on-size GM regressions indicated positive allometry for all four length–vari-

able pairs, because the intercepts were negative (Table III). The coefficient of determination, r^2, was close to 1.0, but the intercept (which determines allometry) was only barely significant.

The linear ratio-on-size (y-on-x) regressions produced positive slopes indicative of positive allometry for all four length–variable pairs (Table III). The coefficient r^2 was close to zero, implying a poor fit. However, the regression parameters (including the slope, which determines the amount of allometry) were statistically significant. The equivalent polynomial regressions had the highest r^2 values, and the regression lines were similar to those of the ratio-on-size regression and identical for PCL–FL (Table III). This indicated that the ratio-on-size regression itself provided good fitting equations. The SEE values of the equivalent polynomial regression were lower (i.e., better) than those of the size-on-size GM regression.

The three-parameter equations (4a and 5a) for the four length–variable pairs produced good fits, but

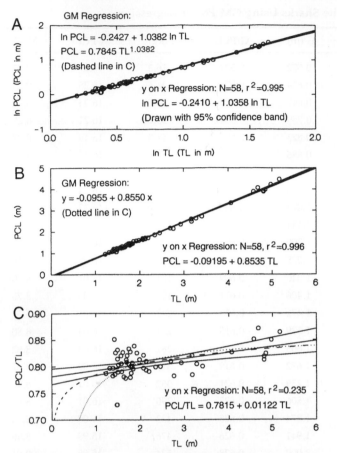

FIGURE 1 Equations for precaudal length–total length regression plots: (A) allometric (size-on-size), (B) linear size-on-size, and (C) linear ratio-on-size. The calculated geometrical mean (GM) regressions in A (dashed line) and B (dotted line) were redrawn in C for comparison. y on x regression in (B) and (C) are drawn with 95% confidence band.

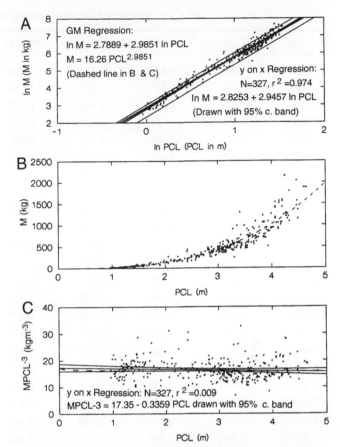

FIGURE 2 Mass (M)–precaudal length (PCL) regression plots, with 95% confidence (c.) bands. (A) Allometric equation (size-on-size; using the prolate spheroid model: top line, calculated M; bottom line, calculated 50% of M). (B) M versus PCL data without a regression line [the calculated geometrical mean (GM) regression in A was redrawn in B (dashed line)]. (C) Linear ratio-on-size equation, with ratio M/PCL^3 [the calculated GM regression in A was redrawn in C (dashed line)].

they were not substantially better than any from the three two-parameter methods. The PCL/TL-versus-TL plot (Fig. 1C) indicated that a curvilinear equation with a minimum between 1.5 and 3 m TL might better fit the data. However, the regression results showed that one or more of the three parameters were not statistically significant (Table III).

The observed variation in M for sharks of similar TL or PCL was relatively large (Fig. 2 for PCL; the plots for TL are similar). A direct nonlinear fit is not shown in Fig. 2B, following the recommendation of Ricker (1979). The M–TL GM power regression indicated that *b* was significantly larger than 3.0 (Table III). Using this equation ($M = 7.914\ TL^{3.0958}$), the median M of a 1 m TL prenatal white shark is predicted to be 7.9 kg (Table III). The M–PCL GM power regression indicated an isometric relationship because *b* was not significantly different from 3.0. Using this equation ($M = 16.26\ PCL^{2.9851}$), the median M of a 1 m

PCL (about PCL_o) white shark would be 16.3 kg (Table III).

The ratio-on-size plots for the M–length variables provided good graphical presentation of the raw data and pointed to possible outliers (Fig. 2C for PCL). The same plot also indicated that the redrawn isometric GM regression is well within the 95% CB for the line of the ratio-on-size regression, thus verifying the latter.

The theoretical M calculations based on the prolate spheroid model provided an approximate upper limit for the observed M (upper line in Fig. 2A). Half of the calculated M agreed with an approximate lower limit for the observed M (lower line in Fig. 2A).

When Casey and Pratt's (1985) idea of using both G and TL together was applied, it provided more accurate M predictions. We used an allometric equation instead of a linear equation without a constant, and

TABLE IV Predicted Dimensions of White Sharks Using GM Power Regression

PCL (m)	FL (m)[a]	PCL/FL	TL (m)[a]	PCL/TL	G (m)[a]	G/PCL	M (kg)[a]	M/PCL³	M/TL³
1.000	1.140	0.877	1.263	0.792	0.572	0.572	16.3	16.27	8.07
1.100	1.252	0.879	1.385	0.794	0.634	0.576	21.6	16.25	8.14
1.200	1.364	0.880	1.506	0.797	0.697	0.580	28.0	16.23	8.21
1.300	1.476	0.881	1.627	0.799	0.759	0.584	35.6	16.21	8.27
1.400	1.587	0.882	1.747	0.801	0.823	0.588	44.4	16.19	8.33
1.500	1.698	0.883	1.867	0.803	0.886	0.591	54.6	16.17	8.39
1.600	1.810	0.884	1.987	0.805	0.950	0.594	66.2	16.15	8.44
1.700	1.921	0.885	2.106	0.807	1.015	0.597	79.3	16.14	8.49
1.800	2.032	0.886	2.226	0.809	1.079	0.600	94.0	16.12	8.53
1.900	2.143	0.887	2.344	0.810	1.144	0.602	111	16.11	8.58
2.000	2.254	0.887	2.463	0.812	1.209	0.605	129	16.10	8.62
2.100	2.365	0.888	2.582	0.813	1.275	0.607	149	16.09	8.66
2.200	2.475	0.889	2.700	0.815	1.340	0.609	171	16.07	8.70
2.300	2.586	0.889	2.818	0.816	1.406	0.611	195	16.06	8.73
2.400	2.696	0.890	2.936	0.817	1.472	0.613	222	16.05	8.77
2.500	2.807	0.891	3.054	0.819	1.539	0.615	251	16.04	8.80
2.600	2.917	0.891	3.171	0.820	1.605	0.617	282	16.03	8.83
2.700	3.027	0.892	3.289	0.821	1.672	0.619	315	16.02	8.87
2.800	3.138	0.892	3.406	0.822	1.739	0.621	352	16.01	8.90
2.900	3.248	0.893	3.523	0.823	1.806	0.623	390	16.01	8.93
3.000	3.358	0.893	3.640	0.824	1.874	0.625	432	16.00	8.95
3.100	3.468	0.894	3.757	0.825	1.941	0.626	476	15.99	8.98
3.200	3.578	0.894	3.874	0.826	2.009	0.628	524	15.98	9.01
3.300	3.688	0.895	3.990	0.827	2.077	0.629	574	15.97	9.04
3.400	3.798	0.895	4.107	0.828	2.145	0.631	628	15.97	9.06
3.500	3.908	0.896	4.223	0.829	2.213	0.632	684	15.96	9.09
3.600	4.017	0.896	4.339	0.830	2.281	0.634	744	15.95	9.11
3.700	4.127	0.897	4.455	0.831	2.350	0.635	808	15.95	9.13
3.800	4.237	0.897	4.571	0.831	2.418	0.636	875	15.94	9.16
3.900	4.346	0.897	4.687	0.832	2.487	0.638	945	15.93	9.18
4.000	4.456	0.898	4.802	0.833	2.556	0.639	1019	15.93	9.20
4.100	4.565	0.898	4.918	0.834	2.625	0.640	1097	15.92	9.22
4.200	4.675	0.898	5.034	0.834	2.694	0.642	1179	15.91	9.25
4.300	4.784	0.899	5.149	0.835	2.764	0.643	1265	15.91	9.27
4.400	4.894	0.899	5.264	0.836	2.833	0.644	1355	15.90	9.29
4.500	5.003	0.899	5.379	0.837	2.903	0.645	1449	15.90	9.31
4.600	5.112	0.900	5.495	0.837	2.972	0.646	1547	15.89	9.33
4.700	5.222	0.900	5.610	0.838	3.042	0.647	1649	15.89	9.34
4.800	5.331	0.900	5.724	0.839	3.112	0.648	1756	15.88	9.36
4.900	5.440	0.901	5.839	0.839	3.182	0.649	1868	15.88	9.38
5.000	5.549	0.901	5.954	0.840	3.253	0.651	1984	15.87	9.40
5.100	5.658	0.901	6.069	0.840	3.323	0.652	2105	15.87	9.42
5.200	5.767	0.902	6.183	0.841	3.393	0.653	2230	15.86	9.43
5.300	5.876	0.902	6.298	0.842	3.464	0.654	2361	15.86	9.45
5.400	5.985	0.902	6.412	0.842	3.534	0.655	2496	15.85	9.47

(continues)

TABLE IV   *(Continued)*

PCL (m)	FL (m)[a]	PCL/FL	TL (m)[a]	PCL/TL	G (m)[a]	G/PCL	M (kg)[a]	M/PCL³	M/TL³
5.500	6.094	0.902	6.527	0.843	3.605	0.655	2637	15.85	9.48
5.600	6.203	0.903	6.641	0.843	3.676	0.656	2782	15.84	9.50
5.700	6.312	0.903	6.755	0.844	3.747	0.657	2933	15.84	9.52
5.800	6.421	0.903	6.869	0.844	3.818	0.658	3090	15.84	9.53
5.900	6.530	0.904	6.983	0.845	3.889	0.659	3251	15.83	9.55
6.000	6.639	0.904	7.097	0.845	3.960	0.660	3419	15.83	9.56

PCL, precaudal length; FL, fork length; TL, total length; G, girth; M, body mass.

[a] 95% confidence bands for predictions of FL, TL, G, and M were approximated by multiplying and dividing table entries by $e^{2SEE} = 1.06$, 1.04, 1.25, and 1.54, respectively.

this produced a better residual distribution: ln M = 3.65 + 0.924 ln (G²TL, N = 140, SEE = 0.118, r^2 = 0.992. For comparison purposes, we also calculated the allometric regression of M on TL with the same N: ln M = 2.14 + 2.93 ln TL, N = 140, SEE = 0.178, r^2 = 0.981. Thus, the use of G and TL apparently produced a substantially smaller SEE. The back-transformed GM power regression using PCL was calculated for reporting purposes: M = 45.98 (G²PCL)$^{0.9267}$, N = 140, e^{2SEE} = 1.264, r^2 = 0.992.

As a result of our analysis, we can now predict one measurement from another. To provide field researchers with handy guidelines for converting or evaluating relative measurements, we have listed these calculations (Table IV). Note that there is a fairly broad CB around these predictions, especially for G and M.

Discussion

Methods

The use of the GM regression (Ricker, 1973) to calculate certain functional relationships has been criticized. Kuhry and Marcus (1977) criticized the GM regression and suggested the major axis regression to be superior. More recently, in an extensive review, McArdle (1988) compared ordinary, major axis, and reduced major axis (identical to GM) regressions and concluded that the GM regression was more efficient and less biased than the major axis regression. The GM regression was not appropriate for the ratio-on-size data, because the error of the y variable (ratio) was much larger than that of the x variable (Ricker, 1973; McArdle, 1988).

We were tempted to use the ordinary regression in order to avoid the cumbersome GM regression, be-

cause our r values were close to 1.0. However, the ordinary regression produced biased and, in some cases, unreasonable results. For example, our GM regression results indicated that the M–PCL functional relationship was isometric, whereas that for M–TL was allometric, both of which were empirically expected. The ordinary regression results indicated the reverse.

The distinction between GM and ordinary regression was not as critical for prediction purposes as it was for functional analyses. We used the GM regression for all of our functional relationships and for predictions, following Ricker's (1973) recommendations based on similar examples from the fishery literature. The ordinary regression is the proper one to use for prediction in most other cases of model II regressions (Ricker, 1973; Sokal and Rohlf, 1981; McArdle, 1988). The CB for a new observation (prediction) is much wider than that for the line (functional relationship). In our study, the GM regressions produced predictions similar to those of the ordinary regressions.

We are not convinced that a correction for back-transformation bias is necessary. Sokal and Rohlf (1981) stated that the uncorrected back-transformed response (i.e., median response) is the preferred one for biological systems with a skewed M distribution. Our corrections for bias from back-transformations, following the methods of Sprugel (1983) and Miller (1984), were negligible; the correction factor was 1.024 for M predictions from PCL, compared with a prediction CB which is approximately M/1.54 to 1.54M wide.

From the practical point of view of length conversions, all three methods tested were adequate. The available white shark data were not suitable to demonstrate the advantages of the allometric equation due to insufficient PCL, FL, and TL data for subadults and adults, the use of nonconsistent TL measurements, and the variability or lack of accuracy of the data used. In our study of relative growth, the allometric equation produced an adequate statistical fit in most cases. It is also favored because it is simple and easily interpreted (von Bertalanffy, 1960; Gould, 1966; Ricker, 1979). We preferred the allometric equation to relate length pairs, because it allowed easy mathematical conversion between different length variables in the M–length equation.

The linear size-on-size equation is the least suitable for use with embryonic sharks (Parr, 1956). It has the mathematical property that allometry is determined by the y axis intercept, which is not reasonable from a biological point of view. Accordingly, y is not zero when x is zero, unless the relationship between y and x is isometric.

We concur with Bass (1973) and Bass *et al.* (1975) that the equation $y/x = a + bx$ successfully describes the variation of many of the proportional changes in sharks, including the white shark. Our two-parameter ratio-on-size regressions all produced normal and homoscedastic residual distributions, and we were unable to find any other possible serious drawbacks. Statistical textbooks (e.g., those by Simpson *et al.*, 1960; Sokal and Rohlf, 1981) have discussed the serious drawbacks of using ratios in statistical work. However, such ratios have proved to be most applicable in studies of the taxonomy and morphology of elasmobranchs.

Our results of large-scale morphometrics (PCL, FL, TL, and G) could be satisfactorily described with two-parameter regressions, although the need for three-parameter equations was substantiated for other white shark morphometrics (Bass *et al.*, 1975). The ratio-on-size plot is indeed excellent for a graphical presentation of the raw data and as a check for the possibility of maxima or minima. Our PCL–TL data suggested a minimum, but this could not be substantiated statistically. Two outliers with a very low PCL/TL ratio of 0.73 (Fig. 1C) were retained in the regression because they did not have high leverage. We could, however, substantiate a maximum in the first dorsal fin height versus TL ratio-on-size plot, which proved to be essential for TL validation in large white sharks (Mollet *et al.*, Chapter 10).

Regressions

Our results were in good agreement with all previously reported or updated results. Bass *et al.* (1975) reported positive simple (linear) allometry of PCL/TL versus TL: PCL/TL = 0.803 + 0.00588 TL (calculated from the given end points). The intercept was slightly higher and the allometry (slope b) was smaller, but fell within the 95% CB of our results. Casey and Pratt (1985) reported FL = −0.068359 + 0.9517 TL (N = 79, r^2 = 0.996). Their results agreed well with our y-on-x results: FL = −0.055(0.012) + 0.942(0.005) TL, N = 69, r^2 = 0.998. Our results also agreed with the updated GM regression reported by G. Cliff and S. Dudley (personal communication): PCL = −0.03231 + 0.9091FL, N = 142, r^2 = 0.9943.

Cliff *et al.* (Chapter 32) also reported an updated isometric GM regression, M = 16.504 (kgm$^{-2.944}$) PCL$^{2.944}$ (N = 383, r^2 = 0.9534, 95% CB for b = 2.880–3.009), which agreed well with our N = 327 results: a = 16.3, CB of b = 2.932–3.038. The use of the PCL–TL exponent, b = 1.038, allowed conversion of reported M–TL (b range 3.15–3.20) power regression results (Compagno, 1984a; Tricas and McCosker, 1984; Casey and Pratt, 1985) to M–PCL (b range 3.03–3.08) power regression results. No CBs for b could be calculated, but we suggest that they would include b = 3.0, that is, isometry.

Compagno (1984a) reported M = 8.27 (kgm$^{-3.14}$) TL$^{3.14}$ (from 98 specimens, mostly from California, and with TLs ranging from 1.27 to 5.54 m). Assuming r = 0.98, b(GM) was estimated to be 3.20, in agreement with our N = 70 California GM results: CB of b = 3.111–3.338. Tricas and McCosker (1984) reported M = 7.66 (kgm$^{-3.15}$) TL$^{3.15}$ (GM regression from 127 worldwide specimens, with TLs ranging from 1.25 to 6.4 m), and Casey and Pratt (1985) reported M = 7.44 (kgm$^{-3.095}$) TL$^{3.095}$ (y-on-x regression based on 200 worldwide sharks, and we estimated b(GM) = 3.15). These results were in agreement with our N = 327 GM results: a = 7.9 (7.4 < 95% < 8.5), b = 3.10 (3.04 < 95% < 3.15).

Prolate Spheroid Approximation

The calculated M (M = 20.5 PCL^3) based on the prolate spheroid approximation, using the observed mean G/PCL ratio 0.621, was 25–30% larger in the size range of 1–5 m PCL compared to the GM power regression (M = 16.3 PCL$^{2.99}$), using our data from 327 white sharks. Considering the number of approximations involved (circular vertical cross sections, elliptical longitudinal cross section with maximum G at 0.5 PCL, isometry of all dimensions, a weightless caudal fin, and a specific gravity of 1.0 gcm^{-3}), agreement to within 20–30% should be considered satisfactory. Larger calculated M was an indication that the utilized observed mean G/PCL ratio (0.621) was too large. The need of a lower effective G/PCL ratio (90%

of the observed ratio) very likely is an indication that a white shark has more taper in the back half of the body.

Mass Based on Girth and Total Length (or Precaudal Length)

Casey and Pratt (1985) indicated that M estimates for individual sharks based solely on length are questionable because of considerable G variations. Accordingly, they suggested that a more accurate M could be calculated using $M = kG^2TL$ instead of $M = aTL^b$. Initially, we thought that the use of PCL instead of TL would allow more accurate M prediction based solely on length, but the improvement was only marginal. The only improvement we can suggest is the use of the allometric equation $M = a(G^2TL)^b$ [or $M = a(G^2PCL)^b$]. These equations produced a better residual distribution, and a single equation [instead of the three equations given by Casey and Pratt (1985)] was suitable for the entire size range. The equations $M = 37.73 (G^2TL)^{0.9334}$ and $M = 45.98 (G^2PCL)^{0.9267}$ based on 140 M, G, and TL (or PCL) data points produced good (i.e., better than 20%) agreement for over 90% of the data between observed and calculated M. This was a substantial improvement over predictions based on TL (or PCL) only.

Summary

In large and relatively rare organisms such as the white shark *C. carcharias*, statistically tested conversion equations are necessary for relating different length measurements and for making M predictions. Five equations were evaluated, and we propose that the allometric equation (power function) is the most convenient for analyzing relative growth of length variables. The ratio-on-size plot is the most suitable for a graphical presentation of raw morphometric data. The corresponding ratio-on-size regression presented no serious statistical problems. Along with M (in kilograms), data from four general sources were used: PCL, FL, TL, and G (all in meters). Parameters in the power function $y = ax^b$ were determined using a GM power regression, $\ln y = \ln a + b \ln x$, because ordinary regression produced biased results. The b's were all significantly larger than 1.0, indicating positive allometry for PCL–FL (1.017), FL–TL (1.020), and PCL–TL (1.038). Using the PCL–TL power function, reported M–TL (b range 3.15–3.20) GM power regression results were easily converted to M–PCL (b range 3.03–3.08). Our analysis also produced M–PCL b's close to 3.0, indicating isometry. Therefore, a constant G/PCL ratio ($x = 0.621$, $N = 141$) was used to calculate white shark M from PCL, assuming a prolate spheroid model. The results ($M = 20.5 PCL^3$) were 25–30% larger (PCL range 1–5 m) than those based on the GM power regression ($M = 16.3 PCL^{2.99}$, $N = 327$). More accurate M prediction ($\leq 20\%$ for individual sharks) required the use of G and PCL according to $M = 46.0 (G^2PCL)^{0.927}$ ($N = 140$).

Acknowledgments

We thank three anonymous referees for helping to focus and strengthen our findings. We are grateful to N. E. Kohler, H. L. Pratt, and J. G. Casey, National Marine Fisheries Service, Narragansett; A. P. Klimley, Bodega Marine Laboratory; T. C. Tricas, Florida Institute of Technology; S. F. J. Dudley, Natal Sharks Board; and D. A. Ebert, Ocean Resource Consulting Associates, for providing data and many helpful discussions. Thanks go to D. C. Powell, C. J. Farwell, P. S. Nygren, M. A. Ferguson, and G. Van Dykhuizen (Husbandry Division) and C. Harrold, J. M. Watanabe, R. E. Phillips, N. H. Allen, and W. A. Anderson (Research Division) of Monterey Bay Aquarium for their support of this study. We appreciate the help given by librarians A. Baldridge, Hopkins Marine Station; S. Baldridge, Moss Landing Marine Laboratory; and F. Wolfe, Monterey Bay Aquarium.

A Review of Length Validation Methods and Protocols to Measure Large White Sharks

HENRY F. MOLLET
Monterey Bay Aquarium
Monterey, California

GREGOR M. CAILLIET
Moss Landing Marine
Laboratory
Moss Landing, California

A. PETER KLIMLEY
Bodega Marine Laboratory
University of California, Davis
Bodega Bay, California

DAVID A. EBERT
US Abalone
Davenport, California

ANTONIO D. TESTI
Italian Shark Research Project
Milan, Italy

LEONARD J. V.
COMPAGNO
Shark Research Centre
Cape Town, South Africa

Introduction

The total length (TL) and mass of large (TL >6 m) white sharks *Carcharodon carcharias* are often difficult to measure, leading to much speculation (Randall, 1973, 1987; Cappo, 1988; Ellis and McCosker, 1991). Reported TLs are often rough estimates or speculations. For many large white sharks, only the jaw and teeth have been preserved, providing the only specimens for length validation. Less frequently, fins are preserved. Their dimensions might be suitable for length validation if measured when fresh or if a conversion for shrinkage were possible.

A white shark with an estimated TL of >7 m was captured in a gill net near Kangaroo Island, South Australia, on April 1, 1987 (Jury, 1987; Cappo, 1988). A photograph of the head and the pectoral fin (as well as the caudal fin in the background) of this white shark (designated KANGA) was included in the book by Ellis and McCosker (1991, p. 55). Following the method of Randall (1973), TL was estimated by Cappo (1988, personal communication) to be 6.4 and 5.8

m from the dried upper jaw perimeter (DUJP) and the enameloid height of the first upper tooth, respectively. This was about 1 m less than the estimated TL by fisherman Peter Riseley.

Sixteen days later, another large white shark of disputed 7 m TL was caught on an 8-mm steel line baited for tuna and swordfish 1.8 km southeast of Fifla Farallon, Malta, by Alfred Cutajar (Abela, 1989; Fergusson, Chapter 30). Photographs of this white shark (MALTA) were shown by Ellis and McCosker (1991, p. 56) and Fergusson (Chapter 30). From upper jaw perimeter and tooth enameloid height (UA1E2), following the technique of Randall (1973), TL was estimated to be 5.7 and 5.3 m, respectively. This was 1.5 m less than Abela's (1989) measurement. The relatively small dorsal and pectoral fins indicated that the shark was much shorter than 7 m, again contradicting the TL measurement. These low TL estimates also were in conflict with photographs that indicated a TL of at least 6–6.5 m (I. K. Fergusson, personal communication).

In the past, extrapolations from upper jaw perimeters and UA1E2's, mass, vertebrae dimensions, and number of vertebral bands of smaller sharks (TL <5 m) were used to evaluate TL estimates >6 m (Randall, 1973, 1987). Mass was expected to be too variable for length validation (see Chapter 9, by Mollet and Cailliet). Jaw and teeth morphometrics of more recently captured white sharks, 5–7 m TL, indicated large variations for sharks of similar length (G. Hubbell and M. Cappo, personal communication; L. J. V. Compagno, unpublished data). This suggested to us that TL estimates from jaw and teeth data might not be suitable. At least, TL estimates had to be based on the regression 95% confidence band for a new observation rather than the regression line.

The preliminary analysis of our white shark morphometric data and the summary of proportional dimensions by Bass *et al.* (1975) indicated that the size of fins might yield more accurate TL estimates than would size of the jaw perimeter, UA1E2, or mass. The purpose of this chapter is to validate indirect TL estimates and evaluate extrapolated TL estimates for the Kangaroo Island and Malta white sharks.

Materials and Methods

Morphometrics for Existing Data

Measurements have been taken from more than 70 white shark specimens (Appendices 1 and 2), but many were not measured by us. Most of the TL measurements were probably taken with the caudal fin in the natural position. TL of four sharks was certainly measured with the caudal fin in a depressed (i.e., extended) position (Appendix 3) (Compagno, 1984a). We assumed that the jaws of each shark were prepared in an identical manner. We further assumed that the upper jaw perimeter and the UA1E2 of the first upper functional tooth (UA1) were measured according to the following definitions, both being modifications of the method of Randall (1973) (see illustrations in Appendix 3, but note that the tooth sketched and measured carried labels of the second upper tooth, UA2). (1) DUJP is measured, with a string, from one extreme corner of the mouth (i.e., the medial area of the widest part where the upper and lower jaws join), along the curvature of the upper jaw just above the base of the teeth, all the way around, to the other extreme corner. (2) UA1E2 is the vertical measurement on the labial surface of the first upper functional tooth, along the medial axis of the tooth from the apex of the tooth cusp to a line between the mesial and distal proximal edges of the enameloid.

The dried jaw and tooth data reported by Randall

(1973) were reconstructed from 8-in. × 10-in. prints made from the negatives of the original figures (Appendix 2). Some reported jaw or tooth measurements required conversion. Data for fresh upper jaw perimeters were converted to DUJP, assuming a shrinkage of 4%, on the basis of mean shrinkage from three upper jaws (G. Hubbell and M. Cappo, personal communication). UA1E2 was calculated from the medial enameloid height (UA1E1), enameloid mesial (or distal) edge length (UA1EM or UA1ED), or tooth total height (UA1H), using the mean ratios of all the Hubbell data (Appendix 2): UA1E2/UA1E1 = 1.154 ± 0.018, N = 9; UA1E2/UA1EM = 0.908 ± 0.026, N = 9; and UA1E2/UA1H = 0.804 ± 0.010, N = 9. Here and throughout this chapter, means are reported ±1 standard error.

Morphometrics for KANGA and MALTA

The morphometrics on KANGA, reported initially by Jury (1987) and Cappo (1988, personal communication), are summarized in Table I. The fisherman who had collected KANGA, Peter Riseley, estimated a TL >7 m, by comparing the shark to the distance between fore and aft portions of his boat (Cappo, 1988). KANGA was too large to be brought aboard; mass was not estimated. Only the pectoral fin height, head size, and girth at the first gill slit were estimated.

More morphometric data and a mass estimate are available for MALTA (Abela, 1989, personal communication) (Table I). Additional pectoral fin, jaw, and tooth measurements were taken by one of us (A.D.T.) in Malta in March 1990. Abela distributed a table that contained tooth edge–length measurements, which are larger than the vertical measurements used by Randall (1973) and by us. The TL of MALTA (Abela, 1989, personal communication) was treated as an estimate after we thoroughly investigated all available correspondence (approximately 40 documents; see Acknowledgments). We concluded that there were too many inconsistencies and treated the reported MALTA TLs of 7 m (straight line) to 8.85 m (over the curve) as estimates. In fact, Ian Fergusson (personal communication) interviewed John Abela and Alfred Cutajar in autumn 1994 and concluded, without reservation, that Abela had measured the fish lying flat on the floor. His 7.14 m TL and 7.77 m over-the-curve TL should, accordingly, be considered credible. The 8.85 m over-the-curve TL quoted in *AQUA* (Abela, 1989) was a misprint or error; the 7.77 over-the-curve TL is consistent with a white shark measuring 7.14 m TL (H. F. Mollet, unpublished data).

Only a rough estimate of the pectoral fin height

TABLE I **Kangaroo Island (KANGA) and Malta (MALTA) White Shark Morphometric Measurements for Length Validation**

Measurement	FAO abbreviation	Other sharks, CV (N) (%)[a]	KANGA m	KANGA %TL	MALTA m	MALTA %TL
Total length	TL		<7.0	100	7.0	100
Prebranchial length	PG1	6.0 (37)	1.346*	19.2*		
Head height	HDH	20.0 (13)	1.219*	17.4*		
Head width	HDW	3.1 (10)	1.219*	17.4*		
Trunk height	TRH	21.0 (17)			1.0–1.2*	14–17
Girth at first gill slit	GIR(GS1)		3.830*	54.6*		
Girth at first dorsal fin (gutted)	GIR(D1)	14.0 (21)			3.2–3.5*	45–50
Pectoral fin anterior margin[b]	P1A	5.0 (49)	1.128*,**	16.1*,**	0.938**	13.4
Pectoral fin base	P1B	10.0 (33)			0.392**	5.6
Pectoral fin inner margin	P1I	15.0 (29)			0.228**	3.2
Pectoral fin posterior margin	P1P	8.4 (23)			0.742**	10.6
Pectoral fin height	P1H	9.1 (10)	1.0*	14.3*	0.833**	11.9
Pectoral fin length	P1L	5.6 (5)			0.536**	8.5
Caudal fin, tip to tip (span)	CFTT	8.3 (8)			1.372	19.6
Caudal fin dorsal margin	CDM	5.9 (42)			1.021**	14.6**
Caudal fin preventral margin	CPV	8.3 (42)			0.893**	12.7**
First dorsal fin height[c]	D1H	10.7 (34)			0.502	7.16
First dorsal fin height (dried)[d]	D1H(D)				0.457	6.62
Dried upper jaw perimeter	DUJP	9.2 (30)	1.250	17.8	1.120	16.0
Fresh upper jaw perimeter	FUJP		1.297	18.5	1.220	17.4
Dried lower jaw perimeter	DLJP				0.965	13.8
Fresh lower jaw perimeter	FUJP				1.000	14.3
Dried jaw, corner to corner	ca. MOW	13.0 (36)	0.829	11.8	0.615	8.77
Dried jaw, symphysis to symphysis	ca. 2×MOL	14.0 (28)	0.706	9.93	0.590	8.42
First upper tooth enameloid height (larger)	UA1E1	8.2 (48)	0.0516	0.736	0.0469[e]	0.669
First upper tooth enameloid height (smaller)	UA1E1		0.0501	0.715	0.0445[e]	0.792
First upper tooth total height (larger)	UA1H	13.0 (28)			0.0555[e]	0.792
Mass (kg)[f]	M				2400–3600	
Condition factor (kgm⁻³)	W/TL³	23.0 (327)			7.0–10.5	

FAO, Food and Agriculture Organization; CV, coefficient of variation.

*, estimated value; **, calculated value.

[a]CV (%) = 100 (SEE)/mean y of the sample, where SEE is the standard error of estimate of the regression.

[b]Pectoral fin measurements for MALTA were calculated assuming 10% shrinkage.

[c]A linear ratio-on-size regression is not suitable for first dorsal fin height (see text).

[d]The first dorsal fin height decreased by 8.9% in 3 months due to shrinkage (J. Abela, personal communication).

[e]Calculated from measurements made by A. D. Testi with calipers.

[f]Mass allows for removal of 1600–2400 kg of muscle; 300–600 kg of liver; 200 kg of stomach contents (1.8 m blue shark, 2.4 m dolphin, and 0.6 m turtle); a 180–250 kg head; 70 kg of pectoral, dorsal, and caudal fins; and 50 kg of body fluids.

was available for KANGA, and the anterior margin was calculated from the height estimate, using the mean margin/height ratio of 0.878 ± 0.014 (N = 11). The pectoral fin of MALTA became extremely desiccated from display in the Maltese bar where it was measured, and we considered shrinkages between 10% and 15%. Only the nonconventional caudal fin span of MALTA was available. The caudal dorsal margin was calculated from the margin/span ratio of 0.744, determined from a photograph of a 5.2-m female (Postel, 1958).

Regression Analysis and 95% Confidence Band for a New Observation

The linear ratio-on-size regression [y/TL $\times$ 100 = y (%TL) = $a + b$ TL] was selected as the most suitable equation for our purpose (Parr, 1956; Bass, 1973; Compagno, 1984a; Mollet and Cailliet, Chapter 9). For all body morphometrics available for the sharks in Appendix 1, and the jaw and tooth data in Appendix 2, we regressed the morphometric data (expressed as %TL) on TL. Outliers with large studentized residuals and/or additional evidence of data error were deleted. KANGA and MALTA data were excluded in the calculation of regression lines and their confidence band. Female and male morphometrics were combined in all regressions after appropriate statistical tests (Neter et al., 1983). In the jaw perimeter and tooth enameloid plots, we used different symbols for our data and those reconstructed from Randall (1973). However, the regression lines pertain to the combined data. We used the condition factor (MTL^{-3}) as the dependent variable for the analysis of the mass–TL relationship. We relied on the mass data (N = 327) of Mollet and Cailliet (Chapter 9, with the confidence band for the line) and calculated the 95% confidence band for a new observation. We checked that the calculated TL ranges were similar to the ones based on a geometrical mean (GM) power (log W versus log TL) regression. We combined the mass data of the white sharks <5.5 m TL of known sex used by Mollet and Cailliet (Chapter 9), Cliff et al. (Chapter 32), and Uchida et al. (Chapter 14) to determine the lack of a significant difference between the condition factor versus TL regressions for males (N = 218) and females (N = 281).

A regression coefficient of variation (CV, expressed as a percentage) was calculated for the relationship of all morphometrics with TL. It is given by CV (%) = 100 (SEE)/mean y of the sample (Bass, 1973), where SEE is the standard "error" of estimate (a standard deviation). The regression CV provided an indication of the suitability of a morphometric size for TL esti-

mation. A regression with a CV of >10%, or a regression with CV <10% but based on a small sample (say, N < 20), is of questionable usefulness to reasonably estimate TL. To conserve space, the CVs are reported only for the morphometrics measured or estimated for KANGA or MALTA and were added to Table III.

The three morphometrics available for both MALTA and KANGA, and three additional ones available for MALTA only, were examined in more detail to evaluate their usefulness for validating TL (Table II). In these cases, the analytical regression equation with the standard error of the parameters and the standard deviation of estimate of the regression were calculated using the MGLH module of SYSTAT (Wilkinson, 1988a). For the dorsal fin height, we also considered a three-parameter fitting equation following the method of Bass (1973), and used the NONLIN module of SYSTAT (Wilkinson, 1988a) to calculate the fitting parameters and their asymptotic standard errors. In this case, we also calculated the constant (intercept) of the one-parameter fitting equation for comparison with the results of Bass et al. (1975). One of these morphometrics was mass, and we used the results of Mollet and Cailliet (Chapter 9).

The procedure used to determine the possible TL range for a large shark of unknown TL from a measured morphometric size is best understood with reference to, for example, Fig. 1B. We used the 95% confidence band for a new observation, also known as the prediction confidence band (Sokal and Rohlf, 1981), to determine the possible TL range from a morphometric measurement of a shark with disputed TL. For example (see Fig. 1B), an observed morphometric datum [i.e., caudal fin dorsal margin (CDM) = 1.021 m] of a shark with unknown TL produces a hyperbola ($y = 102.1/x$) in a ratio-on-size plot, rather than the horizontal line that would occur in a size-on-size plot. Therefore, the possible TL range is determined by the intersection of this hyperbola with the confidence limits defining the 95% confidence band for a new observation. In this example, the TL of MALTA is between 4.6 and 7.0 m, predicted from the CDM of 1.021 m. Assuming that MALTA was, indeed, 7 m TL produces a data point (M) for the CDM (=14.6%) that falls on the lower limit of the 95% confidence band.

The graphical program used by us (SYGRAPH) (Wilkinson, 1988b) provided the 95% confidence band for the regression line of existing observations based on W (i.e., "Working and Hotelling") statistics (Neter et al., 1983; Mollet and Cailliet, Chapter 9). Therefore, we had to calculate the required constants to produce the confidence band of a new observation based on t

TABLE II Ratio-on-Size Regression Parameters for Morphometric Measurements of White Sharks

| Measurement | FAO abbreviation | N | Estimated TL Range (m) | Regression parameters[a] | | | | Reference |
				a (SE)	b (SE)	c (SE)	SEE	
Pectoral fin anterior margin	P1A (%TL)	49	1.29–6.0	23.1 (0.3)	−0.902 (0.100)		1.03	This study
		34	1.70–3.91	25.0	−0.950		0.9	Bass et al. (1975)[b]
		90	1.3–6.0	23.1	−0.69			L. J. V. Compagno, unpublished data
Caudal fin dorsal margin	CDM (%TL)	42	1.26–6.00	23.9 (0.4)	−0.943 (0.134)		1.24	This study
		34	1.70–3.91	24.8	−0.769		0.55	Bass et al. (1975)[b]
First dorsal fin height	D1H (%TL)	34	1.31–6.0	14.8 (2.0)	−0.73 (0.32)	−7.2 (2.5)	0.92	This study
		34	1.31–6.0	9.2 (0.4)	0.15 (0.13)*		1.02	This study
		34	1.31–6.0	9.7 (0.2)			1.03	This study
		33	1.70–3.91	9.8			1.6	Bass et al. (1975)[b]
Dried upper jaw perimeter	DUJP (%TL)	31	1.98–6.1	19.7 (1.1)	0.21 (0.25)*		1.89	This study
		16	1.98–5.4	20.7 (0.5)	−0.17 (0.14)*		0.64	Randall (1973)[c]
Tooth enameloid height	UA1E2 (%TL)	49	1.96–6.4	0.97 (0.03)	−0.026 (0.007)		0.071	This study
		17	1.98–5.4	0.93 (0.04)	−0.009 (0.011)*		0.052	Randall (1973)[c]
Condition factor	M/TL³ (kgm⁻³)	327	1.22–5.55	8.39 (0.33)	0.193 (0.093)		2.06	Mollet and Cailliet (Chapter 9)

FAO, Food and Agriculture Organization; SE, standard error; SEE, standard error of estimate.
*Not significant
[a] $y(\%TL) = a + b \ TL \ (+ \ c/TL \ \text{for D1H}); \ y \ (\%TL) = a$ also included (see text).
[b] Parameters calculated with TL = 1.70 and 3.91 m.
[c] Parameters calculated from original figures.
*Not significant.

statistics. Details on this procedure, based on the method of Sokal and Rohlf (1981) or Neter *et al.* (1983) are available from H. F. Mollet upon request.

The calculation of hyperbolic confidence limits for a new observation (Sokal and Rohlf, 1981; Neter *et al.*, 1983) is tedious and not covered by the computer programs available to us. Therefore, we calculated confidence bands using Bass' (1973) simplified linear method, which defined the 95% confidence limits as ±d (±2 standard deviations about the regression line). Bass' (1973) confidence band requires only the calculation of the standard deviation of the regression, which is standard output of a regression calculation in SYSTAT and most other statistical programs, and is called SEE (Wilkinson, 1988a). A comparison of the two confidence bands indicated that the differences were acceptably small for all linear relations (a factor of 1.15 maximum at TL = 7 m, if N > 20). In the future, we are planning to use Bass' (1973) method to calculate the 95% confidence band for a new observation. This method also justified the use of Bass' (1973) confidence band for a three-parameter (nonlinear) fitting curve, which provided a better fit for the existing first dorsal fin height (D1H) data for the white shark (Fig. 1C). The calculation of more accurate confidence

limits for a three-parameter fit were beyond our statistical capabilities.

Results

Morphometrics for Existing Data

Our morphometric data, excluding those from KANGA and MALTA, were composed of measurements from more than 70 white sharks worldwide, spanning 100 years and 1.261–6.408 m TL (Appendices 1 and 2). In our database, all white sharks >4.75 m TL were females. The database included two embryos that were larger than five free-swimming pups.

The analysis of existing morphometric data indicated that variability, measured as CV, was high (Table I), thus decreasing the likelihood of validating a TL measurement using other variables. To predict TL within 10% of 7 m, a CV of <2–3% would be required (see Simpson *et al.*, 1960). Few variables [e.g., fork length, precaudal length (PRC), and presecond dorsal fin length) produced CV values this low. However, these morphometrics were not available for MALTA or KANGA, and therefore are not

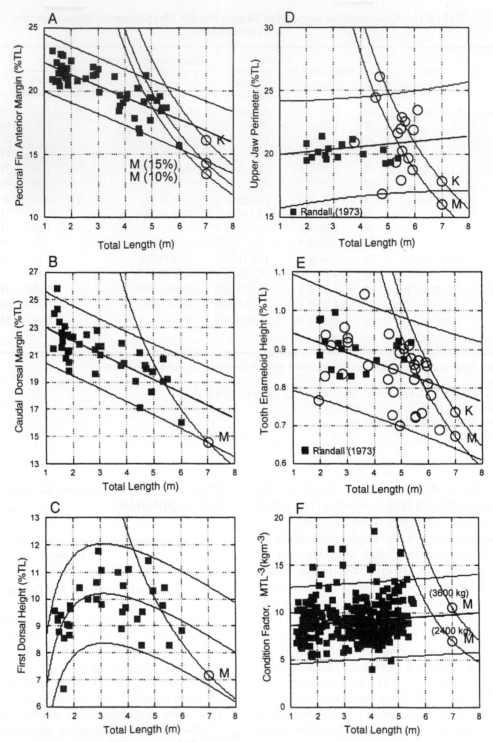

FIGURE 1 Ratio-on-size regression plots with a 95% confidence band for a new observation: (A) pectoral fin anterior margin (P1A), (B) caudal fin dorsal margin (CDM), (C) first dorsal fin height (D1H), (D) dried upper jaw perimeter (DUJP), (E) tooth enameloid height (UA1E2), and (F) condition factor (MTL^{-3}). The following apply: data points for KANGA and MALTA, assuming 7 m TL, are labeled K and M, respectively; 10% and 15% shrinkages are assumed for the pectoral fin anterior margin of MALTA (A); D and E identify the Randall (1973) data, but regressions are for all data; a mass range of 2400–3600 kg was used for MALTA (F), and intersections of the confidence band with the more vertical lines defines the TL ranges for KANGA and MALTA (see text).

listed in Table I. Only the pectoral anterior margin and the caudal dorsal margin had sufficiently low CV values and reasonable sample sizes to potentially predict TL. All other variables had CV values exceeding 6% or had small sample size. The regression from many morphometrics showed large variability, as indicated by a CV of over 10% (Table I).

Using various regression techniques, the relationships between the six morphometric variables available for KANGA or MALTA and TL ranged from negative and positive allometry to isometry (Table II). The pectoral fin anterior margin, CDM, and UA1E2 all showed significant negative allometry using a two-parameter fitting equation. D1H, however, was best described with a three-parameter fitting equation with the maximum of the curve at 3.1 m (Fig. 1C). This indicated positive allometry for small sharks and negative allometry for large sharks. In contrast, the standard two-parameter fit produced a nonsignificant slope (isometry) and was not useful. The two-parameter fit was no better than assuming that the first dorsal height was constant. The upper jaw perimeter regression also yielded a nonsignificant slope, indicating that isometry would be a good first approximation. The condition factor indicated barely significant positive allometry $[T = 0.193/0.093] = 2.08$ (see Chapter 9, by Mollet and Cailliet).

Most of the data for upper jaw perimeter and tooth enameloid for the smaller white sharks came from Randall (1973) and were less variable (Fig. 1D and E and Table II). The upper jaw perimeter plot showed a pronounced increase in variability for sharks larger than 4 m TL, and this could be substantiated statistically. The variance F test $[F^* = 15.2 >> F(0.025,17,10) = 0.34$ to $F(0.975,17,10) = 2.94]$ confirmed highly significant heteroscedasticity for the jaw regression of large sharks. UA1E2 data produced negative allometry, because the larger sharks had relatively smaller teeth, a result that differs from that of Randall (1973), who had data only from small sharks.

We found no difference between males and females for the six morphometrics of interest. This was confirmed by statistical tests. Thus, we combined all data regardless of sex.

Morphometrics for KANGA and MALTA

We confirmed that both KANGA and MALTA could be 7.0 m TL. The pectoral fin, jaw, and teeth indicated that KANGA was bigger than MALTA (Table III). The other morphometrics listed were, unfortunately, not available for both sharks. All three reported KANGA morphometrics, assuming the 7 m TL to be correct, fell within the 95% confidence band for a new observation (Fig. 1A, D, and E). However, the TL ranges determined from these morphometrics were broad, ranging from 2.2 to 3.6 m (Table III). Therefore, based on the mean TL range, KANGA

TABLE III **Summary of Estimated Total Length Ranges of Large White Sharks**

Measurement	FAO abbreviation	Observed		Estimated TL range		Comments
		KANGA (m)	MALTA (m)	KANGA (m)	MALTA (m)	
Jaw perimeter	DUJP	1.250	1.120	5.1–7.3 (2.2)*	4.5–6.5 (2.0)*	
Tooth enameloid height	UA1E2	0.0516	0.0469	5.3–8.9 (3.6)*	4.7–7.5 (2.8)*	
Pectoral fin anterior margin	P1A	1.128	0.938	5.5–8.3 (2.8)*	4.3–6.1 (1.7)	10% shrinkage for MALTA
Pectoral fin anterior margin	P1A		0.993		4.7–7.0 (2.3)*	15% shrinkage for MALTA
Pectoral fin base	P1B		0.392		4.5–7.3 (2.8)	10% shrinkage for MALTA
Pectoral fin inner margin	P1I		0.228		4.0–>10	10% shrinkage for MALTA
Pectoral fin posterior margin	P1P		0.742		3.9–6.5 (2.6)	10% shrinkage for MALTA
Pectoral fin height	P1H	1.0	0.833	5.0–>10	4.0–>10	10% shrinkage for MALTA
Caudal fin	CDM		1.021		4.6–7.0 (2.4)	
Dorsal fin	D1H		0.502		4.2–7.5 (3.3)	
Body mass (kg)	BM		2400		5.6–7.3 (1.7)	
			3600		6.3–8.3 (2.0)	
Mean total length[a]	TL			5.3–8.2	4.6–7.0	

FAO, Food and Agriculture Organization; TL, total length.

[a]The upper and lower limits of the ranges indicated by * were used to estimate a mean range.

could have been 5.3–8.2 m long, consistent with the >7 m TL estimate of Riseley.

The majority of the MALTA morphometrics, assuming the 7 m TL to be correct, also fell within the 95% confidence band for a new observation (Fig. 1). This is true for all but the upper jaw perimeter (Fig. 1D). Also, only the pectoral fin anterior margin, using a 15% shrinkage rate, fit near the confidence band (Fig. 1A). The TL estimates from these morphometrics were also broad (Table III). Based on the mean of the same three morphometrics used for KANGA,-MALTA could have been 4.6–7.0 m long, making the 7 m TL estimate by Abela possible. The TL estimates for MALTA, using the estimated mass range limits of 2400 and 3600 kg, were also broad and in the range of 5.6–7.3 and 6.3–8.3 m, respectively. This demonstrates that even the lower mass estimate was compatible with a 7 m TL shark (Fig. 1F).

None of the other pectoral fin morphometrics (Table III), nor any remaining morphometrics (Table I) for KANGA or MALTA, ruled out a length of 7 m for either shark. In fact, some even supported it. Again, a shrinkage between 10% and 15% for the MALTA pectoral fin was sufficient to produce a TL range that included 7 m. The pectoral fin inner margin and height produced TL ranges with indeterminate upper limits for both KANGA and MALTA, because insufficient existing data were available or the regression CV was >10%. The pectoral fin anterior margin is easiest to measure consistently, and the consideration of additional pectoral fin morphometrics probably amounts to duplication. Prebranchial length, head size, girth, and additional jaw morphometrics for KANGA, and trunk height, girth, and additional caudal fin and jaw morphometrics for MALTA (Table I), also supported a TL of 7 m for either. We did not expect these morphometrics to produce precise TL ranges, so no plots are shown, nor were TL ranges estimated.

Discussion

Morphometrics for Existing Data

Our white shark data, after adjusting for the lack of standards, proved to be remarkably useful and agreed in general with the results of Bass et al. (1975). The regressions for the pectoral fin anterior margin and CDM were in good agreement, although CDM variability in our data was twice as large. We suggest that approximately 0.5% of the difference for small white sharks was due to the incorrect assumption of a constant angle between the sagittal (median) plane and the caudal fin by Bass et al. (1975). They calculated TL as PRC + 0.8 CDM, with 0.8 equaling the cosine of 37° (an approximation of the heterocercal angle). Accordingly, our pectoral fin anterior margin results for small sharks were in better agreement with results from 90 white sharks of the eastern North Pacific, the North Atlantic, South Africa, and Australia (L. J. V. Compagno, unpublished data).

Our data for height of the first dorsal fin was fit by a three-parameter regression, with a maximum at 3.1 m (Table II and Fig. 1C). This was a refinement of the method of Bass et al. (1975) and explained the observed small dorsal fin height of MALTA. We suggest that such a maximum for fin height is expected for any shark and any fin that becomes more erect after birth. Following birth, the dorsal fin becomes more erect, producing *apparent* accelerated growth for D1H. Once the form of the fin becomes stable, we saw that growth decelerates. Bass et al. (1975) reported a constant dorsal fin height for the white shark (N = 33) but large positive allometry for the shortfin mako *Isurus oxyrinchus* (N = 18). Garrick (1967) demonstrated the importance of dorsal fin shape in the systematics of the shortfin mako, and this led to synonymy of the three shortfin mako species. Bass (1973: Fig. 4) observed a dorsal fin maximum for the blacktip shark *Carcharhinus limbatus* (N = 128) and, among carcharinid sharks in general, an initial accelerated growth rate (positive allometry) in the first dorsal fin followed by decelerated growth (negative allometry) as the fish grew.

We used TL as the reference length in the analysis because we were unable to convert our data to the less ambiguous PRC. In principle, if the PRC and CDM were measured, in addition to TL, one can determine how TL was measured and then use a consistent TL. Unfortunately, for almost 50% of the white sharks in our database, the PRC had not been measured. In addition, TL for several sharks was allegedly measured with the caudal fin in the natural position, but the sum of PRC and CDM was <100%, instead of the expected 102–106%. Francis (Chapter 15) reported PRC + CDM = 101.8% (heterocercal angle 23°) for a 1.521 m TL female pup. We estimated PRC + CDM = 105.9% (heterocercal angle 49°) for the Gans Bay female with 6.00 m TOT (total length with the caudal fin in the depressed position) to 5.67 m TL (caudal fin in the natural position).

Our morphometric regressions indicated a generally larger variation about the regression line compared to the findings of Bass et al. (1975) (Table I, CV column). Some morphometrics showed much larger variation than was expected from natural variability (Simpson *et al.*, 1960), indicating that methodological uncertainties were apparent: (1) the use of ill-defined TL (with the caudal fin in the natural or depressed position), instead of PRC, as proposed by Parr (1956), Springer (1964), Van Dykhuizen and Mollet (1992), and Mollet and Cailliet (Chapter 9), and as used for lamnoids by Gruber and Compagno (1982), Gilmore (1983), and Cliff *et al.* (1989); (2) the guess about the exact location of the origin of the fins (M. P. Francis and R. H. Rosenblatt, personal communication); (3) the lack of standards before the work of Compagno (1984a) [e.g., the height of the first dorsal fin was used to designate the anterior margin (Tortonese, 1956)] and the confusion about fin origin and insertion (M. P. Francis, R. H. Rosenblatt, and J. Seigel, personal communication); (4) the distortions due to removing the shark from the water, gutting, freezing, or preservation (Olson, 1954; Springer, 1964; Bass, 1973); and (5) the absence of experienced scientists before the press had arrived and possible interference by onlookers.

Better definitions for certain morphometrics, along with a standard protocol for measuring white sharks, are presented in Appendix 3. Use of the protocol will ensure consistent measurements (see also Cappo, 1988). For example, in our database, we assumed that the jaws were always prepared in the same manner and shrank by the same amount during the drying process. However, the jaw of the Gans Bay shark had an upper perimeter of 1.375 m but appeared to be subjectively smaller than the KANGA jaw (1.250-m perimeter), suggesting different methodologies for preparation or measurement. We assumed that the first upper tooth was the largest, a point supported by the data we were able to check. However, the term *largest tooth* (Randall, 1973) is ambiguous. The first upper tooth (UA1) can be slightly larger, about equal, or slightly less than the second upper tooth (UA2) (L. J. V. Compagno, unpublished data on 80 white sharks). We assumed that reported UA1E2's were measured vertically and consistently, following the method of Randall (1973), and that the shape of the first upper tooth from different sharks was the same, thus allowing calculations from ratios if necessary. Often, the easier-to-measure tooth enameloid edge (UA1EM and UA1ED; see Appendix 3) is reported without stating what was measured. All these questionable assumptions may have contributed to the large variation of jaw and tooth data of larger sharks. However, most of the jaw and tooth data of large sharks used in this analysis were measured by Gordon Hubbell.

Morphometrics for KANGA and MALTA

The most valid TL estimates for KANGA and MALTA are still the original ones provided by Riseley and Abela, respectively. The TL of large white sharks has to be directly measured if a precision >10% (0.7 m range at 7 m TL) is required. Our TL range estimates from fin, jaw, tooth, and mass regressions from white sharks of known TL varied from 25% to 50%. The TL ranges from fin morphometrics were disappointingly large because the regression CVs were so large. The calculated ranges in TL estimated from jaw and tooth morphometrics were large because white sharks >4 m TL varied more than those of smaller sharks (see Randall, 1973). We did not expect mass, which varies greatly, to reasonably validate TL estimates (see Chapter 9, by Mollet and Cailliet). This was confirmed by the large CV (23%) for the condition factor, M/TL^3 (Table I). The mass of 167 males and 223 females from the Natal Shark Board (see Chapter 32, by Cliff *et al.*) had fewer outliers and, correspondingly, a smaller CV (12%). However, the variations were still too large for precise TL estimation.

More accurate morphometrics would be required to produce smaller TL ranges from body measurements. Body morphometrics >65% TL had CVs smaller than 2.5%, in agreement with the results of Bass *et al.* (1975), and would yield smaller TL ranges. However, if a large fraction of the body of a large white shark can be measured, then TL (or better, PRC) likely could be measured as well.

Fin morphometrics should be taken on fresh material, or the shrinkage of fins should be determined experimentally. The estimate of the fresh pectoral anterior margin (1.128 m, 16.2%) for KANGA was within the limits expected for a 7 m TL white shark. The MALTA pectoral fin measurements were taken on a fin that had dried for 2 years in a bar. The anterior margin (0.938 m, 13.4%, assuming 10% shrinkage) was less than expected (Fig. 1A). A shrinkage of 15% (certainly possible) produced an upper limit of TL at about 7 m. R. Smera (personal communication) observed fin shrinkage of up to 17% and 26% for pectoral and dorsal fin morphometrics, respectively, for a 5.3-m female caught in Cananéia, Brazil (Appendix 1). However, it probably is not valid to compare shrinkages of the fins of a fish prepared by a taxidermist with those of air-dried fins.

The use of improved methods for measurement and the inclusion of more recent data from large white sharks reduced the difference between estimated TL and calculated TL from jaw and tooth morphometrics. Randall (1973) did not use a confidence band for a new observation, and his data came from small sharks. The variability of upper jaw perimeter and UA1E2 of large white sharks was surprisingly large (Fig. 1D and E). We concur with Ellis and McCosker (1991), who found that tooth size may reach an upper limit in white sharks and may not accurately reflect age or overall size. We conclude that the sizes of the jaw and teeth of KANGA are compatible with the original estimate of 7 m TL, whereas Cappo (1988), following the method of Randall (1973), estimated a TL of 6.4 and 5.8 m from jaw and teeth, respectively. Our TL range calculated from the jaw data for MALTA did not include 7 m. We suggest that the upper range in our data is an underestimate, because we included data from small sharks. If sharks <4 m TL are excluded from the regression, the upper limit becomes indeterminate. The upper jaw perimeter of 1.120 (16%) becomes compatible with a 7-m shark.

The number of bands in the central portion of a vertebra (Cailliet *et al.*, 1985), determined by the proper laboratory techniques (Cailliet *et al.*, 1983b), may provide more precise TL estimates. At least, such measurements should provide an accurate estimate of the age of a large white shark. A visual band count of 15 actually was reported by Abela for a vertebra of uncertain location in the vertebral column of MALTA. This number of bands was much lower than expected, because a 15-year-old white shark should be about 5 m long (Cailliet *et al.*, 1985). This TL would be considerably less than the minimum estimate of 6–6.5 m from photographic evidence (see Chapter 30, by Fergusson). The 5.67–6.00 m TL Gans Bay and 5.36–5.7 m TL North Cape females had 23 and 22 ± 1 bands, respectively (see Chapter 15, by Francis). We suggest that use of the proper techniques would have shown additional bands in the MALTA vertebra.

It is problematic to extrapolate beyond the range of the available data. The ratio-on-size regression results showed that KANGA and MALTA could have been 7 m TL, as reported by Riseley and Abela, respectively, and that our methods and data were not adequate for a reasonably precise TL validation. The use of a ratio-on-size instead of a size-on-size regression has been questioned by many (e.g., Simpson *et al.*, 1960; Sokal and Rohlf, 1981), but we suggest ratio-on-size to be the more appropriate for length validation of large white sharks. The ratio-on-size data used in morphometrics should be considered as a transformation, because all the measured data are divided by the same number (i.e., TL). We suggest that ratio-on-size is not prone to type II errors (see McArdle, 1988) and agree that, because of large type II errors, size-on-size could

be particularly inappropriate for the determination of TL ranges. However, we found that size-on-size regressions produced very small type II errors (0.1–0.2 m); Mollet and Cailliet (Chapter 9) further discuss the advantages of the ratio-on-size compared to the size-on-size regression. The TL ranges from the condition factor (a mass/TL^3 ratio) versus TL regression were very similar to those from a GM power regression (log W versus log TL). Indeed, it is true that the size-on-size regressions would yield narrower 95% confidence bands for large TL and wider ones for small TL compared to those of the ratio-on-size regression. Along with Simpson *et al.* (1960), Sokal and Rohlf (1969), and Cailliet *et al.* (1986), we suggest that the cautious use of a two-parameter ratio-on-size regression is more appropriate for morphometrics that show large variations or reach an upper absolute limit for large fish. The use of a size-on-size regression requires a three-parameter curve to obtain equally good fits. We found that the ratio-on-size regression, like the log–log transformation, produced more homoscedastic and normal residual distributions for many morphometrics compared to the size-on-size regression. Mollet and Cailliet (Chapter 9) could not substantiate such an improvement, but they used a different set of white shark data comprising only large-scale morphometrics.

Summary

Morphometrics of over 70 white sharks *C. carcharias* were used to validate estimations of TL. The TLs of two large white sharks were estimated from upper jaw perimeter, UA1E2, and pectoral fin size. KANGA was estimated to be 5.3–8.2 m TL, consistent with the >7 m TL estimate of Peter Riseley. MALTA was estimated to be 4.6–7.0 m TL, making the disputed 7 m TL measurement by John Abela possible. None of the additional morphometrics, nor the estimated mass, of MALTA ruled out a TL of 7 m. The most solid TL estimates for KANGA and MALTA are still the original ones provided by Riseley and Abela, respectively. The TL of large white sharks has to be directly measured if a precision range of >0.7 m (10% at TL = 7 m) is required. For these two sharks, the TL range estimates from the 95% confidence band for new observations of upper jaw perimeter, tooth enameloid height, fin morphometrics, and mass from sharks of known TL varied between 25% and 50%. Better definitions for certain morphometrics, to allow for more consistent measurement, and a proper protocol for measuring large white sharks are needed. Such a protocol was presented here.

Acknowledgments

J. Abela brought the MALTA shark to the attention of elasmophiles, and we are grateful for all the data and information he shared with us. I. K. Fergusson provided the contacts and data that initiated this chapter. M. Cappo, J. E. McCosker, J. E. Randall, R. H. Rosenblatt, and G. L. Wood provided us with copies of their correspondence with J. Abela. G. Hubbell contributed the bulk of the jaw and tooth data for large white sharks. J. E. Randall provided negatives for the reconstruction of data in his 1973 article.

Morphometrics of white sharks were made available by A. F. Amorim and C. A. Arfelli, G. Burgess, M. Cappo, J. Castro and G. Hubbell, D. Catania, G. Cliff and S. F. J. Dudley, M. P. Francis, R. Hueter, R. Menni, J. Seigel, R. Smera, H. J. Walker and R. H. Rosenblatt, and R. Warneke. M. P. Francis' help with the evaluation of early morphometrics was much appreciated. S. Baldridge provided hard-to-find references crucial to enlarging the database. A. Mollet, P. Stipa, and M.-E. Genchi translated Spanish and Italian papers. Finally, we acknowledge the comments of an anonymous reviewer, who improved our manuscript.

APPENDIX 1   General Information for White Sharks Used in this Study

No.[a]	TL (m)	Sex	Location	Date	Collection no.	Reference
1	1.261	M	Los Angeles, California	June 18, 1985	LACM 43804-1	J. Seigel (personal communication)
2	1.290	M	Baja California, Mexico	June 1981		Cailliet *et al.* (1985); this study
3	1.307	M	Newport Pier, California	June 12, 1985	LACM 43805-1	J. Seigel (personal communication)
4	1.321	F	Redondo Beach, California	May 31, 1984	LACM 43638-1	J. Seigel (personal communication)
5	1.400	M	Algoa Bay, South Africa	1950		Smith (1951)
6[b]	1.430	F	North Cape, New Zealand	November 13, 1991		Francis (Chapter 15)
7[b]	1.449	F	North Cape, New Zealand	November 13, 1991		Francis (Chapter 15)
8	1.521	F	Kaipara Harbor, New Zealand	January 19, 1993		Francis (Chapter 15)
9	1.540	M	Bodega Bay, California	September 10, 1984	MBA1, MLML31	This study
10	1.543	F	La Jolla Pier, California	July 21, 1948		Fitch (1949)
11	1.590	M	Los Angeles, California	January 1, 1990	LACM	J. Seigel (personal communication)
12	1.594	F	San Diego, California	November 9, 1955	SIO 55-95f	A. Flechsig (personal communication)
13	1.626	M	La Jolla Pier, California	November 6, 1955	SIO 55-95e	Klimley (1985b)
14	1.629	F	La Jolla Pier, California	October 31, 1955	SIO 55-95b	Klimley (1985b)
15	1.632	M	Southern California	October 31, 1955	SIO 55-95b	Klimley (1985b)
16	1.637	F	Port Hueneme, California	June 1993	LACM	J. Seigel (personal communication)
17	1.696	F	San Diego, California	October 1955	SIO 55-95g	A. Flechsig (personal communication)
18	1.764	F	Los Angeles, California	December 31, 1989	LACM	J. Seigel (personal communication)
19	1.806	F	La Jolla Pier, California	October 30, 1955	SIO 55-95a	Klimley (1985b)
20	1.807	F	San Diego, California	November 12, 1955	SIO 55-95g	A. Flechsig (personal communication)
	1.831	M	Darien, Georgia	January 6, 1994		G. Burgess (personal communication)
21	1.837	F	Los Angeles, California	January 4, 1990	LACM	J. Seigel (personal communication)
22	1.843	F	Santa Monica Bay, California	January 13, 1990	LACM 44842-1	J. Seigel (personal communication)
23	1.877	M	Los Angeles, California	January 23, 1990	LACM	J. Seigel (personal communication)
24	1.877	F	Santa Monica Bay, California	January 13, 1990	LACM	J. Seigel (personal communication)
	1.943	M	Soquel Point, California	August 10, 1959	CAS 26376	W. I. Follett (personal communication)
25	1.960	M	Playa Sur, Mazatlán, Mexico	January 25, 1964		Kato (1965)
	1.967	F	Cape Lookout, North Carolina	April 18, 1974		G. Burgess (personal communication)
26	2.000	F	La Jolla Cove, California	November 4, 1971	SIO 71-196	I. Taylor (personal communication)
27	2.170	F	Puerto Quequén, Argentina	December 11, 1956		Siccardi *et al.* (1981)
	2.184	M	Indian Beach, California	July 30, 1959	CAS 26367	W. I. Follett (personal communication)
28	2.240	M	Puerto Quequén, Argentina	July, 12, 1953		Siccardi *et al.* (1981)
	2.254	M	Stinson Beach, California	November 9, 1959	CAS 26694	W. I. Follett (personal communication)
	2.419	F	Stinson Beach, California	December 27, 1960	CAS 27013	W. I. Follett (personal communication)
29	2.420		Sète, Gulf of Lions, France	1876		Moreau (1881)

(continues)

APPENDIX 1 *(Continued)*

No.[a]	TL (m)	Sex	Location	Date	Collection no.	Reference
30	2.540	F	Gulf of Maine	October 1937		Schroeder (1939)
	2.650	M	Stinson Beach, California	December 28, 1960	CAS 27015	W. I. Follett (personal communication)
	2.673	M	Bolinas Bay, California	November 13, 1959	CAS 26695	W. I. Follett (personal communication)
	2.774	F	Indian Beach, California	July 29, 1959	CAS 26366	W. I. Follett (personal communication)
31	2.830[c]	M	Natal, South Africa	October 1987	DAE-871111-01	This study
32	2.830	M	Puerto Quequén, Argentina	January 29, 1960		Siccardi *et al.* (1981)
33	2.896	F	Vero Beach, Florida	March 27, 1985		G. Hubbell (personal communication)
34	2.935[c]	M	Natal, South Africa	October 1987	DAE-871111-02	This study
	2.959	F	Tomales Bay, California	July 2, 1960	CAS 26781	W. I. Follett (personal communication)
35	3.025[c]	F	Brighton Beach, South Africa	April 1987	NSB-BRI-873	This study
36	3.050	M	Puerto Quequén, Argentina	December 29, 1952		Siccardi *et al.* (1981)
	3.251	M	Seal Rocks, Victoria, Australia	January 19, 1969	RW 376	R. Warneke (personal communication)
	3.270	F	Stinson Beach, California	January 2, 1961	CAS 27014	W. I. Follett (personal communication)
37	3.643	F	Balboa, California	February 1–3, 1960	SIO	C. Limbaugh and R. H. Rosenblatt (personal communication)
	3.750	M	Seal Rocks, Victoria, Australia	November 9, 1969	RW 459	R. Warneke (personal communication)
38	3.790	M	Marathon, Florida	April 23, 1992		J. Castro and G. Hubbell (personal communication)
39	3.826[c]	M	Cape Town, South Africa	August 26, 1987	LJVC-870830	This study
40	3.830		Mediterranean Sea	Before 1865		Duméril (1865)
41	3.870	F	Gulf of Maine	August 24, 1949		Scattergood *et al.* (1951)
42	3.930	M	Monterey Bay, California	September 25, 1978		Cailliet *et al.* (1985); this study
43	4.064	F	Amagansett, New York	October 5, 1964		Pratt *et al.* (1982)
	4.064	F	Seal Rocks, Victoria, Australia	March 4, 1969	RW 383	R. Warneke (personal communication)
44	4.100	M	Cap Bon, Tunisia	May 22, 1956		Postel (1958)
45	4.353	M	Santa Catalina Island, California	May 17, 1985	LACM CCS85-2	J. Seigel (personal communication)
46	4.500	F	Muscongus Bay, Maine	July 20, 1961		Scud (1962)
47	4.510	M	Santa Catalina Island, California	May 17, 1985	LACM CCS85-2	J. Seigel (personal communication)
48	4.570	M	Moriches Inlet, New York	June 29, 1979		Pratt *et al.* (1982)
49	4.572	M	Bornholm Beach, Australia	August 22, 1982	F9682	G. Hubbell (personal communication)
50	4.674		Craig, Alaska	October 1961		Royce (1963)
51	4.700	F	Englewood, Florida	February 1, 1939		Springer (1939), Bigelow and Schroeder (1948)
52	4.712	M	Anacapa Islands, California	April 1985	S41285	G. Hubbell (personal communication)
53	4.750	M	Tossa de Mar, Spain	November 17, 1992	MSP-0492M	I. K. Fergusson (personal communication)
54	4.788	F	Davenport, California	October 6, 1989	MLML #32	This study
55	4.942	F	Point Dume, California	August 30, 1982	LACM 42894	J. Seigel (personal communication); Cailliet *et al.* (1985)
56	4.960[d]	F	San Miguel Island, California	March 18, 1958		Kenyon (1959)
57	5.105	F	Monterey Bay, California	January 23, 1957		Follett (1966)
58	5.200	F	Cap Bon, Tunisia	May 16, 1956		Postel (1958)
	5.234	F	Santa Barbara Island, California	November 8, 1993		C. Winkler (personal communication); J. Seigel (personal communication)
59	5.300	F	Todohokke, Japan	May 30, 1985		Nakano and Nakaya (1987)
60	5.300	F	Cananéia, Brazil	December 8, 1992		Arfelli and Amorim (1993)
61	5.350	F	Egadi Islands, Sicily	May 8, 1987	MSI/0287F	I. K. Fergusson (personal communication)
62	5.360	F	North Cape, New Zealand	November 13, 1991		Francis (Chapter 15)

(continues)

APPENDIX 1 (*Continued*)

No.[a]	TL (m)	Sex	Location	Date	Collection no.	Reference
63	5.368	F	Point Vincente, California	September 18, 1985	LACM CCS85-9	J. Seigel (personal communication); G. Hubbel (personal communication)
64	5.474	F	Port Pirie, Australia	January 1984	X11384B	G. Hubbell (personal communication)
65	5.486	F	Malta	1973		J. Abela (personal communication)
66	5.520	F	Dunedin, New Zealand	January 1886		Parker (1887)
67	5.537	F	Israelite Bay, Australia	July 20, 1980	F82380	G. Hubbell (personal communication)
68	5.537	F	Whyalla, Australia	January 1984	X1138C	G. Hubbell (personal communication)
69	5.633	F	Anacapa Islands, California	November 6, 1985	L11985	G. Hubbell (personal communication)
70	5.740	F	Bunbury, Australia	July 2, 1991	H10991	G. Hubbell (personal communication)
71	5.944	F	Ledge Point, Australia	March 22, 1984	H5384	Randall (1987); G. Hubbell (personal communication)
72	5.67–6.00	F	Gans Bay, South Africa	January 17, 1987	LJVC-870303	This study
73	6.096	F	Alberton, PEI, Canada	August 4, 1983	M91683	G. Hubbell (personal communication)
74	6.408	F	Cojímar, Cuba	1945		Guitard and Milera (1974); Randall (1987)
75	~7.0	F	Fifla Farallon, Malta	April 17, 1987	MMA-0187F	Fergusson (Chapter 30); this study
76	>7.0	F	Kangaroo Island, Australia	April 1, 1987		Jury (1987); Cappo (1988)

TL, Total length.
[a]Only numbered individuals were included in the analysis.
[b]Embryo.
[c]TL (actually TOT) was measured with the caudal fin in the depressed position.
[d]TL was calculated from the fork length.

APPENDIX 2 Measurements of the Jaw and Dentition of White Sharks

Measurements						General information	
TL (m)	FUJP (m)	DUJP (m)	UA1E2 (mm)	UA1H (mm)	UA1W (mm)	Collection no.	Reference
1.960			15.0		12.0		Kato (1965)
1.975		0.385	17.5				Randall (1973)
1.975		0.410	19.3				Randall (1973)
2.040			20.0				Randall (1973)
2.170			18.0		13.5		Siccardi *et al.* (1981)
2.195		0.433	20.2				Randall (1973)
2.420			21.0		17.0		Siccardi *et al.* (1981)
2.420			22.0	29.0	21.0		Moreau (1881)
2.420		0.487	20.5				Randall (1973)
2.560		0.547	25.5				Randall (1973)
2.675		0.533	24.5				Randall (1973)
2.715		0.530	22.5				Randall (1973)
2.795		0.570	25.2				Randall (1973)
2.830			23.6	33.3	22.3	DAE-871111-01	This study
2.935			28.1	33.5	24.3	DAE-871111-02	This study
3.025			28.1*	35.0	28.0	NSB-BRI-873	This study
3.050			28.0		25.0		Siccardi *et al.* (1981)
3.140		0.618	26.0				Randall (1973)

(*continues*)

APPENDIX 2 *(Continued)*

	Measurements					General information	
TL (m)	FUJP (m)	DUJP (m)	UA1E2 (mm)	UA1H (mm)	UA1W (mm)	Collection no.	Reference
3.310		0.685	29.9				Randall (1973)
3.643			38.0	38.0	36.0		C. Limbaugh and R. H. Rosenblatt (personal communication)
3.710		0.785	31.0				Randall (1973)
3.790		0.790	32.5	41.0	26.0		J. Castro and G. Hubbell (personal communication)
4.075		0.815	35.5				Randall (1973)
4.500			26.0[a]	36.0	29.0		Scud (1962)
4.572		1.120	43.0	54.0		F9682	G. Hubbell (personal communication)
4.674			34.0		37.0		Royce (1963)
4.700			37.0	52.0			Springer (1939)
4.712		1.230	40.0	49.0		S41285	G. Hubbell (personal communication)
4.788	0.838	0.805*				MLML #32	This study
4.825		0.978	43.9				Randall (1973)
4.885		0.938	45.0				Randall (1973)
4.942			34.6*	43.0	33.0	LACM 42894	J. Seigel (personal communication)
4.960			44.2*	55.0	43.1		Kenyon (1959)
5.105			46.0				Follett (1966)
5.110		0.985	44.7				Randall (1973)
5.300	1.065	1.022*	48.3[b]				Arfelli and Amorim (1993); R. Smera (personal communication)
5.360[c]	1.150	1.104*	51.0				Francis (Chapter 15)
5.368	1.220	1.160	47.0	58.0	39.9	LACM CCS85-9	G. Hubbell (personal communication)
5.370		1.055	49.3				Randall (1973)
5.474		1.200	45.0	56.0		X11384B	G. Hubbell (personal communication)
5.486		0.980	47.1[d]	63.8[d]	44.0[d]		J. Abela (personal communication); this study
5.520			40.0		37.0		Parker (1887)
5.537		1.120	40.0	49.5		F82380	G. Hubbell (personal communication)
5.537		1.265	47.0	60.0		X1138C	G. Hubbell (personal communication)
5.633	1.310	1.270	49.0	61.0		L11985	G. Hubbell (personal communication)
5.740		1.128	42.0	56.0		H10991	G. Hubbell (personal communication)
5.944		1.300	51.0	63.0		H5384	Randall (1987); G. Hubbell (personal communication)
5.67–6.00			48.6[b]	62.0	50.0	LJVC-870303	This study
6.096		1.430	47.5	59.0		M91683	G. Hubbell (personal communication)
6.401			44.0	57.0			Guitard and Milera (1974); Randall (1987)
7.010	1.220	1.120	46.9	55.5	38.4		J. Abela (personal communication); this study
7.010	1.297	1.250	51.6				Cappo (1988)

TL, Total length; FUJP, fresh upper jaw perimeter; DUJP, dried upper jaw perimeter; UA1E2, tooth enameloid height; UA1H, tooth total height; UA1W, tooth total width of the larger first upper tooth of white sharks; *, calculated value (see text).

[a] Scud's (1962) results were not included in the final regression analysis because this tooth was likely the third upper.

[b] The left first upper replacement tooth and the right first upper functional tooth were 46 mm.

[c] Used TOT (total length with the caudal fin in the depressed position) = 5.9 m, based on a 5 m precaudal length expected length of a shark, consistent with its 22 vertebrae bands.

[d] Determined from a photograph of John Abela, placing a ruler along the enameloid edge of the tooth.

APPENDIX 3 **Protocol for Verification of Large White Shark Records and International Specimen Data Form**

A. If you catch a large white shark and cannot release it, bring it in intact to experts so that its size can be verified. If possible, do not cut up the shark. DO NOT REMOVE THE JAWS and DO WATCH THE SHARK before and during the necropsy to prevent spectators from stealing individual teeth or other parts.

B. Call scientists or instructors from local museums, fisheries institutes, universities, or high schools or local physicians to take the measurements or assist the captors in doing so, as well as conducting a necropsy of the captured animal. Also, call in local officials for the purpose of verifying the measurements.

C. In most places where white sharks are captured, they can be hauled onto piers, placed in the shade, and iced at least until scientists arrive to conduct the necropsy. If lifting the shark out of the water onto a boat or pier, use a sling to lift horizontally. DO NOT tie a rope around the tail and lift vertically, as the animal will be distorted and stretched.

D. Obtain the precise locality data for the shark when captured, including coordinates, as well as capture date, captor, vessel, gear, distance from shore, bottom and capture depth, habitat, and anything else of interest from the fishers that captured the shark, including behavioral notes.

E. Obtain the following basic morphometrics using straight-line measurements between parallels and not over the curve, with standardized methodology (see Fig. A1A and B):
   1. TOT (total length from the snout tip to the extended caudal tip, i.e., with the caudal fin in the depressed position)
   2. TLn (natural total length from the snout tip to the caudal tip, with the caudal fin in the natural position)
   3. FOR (fork length from the snout tip to the notch at the juncture of the upper and lower postventral caudal margins)
   4. PRC (precaudal length from the snout tip to the anterior edge of the bottom of the upper precaudal pit)
   5. HDL (head length from the snout tip to the upper end of the fifth gill slit)
   6. PD1 (predorsal length from the snout tip to the origin of the first dorsal fin)
   7. SVL (snout–vent length from the snout tip to the origin of the vent, (i.e., cloacal aperture)
   8. GIR (girth, or circumference around the body at the posterior end of the pectoral bases)
   9. P1A (pectoral fin anterior margin)
   10. P2A (pelvic fin anterior margin)
   11. CLO (clasper outer length from the pelvic fin base to the clasper tip)
   12. CLI (clasper inner length from the front of the vent to the clasper tip)
   13. CLB (clasper base width, measured across the clasper base)
   14. D1H (first dorsal fin height, perpendicular to the body axis from the base to the apex)
   15. CDM (caudal fin dorsal margin from the upper precaudal pit to the tip of the caudal fin)
   16. CPV (caudal fin preventral margin from the lower precaudal pit to the tip of the lower caudal lobe)

F. Have the local news media or a photographer photograph the following items with a *normal* or short *telephoto* lens, *not* a wide-angle lens, in all cases with a SCALE BAR (meter stick or equivalent) provided in the photo: (1) lateral view of the shark; (2) lateral view of the shark's head; (3) first dorsal fin; (4–5) pectoral fin (state which side), dorsal and ventral surface; (6) abdomen between pectoral and pelvic fins; (7–8) pelvic fin (state which side), dorsal and ventral surface; and (9) caudal fin.

G. If possible sketch or prepare a diagram to note the peculiarities of the shark, including color pattern on the side of the head, fins, flanks, and tail. Note all scratches, bite marks, damage to fins, and other injuries.

H. Weigh the shark AFTER measuring it, but BEFORE cutting into it. Weigh with the stomach contents intact. In many places, sharks can be weighed intact in fish-processing plants. If a truck scale is available, weigh the truck (tare it), load the shark, weigh the truck with the shark, and subtract the former. If the truck scale is some distance from the shark, weigh the shark with the truck's fuel tank at the same level each time, or add or subtract any fuel difference between the two weighings. If it is necessary to dismember the shark to weigh it, carefully keep as much of the body fluids as possible and weigh them, too.

I. Have the total mass and various measurements and a description of the methods used in obtaining them, as well as any photographs taken, validated by local officials (notaries, police, judges, the mayor, etc.) with signed affidavits or notarized statements, with the scientific or medicinal staff, as well as the captors, signing them in public.

J. Examine the animal for stomach content. Weigh and sort ALL contents, looking for the remains of small animals, such as squid beaks, fish otoliths, fish bones, etc., as well as those of large ones. A sieve and water for flushing will help the process. Photograph the intact stomach contents.

K. Note the condition and size of the ovaries, oviducts, and any embryos or fetuses of females, and the testes, epigonal organs, sperm ducts, sperm sacs, and claspers of males. Photograph the process of examining them. In males, check the sperm sacs for spermatophores, and check whether the urogenital duct will project seminal fluid when the sperm sacs are palpitated. In females, check the ovaries for ovarian follicles, and measure a small sample of them. Photograph, weigh, and measure ALL fetuses, and if possible, freeze them for further examination or fix them in 10% formalin solution (1 part concentrated formaldehyde to 9 parts water).

L. Count the number of intestinal valves in the valvular intestine. Save any tapeworms or other parasites from the intestine or other body organs for a parasitologist (freeze them or put them in 10% formaldehyde).

M. Weigh (to the nearest gram) the liver, ovaries, or testes, heart, and brain, with the smaller organs weighed on a small sensitive scale.

(continues)

IWS SPECIMEN REGISTER NO.: | c | | | | .

Shark Research Center, South African Museum | **WHITE SHARK PROJECT BASIC DATA**

International White Shark Specimen Data Sheet
Carcharodon carcharias (Linnaeus, 1758) Family Lamnidae

Version V

Field No.:_____. Accession No.:_____.
Station No.: _____. Catalog No.:_____.
Photos:_____. Radiographs:_____.

MATERIAL SAVED: jaws - tooth sets - cranium - vertebral column - hyobranchial
skeleton - fin skeletons and girdles - vertebrae - denticles - entire skeleton -
entire shark - other _____.

Locality:_____.
Area: _____.
Province/State: _____ Country: _____.
Latitude: _____°____'____" N S Collector: _____.
Longitude: _____°____'____" E W Vessel:_____.
Date: Y: _____ M:_____ D:_____. Gear: _____.
Time: _____ AM PM. Distance from shore:_____km/m.
Habitat:_____.
_____ Depth: _____m.
Data from: _____.
Other:

Sex: F M ? Maturation Stage: 1 2 3 4 ? Other: _____.

Basic MORPHOMETRICS:
TOT:_____. P1A:_____.
TLn:_____. P2A:_____.
FOR:_____. CLO:_____.
PRC:_____. CLI:_____.
HDL:_____. CLB:_____.
PD1:_____. D1H:_____.
SVL:_____. CDM:_____.
GIR:_____. CPV:_____.

WEIGHTS:
Body: _____kg
Liver: _____kg
Gonads: _____kg
Heart: _____gm
Brain: _____gm

SPIRAL Valve #: _____.
HEART Valve #: _____.

Tooth measurements

TOOTH UL P____L____I____A____A____I____L____P____UR
COUNTS: LL P____L____A____A____A____L____P____LR

VERTEBRAE:
MP_____ LM_____
DP_____ LD_____
DC_____ HM_____
MP Sample? Y N

TOOTH Measurements/Serration #
UA2H:_____mm UA2ED_____mm
UA2W:_____mm UA2RD_____mm
UA2E1:_____mm
UA2E2:_____mm UA2SM_____#
UA2EM:_____mm UA2SD_____#

COLOR PATTERN:
Axillary spot: y n ?
Black P1 tip: y n ?
Small side spots: y n ?
Caudal base pattern: y n ?
Dorsolateral
Color?:

BIOLOGICAL DATA:

OTHER DATA:

DUJP:_____. DLJP:_____.

DUJP
DLJP

Jaw Measurements
(others optional)

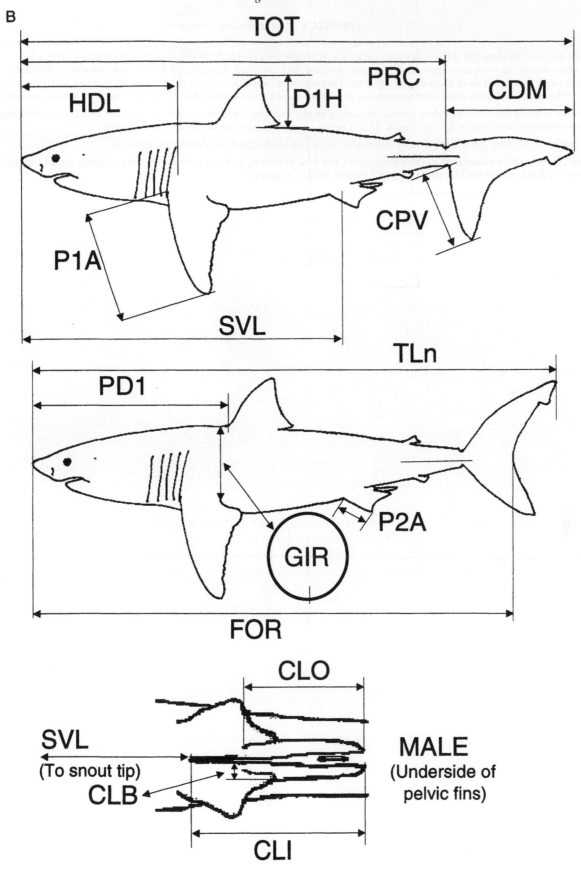

Basic white shark morphometrics

APPENDIX 3 (*Continued*)

N. Remove the jaws, photograph them intact with a scale bar, measure the upper and lower jaw perimeters fresh according to Fig. A1A, and if possible, repeat the measurements some time later after the jaws are dried. Measure (to the nearest millimeter) the first and second upper anterior functional teeth according to Fig. A1A. Determine the dental formula, and count the tooth rows (see Welton and Farish, 1993). If the functional teeth (first row) are missing or broken, record and measure the replacement teeth in the second row.

O. If possible, remove the entire vertebral column (including caudal vertebrae), and freeze, preserve, or dry intact; properly label sections. Remove vertebrae from below the first dorsal fin if it is not possible to save the whole vertebral column.

P. Send copies of the data, photos, and other information collected on the shark to a shark researcher(s).

Q. Deposit representative material of the shark (i.e., jaws and fins, vertebrae, pelvic fins with claspers for males; dermal denticle samples from the midback under the first dorsal fin) in a natural history museum.

PART

IV

Physiology

11

Temperature, Swimming Depth, and Movements of a White Shark at the South Farallon Islands, California

KENNETH J. GOLDMAN[1]
Department of Biology
San Francisco State University
San Francisco, California

JOHN E. McCOSKER
California Academy of Sciences
San Francisco, California

SCOT D. ANDERSON
Inverness, California

A. PETER KLIMLEY
Bodega Marine Laboratory
University of California, Davis
Bodega Bay, California

Introduction

Like other members of the family Lamnidae, white sharks *Carcharodon carcharias* have countercurrent heat exchangers in the circulatory system. These exchangers enable them to achieve a body temperature warmer than that of ambient seawater (Carey and Teal, 1969; Carey *et al.*, 1981, 1982; Smith and Rhodes, 1983). The heat exchangers are formed from masses of parallel arteries and veins, the retia mirabilia (Burne, 1923). By retaining metabolic heat within the body, they raise the temperature of the muscle, viscera, brain, and eyes (Tricas and McCosker, 1984; Carey *et al.*, 1985; Block and Carey, 1985; Wolf *et al.*, 1988).

We are beginning to learn more about thermal physiology and its relationship to ecology in this species. Using telemetry, McCosker (1987) demonstrated elevated stomach temperature in a free-swimming white shark, and Carey *et al.* (1982) and Strong *et al.*

(1992) described white shark swimming depth and movements. In this chapter, we report data, collected by telemetry, on stomach temperature, combined with seawater temperature, swimming depth, and movements of a white shark off the west coast of North America. We discuss elevated organ temperatures as they relate to thermal physiology.

Materials and Methods

Transmitters were manufactured by VEMCO Ltd. of Nova Scotia, Canada (model V4TP-8H). They operated at 32.768 kHz, and had a battery life of 57–91 days. Each transmitter possessed a thermistor and a depth sensor having ranges of 0–30°C and 0–200 m, respectively. All transmitters were calibrated by the manufacturer; we verified thermistor readings on four separate occasions in a digital-readout water bath using a thermometer (Fluke K/J 51). The two devices always registered within 0.1°C. The VEMCO calibrations and our water bath calibrations were within 0.2°C.

The shark was fed a 3- to 4-kg piece of prepared

---

[1]Current address: Virginia Institute of Marine Science, School of Marine Science, College of William and Mary, Gloucester Pt., VA 23602.

pinniped flesh with the transmitter attached. We did not chum, in order to avoid leaving a large residual odor corridor that might alter the subsequent behavior of the shark. A large float was used as a visual decoy to attract a shark to the surface in an area where we knew they had been present (see Chapter 20, by Anderson *et al.*). When a shark appeared, the bait (with the transmitter) was attached to the float line and placed in the water.

The shark was tracked, and data were accumulated, using an ultrasonic telemetry receiver (Dukane N30A5A). Thermal and swimming depth data were passed through an analog-to-digital converter (Ultrasonic Telemetry Systems, Brea, California) and were decoded with a microcomputer interface and a laptop computer (Zenith Z-171; for a description, see Cigas and Klimley, 1987). The tracking vessel used for this study was a 5.4-m Boston whaler.

Temperature–depth profiles (TDPs) of the water column were obtained by lowering a second transmitter while the shark was being tracked. These profiles were obtained at the buoy located on the southeast side of South Farallon (East Landing), where conditions were sufficiently calm. These profiles are considered to be estimates of water temperature (T_w) at the swimming depth and location of the shark, because sea temperature was not measured below 20 m (bottom depth at the buoy), and we could not account for water movement around the island. Off central California, fall is a period of reduced upwelling, resulting in uniform temperatures at depth in the vicinity of the islands. Temperature at the surface was taken (with the Fluke K/J 51 thermometer) at various times and places around the island as the shark was tracked. These surface temperatures never differed

from those at the profile site by more than 0.1°C. This validates the use of these data as estimates of T_w (T_w) at swimming depth. Maximum and minimum stomach temperatures were compared to T_w at the surface and T_w at the mean swimming depth for each day (Table I).

The position of the shark was determined by taking positions of the tracking vessel using a theodolite from atop Lighthouse Hill on Southeast Farallon (for a further explanation of technique, see Klimley *et al.*, 1992). These positions were taken frequently, no more than 15 minutes apart. After the first day of tracking, we realized that theodolite positions alone did not give the best indication of the shark's location. Thus, for the rest of the study, compass bearings were taken from the boat to the strongest acoustic signal as well, thereby estimating the position of the shark. These data appear as vector arrows on the tracking map (Fig. 3). According to manufacturer specifications, the typical range of the transmitters used, under ocean conditions and with the marine life existing at the South Farallon Islands, is approximately 2 km. Range was reduced in the near-vicinity of the island by acoustic "background" noise. The signal could be heard at times when data could not be recorded. Therefore, shark positions could be taken at times when temperature and swimming depth data could not.

Results

Stomach temperature (T_s), swimming depth, and movements of a free-swimming adult male white shark were obtained, by means of acoustic telemetry,

TABLE I **Measurements Telemetered from a White Shark at the Farallon Islands**

Date	T_s (mean ± SD)	T_wSS	T_wSD	$T_s - T_w$SS (max)	$T_s - T_w$SS (min)	$T_s - T_w$SD (max)	$T_s - T_w$SD (min)	D_s(m) (mean ± SD)
October 17, 1991	24.6 ± 0.3	14.7		11.2[a]	8.7[a,b]			23.5 ± 3.9
October 18, 1991	26.2 ± 0.2	14.6	13.2	12.2[a]	11.2[a]	13.6[a]	12.6	34.1 ± 1.0
October 20, 1991	26.5 ± 0.2	16.4	13.8	11.0[a]	9.0[a]	13.6[a]	11.6	32.2 ± 0.7
October 22, 1991	27.3 ± 0.2	14.4	13.9[c]	13.2[a]	12.4[a]	13.7[a]	12.9	13.9 ± 0.9
October 24, 1991	26.4 ± 0.1	13.3		13.2[a]	12.9[a]			28.3 ± 1.4

T_s, Mean stomach temperature; T_w, sea-surface temperature (°C); T_w, estimated water temperature; SD, swimming depth (except where it indicates standard deviation); SS, sea surface; D_s, mean swimming depth.

[a]Statistically significant (*t* test; *p* < 0.0005) between maximum (max) and minimum (min) differences between T_s and T_w at either SS or SD.

[b]Low value due to ingestion of seawater with the bait. The value was 9.5°C once the mean T_s was reached.

[c]Measured, not estimated, value.

at intermittent periods over the course of 8 days (October 17–24, 1991) (Figs. 1 and 2). During the first 7 minutes following ingestion of the bait, T_s rose 1.3°C, from 23.4°C to 24.7°C (Fig. 1A). Surface T_w was 14.7°C (Table I), and a depth profile was not obtained. Mean T_s was 24.6°C, and the maximum elevation of T_s over T_w was 11.2°C, with a minimum elevation of 8.7°C.

The animal's mean swimming depth was 23.5 m. Theodolite positions of the tracking vessel were taken over a 90-minute period (solid circles, Fig. 3), and compass bearings were used on each subsequent day. The shark's general location was always between the boat and the island.

The shark was relocated on day 2 (October 18), and

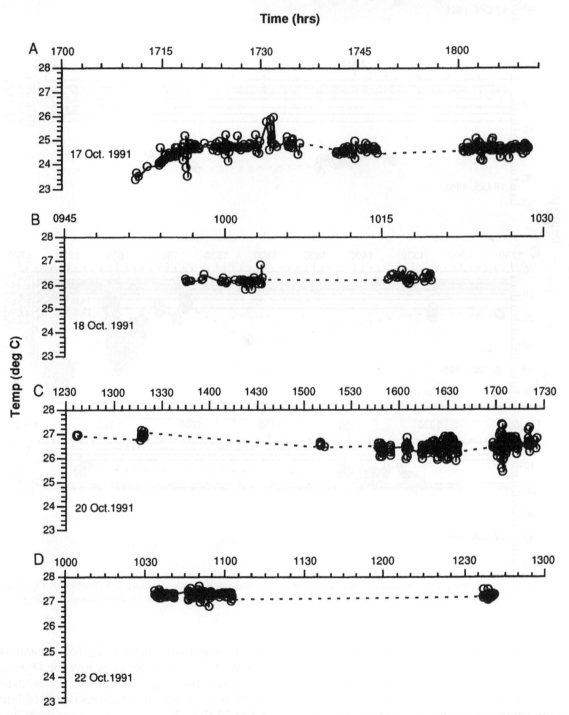

FIGURE 1 Stomach temperature record of *Carcharodon carcharias* at the South Farallon Islands: (A) October 17, (B) October 18, (C) October 20, and (D) October 22, 1991.

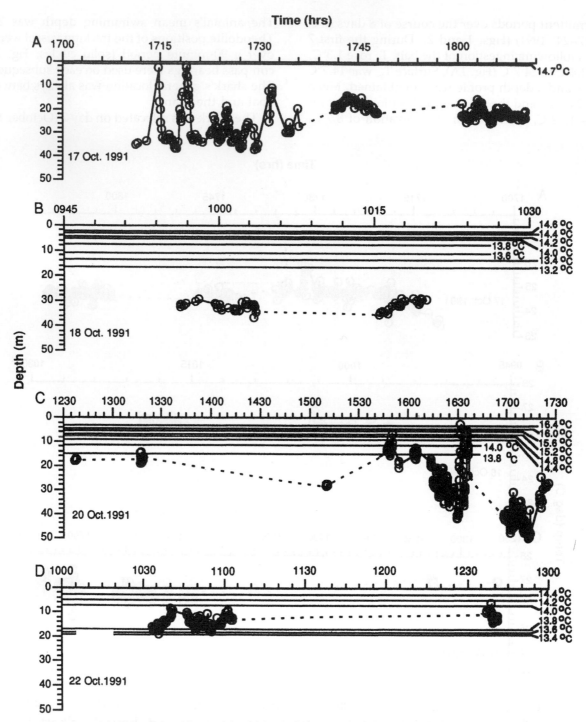

FIGURE 2 Swimming depth record of *Carcharodon carcharias* at the South Farallon Islands: (A) October 17, (B) October 18, (C) October 20, and (D) October 22, 1991. Temperature depth profiles are shown along the right margin.

data were acquired from 0956 to 1021 hours. T_s ranged from 25.8°C to 26.8°C (Fig. 1B). A TDP was obtained to a depth of 16.1 m. T_w was 13.2°C (Fig. 2B), compared to a mean T_s of 26.2°C. The maximum and minimum elevations of T_s over surface T_w were 12.2°C and 11.2°C, respectively, or 13.6°C and

12.6°C when compared to T_w for a mean swimming depth of 32.1 m (Table I). A low fog layer prevented theodolite tracking, but our location and that of the shark were similar to positions recorded later, on October 30 (Fig. 3).

The animal was then relocated 2 days later (Octo-

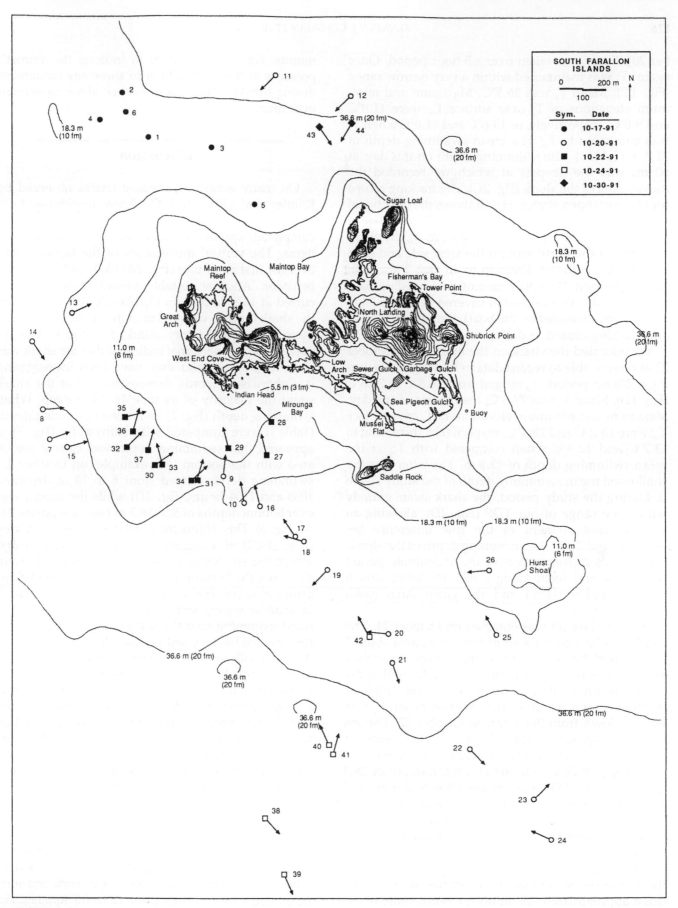

FIGURE 3 Intermittently recorded horizontal movements of tagged *Carcharodon carcharias*, October 17–24, 1991. Symbols indicate the position of the tracking vessel; vectors indicate the direction of the shark from the boat.

ber 20). Data were taken over a 5-hour period. Once again, T_s was maintained within a very narrow range (Fig. 1C). Mean T_s was 26.5°C. Maximum and minimum elevations of T_s over surface T_w were 11.0°C and 9.0°C, respectively, or 13.6°C and 11.6°C when T_s was compared to T_w at a mean swimming depth of 31.3 m (Table I). The swimming depth on this day, at 50 m, was the deepest at which we recorded the movements of the shark (Fig. 2C). The tracking record for this day (open circles, Fig. 3) shows the directional compass bearings. During the last 135 minutes of tracking (theodolite positions 11–26, Fig. 3), the shark swam from the north around the west side and then in a south–southeast direction to an area 2 km from shore (position 24) before turning and heading back to the islands. This excursion covered at least 8.2 km (the distance traveled by the boat). Data accumulation and tracking ceased at dark.

We relocated the shark on the morning of October 22 and were able to record data in two segments during a 2-hour period. T_s ranged from 26.8°C to 27.6°C (Fig. 1D). Mean T_s was 27.3°C, the highest of any day. Maximum and minimum elevations of T_s over surface T_w were 13.2°C and 12.4°C, respectively, increasing to 13.7°C and 12.9°C when compared with T_w at the mean swimming depth of 13.9 m. The latter was the shallowest mean swimming depth of the study (Table I). During the study period, the shark swam entirely within the range of our TDP (Fig. 2D), allowing an accurate measurement of the true difference between T_s and T_w at the swimming depth of the shark. The shark's movements over the 150-minute period were back and forth along a 600- to 700-m stretch between Indian Head and the Great Arch (solid squares, Fig. 3).

A small data set was obtained on October 24. The shark was located for a short time in an area about 2 km south of the islands. The distance from the islands and the direction of movement may indicate that the shark was on another excursion away from and back to shore (open squares, Fig. 3). The mean T_s was 26.4°C, down from the mean on October 22. The sea surface temperature was 13.3°C, giving maximum and minimum elevations of 13.2°C and 12.9°C, respectively, while swimming at a mean depth of 28.3 m (Table I). No TDP was obtained due to rough seas. Therefore, comparison of T_s was with surface T_w only. While the signal was strong enough to track the shark for 45 minutes, we could record data for only an 8-minute period.

After 5 days of stormy weather, which prevented the use of our small tracking boat, the shark was located approximately 350 m off the north shore of the islands. The two theodolite positions (solid diamonds, Fig. 3) serve more to indicate the animal's presence at the islands than to show any movement during the day. They do, however, show movement over time.

Discussion

On many occasions, feeding sharks observed by Klimley et al. (1992; K. J. Goldman, unpublished observation) remained at an attack site after feeding was completed, apparently searching for remaining food items. The vertical movements of the tagged shark over the first 8 minutes (Fig. 2A) likely indicated such behavior. Another notable vertical movement occurred at 1630 hours, on October 20 (Fig. 2C), when the shark came to within 3 m of the surface. Although there was no predatory attack, feeding, or bait involved, the strong signal indicated that the shark was close to the boat and may have been investigating it. Otherwise, records demonstrated that the shark spent the majority of its time 16–34 m deep. When swimming depth (Fig. 2) and mean swimming depth (Table I) were compared with bathymetry (Fig. 3), it appeared that swimming depth was closely associated with the bottom. For example, on October 22, swimming depth ranged from 8 to 18 m (between 1033 and 1104 hours, Fig. 2D) while the animal was over bottom depths of 5.5–18.3 m (see data points 30–34, Fig. 3). This differs from the observation by Carey et al. (1982) of a tagged shark in very deep water, where the shark's association was with the thermocline, not the bottom, which lay many meters below. Strong et al. (1992) also observed a tagged white shark in shallow water, and noted a strong tendency toward swimming near the surface or along the bottom (but not midwater), with mean daytime swimming depths shallower than those at night. Swimming depth is probably an important factor in the dynamics of predation on pinnipeds at the Farallones. Another is that the dark gray dorsum of white sharks allows them to be cryptic when swimming over a rocky bottom (Klimley, 1994) and in water of high turbidity (K. J. Goldman, unpublished observation). With its dark dorsum, swimming closer to the bottom when underwater visibility is good may enable a shark to get closer to its prey while remaining unseen, thus increasing the chance of successful predation (see Klimley, 1994; Pyle et al., Chapter 34).

The overall tracking record (Fig. 3) indicated that different types of search patterns may be used by white sharks for locating prey. The back-and-forth nearshore sweeps on October 22 (solid squares, Fig. 3) and the long excursion on October 20, and possibly again on October 24 (open circles and open squares, Fig. 3), may indicate two search patterns, similar to those of white sharks observed by Strong *et al.* (1992; see also Chapter 37) in South Australia. Obviously, recorded movements of more sharks are needed to determine the commonality of these patterns.

During most of our tracking periods, the shark swam at and along sites where large numbers of predatory attacks on pinnipeds have been recorded over the years (Fig. 4). This indicates that white sharks stay close to where more seals and sea lions are likely to be in the water. The nearshore area (<450 m of the islands) has been defined as a "high-risk zone" for pinnipeds (Klimley *et al.*, 1992). At high tide, more seals are forced into the water due to a loss of haul-out space (see Chapter 25, by Anderson *et al.*) but remain near the island. Interestingly, the shark was more active on days when tidal fluctuations were large. The back-and-forth nearshore sweeps and the excursions away from and back to the islands occurred on days with large tidal ranges (1–2 m), but not on days with small ranges (<1 m; Fig. 3).

The possession of warm muscle, viscera, and stomach, as well as brain and eyes, probably allows an increase in neural, digestive, and muscle activity similar to those patterns apparent in other animals. We recorded maximum elevations of 13.2°C over surface T_w and 13.7°C over T_w at the mean swimming depth of the shark. These elevations are almost twice that observed by McCosker (1987), but 70% of the difference in the maximum elevation of T_s over surface T_w between the two studies can be accounted for by differences in surface T_w. In the McCosker study, surface T_w was 18.1°C, and T_s ranged from 24.0°C to 25.5°C. In this study, surface T_w ranged from 13.3°C to 14.7°C, while T_s ranged from 24.5°C to 27.5°C. Most of the 30% difference due to T_s derived from the 2.0°C difference between the upper limits of T_s at the two study sites (25.5°C and 27.5°C, respectively). These preliminary data may indicate that the white shark, in colder water, possesses a slightly higher mean T_s, which may equalize the effects of colder water (i.e., maintaining a similar thermal inertia to the shark in the warmer environment).

Maximum and minimum differences between T_s and surface T_w (Table I) most certainly underestimate the true differences ($T_s - T_w$ at the swimming depth of the shark), while those between T_s and T_w, at mean swimming depth, give a more precise representation (even though TDPs did not measure deeper than 20 m). This statement is supported by data from October 22, when the shark swam within the boundaries of our TDP during the entire tracking (Fig. 2D), thus allowing a more accurate measurement of T_w at

the swimming depth of the shark. The estimated maximum and minimum differences ($T_s - T_w$) on other days were virtually identical (Table I). This also indicates that T_w changed very little between 20 m and the shark's mean swimming depth.

Over the 8-day period, mean T_s slowly increased and then decreased (Table I). This may have been due to the digestion of food eaten prior to the ingestion of our bait, the digestion of our bait, or both. It could also simply reflect natural temperature fluctuations in the stomach, or relate to the increase and decrease of ambient T_w during that time (Table I). Bluefin tuna *Thunnus thynnus* elevate their stomach temperature after ingestion of food by as much as 15°C over a 12- to 20-hour period, before returning to the 5–6°C elevation otherwise maintained (Carey *et al.*, 1984). We cannot discount the presence of our bait in the stomach as the cause of the rise in T_s, nor can we prove it to be the cause. It is possible that our shark had fed just prior to our study. There were several predatory attacks on pinnipeds during the study period, but we know that this shark did not participate in them. However, we are not able to discount participation in two attacks during that week or in an attack that occurred in the same area as our tagging, 9 hours earlier. And, of course, attacks occur that go unobserved by researchers at the Farallones. The shark's continued presence at the island through October 30 may also indicate incomplete or unsuccessful feeding (see Klimley *et al.*, 1992).

A shark's stomach would probably experience the least amount of thermal change due to changes in ambient T_w, as compared to body muscle (Carey and Teal, 1969; Carey *et al.*, 1982), because it is centrally located within the body cavity. It lies ventral to the subcutaneous lateral rete and dorsal (and slightly posterior) to the suprahepatic rete, thus rendering it a good indicator of body core temperature, T_b. This decreases the likelihood that the changes in T_s that we observed in our shark were due to changes in T_w.

An elevated visceral and stomach temperature has been shown to increase the digestive capabilities of bluefin tuna, allowing them to process roughly three times more food (particularly proteins). This is due to the increased activity rates of digestive enzymes, such as trypsin and chymotrypsin (Stevens and McLeese, 1984). The evidence to this point does not confirm that white sharks warm their stomachs during digestion. If they do warm their stomachs then, on the basis of our data, it appears to be a slow process with a small increase (maybe 2–3°C) and subsequent decrease over many days, unlike the case of the bluefin tuna. The primary prey of white sharks at the South Farallon Islands are juvenile northern elephant

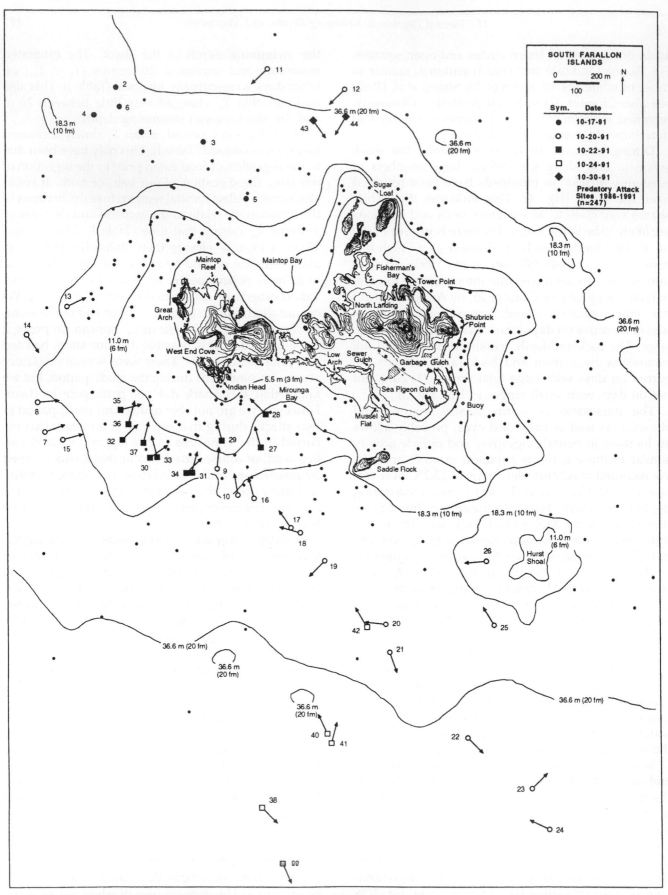

FIGURE 4 Recorded predatory attacks on pinnipeds at the South Farallon Islands, 1986–1991 (updated from the work of Klimley *et al.*, 1992), along with the tracking record.

seals *Mirounga angustirostris* (Klimley *et al.*, 1992), which can weigh 100–300 kg, constituting a large meal. An increase in T_s would aid in the digestion of food material, particularly lipids (e.g., seal blubber), which are rich in energy but somewhat slow to digest (i.e., not easily hydrolyzed). Even if they do not raise their T_s during digestion, it would seem likely that they achieve some increase in their digestive rates simply due to the maintenance of a warm stomach (and viscera). The digestion and the assimilation of large amounts of high-energy lipids could also be important as reserves to white sharks, if they spend long periods without feeding (Carey *et al.*, 1982).

If the white shark indeed does achieve benefit from being warm-bodied, these effects may be secondary to those achieved millions of years ago; that is, the development of the retia allowed white sharks to inhabit cool waters (see Chapter 8, by Purdy). The selection for and development of the orbital rete to warm the brain and eyes, the suprahepatic rete to warm the viscera and stomach, and the subcutaneous lateral rete to warm the muscle appear to be important driving aspects in the evolution and life history of lamnid sharks.

The majority of extant shark species live in warm and tropical waters. Of the cold-water species, some attain size equal to or greater than that of the white shark. Animals such as basking sharks *Cetorhinus maximus* and prickly sharks *Echinorhinus cookei* occur in the waters off central and northern California. These are slow-moving animals probably possessing a low metabolic rate, and neither is an active or aggressive predator. *Cetorhinus* has been shown to have a muscle temperature close to that of ambient T_w (Carey *et al.*, 1971). By contrast, *Carcharodon* is an active animal and an aggressive predator with a metabolic rate probably close to that of birds and mammals (Emery, 1985). Moreover, these sharks possess a body temperature significantly higher than that of the ambient water. Being warm-bodied most likely allowed white sharks to maintain their place as effective hunters of large swift-moving prey in cool and cold environments (see Chapter 8, by Purdy).

Our temperature data are comparable to those of Carey *et al.* (1985), Emery (1985), and McCosker (1987). Regardless of the ambient T_w, T_s of the white shark appears to be maintained within a very narrow range. With T_s being a good indicator of body core temperature, this suggests thermoregulation in this species. One possibility is that, through the evolution of the retia, they have developed an optimal physiological operating temperature maintained with minimal effect from ambient T_w. This would allow their physiological needs to be met with the highest possible efficiency.

The thermal tolerance of this species is demonstrated by its latitudinal distribution (see Compagno, 1984b); the role played by ambient T_w in affecting this distribution might be even more important to smaller sharks. It is possible that small white sharks have a more difficult time maintaining optimal body temperature in colder waters due to the potential for greater heat loss (by conduction and convection) from a small body. If this is true, it might explain the geographic segregation between small young white sharks and larger, older ones. Off the east coast of North America, white sharks <200 cm total length are almost exclusively found in the Mid-Atlantic Bight during the warm months between June and September (Casey and Pratt, 1985). Along the North American west coast, smaller white sharks are found with greater frequency in the warmer waters south of Point Conception, whereas large individuals are found north of there (Klimley, 1985b). In the Mediterranean, too, small white sharks are more frequently found in the warmer areas (see Chapter 30, by Fergusson).

Summary

T_s, swimming depth, and movements of an adult male white shark *C. carcharias* were intermittently recorded by acoustic telemetry over an 8-day period at the South Farallon Islands, California. The shark remained near sites where large numbers of attacks on pinnipeds, the species' primary prey, have occurred. Temperature profiles of the water column (T_w) were concurrently obtained. Maximum elevation of T_s over T_w was 13.7°C, and daily mean swimming depth ranged from 13.9 to 32.1 m. We commented here on the role of temperature regulation in the ecology of this species.

Acknowledgments

Several organizations and many individuals helped with different aspects of this study. We thank the U.S. Fish and Wildlife Service and Point Reyes Bird Observatory (PRBO) for access to the South Farallon Islands, as well as T. Charkins (vessel *Kumbaya*) for transporting us and our equipment to and from the islands. We are grateful for the assistance provided by L.V.A.O., and for funding provided by the Samuel S. Johnson Foundation and the California Academy of Sciences (CAS). Many people at San Francisco State University provided support in various forms, including R. Larson, T. Niesen, A. Arp, and E. Gartside. Special thanks go to M. Fountain for assistance in the field; to D. Nelson, S. Gruber, J. McKibben, and A. Stull for assistance with telemetry; and to the staff of the Steinhart Aquarium, especially T. Tucker and J. Rampley, and the director of the CAS, R. Eisenhardt. The following PRBO biologists and volunteers helped in taking theodolite positions and in deploying our boat: P. Pyle, J. Walsh, B. Sydeman, D. Sibley, P. Vernon, C. Alexander, and J. Feldman. J. Worobieff (vessel *Ranger 85*) and a British Broadcasting Corporation film crew (vessel *Sea Biscuit*) helped transport gasoline to the island. Pin-

niped bait was obtained through PRBO (National Marine Fisheries Service permit 667). D. G. Ainley and several reviewers greatly improved the manuscript. Finally, a special acknowledgment goes to P. and S. Goldman for their tremendous personal support. The tagged shark was given the name Boots, in fond memory of Sam J. "Boots" Cohen (uncle of K.J.G.). This chapter is dedicated to his memory.

12

The Brain and Cranial Nerves of the White Shark: An Evolutionary Perspective

LEO S. DEMSKI
Division of Natural Sciences
New College
University of South Florida
Sarasota, Florida

R. GLENN NORTHCUTT
Department of Neurosciences and Neurobiology
Unit of Scripps Institution of Oceanography
University of California, San Diego
La Jolla, California

Introduction

The central nervous and sensory systems of elasmobranchs have been studied in some detail, especially with respect to three features: brain–mass to body–mass ratios; relative brain area development, particularly as it may involve sensory mediation of behavior; and identification of neural pathways and cell groups and their probable evolution (see the review by Northcutt, 1989). Studies support several generalizations. First, sharks, skates, and rays generally have larger brains than other ectothermic vertebrates. Indeed, certain sharks and batoids have brain–body mass ratios that overlap the range for mammal and birds; a striking degree of development, based on both relative size and structural complexity, of the cerebrum and the cerebellum has been well documented for certain carcharhinoids, stingrays, and mantas. Second, relative development (i.e., the size and/or complexity) of sensory areas can be used to estimate behavioral capabilities in sharks. The enlarged dorsal nucleus of the medulla is correlated with an animal's capacity to analyze electrosensory information, whereas the relative thickness and position of cell layers in the optic lobes or the tectum are probably indicative of visual abilities. Finally, within

major taxonomic groups (e.g., Carcharhiniformes and Batoidea), there has been an independent tendency for brain enlargement in certain species. Within the carcharhiniforms, for example, scyliorhinids and triakids most often have less well-developed brains than do carcharhinids and sphyrnids. Presumably, selection pressures are driving the brain development, but with a paucity of behavioral information, little can be said about the nature of the underlying causes.

The database for relative brain size is still small; large taxa are represented by only a few observations, some of which are suspect because of methodology. Even less is known concerning the contributions of specific parts of the brain. We therefore saw a need to document the brain size and relative proportions in the Lamniformes, as most measurements and observations have been of the brains of squatinomorphs, squalomorphs, and other galeomorphs, especially the carcharhinids. Previous information on lamnoid sharks is limited: recent data on one basking shark *Cetorhinus maximus* [brain mass and description of brain areas (Kruska, 1988)]; one report of the brain–body mass ratio in the shortfin mako *Isurus oxyrinchus* (Myagkov, 1991); photographs of dissections of heads showing the dorsal aspect of the brain and most of the cranial nerves in small specimens of *I. oxyrinchus* and a white shark *Carcharodon carcharias* (Gilbert, 1963, personal communication); an old report of brain–body mass ratio on one sand tiger shark *Odontaspis* (=*Eugomphodus*) *taurus* (Crile and Quiring, 1940); superficial descriptions, with photographs, brain mass, and body lengths of one specimen each of the bigeye thresher *Alopias superciliosus*, the thresher *Alopias vulpinus*, *I. oxyrinchus*, the longfin mako *Isurus glaucus* (=*paucus*), and *Odontaspis* (=*Eugomphodus*) sp. (Okada *et al.*, 1969); and relative ratios of cerebral and cerebellar areas based on two-dimensional measurements on photographs of dorsal and lateral aspects of the dissected brain of one each of *I. glaucus* (=*paucus*), *I. oxyrinchus*, *A. superciliosus*, and *A. vulpinus* (Sato *et al.*, 1983).

This chapter represents the addition of data on the brain of a male (presumed adult) white shark. Estimated ratios of brain–body mass and relative brain area mass are provided, as is a preliminary description of the internal configuration of the telencephalon, the proximal portion of the cranial nerves, and certain gross features of the eye, nose, and postorbital vascular rete. The anatomical observations are compared with the results of similar studies on other taxa and discussed in the context of likely sensory and behavioral capabilities of the white shark, as well as possible trends in brain evolution in the Lamniformes.

Materials and Methods

The white shark on which this chapter is based was caught near Jacksonville, Florida, on December 7, 1992, by commercial fisherman Martin Fisher of Sarasota, Florida. The fish had well-developed claspers which were associated with some milky fluid (presumed to be milt). The length of the animal was reported to be 358 cm (11 ft 9 in). The measurement is assumed to be total length but was completed in an unknown manner. No other measurements were taken on the whole animal. The flesh, removed for sale, weighed 490 lb (222.3 kg), and the width of the head (257 mm) is documented in Fig. 1. The body remained on the dock for at least several hours for weighing. With the help of Gordon Hubbell, we contacted Fisher, who agreed to remove the head, freeze it, and later transport it in ice to one of us (L.S.D.) in Sarasota. The jaws were removed for sale, and most of the skin of the head was retained for tanning. The head arrived cold but not frozen; it had sustained damage by knife cuts into the otic areas and one of

the nares. Within 1 hour, the specimen was immersed in freshly prepared alcohol–formalin–acetic acid (AFA) fixative, in which it remained until the brain was firm, with a chalky white color (about 1 week). During the period of dissection (about 2 weeks), the head was stored in weak fixative.

A dorsal exposure of the brain and cranial nerves was chosen, as this approach has been standardized in studies of other species and corresponds to the only published photograph of a white shark brain (Gilbert, 1963a). As the ear structures were already damaged, we did not preserve them, but rather made maximum exposure of the related cranial nerves. The orbital contents were intact, permitting clear identification of the extraocular muscles and their innervation and isolation of the trigeminal and anterior lateral-line nerves associated with the area (Fig. 1). Ampullae of Lorenzini were intact on the rostrum and nearby regions. We did not map the electrosensory system, as such a study would be grossly incomplete without the jaws.

Following the initial dissection, the specimen was sent to La Jolla for completion of dissection, photography, and drawing of the brain and cranial nerves (Fig. 2). The brain, including the olfactory bulbs, was then removed and weighed; following this, representative areas were weighed separately. The procedures used have been standardized in studies on other species (Northcutt, 1977, 1978). The whole brain mass (34.2 g) was used to compute an approximate brain–body mass ratio (Table I and Fig. 3), while mass of the individual areas was used to determine their contribution to the total brain mass (Fig. 4). The telencephalic hemispheres (lobes) were further cut transversely into several thick slices in order to reveal features of their internal structure, including the extent and configuration of the lateral ventricles (Fig. 5). No attempt was made to study the microscopic anatomy of the brain or nerves, since, in our experience, elasmobranch neural tissue that has not been fixed soon after death and/or has been frozen without cryoprotection is unsuitable for sectioning.

To compare the relative brain size in *Carcharodon* with that in other cartilaginous fishes, it is necessary to have both a brain and body mass. Lacking the latter, we estimated body mass based on the animal's measured length. The estimated mass of 430 ± 150 kg (95% confidence level) was determined on the basis of length–mass allometry (see Chapter 9, by Mollet and Cailliet). The elasmobranch brain–body mass ratios were plotted on log–log coordinates (Fig. 3) and, with one exception, were within an equal-frequency ellipse calculated to enclose 95% of the data points

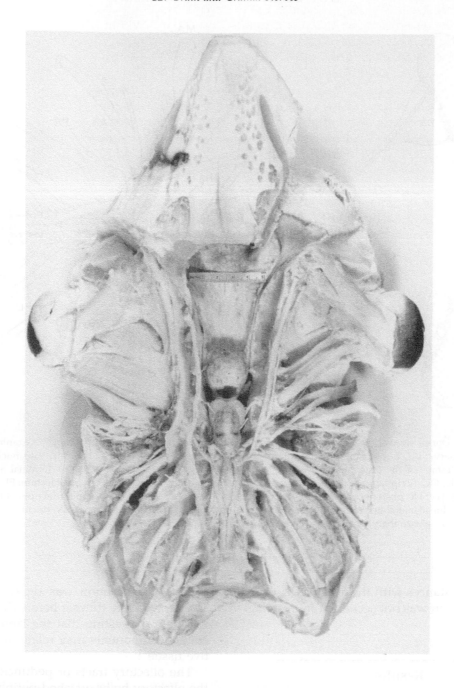

FIGURE 1 Photograph of the dorsal aspect of the brain, cranial nerves, and associated sensory structures (e.g., eyes, extraocular muscles, and ampulla of Lorenzini on the rostrum).

(Sokal and Rohlf, 1981). The slope of the line in Fig. 3 represents the coefficient of allometry for the elasmobranch data.

In order to initiate a study of the general cutaneous innervation, a strip of skin (approximately 15 mm wide) extending from near the ventral to dorsal mid-lines was obtained from the posterior part of the head, and representative samples were fixed in AFA, 10% formalin, or 2% glutaraldehyde. Some pieces were later treated with 2% osmium tetroxide to blacken myelinated cutaneous nerves. Several nerve bundles were teased free of the connective tissue and

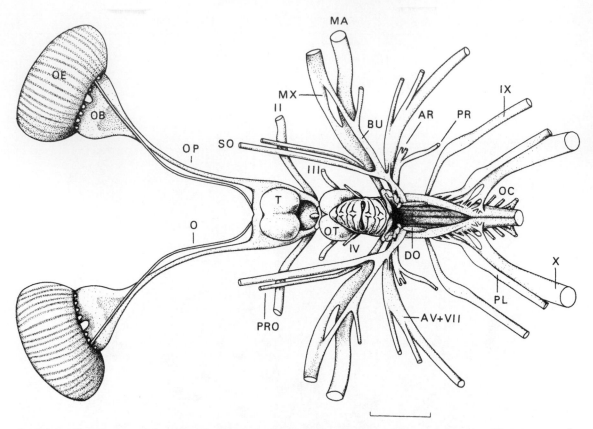

FIGURE 2 Dorsal view of the brain of a white shark. AR, Anterior ramus of the octaval nerve; AV, anteroventral lateral-line nerve; BU, buccal ramus of the anterodorsal lateral-line nerve; DO, dorsal octavolateralis nucleus; MA, mandibular ramus of the trigeminal nerve; MX, maxillary ramus of the trigeminal nerve; O, terminal nerve; OB, olfactory bulb; OC, occipital nerves; OE, olfactory epithelium; OP, olfactory peduncle; OT, optic tectum; PL, posterior lateral-line nerve; PR, posterior ramus of the octaval nerve; PRO, profundal nerve; SO, superficial ophthalmic ramus of the anterodorsal lateral-line nerve; T, telencephalon; II, optic nerve; III, oculomotor nerve; IV, trochlear nerve; VII, facial nerve; IX, glossopharyngeal nerve; X, vagal nerve. Bar, 3 cm.

traced for short distances with the aid of a dissection microscope. The skin was not sectioned for histological study.

Results

Gross Brain Features

Given the size of the head, the first general impression of the brain was that of relative smallness. We had expected that an adult white shark would have a large brain with a greater relative size of the telencephalon and the cerebellum, as is typical of requiem sharks (Carcharhinidae) and hammerheads (Sphyrnidae). On the other hand, the peripheral cranial nerves were larger than those usually encountered in elasmobranchs (Figs. 1 and 2). Indeed, some branches were initially mistaken for muscles. The gross pattern

of nerve distribution was typical for galeomorphs. The nerves were thinner before they exited the cranial cavity, suggesting that the large size of the extracranial components may relate to thickened connective tissue.

The olfactory tracts or peduncles were long, with the olfactory bulbs attached peripherally to the olfactory epithelium or sac. The telencephalon did not have an enlarged dorsoposterior region related to an enlarged central nucleus of the dorsal pallium (i.e., a characteristic of carcharhinids and sphyrnids), but was represented by a thin membrane (Figs. 1 and 2). In addition, the thick transverse slices of the telencephalon revealed a large ventricle and only a moderately developed central nucleus (Fig. 5). The situation was similar to that observed in the basking shark (cf Fig. 4) (Kruska, 1988). The development of the dorsal pallium, and hence the reduction in the ventricle in the white shark, was intermediate between the condi-

TABLE I **Brain and Body Mass of Various Sharks**

Common name	Species	Brain mass (g)	Body mass (kg)	Reference
	Squalomorphii			
Frilled shark	*Chlamydoselachus anguineus*	3.39	3.55	Myagkov (1991)
Sevengill shark	*Heptranchias perlo*	1.94	2.12	Myagkov (1991)
Dumb gulper shark	*Atractophorus armatus* (=*Centrophorus harrissoni*)	4.31	2.82	Myagkov (1991)
Gulper shark	*Centrophorus granulosus*	3.16	1.36	Myagkov (1991)
Longsnout dogfish	*Deania quadrispinosa*	3.47	1.75	Myagkov (1991)
Blackbelly lanternshark	*Etmopterus lucifer*	0.48	0.06	Myagkov (1991)
Greenland shark	*Somniosus microcephalus*	10.29	280.00	Myagkov (1991)
Spined pigmy shark	*Squaliolus laticaudus*	0.44	0.06	Myagkov (1991)
Piked dogfish	*Squalus acanthias*	3.87	4.20	Northcutt (1978)
	Squatinomorphii			
Angel shark	*Squatina squatina*	2.06	6.00	Bauchot *et al.* (1976)
	Galeomorphii			
Horn shark	*Heterodontus francisci*	4.30	2.93	Northcutt (1978)
Nurse shark	*Ginglymostoma cirratum*	31.65	45.30	Northcutt (1978)
White shark	*Carcharodon carcharias*	34.21	280–580	This study
Basking shark	*Cetorhinus maximus*	20.70	385	Kruska (1988)
Shortfin mako shark	*Isurus oxyrinchus*	12.13	40.50	Myagkov (1991)
Sand tiger shark	*Odontaspis* (=*Eugomphodus*) *taurus*	82.55	123.00	Crile and Quiring (1940)
Dusky smooth hound	*Mustelus canis*	8.31	6.50	Northcutt (1978)
Small-spotted cat shark	*Scyliorhinus canicula*	1.38	0.57	Ridet *et al.* (1973)
Finetooth shark	*Aprionodon* (=*Carcharhinus*) *isodon*	18.75	10.87	Northcutt (1978)
Silky shark	*Carcharhinus falciformis*	43.32	36.24	Northcutt (1978)
Bull shark	*Carcharhinus leucas*	54.36	83.80	Northcutt (1978)
Oceanic whitetip shark	*Carcharhinus longimanus*	35.40	40.00	Northcutt (1978)
Dusky shark	*Carcharhinus obscurus*[a]	20.76	12.00	Bauchot *et al.* (1976)
Tiger shark	*Galeocerdo cuvier*	29.44	200.50	Myagkov (1991)
Bull shark	*Prionace glauca*	21.21	36.10	Myagkov (1991)
Sharpnose shark	*Rhizoprionodon porosus*	7.18	3.75	Myagkov (1991)
Scalloped hammerhead shark	*Sphyrna lewini*	59.88	55.71	Northcutt (1978)

[a]Juvenile.

tion typical of squalimorphs (large ventricles) and that of carcharhiniforms [greatly reduced ventricles (cf. Fig. 5) (Northcutt, 1977)].

The attachment to the telencephalon of the non-nervous tela choroidea of the third ventricle occurred in a more rostral and dorsal position than in sharks of other orders (Figs. 1 and 2) (R. G. Northcutt, unpublished observation). Gilbert's (1963a) photograph of the dissected brain of *I. oxyrinchus* suggests that makos also have the more dorsorostral attachment.

Embryological studies are needed to further assess the importance of these differences.

Grossly, the diencephalon is similar to that in most sharks; however, conclusions about its relative development, on the basis of gross size without histological analysis of cell and axon distribution patterns, could only be tentative. The midbrain dorsal surface or optic lobe (tectum) was of moderate size and likely not as large as that of advanced carcharhiniforms (for a comparative summary, see Northcutt, 1977, 1978).

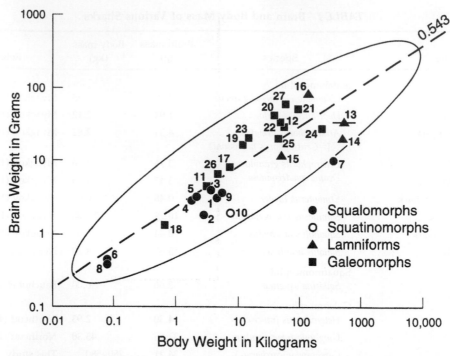

FIGURE 3 Distribution of brain and body mass for the shark taxa listed in Table I. The brain–body data are enclosed by an equal-frequency ellipse with a 95% confidence region; the major axis of the ellipse has a slope of 0.543.

The body of the cerebellum had anterior and posterior lobules and the foliation typical of other lamniforms (for the shortfin mako, see Northcutt, 1977, 1978; for the basking shark, see Kruska, 1988). It was certainly larger with more foliation than the cerebellum of squaliforms, but did not reach the level of these characteristics attained by the myliobatiforms, sphyrnids, and some carcharhinids (for comparisons, see Northcutt 1977, 1978, 1989). The medulla was large and open dorsally, a situation more typical of squalimorphs than galeomorphs (see references cited above). The dorsal nucleus, the area that receives electrosensory input from the anterior lateral-line nerve, and the medial nucleus, the region of termina-

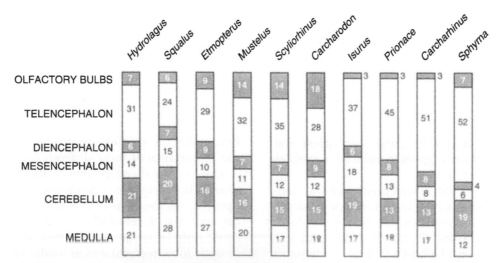

FIGURE 4 Relative mass of the major brain divisions as a percentage of total brain mass in a number of cartilaginous fishes, including the white shark.

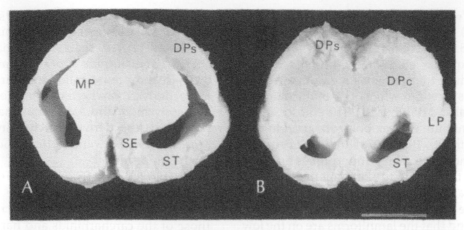

FIGURE 5 Transverse sections through (A) rostral and (B) caudal portions of the cerebral hemispheres of a white shark. Note the large lateral ventricles in A (see text for details). DPc, pars centralis of the dorsal pallium; DPs, pars superficialis of the dorsal pallium; LP, lateral pallium; MP, medial pallium; SE, septal nuclei; ST, corpus striatum. Bar, 1 cm.

tion of primary lateral-line mechanosensory input, were both relatively small. The rostral spinal cord was typical in gross appearance and size to that of other large elasmobranchs.

Sensory Areas

The olfactory sac or epithelium was large, as were the bilobed olfactory bulbs (Figs. 1 and 2; see below). The terminal nerve extended across the lateral lobe of the olfactory bulb into the fissure between the olfactory bulb and epithelium (Fig. 2) (for a review of elasmobranch terminal nerve anatomy and function, see Demski, 1993). The long olfactory tracts merged with the telencephalic hemispheres in the lateral position typical of elasmobranchs.

The eyes were large, as were the extraocular muscles (Fig. 1). A pad of thick connective tissue was present above the orbital contents. Its function is unknown, but it may cushion the orbital contents when the head contacts prey or other objects. We did not examine the eye or the retina (see Gruber and Cohen, 1985).

The semicircular canals and otolith organs were not studied in detail; however, casual observation confirms the size relationships of the canals as illustrated by Gilbert (1963a). The peripheral lateralis system was not studied in detail. Based on the reduced nature of the anterior lateral-line lobe (see above), the electrosensory system is not likely to be exceptional in the white shark.

We made preliminary observations on the innervation of skin. Bundles of myelinated nerve fibers coursed through the thick dermal connective tissue. Their terminations could not be determined at the gross level.

Quantitative Brain Size Estimation

The allometric equation $E = kP^a$ compares brain development among elasmobranchs of different body sizes, where E and P are brain and body mass, respectively, and a and k are constants. A log transformation of the relationship becomes a linear equation $\log E = \log k + a \log P$, with $\log k$ (the y intercept) known as the index of cephalization and a (the slope), the coefficient of allometry. Figure 3 illustrates a plot of brain and body mass from data in Table I and includes the present findings on *C. carcharias*. The horizontal line through point 13 for the white shark represents the loci of positions, if the range of estimated body mass (at the 95% confidence level) is used: that is, 430 ± 150 kg, or 280–580 kg. The coefficient of allometry, or the slope of the main axis of the ellipse, is 0.543. This is lower than earlier estimates (Northcutt, 1977, 1978). Points above or below the line for the equation represent animals with relatively large or small brains, respectively. With the exception of point 16 for the sand tiger shark (see Discussion for comments concerning the validity of this determination), the lamniforms are below the line. In contrast, with the exception of point 18 for *Scyliorhinus* and point 24 for the tiger shark *O. (=E.) taurus*, the other galeomorphs are clearly above the line. The tiger shark *Galeocerdo cuvier* data point, however, is suspect, since earlier estimates give brain–mass to body–

mass ratios that would plot above the line, for example, 107.5 g/200 kg (Crile and Quiring, 1940) and 20.7 g/13.5 kg (Bauchot et al., 1976).

The equal-frequency ellipse (Fig. 3) is based on a statistical procedure that provides a graphical representation of the space into which 95% of the data can be mapped (Sokal and Rohlf, 1981). The ellipse is preferable to the minimum convex polygons used in earlier studies (Northcutt, 1977, 1978). In the ellipse, outlying points do not have as disproportionate an effect on data representation as in the latter method. Results indicate that, as a group, squalomorphs and squatinomorphs have the smallest brains among the superorders of sharks, and that the lamniforms are on the low end of the range for galeomorphs, overlapping some members of the other taxa (Table I and Fig. 3).

We assessed the relative contribution of six major brain areas to the total brain mass for 10 cartilaginous fishes (Fig. 4). The measurements on all species were carried out in a similar fashion (see Materials and Methods). From a comparative standpoint, the most striking feature of the white shark brain was the unparalleled relative contribution of the olfactory bulbs [18%, which is the highest of all chondrichthyans studied (cf. Fig. 1) (Northcutt, 1978)]. The relative mass of the telencephalon (28%) for C. carcharias is more typical of squaliforms (24–29%) than of other galeomorphs (32–52%). The diencephalon and midbrain masses for the white shark place it in a relative size range common to most sharks, with the exceptions of carcharhinids and sphyrnids, in which the areas are relatively smaller. Regarding the cerebellum, C. carcharias has a relative mass (15%) slightly below average (13–20%) (Fig. 4). Interpretation of the relative mass must not be assumed to mean lack of development. For example, carcharhinids generally have a lower relative cerebellar mass than do squaloids; however, from a structural and absolute mass standpoint, their cerebella are clearly better developed. The relative mass of the medulla of the white shark (18%) is within the range for other galeomorphs (17–20%), except Sphyrna (12%), which is low because of its massive telencephalon. Squalimorphs have the greatest relative mass for this part of the brain (27% and 28%), a finding related in part to the small size of their forebrain. Although not studied specifically, the rostral spinal cord of C. carcharias is similar to that of other large sharks.

Discussion

On the basis of brain-to-body size mass ratios and the relative size—and, in some cases, complexity—of certain brain areas, an adult white shark has a moderately sized, fairly generalized brain similar to those in closely related sharks (i.e., makos and basking sharks). The lamnids appear to have smaller relative brain-to-body mass ratios than are characteristic of more derived carcharhiniforms (e.g., carcharhinids and sphyrnids) and have ratios that overlap those of presumed less derived carcharhiniforms, including Mustelus, Prionace, and perhaps Galeocerdo (see comments in Results).

The plotted data (Fig. 3, point 16, and Table I) suggest that sand tiger sharks have the largest brains within the Lamniformes, with ratios overlapping those of the carcharhinids and hammerheads. These data are suspect, however, as the original methodology was not clearly outlined (Crile and Quiring, 1940); more recent data from Okada et al. (1969) indicate that the brain mass-to-body length ratio of an Odontapsis sp. is small (5 g/1 m) compared to that of a derived carcharhinid (Carcharhinus sp.: 51 g/1.8 m, or 28.2 g when normalized to 1 m) and that of a sphyrnid (Sphyrna zygaena, 29 g/1.2 m, or 24.2 g/1 m). Rather, it is in the same range as that of some squaliforms (e.g., Deania rostrata: 5 g/0.94 m, or 5.3 g/1 m) and probably that of the shortfin mako I. oxyrinchus (16 g/2 m, or 8 g/1 m). Thus, it is likely that sand tigers have brain development typical of the other lamnids.

Okada et al. also measured the brain-to-body length ratios in thresher sharks A. superciliosus (30 g/1.5 m, or 20 g/1 m) and A. vulpinus (20 g/1.8 m, or 11.1 g/1 m). The data suggest that the bigeye thresher may have the largest brain among lamniforms. Indeed, the thresher's brain–body ratio may overlap that of some derived carcharhinids. In fact, Okada et al. included photographs suggesting that the dorsoposterior telencephalon and cerebellum are larger and more complex in A. superciliosus than in I. oxyrinchus, I. glaucus (=paucus), or Odontaspis (=Eugomphodus) sp. The presence of enlarged brains in the threshers could be related to aspects of their behavior that differ from that of other lamniforms, for example, use of the long tail in prey capture or, possibly, heavy reliance on vision based on the large size of the eyes in A. superciliosus. It seems likely that the greatest brain development in the order, on the basis of size and complexity of the telencephalon and the cerebellum, has taken place in the threshers rather than in the lamnids. More accurate data on brain–body mass are needed.

The dorsal central area of the telencephalon, or forebrain, is not as large in white sharks as it is in many carcharhinids and sphyrnids. This statement is based on two observations: (1) the lack of a caudal telencephalic protuberance, which, in other sharks, represents the superficial aspect of the central nucleus of the dorsal pallium [this also appears to be the case in makos (cf. figures in Gilbert, 1963a; Okada *et al.*, 1969)]; and (2) the relatively large ventricular spaces in the hemispheres (as observed in the telencephalic slices), typical of species with a small central nucleus.

Neuroanatomical, physiological, and behavioral studies support the idea that the central nucleus is involved in multimodal sensory integration and the control of complex behaviors (Cohen *et al.*, 1973; Platt *et al.*, 1974; Schroeder and Ebbesson, 1974; Graeber, 1978; Graeber *et al.*, 1978; Luiten, 1981; Demski, 1991a). Enlargement of the area in the carcharhinids and sphyrnids seems to be correlated with elaborate social and territorial behaviors that may require learning of individuals and landmarks. These behaviors are especially elaborate in the hammerheads, in which social systems include dominance hierarchies and aggregations subdivided according to sex and size (Myrberg and Gruber, 1974; Klimley, 1981, 1985a, 1987b). If, indeed, the central forebrain controls these activities, such behaviors may not be possible for the white sharks.

The cerebellum, which probably coordinates motor activities, especially as they relate to sensory input (Northcutt, 1978; Roberts, 1978), is well developed in the white shark; that is, it is fairly large, with some lobulation and foliation, but, again, not to the degree seen in some other sharks (see above; for illustrations, see Northcutt, 1977, 1978). Generally, this structure in *Carcharodon* seems comparable to that in other active pelagic species.

In the white shark, all of the major sensory systems known to be involved with prey detection in sharks (e.g., hearing, water movement detection, electric field detection, olfaction, and vision) (see Hodgson and Mathewson, 1978) seem to be well represented based on the size of the relevant cranial nerves and associated brain structures. However, from a comparative standpoint, only olfaction and perhaps vision seem to be exceptional, based on either relative size or numbers of specialized features.

The white shark has the largest relative olfactory bulb size among cartilaginous fishes studied, suggesting that chemical stimuli may be important in guiding feeding, sexual–social, and perhaps other types of behavior. Studies on the sexual behavior of *C. carcharias* are rare (see Chapter 15, by Francis); however, sex pheromones have been implicated in the mating of other galeomorphs (Johnson and Nelson, 1978; Demski, 1991b).

Regarding vision, little definitive information on

extraretinal processing can be gained without a microscopic analysis of the brain. Our gross observations suggest that the midbrain optic lobe is of moderate size, but, as described above, the central area of the dorsal pallium, which is known to be involved in mediating visual discriminations in nurse sharks *Ginglymostoma cirratum* (Graeber, 1978; Graeber *et al.*, 1978), is small. Histological analysis of the retina by Gruber and Cohen (1985) indicates that white sharks have photoreceptors that should permit them to see well under daylight conditions. Our observations of the peripheral visual apparatus also support the probable importance of vision in controlling their behavior. The eyes and the extraocular muscles are massive. In addition, we have confirmed the presence of an orbital vascular rete, which, in lamnids and alopids, is considered a substrate for increasing the temperature of the eye and the brain above the ambient temperature (see Block and Carey, 1985; Goldman *et al.*, Chapter 11). Such brain–eye heaters may allow for faster, and hence more efficient, processing of visual and other types of neural activity.

We can say little concerning the size and complexity of the peripheral lateralis systems, which mediate electroreception and mechanoreception, because of our incomplete specimen. Judging from the lack of thickening of both the anterior lateral-line lobe (and its underlying dorsal nucleus; electrosensory) and the posterior lateral-line lobe (and its underlying nucleus intermedius; mechanosensory), it seems doubtful that the senses are as highly developed as they appear to be in carcharhinids, sphyrnids, and certain batoids, in which both lobes cover most of the floor of the ventricle (see Northcutt, 1978: Fig. 26B). The situation in *Carcharodon* is closer to that typical of squalomorphs, in which the lobes cover only the lateral aspect of the ventricle (cf. Northcutt, 1978: Figs. 1, 2, 13, and 26A).

We also cannot add much information regarding eighth nerve or statoacoustical functions, since the peripheral ear structures were not studied. Moreover, the central nuclei that receive primary eighth nerve afferents are too deep in the medulla to assess relative size through the gross examination of surface structures. The diameter of the eighth nerve roots was in the range typical of other large sharks.

While somatic sensation was not studied in detail, we made preliminary observations on the general cutaneous innervation. Bundles of myelinated nerve fibers were observed in the thick dermal connective tissue; however, their terminations could not be identified. With the exception of the corpuscles of Wunderer, which function as stretch and pressure receptors, specialized general cutaneous receptors have not been described in sharks (Roberts, 1978). The corpuscles of Wunderer are thought to mediate feedback signals involved in controlling locomotion, whereas free nerve endings probably carry information on temperature, touch, and possibly pain (Roberts, 1978). Our initial observations suggest that the white shark most likely has a cutaneous innervation similar to that in other sharks, but additional histological investigations are needed.

We conclude that white sharks do not have the most highly developed brains among sharks. This is especially evident in the small size of the central area of the forebrain compared to that of hammerheads and requiem sharks. *Carcharodon* has a generalized brain with moderate cerebellar size and foliation, an exceptional olfactory bulb, and a retina specialized for diurnal vision. Brain–eye warming may enhance the function of certain neural systems. The cerebellum and the retina seem particularly well adapted for controlling behaviors such as ambushes on seals from below (see Klimley, 1994). The enlarged olfactory bulb suggests that chemical stimuli may be very important for detection and identification of prey, other individual white sharks, including potential mates, and/or various nonbiotic environmental markers. The presence of a relatively small anterior lateral-line area of the medulla suggests that electroreception is probably not a major factor in prey location. Perhaps the system is used mostly after contact with prey, when vision may no longer be effective in guiding an attack. The lack of an exceptional central nucleus of the dorsal pallium of the telencephalon casts doubt on the existence of social hierarchies and perhaps territorial behavior in the white shark.

Summary

The gross appearances of the brain and cranial nerves and total brain and regional brain mass have been described for a 358-cm male white shark *C. carcharias*. An estimated brain–body mass ratio places *Carcharodon* in the range of the lamnids, which generally have smaller brains than do carcharhinids or sphyrnids. The most striking features of the white shark brain are (1) the largest olfactory bulbs relative to total brain mass of any cartilaginous fish measured so far; (2) a fairly small central nucleus of the dorsal pallium, a structure thought to be involved in visual discrimination and perhaps social activities; (3) relatively small dorsal and medial nuclei of the lateral-line lobes, suggesting that electroreception and lateral-line mechanoreception are not well developed compared to the case of many other elasmobranchs; and (4) large eyes and extraocular muscles and the presence of an orbital vascular rete, suggesting that certain visual functions are well developed. The anatomical observations were discussed with respect to sensory–neural control of behavior and to the broader aspects of patterns and processes involved in neural development within elasmobranchs.

Acknowledgments

We express our sincere appreciation to M. Fisher and G. Hubbell for providing the white shark head. Financial support has been provided by National Science Foundation and National Institutes of Health grants (to L.S.D. and R.G.N., respectively) and by the Leonard S. Florsheim, Sr., endowment to the New College Foundation.

13

Reproduction in the
Male White Shark

HAROLD L. PRATT, JR.

Narragansett Laboratory
National Oceanic and Atmospheric Administration–
National Marine Fisheries Service
Narragansett, Rhode Island

Introduction

Knowledge of the reproductive life cycle of white sharks *Carcharodon carcharias* is limited, due to the inaccessibility of this species. Adult white sharks are so large and relatively uncommon that they are not captured frequently anywhere by any method. When they are taken, carcasses are so unmanageable and, except for the jaws and fins, of such low commercial value that they are usually disposed of at sea.

Minimal research has been conducted on the anatomy of white sharks. Ellis and McCosker (1991) illustrated a generalized overview of the placement of organs in the body cavity of an immature specimen, and Matthews (1950) investigated the reproductive system of the male basking shark *Cetorhinus maximus*, a close ally to the family Lamnidae and very similar anatomically to *Carcharodon*.

In the past 12 years, enough large male white sharks have been landed in the western North Atlantic to provide a tentative maturity profile. I present this chapter as a foundation for future efforts.

Materials and Methods

White sharks have been sampled routinely and continuously since 1961 as part of a survey by the

National Marine Fisheries Service, Apex Predator Investigation. Our total database of 380 white sharks for the western North Atlantic was reported by Casey and Pratt (1985). One hundred forty white sharks, mostly juveniles, were examined by the author. Many of the sharks were landed by recreational fishermen using rod and reel, harpoon, or both, from small (<15-m) sportfishing boats. A few were taken by commercial fishermen using pelagic longline gear or caught accidentally in bottom trawls; the remainder were taken by research vessels. The area intensively sampled extended from Cape Hatteras to east of Georges Bank, both on the continental shelf of North America and in the Gulf Stream. Also included were several white sharks collected opportunistically in Florida and California waters. Thirty-eight male white sharks, ranging from 130 to 517 cm in total length (TL) and from 16 to 1558 kg in mass, were caught between June 1971 and July 1991 and dissected to determine maturity as well as organ structure and function.

To determine TL, a caliper-style measurement was made along the body axis from the tip of the snout to a perpendicular line extended from the upper caudal fin (Bigelow and Schroeder, 1948). Because of the rigidity of the tail of *C. carcharias*, this measurement was within a few centimeters of the upper caudal lobe tip measurement used by Compagno (1984a).

Clasper lengths (posterior free tip to the free trailing edge of the pelvic fin lateral to each clasper) and the internal organs were measured to the nearest millimeter as described by Pratt (1979). A few claspers were measured by other workers using cloacal apex or point of rotation as a proximal point. These were converted to the clasper length described above. The size and condition of internal organs were examined through a lateral or ventral incision that exposed the peritoneal cavity from the cloaca to above the pelvic girdle. Testes, epididymis, and spermatophores were preserved in Stieve's fixative (Humason, 1979) or 10% buffered formalin. Tissues were prepared by the paraffin method and sectioned to 7–12 μm. Sexual maturity was determined using the methodology of Clark and von Schmidt (1965) and Pratt (1979). Three conditions of male maturity were defined: (1) juvenile, in which the reproductive organs are relatively quiescent, growing isometrically with body length; (2) subadult, a transitional stage, in which growth of the reproductive organs becomes allometric and signs of incipient activity, such as sperm production and clasper calcification, appear although fertilization is not yet possible; and (3) adult, in which both copulation and fertilization are possible.

Results

Anatomy

Testes

The male white shark has two testes of similar size located in the forward end of the peritoneal cavity. Each is embedded in the anterior portion of a large pale yellow epigonal organ that extends the length of the body cavity (Fig. 1). Testes of juveniles are completely encased in the epigonal organ. Sectioning reveals the testis to be initially composed of one or two lobes, which increase in number as the shark grows. When mature, the edge of the testis erupts from the lateral walls of the epigonal organ. The mature testis is an irregular mass composed of hundreds of lobes, all fitting closely together. Each 10- to 20-mm lobe surrounds a common central germinal zone, around which spermatocysts develop radially (Pratt, 1988). Like all other elasmobranchs whose testes have been investigated, spermatozoa develop in 250-μm spherical spermatocysts, also called seminiferous ampullae or follicles (Parsons and Grier, 1992). The mature spermatocysts are at the outer circumference of the lobe (Stanley, 1963), across which development takes place (Pratt, 1988). Ducts join each mature spermatocyst and coalesce into the ductus efferens, which traverses the epigonal organ. Microscopy shows that sperm travel

singly through the efferent ducts. Paired, broad, double-walled mesorchia, originating under the spine, support each epididymis and testis and enclose the efferent ducts and blood vessels within thin membranous walls. The efferent ducts travel across the anterior edge of the mesorchia and communicate with the head of each epididymis.

Ductus Epididymis

The convoluted tubules of the ductus, with its accessory glands, make up the paired straplike epididymides (Jones and Jones, 1982). The epididymides lie forward and above the testes on the dorsal abdominal wall on either side of the dorsal aorta. The head is a distinct dense node, separated from the main body by a short isthmus. The epididymis is a single highly coiled tubule that gradually increases in diameter as it winds caudally along the hollow of each ventral vertebral arch. For much of its length, the epididymis is embedded in the larger accessory gland (Leydig's gland), which probably secretes material for sperm maturation and support (Fig. 1, ag1). In the adult white shark, the combined ductus epididymis is approximately 35 mm wide and over 1 m long between the head and the ampulla. The head is about 250 mm long, the main body of the ductus epididymis is about 600–700 mm, and the terminal ampulla is 350–400 mm in length. The duct itself varies in diameter from 1 mm in the head to about 25 mm where it joins the ampulla (Fig. 1).

Ampulla Epididymis

The symmetrically paired ampulla is a nearly straight, evenly expanded region of the posterior epididymis in which spermatophores are formed and stored. The ampulla lies along the posterior dorsal body wall under each kidney, adjacent to the ureter and the sperm sac (Fig. 2) and behind a thick layer of peritoneum. Each ampulla is partitioned with large septa of evenly varying height. The septa divide the ampulla into narrow compartments connected in the center by a common lumen. Each ampulla is 33–40 mm in diameter and 500–600 mm long. The ampulla epididymis ends in a thick-walled papilla, which projects into the single urogenital sinus and is the common collection point of both ureters and both ampullae. A single urogenital papilla transfers seminal products to the proximal opening of the clasper groove when the clasper is flexed during insemination.

Sperm Sac

The paired sperm sacs are thin-walled tubes that open on the urogenital sinus. Their relatively straight

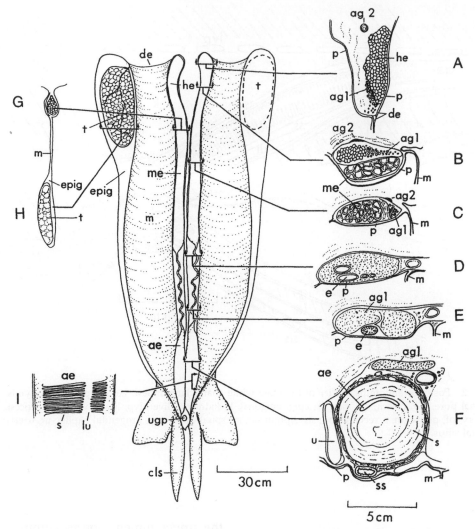

FIGURE 1 Male white shark reproductive system, ventral view. ae, Ampulla epididymis; ag1, accessory gland 1; ag2, accessory gland 2; cls, clasper; de, ductus efferens; e, epididymis; epig, epigonal; he, head of the epididymis; lu, lumen; m, mesentery; me, midepididymis; p, peritoneum; s, septa; ss, sperm sac; t, testis; u, ureter; ugp, urogenital papilla.

course parallels the ampulla and courses forward along its lateral surface to end blindly.

Clasper and Clasper Spurs

In the adult white shark, the claspers are scroll-shaped heavily calcified appendages that protrude 35–40 cm from the medial margin of the pelvic fin. They are robust structures, 5–7 cm in diameter (Fig. 1). Each bears a sharp, conical, calcified, and dentine-covered spur or claw that folds out of the distal surface as the terminal cartilages open. Apparently, the spine serves to lock and hold the clasper in the vagina (Fig. 3). The condition of clasper spurs varies from individual to individual. Juveniles have an undiffer-

entiated soft spur that does not protrude from the clasper when the terminal cartilages are pulled back. The spurs of one or both adult claspers can be completely enveloped in a tough epidermal sheath, as was seen in a 497 cm TL male. Dissection of the thick sheath of large white sharks exposes a sharp, hard, conical spur. The right spur of a 517 cm TL male had 8 mm of hard pointed tip exposed, but the left spur was blunt, naked to the base, and encrusted with irregular concretions of calcium. However, it was not smaller or obviously broken (Fig. 3). A similar irregular spur was noted by Matthews (1950) on an adult basking shark, whose claspers are very similar in form to those of the white shark.

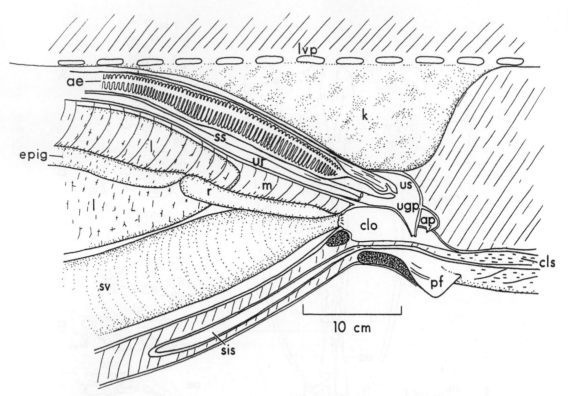

FIGURE 2 Posterior abdominal cavity of the male white shark, lateral view. ae, Ampulla epididymis; ap, abdominal pore; clo, cloaca; cls, clasper; epig, epigonal; k, kidney; l, liver; lvp, lower vertebral process; m, mesentery; pf, pelvic fin; r, rectal gland; sis, siphon sac; sv, spiral valve; ugp, urogenital papilla; ur, ureter; us, urogenital sinus.

Siphon Sacs

The siphon sacs are wide and relatively short (70 cm) compared to those of the mako *Isurus oxyrinchus* and some of the carcharhinids, such as the blue shark *Prionace glauca*. In both of these species, mature siphon sacs extend from the pelvic girdle to the pectoral girdle. In white sharks, they are paired, thick-walled, blind sacs found beneath the dermis on the ventral body wall and anterior to the claspers. They communicate with the clasper groove through a relatively wide passage (Fig. 2).

Spermatophores

White sharks produce typical elasmobranch spermatozoa having helical heads and a total length of about 250 μm. The generation and development of spermatozoa in the testes seem to follow the same stages as described by Stanley (1963) for other elasmobranchs. Transport of the sperm is similar to that observed in *P. glauca* (see Pratt, 1979). Sections of the anterior lumen of the ampulla reveal that, while most sperm enter singly, some enter with heads aligned. In the upper lumen, sperm clump together in directionally aligned masses of 60–200, adhered to and embedded in a cohesive matrix substance. Clusters of sperm settle in random orientation until an aggregate 5–6 mm in diameter is formed. Aggregates in the posterior ampullae continue to accrete a covering of the same substance found inside the structure. Since the aggregate is completely embedded and covered in the matrix, it is termed a spermatophore (Pratt and Tanaka, 1994). Spermatophores were usually abundant in the lower ampullae of adult white sharks and flowed freely from the urogenital papilla of freshly landed specimens. They range in shape from irregular masses and small spheres to large smooth saucers 5–10 mm in diameter.

Sexual Maturity

Among the males examined, 10 larger than 460 cm TL were found to be clearly mature (Fig. 4) on the basis of the clasper criteria of Clark and von Schmidt (1965). Males smaller than 230 cm TL, having short soft

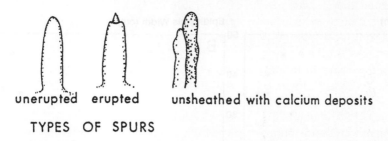

unerupted   erupted   unsheathed with calcium deposits

TYPES OF SPURS

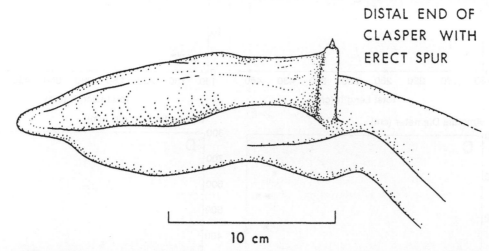

DISTAL END OF
CLASPER WITH
ERECT SPUR

10 cm

FIGURE 3 Distal end of the adult right clasper. The terminal cartilage is splayed open, and the spur is erect.

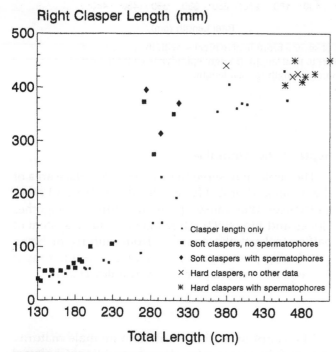

FIGURE 4 Right clasper length graphed against total length, with clasper hardness and presence of spermatophores as indicated.

claspers and thin straight epididymides, were obviously juveniles, but it is difficult to ascertain 50% maturity from this small sample. Males between 230 and 460 cm have developing reproductive systems and display a variety of conditions showing the onset of maturity. For example, a 317-cm shark had long well-developed claspers that were partially calcified. A copious number of sperm aggregates were expressed from the urogenital papilla, but on histological sectioning, these were found to contain very few sperm. The epididymis was strongly convoluted with septa formed in the enlarged ampullae. Various organ diameters and other measurements were taken in the hope of finding another index of maturity (Fig. 5A–D). Allometric growth was noted in epididymis width and siphon sac length when compared to body length. None of these relationships showed a transitional inflection as clearly as did clasper length to total length. Therefore, the robust well-calcified claspers of the 379-cm specimen, which rotate at the base, determine the size at maturity for the white shark in these data.

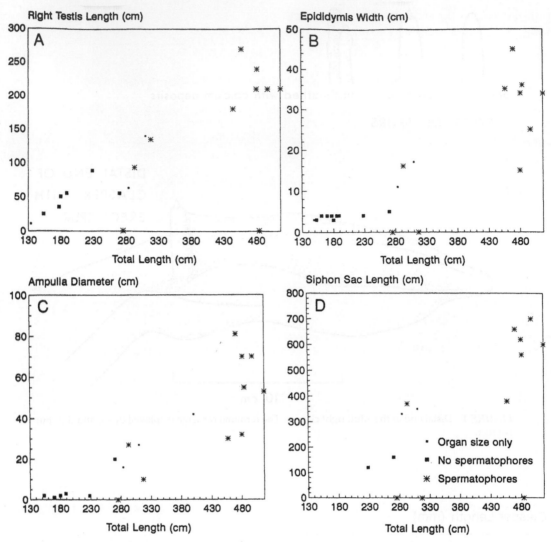

FIGURE 5 Relationships between total length and length of several reproductive organs, with the presence or absence of spermatophores indicated: (A) right testis diameter to total length, (B) right epididymis width to total length, (C) ampulla diameter to total length, and (D) siphon sac length to total length.

Discussion

Anatomy

Anatomically, the reproductive system of the white shark is grossly similar to that of the blue shark (Pratt, 1979). The white shark even more closely resembles the basking shark and the other members of the family Lamnidae in the structure of the testes and the ampullae of the epididymis. The white shark testis is typical of the radial lobed lamnid and alopiid testes described by Pratt (1988; see Figs. 1B and 2) and closely resembles that of *Isurus*, except that it is irregular in shape, rather than a rounded cylinder.

Septa of the Ampullae

The septa may serve to increase the surface area of the organ and probably secretes the matrix and other substances that cause sperm to aggregate together (Jones and Lin, 1992). They may provide a system of baffles to keep the contents from shifting or being lost during prey capture, breaching, and other rapid movement (see Chapter 16, by Klimley *et al.*).

Maturity

The simplest method to determine male maturity is a careful assessment of clasper condition. Clark and von Schmidt (1965; see also Aasen, 1963; Springer,

1960) list three criteria for determining sexual maturity on the basis of clasper condition: (1) the distal end of the clasper and the rhipidion are fully formed and can be spread open on a fresh specimen; (2) the clasper, proximal to the head, is rigid due to calcification of the supporting cartilages; and (3) the clasper rotates around the base easily to allow it to be directed anteriorly. Subadults of some species, however, are not separated easily into these categories (Pratt, 1979; Parsons, 1982). Claspers of the blue shark, for example, calcify slowly, and articulating areas undergo a slow transition as body length increases. To define maturity in male blue sharks, Pratt (1979) used another criterion: the presence of mature sex products such as spermatozeugmata (which he misidentified in the blue shark as spermatophores); spermatozeugmata are now known to be a characteristic of carcharhinid sharks (Pratt and Tanaka, 1994).

Dissection of the ampullae of the epididymides revealed the presence or absence of sperm aggregates (Fig. 4). The presence of spermatophores was not a definitive test of maturity, however. Many of the aggregates in mature white sharks and other lamnoids are sparsely populated with spermatozoa, and often only a minority (i.e., the white to cream-colored aggregates) are densely laden with sperm. Perhaps the empty aggregates constitute a process that continues whether or not sperm are produced. An adolescent shark, 370–390 cm TL having semirigid claspers, may have a preponderance of these empty aggregates, but a microscopic examination of spermatophores is needed to determine whether it is capable of successfully mating. Spermatophores that are nearly devoid of sperm in sharks having flexible (immature) claspers do not support my earlier contention that the presence of spermatophores is a more reliable index of maturity than is clasper condition alone (Pratt, 1979). The criterion may be true, however, for individual species, such as the blue shark.

Estimates of male white shark maturity vary widely in the literature (Table I). This is due in part to differences in criteria and in the interpretation of those criteria without comparative material at hand. For example, a semirigid 30-cm clasper on a 3-m 300-kg white shark is a visually impressive organ, but by no measure is it mature. Degree of calcification is yet another difficult quality to discern without expertise and comparative material. To a worker field-sampling 1–2 m TL carcharhinid sharks who has never examined a white shark before, a semirigid 30-cm clasper could easily be mistaken as mature. Thus, a failproof method of determining sexual maturity in maturing white sharks has yet to be identified.

TABLE I Estimates of Male White Shark Total Length at Maturity

Immature (cm)[a]	Adolescent (cm)	Mature (cm)	Reference
269	268–275	304–339	Bass *et al.* (1975)
		400–430	Castro (1983)
		426	Casey and Pratt (1985)
366–381		442	Bigelow and Schroeder (1948)
	240	550[b]	Compagno (1984)
	275–317	379	This study

[a]Largest measured.
[b]Maximum size.

Determining the body length at which 50% of male white sharks attain sexual maturity is impossible at this time due to the small sample size. Evidence for transition to maturity is found in six sharks ranging from 270 to 317 cm TL (Fig. 4). By strict definition, they are not mature, but the 275- and 317-cm males are approaching that stage. Although doubtful, some of these subadults may even be capable of copulation. Since females mature at a large size (see Chapter 15, by Francis), it is probable that only a proportionately large male with well-calcified claspers is capable of copulation.

Sexual Cycle

The distribution of TL in the sample may be due to aspects of the sexual cycle. The disproportionate presence of very small and very large white sharks in the New York Bight (the Atlantic Ocean between Cape May, New Jersey, and Cape Cod, Massachusetts) cannot be explained at this time, but the TL range indicates that pupping and mating may occur here during the summer, when most of the samples were taken. Some of the males were harpooned while swimming within one body length of a larger shark. The presence here of some of the smallest white sharks ever captured may indicate proximity to a birthing area. Apparent young-of-the-year white sharks exhibited basal cusps on most of the teeth, nearly transparent trailing fin edges, and a yolk sac scar or fold between the pectoral fins. Such neonate sharks from the New York Bight appear in Fig. 4 at 130–140 cm TL.

Summary

Thirty-eight male white sharks *C. carcharias*, ranging from 130 to 517 cm TL and from 16 to 1558 kg body mass, were caught in the western North Atlantic between June 1971 and July 1991 and dissected to determine maturity as well as reproductive organ structure and function. The reproductive anatomy was found to be typical of lamnoid species. In a paired symmetrical system, radial lobed testes deliver spermatozoa to the heads of each highly coiled glandular epididymis. The terminal ampulla of the epididymis is a large, relatively straight storage organ partitioned with thick septa and, in the adult, usually filled with spermatophores. As in other elasmobranchs, sperm are transferred to the female through paired intromittent organs (claspers), which are derivations of the inner margins of the pelvic fins. Although most adults over 450 cm TL had ampullae turgid with spermatophores, no direct evidence of recent copulation was apparent in reproductive tract contents or from clasper condition. The largest immature male in the sample was 317.5 cm TL; the smallest mature male was 379 cm TL, which is estimated to be the size at maturity. Clasper

hardness is the best indicator of male white shark sexual maturity; mature claspers are robust and calcified, each bearing a retractile spur. Circumstantial evidence suggests that the New York Bight, between Cape May and Cape Cod, may be a mating area for the white shark.

Acknowledgments

I thank the many sport and commercial fishermen, tournament officials, and colleagues who have provided samples and data over many years. Particularly helpful were D. Adams, D. Braddick, L. Bullock, R. Conklin, C. Darenburg, T. Edwards, R. Ellis, M. Francis, K. Killam, J. McCosker, F. Mundus, M. Pagano, and G. Parsons. J. Castro provided information on the 379 cm TL shark, and L. Compagno supplied many clasper-to-TL measurements. My thanks also, for help in fieldwork, to the staff of the Apex Predator Investigation, Northeast Fisheries Science Center, National Oceanic and Atmospheric Administration; to J. Casey for support and suggestions; and to G. Skomal, A. Lintala, and L. Natanson, who prepared much of the histological material.

14

Pregnant White Sharks and Full-Term Embryos from Japan

**SENZO UCHIDA and
MINORU TODA**
Okinawa Expo Aquarium
Motobu-cho
Okinawa, Japan

KAZUYUKY TESHIMA
Tohoku National Fisheries Research Institute
Shinhama-cho, Shiogama
Miyagi, Japan

KAZUNARI YANO
Seikai National Fisheries Research Institute
Ishigaki Tropical Station
Fukai Ota, Ishigaki
Okinawa, Japan

Introduction

Little has been reported about embryonic development and birth in the white shark *Carcharodon carcharias*, except for scant information on nine embryos from a female captured in the Mediterranean (Norman and Fraser, 1937). Paterson (1986) gave the dates and locations of capture for four pregnant females caught off Australia in 1981–1982, as well as the water temperatures at the time of capture and the litter sizes. In 1985, we inspected a female, 555 cm total length (TL) (incorrectly reported as 550 cm TL in Uchida *et al.*, 1987), caught off Okinawa and found a number of egg capsules in her uteri. In 1986, one of us (S.U.) obtained photographs of a pregnant female and her seven embryos caught off Wakayama, Japan. Francis (Chapter 15) describes embryos from a female caught off New Zealand on November 13, 1991. Two pregnant white sharks were captured off Japan in May 1992, and we obtained eight embryos from one of these. Bruce (1992) reported the capture of three females pregnant with small embryos off South Australia. In this chapter, we present a description of the pregnant females and embryos caught off Japan and clarify certain aspects of white shark reproduction.

Materials and Methods

A 515 cm TL female white shark was caught on May 22, 1992, in a net off Toyo, Japan (33°31′ N,

134°17′ E; Fig. 1). The net was set 2 km from shore in a water depth of 67 m. The sea-surface temperature at the site and time of capture was 20°C. The female was carrying 10 embryos, of which we were able to obtain eight (Fig. 2A). These were brought to Okinawa by air and preserved in a −30°C freezer in the Okinawa Expo Aquarium. We measured the length of three embryos, a female [140 cm TL, 24.7 kg body mass (BM)] and two males (150 cm TL, 32.4 kg BM, and 135 cm TL, 21.3 kg BM, respectively). The embryos were then dissected in order to observe the internal organs. Minute substances found in the embryos' intestines were identified by viewing under a light microscope. We made 54 morphometric measurements, defined in Fig. 3, for each of the three embryos.

Results

Pregnant Females Previously Caught off Japan

We know of three pregnant females, caught off Japan since 1985, that we have not been able to examine directly except the first. The first was a 555 cm TL individual (Fig. 2B) captured on February 16, 1985, near Kin on the island of Okinawa. (See end of chapter for Tables I–VII.) Her uteri contained egg capsules. A second female was captured on April 2, 1986, close to Taiji on the island of Honshu. She was esti-

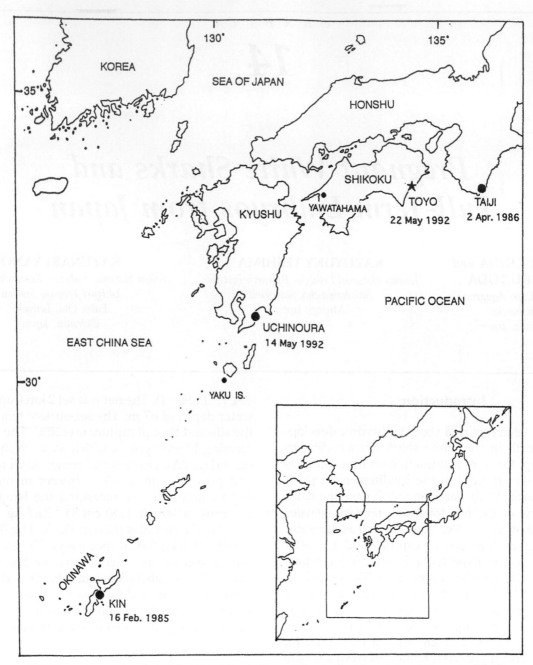

FIGURE 1 Capture location of female white sharks and the smallest free-swimming young. The Toyo specimen of the present study is indicated by a star.

mated from photographs to be 470 cm TL (Fig. 2C). Her seven embryos ranged in size from 100 to 110 cm TL (Table II). More information is available about these two females in the report by Uchida *et al.* (1987). A third female, 480 cm TL, was caught on May 14, 1992, near Uchinoura on the island of Kyushu. She contained five embryos (Fig. 2D), but only one was measured, at 130 cm TL (Table II).

The Toyo Female and Her Embryos

Fishermen observed one embryo exit the mother's cloaca and sink without swimming while they towed the mother to port. At the port, photographs were taken of the mother and her nine remaining embryos (Fig. 2A). We were able to obtain only eight of the embryos. The four females and four males ranged from 135 to 151 cm TL and from 21 to 32 kg BM (Table II). The ovary mass (including the epigonal organ) was 6.75 kg. The ovary was partitioned and contained numerous ova, ranging from 1 to 4 mm in diameter. Each ridge of the partitioned inner surface had numerous yolk-filled ova, ranging from 2 to 4 mm in diameter, and numerous degenerated ova, about 1 mm in diameter.

Morphology of the Embryos

The three embryos from the Toyo specimen had "yolk sac scars" on their throats (Fig. 4B). We give proportional dimensions (in percentage of TL) for a 140 cm TL female (Fig. 4A), a 150 cm TL male (Fig. 4B), and a 135 TL cm male for comparison with 10 free-swimming specimens, five males and five females ranging from 301 to 555 cm TL that were captured around the Okinawa Islands, and 14 males and 19 females ranging from 170 to 391 cm TL that were caught off South Africa (Bass *et al.*, 1975) (Tables III–V). The eye diameter of the embryos was greater than that of larger sharks from Japan and Africa; mouth width of the embryos was shorter than that of the latter. Body height at first dorsal fin origin and pelvic fin origin in the embryos was much lower than in the big Japanese adults. Girth at pectoral fin origin and first dorsal fin origin was smaller than in the adults as well (Table III). These measurements indicate that the near-term embryos are more slender than free-swimming big adults. Length from snout to pelvic fin in the embryos was shorter than in the large adults and African specimens [which include some juveniles (Bass *et al.*, 1975)]. Length from pelvic to anal fins and from anal to lower caudal fins was the same among the embryos and free-living sharks. Length of the upper caudal fin of the embryos was a little longer than in records for other individuals (Table III). We offer additional measurements of the embryos and stomach and intestine contents. The number of laminae in the spiral valve of the 150- and 135-cm embryos was 52 and 51, respectively. Several visceral organs were weighed. The percentage masses of liver, stomach, intestine, spleen, pancreas, rectal gland, and testes to body mass of the 150- and 135-cm male embryos were 18.56% and 13.53%, 1.12% and 1.97%, 0.96% and 1.18%, 0.30% and 0.23%, 0.11% and 0.08%, 0.03% and 0.05%, and 0.23% and 0.35%, respectively. Embryonic stomach and intestine contents (e.g., teeth, yolk material, and fragments of egg capsules) were found in the stomachs of two embryos together with mucuslike substances (Table VI and Figs. 5 and 6). Twenty-three teeth, as well as yolk material weighing 1120 g, were obtained from the stomach contents

of the 150-cm embryo (No. 6 in Table VI; see also Fig. 5). In addition, fragments of amber-colored egg capsules were found in the yolk material. Teeth were also found in the stomach of the 135-cm (No. 1) and 140-cm (No. 8) embryos, and they were similar in shape but slightly smaller than those of functional teeth in both jaws of these embryos (Fig. 6). Intestines of the three embryos were filled with greenish brown material (Fig. 7A), which contained teeth, dermal denticles, and crystal-like pieces (Table VI). At least 139 and 143 teeth of varying sizes were contained in the intestinal fluid of the 150- and 135-cm embryos, respectively (Fig. 7B). Dermal denticles in the intestinal contents were similar in shape to those covering the bodies of embryos (Fig. 7C). Crystal-like pieces present throughout the material in the valvular intestine were minute and sandy to the touch (Fig. 7D).

Other Adult White Sharks Caught off Japan

Thirteen additional white sharks were caught around the island of Okinawa, during all seasons of the year, 1975–1990 (Table VII). The sharks ranged in size from 301 to 555 cm TL and in mass from 270 to 1970 kg. The largest shark was captured near Kin on February 16, 1985. All of the sharks were caught on long lines except the largest. Sea-surface temperature ranged from 21.2°C to 27.4°C at the time of capture.

Discussion

Litter Size

Seven embryos were carried by a Taiji female, five by a Uchinoura shark, and 10 by the Toyo female, one of which exited from her cloaca during towing to the port (Table I). In Australia, 4, 7, 11, 13, and 14 embryos were reported among seven females (Paterson, 1986; Bruce, 1992). Francis (Chapter 15) observed seven embryos in the female inspected in New Zealand.

Body Size at Birth

The two smallest free-swimming white sharks reported from the Pacific and Atlantic oceans, respectively, were 122 cm TL (16 kg BM) and 122 cm TL (12 kg BM) (Casey and Pratt, 1985). The eight embryos in the present study ranged from 135 to 151 cm TL and averaged 144 cm. We therefore suggest, based on the sizes of the smallest free-swimming sharks

FIGURE 2 (A) A 515 cm TL female white shark and nine near-term embryos at Toyo fish market; (B) a 480 cm TL female white shark and her five near-term embryos at Uchinoura fish market; (C) an approximately 470 cm TL female white shark with six of seven embryos at Taiji fish market; and (D) a 555 cm TL female white shark with many egg capsules at Kin. The caption to photograph (C) as published in the work of Ellis and McCosker (1991), contained three errors. (1) the specimen was identified as "Kin-town shark," but this is the Taiji specimen; (2) each embryo was said to weigh 3 lb, but none was weighed; and (3) the yolk material spilling from the stomach was described as "embryonic membrane," but no embryonic membranes were found.

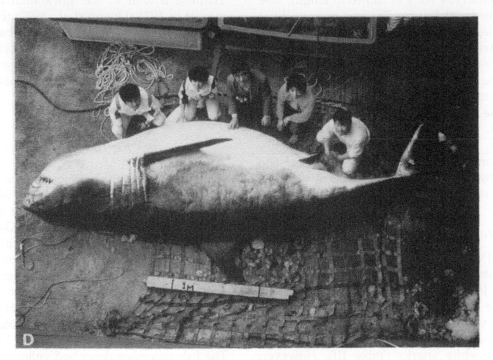

FIGURE 2 (*Continued*)

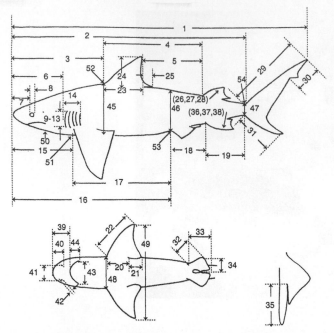

FIGURE 3   Measurements taken in this study.

that the Toyo embryos were at full-term length. The Uchinoura embryo, at 130 cm TL, was apparently at the full-term stage as well. Based on the sizes of the smallest free-swimming white sharks and the sizes of the Toyo embryos, size at birth for the white shark is estimated to be 120–150 cm TL and 12–32 kg BM. Among 26 white sharks smaller than 160 cm TL and reported in the literature, individuals between 120 and 140 cm TL represent 76% of the total (Bigelow and Schroeder, 1948; Bass et al., 1975; Compagno, 1984b; Casey and Pratt, 1985; Klimley, 1985b). The individuals between 120 and 140 cm TL are within the estimated size range at birth.

Proportional Dimensions

Proportional values of mouth widths ranked as follows: Toyo embryos < African specimens (juvenile and adults) < Japanese large adults (Table III). This means that the mouth of the white shark becomes proportionally wider with body size. The near-term embryos showed a more slender body than did other free-swimming white sharks (Table III). However, embryos of younger, 100–110 cm TL, individuals had a fatter body with a bulging belly (Fig. 2C). Thus, just-born pups of white sharks are presumed to show the most slender body in their growing stages, at least after 100 cm TL. The proportional dimensions of body heights and girths at first dorsal and pelvic ori-

gins show females to be a little fatter than males among large white sharks of Japan (Tables IV and V). Although the number of measured samples was low, two males and a female of the near-term embryos indicated about the same percentages in the points, that is, fatness. The embryos had a shorter head and trunk and a slightly longer upper caudal fin than did free-swimming larger sharks, but had the same tail stock (pelvic to lower caudal). These characteristics indicate the growing rate after birth as follows: head and trunk > tail stock > upper lobe of the caudal fin.

Pupping Ground and Season

As described above, a large white shark caught in Okinawa was presumed to be pregnant; pregnant females at Uchinoura and Toyo-cho, caught in May, were thought to have had full-term embryos; and the Taiji embryos, collected early in April, were estimated to be less than full-term. Only two reliable records of small free-swimming white sharks have been reported: a 140 cm TL male captured off Yaku Island in 1978 (Mochizuki, personal communication) and a 170 cm TL male landed at the Yawatahama, Shikoku, fish market in 1974 (Uyeno et al., 1979) (Fig. 1). All came from southwestern Japan. On the basis of these records, we conclude that the coastal waters of this area are an important pupping ground and that parturition occurs during spring. Additional records would be helpful for confirmation.

Reproductive Mode

Embryos of oophagous sharks consume and store yolk in their stomachs (Springer, 1948; Fujita, 1981; Stevens, 1983; Gilmore, 1983). The uteri of a large female white shark caught at Kin, Okinawa, in 1985 had egg capsules that were filled with yolk material. Yolk material contained in egg capsules of the left uterus weighed 9 kg. Unfortunately, no embryos were obtained, and these were presumed to have been aborted during capture. Embryos found in a pregnant female caught at Taiji, Wakayama, in 1986 had expanded bellies, as indicated by photographs. The fish processor described yellowish stomach contents, which we believe to have been a yolk-like substance. In the present study, yolk material and fragments of egg capsules were found in the stomach contents of the 135- and 150-cm embryos. These materials suggest that the white shark embryo is oophagous.

FIGURE 4   (A) The No. 8 embryo from the Toyo specimen, a 140 cm TL female. Bulging in the belly is much less than in embryos from the Taiji specimen (Fig. 2C), which were smaller and younger. (B) The No. 6 embryo from the Toyo specimen, a 150 cm TL male, showing a yolk sac scar in the center of the throat.

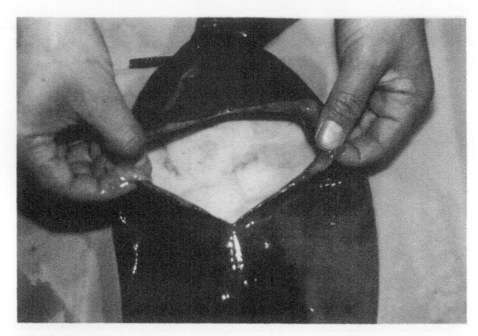

FIGURE 5 The stomach of the No. 6 embryo, a 150 cm TL male, showing yolk material and dark fragments of egg capsules.

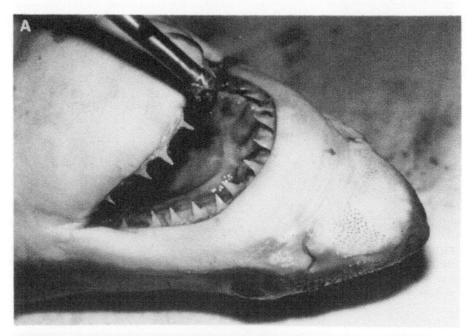

FIGURE 6 (A) The teeth of the No. 8 embryo, a 140 TL cm female, are erect and seem fully functional; (B) teeth found in the stomach of the No. 8 embryo; (C) first tooth on the left lower jaw (J) and a slightly shorter tooth from the stomach (S) of the No. 8 embryo (in both teeth, basal cusps are visible).

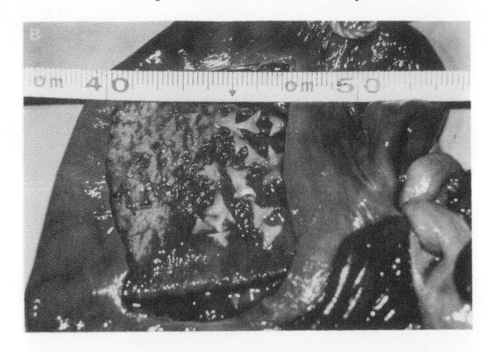

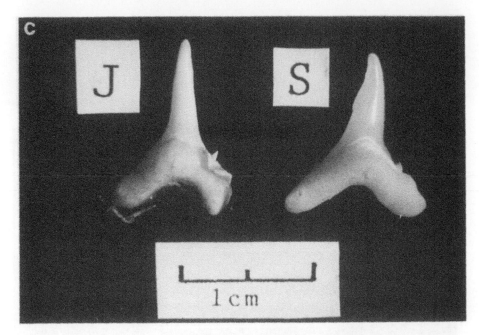

FIGURE 6   (*Continued*)

Teeth and Dermal Denticles in the Stomach

Many teeth and dermal denticles were found in the embryo stomach contents. These contents probably were not a result of fetal cannibalism, because no fragments of cartilage or other tissues were found. During development, the embryo probably swallowed its own teeth and denticles directly, or ate the teeth and dermal denticles of itself and siblings that had dropped off in the uterus. Varying sizes of teeth were found in the embryonic intestines, indicating

FIGURE 7 (A) Longitudinal section, posterior part of the intestine of No. 6 embryo showing greenish-brown material that contained teeth, dermal denticles, and crystal-like pieces; (B) 139 pieces of small teeth found in the intestine of No. 6 embryo; (C) dermal denticles from the intestine of No. 6 embryo, similar in shape to those on the dermis of No. 6; and (D) crystal-like pieces from the intestine of No. 6 embryo, minute and sandy to the touch.

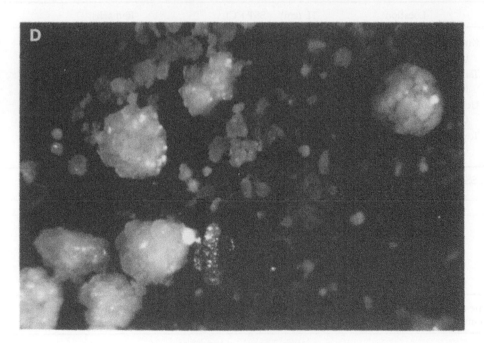

FIGURE 7 (*Continued*)

that tooth replacement occurs several times during embryonic development and that teeth are functional even before the embryo is born.

Summary

A 515 cm TL white shark *C. carcharias*, pregnant with 10 embryos, was caught in a net set off Toyo, Japan (33°31' N, 134°17' E), on May 22, 1992. Fishermen observed one embryo to exit the mother's cloaca and sink without swimming as the mother was towed to port. Another embryo was lost at the dock. We obtained the remaining four male and four female embryos. Sizes ranged from 135 to 151 cm, and masses, from 21.3 to 32.4 kg. Judging from dimensions of the largest embryos and smallest free-swimming juveniles reported in the literature, these embryos had reached maximum length. Three embryos were dissected. Their digestive tracts contained yolk, parts of egg capsules, teeth, and dermal denticles. The presence of yolk and egg material in the embryonic guts indicated that the developing embryos had fed on oocytes (termed oophagy).

Acknowledgments

We are indebted to the Kochi Branch of the *Mainichi Newspapers* and its reporter, T. Imanishi, for informing us and offering photographs of the female white shark and embryos captured at Toyocho. I. Sakurai helped to obtain and transport the Toyo embryos. K. Nishizono and M. Okano helped with our work at Uchinoura and provided photos of the Uchinoura specimens; H. Yamamori provided video tape recordings; and I. Wakabayashi photographed the Taiji specimen. We also express our sincere thanks to Y. Kamei, H. Teruya, N. Tanaka, and M. Nonaka for preparing data; K. Muzik and M. Francis for reviews of the manuscript; T. Nagasaki and the staff of Okinawa Expo Aquarium for their work on the specimens; and M. Taira, H. Kyan, and K. Uema for typing the manuscript.

TABLE I   Details of Female White Sharks Having Egg Capsules or Embryos

Capture date	Location	TL (cm)	SST (°C)	Remarks	Reference
February 16, 1985	Kin, Okinawa, Japan (26°26' N, 128°59' E)	555	21.8	Many egg capsules in uteri; embryo aborted?	Uchida *et al.* (1987)
April 2, 1986	Taiji, Wakayama, Japan (33°35' N, 135°56' E)	~470	16.2	Seven embryos discarded; photograph taken	Uchida *et al.* (1987)
May 14, 1992	Uchinoura, Kagoshima, Japan (31°18' N, 131°8' E)	480	19.0	Five embryos processed into food and manure; photographs and video taken	This study
May 22, 1992	Toyo, Kochi, Japan (33°31' N, 134°17' E)	515	20.0	Eight of 10 embryos obtained by authors	This study

TL, Total length; SST, sea-surface temperature.

TABLE II   Size and Mass of White Shark Embryos

Name of specimen	No.	Sex	Total length (cm)	Body mass (kg)	Remarks
Taiji			100–110		Calculated from photograph
Uchinoura			130		One of the five embryos measured by fishermen
Toyo	1	M	135	21.3	Eight of 10 embryos measured by authors
	2	M	145	24.1	
	3	F	143	26.2	
	4	M	140	23.8	
	5	F	151	31.3	
	6	M	150	32.4	
	7	F	144	26.9	
	8	F	140	24.7	
Mean ± SD			143.5 ± 5.3	26.3 ± 3.8	

TABLE III **External Measurements of Embryos and Free-Swimming White Sharks**

		Embryonic specimen[a]						Free-swimming sharks (cm)	
		No. 1		No. 6		No. 8		Okinawa, Japan[b]—	South Africa[c]—
No.	Measurement	cm	%	cm	%	cm	%	Mean ± SD (%)	Mean ± SD (%) Mean ± SD (%)
1	Total length	135.0	100.0	150.0	100.0	140.0	100.0		
2	Standard length	105.0	77.8	119.0	79.3	109.0	77.9	78.3 ± 0.7	81.8 ± 1.5	82.6 ± 0.8
3	Snout to first dorsal	47.2	35.0	51.0	34.0	48.4	34.6	34.5 ± 0.4	37.4 ± 0.8	38.2 ± 1.6
4	First to second dorsal	42.3	31.3	48.4	32.3	45.0	32.1	31.9 ± 0.4	34.5 ± 1.0	34.2 ± 1.1
5	Between dorsal bases	29.4	21.8	32.6	21.7	32.0	22.9	22.1 ± 0.5	23.9 ± 0.9	24.8 ± 1.2
6	Snout to first gill slit	27.6	20.4	26.5	17.7	26.0	18.6	18.9 ± 1.2	20.4 ± 0.7	21.9 ± 1.0
7	Snout to orbit	5.7	4.2	6.3	4.2	5.4	3.9	4.1 ± 0.2	5.0 ± 0.6	5.7 ± 0.5
8	Eye diameter	2.0	1.5	2.2	1.5	2.4	1.7	1.5 ± 0.1	0.9 ± 0.1	1.0 ± 0.2
9	Gill slit length, first	12.0	8.9	13.2	8.8	12.0	8.6	8.8 ± 0.1	10.1 ± 1.5	10.3 ± 2.1
10	Gill slit length, second	12.8	9.5	13.7	9.1	12.4	8.9	9.2 ± 0.3	9.8 ± 0.9	
11	Gill slit length, third	12.6	9.3	13.7	9.1	12.4	8.9	9.1 ± 0.2	9.8 ± 0.9	9.2 ± 1.5
12	Gill slit length, fourth	13.0	9.6	13.1	8.7	12.6	9.0	9.1 ± 0.4	9.7 ± 0.6	
13	Gill slit length, fifth	13.0	9.6	13.4	8.9	13.6	9.7	9.4 ± 0.3	10.5 ± 0.6	10.2 ± 0.9
14	First to fifth gill slits	7.0	5.2	8.6	5.7	8.0	5.7	9.5 ± 0.3	6.5 ± 1.0	8.2 ± 0.7
15	Snout to pectoral	33.2	24.6	34.0	22.7	2.0	22.9	23.4 ± 0.9	25.0 ± 2.0	26.2 ± 1.9
16	Snout to pelvic	71.7	53.1	80.0	53.3	77.0	55.0	53.8 ± 0.8	58.0 ± 0.9	59.3 ± 2.3
17	Pectoral to pelvic	38.5	28.5	46.5	31.0	43.0	30.7	30.1 ± 1.1	32.9 ± 2.4	30.4 ± 2.6
18	Pelvic to anal	23.4	17.3	24.5	16.3	21.0	15.0	16.2 ± 1.0	16.0 ± 0.7	16.4 ± 1.8
19	Anal to lower caudal	11.2	8.3	13.4	8.9	14.0	10.0	9.1 ± 0.7	8.7 ± 0.5	9.5 ± 1.1
20	Pectoral base	9.0	6.7	9.2	6.1	10.0	7.1	6.6 ± 0.4	7.3 ± 0.6	6.9 ± 0.8
21	Pectoral inner edge	7.7	5.7	8.5	5.7	7.0	5.0	5.5 ± 0.3	4.5 ± 0.9	5.5 ± 0.9
22	Pectoral length	30.0	22.2	32.8	21.9	31.0	22.1	22.1 ± 0.2	19.3 ± 1.3	21.3 ± 1.8
23	First dorsal, base	12.8	9.5	14.6	9.7	13.1	9.4	9.5 ± 0.2	10.2 ± 0.5	9.8 ± 3.2
24	First dorsal, height	12.8	9.5	13.7	9.1	12.6	9.0	9.2 ± 0.2	9.8 ± 0.7	10.5 ± 1.5
25	First dorsal, lobe	3.3	2.4	4.4	2.9	3.7	2.6	2.7 ± 0.2	2.6 ± 0.4	2.7 ± 0.6
26	Second dorsal, base	2.1	1.6	2.5	1.7	2.5	1.8	1.7 ± 0.1	1.3 ± 0.2	1.5 ± 0.3
27	Second dorsal, height	2.0	1.5	2.5	1.7	2.5	1.8	1.6 ± 0.1	1.5 ± 0.1	1.5 ± 0.3
28	Second dorsal, lobe	2.5	1.9	3.1	2.1	1.6	1.1	1.7 ± 0.4	1.9 ± 0.4	2.0 ± 0.6
29	Upper caudal	32.6	24.1	35.0	23.3	34.0	24.3	23.9 ± 0.4	20.9 ± 1.4	21.8 ± 1.1
30	Base of notch to tip	6.1	4.5	7.7	5.1	6.0	4.3	4.6 ± 0.4	4.5 ± 0.4	4.1 ± 0.6
31	Lower caudal	24.2	17.9	26.0	17.3	24.0	17.1	17.5 ± 0.3	17.0 ± 1.5	18.3 ± 1.4
32	Pelvic, origin to lateral lobe	8.8	6.5	9.4	6.3	10.0	7.1	6.6 ± 0.4	5.9 ± 0.8	6.1 ± 0.8
33	Pelvic, origin to median tip	12.5	9.3	13.3	8.9	12.8	9.1	9.1 ± 0.2	9.3 ± 0.6	9.3 ± 1.0
34	Pelvic, width at lateral lobe	7.7	5.7	9.5	6.3	9.0	6.4	6.2 ± 0.3	5.8 ± 1.2	
35	Clasper length	4.6	3.4	4.8	3.2			3.3 ± 0.1	9.5 ± 0.8	
36	Anal fin, base	2.1	1.6	2.2	1.5	2.3	1.6	1.6 ± 0.1	1.5 ± 0.2	1.5 ± 0.3
37	Anal fin, height	1.8	1.3	2.1	1.4	2.5	1.8	1.5 ± 0.2	1.7 ± 0.4	1.6 ± 0.3
38	Anal fin, lobe	2.5	1.9	3.0	2.0	3.0	2.1	2.0 ± 0.1	2.0 ± 0.3	1.9 ± 0.5
39	Snout to mouth	6.9	5.1	8.6	5.7	8.0	5.7	5.5 ± 0.3	5.1 ± 1.6	
40	Snout to nostrils	4.6	3.4	5.7	3.8	5.0	3.6	3.6 ± 0.2	3.3 ± 0.8	4.3 ± 1.4

(continues)

TABLE III (Continued)

No.	Measurement	Embryonic specimen[a] No. 1 cm	%	No. 6 cm	%	No. 8 cm	%	Mean ± SD (%)	Free-swimming sharks (cm) Okinawa, Japan[b]— Mean ± SD (%)	South Africa[c]— Mean ± SD (%)
41	Internasal distance	5.3	4.0	6.0	4.0	5.5	3.9	4.0 ± 0.0	4.1 ± 0.4	4.0 ± 0.3
42	Nostril length	1.8	1.4	1.9	1.3	1.6	1.1	1.3 ± 0.1	1.4 ± 0.1	1.6 ± 0.3
43	Mouth width	10.7	7.9	14.5	9.7	11.6	8.3	8.6 ± 0.8	11.5 ± 1.6	10.4 ± 1.3
44	Mouth length	6.5	4.8	6.4	4.3	6.5	4.6	4.6 ± 0.2	4.6 ± 0.6	4.8 ± 1.0
45	Body height at first dorsal origin	15.5	11.5	20.0	13.3	18.0	12.9	12.6 ± 0.8	20.2 ± 4.6	
46	Body height at pelvic origin	11.5	8.5	15.0	10.0	13.5	9.6	9.4 ± 0.6	13.6 ± 2.8	
47	Body height at precaudal pit	4.0	3.0	4.3	2.9	3.7	2.6	2.8 ± 0.1	2.5 ± 0.5	
48	Body width at pectoral origin	20.1	14.9	24.0	16.0	21.0	15.0	15.3 ± 0.5	15.4 ± 1.7	
49	Distance between pectoral tips	56.5	41.9	58.0	38.7	58.0	41.4	40.6 ± 1.4	42.6 ± 2.8	
50	Girth at mouth	44.5	33.0	55.0	36.7	47.6	34.0	34.5 ± 1.6	37.3 ± 2.3	
51	Girth at pectoral origin	60.5	44.8	67.6	45.1	62.5	44.6	44.8 ± 0.2	50.6 ± 3.1	
52	Girth at first dorsal origin	66.0	48.9	77.8	51.9	69.2	49.4	50.1 ± 1.3	56.3 ± 4.7	
53	Girth at pelvic origin	46.8	34.7	54.0	36.0	49.0	35.0	35.2 ± 0.6	41.2 ± 2.7	
54	Girth at precaudal pit	18.6	13.8	21.6	14.4	18.0	12.9	13.7 ± 0.6	12.9 ± 2.6	

[a]No. 1, a 135 cm TL male; No. 6, a 150 cm TL male; No. 8, a 140 cm TL female.

[b]Five males and five females, ranging from 301 to 555 cm TL, caught off Okinawa, Japan. Individual dimensions are given in Tables IV and V.

[c]Fourteen males and 19 females, ranging from 170 to 391 cm TL, caught off South Africa (Bass et al., 1975).

TABLE IV **External Measurements of Free-Swimming Female White Sharks in Okinawa, Japan**

No.	Measurement	Specimen No. 14[a] cm	%	No. 12 cm	%	No. 10 cm	%	No. 7 cm	%	No. 5 cm	%	Mean ± SD (%)
1	Total length	515.0	100.0	436.0	100.0	508.0	100.0	555.0	100.0	506.0	100.0	
2	Standard length	402.0	78.1	358.0	82.1	418.0	82.3	445.0	80.2	419.0	82.8	81.1 ± 1.8
3	Snout to first dorsal	183.0	35.5	161.0	36.9	192.0	37.8	206.0	37.1	190.0	37.5	37.0 ± 0.8
4	First to second dorsal	170.0	33.0	145.0	33.3	176.0	34.6	189.0	34.1	180.0	35.6	34.1 ± 0.9
5	Between dorsal bases	119.0	23.1	97.0	22.2	126.0	24.8	129.0	23.2	125.0	24.7	23.6 ± 1.0
6	Snout to first gill slit	101.0	19.6	92.0	21.1	103.0	20.3	114.0	20.5	99.5	19.7	20.2 ± 0.6
7	Snout to orbit	26.0	5.0	28.0	6.4	26.0	5.1	23.0	4.1	27.0	5.3	5.2 ± 0.7
8	Eye diameter	4.0	0.8	4.0	0.9	4.5	0.9	4.5	0.8	3.8	0.8	0.8 ± 0.1
9	Gill slit length, first	43.0	8.3	48.0	11.0	47.0	9.3	54.0	9.7	64.0	12.6	10.2 ± 1.5
10	Gill slit length, second	44.0	8.5	46.0	10.6	48.0	9.4	54.0	9.7	52.0	10.3	9.7 ± 0.7
11	Gill slit length, third	44.0	8.5	44.0	10.1	52.0	10.2	53.0	9.5	51.0	10.1	9.7 ± 0.6
12	Gill slit length, fourth	43.0	8.3	44.0	10.1	50.0	9.8	52.0	9.4	52.0	10.3	9.6 ± 0.7

(continues)

TABLE IV (*Continued*)

No.	Measurement	No. 14[a] cm	%	No. 12 cm	%	No. 10 cm	%	No. 7 cm	%	No. 5 cm	%	Mean ± SD (%)
13	Gill slit length, fifth			50.0	11.5	53.0	10.4	56.0	10.1	56.0	11.1	10.8 ± 0.5
14	First to fifth gill slits	25.0	4.9	23.0	5.3	26.0	5.1	34.0	6.1	40.0	7.9	5.9 ± 1.1
15	Snout to pectoral			102.0	23.4	126.0	24.8	154.0	27.7	112.0	22.1	24.5 ± 2.1
16	Snout to pelvic	304.0	59.0	250.0	57.3	290.0	57.1	325.0	58.6	293.0	57.9	58.0 ± 0.7
17	Pectoral to pelvic			135.0	31.0	167.0	32.9	177.0	31.9	188.0	37.2	33.2 ± 2.4
18	Pelvic to anal			74.0	17.0	80.0	15.7	80.0	14.4	80.0	15.8	15.7 ± 0.9
19	Anal to lower caudal			38.0	8.7	46.0	9.1	45.0	8.1	42.0	8.3	8.5 ± 0.4
20	Pectoral base	41.0	8.0	34.0	7.8	38.0	7.5	36.0	6.5	43.0	8.5	7.6 ± 0.7
21	Pectoral inner edge	23.0	4.5	26.0	6.0	24.0	4.7	22.0	4.0	18.4	3.6	4.6 ± 0.8
22	Pectoral length	98.0	19.0	92.0	21.1	99.0	19.5	103.0	18.6	103.0	20.4	19.7 ± 0.9
23	First dorsal, base	52.0	10.1	46.0	10.6	48.0	9.4	59.0	10.6	50.0	9.9	10.1 ± 0.4
24	First dorsal, height	51.0	9.9	45.0	10.3	55.0	10.8	48.0	8.6	52.0	10.3	10.0 ± 0.7
25	First dorsal, lobe	15.0	2.9	13.0	3.0	14.0	2.8	11.0	2.0	14.0	2.8	2.7 ± 0.4
26	Second dorsal, base	7.0	1.4	7.0	1.6	6.0	1.2	6.0	1.1	6.0	1.2	1.3 ± 0.2
27	Second dorsal, height	8.5	1.7	6.0	1.4	7.0	1.4	8.5	1.5	8.0	1.6	1.5 ± 0.1
28	Second dorsal, lobe	11.0	2.1	10.0	2.3	11.0	2.2	10.0	1.8	11.0	2.2	2.1 ± 0.2
29	Upper caudal	111.0	21.6	93.0	21.3	107.0	21.1	113.0	20.4	105.0	20.8	21.0 ± 0.4
30	Base of notch to tip	25.0	4.9	22.0	5.0	21.0	4.1	24.5	4.4	24.0	4.7	4.6 ± 0.3
31	Lower caudal	85.0	16.5	80.0	18.3	88.0	17.3	85.0	15.3	92.0	18.2	17.1 ± 1.1
32	Pelvic, origin to lateral lobe	30.0	5.8	36.0	8.3	30.0	5.9	31.0	5.6	31.0	6.1	6.3 ± 1.0
33	Pelvic, origin to median tip	46.0	8.9	45.0	10.3	42.0	8.3	53.0	9.5	45.0	8.9	9.2 ± 0.7
34	Pelvic, width at lateral lobe	24.0	4.7	30.0	6.9	31.0	6.1	21.0	3.8	33.0	6.5	5.6 ± 1.2
35	Clasper length											
36	Anal fin, base	8.0	1.6	7.0	1.6	7.0	1.4	6.5	1.2	6.5	1.3	1.4 ± 0.2
37	Anal fin, height	10.0	1.9	7.5	1.7	11.0	2.2	13.0	2.3	9.5	1.9	2.0 ± 0.2
38	Anal fin, lobe	12.0	2.3	10.0	2.3	12.0	2.4	12.0	2.2	11.0	2.2	2.3 ± 0.1
39	Snout to mouth	29.0	5.6	26.0	6.0	23.0	4.5	13.0	2.3	28.0	5.5	4.8 ± 1.3
40	Snout to nostrils	18.0	3.5	18.0	4.1	14.0	2.8	15.5	2.8	21.0	4.2	3.5 ± 0.6
41	Internasal distance	22.0	4.3	20.0	4.6	20.0	3.9	27.0	4.9	22.0	4.3	4.4 ± 0.3
42	Nostril length	7.0	1.4	5.0	1.1	7.0	1.4	8.0	1.4	7.0	1.4	1.3 ± 0.1
43	Mouth width	69.0	13.4	52.0	11.9	67.0	13.2	52.0	9.4	45.0	8.9	11.4 ± 1.9
44	Mouth length	31.0	6.0	21.0	4.8	25.0	4.9	26.0	4.7	21.0	4.2	4.9 ± 0.6
45	Body height at first dorsal origin			80.0	18.3	102.0	20.1	150.0	27.0	125.0	24.7	22.5 ± 3.5
46	Body height at pelvic origin			53.0	12.2	75.0	14.8	102.0	18.4	83.0	16.4	15.4 ± 2.3
47	Body height at precaudal pit			16.0	3.7	11.0	2.2	12.5	2.3	11.0	2.2	2.6 ± 0.6
48	Body width at pectoral origin			71.0	16.3	70.0	13.8	71.0	12.8	96.0	19.0	15.5 ± 2.4
49	Distance between pectoral tips			196.0	45.0	225.0	44.3	240.0	43.2	238.0	47.0	44.9 ± 1.4
50	Girth at mouth	175.0	34.0	160.0	36.7	193.0	38.0	203.0	36.6	191.0	37.7	36.6 ± 1.4
51	Girth at pectoral origin					250.0	49.2	284.0	51.2	277.0	54.7	51.7 ± 2.3
52	Girth at first dorsal origin			240.0	55.0	287.0	56.5	319.0	57.5	335.0	66.2	58.8 ± 4.4
53	Girth at pelvic origin			190.0	43.6	218.0	42.9	227.0	40.9	232.0	45.8	43.3 ± 1.8
54	Girth at precaudal pit			78.0	17.9	55.0	10.8	69.0	12.4	55.0	10.9	13.0 ± 2.9

[a]No. 14 was not an Okinawa specimen but was the pregnant Toyo specimen measured with the internal organs removed.

TABLE V External Measurements of Free-Swimming Male White Sharks in Okinawa, Japan

| | | Specimen | | | | | | | | | |
| | | No. 13 | | No. 11 | | No. 9 | | No. 8 | | No. 4 | | |
No.	Measurement	cm	%	cm	%	cm	%	cm	%	cm	%	Mean ± SD (%)
1	Total length	460.0	100.0	435.0	100.0	301.0	100.0	380.0	100.0	471.0	100.0	
2	Standard length	380.0	82.6	355.0	81.6	249.0	82.7	312.0	82.1	391.0	83.0	82.4 ± 0.5
3	Snout to first dorsal	180.0	39.1	163.0	37.5	113.0	37.5	143.0	37.6	176.0	37.4	37.8 ± 0.7
4	First to second dorsal	160.0	34.8	157.0	36.1	101.0	33.6	135.0	35.5	162.0	34.4	34.9 ± 0.9
5	Between dorsal bases	116.0	25.2	100.0	23.0	71.0	23.6	93.0	24.5	116.0	24.6	24.2 ± 0.8
6	Snout to first gill slit	100.0	21.7	90.0	20.7	60.0	19.9	80.0	21.1	91.0	19.3	20.5 ± 0.8
7	Snout to orbit	25.0	5.4	20.0	4.6	13.0	4.3	18.0	4.7	22.0	4.7	4.8 ± 0.4
8	Eye diameter	4.6	1.0	4.5	1.0	4.0	1.2	4.0	1.1	4.0	0.8	1.0 ± 0.1
9	Gill slit length, first	58.0	12.6	42.0	9.7	26.0	8.6	33.0	8.7	48.0	10.2	10.0 ± 1.5
10	Gill slit length, second	52.0	11.3	44.0	10.1	27.0	8.8	33.0	8.7	48.5	10.3	9.8 ± 1.0
11	Gill slit length, third	55.0	12.0	42.0	9.7	28.0	9.1	34.0	8.9	47.0	10.0	9.9 ± 1.1
12	Gill slit length, fourth	49.0	10.7	44.0	10.1	28.0	9.1	35.0	9.2	46.0	9.8	9.8 ± 0.6
13	Gill slit length, fifth	50.0	10.9	44.0	10.1	29.0	9.5	39.0	10.3	50.0	10.6	10.3 ± 0.5
14	First to fifth gill slits	35.0	7.6	29.0	6.7	23.0	7.6	26.0	6.8	32.0	6.8	7.1 ± 0.4
15	Snout to pectoral	111.0	24.1	120.0	27.6	74.0	24.6	105.0	27.6	110.0	23.4	25.5 ± 1.8
16	Snout to pelvic	271.0	58.9	255.0	58.6	169.0	56.1	221.0	58.2	272.0	57.7	57.9 ± 1.0
17	Pectoral to pelvic	165.0	35.9	135.0	31.0	95.0	31.6	113.0	29.7	166.0	35.2	32.7 ± 2.4
18	Pelvic to anal	75.0	16.3	70.0	16.1	47.0	15.6	65.0	17.1	75.0	15.9	16.2 ± 0.5
19	Anal to lower caudal	45.0	9.8	36.0	8.3	26.0	8.6	32.0	8.4	44.0	9.3	8.9 ± 0.6
20	Pectoral base	33.0	7.2	30.0	6.9	19.0	6.3	28.0	7.4	34.0	7.2	7.0 ± 0.4
21	Pectoral inner edge	21.0	4.6	22.0	5.1	16.0	5.3	17.0	4.5	13.0	2.8	4.4 ± 0.9
22	Pectoral length	80.0	17.4	82.0	18.9	63.0	20.9	77.0	20.3	80.0	17.0	18.9 ± 1.5
23	First dorsal, base	45.0	9.8	45.0	10.3	29.0	9.6	42.0	11.1	48.0	10.2	10.2 ± 0.5
24	First dorsal, height	41.0	8.9	41.0	9.4	30.0	10.0	41.0	10.8	42.0	8.9	9.6 ± 0.7
25	First dorsal, lobe	9.0	2.0	11.0	2.5	10.0	3.3	10.0	2.6	10.0	2.1	2.5 ± 0.5
26	Second dorsal, base	6.0	1.3	6.0	1.4	4.0	1.3	6.0	1.6	7.0	1.5	1.4 ± 0.1
27	Second dorsal, height	7.0	1.5	6.0	1.4	4.0	1.3	6.0	1.6	6.0	1.3	1.4 ± 0.1
28	Second dorsal, lobe	7.2	1.6	4.0	0.9	7.0	2.3	8.0	2.1	8.0	1.7	1.7 ± 0.5
29	Upper caudal	80.0	17.4	92.0	21.1	70.0	23.3	81.0	21.3	100.0	21.2	20.9 ± 1.9
30	Base of notch to tip	17.0	3.7	19.0	4.4	15.0	5.0	15.5	4.1	22.0	4.7	4.4 ± 0.4
31	Lower caudal	66.0	14.3	73.0	16.8	55.0	18.3	72.0	18.9	73.0	15.5	16.8 ± 1.7
32	Pelvic, origin to lateral lobe	24.0	5.2	23.0	5.3	17.0	5.5	21.0	5.5	25.0	5.3	5.4 ± 0.1
33	Pelvic, origin to median tip	44.0	9.6	41.0	9.4	29.0	9.6	37.0	9.7	40.0	8.5	9.4 ± 0.5
34	Pelvic, width at lateral lobe	30.0	6.5	30.0	6.9	20.0	6.6	15.2	4.0			6.0 ± 1.2
35	Clasper length	40.0	8.7	40.0	9.2	31.0	10.3	40.0	10.5	41.0	8.7	9.5 ± 0.8
36	Anal fin, base	6.0	1.3	6.0	1.4	5.0	1.7	6.0	1.6	8.0	1.7	1.5 ± 0.2
37	Anal fin, height	5.0	1.1	6.0	1.4	4.0	1.3	6.0	1.6	6.5	1.4	1.4 ± 0.2
38	Anal fin, lobe	6.0	1.3	9.0	2.1	7.0	2.3	7.5	2.0	7.0	1.5	1.8 ± 0.4
39	Snout to mouth	33.5	7.3	19.0	4.4	17.0	5.6	27.5	7.2	11.5	2.4	5.4 ± 1.8
40	Snout to nostrils	17.0	3.7	12.0	2.8	11.0	3.7	15.5	4.1	8.0	1.7	3.2 ± 0.9
41	Internasal distance	19.0	4.1	18.0	4.1	11.0	3.7	15.0	3.9	16.0	3.4	3.9 ± 0.3
42	Nostril length	7.0	1.5	6.0	1.4	5.0	1.5	5.6	1.5	7.0	1.5	1.5 ± 0.0

(continues)

TABLE V (Continued)

| | | No. 13 | | No. 11 | | No. 9 | | No. 8 | | No. 4 | | |
		Specimen										
No.	Measurement	cm	%	cm	%	cm	%	cm	%	cm	%	Mean ± SD (%)
43	Mouth width	54.0	11.7	55.0	12.6	35.0	11.6	48.0	12.6	43.0	9.1	11.6 ± 1.3
44	Mouth length	21.0	4.6	15.0	3.4	13.0	4.2	19.0	5.0	20.0	4.2	4.3 ± 0.5
45	Body height at first dorsal origin	58.0	12.6	98.0	22.5	44.0	14.6	81.0	21.3			17.8 ± 4.2
46	Body height at pelvic origin	47.0	10.2	60.0	13.8	29.0	9.6	52.0	13.7			11.8 ± 1.9
47	Body height at precaudal pit	10.0	2.2		7.0	2.3	10.0	2.6				2.4 ± 0.2
48	Body width at pectoral origin	69.0	15.0	65.0	14.9	50.0	16.6	60.0	15.8	70.0	14.9	15.4 ± 0.7
49	Distance between pectoral tips	177.0	38.5	175.0	40.2	131.0	43.5	166.0	43.7	180.0	38.2	40.8 ± 2.4
50	Girth at mouth	190.0	41.3	163.0	37.5	113.0	37.5	126.0	33.2	190.0	40.3	38.0 ± 2.8
51	Girth at pectoral origin	260.0	56.5	205.0	47.1	146.0	48.3	189.0	49.7	226.0	48.0	49.9 ± 3.4
52	Girth at first dorsal origin	280.0	60.9	220.0	50.6	163.0	54.2	191.0	50.3	260.0	55.2	54.2 ± 3.9
53	Girth at pelvic origin	200.0	43.5	166.0	38.2	119.0	39.5	144.0	37.9	183.0	38.9	39.6 ± 2.0
54	Girth at precaudal pit	52.0	11.3	72.0	16.6	40.0	13.3	51.0	13.4	44.0	9.3	12.8 ± 2.4

TABLE VI **Stomach and Intestine Contents of White Shark Embryos**

Embryo No.	Stomach items	Mass (g)	Volume (cc)	Intestine items	Mass (g)	Volume (cc)
1	80 teeth	4.9		143 teeth	2.8	
	Yolk and egg capsules[a]	654		Dermal denticles, many crystal-like pieces[b], greenish brown fluid	524	
6	23 teeth	0.43		139 teeth, dermal denticles, many crystal-like pieces[b], greenish brown fluid	824	
	Yolk and egg capsules[a]	1120	1000			~800
8	Teeth and egg capsules[a]			Teeth, dermal denticles, crystal-like pieces[b], greenish brown fluid		

[a]Thin fragments.
[b]Smaller than 1 mm in size, sandy texture.

TABLE VII **Capture Records of White Sharks in Okinawa, Japan**

No.	TL (cm)	BM (kg)	Sex	Location	Date	Method	SST (°C)
1	390	~ 400	M	Motobu	November 21, 1975	SL	24.9
2	360	~ 400	M	Motobu	January 18, 1977	SL	21.2
3	400	~ 500	M	Motobu	January 18, 1977	SL	21.2
4	471	1000	M	Motobu	March 5, 1980	SL	21.3
5	506	1790	F	IE Island	June 26, 1981	SL	26.1
6	400	680	M	IE Island	November 15, 1984	SL	24.6
7	555	1970	F	Kin	February 16, 1985	SN	21.8
8	380	543	M	Nago Bay	Feburary 14, 1988	SL	21.1
9	301	270	M	IE Island	August 7, 1989	SL	26.1
10	508	1325	F	IE Island	August 15, 1989	SL	26.6
11	435	850	M	IE Island	August 19, 1989	SL	27.4
12	436	830	F	IE Island	November 2, 1990	SL	24.6
13	460	~ 1100	M	IE Island	November 2, 1990	SL	24.6

TL, Total length; BM, body mass; SST, sea-surface temperature; SL, set line; SN, set net.

15

Observations on a Pregnant White Shark with a Review of Reproductive Biology

MALCOLM P. FRANCIS

National Institute of Water and Atmospheric Research Ltd.
Wellington, New Zealand

Introduction

Lamnoid sharks (order Lamniformes) are typically large and uncommon, making them difficult to study. White sharks *Carcharodon carcharias* are no exception, and they pose the additional problem of being potentially dangerous to researchers. As a result, little is known about their biology, ecology, and behavior. In particular, our knowledge of reproduction in white sharks is rudimentary and is based partly on inferences drawn from other lamnoids. The reproductive mode is thought to be aplacental viviparity, with embryos being nourished by oophagy (Compagno, 1984b; Uchida *et al.*, 1987; Bruce, 1992; Gilmore, 1993).

Review of the Literature

Few reports exist describing pregnant female or embryonic white sharks. Most were based on limited secondhand observations and many of the accounts are inaccurate. The reports are discussed briefly below (Table I).

In a paper describing the anatomy of white sharks, Parker (1887) provided diagrams of parts of the skeleton and brain of a 55-cm embryo that came from the Australian Museum, Sydney. However, L. J. V. Compagno of the South African Museum (personal communication) has determined from the shape of the chondrocranium that the embryo came from a species of *Carcharhinus*, not *Carcharodon*.

Sanzo (1912) described and illustrated a 36-cm lamnid embryo from a shark caught between Italy and Sicily. By a process of elimination, he concluded that the embryo was from a white shark. Sanzo's embryo, which is in the Museum of Zoology in Florence, Italy, has recently been reexamined in detail and identified as *Isurus oxyrinchus* (A. D. Testi, H. F. Mollet, L. J. V. Compagno, and G. Bernardi, unpublished data).

A white shark caught in Egypt contained nine embryos, each 61 cm long (Norman and Fraser, 1937). Each embryo was reported to weigh 49 kg (108 lb), which is obviously erroneous (see also Bigelow and Schroeder, 1948; Randall, 1973). Tricas and McCosker (1984) suggested that the reported weight referred to all of the embryos combined, indicating a mean weight of 5.4 kg per embryo. This seems plausible, although *I. oxyrinchus* embryos 58–60 cm long weigh only 1.4–1.7 kg each (Stevens, 1983).

Bigelow and Schroeder (1948) noted reports of embryos measuring 20–61.6 cm, but gave no details or sources. Their figure of 61.6 cm may refer to the embryos from Norman and Fraser's Egyptian shark. Bigelow and Schroeder (1948, p. 138, footnote 14) also cited a report by Doderlein (1881, p. 69) of a 63-cm specimen and suggested that it may have been an embryo. However, inspection of Doderlein's original

TABLE I Pregnant Female and Embryonic White Sharks

No.	Date	Location	Female TL (m)	Embryo Observed (N)	Embryo TL (cm)	Reference
1	Summer 1934	Alexandria, Egypt	4.30	9	61	Norman and Fraser (1937)
2					20–61.6	Bigelow and Schroeder (1948)
3	November 17, 1981	Queensland, Australia	3.20	4		Paterson (1986); J. D. Stevens (personal communication)
	November 26, 1982		4.00	11		
	November 26, 1982		4.20	14		
				7		
4	February 16, 1985	Kin, Okinawa, Japan	5.55	0[a]	≈100–110	Uchida et al. (1987); Ellis and McCosker (1991)
	April 2, 1986	Taiji, Wakayama, Japan	≈4.70	7		
5	February–March 1988	Taiwan		3	≈100	D. A. Ebert (personal communication)
6	October–November	South Australia	≈4.20	11	≈60	Bruce (1992)
			≈4.70	13	≈5	
			≈5.20	6–7	≈30	
7	November 13, 1991	North Cape, New Zealand	≈5.36	7	143–145	This study
8	May 14, 1992	Uchinoura, Japan	4.80	5	130	Uchida et al. (Chapter 14)
	May 22, 1992	Toyo-cho, Kochi, Japan	5.15	10	135–151	
9	September 1992	Cape Bon, Tunisia	≈5.30	2		Fergusson (Chapter 30)
10	March 1994	South Australia		2[b]	127	J. D. Stevens (personal communication)

TL, Total length; ≈, estimated length.

[a]Numerous ova were present in the uteri, but no embryos were visible.

[b]Two aborted embryos were taken from a litter of unknown size.

account (written in Italian) revealed that a mistake had been made in the translation: the 63-cm measurement referred to the width of a set of jaws in the Palermo Museum.

In an account of the beach-meshing program in Queensland, Australia, Paterson (1986) recorded four pregnant females of a total of 480 white sharks caught. J. D. Stevens (personal communication) investigated this report and obtained copies of the fishing contractor's log entries for three of the pregnant sharks. The smallest of the three was reportedly 3.2 m total length (TL), but it is unlikely that a shark of this length would be mature (see Discussion). In fact, the three Queensland white sharks for which measurements are available are the smallest mature females on record. It is hard to imagine that these sharks were misidentified by an experienced meshing contractor. Perhaps they were incorrectly measured, or perhaps the measurements were not TLs. Unfortunately, the uncertainty over the lengths also casts doubt on the accuracy of the litter size information (Table I). The embryos were not measured.

Uchida et al. (1987) briefly described two pregnant white sharks from Japanese waters. One contained a large number of uterine eggs (192 were counted in the left uterus) but no embryos. The uteri were enlarged, suggesting that embryos had recently been aborted or born (S. Uchida, personal communication). The uterine eggs were thought to have been remnants of ova produced as food for the embryos. The second Japanese shark was not examined, but from photographs and interviews with people involved, Uchida et al. (1987) deduced that she contained seven embryos about 100–110 cm long. Their abdomens were distended by large volumes of yolk, seen spilling from their slashed bellies in the photograph provided by Ellis and McCosker (1991, p. 89). The illustration caption incorrectly states that the pools of yolk were embryonic membranes (S. Uchida, personal communication).

D. A. Ebert (personal communication) was shown photographs of three white shark embryos during a visit to Taiwan in 1988. The embryos came from a large female caught in February or March of that year. Neither the mother nor the embryos were measured, but Ebert estimated the embryos to be 100 cm TL. The abdomens of the embryos were distended.

Bruce (1992) reported a fisherman's account of three pregnant white sharks caught off South Australia in October and November (years not given). The lengths of the sharks and their embryos were apparently estimated rather than measured. Litter sizes were 6–13, and embryo sizes were 5–60 cm.

Two pregnant white sharks carrying large embryos were caught in Japan in 1992, and embryos from one of these were examined in detail (see Chapter 14, by Uchida *et al.*). Fergusson (Chapter 30) discusses the capture of a pregnant female white shark off Tunisia in 1992. She contained two embryos when hauled ashore, but no measurements were taken on the mother or the embryos.

In March 1994, a South Australian fisherman caught two small white sharks in a set net. The net also had a large hole in it, suggesting that the small sharks were embryos aborted by the mother while she struggled to escape (J. D. Stevens, personal communication). One of the embryos was examined in a decayed state, several weeks after capture. It measured 127 cm TL and weighed 14.5 kg, although the capture weight was thought to have been about 18–20 kg (J. D. Stevens, personal communication).

The capture of a pregnant female white shark carrying seven full-term embryos in New Zealand in November 1991 provided an opportunity to improve our knowledge of reproduction in the species. The female was butchered before she could be examined, but two of her embryos, and the jaws from a third, were obtained for study, along with two videotapes and a number of photographs of the mother and the embryos. In this chapter, I report observations made on the female and her embryos.

The new observations presented here and those by Uchida *et al.* (Chapter 14) bring the number of reported pregnant female white sharks or litters to 15 (Table I). I review here the available data on reproductive mode, parturition, litter size, size at birth, female length at maturity, and mating. However, crucial reproductive parameters such as the length of the gestation period and the average annual fecundity remain unknown. Until such data are available, it will be impossible to estimate population replacement rates and to determine the status of white shark populations.

Materials and Methods

Capture, Landing, and Disposal

A female white shark was caught by C. Garrett on the afternoon of November 13, 1991, at North Cape, New Zealand (34°25′ S, 173°03′ E). She was caught in a 140-mm mesh set net in 8 m of water, and towed by Garrett and another fisherman to Houhora Harbour, a distance of 48 km. The shark was dragged ashore on November 14 and winched onto the tray of a tow

truck. During this process, five embryos were aborted through the combined effects of gravity, compression of the abdomen, and direct human assistance. The female and her embryos remained on the truck overnight and were weighed on a log-weighing machine in Kaitaia on November 15. That afternoon, the female was cut up for bait at Awanui Wharf, and two more embryos were discovered inside her. The jaws were removed from an embryo of unknown length and sex by G. Kinnear, and the offal and the seven embryos were discarded over the edge of the wharf.

Garrett asked a taxidermist, K. Flutey, to preserve and mount the jaws of the female. Flutey also requested some of the embryos so that he could make casts from them. On November 16, 3 days after the female had been caught, Garrett retrieved two of the embryos from beneath Awanui Wharf at low tide. The embryos were then frozen and delivered to Flutey.

I eventually obtained from Flutey one intact embryo and one embryo from which the jaws and the pectoral and pelvic fins had been removed. In January 1994, I examined the embryo jaws in the possession of Kinnear. Photographs were obtained from the *Northern Advocate* newspaper, as were videotapes from two private citizens. One of the tapes spanned the period from the landing of the shark at Houhora Harbour to its butchering at Awanui Wharf. This tape also covered the abortion of five embryos and the measuring and weighing of the female.

Total Length Measurements

TL measurements made from the snout to a perpendicular dropped from the posterior tip of the upper caudal lobe while the latter is in its natural position (Bigelow and Schroeder, 1948) are referred to as TL_{nat}. TL measured with the upper caudal lobe flexed down to lie parallel with the body midline (Compagno, 1984b) are referred to as TL_{flex}. For small white sharks, TL_{nat} may be converted to TL_{flex} by multiplying by 1.025. This conversion factor is based on a free-living New Zealand white shark that measured 1521 mm TL_{nat} and 1559 mm TL_{flex}. When the measurement method is uncertain, TL is not subscripted. Other morphometric measurements were taken using the methods of Compagno (1984b).

Results

Observations on the Pregnant Female

The length of the shark cannot be determined precisely. One of the videotapes showed it being measured in a straight line with a measuring tape. The posterior reference point was probably the caudal fork, but this is not certain, because of the angle of the camera and the movement of people around the shark. One of the measurers called out the length as "5 m," but this must be regarded as approximate. A fork length (FL) of 5 m equates to a TL of 5.36 m using the linear geometrical mean regression of Mollet and Cailliet (Chapter 9). Photos taken of the shark lying on the truck showed that its FL was longer than the length of the tray (4.42 m), confirming a TL >4.75 m. The half-girth, measured just behind the pectoral fins, was reported on a videotape as 1.8 m, giving a girth of about 3.6 m.

The first upper right tooth had an enamel height of 51 mm, and the perimeter of the upper jaw measured 115 cm after freezing [both measurements were made by me using the method of Randall (1973)]. These dimensions are greater than any presented by Randall (1973). Furthermore, Randall did not give regression equations for his plots of enamel height and jaw perimeter versus TL, so mathematical extrapolation is impossible. Extrapolations by eye from both plots suggest a TL of 5.5–6 m.

The shark plus her seven embryos weighed 1360 kg. This is an underestimate of weight at capture, because she had lost significant quantities of uterine fluid, as well as other body fluids through dehydration, during the 18 hours she had spent out of the water before weighing. The intact embryo weighed 26.1 kg, indicating a litter weight of about 180 kg.

Four of the five aborted embryos emerged headfirst. The fifth was dragged out tailfirst, after considerable pulling and sideways rocking, by a man who reached into the dilated cloaca. These observations suggest that birth normally occurs headfirst in white sharks. Each embryo was accompanied by a gush of uterine fluid as it emerged from the cloaca. The uterine fluid released by the female during the abortion of her embryos was mostly clear, but during the abortion of one embryo, it was stained yellow–brown, suggesting that some of that embryo's intestinal contents (see below) had been released *in utero*. This presumably resulted from compression of the embryo before or during the abortion.

On December 31, 1991, 46 days after the disposal of the offal and the embryos, I dived under Awanui Wharf to search for vertebrae from the female shark. The wharf is situated on a muddy tidal river, and underwater visibility was zero. Searching was further hampered by rubbish dumped from the wharf and mounds of Pacific oysters *Crassostrea gigas*. In 1 hour of searching, I found a single vertebra that still had a plug of cartilage protruding from the ventral edge. The vertebra had clear black and white bands (Fig. 1),

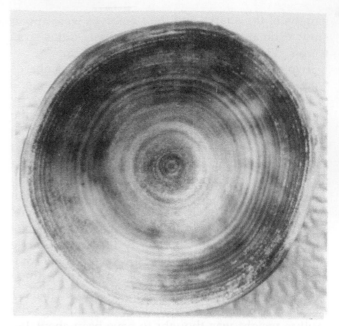

FIGURE 1 Vertebra from an approximately 5.36-m pregnant female white shark caught at North Cape, New Zealand.

presumably as a result of a sulfide reaction with the anaerobic mud. [A technique developed by Hoenig and Brown (1988) for aging sharks from their vertebrae uses ammonium sulfide to stain annual bands.] The vertebra had 22 (±1) black bands and was 63 mm in diameter. The black bands disappeared after 2 days of air-drying and exposure to daylight.

Measurements of the Embryos

Despite the long period between capture and freezing, the two retrieved embryos were in excellent condition. The intact embryo (NMNZ P.27570; see Fig. 2A) measured 1449 mm TL_{nat} and weighed 26.1 kg (Table II). The embryo from which the jaws and fins had been removed measured 1430 mm TL_{nat} and weighed 23.5 kg, suggesting a whole mass similar to that of the intact embryo. The embryos were externally similar to a 1521 mm TL_{nat} free-living female white shark caught in Kaipara Harbour, New Zealand (36°23′ S, 174°15′ E), on January 19, 1993 (Fig. 2B and Table II). Visible differences include a larger abdomen and girth and a more lunate caudal fin in the embryos than in the free-living shark.

Other External Characters of the Embryos

An umbilical cord was not visible in videotapes of the five aborted embryos, nor were there any umbilical cord remnants in the two embryos examined.

FIGURE 2 (A) A 1449-mm female embryo (NMNZ P.27570) from an approximately 5.36-m pregnant female white shark caught at North Cape, New Zealand. (B) A 1521-mm female free-living white shark caught in Kaipara Harbour, New Zealand.

Both embryos and the Kaipara shark had small (about 2–5 mm long), faint, healed scars on the throat between the bases of the second gill slits.

The embryos and the Kaipara shark all had color patterns resembling those of larger juveniles and adults (Fig. 2), including a black blotch in the pectoral axil of all three animals. Minute spiracles were found in the 1449-mm embryo and the Kaipara shark, but not in the 1430-mm embryo. The first dorsal fins of both embryos and the Kaipara shark were rounded at the apex (Fig. 2).

Teeth of the Embryos

The jaws of the 1449-mm embryo were examined *in situ*. The upper teeth were not erect, being oriented posteriorly or obliquely upward (toward the roof of the mouth). The medial margins of left and right teeth 1–5 (counted from the symphysis) of the first (outermost) series had basal cusps, as did the lateral margins of teeth 1–10. The first upper right tooth (Fig. 3) had an enamel height of 11.0 mm, and the width at the enamel base was 8.4 mm. In the lower jaw, some of the small posterior teeth were covered by a denticle-covered layer of skin. The remainder were oriented upward and posteromedially, and most were presumably functional. On the left side, the first nine teeth were uncovered, apart from the fifth, which was not erect and was held by a small sliver of skin. On the right side, the first four teeth were uncovered, teeth 5 and 6 had their points protruding through the skin, and the remainder were completely covered by skin. All visible bottom teeth had two basal cusps. In both jaws, there were tooth scars outside most of the upper and lower teeth, indicating that an earlier series of teeth had been shed.

TABLE II Morphometric Measurements of Embryonic and Free-Living White Sharks

	Smith (1951)	This study	This study	This study	Fitch (1949)	Bass et al. (1975)
Free-living (F) or embryo (E)	F	E	E[a]	F	F	F (N = 33)
Male (M) or female (F)	M	F	F	F		M and F
Total length (mm)	1400	1430	1449	1521	1543	1700–3910
Body weight (kg)	20.0	>23.5	26.1	24.8		
Liver weight (kg)	3.4	4.3		4.0		
Liver weight (%)	17.0	16.5[b]		16.1		
Proportional measurements (%)						
Fork length	89.3	88.1	88.1	88.4	88.5	
Precaudal length	81.5	**76.6**[b]	**78.1**	**78.4**		**81.3**▲**82.6**[b]
Preanal length		69.3	68.4	68.4		
Pre-second dorsal length		**65.5**	**66.8**	**67.0**		**71.1**▲**72.4**
Prepelvic length	56.2	54.5	55.9	54.4	56.8	56.9▼56.0▲59.3
Pre-first dorsal length	38.6	**34.5**	**35.9**	**36.6**	37.0	**36.9**▲**38.2**
Head length		**24.8**	**26.6**	**27.0**		
Prepectoral length	26.4	24.5	24.2	25.4	28.2	26.2
Prebranchial length		19.7	20.6	20.4		22.4▼21.4▲21.9
Prespiracular length			11.3	11.8		11.8
Preoral length	5.9		6.3	6.2		8.2▼7.2▲8.2
Preorbital length		4.5	4.2	4.2	6.2	5.7▼5.3▲5.7
Prenarial length		**3.6**	**3.7**	**4.1**	**4.3**	**4.3**
Snout–vent length		55.9	57.0	55.9		
Vent–caudal length		43.4	43.3	44.7		
Interdorsal space	21.3	21.3	22.1	22.2	21.8	23.7▲24.8
Dorsal–caudal space		10.6	10.8	10.3		
Pectoral–pelvic space			25.9	23.0		
Pelvic–anal space			9.0	10.8		
Anal–caudal space		7.8	8.6	7.7		
Pelvic–caudal space		**18.5**	**19.0**	**19.8**		
Eye length		1.5	1.4	1.4	1.6	1.4▼1.0
Eye height		1.5	1.6	1.4	1.5	
Mouth length	5.4		4.8	5.3		4.8
Mouth width	9.8		10.7	8.2	8.2	9.6▼8.8▲10.4
Nostril width		1.5	1.5	1.4	1.4	1.7▼1.5▲1.6
Internarial space	4.0	4.0	4.1	3.9		4.0▼3.8▲4.0
Interorbital space			7.7	6.9		
Eye–spiracle space			6.1	5.8		
Intergill length		**6.2**	**6.3**	**6.4**		**7.2**▲**8.2**
First gill slit height	8.5			8.5	9.1	9.4▼8.3▲10.3
Fifth gill slit height	8.8			8.2	10.0	9.9▼9.5▲10.2
First dorsal length		11.7	12.4	11.6		
First dorsal anterior margin		13.6	13.6	13.5	13.0	
First dorsal base	9.3	9.6	9.9	9.7	9.7	9.8
First dorsal height	9.3	9.2	8.8	9.4		10.5
First dorsal inner margin		2.7	2.5	2.4		2.7
First dorsal posterior margin		7.7	8.6	7.8		
Second dorsal length		3.5	3.2	3.6		
Second dorsal anterior margin		2.9	2.6	3.0		
Second dorsal base		1.5	1.6	1.6	1.3	1.5
Second dorsal height		1.3	1.5	1.9		1.5
Second dorsal inner margin		1.4	2.1	1.9		2.0
Pectoral anterior margin	20.7		22.2	21.4	21.4	
Pectoral base			7.5	6.2	7.0	6.9
Pectoral inner margin			4.1	5.0		5.5
Pectoral posterior margin			19.0	17.0		
Pectoral height			19.7	19.5		
Pelvic length			8.1	8.7		

(continues)

TABLE II (Continued)

	Smith (1951)	This study	This study	This study	Fitch (1949)	Bass *et al.* (1975)
Pelvic anterior margin			5.9	6.6		
Pelvic base			5.9	5.6	5.8	
Pelvic height			5.1	*5.1*		
Pelvic inner margin			3.7	3.1		
Pelvic posterior margin			5.8	*5.8*		
Anal length		3.1	3.4	3.3		
Anal anterior margin		3.0	2.9	2.9		
Anal base		1.4	1.4	1.6	1.3	1.5
Anal height		1.4	1.5	*1.4*		1.6
Anal inner margin		1.9	2.0	2.1		1.9
Anal posterior margin		*1.3*	1.1	*1.1*		
Dorsal caudal margin		**25.8**	**24.3**	**23.4**		**23.5▼21.8**
Preventral caudal margin		19.2	19.0	15.6		18.3
Upper postventral caudal margin		12.8	11.9	13.0		
Lower postventral caudal margin		9.7	10.4	9.9		
Caudal fork width		7.9	8.2	8.2		
Caudal fork length		10.9	10.8	10.1		
Subterminal caudal margin		1.3	1.4	1.4		
Subterminal caudal width		2.7	2.8	2.6		
Terminal caudal margin		4.1	4.2	3.6		
Terminal caudal lobe		4.7	5.1	4.4		5.2▼4.1
Trunk height		*19.2*	*22.8*	*17.4*		
Abdomen height		*16.8*	*21.7*	*16.4*		
Tail height		*9.8*	*10.7*	*9.2*		
Caudal peduncle height		2.9	2.8	2.8		
Caudal peduncle width			6.2	5.8		
Girth	**60.0**	*50.3*	*51.1*	*44.0*		
Teeth	13 + 13 / 12 + 12	14 + 12 / 12 + 12	12 + 13 / 12 + 12	12 + 13 / 12 + 12		13 + 13[c] / 12 + 12

*a*Specimen NMNZ P.27570.

*b*Measurements showing a consistent trend are in boldface; approximate values are in italics; arrowheads (▼ and ▲) indicate decreasing and increasing trends, respectively, with increasing total length; both symbols designate minimum at intermediate shark lengths.

*c*Usual tooth formula for South African white sharks (Bass *et al.*, 1975); actual total tooth counts ranged from 23 to 28 for upper teeth and from 21 to 25 for lower teeth of 32 specimens (Bass *et al.*, 1975: Table 7).

The jaws of the 1430-mm embryo were not examined in detail, but a number of photographs taken after their removal from the embryo, but before cleaning and drying, were available for study. In the right upper jaw, none of the teeth was erect. Only the third tooth of the first series remained, and it pointed posteriorly (Fig. 4A). Most of the second series of teeth had erupted from the lining of the mouth, but they were not yet functional (Fig. 4A and B). In the lower jaw, most teeth were erect and functional (Fig. 4C). Tooth scars in both jaws indicated that an earlier series of teeth had been shed (Fig. 4C).

The jaws removed by Kinnear were in a dried state, so it was impossible to determine how many teeth had originally been uncovered. The first upper left tooth had an enamel height of 12 mm. In the upper right jaw, the medial margins of teeth 1–8 had basal cusps. In the upper left jaw, only the first five teeth were uncovered, and all had medial cusps. In both the left and right sides of the upper jaw, teeth 1–10 had lateral cusps. In the lower jaw, on both the left and right sides, the first eight teeth had medial cusps, and the first nine teeth had lateral cusps.

Internal Organs of the 1430-mm Embryo

The visceral cavity of the 1430-mm embryo was dissected. Much of the distended abdomen was occupied by a liver weighing 4.3 kg. Assuming that this embryo weighed about the same as the 1449-mm embryo (i.e., 26.1 kg), the liver would have represented 16.5% of its total body weight.

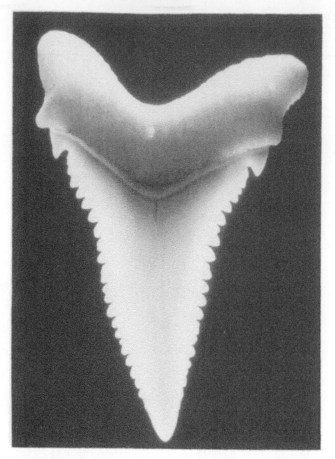

FIGURE 3 First upper right tooth from a 1449-mm female embryo white shark (NMNZ P.27570). (Photo courtesy of A. Blacklock.)

The complete digestive tract weighed 1.2 kg (4.6% of the estimated total weight), but the gut contents weighed only 0.4 kg (1.5%). No yolk or egg membranes were found in the gut, although the intestine (including the spiral valve) was packed with a viscous brown–green material that may have been a waste product of yolk digestion. Similar material was described by Springer (1948) from the spiral valve of embryonic *Carcharias taurus*.

The stomach contained a small quantity of bloody fluid, 81 teeth, and numerous denticles. The teeth were white and displayed no effects of digestion. The largest tooth had an enamel height of 10.5 mm. Using the jaws of the Kaipara shark as a model, I attempted to reconstruct tooth series from the upper teeth removed from the stomach. Most of the teeth appear to have come from a single series, with the addition of a few teeth from an earlier series and one tooth from the next series (Fig. 5). The absence of the third upper right tooth from the main tooth series is consistent with the observation that the third upper right tooth

was the only tooth that remained in the first series of the embryo's jaw (Fig. 4A). The fourth upper left tooth was also missing from the stomach series, but unfortunately, no photos of the upper left jaw were available to determine whether that tooth was still present in the jaw. In a plot of enamel height versus tooth width at the enamel base, most of the upper teeth in Fig. 5 fell on a straight line (Fig. 6). The exceptions were the two teeth from either side of the symphysis; they were markedly more slender than the others.

The intestine contained at least 110 teeth and numerous denticles. Most of the teeth were partly eroded and stained brown. They were smaller than those found in the stomach; the largest had an enamel height of 4 mm, but most had enamel heights <2 mm.

Since a single tooth series in a white shark's jaws comprises about 50 teeth (Table II), the 191 teeth found in the stomach and intestine of the embryo represent almost four full series.

Discussion

Observations on the Pregnant Female

The size of the vertebra found under Awanui Wharf and the number of visible bands (22 ± 1) agree closely with a 65-mm diplospondylous precaudal vertebra having 21 bands that was removed from a 6-m white shark caught at Gans Bay, South Africa (L. J. V. Compagno, personal communication). An age of 21–23 years for white sharks 5–6 m TL is consistent with an extrapolation of the growth curve for eastern North Pacific white sharks given by Cailliet *et al.* (1985).

Measurements of the Embryos

Smith's (1951) proportional measurements of FL, precaudal length, prepelvic length, pre-first dorsal length, and prepectoral length for a 1400 mm TL free-living white shark are all greater than those for the other small white sharks shown in Table II. The reason is unknown, but may relate to the use of a different method for measuring TL. The girth measurement given by Smith (60%) is also considerably larger than that of the two North Cape embryos, but it appears that he reported maximum girth (at the level of the first dorsal fin) rather than the girth behind the pectoral fins, as was used for the embryos and the Kaipara shark.

If Smith's (1951) measurements are ignored, some

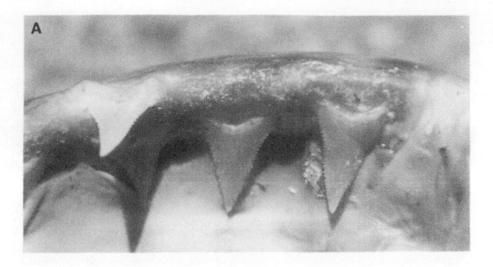

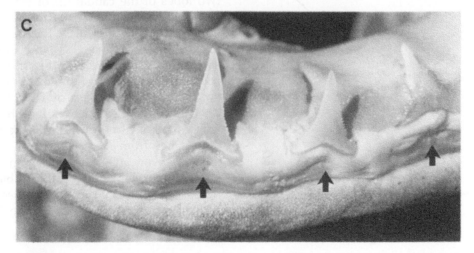

FIGURE 4 (A) First three upper right teeth from a 1430-mm female embryo white shark (viewed from inside the jaw; symphysis to the right). (B) Teeth 4–11 from the upper right jaw of a 1430-mm female embryo white shark (viewed from inside the jaw). (C) First four lower left teeth from a 1430-mm female embryo white shark (symphysis on the left). Arrows indicate scars from a previous series of teeth.

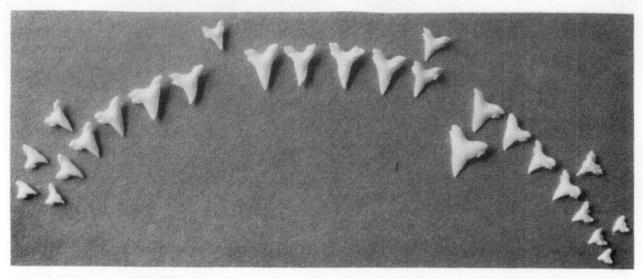

FIGURE 5 Upper jaw tooth series reconstructed from teeth found in the stomach of a 1430-mm female embryo white shark. The teeth are arranged with their anterior faces visible (teeth on the left of the photo are from the right half of the jaw, and vice versa).

consistent allometric patterns are apparent. Precaudal length, pre-second dorsal length, pre-first dorsal length, head length, prenarial length, and intergill length all increased (as a proportion of TL) as TL increased. For the two embryos and the Kaipara shark, the pelvic–caudal space also increased with TL. Conversely, the dorsal caudal margin decreased

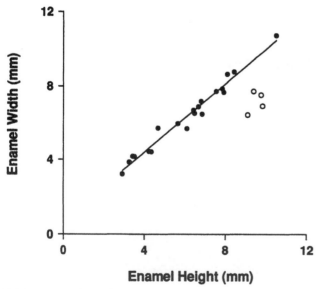

FIGURE 6 Relationship between enamel height and enamel width for the upper jaw teeth shown in Fig. 5. The first two teeth from both the left and right sides of the jaw are indicated by open circles and were omitted when fitting the regression line (enamel width = 0.645 + 0.935 enamel height; N = 20, r = 0.985).

proportionally with increased TL. A number of other measurements showed no consistent trends among the embryos and small free-living sharks, but did appear to increase (prebranchial length, preoral length, and interdorsal space) or decrease (eye length) between small and large free-living sharks. Some of the allometric trends reported by Bass *et al.* (1975), particularly those with minima or maxima at intermediate lengths, are not supported by the new data presented in Table II.

The proportional decrease in the dorsal caudal margin appears to be due to two processes. First, the two lobes of the caudal fin of embryos are presumably folded together in the uterus and are strongly lunate at birth. After birth, the lobes diverge, resulting in "shrinkage" of TL_{nat} and an increase in measurements expressed as proportions of TL_{nat}. This problem is an artifact of the TL_{nat} measurement method and supports the use of TL_{flex} for measuring the TL of sharks. Second, the length of the body increases relative to the length of the tail. Evidence for this comes from the fact that (1) body proportions increase even after the caudal fin has attained the adult shape (Bass *et al.*, 1975) (Table II), and (2) the dorsal caudal margin decreases proportionally throughout life from about 26% to 22% of TL (Table II).

Other External Characters of the Embryos

When referring to black axillary blotches, Bass *et al.* (1975) stated that "we have not seen such marks on specimens from [South Africa]." However, axillary blotches have been found by other South African researchers (Smith, 1951; L. J. V. Compagno, personal communication). The occurrence of axillary blotches is also variable in the Mediterranean (see Chapter 30, by Fergusson).

Smith (1951) did not find spiracles in a 1400 mm TL South African white shark, and Bass *et al.* (1975) reported that spiracles were not always present in white sharks >1700 mm TL. A rounded first dorsal fin was also reported for a 1400-mm South African white shark (Smith, 1951).

Teeth of the Embryos

The teeth found in the jaws of the three North Cape embryos closely resembled those of small free-living white sharks (Smith, 1951; my unpublished observation by comparison with the 1521-mm Kaipara shark). The lower jaw teeth were mainly functional, but the upper jaw teeth were not. Several species of lamnoid sharks have embryonic dentitions that differ from those of postpartum individuals (Moreno *et al.*, 1989; Moreno and Morón, 1992b; Gilmore, 1993). Distinct embryonic dentitions may be an adaptation for oophagy (Gilmore, 1993). However, egg capsules may be swallowed whole in *Alopias superciliosus* (Moreno and Morón, 1992b), indicating that functional teeth are not a prerequisite for oophagy. More detailed analysis of the morphology of the smaller teeth taken from the guts of white shark embryos might elucidate the ontogeny of tooth structure in this species.

Internal Organs of the 1430-mm Embryo

The proportional liver weight of about 16.5% is slightly lower than that reported for a 1500 mm TL white shark embryo from Japan (18.56%; see Chapter 14, by Uchida *et al.*). These values for embryonic liver weight are at the upper end of the range for free-swimming juvenile white sharks: comparative liver weights include 17.0% for a 1400-mm shark (Smith, 1951), 16.1% for the 1521-mm Kaipara shark (Table II), and 5–22% for juveniles <2 m TL (Cliff *et al.*, 1989). Adults may have a liver weight as high as 24% (Uchida, 1983, personal communication; Cliff *et al.*, 1989). The lack of yolk in the stomach of the 1430-mm embryo, and the large liver, are consistent with suggestions that lamnoid embryos consume all intrauterine yolk supplies and store the energy in an enlarged liver before birth (Gilmore, 1993).

The presence of teeth in the guts of white shark embryos may be a general phenomenon. Teeth were also found recently in embryos from Japan and South Australia (Uchida *et al.*, Chapter 14; J. D. Stevens, personal communication). The South Australian embryo had 69 teeth in its stomach, the largest of which were 10.3 and 10.4 mm enamel height. The intestine was not examined.

There are three possible explanations for the presence of teeth and denticles in the gut of white shark embryos. First, and most likely, is that the embryo had swallowed its own shed teeth and denticles. This explanation is supported, for the teeth at least, by (1) the presence of most of a single series of teeth in the stomach, (2) the absence in the stomach of the third upper right tooth and the presence of a matching tooth in the jaw of the embryo, and (3) the presence of tooth scars in the jaws. Second, embryos may shed their teeth and denticles into the uterus, from which they are ingested incidentally during feeding. Although this is possible, such behavior would probably lead to a more random assortment of upper jaw teeth in the gut than was actually found. Third, the embryo had eaten one (or more) of its siblings. The teeth in the stomachs of the 1430-mm North Cape embryo and the 1270-mm South Australian embryo were large, and must have come from embryos of similar size. It is difficult to imagine embryos devouring similar-sized siblings.

Teeth have also been found in the guts of two embryos of *C. taurus* from South Africa (G. Cliff, personal communication) and in a New Zealand free-living 1280 mm TL specimen of *Mitsukurina owstoni* (C. Duffy, personal communication). Teeth were not found in the guts of embryonic *A. superciliosus* by Moreno and Morón (1992b). The ingestion of teeth may not be restricted to lamnoid sharks, and may simply have been overlooked by previous workers.

Reproductive Mode

Intermediate-stage white shark embryos exhibit the enormously distended abdomens that are characteristic of oophagous species (Uchida *et al.*, 1987; Ellis and McCosker, 1991). The 1430-mm late-stage embryo in this study had no yolk in its gut, possibly because it was near birth. However, Uchida *et al.* (Chapter 14) found yolk and egg membranes in the stomachs of similar-sized embryos removed from the Toyo-cho shark. White sharks are therefore clearly oophagous, but are they also embryophagous? If *embryophagy* is defined as the consumption of one's siblings, then any embryo that consumes fertilized eggs is technically practicing embryophagy. However, the consumption of well-developed embryos has been documented only in *C. taurus* (Gilmore *et al.*, 1983;

Gilmore, 1993). It therefore seems sensible to restrict the term *embryophagy* to those embryos that eat other well-developed embryos.

Evidence is lacking to indicate that white shark embryos are embryophagic. In the embryophagous *C. taurus*, only one embryo survives in each uterus, so litter size is never more than two embryos (Bass *et al.*, 1975; Gilmore *et al.*, 1983; Gilmore, 1993). In white sharks, maximum litter size is at least 10 (see Chapter 14, by Uchida *et al.*), and perhaps as high as 14 (Table I). The possession of relatively large litters provides strong evidence that embryophagy does not occur in white sharks (see also Gilmore, 1993).

Maximum litter sizes in lamnoid sharks vary from 2 to 18 embryos, with most species at the lower end of the range (Table III). Only *I. oxyrinchus*, with up to 18 embryos, has a litter size comparable to that of white sharks. Why is litter size typically 4 or fewer embryos in most lamnoids, when embryophagy is known only in *C. taurus*? Pregnant females of most species are seldom caught, and the causes of differential fecundity are unknown, although Gilmore (1993) discussed several hypotheses.

Placental attachments do not form between white shark embryos and their mother. Uchida *et al.* (1987) found no evidence of umbilical cords in photos taken of intermediate-stage embryos. Late-stage embryos in this study and intermediate- and late-stage embryos reported by Uchida *et al.* (1987; see also Chapter 14) had small well-healed scars on their throats. These scars persist for an unknown period after birth (Ellis, 1975; Stevens, 1983; Klimley, 1985b), but they are not the site of umbilical attachments; they may represent the site of absorption of the yolk sac and its stalk (see also Pratt and Casey, 1990).

The available evidence indicates that the reproductive mode in white sharks is aplacental viviparity, with embryos being nourished by oophagy.

Parturition

Embryos >100 cm TL have been caught from late winter to summer (Table I). Embryos ≥127 cm were almost certainly near birth, indicating that parturition occurs in spring or summer (Table I). Further support for this timing is provided by a 5.2-m female caught off Tasmania, Australia, on January 30, 1993. She had large (40 cm wide), flaccid, empty uteri, indicating that she had recently given birth (B. D. Bruce and J. D. Stevens, personal communication). Another 5.2-m female caught off South Australia in April 1990 had a large ovary and large but empty uteri (Bruce, 1992) and may also have been postpartum. Most neonate white sharks (<155 cm TL) have also been caught in spring–summer (Casey and Pratt, 1985; Klimley, 1985b; Fergusson, Chapter 30) (Table IV).

Pregnant white sharks reputedly carrying small embryos have been caught in spring or summer (Norman and Fraser, 1937; Bruce, 1992) (Table I). The embryo lengths of 5–60 cm reported by Bruce (1992) for spring-caught white sharks were estimated by a fisherman, and are thus almost certainly imprecise and inaccurate. Nevertheless, the embryos were clearly at an early or intermediate developmental stage, rather than being close to birth. The Kin white shark reported by Uchida *et al.* (1987) had large empty uteri, suggestive of recent birth, in winter (February). There are several possible explanations for these observations: (1) the reported embryo lengths and/or capture dates were incorrect; (2) the reproductive cycle is non-

TABLE III Maximum Litter Sizes Reported for Lamnoid Sharks

Family	Common name	Species	Litter size (max)	Reference
Odontaspididae	Sand tiger shark	*Carcharias taurus* (=*Eugomphodus*)	2[a]	Bass *et al.* (1975); Gilmore *et al.* (1983)
Pseudocarchariidae	Crocodile shark	*Pseudocarcharias kamoharai*	4	Bass *et al.* (1975); Fujita (1981)
Alopiidae	Pelagic thresher	*Alopias pelagicus*	2	Otake and Mizue (1981)
	Bigeye thresher	*Alopias superciliosus*	4	Guitart Manday (1975); Moreno and Morón (1992b)
	Thresher	*Alopias vulpinus*	7	Moreno *et al.* (1989)
Lamnidae	White shark	*Carcharodon carcharias*	14	This study (Table II)
	Shortfin mako	*Isurus oxyrinchus*	18	Branstetter (1981)
	Longfin mako	*Isurus paucus*	4	Gilmore (1993)
	Salmon shark	*Lamna ditropis*	4	Paust and Smith (1986)
	Porbeagle	*Lamna nasus*	5[b]	Bigelow and Schroeder (1948); Templeman (1963); Gauld (1989)

[a]Litter sizes often exceed 2 in early gestation, but only two embryos survive to birth (Gilmore, 1993).
[b]The number is rarely greater than 4.

TABLE IV Small Free-Living White Sharks from the Southern Hemisphere

No.	Date	Location	TL (cm)	Sex (M/F)	Mass (kg)	Reference
1	March 1967	Eden, New South Wales, Australia	139			Kemp (1991)
2	1950	Algoa Bay, South Africa	140	M	20	Smith (1951), L. J. V. Compagno (personal communication)
3	December 1984	Cronulla, New South Wales, Australia	146	M	31	J. D. Stevens (personal communication)
4	January 18, 1992	Adelaide, South Australia, Australia	147		27	B. D. Bruce (personal communication)
5	May 1, 1986	Ciskei, South Africa	151	M	30	L. J. V. Compagno (personal communication)
6		Durban, South Africa	152		25	Randall (1973)
7		Durban, South Africa	152		25	Randall (1973)
8	January 19, 1993	Kaipara Harbour, New Zealand	152	F	25	This study
9	March 1, 1981	Port Stephens, New South Wales, Australia	153	F	31	J. D. Stevens (personal communication)
10	September 14, 1991	Bayly's Beach, New Zealand	≈155	M	37	This study
11	January 1985	Bird Island, South Africa	159	M	31	L. J. V. Compagno (personal communication)
12	December 15, 1989	Bird Island, South Africa	160	M	40	D. A. Ebert (personal communication)
13	September 27, 1964	Natal, South Africa	170	F	47	Bass *et al.* (1975), L. J. V. Compagno (personal communication)
14	January 15, 1992	Manukau Harbor, New Zealand	174	F	74	This study

synchronous, with females carrying embryos at different stages of development during spring–summer; or (3) the gestation period is longer than 1 year, resulting in two (or more) cohorts of embryos being present in the population at any given time. The second and third explanations seem more likely than the first.

Birth in white sharks is probably headfirst. In most viviparous shark species, young are born tailfirst (Uchida *et al.*, 1990; Pratt and Castro, 1991), but headfirst birth occurs in *C. taurus* (Gilmore *et al.*, 1983).

Embryos and pregnant or postpartum white sharks have been reported from New Zealand, Australia, Taiwan, Japan, and the Mediterranean Sea (Table I). Newborn and 0+ young [i.e., <176 cm TL, based on the growth curve provided by Cailliet *et al.* (1985)] have been reported from New Zealand, Australia, South Africa, the eastern North Pacific, the western North Atlantic, and the Mediterranean (Casey and Pratt, 1985; Klimley, 1985b; Fergusson, Chapter 30) (Table IV). Therefore, parturition probably occurs in many distinct, mostly temperate, locations worldwide.

Size at Birth

Estimation of the length of white sharks at birth has been hampered by the rarity of pregnant females and small free-living young. Unfortunately, the lengths of the three "smallest" free-living white sharks reported in the literature are erroneous. Bigelow and Schroeder (1953) reported a 91-cm (3-ft) white shark, but later corrected this to 145 cm (Bigelow and Schroeder, 1958). Ellis and McCosker (1991, p. 65) illustrated a small white shark with a caption stating that it was 104 cm (41 in.) long. However, the photograph is of a specimen measured and dissected by H. L. Pratt, who informed me (personal communication) that it was actually 122 cm long. Klimley (1985b: Appendix 1, No. 39) cited a record of a 1085-mm white shark provided by C. Swift. The specimen is in the collection of the Scripps Institution of Oceanography, La Jolla, California, and the catalogue entry states that it was 2000 mm long. R. H. Rosenblatt (personal communication) has examined the specimen and confirmed that it is much larger than 1 m. He suggested that the data sheet measurement of 1085 mm was a transcription error for 1850 mm.

The smallest reliably measured free-living white sharks appear to be three 122-cm North American specimens (Casey and Pratt, 1985; Klimley, 1985b). There have been unconfirmed reports of free-living white sharks <122 cm (Casey and Pratt, 1985; L. J. V. Compagno, personal communication), but to my knowledge, none has been accurately measured.

A number of sharks in the size range 125–140 cm have also been caught (Smith, 1951; Casey and Pratt,

1985; Klimley, 1985b; Ellis and McCosker, 1991; Fergusson, Chapter 30). The smallest Southern Hemisphere white shark reported so far was 139 cm (Table IV), but this may simply reflect the fact that fewer juvenile white sharks have been caught there than in the Northern Hemisphere.

Length at birth is therefore about 120–150 cm. This range will probably be extended at both ends as further information is obtained. Data are insufficient to determine whether length at birth varies regionally.. A wide range in length at birth is typical of many shark species, but the range in *weight* at birth in white sharks is remarkable: the 122-cm free-living sharks weighed only 12–16 kg (Casey and Pratt, 1985), whereas large embryos weighed 26–32 kg (Uchida *et al.*, Chapter 14; this study). However, newborn white sharks may lose weight initially while they are learning to feed, so the real difference in weight at birth may be less than this.

Female Length at Maturity

For the purposes of this chapter, female white sharks were judged to be mature if they were pregnant or if either the ovary **or** the uteri were large and well developed (Fig. 7). The largest immature females were 4.72 m (Springer, 1939) and about 5 m (Parker, 1887). [Parker reported the length of this shark as 17 ft (5.18 m), but it was probably measured over the curve of the body, as was another larger white shark discussed in the same paper. From measurements given by Parker for the latter shark, the straight-line length is 95.7% of the length measured over the curve. Applying this conversion factor to a length of 5.18 m gives an estimated 4.96 m TL for the immature shark.] In both of these sharks, the ovaries were small and the oviducts were not well developed.

The three smallest mature females reported (3.2-, 4-, and 4.2-m pregnant sharks) were all caught and measured by the same Queensland beach-meshing contractor (Paterson, 1986; J. D. Stevens, personal communication). Maturity at 3.2 m seems highly unlikely, given the size range of other immature and mature females (Fig. 7). This also raises concern about the accuracy of the other two Queensland measurements. A pregnant female of about 4.2 m, reported by Bruce (1992), was apparently not measured. The next smallest mature female was the 4.27 m (14 foot) pregnant Egyptian shark (Norman and Fraser, 1937), but this measurement is suspect because of other errors in the account. The 4.45 m mature female reported by Bass *et al.* (1975) was probably slightly longer than this, because their computational method underestimates TL.

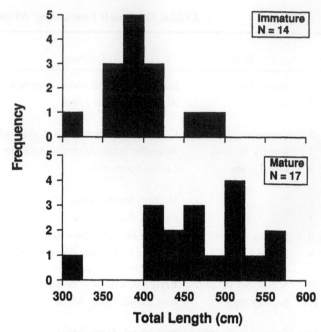

FIGURE 7 Length–frequency distributions of immature and mature female white sharks (N = sample size). Sources were Parker (1887), Norman and Fraser (1937), Springer (1939), Bigelow and Schroeder (1948), Scattergood *et al.* (1951), Bass *et al.* (1975), Casey and Pratt (1985), Paterson (1986; length data provided by J. D. Stevens, personal communication), Uchida *et al.* (1987; see also Chapter 14), Cliff *et al.* (1989), Bruce (1992, personal communication), Nakaya (1994), P. Coutin (personal communication), J. D. Stevens (personal communication), and this study.

I conclude that most female white sharks mature within the size range 4.5–5 m (Fig. 7). Some females may mature at <4.5 m, but this remains to be confirmed. Male white sharks mature at about 3.6 m (Nakaya, 1994; Pratt, Chapter 13), but the number of males examined over the critical size range is small.

Female white sharks grow larger than males. No male is known to exceed 5.5 m TL (Compagno, 1984b), whereas females may exceed 7 m (see Chapter 10, by Mollet *et al.*). The morphometric database compiled by Mollet *et al.* contained 20 females >5 m but no males in this range.

Mating

Mating has rarely been observed in wild sharks of any species, so it is not surprising that there are no published accounts of mating in white sharks. In other shark species, a number of physical signs have been used to infer recent mating, including semen or spermatophores flowing from the claspers, swollen siphon sacs, chafed claspers, and bite marks on females (Clark and von Schmidt, 1965; Springer, 1967; Pratt, 1979; Gilmore *et al.*, 1983). Published and un-

TABLE V Evidence of Mating by White Sharks in Australia and New Zealand

Date	Location	Evidence	Reference
November 9, 1969	Westernport, Victoria, Australia	Seminal fluid in the claspers of a 3.75-m male	R. M. Warneke (personal communication)
February 1981	New South Wales, Australia	Spermatophores oozing from the genital papilla of a 4.5-m male	Stevens (1984)
April 26, 1990	Streaky Bay, South Australia, Australia	Semicircular series of healed tooth marks on the left flank below the first dorsal fin of a 5.2-m female	Bruce (1992)
November 1990	Port Welshpool, Victoria, Australia	Bite marks on the pectoral fin of a 3.5-m immature female	P. Coutin (personal communication)
ca. November 1991	Nugget Point, Otago, New Zealand	Observation of mating between two individuals	A. Strachan (personal communication)
January 30, 1993	Southern Tasmania, Australia	Numerous bite marks on the left pectoral fin of a 5.2-m mature female	B. D. Bruce and J. D. Stevens (personal communication)

published accounts of such mating "indicators," and one observation of mating in white sharks of Australia and New Zealand, are summarized in Table V.

In November 1969, R. M. Warneke (personal communication) caught and examined a 3.75-m male white shark and observed the following:

Claspers chafed at base, at angle with outer flange of pelvic fin, indicating recent sexual activity? Several times, when the carcass was on its side and being maneuvered for measuring, the uppermost clasper momentarily rotated 90°. Adjacent to the base of each clasper was a swollen chambered sac filled with a fluid containing semi-transparent oblong–ovoid capsules (from memory about 7–8 mm long), which looked remarkably like over-cooked, boiled rice. This semen was running from the groove in each clasper.

Similarly, Stevens (1984) found spermatophores oozing from the genital papilla of a 4.5-m male caught in February.

Fresh bite marks were found on an immature female caught in November (P. Coutin, personal communication) and a mature female caught in January (B. D. Bruce and J. D. Stevens, personal communication). Healed bite marks were found on a mature female in April (Bruce, 1992). Because bite marks have been found in immature females and mature white sharks of both sexes, some bites may result from nonsexual intraspecific aggression rather than mating activity (Pratt *et al.*, 1982; Casey and Pratt, 1985; Bruce, 1992). Bite marks on immature females might be explained by premature mating, as has been documented in subadult female *Prionace glauca* (Pratt, 1979). However, Pratt (1993) argued that female lamnoids are incapable of long-term sperm storage, raising doubts that premature mating is effective in white sharks.

An account of two white sharks mating in southern New Zealand is contained in a letter written to the New Zealand Department of Conservation by a temporary employee. A. Strachan was employed to count and monitor New Zealand fur seals *Arctocephalus forsteri*, at a colony at Nugget Point, Otago (46°27′ S, 169°49′ E). White sharks were frequently seen from the clifftop vantage point, and at an unspecified date before December 22, 1991 (probably in November), she made the following observation:

I have unwittingly been fortunate to witness a mating [between two white sharks]. I had thought at the beginning they were fighting as one animal appeared to be attempting to grasp the other with its great mouth, making great gouges in its side. However, they had eventually become motionless, one under the other, turning over from time to time belly to belly. This obvious copulation lasted some forty minutes before the animals finally parted and glided off in opposite directions.

This account sounds highly plausible, based on what is known about mating behavior in other sharks, but I have been unable to contact Strachan to confirm her observations or to obtain further details.

Despite reservations about whether all of the above observations indicate mating activity, it is probably significant that all were made during the austral spring–summer, except for the observation by Bruce (1992). The latter occurred in autumn (April) and involved *healed* bite marks. Because parturition is also thought to occur in spring–summer, mating may occur soon after parturition, and females may carry successive litters of embryos with little or no resting period in between. This, however, remains to be demonstrated.

Summary

A 5.36-m pregnant female white shark *C. carcharias* was caught at North Cape, New Zealand, on November 13, 1991. Her first upper right tooth had an enamel height of 51 mm, and vertebral bands indicated an age of about 22 years. She was carrying seven full-term embryos. Two of these were obtained for study, along with the jaws of a third. The two embryos measured 143 and 145 cm TL, and the larger one weighed 26.1 kg. Morphometric data from the embryos were compared with data from free-living white sharks. One embryo was dissected and found to have a large liver (16.5% of body weight). There was no yolk or egg membranes in the gut, but the stomach and intestine contained 191 teeth. It is thought that the embryo had ingested its own shed teeth.

A review of the reproductive biology of white sharks revealed their reproductive mode to be aplacental viviparity, with embryos nourished by oophagy. No evidence of embryophagy was found. Maximum litter size is at least 10, and perhaps as high as 14. Parturition probably occurs in temperate locations worldwide during spring–summer. The gestation period is unknown. Length at birth is 120–150 cm, and female length at maturity is about 4.5–5 m. Mating probably occurs in spring–summer. A mating event observed in New Zealand was described.

Acknowledgments

This chapter would not have been possible without the foresight and help of C. Garrett, who expended much time and energy in bringing the North Cape shark ashore, and K. Flutey, who was responsible for saving the two embryos. Valuable information was also obtained from videotapes taken by J. Bradley and R. Grange and photographs provided by the *Northern Advocate* newspaper and R. Grange. G. McGregor provided the Kaipara Harbour white shark. For permission to use their unpublished data, I am indebted to B. D. Bruce, G. Cliff, L. J. V. Compagno, P. Coutin, D. A. Ebert, H. L. Pratt, J. D. Stevens, and S. Uchida. For other information, assistance, advice, and comments on the manuscript, I thank the above as well as D. G. Ainley, S. Ballara, C. Bradley, G. Cuthbert, R. Ellis, I. K. Fergusson, R. Forlong, M. A. Fraser, R. G. Gilmore, I. Gordon, M. Johnson, A. P. Klimley, J. E. McCosker, M. McGrouther, H. F. Mollet, J. Pepperell, R. H. Rosenblatt, P. Saul, C. Sherman, P. Stipa, A. Strachan, C. Thorburn, and R. M. Warneke.

Behavior

16

The Behavior of White Sharks and Their Pinniped Prey during Predatory Attacks

A. PETER KLIMLEY
Bodega Marine Laboratory
University of California, Davis
Bodega Bay, California

PETER PYLE
Point Reyes Bird Observatory
Stinson Beach, California

SCOT D. ANDERSON
Inverness, California

Introduction

Humans have a keen and morbid fascination with the predatory behavior of the white shark *Carcharodon carcharias* because it occasionally attacks humans (Baldridge, 1974). However, little is known about the actual feeding behavior of this species under natural conditions. To date, the white shark's mode of prey capture and handling has been inferred from (1) bite scars on pinnipeds surviving attacks (Tricas and McCosker, 1984; Ainley *et al.*, 1985; McCosker, 1985), (2) injuries sustained by humans during attacks (Egaña and McCosker, 1984), (3) human accounts of the behavior of sharks during attacks on humans (Egaña and McCosker, 1984; Tricas and McCosker, 1984; McCosker, 1985), and (4) observations of sharks scavenging on bait (Tricas and McCosker, 1984; Tricas, 1985). Bites found on surviving pinnipeds and human reports of white sharks leaving after seizing and releasing them led Tricas and McCosker (1984) to propose the "bite, spit, and wait" hypothesis of pinniped capture. This hypothesis has three elements. First, sharks seize the prey and then release it intact but bleeding. Second, sharks withdraw and wait until the prey lapses into a state of shock or bleeds to death. Finally, sharks then return to feed on the mori-

bund prey, which no longer can struggle and potentially injure the predator.

Other explanations can account for these observations. For one, scarred pinnipeds observed on shore could only be survivors of poorly executed attacks during which the pinnipeds (especially otariids; see Chapter 24, by Long *et al.*) were able to escape from the attacker (Ainley *et al.*, 1985). Pinnipeds seized in a normal lethal manner would not reach shore because they would be consumed. For another, sharks may not release potential prey to permit them to die but, rather, let them go in response to their defensive behavior or unpalatability (see Chapter 19, by Collier *et al.*).

Descriptions of predatory behavior of white sharks have so far been few. Ainley et al. (1985) gave accounts of the behavior of sharks during four attacks on seals at the South Farallon Islands. In this chapter, we quantitatively describe the feeding behavior of sharks and the response of phocids (the northern elephant seal *Mirounga angustirostris*) and otariids (the California sea lion *Zalophus californianus*) during 129 predatory attacks recorded on video at the Farallones. We also contrast the behavioral sequences of sharks during feeding bouts on pinnipeds and attacks on prey that were rejected (a decomposed pinniped, bird, and human).

Materials and Methods

We videotaped white sharks during 129 of the 247 predatory attacks documented from 1988 to 1992. These video records were transcribed into diagrams (Fig. 1). Each behavioral pattern of the shark and pinniped was recorded as either an "event" or a "state" (Lehner, 1979). We defined discrete actions of short durations as events and represented these by alphabetical codes in the diagrams here. We defined continuous actions with longer durations as states and represented these in the diagrams as horizontal bars. Most of the behavioral patterns were directly observed acts, such as the prey being seized by the shark while in either a horizontal or vertical position or while leaping from the water. A few behavioral patterns were indirect indicators of the shark's seizing the prey, such as an explosive upward splash of water, the sudden appearance of blood at the sea surface, or splashing at the anterior of the shark.

Analysis of the video records led to the creation of an ethogram, a catalog of species-specific behavior, for sharks and pinnipeds, distinguishing among pho-cids, otariids, and unidentified prey. Each predatory attack was described in sequence. In order to infer the shark's method of prey capture, we calculated for each behavioral pattern its relative frequency of occurrence in the first six positions of the sequence of acts in each predatory attack. The onset of a behavioral state was scored as an event. For example, the appearance of a red area of water and the appearance of a seal, floating motionless at the surface without bleeding, were the first and second events in predation P-120 (event B and state MOT, identified by 1 and 2 in Fig. 1A). The four very common swimming states were excluded from this analysis. We used only attacks in which the prey was identified.

Using a theodolite, we took bearings from our position at the peak of Lighthouse Hill (elevation, 102 m) to the shark or prey at the start and end of feeding bouts (see Fig. 1, with start locations, in Klimley et al., 1992). Additional bearings were taken during long attacks. The distance moved by the prey during each attack was calculated by summing distance intervals between successively plotted positions. We ascertained whether the prey moved farther or closer to shore by subtracting the distance to

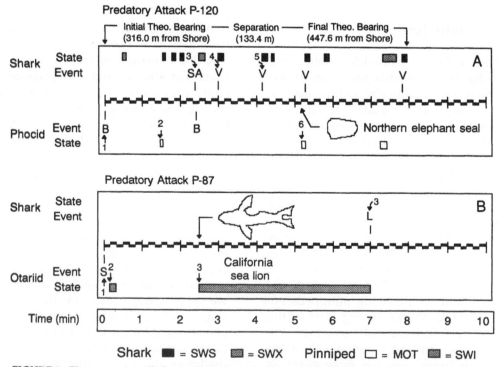

FIGURE 1 The sequences of behavioral events and states in the feeding bouts of white sharks on phocids and otariids. (A) Predatory attack P-120 on a northern elephant seal, October 15, 1989; and (B) predation P-87 on a California sea lion, November 13, 1988. The activity of the white shark is given above the abscissa; that of the prey, below. Behavioral events are indicated by alphabetical codes above the lines leading from the abscissa; behavioral states are denoted by horizontal bars. Numbers indicate the order of events (or the behavioral state onset) in the first six positions in the attack sequence.

shore from the initial position (identified by the arrow over event B in Fig. 1A) and the final position [arrow above the end of the solid bar for state SWS (slow swimming)].

Bite information was recorded for all 247 predatory attacks. Crucial to our understanding of the shark's mode of prey capture was an awareness of which part of the prey's body was first bitten by the shark. Outlines of the initial wounds were sketched on lateral and dorsal–ventral diagrams of phocids and otariids after the prey was first seen at the surface. The time when they were first observed was also recorded (see views of predations P-109 and P-87 in Fig. 1). We calculated the percentage of bites overlapping 10 body zones, each comprising 10% of the pinniped's length.

Results

Ethogram of White Shark Behavior

We recorded 1152 behavioral acts by white sharks during the 129 video records of predatory attacks on pinnipeds (Table I). Sharks performed 749 acts during 56 feeding bouts on phocids, 81 acts during 13 bouts

on otariids, and 322 acts during 62 bouts on unidentified prey. The following behavioral events and states were performed by white sharks. The behavioral patterns are described in decreasing order of frequency.

Slow swimming (state SWS): The shark swam slowly (i.e., <1 m/sec) at the surface of the water.

Exaggerated swimming (state SWX): The shark swam by moving its tail back and forth over a wide arc, often displacing water with each beat. When swimming in this manner, the shark was believed to carry a pinniped in its mouth. In form, exaggerated swimming resembled carrying, a behavior during which the prey was actually observed in the shark's mouth. The exaggerated tail beats appeared necessary to propel the shark forward when carrying a bulky carcass.

Tail slap (event TS): The shark lifted its caudal fin out of the water and slapped the sea surface, propelling water usually in the direction of a second shark. This appeared to be a signal, functioning to discourage another shark from feeding on the prey (see Chapter 22, by Klimley *et al.*).

Horizontal bite (event H): While swimming horizontally at the sea surface, the shark seized the pinniped (Fig. 2A).

TABLE I Ethogram of White Shark Behavioral Patterns Based on 131 Videotape Records of Predatory Attacks

No.	Behavioral pattern	Type	Code	Total pinnipeds N	% Total	R	Phocids N	% Total	R	Otariids N	% Total	R	Unidentified N	% Total	R
1	Slow swimming	State	SWS	408	35.4	1	227	29.9	1	44	54.3	1	137	42.5	1
2	Exaggerated swimming	State	SWX	314	27.3	2	220	29.4	2	15	18.5	2	79	24.5	2
3	Tail slap	Event	TS	98	8.5	3	72	9.6	4				26	8.1	4
4	Horizontal bite	Event	H	95	8.2	4	76	10.1	3	3	3.7	5	16	5.0	5
5	Vertical bite	Event	V	83	7.2	5	48	6.4	5	5	6.2	4	30	9.3	3
6	Shake torso	Event	T	38	3.3	6	29	3.9	6	2	2.5	8	7	2.2	6
7	Carrying	State	CAR	27	2.3	7	18	2.4	7	3	3.7	5	6	1.9	7
8	Splash at anterior torso	Event	SA	16	1.4	8	13	1.7	8				3	0.9	11
9	Splashing	State	SPL	14	1.2	9	12	1.6	9				2	0.6	12
10	Swirl of water	Event	W	9	0.8	10	4	0.5	11				5	1.6	8
11	Side by side	Event	SS	8	0.7	11	4	0.5	11				4	1.2	9
12	Push	Event	P	7	0.6	12	3	0.4	13				4	1.2	9
13	Mouth	Event	M	6	0.5	13				6	7.4	3			
14	Leap	Event	L	5	0.4	14	2	0.3	16	3	3.7	5			
15	Tilting	State	TIL	5	0.4	14	5	0.7	10						
16	Chasing	State	CH	4	0.4	16	3	0.4	13				1	0.3	13
17	Dive and bite	Event	DV	3	0.3	17	3	0.4	13						
18	Fast swimming	State	SWF	3	0.3	17	2	0.3	16				1	0.3	13
19	Breach	Event	BR	2	0.2	19	1	0.1	20				1	0.3	13
20	Release	Event	R	2	0.2	19	2	0.3	16						
21	Repetitive bites	Event	RB	2	0.2	19	2	0.3	16						
22	Follow	Event	FOL	1	0.1	22	1	0.1	20				1	0.3	13
23	Head to tail	Event	HDT	1	0.1	22	1	0.1	20						
24	Tail hit	Event	THT	1	0.1	22	1	0.1	20						
	Behavior totals			1152			749			81			322		

FIGURE 2 Two behavioral events performed by white sharks and two behavioral states of pinnipeds: (A) a white shark approaching from the side to seize a northern elephant seal (event H), (B) a white shark seizing a seal from underneath (event V), (C) a dead seal recovered after the initial bite (found floating at the surface motionless but not bleeding; state MOT), and (D) a seal swimming intermittently after being bitten by a white shark (state SWI).

Vertical bite (event V): The shark swam upward from beneath the prey and grasped the prey in its jaws at the surface. The angle of the shark's body to the sea surface ranged from 60° to 90° (Fig. 2B).

Shake torso (event T): Forward movement was momentarily halted as the shark shook its forward body from side to side in the horizontal plane. This movement often dislodged a piece of flesh from a pinniped held in its mouth. The rest of the carcass then floated to the surface.

Carrying (state CAR): The shark swam with an immobile or struggling pinniped held in the jaws. The tail was moved back and forth over a wider arc than when the shark did not hold prey in its mouth to generate sufficient force to propel the body forward with its extra load.

Splash at anterior torso (event SA): Vigorous splashing occurred near the head of the shark and was usually followed by the appearance of blood. This splash-

ing indicated that the pinniped had been seized by the shark.

Splashing (state SPL): Much splashing occurred near the pinniped carcass, yet the shark was not seen at the sea surface.

Swirl of water (event W): The surface of the water near a pinniped carcass was disturbed by a vigorous movement of the shark just below the surface.

Side by side (event SS): Two sharks swam by each other in opposite directions at the sea surface less than a body length apart. When the two sharks were next to each other, they usually slowed and performed tail slaps in the direction of each other (for a further description, see Chapter 22, by Klimley *et al.*).

Push (event P): The shark contacted prey with its snout. When the shark approached rapidly from underneath, it struck the pinniped forcefully and often propelled the prey out of the water. When the shark approached from the side, contact was usually less forceful, and the pinniped was pushed along the sea surface.

Mouth (event M): The shark grasped the prey in its mouth and then quickly released it without removing any flesh. This behavior was recorded six times during a single feeding bout on a highly decomposed otariid from which only a single bite of flesh was removed before the otariid floated out of view.

Leap (event L): The shark propelled two thirds of its entire length from the water (at a 30–60° angle with the sea surface), seized the prey, and held it in its jaws. While the shark was airborne, the anterior torso bent downward and the tail lifted so that the shark reentered the water headfirst.

Tilting (state TIL): The shark rotated its body onto one side and beat its tail to one side in a forceful and exaggerated manner; this behavior may be a display used to threaten other sharks competing to feed on the same carcass (see Chapter 22 Klimley *et al.*).

Chasing (state CH): The shark swam quickly in pursuit of a fleeing pinniped.

Dive and bite (event DIV): The shark bit the prey while forcing it underwater, whereupon the shark usually lifted its slashing tail out of the water as it propelled itself downward.

Fast swimming (state SWF): The shark swam rapidly at the surface of the water.

Breach (event BR): The shark propelled from two thirds to its entire length from the water at a 30–60° angle to the sea surface. In contrast to the leap, the shark usually landed with its body straight and often on its side. The entire body of the shark contacted the surface of the water at one time, displacing water in all directions. On two occasions, breach preceded tail slap and, for that reason, appeared to be a display, perhaps functioning to discourage other sharks from feeding on a carcass (see Chapter 22, by Klimley *et al.*). However, other acts of breach were observed that were not associated with predatory attacks (see Chapter 34, by Pyle *et al.*).

Release (event R) The shark released an intact object from its mouth. This behavior was observed only twice when sharks were feeding on active pinnipeds, but it also occurred with a decomposing sea lion, brown pelican *Pelecanus occidentalis*, and human.

Repetitive bites (event RB): The jaws were closed three to five times in quick succession (<5 seconds) on the pinniped.

Follow (event FOL): One shark followed another shark less than a body length away as the leader moved in a sinuous path.

Head to tail (event HDT): Two sharks swam in a tight circle, with the head of one close behind the tail of the other.

Tail hit (event THT): One shark struck another shark with its caudal fin. Actual contact between sharks was observed only once. However, two sharks may have contacted each other when they alternately directed five tail slaps at each other, but the bodies of the sharks were blocked from our view by splashing water (predation P-143 in Chapter 22, by Klimley *et al.*). A bout of tail hitting may have resulted in an oval white contusion on the peduncle of a shark observed feeding on a phocid on November 12, 1992 (see White Peduncle in Chapter 33, by Klimley and Anderson).

Ethogram of Pinniped Behavior

We observed 474 behavioral acts by prey during the 129 video records (Table II). Phocids performed 297 acts; otariids, 46 acts; and unidentified pinnipeds, 131 acts. The following behavioral events and states were performed by pinnipeds when attacked by white sharks. Some of the behavioral patterns described below were also performed by the brown pelican and the human when attacked by a white shark.

Blood-stained water (event B): An area of water became red from mixing with blood from the bleeding pinniped. Initially, the area was small, circular, and brightly colored. The colored area sometimes elongated toward the site where the pinniped later floated to the surface. When this happened, we believe that the bleeding prey was chased or transported by the shark held in its jaws. If the prey did not move, the area grew and became less bright with greater dilution.

Floating motionless (state MOT): The pinniped, usually with a large bite removed from its body, floated at the sea surface without moving its limbs. The water around the carcass was not colored red by an actively bleeding wound. The pinniped appeared to have lost its blood supply. Some nonbleeding carcasses examined by us at the site of the attack had a deflated appearance, resulting from the missing blood supply (see the northern elephant seal in Fig. 2C).

Explosive splash (event S): A large amount of water was suddenly propelled upward. Although a blood stain was often seen at this time, the pinniped was not.

Floating motionless in blood-stained water (state MTB): The profusely bleeding prey floated immobile at the surface within an area of blood-stained water. The prey usually had a crescent wound where the shark had removed a large amount of flesh.

Slow swimming (state SWS): The pinniped swam slowly at the surface of the water. In contrast to the case of intermittent swimming, the individual swam in a continuous and oriented manner.

TABLE II **Ethogram of Pinniped Behavioral Patterns Based on 131 Videotape Records of Predatory Attacks**

No.	Behavioral pattern	Type	Code	Total pinnipeds			Phocids			Otariids			Unidentified		
				N	% Total	R	N	% Total	R	N	% Total	R	N	% Total	R
1	Blood-stained water	Event	B	209	44.1	1	122	40.4	1	19	19.6	2	80	61.1	1
2	Floating motionless	State	MOT	94	19.8	2	69	23.2	2	6	13.0	4	19	14.5	3
3	Explosive splash	Event	S	52	11.0	3	20	6.7	4	10	21.7	1	22	16.8	2
4	Floating motionless in blood-stained water	State	MTB	48	10.4	4	44	14.8	3	2	4.3	6	6	4.6	4
5	Slow swimming	State	SWS	27	5.7	5	20	6.7	4	6	13.0	4	1	0.8	6
6	Oil slick	Event	SL	23	4.9	6	15	5.1	6	2	4.3	6	6	4.6	4
7	Intermittent swimming	State	SWI	7	1.5	7				7	15.2	3			
8	Moving limbs	State	MLM	4	0.8	8	3	1.0	7	1	2.2	9			
9	Bobbing	State	BOB	3	0.6	9	2	0.7	9	1	2.2	9			
10	Fast swimming	State	SWF	3	0.6	9	3	1.0	7						
11	Rolling	State	ROL	3	0.6	9				2	4.3	6	1	0.8	6
12	Bubbling	State	BUB	1	0.2	12	1	0.3	10						
Behavioral totals				474			297			46			131		

Oil slick (event SL): The sea surface became smooth due to the dampening effect on small waves exerted by oils escaping from the body of the prey. The oil film was more reflective than the surrounding water.

Intermittent swimming (state SWI): The pinniped slowly swam a short distance, stopped, and swam a small distance again. The prey often appeared disoriented, moving in one direction and then in another (Fig. 2D).

Moving limbs (state MLM): The moribund pinniped floated on either its back or its belly and feebly moved its limbs up and down.

Bobbing (state BOB): The pinniped remained in a vertical position, moving its head repeatedly up and down, above and below the sea surface.

Fast swimming (state SWF): The pinniped swam rapidly at the surface in a continuous and oriented manner.

Rolling (state ROL): The pinniped, horizontal at the sea surface, rotated around its long body axis.

Bubbling (state BUB): A large number of small bubbles appeared at the surface near where the shark had initially seized a pinniped. We assumed that the bubbles were released from the pinniped's trachea and lungs as the shark consumed the chest region of the pinniped.

Strike with object (event ST): The pinniped or human struck the shark either with its limbs or with an object held in them. This behavior was never re-corded for a pinniped, but was recorded three times during attack P-99 on a human.

Description of Predation

The sequence of behaviors in predation P-120 is an example of a white shark feeding bout in which the prey is a phocid (Fig. 1A). Observers were alerted to the shark's initial strike by the appearance of a small bright-red area on the sea surface (event B). The shark surfaced and began to swim with exaggerated tail beats 0.3 minute later, possibly with the phocid held in its jaws (state SWX). The phocid surfaced after 1.5 minutes and floated motionless, no longer bleeding, 75–100 m from the site of the initial strike (state MOT). The phocid was no longer inside the area tinted red with blood; the area had by now expanded and elongated toward the site where the dead seal reappeared. The phocid had already lost a massive amount of blood. The carcass remained at the surface for 0.1 minute before the shark seized the carcass and submerged with it. Splashing and a small amount of blood were observed 1 minute later near the head of the shark, indicating that the pinniped had been bitten again (events SA and B in Fig. 1A). Later, the carcass was seized four more times by the shark from underneath, with no loss of blood (event V in Fig. 1A). During the feeding bout, the phocid moved a distance of 133.4 m almost directly offshore of the location where the attack began.

The interval between the appearance of blood (event B, labeled 1, in Fig. 1A) and the appearance of the bloodless and immobile phocid at the surface (event MOT, labeled 2) was 1.5 minutes, almost twice the duration of the 0.8-minute interval between the second and third events (onset of state MOT and event SA), and almost three times the duration of the 0.6-minute interval between the third and fourth events (events SA and V).

The sequence of behavior during predation P-87 is an example of a white shark feeding bout in which the prey is an otariid (Fig. 1B). Observers were alerted to the attack by an explosive splash of water propelled upward 3–5 m (event S). The otariid quickly surfaced and swam in a slow and jerky manner, often changing directions (state SWI). After swimming underwater for 2 minutes, the otariid surfaced and swam toward shore in a slow and jerky fashion for 3.5 minutes; a large amount of flesh was missing from its chest region (see the outline of the wound in Fig. 1B). The shark then jumped entirely out of the water and seized the animal in its jaws (event L). Neither the shark nor the otariid was seen again.

Intervals of time between subsequent events in the otariid predation P-87 increased, while the time intervals decreased during predation P-120 on the phocid. An interval of only 0.2 minute elapsed between the explosive splash (event S, labeled 1, in Fig. 1B) and the appearance of the otariid at the surface swimming intermittently (onset of state SWI, labeled 2). This interval was considerably shorter than the 2.5-minute interval between the second event and the continuation of intermittent swimming by the sea lion (onset of state SWI, labeled 3).

We observed differences in the sequences of behavioral events during feeding bouts on phocids (N = 55) and otariids (N = 13). The initial events were usually different. We were alerted to 76.8% of the predatory attacks on phocids by the appearance of blood at the sea surface and to only 10% of the attacks by an explosive splash at the surface (bars over events B and S, respectively, left histogram, first event, in Fig. 3). In contrast, our attention was first directed to 53.8% of attacks on otariids by an explosive splash and to 20% of the attacks by the appearance of blood (bars over events S and B respectively, right histogram).

Phocids were often observed dead or immobile at the surface early in the predatory sequence. The second behavioral event was most often the phocid floating motionless but no longer bleeding (21% state MOT) or motionless and bleeding (20% state MTB). The high frequency of MOT indicated that some pho-

cids had already lost most of their blood, possibly while struggling alone or being either chased or carried by the shark immediately after the initial strike. By the third event, a much higher percentage of phocids was no longer bleeding (22% state MOT) relative to those that were bleeding (5% state MTB). Immobile bloodless phocids were observed 11%, 5%, and 9% more than bleeding phocids in the fourth, fifth, and sixth positions, respectively, of the feeding sequence. Sharks commonly seized phocids either from horizontal or vertical orientations during the second (19%) and third positions (27%) of the attack sequence (bars over events H and V). Side by side, tail slap, tilting, and breach usually occurred when two sharks were near one another. The displayer's behavior appeared to discourage the recipient from approaching the prey. These acts were common in the fourth, fifth, and sixth positions of the attack sequence (bars over events SS, TS, and BR and state TIL).

Otariids, on the other hand, usually survived the shark's initial strike. We observed them early during an attack: floating at the surface moving only their limbs (5% of bouts), swimming intermittently (14%), swimming slowly (27%), or swimming fast (5%) (bars over states MLM, SWI, SWS, and SWF, right histogram, second event). Otariids initially floated motionless at the surface, either bleeding or not bleeding, in only 5% of the attacks (bars over states MTB and MOT). Even by the third event, otariids continued to swim, either slowly or rapidly, in 6% and 24% of the attacks, respectively. Rarely did sharks seize otariids from either horizontal or vertical orientations in the first six positions of the attack (bars over events H and V). In the few instances when sharks approached otariids soon after the initial strike, sharks attacked them forcibly. They leaped from the water, seized the prey, and submerged with it in 8% of the first events and 22% of the fourth events in the sequence (bars over event L, right histogram). Otariids began to succumb to the predatory aggression of the sharks only by the sixth event. During 25% of feeding bouts, otariids were carried by the sharks in their mouth; during 30% of the bouts, they floated immobile at the surface; and during 23% of the bouts, they swam slowly (bars over states CAR, MOT, and SWS, right histogram, sixth event).

The time intervals between the first and second events in feeding bouts on phocids were longer than those on otariids. Intervals >2 minutes were recorded in 28% of the attacks on phocids (left histogram, Fig. 4); no intervals >2 minutes were recorded for otariids (right histogram). In only 17% of the attacks on phocids, the intervals were <0.5 minute, in contrast to

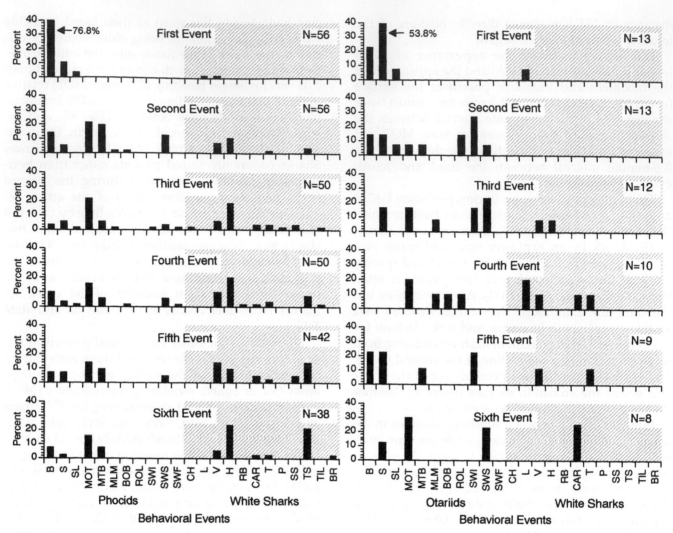

FIGURE 3 Bar diagrams of the relative frequencies of 12 behavioral events performed by white sharks and 11 events by pinnipeds in steps 1–6 of feeding bouts (the first event is at the top). Percentages of events performed by sharks and phocids are shown on the left; percentages of events by sharks and otariids are on the right. The behavior of the sharks is within the stippled areas.

50% for otariids. The frequency of short intervals (<0.5 minute) increased between subsequent consecutive events in feeding bouts on phocids (increasing height to left bars, left histograms for first to fifth Intervals). On the contrary, the frequency of short intervals decreased later in feeding bouts on otariids (lowering height to left bars, right histograms).

The initial bites by sharks were directed at different areas of the prey's body. Sharks bit phocids most often near the base of the hindflippers or on the head (Fig. 5A). The latter bites frequently resulted in decapitation. Otariids were most often bitten slightly posterior of the midbody (Fig. 5B). At times, otariids were decapitated.

The pinnipeds were either chased or carried con-

siderable distances. Distances between the initial and final positions recorded for phocids were commonly as long as 400 m (42% and 30% of distances in 100- and 300-m classes, respectively) (Fig. 6A). Otariids appeared to be transported or pursued even greater distances (38% in the 900-m class and 40% in the >1200-m class). The direction of these movements was most often away from the island. The measurements of distances to shore were skewed toward positive distances (away from shore), with peaks in the 100-m range (Fig. 6B). Otariids moved farther offshore than did phocids (see the peak in the >600-m distance class). By transporting or chasing prey away from the island, the shark would maximize its chance of recapturing the prey after the initial strike.

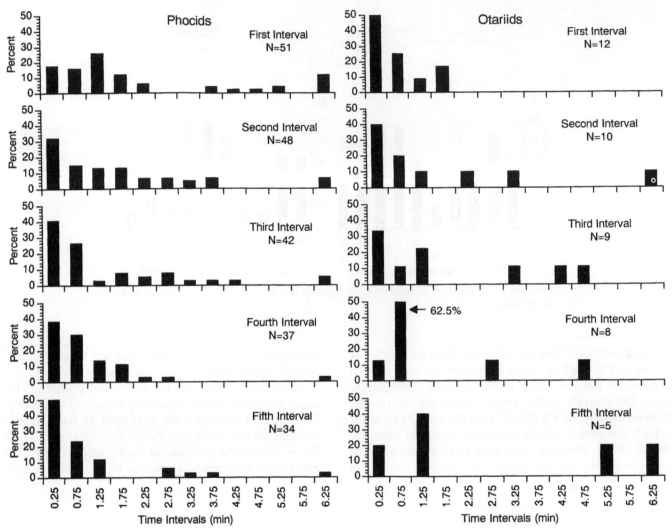

FIGURE 4 The relative frequencies of intervals of increasing duration among the first six events of feeding bouts. The first interval is the period between the first and second events, and so forth. Percentages for attacks on phocids are shown on the left; percentages for attacks on otariids are on the right.

White sharks spent less time handling pinnipeds at the surface during feeding bouts near shore. Sharks spent <1.2 minutes at the surface during feeding bouts on phocids <100 m from shore (Fig. 7A). At greater distances, sharks spent up to 8.2 minutes at the surface. The frequency of surface visits during feeding bouts on phocids <100 m from shore was <9, whereas at greater distances the frequency was usually greater and ranged as high as 20 visits (Fig. 7C). Predatory attacks on otariids usually began at distances >100 m from shore. It is possible that sharks minimize time at the surface near shore due to the risk of alerting other pinnipeds of their arrival; this may not be necessary at greater distances from seal haul-out locations (Fig. 7B and D).

Method of Pinniped Capture

"Bite, Spit, and Wait"

Verification of the "bite, spit, and wait" hypothesis of Tricas and McCosker (1984) would require that we witnessed sharks seizing pinnipeds, releasing them in an intact yet wounded condition, waiting nearby until the prey became immobile, and soon feeding on them. We never observed a predatory attack with this behavioral sequence in the 131 video records. However, such a sequence could have been missed, because we recorded only predation at the surface. For this reason, we examined our data relative to certain questions that relate directly to the above hypothesis.

Did white sharks release seals in a wounded but

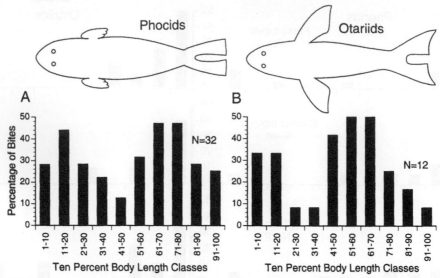

FIGURE 5 Percentages of initial bites overlapping 10% body divisions: (A) phocids and (B) otariids.

intact condition? Sharks removed flesh in 94.4% (34 of 36) of the feeding bouts in which we were able to determine the condition of the phocid immediately after the shark's initial strike. Similarly, sharks removed flesh in 90.9% (10 of 11) of the feeding bouts in which the bodies of otariids were observed after the initial strike. Wounds with little flesh removed were recorded only on two seals and one sea lion. It is

plausible that these wounds were left when the pinnipeds struggled free from the shark's jaws.

Did white sharks always wait until pinnipeds became immobile before returning to complete feeding? Phocids and otariids were first seen as they lay motionless on the surface in 72.2% (N = 54) and 38.5% (N = 13) of the attacks on each, respectively. The absence of movement appeared to be a reliable indi-

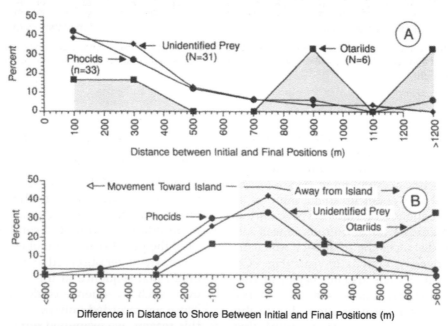

FIGURE 6 (A) The distances between the initial and final positions, and (B) the movements of pinnipeds relative to the island for phocids, otariids, and unidentified prey during predatory attacks. Negative distances indicate movements toward, and positive distances, movements away from, the shore.

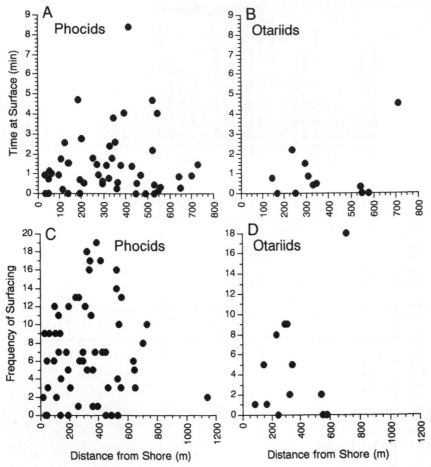

FIGURE 7 The amount of time and frequency of surface visits by sharks when feeding on phocids compared to otariids.

cator of death. Only in one instance did a phocid or otariid begin swimming after floating motionless (see the single act of states SWF and SWS following state MTB in Figs. 8 and 9). This contrasted with how frequently phocids and otariids were observed floating motionless at the surface (102 acts of states MOT and MTB). Sharks may have waited for prey to succumb during the time between appearance of blood-stained water or an explosive splash, thus indicating the initiation of the attack on a phocid and the second attack event, usually the surfacing of motionless prey. However, chasing or carrying their prey underwater would also explain their absence from the surface after the initial strike. Few sharks were seen chasing or carrying phocids when we witnessed events from Lighthouse Hill, but when feeding bouts were witnessed nearby from a boat, sharks seized and transported pinnipeds while the latter bled profusely (e.g., predation P-10: November 8, 1986; P. Pyle, unpublished observation, November 11, 1992).

When pinnipeds survived the initial strike and remained at the sea surface, sharks usually did not wait until these prey were motionless before feeding. Pinnipeds were seen after the initial strike at the surface moving limbs, bobbing, rolling, intermittent swimming, or slow swimming in 27.8% (N = 14) and 61.5% (N = 8) of the attacks on phocids and otariids, respectively. Sharks attacked 64.4.3% (N = 9) of the phocids and 75% (N = 6) of the otariids performing such actions.

As an example of the capture of an active phocid, consider attack P-109. Two sharks chased a subadult elephant seal (estimated weight, 225 kg) that initially fled swimming with its head held out of the water. The seal reversed its direction and swam directly toward the two sharks, abruptly veering off to one side as it moved between them (they were not exactly abreast of each other). The sharks later captured and consumed the seal. As an example of the capture of an active otariid, consider attack P-71. A Steller sea

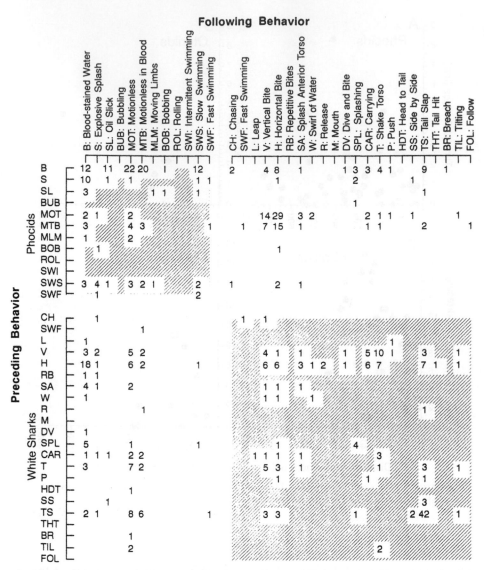

FIGURE 8 The numbers of acts immediately preceding or following each behavioral event and state recorded during white shark predatory attacks on phocids. For example, the totals of vertical and horizontal bites following the appearance of nonbleeding and bleeding yet motionless prey can be found across the rows labeled states MOT and MTB and down the columns labeled events V and H.

lion *Eumetopias jubatus* surfaced with a large bite removed from its midsection 0.8 minute after the appearance of a pool of blood. The sea lion swam intermittently toward shore for 1.3 minutes, until a white shark seized it from the side. The shark carried the sea lion for 1.2 minutes before removing a second bite. The shark then swam slowly at the surface for 0.3 minute, searching for the rest of the carcass. The shark then seized the carcass from underneath and carried it at the surface for 1 minute before consuming the remnants.

Did the attacked pinnipeds struggle at the surface for prolonged periods, as expected if their predators were waiting for them to die (or become motionless) before returning to feed? Surviving pinnipeds swam at the surface for relatively short intervals. The durations of swimming bouts at the surface by the surviving phocids ranged from <0.1 to 3.2 minutes (median, 0.6 minute; mean, 1.1 minutes; N = 10); the activity bouts for the surviving otariids ranged from <0.1 to 17.3 minutes (median, 0.9 minute; mean, 3.7 minutes; N = 7).

Did sharks return to feed on the carcasses soon after the pinnipeds died? The bite, spit, and wait

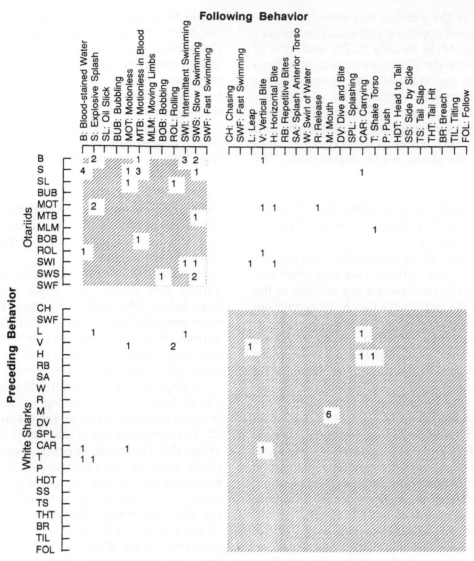

FIGURE 9 The number of acts immediately preceding or following each behavioral event and state recorded during white shark predatory attacks on otariids.

hypothesis predicts a quick return because of the absence of any risk of being bitten or clawed. We observed carcasses of phocids to float for periods ranging from 0.1 to 140 minutes (median, 2.2 minutes; mean, 14.4 minutes; N = 27) and carcasses of otariids to float for intervals ranging from 0.5 to 120 minutes (median, 6.5 minutes; mean, 41 minutes; N = 5). For instance, in attack P-72, a northern elephant seal, with a wound on its pelvis, was permitted to drift for 87 minutes before the white shark fed again. During attack P-80, an explosive splash was followed 1 minute later by the appearance of a dead California sea lion. The beheaded carcass floated at the surface for 76 minutes until we could no longer see it.

Bloodletting, or Exsanguination

In 34% of the attacks, the interval between the initial strike and the appearance of the seal was <1 minute; in 66% of the attacks, the seal appeared ≥1 minute after the initial strike. In the case of shorter intervals, the sharks removed bites quickly and killed the seals. The longer intervals are consistent with the seal's struggling underwater, being chased, or being carried in the mouth of the shark before dying. Dead or immobile pinnipeds often were no longer bleeding when we first observed them after the initial strike. Phocids and otariids appeared at the surface motionless yet not bleeding (state MOT) in 50% (N = 54) and

30.8% (N = 13) of the attacks, respectively. In contrast, the two species appeared at the surface motionless and bleeding (MTB) in only 22.2% and 7.7% of the attacks, respectively. These pinnipeds may have died from hypoxia due to blood loss. Transport of the seal by the shark would explain why phocids appeared at the surface after longer intervals, despite being decapitated or missing their hindflippers and being unable to swim. An even more complex feeding strategy by the shark might explain this. The shark may carry the pinniped in its mouth until the shark no longer sees the pinniped bleeding. Only then would the shark bite to remove flesh. This behavior would ensure that the pinniped could not escape after being released.

It was difficult to evaluate bloodletting as a method of prey capture, because both the attacking shark and its prey were usually underwater out of view at the apparent time of the pinniped's death. However, observations of sharks on the surface feeding on nonbleeding carcasses twice as often as bleeding ones indicated that sharks may indeed assess the condition of the prey based on whether or not it is bleeding. We observed that sharks seized nonbleeding phocid carcasses from the side 29 times and from underneath 14 times, whereas sharks bit bleeding carcasses from the side only 15 times and from underneath only 7 times (states MOT and MTB, followed by events H and V, in Fig. 8). The ratio of the 44 bites (events H and V pooled) on nonbleeding to 21 bites on bleeding carcasses differed from the ratio of the 69 records of nonbleeding carcasses and 44 records of bleeding carcasses in a statistically significant manner (χ^2 test of independence, $p = 0.0001$) (Table II). With regard to otariids, sharks were never observed to bite bleeding carcasses, but on two occasions fed on nonbleeding carcasses (state MOT, followed by events H and V, in Fig. 9). An example of this was the sequence during the only predation (P-195) in which an otariid, surviving the initial strike, was observed to die while on the surface. This individual spent 0.5 minute at the surface rolling before dying (state ROL), floated at the surface within a pool of blood for 11 minutes (state MTB), and then was bitten only 0.5 minute after bleeding stopped.

Item Rejection

The contrast between the behavior of sharks when prey are not readily eaten and that when it is consumed provided insight into factors underlying prey selection. For instance, during attack P-173, the shark swam slowly four times, swam vigorously only twice, and seized the pinniped only once during the 23-minute period when a boat was close to the pinniped (states SWS and SWX and event H in Fig. 10A). After the boat's departure, the shark swam vigorously at the surface eight times and seized the prey three times (states SWS and SWX and Event S). The shark may have hesitated from feeding early during this bout because it was disturbed by the nearby boat or the boat's outboard engine.

During attack P-205, a badly decomposed and gas-filled sea lion carcass floated away from the island (Fig. 10B). A white shark swam slowly near the carcass. The shark then momentarily grasped the rotting carcass in its jaws and quickly released it without removing any flesh six times before removing only a single bite (events M and H). The shark swam away and appeared to reject the carcass due to its advanced state of decomposition.

In attack P-60, a brown pelican sitting on the sea surface was struck by a shark swimming upward from below. The bird and a great amount of water were propelled upward (Fig. 10C). The bird fell to the surface, where it bled profusely and fluttered its wings spasmodically until it died 2 minutes later. The shark never returned to feed on the pelican.

During attack P-99, a white shark seized a commercial abalone diver by the leg while he was pausing in midwater at a depth of 4.6–6.4 m to clear his ears (R. Lea, personal communication). According to the victim (M. Tisserand, personal communication), "the shark swam up from underneath, seized me, carried me down for five to seven seconds, and suddenly let go and swam off" (event V, state CAR, and event R in Fig. 8D). While being carried in the shark's jaws, the victim lost much blood (event B) (R. Lea, personal communication). Similar to the situation during feeding bouts on pinnipeds, the shark carried the diver underwater for some distance. The diver struck the shark with the butt of a bang stick three times before the shark released him (events ST and R) (R. Lea, personal communication).

Discussion

Tricas and McCosker (1984), and later Ainley et al. (1985), hypothesized that sharks approached pinnipeds floating at the surface from behind or beneath. Their hypothesis was based on the concentration of wounds on the posterior region of pinnipeds that survived predatory attacks. We also recorded high percentages of bites on the posterior bodies of phocids and otariids. Neither Ainley et al. (1981, 1985) nor we observed pinnipeds prior to the initial attack. Rather, attention was first drawn to predatory attacks by the appearance of blood, an oil slick, or an explosive

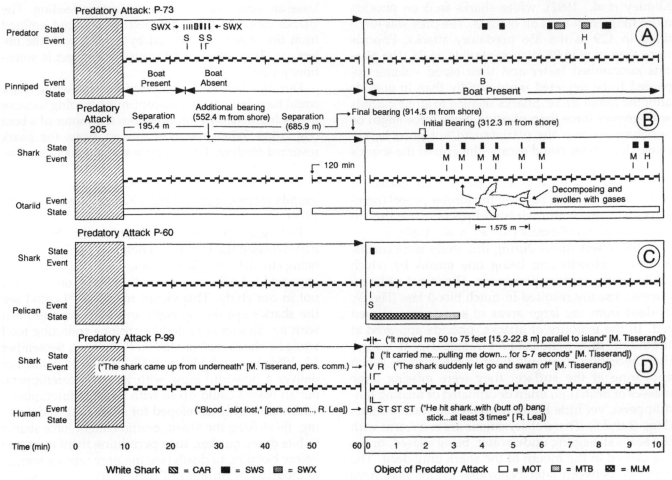

FIGURE 10 Sequences of attacks on items not immediately eaten: (A) attack P-73 on an unidentified pinniped, October 21, 1988; (B) attack P-205 on a decomposing California sea lion, September 10, 1991; (C) attack P-60 on a brown pelican, October 6, 1988; and (D) attack P-99 on a human, September 9, 1989.

splash. The high frequency of attacks on otariids that started with an explosive splash would result if sea lions either swam closer to the surface when seized or were struck with greater force than seals. The rarity with which sharks were seen prior to the predatory strikes was consistent with the behavior of sharks toward baits deployed at the South Farallon Islands (Klimley, 1987) and their response toward models (see Chapter 20, by Anderson *et al.*).

We observed a high percentage of bites on the anterior bodies of phocids and otariids (see Chapter 24, by Long *et al.*). Ainley *et al.* (1985) recorded wounds on the face and neck mainly of phocids, and attributed these to attacks on them while floating solitarily at the surface. We never observed phocids attacked while floating at the surface during our study. The high frequency of head wounds in our study appears inconsistent with a shark's approaching a pinniped at the surface that was breathing with its

head out of the water. Such an anterior attack may be inflicted in the following manner. Seeing a pinniped at the surface, the shark may begin to swim upward from a position near the bottom. By the time the shark arrives at the surface, where it can seize the pinniped, the latter may have begun to dive downward with its head directly in front of the approaching shark. However, bites to the head and the dorsal aspect of the bodies of pinnipeds (see the wound on the phocid in Fig. 2D) may, alternatively, result if sharks at times attacked pinnipeds in midwater by swimming downward from a position near the surface. Supporting the initiation of attacks from a position near the bottom or surface were the depths recorded from sharks tracked at Dangerous Reef, South Australia, which swam most frequently near the surface (1–3 m) and along the bottom (20 m), but rarely in midwater (Strong *et al.*, 1992).

As noted previously (Ainley *et al.*, 1981, 1985;

Klimley et al., 1992), white sharks feed on phocids more frequently than on otariids. The prey was identified in 129 of the 356 predatory attacks. Phocids were they prey in 98 attacks; otariids, in 28. The phocids succumbed faster and their blood volume appeared to be depleted more rapidly than in otariids after the initial strike. Sharks seemed to strike otariids with greater force than phocids. Such force might be necessary to stun the otariids and not phocids, because the former could escape the grip of the shark's jaws by biting or striking sharks with their foreflippers. Phocids are less able to escape than otariids if seized from behind, because the former propel themselves forward with their hindflippers, while the latter use their foreflippers (Ainley et al., 1985).

Some observations during this study were consistent with bloodletting being one means by which white sharks capture pinnipeds. The initial strike on phocids usually resulted in much blood loss (Fig. 3), evident from the large areas of sea surface colored red. In the majority of attacks, phocids appeared at the surface after periods of >1 minute (Fig. 4), greater than necessary to simply float to the surface. When appearing at the surface, they were missing large masses of flesh (i.e., often decapitated or missing hindflippers), yet little blood flowed from their wounds (Figs. 3 and 5). These observations are consistent with the seal's struggling underwater, being chased, or being carried in the mouth of the shark until dead. The intervals between later behavioral events may have been shorter because the sharks simply took a bite and released the carcasses, with no need to kill the phocids (Fig. 4). In contrast, we recorded short intervals between the initial strike and the appearance of the bitten otariids' swimming crippled at the surface. The subsequent intervals between behavioral events may have been much longer because the initial strike was not lethal, and sharks may have had to recapture and then carry the otariids until exsanguinated.

Although white sharks were rarely observed to release pinnipeds captured alive without removing a bite, sharks did spit out a decomposed otariid, a brown pelican, and a human. In the case of the human, we were unable to corroborate whether the shark used the bite, spit, and wait method of prey capture (Tricas and McCosker, 1984; McCosker, 1985), because the diver was promptly rescued. The bite, spit, and wait hypothesis predicts that a shark should wait until the diver becomes immobile or dies before returning to feed. In attack P-60 on a pelican, the shark did not resume feeding once the pelican died (Fig. 10C). On the basis of our observations, there are at least three additional reasons that sharks may release an item and refrain from further feeding. The sharks could be (1) frightened by a sound emitted from the object, (2) harmed by contact with the object, or (3) uninterested because the object is somehow judged unpalatable.

During attack P-73 on a pinniped, the white shark could have temporarily discontinued feeding because it was frightened by sounds from the motor of a boat (Fig. 10A). When the boat moved away, the shark resumed feeding. Other species of sharks (e.g., Carcharhinus falciformis, C. longimanus, and Negaprion brevirostris) withdraw from the sources of sudden loud sounds (Myrberg et al., 1978; Klimley and Myrberg, 1979).

During attack P-99 on a human, the shark may have released the victim as a defensive reflex to avoid being struck by the diver's bang stick (Fig. 10D). This appeared to be the case in another attack on a human not in our study. This victim reported, "I could see the shark's eye on my right and attempted to hit it with my ab iron [a knife-like abalone collecting tool] while he shook me" (attack on P. Parsons, September 30, 1984; Lea and Miller, 1985). A phocid could not deliver a powerful blow with its small foreflippers, but an otariid could do so with its large foreflippers, which are highly developed for swimming and walking. By striking the shark, otariids might force sharks to bite down quicker, thus permitting them to escape before bleeding to death (see the prey capture scenario modeled by Strong in Chapter 21).

Item Rejection Due to Low Energy Value?

Finally, certain prey could be deemed less palatable after initial sampling. In attack P-205, an immature but highly decomposed otariid was seized and released (Fig. 10B). Other species also are seized and released by white sharks. Randall et al. (1988) recovered jackass penguins Spheniscus demersus having shark bite scars. Ames and Morejohn (1980) found sea otters having wounds in which white shark tooth fragments were embedded but from which no flesh was removed. The birds, sea otter, and human are composed mainly of muscle tissue, whereas pinnipeds and cetaceans possess a great deal of superficial fat tissue. Sharks may prefer energy-rich marine mammals in favor of energy-poor prey. Supporting this are observations of sharks selectively feeding on the blubber but not the muscle layers of mysticete whales (Pratt et al. 1982; D. J. Long, personal communication). Klimley (1987) also observed that white sharks consumed seal carcasses, but would only seize and release sheep carcasses. The polar bear Ursus maritimus and the leopard seal *Hydrurga leptonyx* often remove only the fatty layer from phocids when feeding (D. Siniff, personal communication).

Summary

The behavior of white sharks *C. carcharias* and their pinniped prey is described from 247 predatory attacks, 129 of which were recorded on videotape, at the South Farallon Islands, California, 1986–1992. An ethogram was constructed from 24 behavioral patterns performed by the sharks and 12 patterns exhibited by the pinniped prey during attacks. We saw neither shark nor prey before initiation of the attacks, which usually were recognized by the appearance of blood-stained water in the case of phocids (mainly northern elephant seals *M. angustirostris*) or an explosively upward splash in the case of otariids (mostly California sea lions *Z. californianus*). The dead phocids most often reappeared at the surface immobile with a bite removed. In 34% of the attacks, the interval between the initial strike and the appearance of the seal was <1 minute; in 66% of the attacks, the seal appeared after an interval of ≥1 minute. In the case of shorter intervals, the sharks removed bites and killed the seals quickly. The longer intervals are consistent with the seal's struggling underwater, being chased, or being carried in the mouth of the shark until dead. Transport by the shark would explain why phocids appeared at the surface only after longer intervals, despite an inability to swim due to decapitation or removal of the flippers. Bitten otariids, on the other hand, quickly surfaced after the initial strike and began rolling, moving intermittently (i.e., disoriented and stop-and-go locomotion), or swimming slowly. The otariids seized may force the sharks to initially release them by defending themselves using their large foreflippers. On occasions, white sharks chased and seized these wounded otariids. Their carcasses later floated to the surface no longer bleeding, indicating possible "bloodletting," or exsanguination by the shark. Our observations did not support the hypothesized bite, spit, and wait mode of prey capture (i.e., the white shark seizes and releases a pinniped, leaves it intact but bleeding, waits for it to lapse into shock or die, and then returns to feed with little risk of injury).

Acknowledgments

We thank M. Elliot of the U.S. Geological Survey for lending us two Wild T-2 theodolites for use in this study. Members of the Farallon Patrol transported investigators and equipment back and forth from the South Farallon Islands. D. G. Ainley offered guidance and encouragement for our white shark studies. E. Uhlinger read the manuscript and made many helpful suggestions. The U.S. Fish and Wildlife Service permitted us to carry out these studies in the Farallon National Wildlife Refuge. This is contribution 595 of Point Reyes Bird Observatory International Biological Research.

17

Diving Behavior of Elephant Seals: Implications for Predator Avoidance

BURNEY J. LE BOEUF *and* DANIEL E. CROCKER

Department of Biology and Institute of Marine Sciences
University of California
Santa Cruz, California

Introduction

Northern elephant seals *Mirounga angustirostris* are an important prey of white sharks *Carcharodon carcharias* in the coastal waters of California and northwestern Mexico (Ainley *et al.*, 1981, 1985; Le Boeuf *et al.*, 1982; Klimley, 1985b; Klimley *et al.*, 1992). If white shark predation is a significant cause of mortality in elephant seals, we expect the seals to have evolved strategies to reduce predator pressure (see Ainley *et al.*, 1985). The aim of this chapter is to examine the diving pattern and aquatic behavior of elephant seals from this perspective, focusing on diving to and from the rookery over the continental shelf. We attempt to test the hypothesis that the diving pattern of elephant seals not only serves migration and foraging, which has been amply demonstrated (Le Boeuf *et al.*, 1986, 1988, 1989, 1992, 1993; DeLong and Stewart, 1991), but is also, in part, an adaptation to reduce encounters with near-surface predators, such as the white shark and the killer whale *Orcinus orca*.

This study did not begin with the hypothesis and proceed to data collection. Rather, we were asked by the editors of this volume to examine the diving behavior of elephant seals vis-à-vis its predator, the white shark. Post hoc, we used dive data that were already collected and summarized elsewhere and observations incidental to other studies at Año Nuevo State Reserve that bear on the hypothesis. In a study

of this kind, we could not manipulate variables after the fact. Consequently, this chapter is intentionally speculative. Our aim is to stimulate further thought on the subject.

Northern elephant seals dive deeply (modal depth, approximately 500 m), for long periods (mean durations, approximately 20 minutes), and continuously throughout their 2- to 8-month periods at sea. They pause only 2–3 minutes between each dive (Le Boeuf *et al.*, 1988; Naito *et al.*, 1989; DeLong and Stewart, 1991). This pattern is typical of both sexes and all age groups (Le Boeuf, 1994; Thorson and Le Boeuf, 1994).

Le Boeuf *et al.* (1988) proposed three explanations for the diving pattern of elephant seals: feeding, energy conservation, and predator avoidance. Mass gain during the period at sea confirmed that diving serves foraging. Adult females increase their mass by about 1 kg/day during the 2.5 months that they remain at sea postbreeding (Le Boeuf *et al.*, 1988, 1989). The characteristics of female dives suggest foraging on prey in the deep scattering layer well beyond the continental shelf; male dives usually do not follow the diel ascending and descending of the deep scattering layer, but rather indicate benthic foraging or a sit-and-wait strategy (Le Boeuf *et al.*, 1992, 1993). It is not clear what proportion of dives results in prey catching. From a combination of data on dive depth and duration, angle of descent and ascent, horizontal distance traveled, and swim speed we estimate that 60–

65% of the dives of adults represent foraging or foraging attempts, 24–32% indicate transit to foraging areas or transit between foraging bouts, and the remaining fraction reflects rest, sleep, or physiological processing of food (Le Boeuf et al., 1993; Crocker et al., 1996).

Diving also serves migration between island breeding sites and foraging areas (DeLong et al., 1992, Le Boeuf, 1994; Stewart and DeLong, 1993, 1994). Elephant seals exhibit "yo-yo" diving while traveling, averaging 1.2 km per dive in horizontal transit; direct measurement of swim speed shows that they do not swim at the surface as sea lions and fur seals do (Le Boeuf et al., 1992; Davis et al., 1993). Females that breed in central California forage while meandering in the eastern North Pacific Ocean, from the coast to as far west as 150° W, between 44° and 52° N; males forage in areas along the continental margin from Oregon and Washington northward to the eastern Aleutian Islands (Le Boeuf et al., 1993).

Elephant seals may dive continuously while traveling and foraging because, by lowering their metabolic rate, they expend less energy diving than by resting or swimming at the surface (Le Boeuf et al., 1988). Laboratory studies confirm that the metabolic rate is lower during periods that include diving than when the seal is simply resting at the surface (Thorson and Le Boeuf, 1994; Webb, 1994).

Avoidance of near-surface predators may be an added benefit to deep continuous diving. Because elephant seals dive in horizontal transit, rather than swim at the surface, and because of relatively long-duration dives (mean, 18–22 minutes) followed by brief surface intervals between dives (1– 3 minutes), they spend only about 10% of the total time at sea at the surface. This is a smaller percentage of time at the surface to total time in the water than in some whales (Mate et al., 1992; Martin et al., 1994). By remaining submerged at great depths for 90% of the time, elephant seals may reduce the encounter rate with a near-surface predator.

In the discussion to follow, we focus on the diving pattern of elephant seals in waters over the continental shelf when leaving or approaching the rookery during their biannual migrations, comparing this pattern to that in the open sea. We assume that elephant seals are more vulnerable to white shark predation in waters over the shelf, because white shark abundance is high on the continental shelf compared to their occurrence off of it (Strasburg, 1958; Klimley, 1985b; Klimley et al., 1992). Klimley et al. (1992) make a convincing case that white shark attacks on seals occur more frequently in nearshore waters. Based on their observations of white shark predation on seals at the Farallon Islands, they defined the "high-risk zone" of white shark attack on seals to be waters on

or near the surface, over shallow depths (<11 m) near departure and entry points within 450 m of a traditional breeding or resting site. Evidence for the presence of white sharks in pelagic waters is scanty (Carey et al., 1982; Pratt et al., 1982; Klimley, 1985b; Long, 1991a; Long et al., Chapter 24), perhaps due to low search effort.

We make several predictions here about seal behavior that would reduce detection or attack by the predator, and we predict which seals would be most vulnerable to shark attack, based on their life history strategies and experience. These predictions and their logic are as follows.

1. Seals postpone or avoid departing the rookery when freshly wounded or bleeding. Sharks have a well-developed olfactory sense (see Chapter 12, by Demski and Northcutt) and are attracted to chum or a bloody spoor. A wounded seal crossing the shelf would increase the probability of being detected by leaving an inadvertent odor trail.

2. Seals more frequently cross the continental shelf at night than during the day or in a group rather than alone. Crossing the shelf under cover of darkness would reduce the probability of visual detection and attack, assuming from behavior (Klimley et al., 1992) and retinal physiology (Gruber and Cohen, 1985) that white sharks are primarily diurnal predators. Exiting the rookery in a group, as Adélie penguins Pygoscelis adeliae do to avoid leopard seals Hydrurga leptonyx (Penny and Lowry, 1967; Oritsland, 1977; Horning and Fenwick, 1978), might provide some margin of safety to an individual (Hamilton, 1971).

3. Diving is as deep as possible; seals swim along the bottom of the continental shelf. This behavior would reduce the probability of detection and minimize the probability of attack by restricting the angle of attack by a near-surface predator. Moreover, a seal on the ocean bottom restricts the visual field and is in a better position to see the predator and take evasive action before it is itself seen. Due to the spatial geometry, we would further expect descent rates to the aphotic zone to exceed ascent rates, because the seal is back-lit on its way down and is more easily detected and vulnerable. On ascent, it can use back-lighting to its advantage to scan the near surface for predators. If a predator is sighted, the seal can then prolong its ascent or surface elsewhere, some distance away.

4. Time to cross and time at the surface over the shelf are minimized by faster swim speeds and more directional travel than when swimming in deep water. The object is to minimize time in the relatively shallow water near rookeries or to minimize exposure at the surface by reducing surface intervals and/or the

total time spent at the surface during the shelf crossing. The mean swimming depth of white sharks when in the vicinity of rookeries, determined from transmitter tags, is 16–35 m (see Chapter 11, by Goldman *et al.*, and Chapter 37, by Strong *et al.*). Other data also indicate shallow swimming. Carey *et al.* (1982) reported that a white shark tagged near Long Island, New York, remained near the thermocline, at 5–20 m, making occasional excursions to about 50 m; and most sharks caught off the west coast of North America are in waters <15 m deep (Klimley, 1985b). Conversely, records of deep-dwelling white sharks are rare. A white shark has been observed at 200 m by a Remotely Operated Vehicle in Monterey Bay, California (see Chapter 11, by Goldman *et al.*), and one white shark has been caught at 366 m (Klimley, 1985b).

5. Splashing upon surfacing is minimized, and porpoising to dive is avoided. Splashing at the surface provides visual and acoustic cues to a seal's presence and should be avoided.

6. Seals most at risk to white shark attack are young pups entering the water for the first time and breeding-age males during the breeding season. Newly weaned pups, 1–3.5 mo of age, should be easy prey because they are naive. They also are fat (47.6% ± 2.7% adipose tissue), or about twice the blubber content of postbreeding females and breeding-age males (Deutsch *et al.*, 1990; Kretzmann, 1990, Crocker, 1993). If white sharks prefer an energy-rich fat diet (Klimley, 1994), which is suggested by their preference for elephant seals and harbor seals over sea lions and sea otters (Ainley *et al.*, 1981, 1985; Le Boeuf *et al.*, 1982; Ames *et al.*, Chapter 28), and based on the observation that they feed selectively on the blubber rather than the muscle of baleen whales (Pratt *et al.*, 1982; Long and Jones, Chapter 27), young pups would appear to be nutritious prey. Anderson *et al.* (Chapter 25) show that white shark predation on juveniles at the Farallon Islands is highest at high tide, when many seals, for lack of resting space, must enter the nearshore water.

The reproductive strategy of male mammals puts a priority on maximizing copulations with females (Bateman, 1948; Trivers, 1972). This often puts them at high risk of injury or death. Consequently, male mortality exceeds that of females at all stages of life in most mammals, a difference that is attributable to the effect of functional gonads and their products, androgens, on male behavior (Trivers, 1985).

Methods

The data used to test our hypotheses were derived from three sources: (1) observation at Año Nuevo, California, 1968–1993; (2) free-ranging dive records

obtained from 42 seals tagged at the Año Nuevo rookery, 1989–1992; and (3) video recordings obtained by a camera attached to the backs of three juvenile seals in 1993.

Shark-bitten seals were censused and identified daily every breeding season. A record of the sex, age category, date, and location of the animal and a description of the injuries were also recorded (see Le Boeuf *et al.*, 1982).

Dive records were obtained from time–depth recorders (TDRs), one of which could also measure swim speed. The instruments were attached to the backs of seals before they departed for the at-sea portion of the annual cycle—lasting 2–8 months—and were recovered upon return to the rookery (Le Boeuf *et al.*, 1988, 1989, 1992, 1993; Le Boeuf, 1994). Our sample was composed of 10 postbreeding females, six pregnant females, 14 juveniles (15–21 months old, with the experience of two or three trips to sea), two young-of-the-year (3.5 months of age), and 10 adult and subadult males (5–12 years of age).

Direct observation of diving was provided by 2-hour video recordings obtained from a Sony Handycam, enclosed in a waterproof housing and attached to the dorsal midline above the shoulders of three juveniles, 15–21 months old (Davis *et al.*, 1993). The seals were captured at Año Nuevo in May or October. Release sites and respective distances from Año Nuevo were: beaches in Santa Cruz, 30 km; Pacific Grove, 96 km; and a deep-water site in Monterey Bay, California, 60 km.

We define the continental shelf off central California as the inshore area from the shoreline seaward to the 140-m isobath, the depth where the slope becomes significantly steeper (Fig. 1). We assumed that during departure or return to the rookery, a seal was over the shelf if it exhibited dive depths <140 m. This assumption is validated, first, by the monotonic stairstep pattern of increasing or decreasing depth in dive records that reflects the bathymetry of the area (Fig. 2). Second, when the seals depart the rookery and reach a depth of 140 m, the depth of subsequent dives increases abruptly to 300 m or more. Third, dive depths <140 m in the open ocean (false positive) compose only a small fraction of dives for postbreeding (1.1% ± 0.6%, N = 18) and pregnant females (3.5% ± 0.9%, N = 7) and adult males (3.9% ± 2.6%, N = 5). Preliminary data indicate that subadult males, 4–7 years old, may be an exception. Last, the majority of shallow dives in waters beyond the continental shelf occur around midnight (Le Boeuf *et al.*, 1988). We conclude that using the pattern in TDR dive records is a valid method of determining location on and off the continental shelf. Doing this determination by estimating location coordinates from light lev-

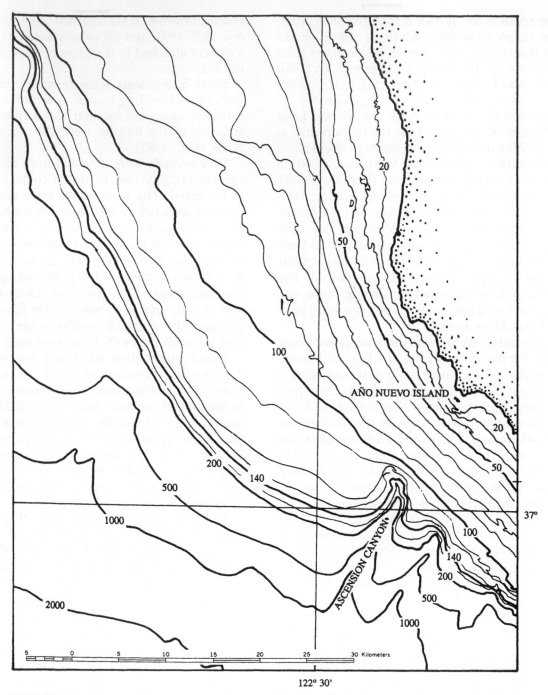

FIGURE 1 Bathymetry near Año Nuevo Island and the adjacent mainland rookery. The continental shelf break is at 140 m.

els (DeLong *et al.*, 1992; Hill, 1994) has a measurement error greater than the width of the continental shelf off California, and thus, is a method inappropriate to the scale of our study.

Results

We consider here each of the predictions discussed above.

Seals Postpone or Avoid Departing the Rookery When Freshly Wounded or Bleeding

Observations at Año Nuevo during the breeding season indicate that wounds are almost exclusively a problem for males. Deutsch *et al.* (1994) report that 90% of adult males monitored at the peak of the breeding season had fresh bleeding wounds obtained during aggressive competition for mates. In contrast,

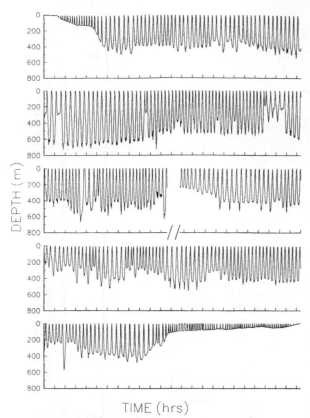

FIGURE 2 Segments of the dive records obtained by a time–depth recorder attached to an adult female at Año Nuevo on her postbreeding trip to sea. (First segment) Departure from the rookery and increasingly deeper dives until the continental shelf break is reached (at 140 m). (Middle segments) Records typical of 99% of dives in the open ocean. (Bottom segment) Transition from deep to shallow diving upon return to the rookery.

only 20% of females had wounds, and these were far less severe than in males. Long fights between similarly ranked males sometimes begin on land and end in the water, with one or both males bleeding profusely. Typically, both males return to shore within about 30 minutes after the end of a fight.

Most males remain on the rookery, sleeping for 2–3 weeks after the last estrous females have departed (B. J. Le Boeuf, unpublished data). This delay in returning to sea may allow rest and healing. If the primary function of the delay is healing, a strong positive correlation should exist between the severity of the injury and the duration of the stay after the females have left. This prediction requires further testing.

Seals More Frequently Cross the Continental Shelf at Night Than during the Day or in a Group Rather Than Alone

We find no support for this prediction. Mesnick and Le Boeuf (1991) found that the frequency of day-

time departures of postbreeding females did not differ significantly from that of nighttime departures (63% versus 78%, respectively). In the thousands of departures we have observed to date, regardless of age, elephant seals left individually, not in groups. Moreover, a departing seal did not trigger departure in others nearby. Similarly, elephant seals arrived on the rookery singly, not in groups.

Diving Is as Deep as Possible; Seals Swim along the Bottom of the Continental Shelf

This prediction is confirmed for adult females leaving the rookery after breeding (Fig. 3). The diving records of departing females showed steadily declining depths, suggesting that dives were bathymetrically constrained by the substrate at the deepest depth of each dive. Once the 140-m shelf break was reached, the diving pattern changed abruptly to one of variable deep diving, indicating that the dives were no longer bottom-limited. A similar pattern, in reverse, was observed in all females returning to the rookery in the spring (Fig. 2).

Dive depths of juveniles and adult males were more variable at departure. Ten of 14 juveniles exhibited an apparent bottom-diving pattern like that of adult females. The four remaining juveniles deviated from this pattern by exhibiting 6–21 shallow dives <20 m deep. One juvenile apparently returned to the shelf twice before exiting. Three juveniles exhibited a pattern of dives that were not flat-bottomed, but rather, "jagged-bottomed." The dive depths of some adult males, although flat-bottomed and suggestive of bottom swimming, decreased and increased again a few times before deep diving commenced, suggesting that the seal did not take a direct route to deep water. Moreover, water entry was not a reliable signal of departure in adult males, as it was in other adult females and juveniles (see below).

The bottom diving inferred from TDR records was confirmed by video records of two homing seals released on beaches in Monterey Bay. In swimming to deep water over the continental shelf on their return to the rookery, the seals spent up to 90% of each dive swimming horizontally over the ocean bottom. The dorsal surface of both seals was always oriented toward the water surface, giving it a view of what was above and ahead. Distance from the substrate appeared to be 1–3 m over a rocky bottom with kelp and about 1 m above a sandy bottom (Fig. 4). In the latter case, the swimming seal was so close to the substrate that it often stirred up sand with the sculling movements of its hind flippers. A dive recorder attached to one of the seals revealed gradually increasing depths as the seal moved away from shore.

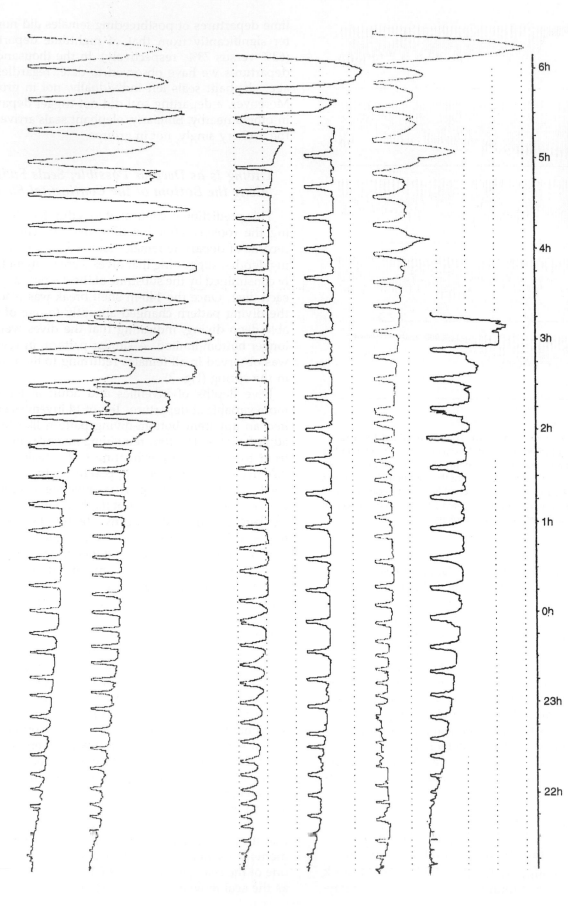

FIGURE 3 Initial portions of the dive records from six postbreeding females, from departure from the Año Nuevo rookery to attainment of the continental shelf break (where dives exceed 140 m depth). Depth, on the ordinate in 100-m intervals, is shown by dotted lines; time, in hours, is shown on the abscissa. The females in the top two records reached the shelf break within 5 hours; the third record shows an extended surface interval of 20 minutes after approximately 5.5 hours.

FIGURE 4 Single frame from a video recording taken by a Sony Handycam attached to the back of a 2-year-old juvenile elephant seal. The seal is seen swimming near the substrate bottom as it crosses the continental shelf to deep water on its return to Año Nuevo. The camera is oriented backward, showing the time–depth recorder, the seal's hind flippers, a following harbor seal, and kelp attached to the substrate bottom.

The diving record of another translocated seal, the only one that dived into the aphotic zone, bears on the prediction that the seal may be more in danger from predators on descent than on ascent. The mean time to reach dark waters (where we could not see the seal) was 24.5 ± 7.5 seconds (N = 17 dives), or 35 m, if the seal was descending at a rate of 1.44 m/sec (see Le Boeuf *et al.*, 1992). In contrast, the mean time to ascend from 35 m to the surface was three times as long, 74.1 ± 72.2 seconds. In 14 of these dives, ascent duration exceeded descent duration by a mean of only 24.8 ± 19.7 seconds. This difference may have been due to the camera attachment with the lens pointing forward, that is, pointing up to light during ascent and down to dark during descent. In three dives, however, the ascending seal appeared out of the darkness but paused three times, for 2, 2.2, and 5.5 minutes, during which it swam horizontally just above the aphotic zone before its final ascent to the surface. Similarly, Le Boeuf *et al.* (1992) found that ascent rates were 16.6% slower than descent rates in a free-ranging adult female. This behavior is consistent with the seal's scanning for a predator while ascending and taking evasive action.

Time to Cross and Time at the Surface Over the Shelf Are Minimized by Faster Swim Speeds and More Directional Travel Than When Swimming in Deep Water

Swim Speed and Number of Dives

A swim-speed recorder attached to a postbreeding adult female showed that she swam faster when on rather than off the shelf. Her mean horizontal swim speed was 1.53 ± 0.59 m/sec (N = 18) on the shelf at departure and 1.23 ± 0.21 m/sec (N = 1661) off the shelf in deep water (Le Boeuf *et al.*, 1992; our unpublished data), a statistically significant difference [t = 2.03, degrees of freedom (df) = 1676, $p < 0.05$]. At this mean swim speed, a seal departing the Año Nuevo rookery would be out of the high-risk zone of shark attack (Klimley *et al.*, 1992) in 5 minutes.

The total time and number of dives required to reach a point where depths were >140 m were calculated for all subjects (Table I). The performances of postbreeding and postmolt females were not significantly different (t dives = 1.51, df = 13, $p > 0.05$; t duration = 1.62, df = 13, $p > 0.05$). Five of nine postbreeding females crossed the shelf in 3.5–

TABLE I   Number and Duration of Dives of Northern Elephant Seals While Crossing the Continental Shelf to and from Año Nuevo, California

	N	Number of dives	Duration of dives (hours)
Postbreeding females	10	38.1 ± 25.1	10.4 ± 9.3
Postmolt females	5	66.2 ± 37.6	19.8 ± 11.2
Juvenile males	6	45.7 ± 17.4	9.2 ± 2.9
Juvenile females	6	27.8 ± 9.7	5.2 ± 1.9
Breeding-age males	10	159.4 ± 180.9	53.2 ± 64.8

Values are expressed as means ± SD.

7.5 hours, three in 11.8–13 hours, and one in 34.8 hours. Assuming a direct route from Año Nuevo to the head of Ascension Canyon, 12 km distant, the fastest time calculates to a minimum swim speed of 0.95 m/sec (3.43 km/hour). This speed is consistent with direct measurements. Assuming a direct route to the shelf edge due west of the Año Nuevo rookery, 30 km distant, the fastest time calculates to a minimum swim speed of 2.38 m/sec (8.57 km/hour). Assuming that all females were traveling at similar speeds, the calculated swim speeds indicate that some took the most direct route to deep water to the southwest, while others, to varying degrees, crossed the shelf in a westerly or northwesterly direction, a more direct route to their final destination (Le Boeuf et al., 1993).

Juvenile females spent less time and exhibited fewer dives on the shelf than did juvenile males ($t = 2.84$ and 2.20, respectively; df = 10, $p < 0.05$) and adults of both sexes, with the exception of postbreeding females ($t = 1.71$ and 1.16, respectively; df = 14, $p > 0.05$). One juvenile female reached the shelf break in 110 minutes. This calculates to swimming at a rate of 1.82 m/sec (6.56 km/hour) if the most direct route to the head of Ascension Canyon were taken.

Breeding-age males spent significantly more time ($t = 5.59$, df = 18, $p < 0.05$) and a greater number of

dives on the shelf ($t = 1.97$, df = 14, $p < 0.05$) than other groups of seals, with the exception of postmolt females, with whom they did not differ ($t = 1.59$ and 1.56, respectively; df = 13, $p > 0.05$). Either males were considerably slower in crossing the shelf than postmolt females and juveniles (the fastest time for an adult male, assuming the shortest distance to the head of the canyon, was 0.44 m/sec, or 1.58 km/hour) or they took a longer route to deep water, crossing the continental shelf in a northwesterly direction. The dive rate data presented below indicate the latter.

Diving Pattern on and off the Shelf

Comparison of TDR records indicated different diving performance on and off the shelf. That is, we compared the first day or two of diving after departure, and the last day or two before return to the rookery, to diving performance in the open ocean when dives were not bottom-limited (Table II).

The diving rate, regardless of age or sex, was significantly faster ($t \geq 4.00$, df = 10–20, $p < 0.05$), and the mean dive duration was significantly shorter in waters over the shelf than off of it ($t \geq 5.78$, df = 10–208, $p < 0.05$). The mean surface interval was significantly shorter on the shelf than off the shelf for postbreeding females ($t = 3.84$, df = 18, $p < 0.05$) and juveniles ($t = 3.7$, df = 20, $p < 0.05$), but not for postmolt females and breeding-age males. The faster swim rate on the shelf indicates a higher oxygen consumption rate, which requires the seals to come to the surface more often and lowers dive durations.

Apparently, when a seal was over the shelf, swimming fast took precedence over reducing total time at the surface. Postbreeding females and juveniles spent more time on the surface when crossing the shelf than when off of it (increases of 14.7% and 15.7%, respectively). There was no difference in this statistic for postmolt females; breeding-age males showed the reverse trend, spending more time at the surface when at sea than when crossing the shelf.

During shelf crossings, extended surface intervals

TABLE II   Dive Parameters of Northern Elephant Seals on and off the Continental Shelf

Location relative to shelf	Postbreeding females (N = 10)		Postmolt females (N = 6)		Breeding-age males (N = 10)		Juveniles (N = 11)	
	On	Off	On	Off	On	Off	On	Off
Dives per hour	4.1 ± 0.7	2.9 ± 0.2	3.4 ± 0.7	2.2 ± 0.2	3.2 ± 0.3	2.4 ± 0.3	5.5 ± 0.6	3.6 ± 0.5
Surface interval (minutes)	1.7 ± 0.4	2.2 ± 0.1	1.9 ± 0.3	2.9 ± 1.3	2.3 ± 1.0	4.3 ± 4.0	1.4 ± 0.3	1.8 ± 0.2
Dive duration (minutes)	13.9 ± 3.0	20.9 ± 2.4	16.4 ± 4.5	27.9 ± 2.0	16.7 ± 1.6	21.6 ± 1.8	9.8 ± 1.4	14.9 ± 2.6

Values are expressed as mean ± SD.

(ESIs), surface intervals >10 minutes, were rare for females and juveniles. One postbreeding female had an ESI lasting 20 minutes and one postmolt female had one lasting 19 minutes. There was only one ESI for a juvenile, and it lasted 36 minutes. Male behavior did not follow the prediction. Three of 10 breeding-age males had a total of 45 ESIs that ranged from 11 to 97 minutes; 83% of these occurred during the first 2 days of departure. Despite spending more unusually long periods at the surface, breeding-age males maintained other characteristics associated with shark avoidance on the shelf: a high dive rate and swimming along the ocean bottom.

Do seals linger over the shelf to feed? The dive pattern suggests that rapid transit is the priority and the seals do not forage until reaching deep water. All categories of seals had a higher diving rate on the shelf than in the open sea (Table II), and all seals, except breeding-age males, crossed the shelf quickly (Table I). No evidence of foraging was observed in the video records of homing seals on or off the shelf. The two-dimensional TDR patterns observed in seals diving over the shelf, regardless of sex, are distinct from those suspected to be pelagic foraging dives (Le Boeuf *et al.*, 1988, 1992, 1993) but are indistinguishable from the putative benthic foraging dives of breeding-age

males (Le Boeuf *et al.*, 1993). Finally, foraging off the shelf is evidently preferred, because all categories of seals, except subadult males, spent most of their time at sea and most of their dives in deep water. Subadult males appear to supplement their diet by foraging on benthic prey over the shelf during migration (Le Boeuf *et al.*, 1993). In summary, foraging on the shelf at departure or return to the rookery is unlikely for adult females and juveniles, but we cannot exclude this possibility for breeding-age males.

Splashing upon Surfacing Is Minimized, and Porpoising to Dive Is Avoided

Observations of seals in the water near the rookery support this prediction. Elephant seals break the water quietly upon surfacing and merely lower the head into the water when diving. This conclusion is reinforced by the video recordings from three homing juveniles to which video cameras had been attached (Davis *et al.*, 1993). Only the snout and the anterior portion of the head break the surface on ascent as the seal floats or treads water in a near-vertical position (Fig. 5); on descent, the head is submerged and sinks as the seal begins to swim downward.

FIGURE 5 Single-frame image taken by a video camera attached to the back of a 2-year-old elephant seal, showing that only the anterior portion of the head is exposed to air during surface intervals between dives.

*Seals Most at Risk to White Shark Attack
Are Young Pups Entering the Water
for the First Time and Breeding-Age
Males during the Breeding Season*

Weaned Pups

The behavior of newly weaned elephant seal pups learning to swim and dive may have been selected to avoid the high-risk zone. Newly weaned pups enter the water for the first time at about 5 weeks of age (Reiter *et al.*, 1978). Initially, their aquatic activities are restricted to dawn and dusk. As they develop, they become more proficient at swimming and diving and spend increasingly more time in the water. By 3.5 months, when ready to depart for sea, they spend all night in the water (12.5 hours/day) and sleep on land during the day. Throughout this learning period, they remain in shallow coves protected from high surf; the mean dive depth increases with age from <2 m at 1.5 months of age to 6 m at 2.5 months (Thorson and Le Boeuf, 1994).

A mean of 46% (range, 35–62.1%) of the young seals that leave the Año Nuevo rookery never return to land but die at sea (Le Boeuf *et al.*, 1994). The causes of this high mortality—the highest during any period at sea for any age or sex group—and the proportion attributable to white sharks are unknown.

Free-ranging dive data exist for two elephant seals on their first trip to sea (Thorson and Le Boeuf, 1994). The diving pattern upon departure was similar to the behavior of 1- to 2-year-old juveniles (Tables I and II). Swimming over the shelf was rapid, with minimal time spent at the surface; the seals appeared to follow the bottom contour in waters over the shelf. Off the shelf, these young seals exhibited a dive pattern similar to that of older juveniles and adults, averaging 10 minutes per dive, 1.6 minutes on the surface between dives, and diving to mean depths of 200 m (maximum, 553 m).

Adult Males

The behavior of males during the breeding season supports the logic that they take higher risks of shark attack than other seals and were more likely to encounter white sharks. For example, (1) breeding-age males took more dives when crossing the shelf and spent more time in waters on the shelf, by a large margin, than did females and juveniles (Table I). (2) Unlike adult females and juveniles, breeding-age males entered and left the water repeatedly during the breeding season, moving long distances along the shoreline of rookeries in search of unguarded females. (3) When leaving the rookery at the close of the breeding season, some adult males spent a great deal of time near the surface or exhibited shallow diving (Fig. 6). Dive records show that some males moved off the shelf only to return to it or the rookery. The latter suggests searching for departing estrous females or virgin females that copulate at sea. (4) Only breeding-age males assumed the vulnerable position of resting at the surface with head and tail out of the water and eyes closed within 0.5 km of rookery beaches.

The increased time spent by males in the danger zone during the breeding season may explain why they are more frequently injured by sharks than are adult females (Le Boeuf *et al.*, 1982). An update of shark bitten seals during the breeding season at Año Nuevo for the years 1981–1993 confirms that injured males outnumbered injured females by >5:1 (Table III), despite females' outnumbering males by 2:1 or more. The majority of males injured (50%) were fully grown adults.

We have no data on males returning to the rookery outside the breeding season, when their testes are less active. We would predict, however, that their behavior at this time, with respect to predator avoidance, would be no different from that of adult females.

Discussion

Much of the diving behavior of elephant seals on the continental shelf is consistent with the hypothesis that selection has acted to minimize the probability of encounters with near-surface predators such as white sharks. Diving in waters over the shelf, relative to the open ocean, is characterized by fast swimming, a high dive rate, short surface intervals and dive durations, and swimming at the bottom. Juveniles and adult females cross the shelf rapidly and do not linger. Newly weaned pups learn to swim and dive at night in shallow water close to shore.

Although behaving similarly in some respects, breeding-age males deviate significantly from this predator-avoidance pattern. They spend considerable time in crossing the continental shelf. They enter waters near the rookery repeatedly and rest on the water's surface during the breeding season. The risk-taking behavior of males may promote their reproductive success, but they evidently pay a high price, incurring more shark bites during the breeding season than do females. It is possible that this sex difference is due to males escaping from the predator with only a wound, while the smaller adult females more readily succumb. It is difficult to test this possibility directly, but the following points indicate it to be

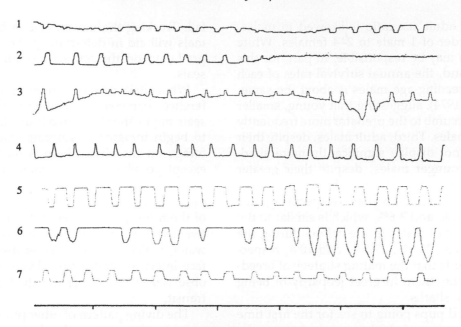

TIME (hrs)

FIGURE 6 Portions of the dive records of seven breeding-age males over the continental shelf during departure from the Año Nuevo rookery, showing behaviors that put them at risk of encountering white sharks. All dives were <140 m, except for the deepest dives, in examples 3 and 6. (1) Shallow diving and near-surface swimming with long interdive surface intervals; (2) a seal returns to the rookery after an excursion on the shelf; (3) a seal moves on and off the shelf; (4 and 7) seals not moving to deeper water but swimming parallel to shore and remaining on the shelf longer than necessary; and (5 and 6) extended surface intervals.

TABLE III **Shark-Bitten Northern Elephant Seals on Año Nuevo Island and Adjacent Mainland during the Breeding Seasons 1981–1990**

Year	Adult males (8+ years)	Subadult males			Total males	Adult females	Juveniles	Grand total
		7 years	6 years	5 years				
1981	1	4	3	2	10	0	0	10
1982	—	—	—	—	14+	2	1	17+
1983	10	1	1	1	13	1	0	14
1984	3+	—	—	—	3+	4	0	7+
1985	6	3	4	0	13	3	1	17
1986	2	2	3	0	7	1	0	8
1987	0	1	2	0	3	2	0	5
1988	5	2	0	0	7	0	0	7
1989	3	2	3	0	8	0	0	8
1990	6	0	1	0	7	2	0	9
1991	4	1	1	0	6	0	0	6
1992	0	2	1	0	3	0	0	3
1993	5	2	0	0	7	0	0	7

—, Missing data for males in 1982 and 1984; males were not classified according to age in 1982.

false. First, the adult sex ratio is skewed to males, being on the order of 1 male to 2–4 females. White shark predation may be responsible, in part, for this difference. Second, the annual survival rates of each age group of breeding-age males is about the same, 55% (Le Boeuf, 1974), suggesting that young, smaller males do not succumb to the predator more frequently than do larger males. Third, adult males, despite their great size, did not display significantly more shark wounds than younger males, despite their greater size. The incidence of shark wounds in adult males and males of age 7, 6, and 5 years, respectively, was 50%, 23.8%, 22.6%, and 3.6%, which is similar to the relative frequency of males in the colony during the study period, 41%, 21.9%, 27.5%, and 10%, respectively. We conclude that the mating strategy of breeding-age males puts them in great jeopardy of being injured by white sharks.

Newly weaned pups going to sea for the first time are the other category of elephant seals that may be especially vulnerable to shark predation. They must learn to swim and dive on the edge of the high-risk zone. To what extent gaining aquatic experience at night or in sheltered coves is effective in lowering predation is unknown. The highest annual mortality rate is evident in young-of-the-year during their first trip to sea, which extends from April to September–October. Observations of shark attacks and shark-bitten juveniles, however, are rare in April on both the Farallones and Año Nuevo (Le Boeuf et al., 1982; Ainley et al., 1981, 1985).

Adult female white sharks appear to move south in spring and breed below Point Conception during summer (Klimley, 1985b). If young elephant seals move north to the Gulf of Alaska and across the North Pacific, as older juveniles do on their third and fourth trips (Le Boeuf et al., 1993; Le Boeuf, 1994), they are doing so at a time when white sharks appear to have left the area. Moreover, the seals are moving into northern latitudes, where white sharks are far less common and are replaced by killer whales as the most significant predator (Jefferson et al., 1991). There are no reports of white sharks north of the Queen Charlotte Islands, Canada (Klimley, 1985b). If white sharks are moving south in spring, then juvenile mortality should be higher at southern California rookeries such as San Miguel Island than at central California rookeries. This is apparently not the case (Le Boeuf and Reiter, 1988; Le Boeuf et al., 1994) and can be explained if the sharks do not feed while breeding. Shark attacks on juveniles, age 1–3 years, are most common in October and November on the Farallones (Klimley et al., 1992), which suggests that the young-of-the-year may be most vulnerable when returning to the rookery in the

fall. Tracking the movements of both groups of animals will aid in determining the role of white shark predation on the annual mortality rate of elephant seals.

What alternative hypotheses might explain the differences between on-shelf and off-shelf diving? The seals might simply be in a hurry to get to deep water to begin foraging. Assuming that the seals do not forage on the shelf, there is no point in lingering, except possibly to take a more direct route to their foraging destination. Males appear to do this, because they migrate farther north than females, some of them for great distances. From the other perspective, the differences may be due to foraging over deep water, which produces slower dive rates, longer surface intervals, and longer dive durations than those observed in dives over the shelf when the seal is in transit.

The diving pattern of other pinnipeds may provide a perspective on the role of predation in shaping the northern elephant seal's diving pattern. The closest relative of the northern elephant seal is the southern elephant seal *Mirounga leonina*, which breeds on subantarctic islands and mainland sites (Laws, 1994). Southern elephant seals have only one major predator, the killer whale (Voisin, 1976; Carrick and Ingham, 1962; Ling and Bryden, 1981; Lopez and Lopez, 1985). Arrival of the whales at some rookeries coincides with breeding and the entry of weaned pups into the sea (Condy, 1978; Condy et al., 1978). Despite different predators and environments, the general diving patterns of both species are similar (Hindell et al., 1991; Le Boeuf and Laws, 1994). Campagna et al. (1995) reported that postbreeding adult females in Patagonia crossed the broad continental shelf quickly at a mean horizontal swim speed of 1.6 ± 0.1 m/sec, suggesting that they were not feeding. Like northern elephant seals, all females had significantly higher diving rates and shorter dive durations on the shelf than off it, and they dived along the substrate bottom on their way to the shelf break.

It is not clear whether other phocids, with or without near-surface predators, exhibit a diving pattern that may have functions other than foraging. The diving patterns of only a few of them have been described (e.g., Weddell seal *Leptonychotes weddelli*) and none have been examined from this perspective.

Otariid species, whose diving behavior has been studied [extensively in lactating females (e.g., Gentry and Kooyman, 1986)], do not exhibit the behaviors associated with avoidance of a near-surface predator. Northern fur seals *Callorhinus ursinus*, whose feeding range overlaps that of the northern elephant seal, spend 74% of the time at sea resting or active at the surface (Gentry *et al.*, 1986). Most otariids swim at the surface during transit, "porpoising" to breathe, for example, California sea lions *Zalophus californianus* (Peterson and Bartholomew, 1967) and South African fur seals *Arctocephalus pusillus* (Kooyman and Gentry, 1986). Sea otters *Enhydra lutris* also spend long periods at the surface, much of it in the high-risk zone. Of these animals, sea otters are attacked and killed (Miller and Collier, 1981; Ames and Morejohn, 1980), but they are not eaten (see Chapter 28, by Ames *et al.*). A similar low preference for fur seals is suggested by the absence of fur seal remains in shark stomachs. California sea lions are consumed by white sharks (Ainley *et al.*, 1981, 1985), but not as frequently as elephant seals and harbor seals.

This analysis, although speculative, underscores gaps in our knowledge of the relationship between white sharks and northern elephant seals and suggests directions for future research. Knowledge of the habits of either animal will enhance our understanding of this predator–prey relationship.

Summary

We hypothesize that the diving pattern of northern elephant seals *M. angustirostris* is, in part, an adaptation for avoiding encounters with near-surface predators such as the white shark *C. carcharias*. We tested this hypothesis using behavioral observations and free-ranging dive records from 42 seals to which instruments were attached at Año Nuevo, California, 1989–1992. Consistent with the hypothesis are the following general patterns in elephant seal diving: (1) "yo-yo" diving with no swimming at the surface, (2) brief interdive surface intervals (i.e., 1–3 minutes), and (3) long-duration dives (mean, 10–22 minutes) in the modal depth range of 200–600 m. Elephant seals also may minimize encounters with white sharks along the routes to and from the rookery over the continental shelf (<140 m deep) by swimming faster in waters over the shelf than off it, surfacing for shorter intervals, and swimming at the bottom. Newly weaned pups learn to swim and dive at night, remaining close to shore in shallow water. Breeding-age males, in their effort to mate, spend more time in the "high-risk zone," as described by others, near islands than do females and juveniles and are more frequently injured by sharks. Pups suffer the highest annual mortality rate during their first year at sea (mean, 46% per year), but the role of white shark predation in causing such low survivorship is unknown. The diving pattern of elephant seals appears to have been molded by constraints of predator avoidance as well as foraging, especially during diving over the continental shelf.

Acknowledgments

This research was funded in part by grants from the National Science Foundation and the Minerals Management Service and by a gift from the G. Macgowan estate, for which we thank George Malloch. We thank D. G. Ainley, S. B. Blackwell, and A. P. Klimley for comments on the manuscript.

Repetitive Aerial Gaping: A Thwart-Induced Behavior in White Sharks

WESLEY R. STRONG, JR.[1]
The Cousteau Society
Chesapeake, Virginia .

Introduction

During an extended study of the behavior and ecology of white sharks *Carcharodon carcharias* in South Australia, six individuals exhibited a heretofore undescribed behavior pattern. In 15 instances, "repetitive aerial gaping" (RAG) followed a series of thwarted attempts to feed on baits positioned at the surface.

Historically, opportunities to closely observe white sharks have been limited to brief, but often intense, encounters. RAG was probably witnessed during some of these earlier excursions, but it was never described, probably because isolated observations were regarded as "anomalies." Unfortunately, behaviors interpreted as abnormal are often merely infrequent. Now, in fact, RAG appears to occur outside the study population; following our initial description of this behavior (Cousteau and Richards, 1992), RAG was reported in South African white sharks (L. V. J. Compagno, personal communication). The form, causation, and function of RAG are discussed here and compared to biting patterns observed during normal feeding.

Methods

In lower Spencer Gulf, South Australia, sharks were attracted to the research vessel using a standardized baiting procedure (see Chapter 37, by Strong *et al.*). Liquefied and particulate fish and mammal tissues formed the basis of a robust odor corridor. In addition, we normally floated two small tethered baits (approximately 1-kg pieces of tuna) 6–10 m downcurrent of our vessel. Crew members were generally able to move the baits out of the sharks' reach by pulling the bait lines, thereby allowing only moderate food intake ($\bar{X} = 1 \pm 1.1$ kg/hr per shark; range, 0–4.5 kg/hr). However, when the collection of other relevant data was concluded, we occasionally used a different baiting method. A larger bait (2–10 kg) was suspended from a stern-mounted crane so that it hung immobile at the sea surface, but could be hoisted rapidly out of the water as a shark tried to seize it (hereafter called the vertical baiting technique).

Using film and video analysis, RAG bouts were compared to normal surface feeding attempts (wherein bites were aimed at nearby baits, but no contact was made). Coefficients of variation were derived for durations of gape cycles (i.e., jaws closed—fully open—closed) by dividing the standard deviation (SD) by the mean ($\bar{X}$). An index of stereotypy (ST) was calculated, as follows (Barlow, 1977; see also Tricas, 1985):

---

[1]Present address: Department of Biological Sciences, University of California, Santa Barbara, California.

$$ST = \frac{100}{(100 \times SD/\bar{X}) + 1}$$

Throughout this chapter, all means are reported ±1 SD.

Results and Discussion

Form

Although there was considerable inter- and intra-individual variation in overall duration and intensity (Table I), RAG generally took the following form: holding its head out of the water, the performer rolled onto its side and opened and closed its mouth in a moderately slow, rhythmic, partial gape while swimming slowly and awkwardly along the surface (Figs. 1 and 2). In every case, the mouth was at or above water level at the time of onset. Individuals that exhibited RAG repeated the pattern an average of 2.6 ± 1.4 times, with subsequent occurrences a short time later ($\bar{X} = 27 \pm 28$ minutes) under similar conditions (Table I).

The most notable difference between RAG and normal surface feeding is that RAG gapes were not oriented toward the food or other possible targets (i.e., not "goal directed"). The entire pattern was distinctive, however. The two primary components of RAG, jaw gaping and tail sweeping, are among the species' most common action patterns. However, the peculiar execution of these basic movements, along with the animal's proximity to the sea surface, made the response highly distinguishable. Jaw gapes, in particular, were executed in a slow sustained series, making them appear more rhythmic than the intermittent gaping associated with normal surface-feeding attempts (Fig. 3). Kruskal–Wallis one-way analyses of variance (ANOVAs) indicated no significant differences between sharks for durations of individual gapes exhibited during RAG ($P = 0.169$) and during normal feeding attempts ($P = 0.158$) (Table II). Data were thus pooled within each of the two categories. Full gapes executed during feeding attempts averaged 0.408 seconds (±0.177; n = 14), while RAG gapes lasted three times as long ($\bar{X} = 1.272 \pm 0.715$ seconds; n = 13). It should be noted that, like the example in Fig. 3, the mouth was seldom completely opened or closed during RAG. Thus, durations given for RAG in the above comparison are not for full gapes, but rather, the fullest observed and covering at least half of the full-gape range (e.g., 0.25–0.75). The fact that the jaws did not open as far during RAG, however, reinforces the observation that the gapes are slow. Pooled ST values averaged 1.7 for RAG and 2.3 for normal surface feeding (Table II). The latter are similar to Tricas' (1985) ST values for gapes recorded during surface feeding (Table II). Reviewing published data for a variety of modal action patterns

---

TABLE I   Bouts of Repetitive Aerial Gaping in Six South Australian White Sharks

Observation no.	Shark no.	Est. TL (m)	Sex	Date	Time	Distance (m)	Duration (sec)	Intensity[a]	Other[d] sharks	Vertical baiting
1	17	4.3	F	March 6, 1990	1715	2	**3**[b]	Low	Yes	Yes
2	17	4.3	F	March 6, 1990	1740	7	**10**	High	Yes	Yes
3	17	4.3	F	March 6, 1990	1800	7	**13**	**High**	Yes	Yes
4	15	3.4	F	March 12, 1990	1345	1	**2**	Low	No	Yes
5	15	3.4	F	March 12, 1990	1420	6	**8**	Medium	No	Yes
6	15	3.4	F	March 12, 1990	1430	12	**15**	High	No	Yes
7	15	3.4	F	March 12, 1990	1607	10	**13**	Medium	No	Yes
8	15	3.4	F	March 12, 1990	1610	19	**24**	**High**	No	Yes
9	43	4.0	M	February 21, 1991	1300	**4**	7	**Low**	Yes	No
10	43	4.0	M	February 21, 1991	1320	7	9	**High**	Yes	No
11	47	4.0	M	February 22, 1991	1621	10	13	**Medium**	Yes	Yes
12	47	4.0	M	February 22, 1991	1630	4	13	**Low**	Yes	Yes
13	60	3.5	F	September 15, 1991	1025	1	6	Low	Yes	Yes
14	65	3.0	M	September 20, 1991	1412	2	5	**Low**	Yes	No
15	65	3.0	M	September 20, 1991	1439	5	8	Medium	Yes	No
		3.7 ± 5.0[c]				6.5 ± 4.8[c]	9.9 ± 5.5[c]			

Est. TL, Estimated total length.

[a] Subjective assessment based on physical activity level.

[b] Boldface values measured or estimated from film or video records; others estimated immediately following observations.

[c] Mean ± SD.

[d] Other sharks, other sharks in area during bout.

(MAPs), Barlow (1977) concluded that ST values are typically low, ranging from 2 to 8. Although the present data lie at the lower extent of this range, RAG gape durations "cluster about a mode" and can be characterized as an MAP.

Total RAG bout duration as well as distance traveled by the shark varied markedly, ranging from 2 to 24 seconds ($\bar{X} = 9.9 \pm 5.5$; n = 15) and from 1 to 19 m ($\bar{X} = 6.5 \pm 4.8$; n = 15), respectively (Table I). Generally, the intensity of the series of surface lunges that preceded each RAG bout reflected that of the bout itself. Body orientation also varied, the sagittal plane of the body axis ranging from tilted slightly off-center to completely inverted. Such variation is not uncommon for MAPs, however—even the most highly stereotyped action patterns are graded to some degree and, if slow enough, are influenced by the environment (Barlow, 1968). In the present case, differences may have been due to several factors, including interactions with the immediate physical environment and simple fatigue. Concerning the former, the shark's approach path relative to the position of the bait influenced its posture at onset, and the motion of surface waters appeared, in some cases, to affect the subsequent demonstration. As to the latter, RAG was initiated abruptly following attempts to grasp baits, but resumption of normal posture and swimming often took approximately 4–8 seconds following its cessation. Also, relative to the frequency of approaches observed before RAG occurred, there was sometimes a long hiatus following RAG bouts before feeding attempts resumed. Empirical efforts to characterize the entire behavior pattern as to degree of stereotypy would presently be unproductive due to the limited number of observations and the aforementioned variability. Nonetheless, the pattern is unique and distinctive, partly because of the temporally stereotyped jaw-gaping component.

Causation

Baits pulled away travelled horizontally, just beneath the surface until they came against the hull at which point the sharks generally turned away. In the vertical baiting configuration, however, the bait could be hoisted directly out of the water as a shark tried to seize it, moving it out of reach more quickly than a bait drawn horizontally. This distinction is important, because 11 of 15 (73%) RAG occurrences were elicited while using the vertical baiting technique, but the technique itself was employed less than 5% of the time.

In each case, RAG followed a series of failed attempts to seize a bait and was immediately preceded by an open-mouthed lunge and miss as a bait was withdrawn. Lack of contact with prey after the feeding MAP was initiated appears to play a role in triggering RAG, but does not solely explain its onset. One factor common to each observation of RAG was that a portion of the anterior buccal cavity was exposed to air before (i.e., during the surface lunge) and during its execution. While failure to obtain food *while submerged* may eventually evoke a similar response, we never observed an underwater bout of RAG during more than 135 hours of underwater observation. Thus, it appears that contact with air is required to initiate the response.

Function

Although RAG could be an artifact (e.g., a manifestation of conflicting "drives") and have no function per se, it is worth considering adaptive explanations for such a pervasive behavior. Five working hypotheses are discussed here. First, however, it should be noted that two terms used below, *thwarting* and *frustration,* are anthropomorphic and should generally be avoided (for a detailed discussion, see Kennedy, 1992). These concepts are in wide use, however, and lacking satisfactory replacements, they are used here with reservation.

Maintenance

RAG may be a simple maintenance behavior associated with ingestion or respiration. As the head reenters the water following a surface lunge, large volumes of air can be seen leaving the mouth and gills. While other maintenance-related explanations are possible, it seems most likely that, in this regard, RAG acts as an internally mediated comfort movement, helping void air that remains trapped in the buccal cavity or that has entered the stomach.

Thwarting

Thwarting is defined as a situation in which some physical constraint or "barrier" prevents an animal from completing a previously reinforced behavior (Archer, 1976). As noted, RAG was observed only in association with surface feeding. The barrier, therefore, appears to have been the air–water interface and the previously reinforced behavior, food consumption. Having successfully "captured" a number of moving baits, frustration emerged when baits repeatedly "escaped" into the air, where the pursuer could not follow.

FIGURE 1 Repetitive aerial gaping by a 400 cm total length male white shark. (A) The shark submerges, having just made a surface lunge at the bait in the foreground. (B) The tail flexes slowly, and the shark rises and moves forward (note that the bait is now aligned with the right pectoral fin). (C) The mouth remains out of the water; the jaws gape slowly as the tail continues to propel the shark forward. (D) The snout lifts slightly as the forward motion slows. (E) The body turns as the response dissipates and the shark begins to right itself. Note that this instance did not involve the vertical baiting technique; see text for details on baiting (observation 12 of Table I).

FIGURE 1 (Continued)

FIGURE 1   *(Continued)*

Aggression Reduction

RAG may immediately reduce frustration, as suggested above, but have most of its adaptive value in lowering intraspecific aggression. Following thwarted feeding attempts that did not lead directly to RAG, a number of sharks were observed to bite nearby objects, including the ship's hull, rudders, and dinghies tied alongside. Under similar conditions, white sharks have been observed to bite conspecifics (R. Taylor and V. Taylor, personal communication; W. R. Strong *et al.*, unpublished data). All aggregations thus far described for this species have occurred in close proximity to food sources. Given that (1) feeding aggregations are relatively common in white sharks, (2) the feeding activities of an individual can be thwarted by the presence of a larger conspecific (W. R. Strong *et al.*, unpublished data), and (3) resultant frustration can give rise to aggression (Dollard *et al.*, 1939), such a mechanism for redirecting and harmlessly discharging frustration may serve to reduce potentially damaging aggression.

Extended Feeding Attempt

RAG could be a feeding response, exaggerated to increase the probability of seizing prey above the air–water interface.

Predator–Prey Communication

It is remotely possible that RAG has been ritualized for communication, perhaps to intimidate and confuse prey escaping at the surface. Killer whales appear to take advantage of the confusion resulting when escape behaviors and alarm signals of pinnipeds exiting the water are misinterpreted by conspecifics as they are transmitted along the shore. Mistaking the cause of the disturbance or avoiding the disturbance

FIGURE 2   Underwater aspect of repetitive aerial gaping, drawn from video (a 300 cm total length male; observation 14 of Table I).

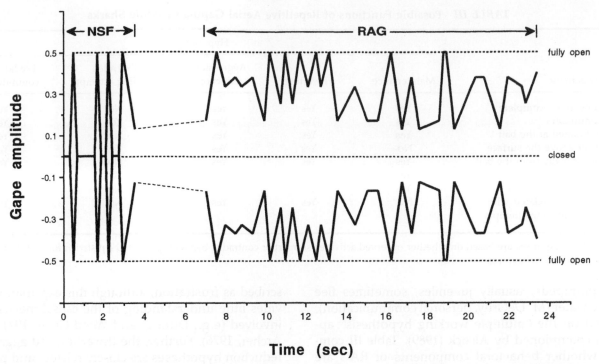

FIGURE 3 Comparison of gape amplitudes recorded during normal surface feeding (NSF) and repetitive aerial gaping (RAG) in a 340 cm total length female white shark. Data are real-time visual estimates of the relative distances between the upper and lower jaws from two videotaped sequences (observation 8 of Table I).

TABLE II Gape Duration Stereotypy during Repetitive Aerial Gaping and Normal Surface Feeding

Behavior pattern	Shark no.	n	Range (seconds)	Mean (seconds)	SD (seconds)	CV (%)	ST
RAG	17	3	0.867–3.300	1.700	1.386	81.5	1.2
	15	3	0.833–1.567	1.189	0.367	30.9	3.1
	43	4	0.733–1.100	0.858	0.164	19.1	5.0
	47	3	0.867–1.800	1.478	0.530	35.8	2.7
	Pooled	13	0.733–3.300	1.272	0.715	56.2	1.7
NSF	42	4	0.222–0.430	0.293	0.095	32.4	3.0
	43	4	0.178–0.652	0.402	0.204	50.7	1.9
	47	6	0.333–0.767	0.489	0.180	36.8	2.6
	Pooled	14	0.178–0.767	0.408	0.177	43.4	2.3
	1[a]	20	0.210–1.540	0.540	0.331	58.5	1.7
	2[a]	7	0.225–0.295	0.264	0.028	10.8	8.5
	3[a]	4	0.542–1.083	0.740	0.239	32.3	3.0

SD, Standard deviation; CV, coefficient of variation; ST, index of stereotypy; RAG, repetitive aerial gaping; NSF, normal surface feeding.

[a]Data from Tricas (1985) for white sharks 3 m (2) and 3.5 m (1 and 3) long.

TABLE III **Possible Functions of Repetitive Aerial Gaping in White Sharks**

Observed activity	Hypothesis				
	Maintenance	Thwarting	Aggression reduction	Extended feeding attempt	Predator–prey communication
Gapes slow and incomplete	Yes	Yes	Yes	No	Yes
Gapes continuous	Yes	Yes	Yes	Yes	Yes
Gapes not aimed at the bait	Yes	Yes	Yes	No	Yes
Mouth at or above the surface	No	Yes	Yes	Yes	Yes
Shark moves away from the bait	Yes	Yes	Yes	No	Yes
Solitary sharks perform RAG	Yes	Yes	No	Yes	Yes
Hiatus often follows RAG	Yes	Yes	Yes	No	No
RAG intensity correlated with preceeding feeding attempts	No	Yes	Yes	No	Yes

Yes or No, responses are based on whether observed activities support or contradict behavior predicted *a posteriori* for each hypothesis. RAG, Repetitive aerial gaping.

itself, pinnipeds, usually juveniles, sometimes flee into the water (M. DeGruy, personal communication).

Based on the "multiple working hypothesis" approach championed by Alcock (1989), Table III compares whether behavioral components of RAG are logically consistent with each of the five hypotheses presented. The predictions inferred are somewhat subjective, and other interpretations are possible. Assuming that the predictions are correct, the extended feeding attempt hypothesis can be essentially eliminated based on the number of inconsistencies with critical predictions. Although the predator–prey communication hypothesis is generally supported in Table III, it is the least parsimonious and seems improbable.

The maintenance hypothesis, on the other hand, is the most parsimonious of all and will be further considered, although it fails two of the predictions in Table III. The postures illustrated in Figs. 1 and 2, along with the pharyngeal pumping associated with jaw gaping, lend support to the idea that the sharks may be engaged in relieving discomfort associated with trapped air. However, both of these were low-intensity RAG bouts (see observations 12 and 14, Table I). One-third (5) of the bouts observed were of "high intensity" wherein the shark, often leaning only slightly to one side, swam at moderate speed across the surface, causing water to flush turbulently into the mouth in a manner that would appear to confound the presumed desired effect.

Based on the evidence in Table III, especially the observation that RAG bout intensity appears positively correlated with that of the preceding feeding attempts, RAG tentatively appears to be a manifestation of frustration incurred during repeated thwarted attempts to feed at the air–water interface. Similar responses to thwarting heretofore have been de-

scribed as frustration, although this explanation provides little understanding of the causal mechanisms involved (e.g., Duncan and Wood-Gush, 1971, 1972; Archer, 1976). Further, the thwarting and aggression reduction hypotheses are closely related, and present evidence provides no compelling basis for preferring one over the other. It seems likely that a response as pervasive as RAG serves some purpose, if only to release frustration, an activity that may be intrinsically beneficial. It logically follows that such a release may reduce intraspecific aggression, the utility of which is particularly apparent, given an individual white shark's potential for mortally wounding a conspecific with a single bite. RAG may, therefore, be functionally similar to the "redirected attacks" described by Tinbergen (1953), Bastock *et al.* (1953), and others, although these are usually aimed at subordinate animals rather than air (although it is true that few other "targets" would be naturally available to a shark). Alternatively, if two incompatible activities are strongly activated (e.g., approaching and avoiding), they may suppress the expression of one another and thereby disinhibit a third, seemingly irrelevant, activity or "displacement" (Armstrong, 1947; Tinbergen and van Iersel, 1947; McFarland, 1966). However, displacement activities are recognizably similar to or derived from motor patterns normal for the species (Tinbergen, 1952). RAG does not resemble other common behaviors, and although ineffective for feeding, the gaping component of RAG does not appear out of context. Thus, while RAG appears to result from the failure of sensory feedback to confirm some "expectation," it does not fit neatly into existing paradigms.

All observations of RAG were made under artificial conditions; most elusive are natural scenarios in which white sharks could be thwarted at the air–

water interface. It is an interesting coincidence, however, that the major prey of these animals, delphinids and pinnipeds, commonly leap >1 m into the air and that the latter often escape onto shore when in danger (Martinez and Klinghammer, 1970). Further, both prey types are gregarious; this may, in some situations, contribute to excited feeding behavior and could give rise to a build-up of frustration if a train of predatory attempts goes unrewarded. Presently, no firsthand observations exist to support or deny such an inference. The same surface-feeding scenarios could, however, cause air to be trapped inside the animal; although this hypothesis is less well supported, it cannot be rejected until more is known. In the meantime, it is intriguing—even disarming—to consider that white sharks may ingest air in the course of surface feeding and actually interrupt their activities to "burp."

RAG is an interesting addition to the white shark's ethogram. Although difficult, it is conceivable, through observation and experimentation, to better define the causative factors of RAG, to relate how variations in external stimuli affect its performance, and to develop a clearer understanding of its purpose in its natural context.

Summary

RAG, the behavior pattern described here, was observed in six bait-attracted white sharks, *Carcharodon carcharias*, that were prevented from feeding on baits positioned at the surface by pulling the bait away each time an advance was made. RAG always followed a series of thwarted feeding attempts and took the following basic form: the shark held its head out of the water, rolled onto its side, and opened and closed its mouth in a series of moderately slow, rhythmic, partial gapes while swimming slowly and awkwardly along the surface. RAG bouts averaged about 10 seconds (± 5.5 SD; n = 15) and were not oriented toward any specific object. Jaw movements recorded during RAG were temporally stereotyped and were three times slower than those associated with normal feeding. Based on a comparison of alternative hypotheses for the function of this novel behavior, it is tentatively suggested that RAG is a manifestation of "frustration" incurred during a series of thwarted feeding attempts and may function to reduce intraspecific aggression.

Acknowledgments

I am grateful to the members and employees of The Cousteau Society for financial and logistic support and am especially indebted to J.-M. Cousteau and the dedicated crew of *Alcyone*: C. Jouet-Pastre, N. Durassoff, M. Deloire, C. Davis, P. Martin, S. Arrington, P. Allioux, M. Blessington, D. Brown, J. Cramer, B. Gicquel, P. Pique, L. Prezlin, M. Richards, A. Rousset, T. Stern, C. Van Alphen, M. Westgate, C. Wilcox, and Y. Zlotnicka. Superb logistical and technical support was provided by J. Brody, Y. Goldman, L. Sullivan, and the rest of the Cousteau team. I am likewise indebted to the South Australian Department of Fisheries for administrative support and for making *Ngerin* and her excellent crew available for our fifth expedition: N. Chigwiddin, K. Branden, B. Davies, C. Fooks, D. Kerr, D. Short, and N. Wigan as well as G. Dubas and I. Gordon. R. Fox, A. Fox, B. Talbot, R. Taylor, and V. Taylor contributed their unique talents and much inspiration to this effort. Sincere thanks to the National Parks and Wildlife Service of South Australia for permission to conduct research in park waters and to the following for equipment loans and donations: New South Wales Fisheries Research Institute (tags), the California Department of Fish and Game (ultrasonic receiver), Ocean Technology Systems (underwater communications), J.B.L. Enterprises (tagging spears), and Pelican Products (waterproof cases). Special thanks to R. C. Murphy, D. R. Nelson, Barry Bruce, A. Ebeling, and T. Tricas for scientific support and to B. Talbot for the photos in Fig. 1. Finally, I thank my anonymous reviewers and the editors of this volume for their many meaningful comments.

White Shark Attacks on Inanimate Objects along the Pacific Coast of North America

RALPH S. COLLIER
Shark Research Committee
Van Nuys, California

MARK MARKS
Shark Research Center
South African Museum
Cape Town, South Africa

RONALD W. WARNER
California Department of Fish and Game
Eureka, California

Introduction

Along the Pacific coast of North America, attacks by the white shark *Carcharodon carcharias* on boats have been described by Gilbert (1963b). Unfortunately, the list provided did not include information other than date, location, and, occasionally, the boat owner's name. Subsequently, Follett (1974) described details of the attack on Henry Tervo's boat, September 10–12, 1959. Otherwise, white shark attacks (strikes) on other inanimate objects have gone unreported.

In various reports, the identification of the attacking shark has often been inaccurate, as demonstrated by the conclusion that a killer whale *Orcinus orca* attacked and sunk a skiff near Bodega Bay, California, in 1952 (Caras, 1964). Some years later, Miller and Collier (1981) examined photographs of the skiff's hull and determined the attacker to be a white shark. Conversely, white shark attacks on humans, even from the same locations, have been well documented (Bolin, 1954; Fast, 1955; Gilbert, 1963b; Collier, 1964, 1992, 1993; Follett, 1974; Baldridge, 1974a; Miller and Collier, 1981; Lea and Miller, 1985; Ellis and McCosker, 1991; Tricas and McCosker, 1984).

In this chapter, we discuss white shark strikes on a number of inanimate objects of varying shapes, sizes, and colors. These events are presented chronologically.

Methods

We review here the available literature and summarize accounts from primary sources (see Acknowledgments). The geographic distribution of events extends from Isla de Guadalupe, Baja California, Mexico (29°07′ N, 118°21′ W), to Esperanza Inlet, Queen Charlotte Islands, British Columbia, Canada (49°48′ N, 127°07′ W) (Fig. 1).

Results

Boats

1. 1955–1957; Guadalupe Island, Mexico (29°7′ N, 118°21′ W). Dr. Carl Hubbs, his wife Laura, and Al Stover were conducting pinniped counts aboard a 4.9-m blue fiberglass boat 50–100 m from shore at West Elephant Seal Beach. They observed a 3- to 4-m white shark motionless at the surface 30–40 m from their location. The shark, without provocation, turned and

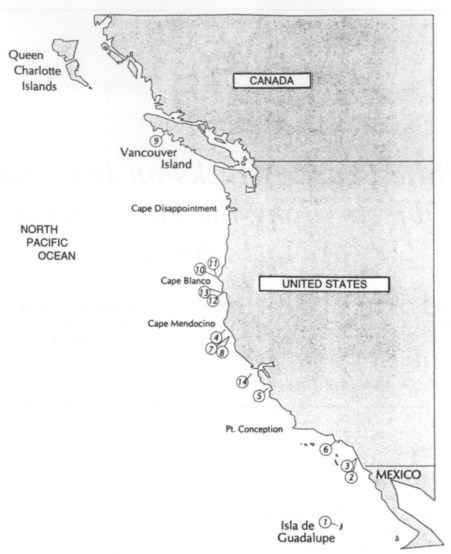

FIGURE 1 Locations of white shark attacks on boats and other inanimate objects in chronological order.

charged at a very high rate of speed, ramming the boat so violently that it nearly threw the Hubbses into the water (A. Stover, personal communication).

2. November 5, 1958; Pacific Beach, California (32°45′ N, 117°11′ W). Bob Shay's 4.3-m skiff was struck twice by a "heavy bodied shark with a head like a mackerel shark *Lamna ditropis*." The shark slashed the skiff's motor, apparently while trying to capture an injured fish tied to the stern. A white shark was presumed to be the attacker (C. Limbaugh, personal communication).

3. July 30, 1959; La Jolla, California (32°50′ N, 117°16′ W). James Randle's 4.3-m skiff was violently struck by a white shark 3–4 m in length. A cloth sack

containing several fish was removed by the shark, which returned 10 minutes later. At that time, several .38-caliber bullets were fired into the shark, which then disappeared in "a small cloud of blood."

4. July 21, 1985; Shelter Cove, California (40°01′ N, 124°01′ W). Jack Siverling was bottom fishing for lingcod *Ophiodon elongatus* about 1.5 km east–southeast of the whistle buoy at Shelter Cove. He hooked a fish, which suddenly took off at a sharp angle to the boat before his 100-lb test line parted. Seconds later, the entire stern of his 4.9-m fiberglass boat, with two outboard motors, was lifted out of the water to such a height that Siverling was nearly thrown into the water. A large swirl was observed astern, followed mo-

mentarily by a 5- to 6-m white shark that swam by 1 m below the boat (J. Siverling, personal communication).

5. April 9, 1989; Monterey Bay, California (36°37′ N, 121°56′ W). At 1035 hours, Jon Capella and friends were aboard the *Xeno*, a 10.7-m boat with a white and blue hull, about 1.5 km south of the whistle buoy at Point Pinos. The depth was 10 fm. They observed a 5- to 7-m white shark (Fig. 2D) feeding on a harbor seal *Phoca vitulina*. After 10 minutes, the shark began to circle the boat, finally ramming the bow with its head. The shark struck four times over a 15- to 20-minute period. Just prior to its departure, while astern of the boat, the shark rolled to one side and began slapping the swimstep and propeller with its tail. The boat had drifted some 20–30 m from the seal at this time, whereupon the shark returned to the seal for several additional bites before it departed (J. Capella, personal communication).

6. September 1989; Palos Verdes, California (33°46′ N, 118°20′ W). Tony DiCristo and Dan Fink were aboard the *Velmar*, a 11.6-m fiberglass boat, about 10 km off Palos Verdes, filming a 3- to 4-m white shark feeding on a dead whale. The shark left the carcass, swam over, and bit the swimstep several times, causing only minimal damage. Later, the shark rammed the boat several times, pushing it through the water. These events took place over a 4- to 5-hour period. When the boat drifted too close (4–6 m) to the whale, the shark would leave the carcass, focusing its attention on the boat (T. DiCristo and D. Fink, personal communication).

Crab Trap and Crab Trap Buoys

7. January 1983; Low Gap (Usal), California (40°06′ N, 124°01′ W), and Jackass Creek, California (39°53′ N, 123°56′ W). The January 1983 marine resources monthly report of the California Department of Fish and Game stated that "two crab fishers brought in three crab trap buoys that were hit by white sharks [Fig. 2C]. One buoy was off Low Gap, about 3 km south of Usal, and two were from Jackass Creek: all in about 9 fathoms of water. Two teeth removed from one buoy measured 42.0 mm and 42.2 mm in height. The other buoys showed serrated tooth patterns. All buoys were x-rayed, but no additional teeth or fragments were found."

8. April 1984; Usal, California (40°06′ N, 124°01′ W). The April 1984 marine resources monthly report of the California Department of Fish and Game described an incident reported by crab fisher Chuck Chernoff. While in the Usal area and retrieving his crab gear in about 20 fm, Chernoff reported that a

crab trap was almost halfway to the surface in calm water when something hit it so hard that the trap jerked past the stern of the boat, snapping the buoy line off the block. He retrieved the trap and brought it to a California Fish and Game office. Whatever hit the pot was large and powerful. Three of the welded 0.75-in. bars on the pot were bent, welds were broken, and the woven stainless steel wire was crushed all the way into the bait jar. Several large white sharks (4.3–5.5 m) had been reported in the area. Additionally, this is the area where lacerated crab trap buoys, having embedded white shark teeth, were found in 1983 (see case 7).

Otter Board and Float Bags

9. August 17, 1961; Esperanza Inlet, British Columbia, Canada (49°48′ N, 127°07′ W). While fishing for salmon *Oncorhyncus* spp. in the Vancouver Island area, Greg Trenholme observed a 4- to 6-m white shark surface astern of his boat, whereupon it charged and seized one of his canvas float bags. After several seconds, the shark released the bag and swam off, descending from sight. Upon examination, a single tooth and several fragments were removed from the damaged bag, identifying the attacker as a white shark (K. Ketchen, personal communication).

10. October 18, 1987; Elk River, Oregon (42°48′ N, 124°32′ W). In waters 3–5 fm deep near Cape Blanco, Nelson Miles of the *Elaine M*, a 9.5-m commercial salmon troller, witnessed a 3- to 4-m white shark charge and repeatedly strike his gray and white float bag. After the shark departed, six tooth punctures were found in the float bag (N. Miles, personal communication).

11. November 27, 1988; Elk River, Oregon (42°48′ N, 124°32′ W). The *Showdown*, a 12-m commercial salmon troller piloted by Dave Tilly, was fishing in calm waters 6–10 fm deep, near the mouth of the Elk River. At 1100 hours, he saw a white shark, 6–7 m in length, surface astern, then charge and strike a single blow to one of the float bags. It then submerged and was not seen again. The ring of tooth scars on the float bag measured 550 mm in diameter (D. Tilly, personal communication).

12. August 28, 1989; Port Orford, Oregon (42°44′ N, 124°30′ W). At 1430 hours, Gerald Moser was trolling for salmon in 10 fm of water, near the Port Orford whistle buoy. The *Tequila*, a 7-m fishing boat, was approached from the stern by a 3- to 4-m white shark. The shark struck the troller's fishing gear, which was partially submerged, biting a white inflatable float bag in half, then seizing and biting one of two wooden otter boards several times (Fig. 2A). The shark

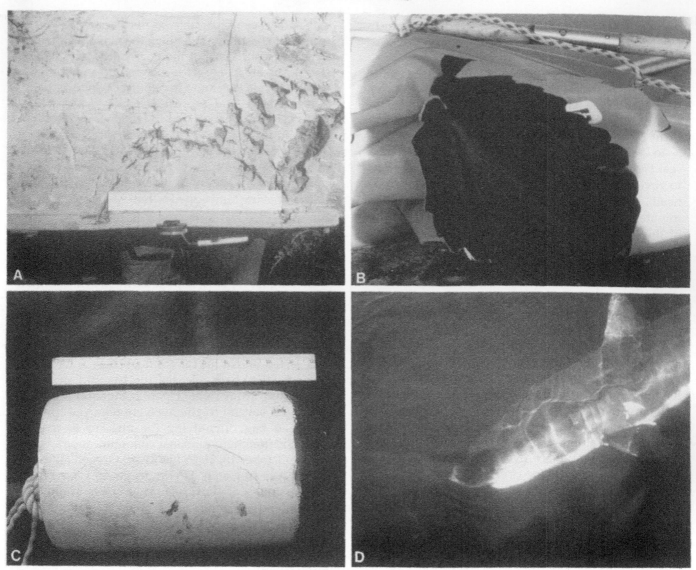

FIGURE 2 (A) The trawler board from Port Orford, Oregon (case 12). (Courtesy of G. Moser.) (B) The inflatable boat from the Farallon Islands, California (case 14). (Courtesy of A. P. Klimley.) (C) The crab pot buoy from Jackass Creek, California (case 7). (Courtesy of K. Collier.) (D) The white shark that rammed the *Xeno* in Monterey Bay, California (case 5). (Courtesy of J. Capella.)

departed after mouthing this equipment for several minutes.

13. October 1990; Port Orford, Oregon (42°45′ N, 124°30′ W). A white shark struck a float bag trolled behind an unidentified fishing vessel near Cape Blanco. The shark severed the float bag from its line and towed the bag into a kelp patch, where it was released. Upon retrieval, many individual tooth punctures were noted in the float bag. This incident was reported to local fisher J. Hassett, who also observed the damaged float bag (J. Hassett, personal communication).

Inflatable Boat

14. November 5, 1985; Farallon Islands, California (42°00′ N, 123°00′ W). Peter Klimley and Jim Wetzel were conducting field studies throughout the day from a 6.7-m boat. Klimley wrote in his field notes, "the inflatable was being used to ferry ourselves back and forth to a research skiff which was kept on a buoy at Fisherman's Bay. The shark attacked the inflatable while it was tied to the buoy with no persons aboard [Fig. 2B]. There was a large swell that day and the skiff was being jerked rhythmically in the water like a

large fishing jig" (A. P. Klimley, personal communication).

Discussion

Available for comparison with the observations reported here is a wealth of published information on attacks by white sharks on humans at nearby sites. This affords us the opportunity to discuss the popular hypothesis that white shark attacks on humans may be the result of mistaken identity.

Following exhaustive analysis of over 1700 cases of shark attacks against humans worldwide, Baldridge and Williams (1968; Baldridge, 1974a) concluded that "50–75% of all reported attacks upon man [worldwide] were motivated by a drive or drives other than feeding." Miller and Collier (1981) reported 47 cases of unprovoked shark attack along the California and Oregon coasts and reached the general conclusion that "*most* of the attacks resembled the feeding behavior of an isolated, large shark that appears to be investigating an object." In their description of the postulated "bite and spit" attack behavior adaptation, Tricas and McCosker (1984) believed that diver and surfer silhouettes, when viewed from below, resembled those of pinnipeds, a natural prey species of the white shark. In this case, attack was likely more than mere investigation.

The events described here do not support the "mistaken identity" hypothesis, because white sharks attacked inanimate objects of a variety of shapes, colors, and sizes, none resembling the shape, size, or color of a marine mammal (Table I). The shapes of inanimate objects included the conical boats (cases 1–6), the circular crab trap buoys (case 7), the rectangular float bags (cases 9–13), and the complex shape of the inflatable boat (case 14). Object colors varied as well: crab trap buoys—blue and yellow; boats—blue and white; float bags—white and gray; otter boards—white and gray; and the inflatable—gray. The sizes of the objects ranged from the small crab trap buoys (35 cm in length) to the large boats (11.6 m in length). The movement of the object did not appear to be a significant contributor to the white sharks' attacking the object. Further, we believe that the object color(s) probably was not important, as any object viewed from below and back-lighted by a bright sun or light sky would appear as only a dark, colorless silhouette (see Chapter 21, by Strong). There appeared to be no correlation between an object's size, color, or shape and a "preferred silhouette" (i.e., a pinniped) for determining the probability of a white shark encounter, whether an inanimate object or a human.

On the basis of the events presented, we suggest that white sharks often strike unfamiliar objects to determine potential food value (see Chapter 20, by Anderson *et al.*, and Chapter 21, by Strong). Some evidence suggests that white sharks decide a prey's palatability while it is lodged in the shark's mouth, whether the shark is moving or stationary. Klimley

TABLE I Characteristics of Inanimate Objects Attacked by White Sharks

Object	No.	Location	Shape	Color	Size (m)	Movement
Crab buoy	1	Low Gap, California	Cylindrical	Yellow and blue	0.35	Minimal
	2	Jackass Creek, California	Cylindrical	Yellow and blue	0.35	Minimal
Boat	1	Pacific Beach, California	Conical		4.30	Minimal
	1	Monterey, California	Conical	White and blue	10.70	Minimal
	1	Palos Verdes, California	Conical	White	11.60	Minimal
	1	Shelter Cove, California	Conical		4.90	Minimal
	1	Guadalupe Island, Mexico	Conical	Light blue	4.90	Minimal
	1	La Jolla, California	Conical		4.30	Minimal
Inflatable	1	Farallon Islands, California		Gray	3.10	Erratic
Float bag	1	Esperanza, British Columbia, Canada	Rectangular		0.60	Constant
	1	Port Orford, Oregon	Circular	White	0.60	Constant
	1	Elk River, Oregon	Rectangular	White and gray	0.60	Constant
	1	Elk River, Oregon	Rectangular	White and gray	0.75	Constant
	1	Port Orford, Oregon	Rectangular	White and gray	0.60	Constant
Otter trawl board	1	Port Orford, Oregon	Rectangular	White and gray	0.75	Constant

(1994) argues that white sharks reject low-fat content (i.e., energy-poor) items, such as birds, sea otters, and humans, to feed on high-fat content pinnipeds and whales. Further, there are several cases of white shark attacks on humans from this same area that support this "high-energy content" hypothesis (Collier, 1993, unpublished data). It appears that white sharks are somewhat indiscriminate in their strikes on surface objects, regardless of whether or not the object resembles a prey species. It would seem that grasping an unfamiliar object would be the only reliable method of determining palatibility.

Alternatively, in some instances, the shark could be eliciting a territorial behavior (i.e., butting a boat in the vicinity of food), or the behavior could be a form of displacement behavior (see Chapter 21, by Strong). Further, cases 2 and 3 could have been the result of the white sharks' attempting to eat the fish that had been tied to the boat. We do not believe the remaining cases (1 and 4–14) are the result of attracting or baiting the shark directly to the object struck. The presence of marine mammals (pinnipeds or whales) was reported in cases 1, 5, 6, and 14. The white shark which so violently struck the boat in case 1 was not feeding prior to its attack. Proximity to pinnipeds might have contributed to this incident. The crab trap (case 8) and buoys (cases 7 and 8) had a small amount, usually <1370 g, of bait present. The buoys were on the surface, 8–10 fm from the bottom (the location of the bait). This distance would reduce the likelihood that the bait was a significant contributor to the strike on the buoy.

Although bait was present in case 8, the shark struck the trap only after it began moving toward the surface. Cases 9–13 demonstrate a white shark's ability to "run down" and effectively attack a moving object, in these cases float bags and an otter board. All of these objects were being towed at a constant speed with no baiting. The inflatable (case 14) was being erratically "jerked" near a pinniped haulout site, without any baiting.

It was not possible to determine whether the white shark uses a search image when hunting. The sizes, colors, and shapes of the objects discussed here indicate that the white shark investigates many foreign objects, in addition to its hunting strategies. We believe that these objects represent only a small proportion of things that white sharks investigate (see Chapter 20, by Anderson *et al.*, and Chapter 21, by Strong).

Summary

White shark strikes on a variety of inanimate objects have been described. Results indicate that the white shark approaches and seizes surface objects, without regard to shape, size, or color. We suggest that the shark is determining suitability as food in most cases.

Acknowledgments

Many individuals contributed to the information we presented: J. Capella, A. P. Klimley, J. Siverling, T. DiCristo, D. Fink, N. Miles, D. Tilly, G. Moser, A. Stover, K. Ketchen, G. McFarlane, V. Springer, W. Cox, R. Rainey, B. Weisner, and W. Pollock. S. Vukas provided assistance with maritime particulars, including event locations.

White Shark Reactions
to Unbaited Decoys

SCOT D. ANDERSON
Inverness, California

R. PHILIP HENDERSON,
PETER PYLE, *and* DAVID G. AINLEY[1]
Point Reyes Bird Observatory
Stinson Beach, California

Introduction

In a paper published in 1985, in order to explain the high frequency of attacks by white sharks *Carcharodon carcharias* on surfers in California, J. E. McCosker hypothesized that a person in normal position on a surfboard duplicates the image of a pinniped swimming at the surface. At that time, it was well known that pinnipeds were the primary prey of white sharks in California (Ainley *et al.*, 1981, 1985; Klimley, 1985b; Le Boeuf *et al.*, 1982). McCosker (1985) also published a convincing photograph (his Fig. 5) comparing the silhouettes, from below, of both a seal and a surfer side by side.

In 1988, we discussed methods to view sharks in order to identify them by natural markings for a study of their residency patterns around the South Farallon Islands (see Chapter 33, by Klimley and Anderson). We had few logistical or monetary resources and we wanted to avoid baiting, because we did not want to artificially supply food, thus biasing our other studies of predation. Moreover, baited sharks injure themselves in the process, which might lead to avoidance of our study area by the sharks. Being fishing enthusiasts, and with McCosker's photo well in mind, we decided to attempt to use decoys to bring the sharks to the surface and decided that the best decoy would be a surfboard "fished" in the high-risk

zone for pinnipeds in nearshore waters around the islands (see Klimley *et al.*, 1992). We were immediately rewarded with spectacular results (see the photograph in Davidson, 1992) (Fig. 1).

We began to conduct our photo identification efforts in 1989. Since then, we have amassed many hours of observations of white sharks lured to the surface by various "decoys." Even though our purpose was other than a controlled experiment on prey discrimination in white sharks (see Chapter 21, by Strong), in the process of our work, becoming apparent were patterns that offered new insights into the predatory behavior of the white shark. This chapter summarizes what we saw, providing more questions than conclusions, but establishing the basis for future experimentation by us or others.

The white shark is known to attack pinnipeds, as noted above, but also nonprey items such as boats, buoys, otter trawl boards, fish bags, logs, birds, human divers, surfboards, and kayaks (Miller and Collier, 1981; Lea and Miller, 1985; Klimley *et al.*, 1992; Collier *et al.*, Chapter 19; McCosker and Lea, Chapter 39). Firsthand, in the high-risk zone at the South Farallon Islands during the period when sharks are present and hunting, we have observed attacks on the largest sea birds sitting on the water (brown pelicans *Pelecanus occidentalis*; 50 cm long), a log (100–150 cm), and an inflatable boat [310 cm (see also Chapter 19, by Collier *et al.*). In contrast, also in the high-risk zone, we have observed other virtually inanimate objects

---

[1]Present address: H. T. Harvey & Associates, Alviso, CA.

FIGURE 1 Photographs showing the two categories of response to decoys. (A) investigatory—the dorsal fin appears as a shark passes the yellow mat; and (B) attack—a shark leaps at a surfboard.

that have never been attacked, even while pinnipeds are being consumed by sharks nearby: ocean sunfish *Mola mola* (50–200 cm), dense tangles of bull kelp *Nereocystis iuetkeana* (200–500 cm), and western gulls *Larus occidentalis* (30 cm) and smaller sea birds. In addition, we have spent hundreds of hours in the high-risk zone in a small boat (approximately 400 cm), observing white sharks, but no shark has ever attacked it. The paucity of responses to nonprey items under these circumstances is quite different from the case of baited sharks, which, with blood in the water, bite almost anything in the area (documented ad nauseam in film documentaries).

On the basis of the various observations of sharks investigating or attacking nonprey items, McCosker (1985) also speculated that vision, electric fields, vi-

bration, and scent may contribute to stimulating the white shark to attack. In presenting our observations, we do so with this hypothesis in mind and ask the questions, To what degree is vision employed by a white shark during its hunt?, Can white sharks discriminate between objects?, and What size range of prey is of most interest to them?

Materials and Methods

The study site, the South Farallon Islands, central California, is the location of a long-term white shark study in which predation on pinnipeds by white sharks has been well documented (Ainley *et al.*, 1980, 1985; Klimley *et al.*, 1992, Chapter 16; Anderson *et al.*, Chapter 25; Pyle *et al.*, Chapter 26). There, the sharks prey on northern elephant seals *Mirounga angustirostris* and California sea lions *Zalophus californianus*.

We present here observations on decoys deployed between 0900 and 1600 hours, September–November 1989–1992, for a total of 159 hours of effort. The decoys tested included inflatable plastic mats and toys, surfboards, and a neoprene wet suit (see Figs. 2 and 3). The decoys, attached by line to a fishing rod, were drifted into the high-risk zone, or the region 400 m wide where most attacks on pinnipeds have occurred (for a discussion of the high-risk zone, see Klimley *et al.*, 1992), from East Landing. Our main purpose was to identify sharks in the vicinity, and therefore, those decoys that proved most effective in eliciting a response were used most often. Once a shark was sighted, observations continued until it disappeared; the decoy was then retrieved. Notes and photographs were taken to document identification of individual white sharks and their behavior. If no sighting occurred within 60 minutes, the decoy was retrieved.

Responses were categorized into two groups: investigatory and attack. Investigatory behavior was characterized by sharks' circling, swimming near, bumping, or gently mouthing the decoy (see Fig. 1A). Attack behavior was characterized by a shark's rushing aggressively to bite the decoy vigorously, often leaping clear or halfway out of the water to "catch" the decoy (see Fig. 1B).

We statistically compared results using the proportions *C* test (Table II). The null hypothesis (H0) tested that there is no significant difference in the proportion of attacks to investigations. An alternate hypothesis (H1) was that there is a significant difference in the proportion of attacks to investigations between any two decoys. Significance was considered at 0.002, determined by the Dunn–Sidak method (Sokal and Rohlf, 1981).

FIGURE 2 The variety of decoys used in this study: (A) a 170-cm brown inflatable seal-shaped cushion, (B) a 160-cm yellow inflatable mat, (C) a 200-cm black wet suit, and (D) a 180-cm tan surfboard.

Results

Sharks attacked the decoys 23 times (27% of responses) and investigated them 62 times (73%), for a total of 85 responses (Table I). Responses per hour of effort, among decoys that were visited, ranged from 0.25 to 0.78 hours and averaged 0.50. The 170-cm oval decoy was the only one that failed to attract a response, but 6 hours' deployment time is insufficient to draw a firm conclusion. Among frequently deployed decoys, the 220-cm green surfboard was the least effective, with 0.25 responses per hour in 27 hours of effort. The 180-cm surfboard and the 200-cm wet suit received the most attention: 0.78 and 0.72 responses per hour of effort, respectively. The 210-cm brown–gray surfboard, which had the video camera attached, was deployed the most (55.5 hours) and elicited 0.61 responses per hour of effort. None of these rates was statistically significant.

We next looked at the proportions of attacks to investigations. The 180-cm tan surfboard and the 200-cm black wet suit had, by far, the highest proportion of attacks to investigations: 1.29 and 1.00, respectively. The next closest proportion of attacks to investigations was 0.50 (a seal-shaped mat and a rectangular mat), but only the comparison between the 180-cm tan surfboard and the 210-cm brown–gray surfboard even approached statistical significance (Table II).

Discussion

Our procedures were not designed with a rigorous experimental test in mind. True to the most effective method used by fishermen to catch fish, we tended to use what proved to be most effective. One weakness in our plan was that we did not know whether any sharks were present when we deployed our decoys. Thus, an even greater sample size of effort, which we have been accumulating, will be necessary to better understand the patterns we observed. In spite of the

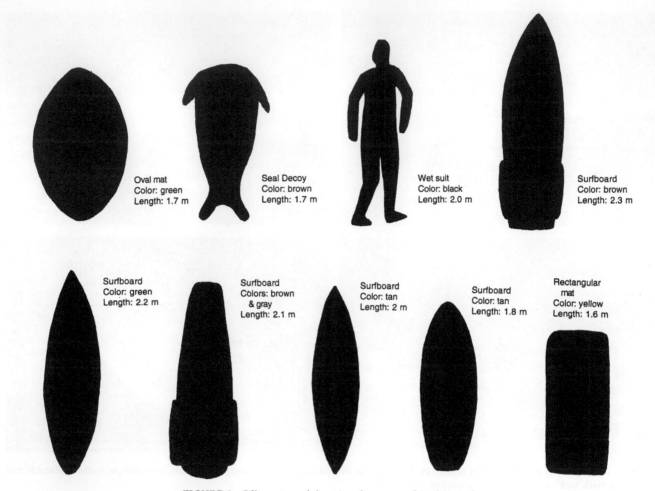

FIGURE 3 Silhouettes of the nine decoys used in this study.

TABLE I White Shark Reaction to Decoys Deployed at the South Farallon Islands

Object	Color	Length (m)	Effort (hr)	Investigations (N)	Attacks (N)	Pooled (N)	Responses/hour of effort	Attacks/ investigation
Surfboard	Brown and gray	2.1	55.5	30	4	34	0.61	0.13
Surfboard	Green	2.2	27.1	5	2	7	0.25	0.40
Surfboard	Tan	1.8	20.5	7	9	16	0.78	1.29
Surfboard	Brown	2.3	13.7	6	1	7	0.51	0.17
Surfboard	Tan	2.0	10.2	5	1	6	0.59	0.20
Seal mat	Brown	1.7	9.9	4	2	6	0.61	0.50
Wet suit	Black	2.0	8.3	3	3	6	0.72	1.00
Rectangular mat	Yellow	1.6	7.8	2	1	3	0.39	0.50
Oval mat	Green	1.7	6.0	0	0		0	
Total			159.0	62	23	85	0.50[a]	0.27

[a]Total hours per sighting, 1.96 ± 0.28 SE (N = 8) represents the mean hours per sighting for each decoy, rather than being based on the total hours and sightings.

TABLE II Values of Proportions C Test Comparing Any Two Decoys

Decoy	2.1-m Brown/gray surfboard	2.2-m Green surfboard	1.8-m Tan surfboard	2.3-m Brown surfboard	2-m Tan surfboard	1.7-m Brown surfboard	2-m Black surfboard
2.2-m green surfboard	0.994 $p > 0.30$						
1.8-m tan surfboard	3.29 $p < 0.01$	1.31 $p > 0.10$					
2.3-m brown surfboard	0.182 $p > 0.80$	0.662 $p > 0.50$	2.31 $p > 0.05$				
2-m tan surfboard	0.308 $p > 0.70$	0.519 $p = 0.60$	2.01 $p < 0.05$	0.119 $p > 0.90$			
1.7-m brown seal mat	1.08 $p > 0.20$	0.183 $p > 0.80$	1.00 $p > 0.30$	0.815 $p > 0.40$	0.648 $p > 0.50$		
2-m black wet suit	1.80 $p > 0.05$	0.801 $p > 0.40$	0.261 $p > 0.70$	1.46 $p > 0.10$	1.30 $p > 0.10$	0.594 $p > 0.50$	
1.6-m yellow rectangular mat	0.779 $p > 0.40$	0.146 $p > 0.80$	0.768 $p > 0.40$	0.629 $p > 0.50$	0.532 $p > 0.50$	0.000 $p > 0.95$	0.491 $p > 0.60$

Statistical significance is at $p = 0.002$ (Sokal and Rohlf, 1981).

varying effort used for the different decoys, can we draw any conclusions from our observations?

The decoys had no scent, gave off no electric fields (because they were plastic, fiberglass, and neoprene rubber), and produced no vibrations or splashing (because they sat softly on the water, and most flexed with waves). Thus, vision can be singled out as the sense used by the sharks to locate, investigate, and attack the decoys (and therefore their prey) in the high-risk zone around the Farallon Islands. This is consistent with other findings: a higher percentage of white shark attacks on humans occurred in clear water (Miller and Collier, 1981), and baited white sharks in South Australia preferred to investigate a sea lion shape over a square (see Chapter 21, by Strong). Moreover, white sharks hunting at the Farallones appear to lie in wait, camouflaged by dark backgrounds, to attack pinnipeds that swim or float above them (Klimley et al., 1992; Klimley, 1994).

On some occasions, sharks investigated or attacked decoys that did not closely resemble pinnipeds (i.e., the rectangular yellow mat). One possible explanation, on the basis that the water around the South Farallon Islands is relatively clean or free of nonprey items or flotsam (e.g., kelp or logs) at this time of year, is that whatever the shark sees within a certain size range (100–300 cm) in the "high-risk zone" is categorized as a pinniped, and thus, is attacked or at least investigated. This behavior may be a key to maximizing their chances of eating (see also Chapter 21, by Strong). The decoys varied through-

out the size range of pinnipeds found on the South Farallon Islands during autumn. Klimley et al. (1992) estimated the lengths of five pinnipeds eaten by sharks to range from 140 to 170 cm; Long et al. (Chapter 24) measured two fatally bitten pinnipeds to be 200 and 250 cm long. Our observations indicating that white sharks do not readily attack the smaller or larger objects in this zone (i.e., numerous small birds and molas <100 cm versus boats >300 cm), and the vigor by which they attacked our decoys, indicate to us that size may be an important factor in the sharks' search image for prey at the Farallon Islands.

Although sharks in our study attacked objects of varying properties, as in the study by Collier et al. (Chapter 19), the greater frequency and vigor of the attacks that we observed lead us to believe that, in some cases, white sharks do use vision to discriminate among prey. As our work at the Farallones continues, we hope to investigate in more detail prey discrimination in this species and the importance of vision in the white shark's predatory strategy.

Summary

Using unbaited decoys, we explored the importance of vision in the predatory behavior of white sharks C. carcharias. Nine different objects, in the approximate size range of pinniped prey, including inflatable plastic mats and toys, surfboards, and a neoprene wet suit, were deployed at the South Farallon

Islands, California, for a total of 159 hours of effort. Positive responses by sharks were either investigatory (62 responses) or attack (23 responses). Attacks were spectacular, indicating that some of the decoys had effectively lured the shark into considering the decoy to be prey. The sharks were attracted visually, because the decoys emitted no scent, electric fields, or vibrations or splashing. Results indicate that further experimental investigation of white shark predatory behavior would be fruitful and would reveal the

degree to which white sharks form a search image when hunting their prey.

Acknowledgments

The research program on the South Farallon Islands, part of the Farallon Islands National Wildlife Refuge, is conducted through the cooperation of the U.S. Fish and Wildlife Service. We also thank A. P. Klimley, K. Goldman, and four anonymous reviewers, whose comments improved this chapter. This is Point Reyes Bird Observatory (PRBO) contribution number 601.

Shape Discrimination and Visual Predatory Tactics in White Sharks

WESLEY R. STRONG, JR.[1]
The Cousteau Society
Chesapeake, Virginia

Introduction

Mechanisms underlying patterns of prey selection can be better understood by dissecting predation events into their component parts, usually given as five primary stages: detection, identification, approach, subjugation, and consumption (cf. Endler, 1986). In white sharks *Carcharodon carcharias*, only the latter two stages have been examined in detail. Tricas and McCosker (1984) conducted kinematic analyses of jaw mechanics manifested during food consumption by baited sharks, and from positions atop Southeast Farallon Island, Klimley *et al.* (1989, 1992, Chapter 16) described spatial and temporal characteristics of white shark attacks on natural prey, mainly pinnipeds, in the surrounding waters.

Far less information is available concerning prey detection, identification, and approach, although two ideas have become quite popular. Despite being countershaded, a fish or a marine mammal seen from directly below will appear as a silhouette against a lighter background of downwelling light (Hobson, 1963). This basic concept, as well as circumstantial evidence from attacks, has led a number of researchers to suggest that white sharks often attack their prey from below and behind (e.g., Miller and Collier, 1981; Tricas and McCosker, 1984; McCosker, 1985;

Myrberg and Nelson, 1991). When critically interpreted, however, this generalization is weak. As Tricas (1985) pointed out, a shark attacking along the surface creates considerable commotion, greatly reducing the possibility of surprise. Any other attack by a submerged fish on a floating object is, by definition, from "below," leaving the term open to a full range of interpretations. Additionally, direct observations by Klimley *et al.* (Chapter 16) of bite placements on pinnipeds indicate a nearly equal tendency to bite the head or the tail and do not, therefore, support the assertion that white sharks attack pinnipeds from "behind." Second, it has been suggested, based on inference only, that visual similarities to silhouetted pinnipeds increase the probability of attacks on human surfers (Tricas and McCosker, 1984). While both of the aforementioned ideas seem logical, there is little direct evidence to support either.

Given the need for information on how and why white shark attacks are initiated, this investigation was aimed at elucidating the detection, identification, and approach phases of the predation sequence. Based on extensive underwater observations, motor patterns used during initial responses toward floating "prey" were examined. Prey preferences, diel patterns of feeding, and the relative importance of vision are discussed here in light of experimental results and existing information. The final section of this chapter contains a series of speculations on the adaptiveness and ultimate causes of behavior patterns described here.

---

[1]Present address: Department of Biological Sciences, University of California, Santa Barbara, California.

Materials and Methods

Predatory responses exhibited under baited conditions were observed during five expeditions, averaging 33 days each, to lower Spencer Gulf, South Australia. Our team identified 67 individual white sharks, 34% of which were resighted on subsequent days (see Chapter 37, by Strong *et al.*). Approximately 900 hours of topside observations were generated over a total of 74 days, during which one or more white sharks were present. Approximately 135 hours of observations were conducted underwater among 1–5 ($\bar{X}$ = 2.3) observers, including 62 hours by W. R. S. A detailed review of the methods used in the attraction and identification of sharks is given in Chapter 37, by Strong *et al*. Other salient methods are as follows. Sharks were attracted via odor corridors consisting of mammal blood and tuna. Measured amounts of each of the attractants were mixed and delivered into the water on a standardized schedule using an automated system. Two approximately 1-kg pieces of tuna or horse meat were floated downcurrent of the research vessel on 6- to 10-m-long tethers, suspended from toy balloons, 40–80 cm below the surface. To increase the probability that sharks would remain in the area, subjects were occasionally allowed to take a bait; the amount consumed by each shark was recorded and averaged 1 kg/hour ($\pm$1.1 SD; range, 0–4.5 kg/hour) for a subset of 14 sharks observed for a total of 114 hours.

Prey preference trials (*preference* being operationally defined as a tendency to inspect or attack one potential prey type versus another) were conducted in waters approximately 20 m deep, adjacent to Dangerous Reef (34°49' S, 136°13' E) and the Neptune Islands (35°14' S, 136°03' E) during the fourth and fifth expeditions, January 19–February 25, 1991, and September 13–24, 1991. Test shapes, a square and a seal lying in the prone position, were constructed from 2.5-cm-thick plywood coated with black epoxy resin. The targets had equal surface areas of 3025 cm² (Fig. 1). Extending vertically from each target was a 65-cm-long PVC tube fitted to a 2-m-long acrylic crossbar; this clear connecting rod did not cast a silhouette when the array was viewed from underwater. With the exception of four small stainless steel screws used to connect the crossbar apparatus to each of the targets, materials used in the array were electrochemically inert. Trials were conducted on clear days between 1100 and 1830 hours. With the arrival of a new shark, baiting was stopped, with the exception of a blood mixture that was allowed to flow continuously so as not to interrupt the odor corridor. The

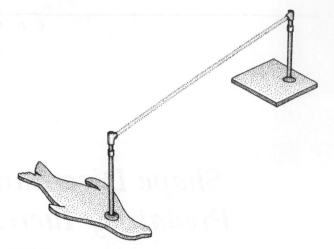

FIGURE 1 Visual discrimination array consisting of two black epoxy resin-coated plywood shapes of equal surface area, held apart by PVC uprights and a 2-m-long clear acrylic crossbar.

array was floated 6–8 m downcurrent of our research vessel. The targets lay flat on the sea surface and were held in position using a pair of 150-lb test monofilament nylon tethers. Positions of the targets, port and starboard, were randomly swapped following each trial series using the quick-disconnect fittings located at each end of the crossbar.

Responses to experimental targets, baits, and other objects were recorded from topside vantage points as well as from underwater antishark cages by divers using audio and video recorders. Underwater observations were confined to a roughly spherical space (hereafter referred to as the visible arena) delimited by water clarity ($\bar{X}$ = 20 m via secchi disk measurements). Divers wearing masks fitted with wireless single-side-band communicators (Ocean Technology Systems, Costa Mesa, CA) recorded the position and identity of each shark at the instant it entered the visible arena and gave ad libitum descriptions of each individual's subsequent activities. Thus, the technique produced real-time audio records of the primary activities of all sharks swimming into view during our dives. Initial and minimum distances of approaches were estimated based on the known dimensions of the target array and on the known depths of stationary observers in the antishark cages.

Arrival and departure times of sharks attracted to the bait station were recorded whenever possible. Nighttime arrivals were noted primarily by regularly checking tethered baits, although sharks were occasionally seen near the surface in the glow of deck lights. A departure was scored as the last time an individual was seen, but only after the shark had not been sighted for at least 6 hours while baiting contin ued. This criterion resulted in the recording of considerably fewer departures than arrivals.

Results

Approach Patterns

An *approach* was defined as an oriented pursuit of a bait or other object. Any traverse of the visible arena, whether or not an approach was made, was called a *pass*. Sometimes the sharks showed no response to the food available, but the majority actually approached objects when they entered the arena. To estimate the actual number of approaches observed during the course of this study, 20 separate timed bouts, during which every pass made by a shark was recorded, were combined into a subset of 27.4 hours of observation. On average, the sharks made 19.5 passes per hour ($\pm$8.35 SD; range, 7–41 passes per hour) through the visible arena. Multiplying the average rate of passes by the total observation time (approximately 900 hours) yielded 17,550 passes. Given that the majority included approaches, it is estimated that roughly 10,000 approaches were observed.

Approaches on floating objects were oriented both horizontally (0–44°) and vertically (45–90°), the great majority being horizontal. As individual sharks became increasingly familiar with the situation, they typically made more frequent passes near or along the surface. Generally, when a shark initiated a horizontal approach on a floating bait, a crew member would haul the attached tether toward the boat. The shark usually responded with a horizontal swim-chase, smoothly adjusting its swimming depth to, or slightly below, the same depth level as that of the bait (Fig. 2).

Newly arrived sharks showed a greater tendency to cruise through the visible arena at deep depths. While possibly serving the cautious tendencies of sharks that were unfamiliar with the situation, this behavior also appeared to facilitate visual detection of surface-borne objects. Upon sighting potential prey at the surface, the sharks oriented their bodies and ascended toward them on steep inclines (Fig. 3). This motor pattern, hereafter called the "vertical approach," was observed in a feeding context only and was exhibited by many of the 67 white sharks observed.

Approach orientation was influenced by the initial depth at which a shark entered the visible arena. In other words, an individual was, to some degree, predisposed to make a horizontal or vertical approach based on its depth at the moment a prey item was detected. Records of the depths at which sharks entered the arena illustrate the spatial readiness of newly arrived sharks to make vertical attacks. A subset of 862 minutes of underwater observations was collected over 7 days. In this sample, 14 individual white

FIGURE 2 A 320 cm total length female white shark exhibiting considerable flexibility while attempting to catch a bait moving horizontally; while the shark approached from just below the surface, this still constituted an "attack from below."

FIGURE 3 Having visually detected an object silhouetted at the surface, a 460 cm total length male white shark exhibits an orienting response and initiates an approach.

sharks, each of which was being observed for the first time, made 296 passes through the visible arena. Initial observation depth was recorded during 192 of these cases; 94 (49%) entered from depths >4 m (42 entered between 5 and 12 m, and 52 between 13 and 20+ m).

No attempt was made to record every observation, but field notes contain more than 160 firsthand accounts in which the vertical approach was used to investigate surface-borne objects, including primarily baits, but also cardboard boxes, birds, and clumps of *Sargassum* sp. as small as 15 cm in diameter. An indicator of the proportion of horizontal to vertical approaches is provided by a subset of event recorder data consisting of 332 minutes of observations collected on February 22, 1991. In this sample, 10 individual white sharks, 7 of which were being observed for the first time, made 109 passes through the visible arena. Vertical approaches were executed during 11 of the passes (10%).

Visual Discrimination Experiment

The estimated total length (TL) of the subjects averaged 354 ± 70 cm (Table I). While there were more male than female subjects, sex had no observable effects on the behaviors observed.

Ten of 11 sharks investigated or attacked the seal shape during their first approaches ($\chi^2 = 5.82$, $P <$

TABLE I Size and Sex of White Sharks in Visual Shape Discrimination Tests

Shark identification[a]	Estimated total length (m)[b]	Sex
15	3.4	F
37	3.7	F
39	4.6	M
42	3.0	M
44	2.3	M
47	4.0	M
49	3.6	M
51	3.7	M
62	3.0	F
65	3.0	M
67	4.6	F
	3.54 ± 0.7[c]	7 M:4 F

[a]Sharks described in the work of Strong *et al.* (1992, Chapter 37).

[b]Converted from estimates or measurements made in feet.

[c]Mean ± SD.

0.025) (Table II). Second and third approaches occurred at mean intervals of approximately 90 seconds (range, 0.5–13 minutes; n = 18). The sharks approached to within mean minimum estimated dis-

TABLE II **Approaches by White Sharks to Target During Shape Discrimination Tests**

Approach no.	Total (N)	Seal		Square		χ^{2a}	P	Estimated minimum approach distance (cm)[b]	Contacts (n)
		%	n	%	n				
1	11	91	10	9	1	5.82	<0.025	90 ± 70	3
2	10	80	8	20	2	2.50	<0.250	90 ± 100	4
3	9	67	6	33	3	0.44	<0.250	60 ± 60	2

[a]χ^2 values computed with Yate's correction (cf. Schefler, 1979).
[b]Mean ± SD.

tances of <1 m and made contact with the targets in 30% of the trials (Table II and Fig. 4). Three of the subjects returned 4, 5, and 7 days later, and upon being presented the targets a second time, each made its first approach toward the seal shape.

Arrivals and Departures

The vast majority of 120 arrivals occurred during the day (Fig. 5a). The largest proportion of 28 recorded departure times occurred around dusk (Fig. 5b).

Discussion

Approach Patterns

Based on observations and film footage taken at Dangerous Reef during January 1980, Tricas (1985) described three general types of horizontal swimming advances on floating baits. Most sharks used an "underwater approach," in which the shark "swam just below the surface until ca. 1 m from the bait and then attacked by deflecting the head upward and emerging out of the water to either swallow or bite

FIGURE 4 A 350 cm total length male white shark closely investigates the seal target. The square target is just out of frame on the left-hand side (the fish in the foreground are tommy roughs *Arripis georgianus*; note the distinct edge of Snell's window).

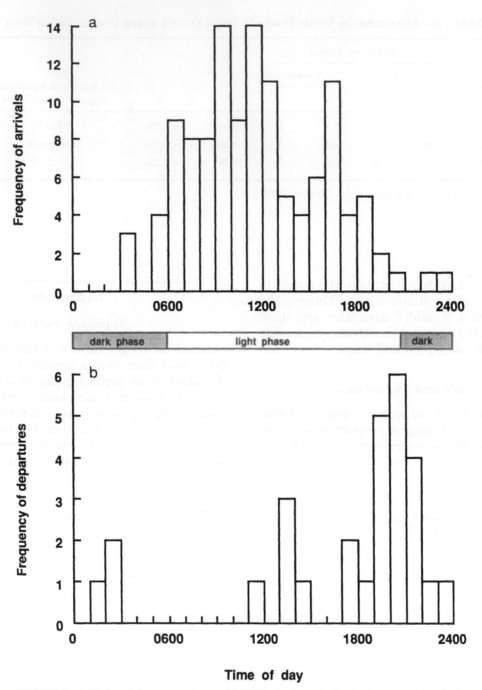

FIGURE 5 Arrival and departure times of individual white sharks during round-the-clock observations under standardized baited conditions; light- and dark-phase transitions are average twilight times for days on which data were collected.

baits." He also described a "surface-charge" involving a rapid rush with the body partially above the surface, and finally, an "inverted approach," in which sharks that had been in the area for extended periods "often rolled over 180 degrees and swam with the ventral side up." Interestingly, our team rarely saw inverted approaches, although the sharks sometimes

inverted immediately prior to or shortly after making contact with floating objects, especially large tethered baits.

As stated above, the majority of approaches observed during the present study were oriented horizontally (Tricas' "underwater approach"). Vertical approaches were, nonetheless, common, with sharks readily engaging in vertically oriented swimming (i.e., 45–90°) during feeding activities, sometimes swimming perpendicular to the surface in direct and rapid pursuit of floating objects. In contrast, depth telemetry from two sharks cruising in open water (i.e., apparently not feeding) revealed a strong tendency to ascend and descend slowly and steadily, at an average rate of about 4 m/min (Strong *et al.*, 1992, Chapter 37).

Interestingly, this propensity for vertical swimming was present in the smallest sharks observed (approximately 220 cm TL), one of which (shark No. 10 in Table I of Chapter 37, by Strong *et al.*) performed this pattern repeatedly, to the general exclusion of horizontal swimming, during a 6.5-hour observation period. Development of this behavior, therefore, appears to precede physical ontogenetic changes believed to be adaptations for feeding on large mammalian prey, including broadening of the teeth as individuals reach 3–4 m TL (Tricas and McCosker, 1984).

Is the pronounced consistency with which white sharks used the vertical approach due more to the limited number of physically possible alternatives or the fact that it constitutes the most efficient approach conceivable under most conditions? Two apparent benefits of using the vertical approach during attempts to capture prey positioned near the surface support the latter—a predator attacking from below is harder for the victim to see as well as to elude. Pinnipeds and cetaceans depend on their own eyesight to avoid predators. Given several predators equidistant from a prey item, as in Fig. 6, the length of an optical pathway entering the prey's eye increases with the depth of the object being viewed. As the pathway length increases, the viewer experiences greater spectral shift and reduction in luminance. Thus, the deepest shark is least visible. In addition, fleeing (i.e., rapid movement away from an approaching predator) is probably the most common escape tactic used by animals (Edmunds, 1974). Predators should therefore use approaches that most likely counteract the prey's re-

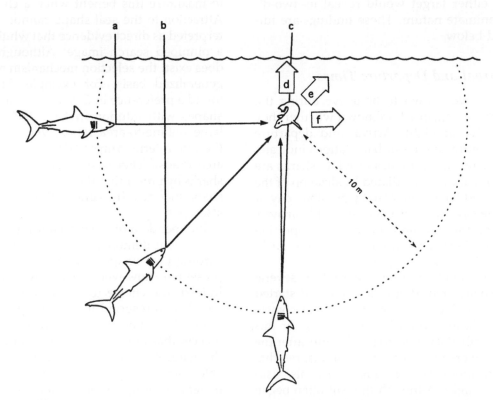

FIGURE 6 Three advantages of attacking surface-borne prey (a sea lion) from below, considering limitations on the prey's vision and possible escape routes. (a–c) Optical pathways—light must travel down to an object and bounce back into the eye of the viewer in order for the object to be seen. In this figure, light attenuation and spectral shift are greatest when viewing the shark attacking from below (c). Conversely, the shark attacking from below probably has the best view of the prey. (d–f) Escape paths—the possibility of fleeing directly away from the shark is ruled out in d, but remains open in f.

sponse (Webb, 1986). For the situation described, extended escape in the direction opposite the vertically approaching shark is impossible (Fig. 6e).

Of course, increased visibility of prey positioned overhead constitutes a third probable benefit. When viewing the surface from below, "the entire 180° dome of land and sky above the water is compressed into . . . a circular window [known as Snell's window] which always subtends a solid angle of 97°; because this angle is constant, the area of surface silhouetted against the sky gets larger as the observer moves deeper" (Lythgoe, 1979). During crepuscular periods when background spacelight is greatly reduced, this overhead window may provide the only adequate light for visual feeding activities (see Fig. 4).

Visual Discrimination Experiment

During their first experiences with the target array, subjects demonstrated a significant preference for the seal shape. This preference decreased gradually during successive passes; it was expected that the sharks would increasingly try the other option, since close inspection of either target would reveal its two-dimensional inanimate nature. These findings are further discussed below.

Arrival and Departure Times

White sharks are known to be active during the daytime, but it has remained unknown whether they are equally active at night. Arrival and departure times of sharks attracted to our bait stations indicate, in fact, that the feeding activities of white sharks are largely diurnal. Olfactory-mediated localization of the bait source should have been equally possible, day or night. Nonetheless, the vast majority of 120 arrivals occurred during the day, and the largest proportion of 28 recorded departures occurred around dusk, even though baiting had continued.

These results could have been biased by several factors. We usually arrived at a new site and started baiting during morning hours, whereupon newcomers often arrived within several hours. Because of this potential bias, only data collected >12 hours after the onset of baiting were used. This criterion insured that the first opportunities for scoring new "arrivals" usually occurred at night. Although the bait often began to attract sharks shortly after it was placed in the water, it is possible that bait dispersal dynamics could bias the rate of arrivals beyond the 12-hour "buffer period"; if, for example, odor corridor robustness reaches an equilibrium at around 24 hours, attractiveness could peak on the following day. While a significant effect of this type seems unlikely, bait dispersal patterns deserve much more attention than they have had to date (bait dispersal is discussed in more detail in Chapter 37, by Strong et al.). Finally, while we did not feed the sharks large quantities of meat relative to what they are known to consume, satiation and other factors probably influenced some of the departures (i.e., some sharks left because their stomachs were getting full, rather than because the sun had gone down). It should be noted that there were exceptions to this diurnal pattern, in that several individuals fed occasionally throughout the night. This appeared to be facilitated by illumination from the vessel's deck lights, however.

Discrimination versus Speculation

Given a choice between the square target and the fusiform seal, white sharks consistently selected the shape that was more common in their environment and, we can assume, more familiar. The utility of this decision is obvious, and the marked consistency with which the seal was chosen is indicative of a tendency to maximize this benefit when a choice is available. Attraction to the seal shape cannot, however, be interpreted as direct evidence that white sharks possess a pinniped search image. Although the possibility does exist, the selection mechanism may have a more generalized basis. For example, Sutherland (1960) found a preference in *Octopus* for "open" shapes (i.e., shapes with interior angles >180°, which result in large outline-to-area ratios) versus "closed" shapes (i.e., no interior angles >180° and smaller outline-to-area ratios). This possibility can be tested in white sharks by presenting the seal along with an inanimate shape that has the same outline-to-area ratio (i.e., is also open).

Independent of the mechanism(s) responsible, this experiment demonstrates an ordering of options regarding prey shape. When only a single object was presented, however, it was invariably investigated. Indeed, the choice to be made in nature is usually whether to respond to a single potential prey item, rather than choosing between two of them. Floating objects that are not acceptable prey are common in the natural environment and may be indistinguishable from edible prey at distances greater than several meters, even up to the point of contact, as indicated by numerous attacks on inedible objects during this study. While it could be argued that inferences drawn from these observations are invalid due to the baited circumstances, it is well known that, under nonbaited conditions, white sharks bite inanimate objects (see, e.g., Klimley et al., 1992; Collier et al., Chapter 19;

Anderson *et al.*, Chapter 20) and animate objects (see, e.g., Ames and Morejohn, 1980; Randall *et al.*, 1988; Klimley *et al.*, 1992, Chapter 16) that apparently are not consumed (i.e., are found beached or adrift and are seldom recorded among gut contents). Along the West Coast of the United States, considerable numbers of sea otters *Enhydra lutris* are found dead from injuries inflicted by white sharks (Ames and Morejohn, 1980; Ames *et al.*, Chapter 28). A similar situation exists in South Africa for jackass penguins *Spheniscus demersus* (Cooper, 1974; Randall *et al.*, 1988). Moreover, attacks on humans rarely proceed beyond the initial strike. The fact that many of these victims were severely injured but not eaten suggests that the shark may have ended the attack after the initial contact.

Further, during two of the ultrasonic trackings reported by Strong *et al.* (Chapter 37), the sharks we were following suddenly surfaced, apparently in response to floating objects. In one instance, a 350 cm TL female shark had been on a relatively straight course for 2 hours, when the animal's audio signal abruptly changed directions; an instant later, approximately 50 m away, the shark broke the surface with a great splash. Two nearby mutton birds *Puffinus* spp. took flight. The shark then rapidly resumed its course. We rode over to inspect the area and found floating a wood plank approximately 125 cm long × 25 cm wide, on or near which the birds had been resting.

Often, at the point of initiating investigatory or attack runs, white sharks were influenced by only a few simple features of the prey candidate. These were visual features that could be perceived at a distance, rather than the variety of other stimulus properties perceivable at closer range or from different positions relative to the prey. Thus, while there is a preference when a choice is available, the key visual stimulus triggering an investigatory approach appears to be relatively simple.

Ewert (1984) and Endler (1986) stated that recognition of an object as prey must precede all other predatory acts. While this idea has limitations, it raises an important question: How sure can a white shark afford to be regarding the identity of an item it has detected before proceeding with a predation event? I suggest that, in many cases, the answer is "not very sure." The capture success of many predaceous fishes depends on speed and the element of surprise (Nyberg, 1971; O'Brien, 1979; Potts, 1980). In roughly a dozen observations by our team and associates (R. Taylor, V. Taylor, and R. Fox, personal communication), white sharks were easily avoided and even harassed by fur seals *Arctocephalus forsteri* and sea lions *Neophoca cinerea*. The fact that these pinnipeds did not

flee upon sighting white sharks indicates that once the predator is detected, the chance of capture drops considerably. Further, the majority of marine mammal species upon which white sharks prey are gregarious; provided that there exists some mechanism for alarming conspecifics, early detection of predators increases directly with increasing numbers of prey and correspondingly higher numbers of peripheral sense organs (Bowen, 1930).

Based on the apparent need for surprise (i.e., achieving a minimum strike distance and speed during the approach), the tendency to strike nonprey objects, and the general consistency with which the vertical approach was executed from deep depths, it appears that white sharks may use a "speculative hunting strategy" that bears an interesting resemblance to one used by octopuses. In describing the predatory tactics of *Octopus cyanea*, Yarnall (1969) reported that while octopuses occasionally sighted and rushed at crabs that were clearly visible, the species' most common hunting method involved an action pattern referred to as the "speculative pounce"; every meter or so during a hunting excursion, the octopus would pounce on and envelop a likely prey shelter site (e.g., coral rocks, algae clumps, and small areas of the substrate) by spreading its interbrachial web over the area, then using the tips of its tentacles to probe beneath the web for crabs and other prey.

In both situations, simple visual cues indicate the possibility of food, and a speculative approach is initiated. During the approach, however, the white shark continues to collect information regarding the object's edibility and may adjust its response accordingly, whereas the octopus sees only the shelter site (see Fig. 7). Although experience likely indicates only a moderate chance of success, both animals follow through to the point of contact, unless, in the case of the white shark, it becomes clear beforehand that the object is not food. Thus, the key feature in Fig. 7 is the interval between the decision points numbered 1 and 2. To maximize the probability of taking prey by complete surprise, this period must be minimized; the opportunity to evaluate a potential prey item is often brief and, in many cases, inadequate to allow identification.

An obvious alternative to the above strategy exists, in which the animal could spend an extended period (i.e., many seconds to many minutes) identifying the potential prey it has detected before making an approach. This would likely entail swimming appropriate patterns to keep the object near the outer boundary of its search space while collecting visual and, less probably, auditory or olfactory cues. If such a careful assessment is made, it could include details such as

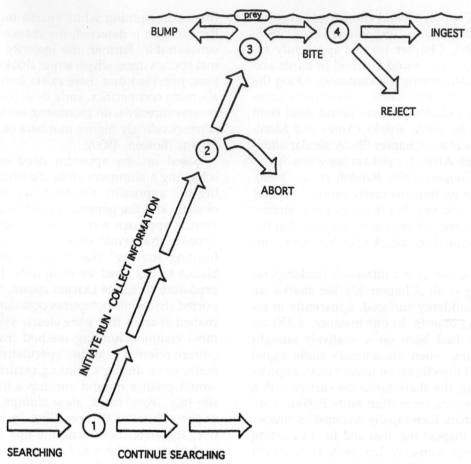

FIGURE 7 Hypothetical points in the decision-making sequence during speculative hunting in white sharks: (1) initiate investigatory/attack run (the predator is aware of potential prey)?; (2) follow through (having approached close, the predator must follow through with attack or abort)?; (3) bite or bump (the predator, approaching the point of contact, must deliver a bite or simply bump the prey)?; and (4) consume (the predator ingests part or all of the prey or spits)?

the prey's orientation or direction of movement, information that the shark might use to adjust its attack. Obtaining this information is, nonetheless, a trade-off with the risk of being detected.

The best feeding strategy is that which maximizes the net rate of energy uptake (McCleery, 1978). While truly indiscriminate foraging would be energetically inefficient, the likelihood of an object's being noticed and attacked should increase as the effective stimulus approaches that of the most strongly reinforced objects within the sharks' realm of experience, particularly if other portentous stimuli have primed the animal (i.e., if stimulus summation occurs). For example, based on movements observed during several ultrasonic trackings, Strong et al (1992) suggested that white sharks cruise temperate waters adjacent to land masses and locate pinniped-dense areas by means

of olfaction, since haul-out sites are rich in odorous stimuli. Such "naturally baited areas" may give rise to a high degree of shape generalization, especially since the costs of attacking nonprey items seem relatively minor for white sharks; because there is little overt risk to the predator, these appear limited to energy expenditures and alarming other potential prey in the area.

Synthesizing the arguments above yields the following tentative model for the initial stages of predation: certain feeding areas may be localized day or night via olfaction, but feeding appears diurnal. Individual prey detection depends heavily on vision (see below). Due to the need for surprise, however, very little time is invested in identification. Approaches on floating prey are often initiated from considerable depths (>5 m), apparently because vertical attackers are harder to detect and elude. An initial strike is often delivered, after which nonprey items are rejected.

Vision in White Sharks

Several lines of evidence indicate that white sharks have the good vision needed to facilitate the strategy suggested above. The depths ($\geq$17 m) from which white sharks oriented toward relatively small surface-borne objects ($\geq$15 cm in diameter) are indeed surprising. Is vision relatively more important to white sharks than to many other shark species, and if so, what selection pressures might account for such specialization? The following paragraphs contain speculations intended solely to stimulate further discovery and should be interpreted cautiously.

The white shark's eyes are larger relative to body size than those of most other large sharks, and the extrinsic ocular muscles are extremely well developed, rotating the heavily constructed eyeballs so as to maintain a constant visual field during maneuvers (Gilbert, 1963a). For the present study, there is little doubt that the sharks initially located the bait via olfaction, but vision was clearly used to orient their actual approaches; fixation of the eyes on surface-borne objects was observed by us on the basis of the contrast between the white sclera and the darker tissues of the eye (i.e., iris, pupil, and surrounding skin). As the eye rolled, the sclera became visible at the base of the eyeball, opposite the pupil, on an axis lined up with the prey. Alignment of the eyeballs toward the stimulus often immediately preceded "orienting responses" (cf. Pavlov, 1927; Lynn, 1966), marked by concomitant changes in the direction of travel and increasing speed.

Arrival and departure times of sharks attracted to our bait stations indicate that feeding is diurnal. Based on characteristics of the white shark's retinal neurons and the spatial organization of the photoreceptors, Gruber and Cohen (1985) predicted such an emphasis on diurnal vision. More specifically, they predicted acute photopic color vision and suggested that the retina is partitioned into areas specialized for nocturnal and diurnal vision, but stressed the latter based on the presence of centralized concentrations of retinal cone receptors. In contrast, organization of the lemon sharks' *Negaprion brevirostris* retina appears better adapted for night vision (Gruber and Cohen, 1985), and correspondingly, lemon sharks feed round-the-clock (Cortes and Gruber, 1990).

Thirty years ago, when sharks were widely thought of as "swimming noses," Perry Gilbert suggested that the importance of vision was generally underestimated. He and others observed that, for many fast active species, attacks on live prey are mainly visually oriented, at least at very close range (see, e.g., Eibl-Eibesfeldt and Hass, 1959; Hobson, 1963; Gilbert, 1963a). Since then, much has been learned about the elasmobranch visual system. Its relative importance to most shark species remains, however, poorly understood. Logically, the extent to which vision is used for predation is linked to the ecomorphological niche that a given species inhabits. Although he does not directly address this issue, Compagno (1990a) places each extant shark species into one of 18 ecomorphs based on body type and prey and habitat affinities. Marked differences between these ecomorphs' lifestyles indirectly indicate differences in the means by which they are carried out.

By virtue of its great size and particular construction, the white shark may experience some unique selection pressures and, at the same time, be free of others. For many fishes, predator–prey interactions have apparently had strong influences on the species' visual physiology (Helfman, 1988). It seems reasonable that the requirements for capturing agile surface-dwelling prey such as pinnipeds and cetaceans differ from those associated with feeding on other marine organisms and that these requirements may influence relevant sensory modalities and behavior patterns. I suggest that vision appears to have been optimized in white sharks and plays a key role in active predation. This hypothesis can be further tested, using advanced histological techniques to better compare the white shark's eyes to those of other species (R. Hueter, personal communication). The means to assess this idea in the field also are developing rapidly. For example, autonomous video systems (e.g., "crittercams") are capable of documenting ever-longer segments of the natural lives of secretive animals such as white sharks.

Relevance to Surfers and Divers

The shape-preference data presented herein support the suggestion by Tricas and McCosker (1984) that surfers lying prone on their boards resemble pinnipeds and may be predisposed to attacks. On the other hand, the species' tendency to launch speculative investigatory attacks on a wide variety of objects that do not resemble pinnipeds suggests the need for a broader awareness. It would be a disservice to surfers and divers to promote the notion that details of appearance (i.e., resembling pinnipeds in form or color) are even remotely as important to their safety as is their behavior based on knowledge of local pinniped and shark distribution patterns and behaviors.

Summary

Predatory responses of bait-attracted white sharks *C. carcharias* were studied during five expeditions to Spencer Gulf, South Australia. During shape discrimination trials, naive sharks showed a significant preference for a seal-shaped target presented simultaneously with a square target. Visually oriented vertical approaches were initiated from depths ≥17 m. Further, arrival and departure data indicate that feeding activities are strongly diurnal. Vision plays a major role in commonly used feeding patterns and appears very well developed, perhaps due to unique selection pressures resulting from the white shark's status as a megacarnivore. Tactics such as a vertical approach pattern appear to be adapted for exploiting a challenging suite of surface-dwelling prey species and may be the basis of a speculative hunting strategy wherein individuals sacrifice much of the possibility of identifying a potential prey item in exchange for an increased chance of capture.

Acknowledgments

I am grateful to the members and employees of The Cousteau Society for financial and logistical support and am especially indebted to J.-M. Cousteau and the dedicated crew of *Alcyone*: C. Jouet-Pastre, N. Durassoff, M. Deloire, C. Davis, P. Martin, S. Arrington, P. Allioux, M. Blessington, D. Brown, J. Cramer, B. Gicquel, P. Pique, L. Prezlin, M. Richards, A. Rousset, T. Stern, C. Van Alphen, M. Westgate, C. Wilcox, and Y. Zlotnicka. Superb logistical and technical support was provided by J. Brody, Y. Goldman, L. Sullivan, and the rest of the Cousteau team. R. C. Murphy, D. R. Nelson, and B. Bruce provided essential support for every phase of this project. Thanks go to R. Fletcher, N. "Chief" Lammer, and R. Foster for helping construct the target array and to S. Arrington and B. Talbot for photos (Figs. 2 and 3, respectively). I am indebted also to the South Australian Department of Fisheries for administrative support and for making *Ngerin* and her excellent crew available for our fifth expedition: N. Chigwiddin, K. Branden, B. Davies, C. Fooks, D. Kerr, D. Short, and N. Wigan as well as G. Dubas and I. Gordon. A. Ebeling, R. Fox, R. Taylor, and V. Taylor contributed their unique talents and much inspiration to this effort. Sincere thanks go to the National Parks and Wildlife Service of South Australia for permission to conduct research in park waters and to the following for equipment loans and donations: New South Wales Fisheries Research Institute (tags), the California Department of Fish and Game (ultrasonic receiver), Ocean Technology Systems (underwater communications), J.B.L. Enterprises (tagging spears), and Pelican Products (waterproof cases). Finally, I thank D. G. Ainley, A. P. Klimley, and M. Larson for their patience and careful review of the manuscript.

22

Tail Slap and Breach: Agonistic Displays among White Sharks?

A. PETER KLIMLEY
Bodega Marine Laboratory
University of California, Davis
Bodega Bay, California

PETER PYLE
Point Reyes Bird Observatory
International Research
Stinson Beach, California

SCOT D. ANDERSON
Inverness, California

Introduction

Intraspecific competition occurs when two or more individuals of the same species simultaneously demand use of a limited resource (Wilson, 1975). Access to this resource is commonly established through agonistic behavior that rarely takes the form of direct fighting. Competitors instead resort to displays—conspicuous and exaggerated motor patterns that demonstrate the displaying individual's ill ease due to the presence of another and its capacity to inflict harm should the competitor remain. The signaler consequently gains an advantage if the recipient heeds the message and withdraws (Burghardt, 1970).

If two white sharks *Carcharodon carcharias* attempt to feed on the same prey, it would be disadvantageous for one to discourage the other from further feeding by biting it and inflicting a wound, because the situation is reciprocal–the individual inflicting the bite this time may receive it next. Such an injury might reduce either shark's future ability to catch prey. For this reason, displays evolve among animals such as sharks. We report here the performance, context, and social consequence of two exaggerated behavior patterns that we hypothesize to be agonistic displays in the white shark. Accordingly, we referred to these patterns as tail slap (TS) and breach. In that quantified behavioral observations are uncommon in

the field study of fish, we closely detail our methods as an aid to other interested researchers.

Materials and Methods

We observed white sharks feeding on northern elephant seals *Mirounga angustirostris* at the South Farallon Islands, central California. Behavior during these feeds was recorded from Lighthouse Hill (elevation, 102 m), using a video camera (RCA CPR350) equipped with a 8:1 zoom lens and a 5.5X tele-extender (Toking VC55) (for further details, see Klimley *et al.*, 1992). We determined distance from the camera to the predatory shark using a theodolite (Wild T-1).

Records of feeding bouts were converted into diagrammatic representations. For example, on the left of Fig. 1 is a 30-minute time scale with 1-minute increments and 10-minute divisions. The scale on the right refers to that 10-minute interval enclosed within the stippled box, and has 10-second increments and 1-minute divisions. Shark and seal behaviors were recorded as either events or states (Lehner, 1979). Each event is indicated by a line leading to a letter; the longer states are denoted by the horizontal bars above or below the letters (for procedural details, see Chapter 16, by Klimley *et al.*).

We determined the magnitude of four components

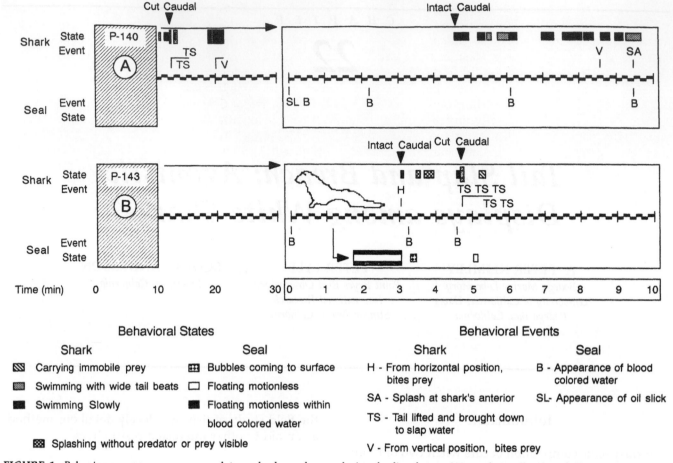

Behavioral States

Shark

- ▨ Carrying immobile prey
- ▨ Swimming with wide tail beats
- ■ Swimming Slowly
- ▨ Splashing without predator or prey visible

Seal

- ▦ Bubbles coming to surface
- ▢ Floating motionless
- ■ Floating motionless within blood colored water

Behavioral Events

Shark

- H - From horizontal position, bites prey
- SA - Splash at shark's anterior
- TS - Tail lifted and brought down to slap water
- V - From vertical position, bites prey

Seal

- B - Appearance of blood colored water
- SL- Appearance of oil slick

FIGURE 1 Behavior sequences among predatory sharks and prey during feeding bouts: (A) predation P-140 and (B) predation P-143. Behavior is described as either continuous states (bars) or discrete events (lines with letters); behavioral states of the shark Intact Caudal (IC) and its behavioral events are designated by narrow bars and upper letters above the time scale, and those of Cut Caudal (CC) are indicated by wide bars and lower letters above the time scale. In predation P-143, the initial wound is indicated by an arrow leading to the time when the wound was first seen.

of TS: (1) length of body lifted out of the water, (2) duration that the body was held above water, (3) distance that water was splashed when the shark's tail contacted the sea surface, and (4) duration that splashed water remained in air. The amount of the shark's body held from the water and the extent of the splash were determined using image analysis software. We measured on a video monitor, in pixels, the separation distance between the first dorsal and caudal fins on each shark. This value and the distance to the shark, the latter obtained from a theodolite reading, were entered into a polynomial equation that converted the measurements to centimeters. The equation was based on measurements, made on videos displayed on the monitor, of black and white bands of known dimensions marked on a pole held perpendicular to the axis of the camera at 200-m intervals to 2.6 km from the island. Fin separation was then used as a reference to measurements of body elevation and

splash extent, as well as to the shark's total length (TL) based on a 1:1.8 ratio (Bigelow and Schroeder, 1948). A more detailed description of this method is given in Chapter 33 (by Klimley and Anderson). The duration of each tail elevation and splash was determined by counting the number of frames in the video recordings.

To index display strength, we used the methodology called "fuzzy logic" to make decisions based on various inputs (McNeill and Freiberger, 1993). The word *fuzzy* refers to an alternative approach to classifying statements or information; rather than two simple states such as true or false, yes or no, or 1 or 0, variables have a scale between 0 and 1. An everyday example of how this logic is applied to problems is as follows. A "fuzzy" washing machine automatically determines the best wash-and-rinse cycle on the basis of information from several inputs. First, the weight (or amount) of clothing is measured by a sensor attached to the receptacle containing the clothes. Based on the range of weights for a number of laundry loads, the sensor is calibrated so that the heaviest weight gives a value of 1; the lightest, 0; and an average, 0.5. Second, the amount of dirt in the clothes is detected by passing through the water a beam of light from a photocell, assuming that light transmission is inversely proportional to the amount of dirt dissolved in the water. The sensor is then calibrated so that the minimum light transmission is 1 and maximum is 0. The machine chooses a particular rinse cycle based on the sum of the values for these sensors.

The four components of TS were treated in a similar manner. For example, we measured the length of tail lifted out of the water for 63 of 83 events. Lengths varied from a minimum of 64.5 to a maximum of 184.3 cm. A nondimensional index of signal strength (SS) was calculated for each tail elevation, ranging from 0 to 1.000, with the following equation:

$$SS = (value - minimum)/(maximum - minimum)$$

An individual shark, identified as Cut Dorsal (CD), elevated 170.2 and 181.6 cm of its body out of the water during the first and second TSs, respectively, on September 30, 1989 (Table I). The index values for these elevations were 0.874 and 0.969, which were summed to give a cumulative index of 1.843 (see the value in the row to the right of *P-104*, under the column labeled "Torso Extension" in Table I). The opponent of CD raised 139.2 cm of its body from the water and had an index value of only 0.618, considerably lower than that of CD. Similar index values were calculated for each of the two sharks from measurements of the duration of tail elevation, splash distance, and splash duration. These values were 0.516, 1.865, and 1.631 for CD and 0.637, 0.214, 0.369 for its opponent (see the values in the next three columns in Table I). Each of the four components of the display were added to give a cumulative index of behavioral vigor for each of the two competitors. In predation P-104, the two TSs of CD had a cumulative SS of 5.855, much greater than the 1.838 cumulative value for the single TS of its opponent. The difference in SS (4.017), normalized again with reference to the minimum and maximum values in that column (see the equation above), was 0.380 in favor of CD.

We also developed an index of "feeding success," because it was not always possible to ascertain which shark fed on the seal after an agonistic encounter. In addition to bites to the prey, two other indicators of feeding success were recorded: (1) the amount of time spent at the surface near the prey and (2) the frequency of visits to the surface. We calculated nondimensional values for the degree of success for each com-

petitor (Table II). For example, CD had index values of 0.148, 0.091, and 0, based on swimming 0.5 minute at the surface during a single visit to the surface, at which time it was not observed to bite the prey (see the values outside and inside the parentheses in the first row of Table II). Each of the three index values for the opponent of CD were 0, because the shark never surfaced again after the bout. The difference between the cumulative index of 0.239 for CD and 0 for its opponent was 0.239, and this value, normalized with regard to the maximum and minimum of the cumulative indices of success in that column, was 0.091.

Results

Description of Tail Slap

We observed 83 tail elevations during 16 of 129 feeding bouts filmed from 1988 to 1991 (Tables I and II). The prey was a northern elephant seal *M. angustirostris* in 13 bouts and was unidentified in the remainder. The behavior pattern consisted of the shark's raising its caudal fin from the water and then rapidly bringing it down. In the usual sequence, the shark rotated onto its side, bent its body at the midsection, and lifted the posterior section of its torso (Fig. 2). The percentage of the body elevated out of the water ranged from 16% to 50% of the shark's TL (median, 0.25; N = 63). The amount of body elevated ranged from 65.4 to 184.3 cm (median, 106.0 cm). The tail was held above water for periods ranging from 466 to 1565 msec (median, 811 msec). The amount of water displaced when a shark's tail slapped the water was considerable. The splash was usually directed laterally toward a second shark. The splashes traveled distances ranging from 132 to 898 cm (median, 400 cm; N = 80). The displaced water remained in the air for periods ranging from 466 to 1998 msec (median, 916 msec).

The intensity of TS depended on the extent of variation among four elements: (1) length of body out of the water, (2) duration of body elevation, (3) splash distance, and (4) splash duration (Fig. 3). Two similar behavioral states were also observed: exaggerated swimming (SWX) and slow swimming (SWS). SWX consisted of the shark's moving its tail back and forth over a wide arc, often displacing considerable water with each beat. The wide tail beats propelled the shark forward at the surface with prey in its mouth. The tail beats resembled those during a less frequently observed behavioral state, carrying, which occurred when a shark held a pinniped in the mouth. SWS consisted of the shark's moving its tail back and forth

TABLE I "Fuzzy Logic" Scores of Signal Strength for Tail Slap Displays

No.	Identification	Date	Time (hours)	Predation (N)	Prey	Name	Total length (cm)	Tail slap (N)	Torso extension	Extension duration	Splash distance	Splash duration	Σ[a]	$\Sigma_A - \Sigma_B$[b]	Normalized[c]
									Indices of signal strength						
1[d]	P-104	September 30, 1989	0632	2		Cut Dorsal	380.6	2 / 1	1.843 / 0.618	0.516 / 0.637	1.865 / 0.214	1.631 / 0.369	5.855 / 1.838	4.017	0.380
2[d]	P-115	October 10, 1989	1530	2	M	Cut Caudal	366.6	1 / 0	0.267	0.303	0.540	0.392	1.502 / 0	1.502	0.126
3	P-136	November 16, 1989	1428	1			453.0	2	1.736	1.152	0.933	0.761	4.582		
4	P-138	November 18, 1989	0931	2, 2	M		432.5	4	0.364 / 0.212	1.242 / 0.114	1.164 / 0.326	1.152	3.922 / 0.652	3.270	0.305
5	P-140	November 21, 1989	1358	2	M	Cut Caudal	368.3 / 368.6	1 / 1	0.356	0.334	0.677 / 0.436	0.217 / 0.587	1.584 / 1.023	0.561	0.030
6	P-143	November 26, 1989	0859	2	M		369.3 / 382.8	3 / 2	1.948 / 1.000	0.515 / 0.364	1.026 / 1.250	0.544 / 0.457	4.033 / 3.071	0.962	0.071
7[d]	P-151	September 18, 1990	1404	2	M		430.2	3 / 1	2.121 / 0.546	1.001 / 0.303	1.583 / 0.333	0.848 / 0.131	5.553 / 1.313	4.240	0.403
8	P-156	September 27, 1990	1030	3	M	Pointed Dorsal	424.0 / 394.4	3 / 2 / 0	0.906 / 1.227	0.728 / 0.697	1.311 / 0.710	1.152 / 0.558	4.097 / 3.192 / 0	0.905 / 3.192	0.065 / 0.297
9	P-162	October 8, 1990	0932	2	M	Cut Dorsal / Cut Caudal	420.5 / 414.7	3 / 2	0.402 / 0.425	0.515 / 0.455	0.682 / 0.372	0.543 / 0.630	2.142 / 1.882	0.260	0.000
10[d]	P-191	November 7, 1990	1146	2	M	Cut Caudal	406.4 / 433.3	14 / 6	4.200 / 1.506	3.942 / 2.093	4.655 / 1.665	4.176 / 1.566	16.973 / 6.830	10.143	1.000
11[d]	P-192	November 7, 1990	1229	2	M	Notched Lobe	444.6	3 / 0	1.224	0.818	1.086	0.805	3.933 / 0	3.933	0.272
12[d]	P-193	November 7, 1990	1448	2	M	Notched Lobe	446.8	2 / 1	0.671 / 0.187	1.091 / 0.212	1.056 / 0.323	1.240 / 0.457	4.058 / 1.179	2.879	0.265
13	P-197	November 12, 1990	0844	2	M		476.7	5 / 1	0.729	1.849	1.849 / 0.329	1.544 / 0.196	5.971 / 0.525	5.446	0.525
14	P-200	November 19, 1990	1244	2	M		409.5	2 / 0	0.344	0.031	0.351	0.392	1.118 / 0	1.118	0.870
15	P-213	October 15, 1991	0720	2	M		345.9	3 / 1	0.821	1.243	0.946 / 0.078	0.588 / 0.131	3.598 / 0.209	3.389	0.317
16	P-221	October 24, 1991	0905	2	M		390.9 / 351.8	8 / 4	1.279 / 0.156	1.033 / 0.091	1.548 / 0.821	1.392 / 0.674	5.252 / 1.742	3.510	0.329

M, Northern elephant seal Mirounga angustirostris.

[a] The values for the four indices of signal strength were summed to give a cumulative strength.

[b] The cumulative signal strength of the tail slaps by the second-appearing shark (B) was subtracted from the signal strength of tail slaps by the first-appearing shark (A).

[c] The signal difference is normalized to a scale of 0 to 1, based on the minimum and maximum values.

[d] Measurements could not be performed for all tail slaps.

TABLE II "Fuzzy Logic" Scores of Combat Success during Tail Slap Displays

No.	Name	Total length (cm)	Tail slap (N)	Surface time		Surface visits		Bites to prey		Σ^b	$\Sigma_A - \Sigma_B{}^c$	Normalized[d]
1[e]	Cut Dorsal	380.6	2	0.148	(0.50)	0.091	(1)	0	(0)	0.239	0.239	0.091
			1	0	(0)	0	(0)	0	(0)	0		
2[e]			0	0.293	(0.99)	0.273	(3)	0.400	(2)	0.966	0.966	0.548
	Cut Caudal	366.6	0									
3		453.0	2									
4		432.5	4	0.645	(2.18)	0.273	(3)	0.400	(2)	1.318	1.318	0.769
			2									
5	Cut Caudal	368.6	1	1.000	(3.38)	0.182	(2)	0.200	(1)	1.382	1.382	0.809
		368.6	1									
6		369.3	3	0.050	(0.17)	0.091	(1)	0	(0)	0.141	0.141	0.029
		382.8	2									
7[e]		430.2	3	0.512	(1.73)	0.273	(3)	0	(0)	0.785	0.785	0.434
			1			0	(0)	0	(0)	0	0	
8	Pointed Dorsal	424.0	3	0.438	(1.48)	1.000	(11)	1.000	(5)	2.438	1.685	1.000
		394.4	2	0.080	(0.27)	0.273	(3)	0.400	(2)	0.753	0.094	0.000
			0	0.095	(0.32)	0.364	(4)	0.200	(1)	0.658		
9	Cut Dorsal	420.5	3	0.444	(1.50)	0.364	(4)	0.600	(3)	1.407	1.407	0.825
	Cut Caudal	414.7	2									
10[e]		406.4	14	0.231	(0.78)	0.364	(4)	0.400	(2)	0.994	0.809	0.449
	Cut Caudal	433.3	6	0.095	(0.32)	0.091	(1)	0	(0)	0.186		
11[e]	Notched Lobe	444.6	3	0.429	(1.45)	0.273	(3)	0.400	(2)	1.102	1.102	0.633
			0									
12[e]	Notched Lobe	446.8	2	0.166	(0.56)	0.455	(5)	0.200	(1)	0.820	0.820	0.456
			1	0	(0)	0	(0)	0	(0)	0		
13		476.7	5									
			1									
14		409.5	2	0.050	(0.17)	0.091	(1)	0.200	(1)	0.341	0.341	0.155
			0	0	(0)	0	(0)	0	(0)	0		
15		345.9	3	0.050	(0.17)	0.091	(1)	0	(0)	0.141	0.141	0.029
			1			0	(0)	0	(0)	0		
16		390.9	8	0.189	(0.64)	0.273	(3)	0.400	(2)	0.862	0.747	0.411
		351.8	4	0.024	(0.08)	0.091	(1)	0	(0)	0.115		

[a] The value for each index of winning within parentheses was normalized to a scale of 0 and 1, based on the minimum and maximum values; within parentheses are surface time in (seconds), visits to the surface (N), and bites to prey (N), summed for all tail slaps.

[b] Values of the three indices of combat success, summed to give overall success.

[c] The cumulative combat success for the second-appearing shark (B) is subtracted from the combat success of the first-appearing shark (A), to obtain the signal strength difference.

[d] Overall success is also normalized, based on minimum to maximum values.

[e] Measurements could not be performed for all tail slaps.

over a narrow arc, displacing little water with each beat. A large difference existed between the four elements of TS and those of SWX and SWS in (1) the medians (see differing heights of the horizontal lines in notches of bars in Fig. 3) and (2) the distributions of measurements [Kruskal–Wallis test with multiple comparisons: N = 94, 99, 112, 112, $p < 0.01$; and N = 106, 111, 123, 123, $p < 0.001$, respectively (see Conover, 1971)].

Behavioral Sequences

We describe here three predatory attacks having TS bouts that increase in complexity.

FIGURE 2 Motor components of tail slap. (A) After rotating onto its side, the white shark is lifting its tail from the water; (B) the shark's tail and posterior torso are poised to slap the water. These photographs depict only the form of the behavioral pattern but not social context, as both sharks performed the behavior after seizing inanimate objects tethered to shore at the South Farallon Islands (photographs by S. Anderson).

Predation P-140 (Fig. 1A)

On November 21, 1989, observers noticed a small bloodstain in an oil slick at 1358 hours, 300 m northwest of Maintop Reef (*P-140* on the map in Fig. 1 of Klimley *et al.*, 1992). Two minutes later, a second bloodstain was observed a small distance away. A 368.3 cm TL white shark, referred to as Intact Caudal (IC), surfaced 2.3 minutes later and swam slowly within the slick. IC then swam vigorously with the prey in its mouth, sweeping its tail back and forth over a wide arc, until it stopped abruptly. Blood appeared again at the surface as IC took another bite of the prey. The shark later seized the prey from below and lifted it from the water. Twelve minutes into the feeding bout, Cut Caudal (CC), a 368.3 cm TL shark (with the tip of its caudal fin missing), appeared,

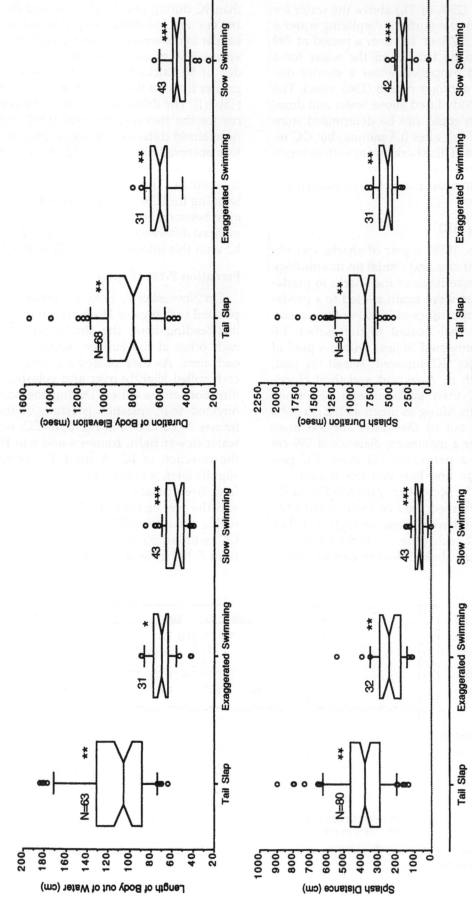

FIGURE 3 Comparison of four potential components of tail slap with two behavioral states, exaggerated swimming and slow swimming, on the basis of minimum, 10th percentile, 25th percentile, median, 75th percentile, 90th percentile, and maximum measurements for each component. Significance levels: *, $p < 0.05$; **, $p < 0.01$; ***, $p < 0.001$.

lifted its tail 107 cm (29% of TL) above the water for 833 msec, and struck the surface, displacing water a distance of 651 cm (177% of TL) over a period of 799 msec. IC then lifted its tail out of the water for a shorter interval and propelled water a shorter distance (466 cm) over a longer period (1365 msec). The exact amount of its body lifted above water and duration of the elevation could not be determined from the video record. IC left after 0.3 minute, but CC remained for 3 minutes. This shark swam with exaggerated tail beats at times—evidence that it carried the prey carcass in its mouth. Later, CC consumed the remainder of the prey.

Predation P-143 (Fig. 1B)

On November 26, 1989, a pair of sharks was observed feeding. Their size and caudal fin morphology duplicated the characteristics of the sharks in predation P-140. Observers were again alerted to a predatory attack by the appearance of blood at the sea surface. An immature seal floated to the surface 1.6 minutes later and remained lifeless within a pool of blood for 1.3 minutes. IC surfaced, seized the seal, and submerged with it. A second and third blood-stain appeared. IC then surfaced and performed three TSs, at one time lifting as much as 182 cm (49% of TL) of its torso out of the water for 566 msec and displacing water a maximum distance of 599 cm (162% of TL) over a period of 833 msec. CC performed two TSs, the first between the second and third TSs of IC and the second after the third TS of IC. The two sharks performed the five TSs in rapid alternating succession and were close enough that they may have contacted each other with their tails. CC was able to lift its tail slightly higher and for longer

than IC during one TS (184 cm and 866 msec versus 182 cm and 566 msec), to produce a splash that was similar in distance and duration (571 cm and 966 msec versus 599 cm and 833 msec), but the cumulative index of SS calculated from the three TSs of IC was greater than for the two TSs of CC: 4.033 versus 3.071 (Table I). The difference between the cumulative indices for the two displays was 0.962, equivalent to a normalized difference in signal intensity of 0.071. IC was observed once afterward at the surface, carrying the carcass of the seal in its mouth for a period of 0.2 minute. The value for the cumulative index of winning for IC was 0.141 versus 0 for CC. The difference between both indices was 0.141, equal to a normalized difference of winning of 0.029. In conclusion, IC won this interaction as well as the prey.

Predation P-191 (Fig. 4)

On November 7, 1990, we observed a more complicated exchange of TSs between IC and CC. During this feeding bout, the sharks repeatedly swam by each other at the surface, approaching within 1 m each time. As they passed a second and third time, each rolled slightly over and splashed water in the direction of the other. During the second pass, IC directed four splashes (starting in the upper left, frames 2–3 and 5–6 in Fig. 5) at CC, which splashed water (lower right, frames 4 and 6 in Fig. 5) twice in the direction of IC. A third TS, performed by CC slightly later, is not shown in the diagram. During the first three passes, IC positioned itself between CC and the seal carcass. IC directed 14 splashes at CC, with a cumulative SS of 16.971; IC directed only six at CC, with an SS of only 6.830 (see the values across from *P-191* in Table I). The difference in SSs, 10.143,

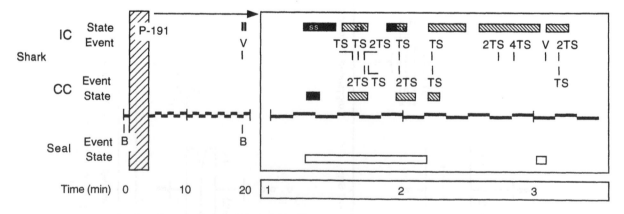

FIGURE 4 Behavior of the Intact Caudal (IC) and Cut Caudal (CC) sharks and seal prey during predation P-191 (see the key to abbreviations in Fig. 1).

FIGURE 5 Depiction of tail slaps performed by (frame 3) the Intact Caudal (IC) and (frame 4) Cut Caudal (CC) sharks during the second passing movement of P-191 (see Fig. 4). Reprinted from Klimley, A. P. (1994) The predatory behavior of the white shark. *American Scientist* **82,** 122–133, with permission.

equaled a normalized value of 1.000. After the bout, IC surfaced four times, for a total of 0.8 minute, and bit the seal carcass twice, for a cumulative index of winning of 0.994. CC appeared only once at the surface, for a period of 0.3 minute, for a value of 0.186. The difference between the two indices was 0.809, normalized to a signal difference of 0.449 in favor of IC. Again, the shark with the higher index of SS had the greater index of winning (and won the carcass).

The individual with the higher SS won the prey in all 14 of 16 predatory attacks for which we could make measurements of SS and winning (Tables I and II). There were many zero values for the cumulative index of winning of eventual losers despite varying positive values for eventual winners (see the lower value for each predation, third column from the right, in Table II). Finally, the winner of the combat was not always the larger shark, but at times was a smaller shark that displayed more vigorously, for example, predations P-140, P-143, and P-191.

A reason other than communication for lifting the tail out of the water might be to redistribute weight in order to push the highly buoyant item held in the mouth underwater more efficiently. To evaluate this hypothesis, we recorded the number of sequences in attack diagrams in which the shark (1) did or did not seize the prey, (2) did or did not perform TS, and (3) swam at the surface or submerged (Fig. 6). In the diagrams of predatory attacks, prey seizure was indicated by vertical lines above the time axis leading to feeding events, abbreviated as H, V, and L (leap into air with prey in mouth). A shark was considered to remain at the surface after performing TS (i.e., "Surface Swimming" in Fig. 6) when the vertical line indicating TS was followed by a horizontal bar denoting SWS or SWX (see the behavioral pattern key in Fig. 1). Alternatively, a shark was considered to dive while performing TS (i.e., "Submergence" in Fig. 6) if the horizontal bar indicating swimming ceased at the vertical line denoting TS. The frequent occurrence of the latter pairing would imply that the sharks used TS to push the prey underwater. We recorded sequences of these three behavioral combinations, distinguishing between them when a single or more than one shark was observed at the surface. We looked for similar

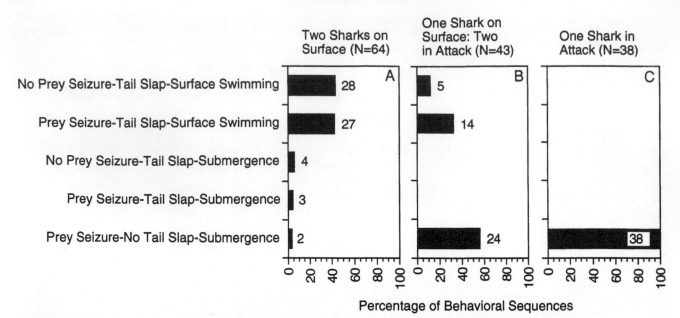

FIGURE 6 The percentage of each of five behavioral sequences in (A) predatory attacks that included two sharks observed on the surface, (B) one shark seen on the surface but two sharks involved in the attack, and (C) one shark in attack. The sample size of each sequence is shown to the right of each bar; the total number of sequences for A–C is indicated by N.

sequences of these combinations in the same number of attacks in which only one shark was observed. These "control" predatory attacks were chosen using a table of random numbers (Zar, 1974: Table D. 45).

TS was recorded only during attacks in which two sharks were identified, never during the control attacks of a single shark (Fig. 6). During the former, TS was observed much more often when both sharks swam at the surface at the same time (N = 62) than when they surfaced at different times during the feeding bout (N = 19) (Fig. 6A and B). This 300% increase in the frequency of TS was in striking contrast to the rarity in which two sharks were simultaneously seen at the surface (i.e., in only 5% of the total duration were sharks seen at the surface). During these infrequent periods, a relatively equal number of TSs were preceded by prey seizure and no prey seizure (32 versus 30) (Fig. 6A). Regardless of whether or not they held prey in their mouths, the sharks did not lift their tails out of the water to submerge the prey, because the sharks usually remained at the surface while performing TS (55 versus 7). Even the 19 TSs performed by a single shark were not associated with the shark's submerging (Fig. 6B). In fact, the sharks did not need to lift their tails out of the water to propel the prey underwater. In 38 instances, a single shark seized the prey, failed to perform TS, but submerged with the prey in its mouth (Fig. 6C). The same sequence of behavioral combinations was ob-

served 26 times during attacks with two sharks (Fig. 6A and B).

Finally, we wondered whether the elements of the display were more exaggerated when sharks held prey in the mouth. Holding the prey in the mouth might be an integral part of the display, enabling the shark to raise its tail higher out of the water and splash water farther. Little difference existed, however, between the medians for the distributions of measurements of length raised from the water (both in centimeters and in percentage of TL), duration of this body elevation, distance the water was propelled (both in centimeters and in percentage of TL), and duration of the splash when prey was or was not in the mouth (see the lines between notches in bars in Fig. 7). No significant difference was detected between the two conditions for the six display elements (Mann–Whitney U test: N = 63, 80, 63, 80, 68, 37; $p > 0.05$). Finally, the absence of any difference in the degree of exaggeration when the shark had or did not have the prey in its mouth further indicates that the behavioral pattern was not used by the shark to propel buoyant prey under the water.

Other Displays

On three occasions, we observed that the shark rolled onto its side and directed exaggerated tail beats in one direction. In one feeding bout, this immediately preceded two TSs. We refer to this behavior as

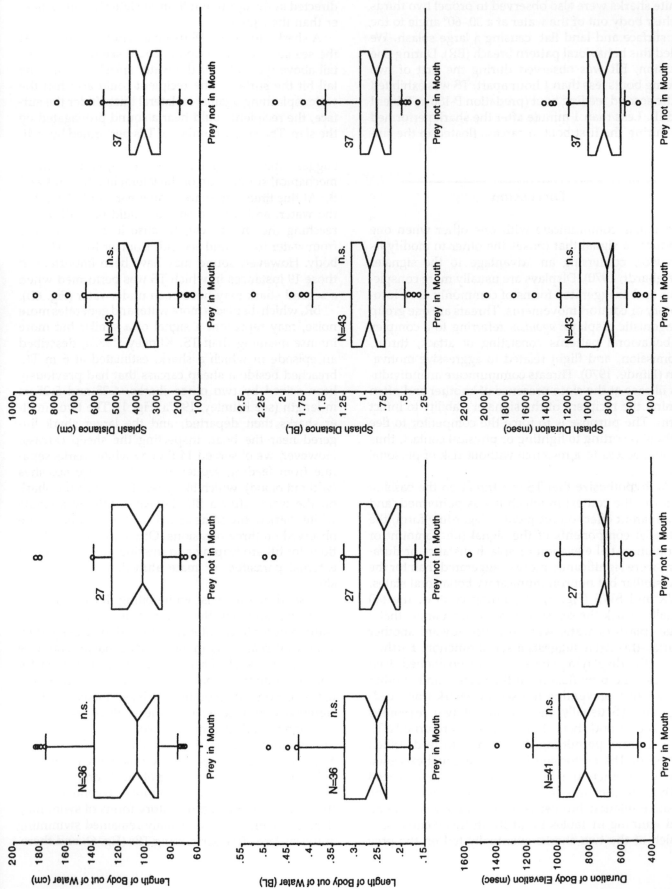

FIGURE 7 Comparison of the measurements of four components of the tail slap when sharks did or did not carry prey in their mouths, on the basis of minimum, 10th percentile, 25th percentile, median, 75th percentile, 90th percentile, and maximum measurements. n.s., Not significant.

tilting, similar to the lateral display in bony fishes. White sharks were also observed to propel two thirds of their body out of the water at a 30–60° angle to the sea surface and land flat, causing a large splash. We called this behavioral pattern breach (BR). During one occasion, BR was observed during the first of two feeding bouts less than 1 hour apart. TS was exhibited in the second feeding bout (predation P-136 in Tables I and II). Less than 1 minute after the shark performed BR during the first bout, a carcass floated to the surface and was quickly consumed by the shark.

Discussion

Animals communicate with one other when one produces a signal that causes the other to modify its behavior, conferring an advantage to the signaler (Burghardt, 1970). Displays are usually more conspicuous and exaggerated forms of commonly used locomotory or comfort movements. Threats are one group of agonistic displays: *agonism* referring to a complex of behavioral patterns consisting of attack, threat, submission, and flight related to aggressive motivation (Hinde, 1970). Threats communicate an individual's ill ease at the close presence of another and often remind the recipient of the signaler's ability to inflict harm. The purpose is to force the competitor to flee without resorting to fighting or physical contact, thus gaining access to a resource without risk of personal injury.

We hypothesize that TS is a threat on the basis of its form, the context in which it was performed, and its apparent effect on recipients. Regarding form, the potential components of the signal (i.e., amount or duration of tail elevation or splash distance or duration) were significantly more exaggerated than for the two similar but noncommunicatory behavioral states, SWX and SWS (Fig. 2). Regarding context, the tail usually struck the sea surface at such an angle that a large splash of water was propelled toward another shark—this itself suggests a social function. Furthermore, the displaying shark at times positioned itself between the prey floating at the surface and another shark, thus preventing the second shark from feeding. The TS usually elicited one of two responses from the second shark. Either the shark returned a TS or withdrew, permitting the signaler to feed. When the SS of TSs performed by an individual exceeded that of another, the former was permitted to feed without further interference from the latter (see the positive relationship between the index values of SS and winning in Tables I and II). In this comparison, which of the two sharks eventually fed on the prey depended more on the frequency and vigor of TSs directed at the opponent than the latter's being smaller than the signaler.

A shark can detect TS using a variety of senses. At the sea surface, the recipient could see the signaler's tail above the water or the water splashed when the tail hit the surface. The recipient could also feel the water splashing against its lateral line. Under the surface, the recipient could hear a sound propagated by the slap. The larger number of TSs performed by individuals at the surface than underwater (62 versus 15) suggests that the signal is perceived by either sight or mechanical stimulation of the lateral line (Fig. 6A and B). At this time, the shark's inner ear would be out of the water, and thus, a sound would have difficulty reaching the shark's ear, because it must first pass from water to air, and second, from air to the shark's body. However, sound may have been important in those 19 instances in which TS was performed when a second shark may have been underwater (Fig. 6B).

BR, which displaces more water and generates more noise, may represent a signal of a similar but more intense meaning than TS. Klimley (1987a) described an episode in which a shark, estimated at 6 m TL, breached beside a sheep carcass that had previously been seized by two other sharks, 4.25 and 5.25 m in length (see Klimley, 1987a: Fig. 1). The two smaller sharks then departed, and the larger shark lingered near the boat, inspecting the sheep carcass. However, we observed 14 BRs by white sharks separate from feeding sequences and 20 large splashes (without blood), which likely resulted when the shark hit the water after a BR. Repeated BRs by a single white shark in the same area 5–20 minutes apart were observed on three occasions. Other possible explanations for BR are leaping and missing prey, removal of external parasites, or mate attraction during courtship.

A shark might twist while biting, lift its tail from the water, and slap the water to propel itself downward in order to counter the upward force exerted by a buoyant seal in its mouth. This behavior could be mistaken for a TS. Displays are easily mistaken for noncommunicatory behavioral patterns because natural selection only modifies the latter, increasing their stereotypy and conspicuousness, thus maximizing the signals' effectiveness in communication (for a discussion of ritualization, see Eibl-Eibesfeldt, 1970). Mistaken identity was not the case with TS, for the following reasons. First, the potential components of the signal were significantly more exaggerated for TS than for two noncommunicatory forms of swimming (Fig. 2). Second, sharks usually remained swimming at the surface after performing TSs (Fig. 6), and therefore did not perform the behavior simply to submerge the prey. Third, no significant difference existed between the distributions of measurements of the elements of the signal when the shark transported the prey versus when the prey was not held in its mouth (Fig. 7).

We are alone in observing white sharks perform TS in the context of natural circumstances, but others have observed TS displayed by sharks attracted by bait to the side of the boat (Table III). In these cases, either the cage protecting human divers or the boat may be considered an opponent by the signaler, instead of another shark. With regard to a shark cage,

Klimley observed at Dangerous Reef, South Australia, that "on two occasions, the shark cage drifted between the shark and baits tied to lines off the stern of the boat. . . . At both times, a shark elevated its tail out of the water and brought it down forcibly to strike the top of the cage." With regard to a boat, Goldman noted the following:

When we arrived, no blood, no sharks, just gulls . . . not for long . . . First one shark, then another . . . bit boat engine twice within 5–10 seconds by two different sharks . . . then boat was bumped on port side and tail slap shot water (two times) into boat (really across it) . . . got soaked good. . . . Shortly thereafter (10–15 seconds) shark on starboard side bumped boat, bit engine and tail

TABLE III **Description of Tail Slaps Taken from Field Notes**

No.	Date	Location	Observer	Description	Reference
1	April 9, 1989	Point Piños, California	J. Capella	"Observed a 6.7 m white shark feeding on a harbor seal. The shark left the seal and began circling the boat, finally hitting the bow with its head. Prior to departing the area, while astern of the boat, the shark rolled on to its side and began slapping the swim-step and propeller with its caudal fin. When the boat had moved 20–30 m from the seal, the shark returned for several additional bites of the carcass before departing."	Collier *et al.* (Chapter 19)
2	November 11, 1992	Farallon Islands, California	L. Gilbert	"Big splash. Second shark surfaces near the first. Much thrashing ensues. Observed some conflict . . . with one exposing its whole head with mouth and teeth (as if trying to bite the other?). Slapped tail on surface five times like a whale."	P. Pyle (unpublished observation)
3	November 12, 1992	Dyer Island, South Africa	L. J. V. Compagno	"Chumming and floating baits. White shark identified as ABW tail-slapped and hit the propeller while taking bait, damaging an engine fin."	L. J. V. Compagno and S. Cook (personal communication)
4	January 11, 1992	Dyer Island, South Africa	L. J. V. Compagno	"Chumming and floating baits. A white shark, possibly ABW, tail-slapped in proximity of another white shark while taking bait."	L. J. V. Compagno (personal communication)
5	May 6, 1993	Dangerous Ree, South Australia	A. P. Klimley	"On two occasions, the shark cage drifted between a shark and the baits tied to lines off the stern of the boat. At both times, a shark elevated its tail out of the water and brought it down forcibly to slap the top of the cage."	A. P. Klimley (unpublished observation)
6	October 15, 1993	Farallon Islands, California	S. D. Anderson and K. J. Goldman	"When we arrived, no blood, no sharks, just gills. Not for long. First one shark, then another. Bit boat engine twice within 5–10 seconds. Then boat was bumped on port side and tail slap shot water (two times) into boat (really across it). Got soaked good. Shortly thereafter (10–15 seconds) shark on starboard side bumped boat, bit engine and tail slapped multiple times. Got completely soaked. It was incredible. Then another shark appeared. These sharks were 'hot' so we headed off."	K. J. Goldman (personal communication)

slapped multiple times . . . got completely soaked . . . it was incredible. . . . Then another shark appeared. . . . These sharks were "hot" so we headed off.

Anderson's comment to Goldman at the time was "You still don't believe in the tail slap?" (S. D. Anderson, personal communication). An alternative to departure by the observers in such a situation would be to perform an experiment. The recipients of the signal could simulate TSs by contacting the water surface with an oar, thus propelling water in the direction of the displaying shark. One would predict that if the two experimenters had performed these artificial TSs more often and vigorously than the white sharks performed their TSs, the latter would have left the vicinity of the seal carcass with the attached transmitter. However, the experiment required a more substantial boat than that available to Anderson and Goldman.

Aggressive displays have been observed in other species of elasmobranchs. The gray reef shark *Carcharhinus amblyrhynchos* performs an aggressive display when confronted by the rapid and direct approach of an intruder (a human diver), particularly if the shark's departure is obstructed by reef topography (Johnson and Nelson, 1973). The display consists of two locomotory elements—laterally exaggerated swimming and spiral looping—and four postural components—elevation of the snout, depression of the pectoral fins, arching of the back, and lateral bending of the body. When a diver continued to pursue the displaying shark, the threat was followed by an attempt by the shark to deliver a slashing bite (Nelson *et al.*, 1986). The authors believed the display was carried out to discourage predation by other, larger sharks.

Scalloped hammerhead sharks *Sphyrna lewini* perform a display, corkscrew swim, while inside a school, apparently to force smaller individuals away (Klimley, 1985). This behavior consists of a rapid burst of swimming into a tight looping trajectory, with the shark rotating its torso almost 360° on its longitudinal axis. The performance of this behavior, sometimes leading to hit (when a hammerhead brings the underside of its jaw into contact with a neighboring shark's anterior torso), induces smaller sharks to withdraw and perform two withdrawal behaviors, head-shake and acceleration. In the former, the hammerhead swings its head back and forth over a wide arc several times, in rapid succession, while abruptly propelling itself forward with rapid tail beats. The latter behavior consists of a rapid burst of swimming with rapid shaking of the head over a small arc. Corkscrew swim was interpreted to be a signal that conveys a shark's intent to actually contact the recipient (hit). The con-

text for these behaviors is competition for a limited resource, the large females being located near the center of the school.

Both TS and corkscrew swim appear to be agonistic displays "requesting" a neighboring shark to withdraw from an area. The hit performed by hammerheads often left light contusions anterior to the first dorsal fin on recipients. White sharks may hit each other with their caudal fins during TS. One of us (P.P.) observed such contact during a predatory attack in 1992, and IC and CC may have alternately contacted each other with TSs during predation P-143. A large light contusion, which could have resulted from such blows, appeared on the caudal peduncle of an individual observed later (see Fig. 2 in Chapter 33, by Klimley and Anderson). Pratt *et al.* (1982) observed that sharks feeding on a fin whale *Balenoptera physalus* possessed either fresh or healed tooth slashes on their flanks between the gills and the caudal peduncle. The authors suggested that these injuries were inflicted by other white sharks, based on (1) the rarity of more than two sharks feeding on the whale at one time, despite the fact that as many as nine sharks visited the whale to feed; (2) the sudden directional change and disappearance without further feeding of a scarred individual; and (3) the presence of lacerations on males and immature females, unlike the mating cuts reported for other shark species (Stevens, 1974; Pratt, 1979; Klimley, 1980).

In the use of surface-generated signals, the white shark is behaviorally convergent with unrelated species that also spend time at the sea surface. In form, TS resembles lobtailing by whales, in which the tail flukes are raised out of the water and slammed against the sea surface (Leatherwood *et al.*, 1982; Silber, 1986). Lobtailing is performed by males competing for social dominance in order to obtain access to females (Silber, 1986). American alligators *Alligator mississippiensis* perform a head-slapping display, in which the jaws are suddenly closed as the undersurface of the head is slapped against the surface of the water (Vliet, 1989). This display, performed primarily by males, is a declaration of presence, eliciting tail slapping by other alligators.

Summary

We have described and interpreted behavioral patterns observed to be common among white sharks *C. carcharias* when more than one attempts to feed on a seal floating at the sea surface. One behavior, TS, consists of lifting the caudal fin and splashing water in the direction of the accompanying shark. We have described behavioral sequences that include TSs during three feeding bouts involving pairs of sharks. The height and duration of tail elevation and the extent and duration of splashing exceeded those of two noncommunicative behavioral states, SWS and SWX, the latter used when sharks swim forward with prey in the mouth. A shark was permitted to feed further on a seal only if the vigor and frequency of its TS were greater than those of its opponent. Thus, we hypothesize that TS is an agonistic display, functioning to ward off potential competitors. We have also commented on another behavior, BR, which is used much less commonly, but which may be a higher-intensity agonistic display.

Acknowledgments

The U.S. Fish and Wildlife Service and Farallon Patrol provided encouragement and logistical support during our stay at the South Farallon Islands. M. Elliot of the U.S. Geological Survey kindly lent us two theodolites for use during the study. W. Clark permitted us to use his image analysis system to photograph the TS from video frames. L. Huff let us use her drawing published in *American Scientist* of the bout of TSs between CC and IC in P-191. D. G. Ainley, G. Barlow, G. M. Cailliet, D. Nelson, and E. Uhlinger read the manuscript and suggested many improvements. This is contribution 594 of the Point Reyes Bird Observatory.

23

Behavior of the White Shark: An Emerging Picture

GEORGE W. BARLOW[1]

Department of Integrative Biology
and Museum of Vertebrate Zoology
University of California
Berkeley, California

Commentary

Why Study the White Shark?

When starting fresh on a research project on animal behavior, investigators commonly follow one of two paths. The biologist may be enticed to march off toward an especially attractive species; *attractive* may mean that the species is shrouded in mystery, is beautiful or powerful, or whatever. The other fork in the path may be chosen because of an intriguing issue, apart from any particular species; in this case, the investigator seeks the species best suited for studying that issue. Either path may lead to productive research, but the attractive-species approach is by far the riskier journey. In either event, however, the prudent biologist should have a short list of requirements the species in question should meet, although, in reality, few species will satisfy all of them.

The species should be small and easily managed in captivity. Thus, it should go through most or all of its life history under controlled conditions; a corollary is that the species should be inexpensive, both to obtain and to keep. Further, it should be accessible in the

field to direct observation and individual tagging, and it should be abundant. Having a long season of availability or breeding is also desirable. Locally occurring species take priority over remote ones, because stocks can be easily replenished and fieldwork does not depend so heavily on external funding. In short, one should undertake projects having a high probability of succeeding, which return a high yield per unit of effort. Avoid low-yield studies.

Seldom considered, too, is the advantage of having a solid literature on the species, providing background knowledge on systematics, morphology, physiology, genetics, ecology, and behavior. Given all these points, an ideal species for many issues, and one found near Bodega Bay, is the three-spine stickleback *Gasterosteus aculeatus*. But when the allure of the species lies in the very fact of our ignorance about it, then the absence of a supporting literature becomes part of the mystery attracting the investigator. Such seems to be the case with the white shark.

The observant reader will have noted that white sharks *Carcharodon carcharias*, other than occurring close to shore, fail to meet any of the criteria for a good research species, so much so that one would have to question the sanity of anyone who undertakes research on them. The obstacles to effective inquiries into their behavior are so great that they seem insurmountable. Fortunately, no one told our speakers that.

[1]Rapporteur Barlow's comments summarize results presented during talks on behavior and predatory strategies given in scientific paper and plenary sessions during the Symposium on the Biology of the White Shark, held March 4–7, 1993, in Bodega Bay, California.

In their ignorance, the speakers have managed to produce a number of pieces of the jigsaw puzzle that portrays the white shark's behavior. This symposium has revealed a great deal about several aspects of its behavior, such as seasonal movements, shifts in feeding behavior between pinnipeds and cetaceans, competitive interactions around a food resource, the mode of attacking prey, and an impressive teasing out of the environmental variables modulating interactions between predator and prey.

Communication among White Sharks

Several papers and comments during discussions touched on communication among white sharks and their cognitive abilities in relation to observed behavior. The white shark presents an intriguing, nearly unique problem in communication, one shared by most, perhaps all, sharks.

The equipment sharks have for emitting signals is limited. They are constrained by their physiology and morphology. Insofar as is known, they cannot make sounds beyond the usual hydrodynamic noises, although they hear well. Unlike teleost fishes, they cannot change color quickly, so acute chromatic signals are ruled out. The fins of sharks, especially the median ones, are nearly immobile, so they are unable to raise and lower them to change their profile. Nor can they erect gill covers or lower the throat to transform the shape of the head, as can teleosts. The vertebral column is relatively rigid, limiting gross body postures (although the gray reef shark has a striking display consisting of a raised head, arched back, and lowered pectoral fins accompanied by a ritualized wobbly mode of swimming) (Johnson and Nelson, 1973). Thus, the white shark is like a mute person running around in a straitjacket. Cetaceans find themselves in a similar situation, but they seem to have compensated through the acoustic channel.

This lack of intricate displays, at least as far as is known, may contribute to the impression that the white shark is simple minded. This view was, to some degree, abetted by the report by Demski and Northcutt (Chapter 12) that the brain of the white shark is small and relatively undifferentiated compared with that of advanced sharks. Those aspects of brain morphology may be informative about the *relative* cognitive structure among sharks, but any judgment about *absolute* cognition based on such information would be risky. Bear in mind that honeybees, with only a few hundred ganglia in the central nervous system, are capable of sophisticated communication and learning.

In my experience with cichlid fishes, at one extreme some species have multiple complex displays, whereas others, at the opposite extreme, have few and relatively simple displays. My impression only is that those with complex displays (e.g., *Etroplus* or *Apistogramma*) are the less intelligent, responding almost like windup toys. In contrast, those with few simple displays (e.g., large *Cichlasoma*) seem more intelligent. They use situational context a great deal, and they rely on subtle aspects of spatial arrangements and possibly on past experience.

The white shark appears to use simple displays, about which there is more later, and spatiotemporal cues. This was especially striking in Strong's (Chapter 21) account and films of white sharks interacting competitively around bait. Displays were lacking, and supplanting through approach–avoid interactions was the rule. I was surprised to see no overt demonstration of intent, such as showing the teeth, other than swimming at the competing shark. Other aspects of the white shark's behavior were obliquely touched on during the symposium, raising the possibility that they learn about their prey and have a geographically extensive cognitive map.

Klimley presented film and data on what seems to be an impressive visual–acoustic display in the white shark (Chapter 23, by Klimley *et al.*). When competing for a stunned or dead prey at or near the water surface, one white shark or the other raised its tail out of the water and slammed it down toward the other. The behavior is noteworthy for the following reasons.

First, if this is a threat display, what is the channel(s) of communication? The percussion of tail on water could send both an acoustic and mechanical signal to the opponent, perhaps even touching the opponent; the action may also be seen by the recipient. Second, is the behavior triggered only in the presence of a competitor? If so, is the tail-slapping behavior primary, or does it merely result from the mechanics of trying to keep possession of the prey, that is, does the shark do something underwater that has the secondary consequence of causing its tail to splash the surface at the other shark? Third, if it is indeed a display, from what is it derived?

Assuming for the moment that tail slapping is a ritualized signal, what was its precursor? Klimley suggested that it parallels tail beating in bony fishes. In most kinds of bony fishes, two individuals stand, usually parallel to one another, and one or both beat their tail at the opponent. To carry out this action, the actor must back water with the pectoral fins, to prevent forward locomotion; some bony fishes also allow the head end to swing as widely as the tail end, so that no propulsive wave passes down the body, but that introduces other problems. If the shark engaged in tail beating in the water, it would have difficulty staying put to continue the interactions, because it cannot use its pectoral fins to resist forward displacement. However, if its tail end were out of the water, then it would derive little propulsion from it and would not be thrust forward by the action. Thus, tail beating would be mechanically liberated, so to speak, if the shark freed its tail of the water, which the white shark does.

Not surprisingly, the audience greeted with mixed responses Klimley's suggestion that tail slapping is a ritualized display. That suggestion, however, should now generate explicit predictions to test the hypothesis. Klimley may already have the data to do this.

Before leaving the discussion of communication among white sharks, I would like to return to ritualized displays in the context of aggressive interactions. That white sharks are capable of inflicting severe injury with their teeth escaped no one at the symposium. Also, persuasive evidence was presented that white sharks engage in competitive encounters. One might naively expect, then, to see dominant sharks take injurious bites out of subordinates. They don't. Bite marks are seen on white sharks, to be sure, but not gaping wounds. Compagno reported an action called "grab–release," in which one shark seizes, say, the pectoral fin of another shark in its jaws, then lets go. In such a circumstance, why isn't the fin bitten off? Kin recognition is not a plausible explanation, because all of the other sharks would have to be kin.

Animals with dangerous weapons often avoid using them in social contests. This can be for a number of reasons, such as damaging one's weapon; this might explain why rattlesnakes entwine necks and push and shove, rather than bite. Whatever the reason, white sharks appear to constrain themselves from injurious attack. This possibility requires more attention because of its theoretical importance. Many predators show no such inhibition, as in fights between wolves from different packs. If white sharks so constrain their attacks on conspecifics, why would they do this?

A tangential issue that arose is whether white sharks risk injury from heterospecifics. They seem so large and formidable that they need not fear any other creatures. Yet, surprisingly, according to Long, they don't attack live animals larger than themselves (Chapter 27, by Long and Jones). And many observers have documented that white sharks withdraw from their smaller prey after slashing them. Why not bite into large whales or basking sharks? White sharks, after all, scavenge huge dead whales. Do large prey present a risk to them, beyond the

scratches thought by some to be inflicted occasionally by pinnipeds?

Another point of potential confusion arose around the issue of whether a particular behavior might indicate one functional system as opposed to another. One example is breaching. Does it indicate courtship or aggression? Inherent in this discussion was the assumption that if it was one, it couldn't be the other. However, we should not be surprised if breaching occurs in both contexts. I say this for two reasons. First, in many bony fishes whose social behavior has been analyzed, courtship starts out like the beginning of a fight. Courtship and aggression, consequently, share many of the same displays, especially initially. Second, sharks (and also many bony fishes) are restricted in the motor patterns they can express.

White sharks have limited common final pathways for the expression of behavior. One expects, therefore, that the same modal action patterns will be used in different settings, and that context will determine the information communicated. Incidently, a similar argument can be made for the gray boundary between aggression and predation. Predators use, in aggressive interactions, the motor patterns and structures that evolved primarily in the service of predation.

Future Directions for Research

That we are in a position to debate the meaning of the behavior of white sharks shows how far we have come. But that very debate also reveals the need to go to the next level, from description to experimentation. The need is apparent in claims emanating from descriptive correlational studies. One paper, for example, reported that white sharks may attack surfboards because they are white, presumably resembling the lighter underside of their prey (Chapter 20, by Anderson *et al.*). In marked contrast, at least two papers suggested that scuba divers are attacked because of their black wet suits, again, because of their resemblance to prey (Chapter 39, by McCosker and Lea; Chapter 42, by Burgess and Callahan). Which is correct, and how can one answer the question based on descriptions of attacks? Obviously, one cannot. But some manipulative studies have begun to appear.

In separate reports, Anderson and Strong described preliminary experiments with floating objects that sharks attack (Chapter 18, by Strong; Chapter 20, by Anderson *et al.*). They offered the sharks "prey" that varied in shape and color. The results are intriguing, but they only scratch the surface. We need more such systematic experiments, testing explicit hypotheses with planned statistical design. This is feasible in the field. Another obvious hypothesis to test is the effect of tail slapping. Is it a signal, and if so, what is the effective component of the signal, mechanical, acoustic, or visual? Workers immersed in the biology of the white shark should feel challenged to employ the experimental method to get answers to questions about the behavior of this shark.

Finally, a cautionary note. This derives from listening to aficionados of white shark biology for 2 days. Beware of the temptation to become provincial. The allure of viewing the world as revolving around the white shark is strong, and that intense involvement has merit. Nonetheless, it should be tempered by placing the biology of the white shark in context. The general theoretical framework that drives the study of the behavior of other kinds of animals has just as much relevance to the white shark. Such theory gives direction and meaning to observations of the white shark, and using that theory will quicken the pace of discovery.

Ecology and Distribution

White Shark Predation on Four Pinniped Species in Central California Waters: Geographic and Temporal Patterns Inferred from Wounded Carcasses

DOUGLAS J. LONG
Department of Ichythyology
California Academy of Sciences
Golden Gate Park
San Francisco, California

KRISTA D. HANNI
The Marine Mammal Center
Marin Headlands
Golden Gate National
Recreation Areas
Sausalito, California

PETER PYLE
Point Reyes Bird Observatory
International Research
Stinson Beach, California

JAN ROLETTO
Gulf of the Farallones National
Marine Sanctuary
Fort Mason
San Francisco, California

ROBERT E. JONES
Museum of Vertebrate Zoology
University of California
Berkeley, California

RAYMOND BANDAR
Department of Ornithology
and Mammalogy
California Academy of Sciences
Golden Gate Park
San Francisco, California

Introduction

The white shark *Carcharodon carcharias* is the primary natural predator on pinnipeds in California waters (Ainley *et al.*, 1981, 1985; Le Boeuf *et al.*, 1982). Evidence of white shark predation on pinnipeds comes from examination of stomach contents from dead sharks (Bonham, 1942; Fitch, 1949; Le Boeuf *et al.*, 1982; Scholl, 1983; Stewart and Yochem, 1985), from observation of white shark attacks on live pinnipeds (Ainley *et al.*, 1981; Klimley *et al.*, 1992), and from fresh bite wounds or healed bite scars on the bodies of live and dead pinnipeds (Ainley *et al.*, 1981, 1985; Le Boeuf *et al.*, 1982; Stroud and Roffe, 1979; Tricas and McCosker, 1984). In California, white sharks prey on harbor seals *Phoca vitulina*, northern elephant seals *Mirounga angustirostris*, California sea lions *Zalophus californianus*, and Steller sea lions *Eumetopias jubatus*. They also attack northern fur seals *Callorhinus ursinus* (Bonnot, 1928) and sea otters *Enhydra lutris* (Orr, 1959; Ames and Morejohn, 1980), but seem not to eat them.

The white shark's predatory strategy usually consists of an attack from behind and/or below an animal at or near the ocean's surface (Ainley *et al.*, 1981, 1985; Tricas and McCosker, 1984; Klimley *et al.*, 1992; Fergusson, Chapter 30). The shark usually inflicts a single deep bite, sometimes followed by a waiting period while the prey bleeds to death or lapses into shock (Tricas and McCosker, 1984; but see Chapter 16, by Klimley *et al.*). Prey may survive an initial hit and escape with wounds that are apparent when the live animal hauls out or when it dies and washes ashore (Le Boeuf *et al.*, 1982; Tricas and McCosker, 1984; Ainley *et al.*, 1985). Although white shark-bitten animals have been noted from Oregon to southern California (Stroud and Roffe, 1979; D. J. Long, unpublished data), previous reports of white shark predation on pinnipeds in coastal California waters are limited to particular sites or species (Le Boeuf *et al.*, 1982; Ainley *et al.*, 1985; Klimley *et al.*, 1992).

Our understanding of the movement patterns and population dynamics of white sharks in response to the population dynamics of pinnipeds off California remains incomplete, because white sharks are difficult to follow or study in the wild. Based on observations of predatory attempts and on shark-bitten pinnipeds, Ainley *et al.* (1985) hypothesized that large white sharks gathered around Southeast Farallon Island (SEFI) primarily in the fall, when the abundance of immature northern elephant seals reached its peak, and then they moved to the adjacent coast in spring, coincident with the annual peak there in the abundance of harbor seals. Another influx of immature elephant seals occurs in the spring at SEFI, when large sharks are not observed at the island (Ainley *et al.*, 1985), and are suspected of breeding south of Point Conception, southern California (Klimley, 1985b). In response to the dramatic increase in the northern elephant seal and California sea lion populations along the central California coast (Lowry *et al.*, 1992; Barlow *et al.*, 1993), it is also possible that white shark movement patterns have changed during the past 20 years. Frequency and distribution of shark-injured pinnipeds may be useful in understanding these seasonal, annual, and geographic fluctuations in white shark abundance, prey choice, and predatory attempts.

This chapter summarizes 548 records of four species of pinnipeds having fresh white shark bites; animals were observed either in the wild, recovered live and brought into captivity, or found dead on beaches, over a 23-year period, 1970–1992. We compare the seasonal patterns of bitten animals recovered along the coast with those observed on SEFI, and we examine interannual variation in numbers of recorded animals during the study period. Although some biases exist in our data set, we believe our results can be used to infer annual, seasonal, and geographic trends and variations in shark–pinniped predatory interactions over a long-term period.

Methods

White shark bite wounds on pinnipeds (Fig. 1) are distinguished by large (10- to 50-cm) ovate or crescentic arcs, triangular punctures, and/or jagged tears on the skin and bones of bitten animals (Ames and Morejohn, 1980; Ainley *et al.*, 1981; Le Boeuf *et al.*, 1982; Long and Jones, Chapter 27). Due to variations in shark sizes, attack directions, and prey avoidance behaviors, characteristics of white shark bites on prey may vary greatly (see Chapter 27, by Long and Jones). Pinniped remains have been recorded in the stomachs of two other species of sharks in California waters, the blue shark *Prionace glauca* and the sevengill shark *Notorynchus cepedianus*, but these sharks are more often scavengers on pinniped carcasses than actual predators (Ebert, 1989, 1991; Long and Spencer, 1995; D. J. Long, unpublished data). The dentitions and bite shapes left by these two shark species are very different from those of the white shark (see Chapter 27, by Long and Jones).

The National Marine Mammal Stranding Network (NMMSN), supervised by the National Marine Fisheries Service, collected data from stranded marine mammals along the central California coast, Del Norte through Monterey counties, from January 1, 1970, through December 31, 1992. Live and dead pinnipeds having white shark bites, and found stranded along the coast, were reported year-round to the NMMSN by state and federal personnel, local and county law enforcement officials, officers from humane societies, academic biologists, and the general public (Seagers and Jozwiak, 1991). For dead pinnipeds, field investigators from the Museum of Vertebrate Zoology, University of California, Berkeley (MVZ), and the Department of Ornithology and Mammalogy, California Academy of Sciences (CAS), recorded data and collected voucher specimens. Since 1975, live shark-bitten pinnipeds from the coast were recovered by The Marine Mammal Center (TMMC) and temporarily housed for observation and medical treatment.

A total of 134 records of shark-bitten pinnipeds from the central California coast were recorded by NMMSN members. Information collected included species, age, sex, stranding location, bite location on the body, cause of death (if determined), and estimated elapsed time since attack. For our summary, only animals with recent wounds, estimated to be ≤1 week old on the basis of granulation of wound and state of healing and scar development were included.

Stranding data from marine vertebrates can be used to monitor changes in biological or oceanic events and their impacts on populations of marine vertebrates (Imber, 1971; Rabalais and Rabalais, 1980; Seagers *et al.*, 1986; Stenzel, 1988; DeGange and Vacca, 1989; Woodhouse, 1991). However, because units of effort are difficult to define (Seagers *et al.*, 1986), these data must be evaluated cautiously. Patterns of currents and tides may also affect observations of strandings (Doroff and DeGange, 1992), but by in-

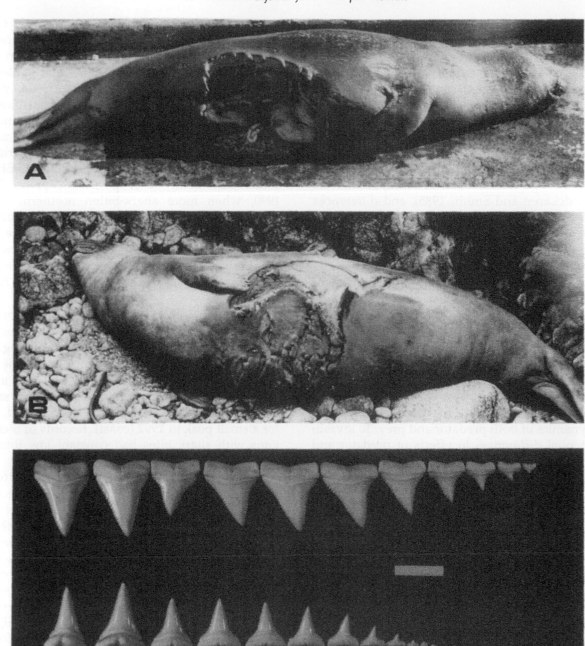

FIGURE 1 Typical bite wounds inflicted by white sharks, showing a large gape, jagged cuts, and differences between the dentitions of the upper and lower jaws. (A) A dead 200-cm juvenile female elephant seal from Southeast Farallon Island with a bite >46 cm wide. (B) A live 251-cm adult female elephant seal from Southeast Farallon Island with a bite >61 cm wide. (C) Upper right dentition of a 296-cm female white shark from Tomales Bay, California (specimen CAS 26781). (Photos by D. J. Long.)

cluding only data from animals with recently inflicted bites, we believe that the interaction with the shark occurred within several kilometers of the stranding location (see Discussion). This would minimize stranding biases due to seasonal and annual variations in current patterns.

At SEFI (37.4° N, 123.0° W), daily observations and weekly censuses of pinnipeds were carried out by volunteers and staff from Point Reyes Bird Observatory (PRBO) (see Huber *et al.*, 1985). All live or dead pinnipeds with shark-related injuries were documented; recorded information usually included the species, age, sex, date, and location of the injury on the body. A total of 414 observations of white shark-bitten pinnipeds were recorded.

For comparative analyses, wounded pinnipeds were placed into three age classes, on the basis of size and sexual maturity. For seals, we classified animals as pups (<1 year old), immatures (1 year to subadult), and adults (sexually mature). For sea lions, we classified animals as immatures (<3 years old, including pups), subadults (3 years to adult), and adults (sexually mature). Age classification of pinnipeds can be difficult (McLaren and Smith, 1985), and differences in source population sizes of each age class may further bias analyses; thus, we interpret our results accordingly. Bite location was categorized into four regions: (1) head and neck (from snout to shoulder), (2) forebody (from shoulder to diaphragm), (3) lower body (from diaphragm to pelvis), and (4) hindquarters (from pelvis to end of rear flippers).

Interannual and seasonal trends in observed shark-bitten pinnipeds were analyzed with linear and polynomial regression (Seber, 1977). Total records by month, for each species and for all pinnipeds, were used as a basis for all analyses. Observed patterns of shark predation likely related to patterns in the population dynamics of both predator and prey. To investigate potential differences in effects of predator and prey behavior, we performed analyses on both the unadjusted numbers of shark-bitten animals, using linear regression, and on these numbers after adjustment for a pinniped population index (PPI). PPI consisted of annual (for annual analysis) or monthly (for seasonal analysis) means of counts of each species, recorded during weekly censuses at SEFI (Huber et al., 1985); these were added as terms to regression analyses. The effect of this procedure was to statistically control, in part, for seasonal and annual variations in pinniped abundance, which would be expected to affect observed numbers of shark-bitten animals. The adjusted analyses should provide a more precise evaluation of patterns in white shark (as opposed to pinniped) dynamics around SEFI and, to a lesser extent, along the coast. To examine nonlinear relationships, we tested for effects of quadratic and cubic terms, as additions to the linear models. Analyses of covariance (ANCOVAs) were used to compare trends (Seber, 1977).

Results

Coastal records of 134 live and dead shark-bitten pinnipeds consisted of 89 California sea lions, 26 harbor seals, 13 elephant seals, and 6 Steller sea lions.

These were distributed from Del Norte County through Monterey County, with a marked concentration of records (N = 73) from southern San Mateo County (Fig. 2). Only four white shark-bitten pinnipeds were recorded from San Francisco Bay.

On SEFI, the total of 414 shark-bitten pinnipeds included 191 California sea lions, 150 elephant seals, 47 Steller sea lions, 17 harbor seals, and 9 unidentified pinnipeds. Throughout the study period, reports of shark-bitten California sea lions at SEFI were greater than for any other species, except for 1981 and 1990, when more shark-bitten northern elephant seals (1981: N = 16; 1990: N = 12) were recorded (Fig. 3). Records of shark-bitten harbor seals were higher on the coast (N = 26) than at SEFI (N = 17), whereas numbers of shark-bitten elephant seals were much higher at SEFI (N = 150) than on the coast (N = 13) (Fig. 4).

Peaks in the number of shark-bitten California sea lions occurred in 1982–1984 (SEFI), 1983 (coast), 1989 (SEFI), 1992 (SEFI), and 1991–1992 (coast) (Fig. 3). At SEFI, there was a small peak in numbers of shark-bitten northern elephant seals in 1981 and a larger peak in 1992. Records of shark-bitten harbor seals remained constant throughout the study period, except for a small peak in 1992 (coast). Records of Steller sea lions with shark bites had constant low totals, but showed a small peak in 1985.

Annual trends in the number of shark-bitten pinnipeds were generally similar between the central California coast and SEFI during the 23-year period (Fig. 3). Few bitten animals were recorded in the early 1970s. Combining totals of shark-bitten animals from the coast and SEFI, the linear increase was significant both before (slope = 2.33 + 0.42, $t = 5.52$, $p < 0.001$) and after (slope = 0.33 + 0.12, $t = 2.75$, $p = 0.013$) adjustment for PPI. The trend was stronger before adjustment, indicating that some, but not all, of the increase in shark-bitten animals resulted from increases in overall pinniped numbers.

Separating the two localities, the unadjusted number of shark-bitten animals increased linearly over time both at SEFI ($t = 5.83$, $p = 0.001$) and on the coast ($t = 3.33$, $p = 0.003$). The increase was significantly greater at SEFI (ANCOVA on year term: $F = 10.44$, df = 46, $p = 0.002$). When looking at bitten animals at SEFI, this increase over time was significant for each species (Table I). For the coast, this increase was significant for all species except Steller sea lions (Table I).

Adjusting for PPI, the increase of shark-bitten pinnipeds at SEFI remained significant ($t = 3.05$, $p = 0.007$), but that of the coast became insignificant ($t = 0.40$, $p = 0.696$). The difference for the coast could

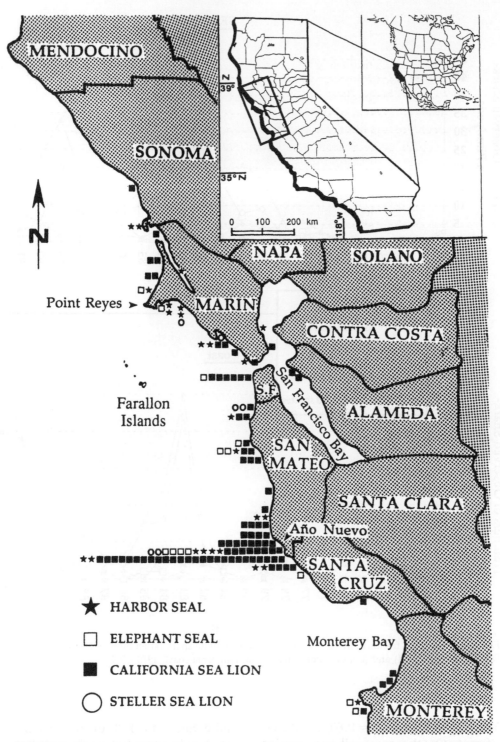

FIGURE 2 The central California coast, showing locations of shark-bitten pinnipeds, 1970–1992, each species indicated by a particular symbol. Attack localities from northern Mendocino, Humboldt, and Del Norte counties to the north, and San Luis Obispo county to the south are not shown.

indicate that annual trends reflected changes in pinniped numbers rather than shark activity, although it is also possible that PPI did not adjust for fluctuation in pinniped numbers on the coast as well as it did at

SEFI. After adjustment for PPI, at SEFI, the increases in shark-bitten elephant seals ($t = 2.12$, $p = 0.047$) and Steller sea lions ($t = 0.40$, $p = 0.696$) remained significant, whereas those of California sea lions ($t = $

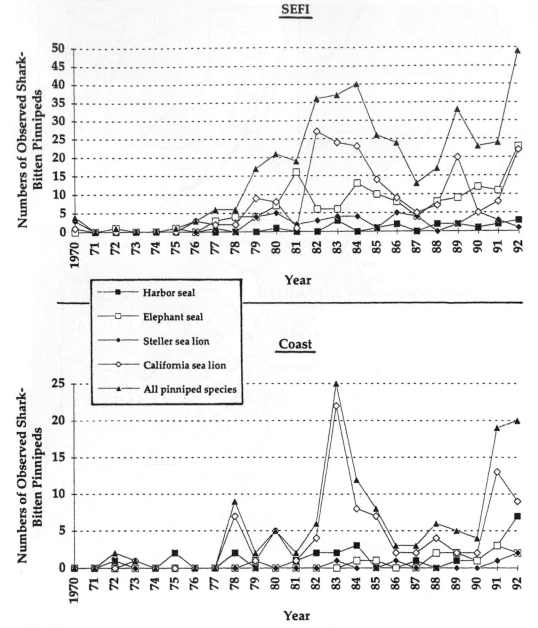

FIGURE 3 Annual trends in the abundance of observed white shark-bitten pinnipeds (top) at Southeast Farallon Island (SEFI) and (bottom) along the central California coast, 1970–1992.

0.99, $p = 0.335$) and harbor seals ($t = 0.60$, $p = 0.555$) became insignificant. On the coast, all four species showed no trend after adjustment for PPI.

Shark-bitten pinnipeds were documented every month of the year, but were recorded more frequently during summer along the coast and during autumn at SEFI (Fig. 4). Using a December–January cutoff (reflecting low numbers at both localities), monthly differences were significant, for the linear and quadratic terms, over the year for SEFI, both before and after

adjustment for PPI; on the coast, they were significant only when using the quadratic monthly term, both before and after index adjustment (Table II). Using ANCOVA, seasonality in shark-bitten animals differed significantly between the two locations, both linearly (month $\times$ location; $F = 5.91$, df $= 23$, $p = 0.025$) and curvilinearly (month^2 $\times$ location; $F = 10.90$, df $= 23$, $p = 0.004$) after adjustment for PPI.

The numbers of shark-bitten pinnipeds, by age class, differed among species (Fig. 5). For phocids,

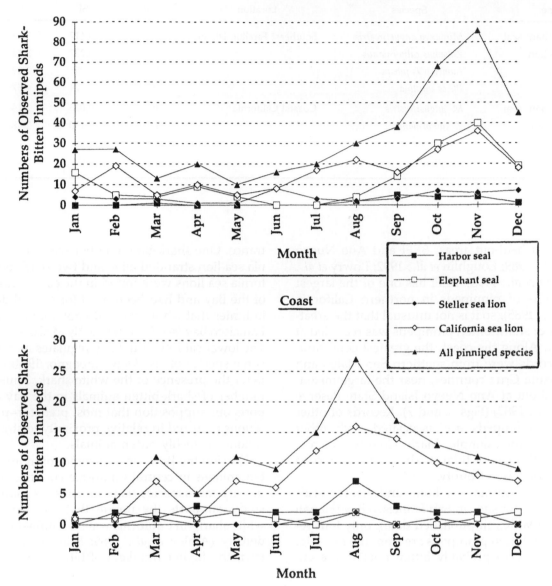

FIGURE 4 Seasonal trends by month of observed white shark-bitten pinnipeds (top) at Southeast Farallon Island (SEFI) and (bottom) along the central California coast.

assuming equal distribution of age classes, pups were observed two times more frequently with wounds than any other age class (Pearson $\chi^2 = 16.05$, $p = 0.007$). For otariids, significantly more immatures than any other age class were observed with bites ($\chi^2 = 8.53$, $p = 0.014$).

The location of shark bites on pinnipeds differed significantly among the four species ($\chi^2 = 46.16$, $p = 0.001$) (Fig. 6). Bites on otariids tended to be in the posterior regions (3 and 4), whereas bites on phocids were distributed evenly across the body. This differ-

ence between otariids and phocids was also significant ($\chi^2 = 27.79$, $p = 0.001$).

Discussion

Geographic Patterns

Pinnipeds having shark bites were reported throughout the study area, but the greatest concentrations of records occurred near the most populated

TABLE I   Regression Analysis of Annual Frequencies of Shark-Bitten Pinnipeds

Common name	Species	Location	β	SE	t	p
Northern elephant seal	*Mirounga angustirostris*	Southeast Farallon Island	0.697	0.117	5.98	0.001
California sea lion	*Zalophus californianus*		0.776	0.232	3.34	0.003
Steller sea lion	*Eumetopias jubatus*		0.126	0.055	2.29	0.030
Harbor seal	*Phoca vitulina*		0.024	0.024	4.57	0.001
Northern elephant seal	*M. angustirostris*	Coastal California	0.048	0.023	2.14	0.044
California sea lion	*Z. californianus*		0.373	0.150	2.48	0.022
Steller sea lion	*E. jubatus*		0.030	0.016	1.84	0.081
Harbor seal	*P. vitulina*		0.109	0.047	2.29	0.032

rookeries and haul-out areas, SEFI and Año Nuevo (Hanan *et al.*, 1989; Loughlin *et al.*, 1992; Lowry *et al.*, 1992; Barlow *et al.*, 1993). SEFI has one of the largest concentrations of pinnipeds in northern California (Huber *et al.*, 1985), so it is not unusual that the greatest number of shark-bitten pinnipeds was recorded at this locality. Along the coast, the greatest concentrations of records were from southern San Mateo and northern Santa Cruz counties, near the large mixed-species haul-out at Año Nuevo Island State Reserve (see Le Boeuf, 1982) (Figs. 2 and 7). Records of other shark-bitten pinnipeds are associated with other haul-out sites, for example, Marin County (Allen *et al.*, 1989), and areas of tidal washes, such as Ocean Beach, San Francisco County.

Although California sea lions and harbor seals both have haul-out sites in San Francisco Bay (Hanan *et al.*, 1989; TMMC, unpublished data), only four records of shark-bitten pinnipeds are from the Bay (Fig. 2), and these were probably bitten outside the en-trance. One shark-bitten harbor seal and one California sea lion stranded alive and two shark-bitten California sea lions were found in the east central section of the Bay and had been dead for several days. This indicates that white sharks do not normally enter San Francisco Bay (see Chapter 39, by McCosker and Lea). The lower salinity of the bay regulates movements of some species of sharks (Compagno, 1984a) and may deter the presence of the white shark. Thus, the low number of shark-bitten animals in the Bay also supports our supposition that most predator–prey interactions occurred in relative proximity to stranding locations of freshly bitten animals.

Outside the Bay, the occurrence of more shark-bitten animals in central California counties near high-density multispecies haul-out sites, as found by this study, supports the notion that the distribution of white sharks in California waters is influenced by prey density (Ainley *et al.*, 1985; Klimley, 1985b). However, the higher numbers of bitten animals at these

TABLE II   Regression Analysis of Monthly Frequencies of Shark-Bitten Pinnipeds

Treatment	Location	Source	β	SE	t	p
Unadjusted	Southeast Farallon Island	Month	4.688	1.459	3.212	0.009
		Month^2	0.945	0.393	2.406	0.040
	Coastal California	Month	0.937	0.502	1.867	0.091
		Month^2	−0.328	0.134	−2.439	0.037
Adjusted	Southeast Farallon Island	Month	4.689	1.539	3.046	0.040
		Month^2	1.002	0.421	2.379	0.045
	Coastal California	Month	0.983	0.529	1.774	0.110
		Month^2	−0.351	0.143	−2.463	0.039

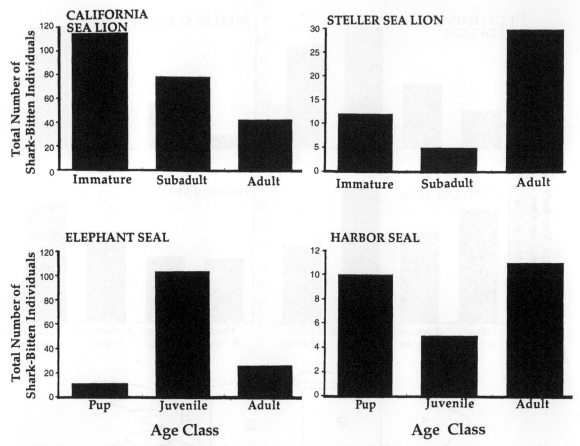

FIGURE 5 Frequency of shark-bitten pinnipeds with respect to species and age class. Sea lions were classified as immature, subadult, and adult; seals were classified as pups, juveniles, and adults.

locations may also reflect the larger population of pinnipeds in the vicinity as well as increased observer awareness at these sites. The fewer records of shark-bitten pinnipeds from Sonoma County northward and from central Monterey County southward, for example, may indicate observer biases, since these coastlines are sparsely populated and/or of difficult access. In these areas, further information is needed to compare with the trends observed along the coast from San Mateo through Marin counties.

Annual Patterns

El Niño/Southern Oscillation (ENSO) is known to affect the marine ecosystem and marine animal distribution and abundance (Hubbs, 1954; Radovich, 1961; Karinen *et al.*, 1985). ENSO occurred in 1976–1977 (mild), 1982–1983 (strong), 1986 (mild), and 1991–1993 (moderate) (Quinn and Neal, 1987; Hayward, 1993). Large influxes of immature California sea lions into central California have been recorded during moderate to strong ENSO (Huber, 1991). The higher numbers of shark-bitten California sea lions recorded during ENSO further support our conclusion that numbers of bitten California sea lions in central California reflect higher prey availability (i.e., more pinnipeds), rather than an increase in shark numbers.

Our results, based on a larger spatial and temporal data set, confirm earlier observations by Le Boeuf *et al.* (1982) and Ainley *et al.* (1985), that the incidence of shark-bitten pinnipeds was very low during the early and mid-1970s, but increased over time (Fig. 2). Our data further indicate a significant increase through 1992 of both shark-bitten animals overall, and of shark-bitten animals at SEFI after adjustment for PPI. Low numbers of shark-bitten pinnipeds before 1978 may be attributable to considerable mortality of juvenile white sharks in gill net fisheries, prevalent before their prohibition in the early 1980s (Compagno, 1984a; see also Chapter 45, by Heneman and Glazer). Now that coastal gill nets are prohibited, increasing records of shark-bitten pinnipeds appear to be correlated with increases in the number of both prey and predators (see Chapter 34, by Pyle *et al.*). The increase in shark-bitten pinnipeds at SEFI, after adjustment for PPI, supports this conclusion.

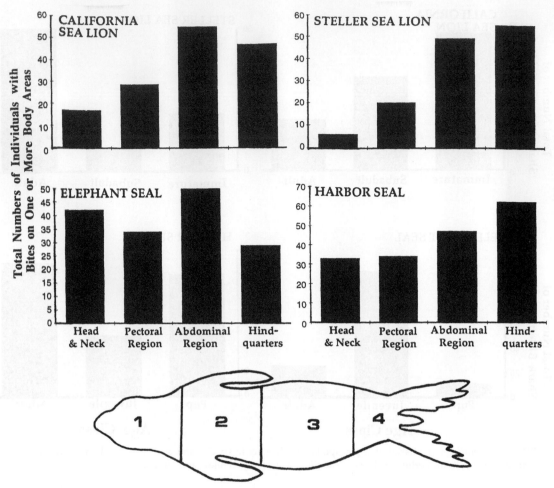

FIGURE 6 Relative locations of white shark bites on pinnipeds: (1) the head and neck, (2) the pectoral region, (3) the abdominal region, and (4) the hindquarters.

The mixed results in trends of bitten animals by species at SEFI, after adjustment for PPI, could be explained by changes in predator and prey populations. Although our results indicate that white sharks in California are polyphagous, results also indicate a degree of specialization, especially among larger white sharks, which can take the larger elephant seals and Steller sea lions. The increase in elephant seals (Huber *et al.*, 1985; Barlow *et al.*, 1993) during the study period may have contributed to an increase in survivorship and fecundity in white sharks (see Chapter 34, by Pyle *et al.*). Finally, the dramatic increase in California sea lions observed through the 23-year period (Huber *et al.*, 1985; Lowry *et al.*, 1992) may have outpaced any increases in white shark numbers that may have occurred, causing the results we observed.

Seasonal Patterns

The seasonal patterns for shark-bitten pinnipeds throughout the year and between the coast and SEFI can be related to both predator and prey dynamics. White sharks live in California waters year-round, with the greatest numbers observed or caught in spring and summer (Squire, 1967; Klimley, 1985b). Our study supports a previously hypothesized (Ainley *et al.*, 1985) correlation between white shark and pinniped population annual cycles off central California. The significantly earlier peak in shark-bitten animals on the coast, followed by the peak at SEFI, may show a general seasonal pattern of sharks relocating from the coast to SEFI (Figs. 4 and 7). This coincides with peaks in prey availability at haul-out sites, particularly California sea lions along the coast in August–

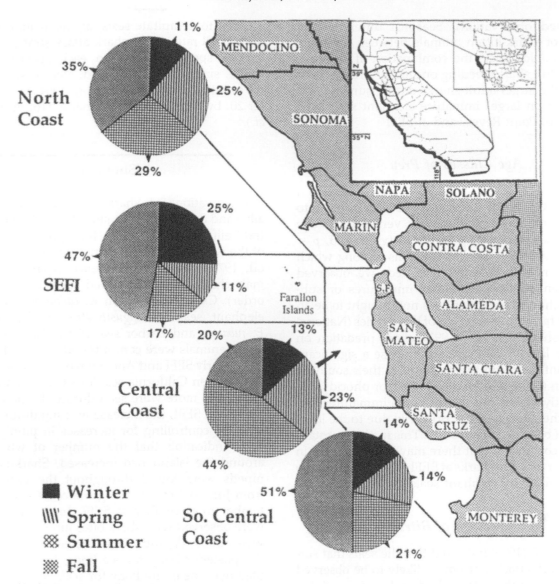

FIGURE 7 Seasonal abundance of recovered shark-bitten pinnipeds along three regions of the California coast and at Southeast Farallon Island (SEFI).

September and northern elephant seals at SEFI in October–November (Ainley *et al.*, 1985; Huber *et al.*, 1985; Lowry *et al.*, 1992; Barlow *et al.*, 1993).

Within the scheme proposed by Klimley (1985b), adult female white sharks may move northward from southern California waters to the central and north coast in the summer, followed by movement out to the Farallon Islands, and then migration back south. However, shark-bitten pinnipeds were recorded throughout the year, especially harbor seals along the coast in spring (see Ainley *et al.*, 1985) and Steller sea lions at SEFI in June (Figs. 4 and 6). We infer that some subadult, adult male, and/or nonbreeding

sharks may live year-round in the waters off central California, feeding mostly on the smaller harbor seals and recently weaned elephant seal pups. The fact that numbers of shark-bitten northern elephant seal and harbor seal pups—which enter coastal waters from March through May—were low during the 23-year period may be explained by the small size of these prey, most of which were probably consumed entirely (see Le Boeuf *et al.*, 1982). Although shark attacks are not usually seen at SEFI in the spring (Ainley *et al.*, 1985), attacks on recently weaned elephant seals by relatively small (estimated 3- to 4-m) sharks were witnessed at SEFI in March 1993 (P. Pyle, unpublished observation). Seasonal differences in numbers of shark-bitten animals between the coast and SEFI could reflect the combination of semiresident subadult sharks feeding on smaller prey and the arrival of the larger adult sharks during summer and fall to prey on larger immature elephant seals near Año Nuevo, Point Reyes, and at SEFI.

Age Classes of Prey

Large influxes of California sea lions were recorded during ENSO. These influxes may have contributed to significant numbers of immature versus other age classes of sea lions being recorded ($\chi^2 = 8.53$, $p = 0.014$). However, for Steller sea lions, among which fewer bitten animals were recorded, we observed more bitten adults than either immatures or subadults. Although predation was not thought to be an important factor in the decline of this species (National Marine Fisheries Service, 1992), shark predation on Steller sea lions in California may pose a significant negative influence on the recovery of their southern population (Long and Hanni, 1993). For phocids, the significantly higher number of bitten immatures versus the other age classes is primarily due to the large number of bitten elephant seals. This further substantiates the conclusion that there may be an increase in the number of white sharks at SEFI coincident with the influx of immature elephant seals to the island.

Location of Bites

Ainley et al. (1985) indicated that the seals that survived a shark attack were more likely to be observed with wounds on the anterior body, whereas surviving sea lions would have bites primarily on the posterior body. In this chapter, although bite wounds for seals were distributed evenly across the body rather than just in the head region, sea lions were bitten significantly more often in the posterior region ($\chi^2 = 27.79$, $p = 0.001$). This could be related to differences in the ability to escape from an attack, as proposed by Ainley et al. (1985). The larger number of lower body bites in sea lions may be due to this species reacting, during a rearbody predatory attempt, using its powerful foreflipper locomotion. Seals, on the other hand, use hindflipper locomotion and are not as able to escape from a posterior attack. This does not explain, however, why more bites are located on the head and neck. One possibility is that white sharks

attempt to decapitate seals, as this would result in a higher kill rate; white shark attack strategies may be species specific, based on each species' different shape, size, swimming speed, alertness, and predator avoidance behavior (Klimley, 1994; see also Chapter 20, by Anderson et al.; Chapter 16, by Klimley et al.).

Summary

Observations of shark-inflicted wounds on 548 alive and dead pinnipeds stranded on beaches in central California were used to infer predatory dynamics of the white shark C. carcharias during a 23-year period, 1970–1992. The prevalence of shark-bitten animals changed with pinniped species (in decreasing order): California sea lion Z. californianus, northern elephant seal M. angustirostris, Steller sea lion E. jubatus, and harbor seal P. vitulina. Most shark-bitten animals were concentrated near haul-out sites, particularly SEFI and Año Nuevo State Reserve. More shark-bitten California sea lions were seen during El Niño, and more were seen during the period of our study. On SEFI, the increase was significant after statistically controlling for increases in pinniped numbers, indicating that the number of white sharks around the island had increased. Shark-bitten pinnipeds were found throughout the year, peaking from July to October along the coast and from September to December at SEFI. Immature pinnipeds were observed with shark bites more frequently than were adults. Wounds were located predominantly on the posterior body for sea lions, but evenly distributed over the entire body for seals. Overall, patterns of temporal and spatial records of shark-bitten pinnipeds could be related to a combination of the geographic distribution of the white shark and the seasonal and annual abundance of pinnipeds in central and northern California waters.

Acknowledgments

This chapter would not have been possible without the assistance of many volunteers and biologists; we especially acknowledge those from NMMSN (C. Keiper, J. Cordaro, and many field volunteers), TMMC (L. Morgan, K. Beckman, and M. Webber), CAS (B. Cutler, I. Szczepaniak, and L. Thomson), MVZ, and PRBO (D. G. Ainley, S. D. Anderson, and R. P. Henderson). The manuscript benefitted from revisions suggested by anonymous reviewers and the editors. This is contribution 589 of PRBO and scientific contribution 118 of TMMC.

25

Tidal Height and White Shark Predation at the Farallon Islands, California

SCOT D. ANDERSON
Inverness, California

A. PETER KLIMLEY
Bodega Marine Laboratory
University of California, Davis
Bodega Bay, California

PETER PYLE *and*
R. PHILIP HENDERSON
Point Reyes Bird Observatory
Stinson Beach, California

Introduction

In California, white sharks *Carcharodon carcharias* prey on both cetaceans and pinnipeds (Ainley *et al.*, 1985; Le Boeuf *et al.*, 1982; Long, 1991a; Long *et al.*, Chapter 24; Long and Jones, Chapter 27). At the South Farallon Islands (SFI), located 30 km off the coast near San Francisco, California, sea lions *Zalophus californianus* and northern elephant seals *Mirounga angustirostris* are preyed upon by white sharks, the northern elephant seal being the preferred prey (Ainley *et al.*, 1981, 1985).

White shark predation at SFI occurs seasonally, September–November (Ainley *et al.*, 1981, 1985; Klimley *et al.*, 1992). Since 1987, systematic watches from the SFI lighthouse have shown that predation occurred most frequently 25–450 m from shore (Klimley *et al.*, 1992), leading these authors to propose that the waters within this perimeter are the "high-risk zone." Predation has been observed during all daylight hours at SFI. Events occurred in bouts and were unevenly distributed, clustered within the same time on the same day or at the same time the following day (Klimley *et al.*, 1992). We hypothesized that tidal height explains the temporal clustering of predation, and we present here an analysis to test this hypothesis.

Methods

Predation watches were conducted during all daylight hours at the lighthouse (elevation, 102 m) from atop Southeast Farallon Island, September 1–November 30, 1988–1992 (Klimley *et al.*, 1992). Observers scanned the water around the islands every 3 minutes during all daylight hours, except when fog or high winds affected observation (see Chapter 26, by Pyle *et al.*). Predatory attacks seen from other parts of the island during days of poor weather conditions were not included in the analysis. We observed 190 attacks, most of which were indicated by the following: (1) a large red bloodstain on the ocean surface and (2) a white shark feeding on a pinniped at the surface, often accompanied by gulls *Larus* hovering over a slick. Events in which sharks bit rotting carcasses or appeared at the surface with no sign of prey or blood were judged as "nonpredatory" and not used in the study. Once an attack was located, the observer quickly recorded the time, a theodolite reading (to position the event), and the prey type.

Elephant seals seen in the water near their haul-out sites around SFI were counted from the lighthouse every 2 hours in the 1989 season.

We determined local tidal height at the time of the attacks and created a control recorded every 2 hours every day from 0700 to 1900 hours during the 1992 season, using the computer program TIDES (Wallner, 1990). We then plotted the fractions of all predations in each of 12 tidal classes, 0.2 m wide, ranging from −0.2 to +2 m relative to mean low tide. We also calculated a normalized predation fraction to express the frequency of predation independent of the frequency of different tidal states, as follows. We obtained tidal

heights, using the TIDES program, at 2-hour intervals during the months of September and October 1990 and the fraction of the total in each of the 12 tidal height classes. This fraction was then normalized by dividing the mean fraction from the 12 classes by the fraction in each class. For a class fraction less than the mean, the calculated value was >1; for a fraction greater than the mean, the resulting value was <1.

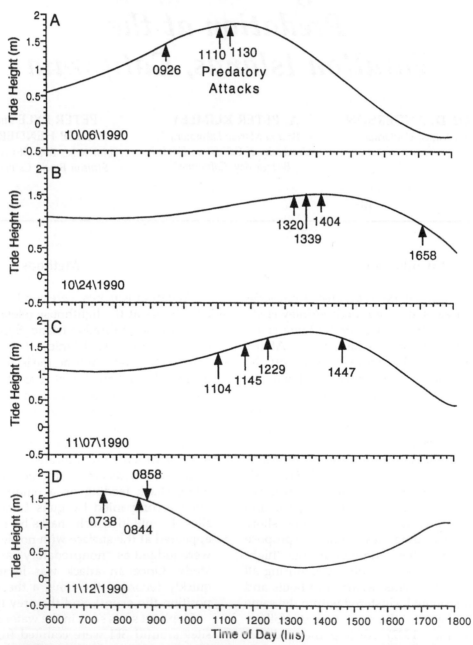

FIGURE 1 Four days in 1990, with multiple predatory attacks (indicated by arrows) of three and four (3 days and 1 day, respectively), plotted on a curve of tidal height for each day at 0600–1800 hours.

The fraction of all predations occurring in each tidal height class was then multiplied by this value.

Results

In 1990, three different attacks were recorded on 4 days, and four were recorded on 2 days. On such days of multiple attacks, most attacks occurred at the highest tide (Fig. 1). For example, on October 24, 1990, three of four attacks occurred between 1320 and 1404 hours, a 44-minute period that coincided with the peak tide. On November 12, 1990, all three attacks occurred between 0738 and 0858 hours, an 80-minute period near the peak tide at 0712 hours. In our plot of the fractions of the total number of attacks at different tidal heights, the frequency of predation increased with tidal height up to those registering 1.6 m; thereafter, frequency declined (Fig. 2). In a comparison of attack frequency with tidal height frequency, predation peaked at tides of 1.6 m, and the frequency of tidal heights was greatest at 1.2 m (Fig. 2). It is likely that the predation tidal height decreased due to the infrequency of extreme tidal states shown by the control.

We plotted a normalized predation fraction against tidal height to determine whether predation frequency was affected by the frequency of various tides (Fig. 3). The slope of the regression, a linear one, was significantly positive ($t = 2.652$, $p = 0.02$). In spite of high tides' being more frequent than low tides, our

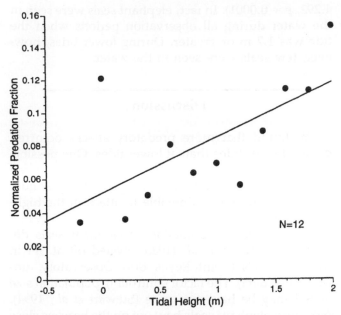

FIGURE 3 Normalized predation fraction plotted against tidal height and fit to a linear regression (slope of the line, $t = 2.652$, $p = 0.02$).

analysis shows that white shark predation is indeed more frequent during high tide.

We plotted the number of elephant seals in the water at different tidal heights to determine whether tidal height changed prey behavior (Fig. 4). Significantly more seals were in the water at high tide ($t =$

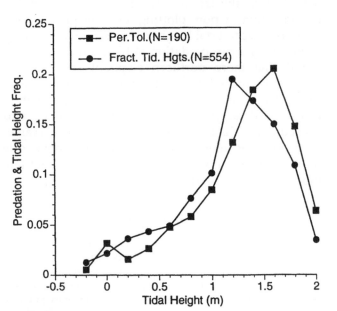

FIGURE 2 Percentage of total predatory attacks (Per.Tol.) and tidal height fraction (Fract. Tid. Hgts.) plotted against tidal height.

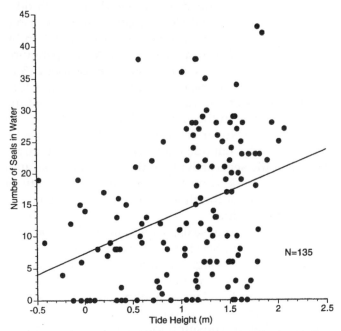

FIGURE 4 Numbers of elephant seals counted in the water, September 1–November 30, 1989, plotted against tidal height (slope of linear regression, $t = 2.652$, $p = 0.0001$).

4.292, $p = 0.0001$). In fact, elephant seals were seen in the water during all observation periods when the tide was 1.7 m or greater. During lower tides, however, few seals were seen in the water.

Discussion

We found that more predatory attacks occurred during higher tides than at lower tides. One possible explanation is that elephant seals move, or are forced to move, due to lack of haul-out space at higher tide, making them more vulnerable to attack in the high-risk zone.

The increasing population of elephant seals described by Ainley et al. (1985) leveled off at SFI in the mid-1980s (Point Reyes Bird Observatory, unpublished data). The factor controlling population size at SFI may be haul-out space (Stewart et al., 1994). Northern elephant seals haul out on the beaches close to the water and, unlike California sea lions, are incapable of climbing the rocky areas above the beaches affected by tide. Thus, elephant seals are forced into the water as beach haul-out space decreases at high tide.

Tidal height also influences predation by killer whales Orcinus orca on southern sea lions Otario flavescens. This situation was found in Patagonia, where sea lions are attacked by killer whales close to shore (sometimes on the beach) at high tide. Parer and Parer-Cook (1991) reported that pups and cows crossed a channel that developed in the reef during high tide, at which time the whales attacked the seals. Tidal height also has been observed to affect predation by leopard seals Hydrurga leptonyx on Adélie penguins Pygoscelis adeliae. Low tide created an ice foot 2–3 m high, and under these conditions the penguins could not jump free of the water. This barrier prolonged the penguins' time in the water and deprived them of an escape route (Muller-Schwarze and Muller-Schwarze, 1975). At SFI, high tide may create a similar situation, in which the seals arriving to the island swim around in the high-risk zone before reaching the safety of shallow water or the shore, due to the lack of haul-out space at high tide.

Newly arriving seals, especially in late October and early November, have yellow marks behind their eyes and other pelagic organisms (i.e., goose barnacles Lepas anatifera) on their fins. These quickly fade or rub off within 1–2 days of hauling out. Although it is difficult to recover or observe the kill of a white shark up close before consumption, we were able to note that four of five seals killed in predatory attacks in

1992 bore marks of newly arrived animals (Point Reyes Bird Observatory, unpublished data). This leads us to believe that seals have difficulty in finding haul-out space, especially when arriving during higher tides, for the first time in the fall season. Conversely, the white shark may be better able to intercept arriving seals at higher tides.

Other factors may explain the tide effect. Most predatory attacks occur in water 5–50 m deep (Klimley et al., 1992). Perhaps the seals are ambushed close to shore, where their concentration is greatest, and the water depth reaches a threshold at which the white shark is undetected near the bottom, but close to shore at high tide. There may be a correlation among water depth, SFI shoreline features, and tidal height. Finally, the northern elephant seal is a deep diver (Le Boeuf et al., 1988) and may immigrate to and away from SFI during higher tides, when they have more water in which to dive (see Chapter 17, by Le Boeuf and Crocker). This seems unlikely, however, since the difference in tidal height is small compared to the range of depths at which most predation is observed.

Summary

Predation by the white shark C. carcharias on pinnipeds has been observed repeatedly at SFI during all daylight hours. Events occurred in bouts and were unevenly distributed, clustered within the same time on the same day or at the same time the following day. We hypothesize that tidal height may affect this phenomenon. Upon plotting tidal height at each event relative to tidal height every 2 hours, we found significantly more predation at higher tides. We also found that more northern elephant seals M. angustirostris, the white shark's preferred prey, are in the water during higher tides. The seal population has reached carrying capacity at the Farallones, and consequently, haul-out space at high tide is scarce. Elephant seals may move, or be forced to move, due to lack of haul-out space at higher tide, making them more vulnerable to attack.

Acknowledgments

We thank the members of the Farallon patrol, who transported biologists and equipment to and from the island. The Farallon Islands are a national wildlife refuge, and we thank the personnel of the San Francisco Bay National Wildlife Refuge for their financial support. The members of Point Reyes Bird Observatory provided financial as well as logistical support, and the following volunteers participated in the watches: C. Alexander, P. Allen, P. Ashman, J. Aycrigg, B. Baez, D. Beadle, B. Champagne, T. Cole, J. Curson, J. de Marignac, L. Gilbert, K. Goldman, K. Hansen, D. Hardesty, D. Huntwork, D. Kapan, N. Karnovsky, F. Lawler, T. Lewandowski, D. Mann, E. McLaren, N. Nencini, P. Otis, P. Rodewald, T. Shantz, H. Smith, J. Sutter, K. Thomas, P. Vernon, O. Williams, D. Wimp-feimer, C. Wise, and E. Van Gelder. We also thank M. Elliot (U.S. Geological Survey) for lending us two Wild T-2 theodolites. This is contribution No. 585 of Point Reyes Bird Observatory.

Environmental Factors Affecting the Occurrence and Behavior of White Sharks at the Farallon Islands, California

PETER PYLE
Point Reyes Bird Observatory
Stinson Beach, California

SCOT D. ANDERSON
Inverness, California

A. PETER KLIMLEY
Bodega Marine Laboratory
University of California, Davis
Bodega Bay, California

R. PHILIP HENDERSON
Point Reyes Bird Observatory
Stinson Beach, California

Introduction

White shark *Carcharodon carcharias* behavior has been linked to environmental variables such as ocean temperature, water clarity, and current patterns (Limbaugh, 1963; Bass *et al.*, 1975; Bass, 1978; Miller and Collier, 1981; Ainley *et al.*, 1981, 1985; Carey *et al.*, 1982; Casey and Pratt, 1985; Cliff *et al.*, 1989; Strong *et al.*, 1992); however, comprehensive analyses of these and other environmental effects have not been performed. The occurrence of large white sharks off the central California coast has been correlated with season (Klimley, 1985b) and prey availability (Ainley *et al.*, 1985), and this information has been used by surfers, divers, and others to assess the risk of encountering this species while involved in commercial or recreational marine activities. A greater understanding of the effects of weather, lunar cycle, current patterns, tidal cycle, and physical oceanographic parameters on white shark behavior could further help to reduce white shark attacks on humans.

Biologists of the Point Reyes Bird Observatory (PRBO) have been monitoring white shark activity and predation on pinnipeds from the South Farallon Islands (SFI; 37°42′ N, 123°00′ W), off San Francisco, California, since 1968 (Ainley *et al.*, 1981, 1985; Pyle *et al.*, Chapter 34). Large white sharks forage in waters around the islands each autumn, coincident with the seasonal occurrence peak of their preferred prey, the northern elephant seal *Mirounga angustirostris*. From 1987 to 1992, we stationed an observer at Southeast Farallon's highest peak in autumn to watch specifically for white shark activity (see Klimley *et al.*, 1992). Here, we use frequency of predatory events and sightings, standardized by observational effort, to assess the effects of environmental variables on the behavior of white sharks at SFI. Multivariate models were developed to assess each environmental effect, after statistical adjustment for other examined effects. Variation in observed activity could relate to (1) behavior of the white sharks, (2) behavior of the prey species, and/or (3) variable observation conditions.

By synthesizing all results and comparing those based on attacks with those based on sightings, we infer which of these three causes most likely account for significant environmental effects.

Study Site, Materials, and Methods

Observations of white shark activity around SFI in 1968–1986 indicated a strong peak of occurrence during autumn (Ainley et al., 1981, 1985). Therefore, we conducted standardized watches from the lighthouse (elevation, 102 m) atop Southeast Farallon Island (Fig. 1) during autumn 1987–1992 (see Klimley et al., 1992). Trained observers rotated 2-hour shifts from dawn to dusk each day, weather and visibility permitting. Observations were terminated if ocean surface visibility in any direction was <1 km, or if wind speeds were >12.4 m/sec (25 knots). Observers continuously scanned the ocean within 1 km of SFI for shark activity. Most predatory events involving pinnipeds (here-

after referred to as "attacks") lasted >5 minutes and were accompanied by blood or a slick on the water, a circling flock of gulls, and/or vigorous thrashing of a shark on the surface (Ainley et al., 1981, 1985; Klimley et al., 1992). Because of the obvious nature of attacks, we believe that we missed few within 1 km of SFI. Nonpredatory events (hereafter called "sightings") often lasted <15 seconds; thus, it is likely that we missed many of these events. We assume, however, that our sightings data represent a randomized sampling of these activities. Observation occurred during the following periods: October 11–November 13, 1987; September 9–November 18, 1988; August 9–December 14, 1989; and September 1–November 30 during each of 1990, 1991, and 1992. A total of 434 observation periods (here defined as those lasting >90 minutes) on 404 days, totaling 3626.5 hours, were logged.

We define an attack (N = 248) as an event including the presence of blood, a victimized pinniped in the water, and/or an actively feeding shark. The prey

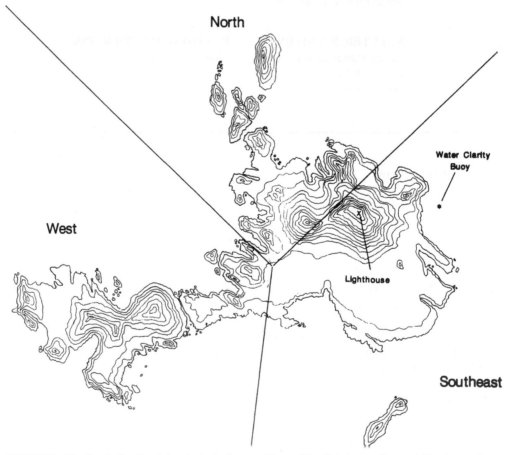

FIGURE 1 The South Farallon Islands, including positions of the lighthouse (from which observations were made), the water clarity buoy and areas of concentrated white shark activity, separated for various analyses.

species was identified during 37.9% of attacks. We separately examined environmental factors associated with attacks on northern elephant seals (N = 87), those on California sea lions *Zalophus californianus* (N = 13), and those on other species of pinnipeds (N = 4). Three attacks on brown pelicans *Pelecanus occidentalis* and one on a human are here considered sightings, rather than attacks. Other sightings were categorized for separate analyses into horizontal surfacings (N = 29) and breaches (N = 32), the latter often including vertical leaps that cleared the water surface (P. Pyle and S. D. Anderson, unpublished observation).

We determined the spatial components of environmental effects by dividing the waters around SFI into three distinct areas—southeast, west, and north (Fig. 1)—on the basis of wind patterns (see below) and concentrations of observed activity (see the map in Klimley *et al.*, 1992); positions of all events, except 15 sightings (which were not sufficiently described as to location), were assigned directional areas. To investigate for lagged effects, we examined correlations of environmental variables recorded on the day of observation and during each of the 3 previous days.

Environmental variables examined included wind direction (see below), surface wind speed (m/sec) and visibility (km), air temperature (°C), barometric pressure (mb), cloud cover (tenths of the sky obscured), swell direction (true; range, 180–320°), swell height (nearest 0.305 m), sea-surface temperature (°C), sea-surface salinity (‰), water clarity (see below), daily upwelling index (see below), and moonlight (see below). The first seven of these variables were recorded daily at SFI at 0600, 0900, 1200, 1500, and 2000 hours PST (see Pyle *et al.*, 1993). To obtain daily values for wind speed, visibility, air pressure, cloud cover, swell direction, and swell height, we averaged data recorded during and within 1 hour of start and end times of observation periods to obtain daily values. We used wind direction recorded at the time closest to the midpoint of the observation period, and categorically scored it for analysis as east (30–140° true), south (150–260°), or northwest (270–20°), reflecting the three prevailing wind directions recorded at SFI (Pyle *et al.*, 1993). We used the highest air temperature recorded during the observation period.

We collected a water sample from the southeast end of SFI to determine sea-surface temperature and salinity at 1200–1300 PST each day (see Walker *et al.*, 1992). Water clarity was scored during the 1990–1992 seasons by examining white disks, suspended 1.2 m apart, beneath a buoy anchored in a sheltered area, approximately 50 m east of SFI (Fig. 1). The clarity score represented the number of disks visible through a 25× scope from the lighthouse between 1200 and 1400 PST. Comparison of observations indicated that underwater visibility (facing downward) was about 1.75 times the number of disks visible (N); thus, underwater visibility $\sim$N*2.1 m, although this relationship may not be linear. From a model that included the terms Julian date, wind direction, wind speed, swell height, sea-surface temperature, and upwelling index as independent variables [both linear and quadratic terms; $F_{(13,215)}$ = 8.74, p < 0.000, adjusted r^2 = 0.3305], we calculated daily clarity scores for 1987–1989 to use in multivariate analyses.

Upwelling indices were provided by the National Oceanographic and Atmospheric Administration (see Bakun, 1973) for both 36° and 39° N, bracketing SFI. We found similar effects of upwelling using indices from both latitudes; our analyses included indices from 39° N. Moonlight was scored 0–100 relative to the proportion of the moon that was complete at midnight prior to the observation day.

White shark activity at SFI is known to be affected by season, number of immature elephant seals present (Ainley *et al.*, 1985), and tidal cycle (see Chapter 25, by Anderson *et al.*); we therefore adjusted for these effects in analyses. Date and date^2, represented as the Julian date within each season, were included as factors in all daily analyses. This "date adjustment" procedure controlled for linear and curvilinear seasonal cycles in both weather and oceanography and in behavioral phenology and occurrence patterns of pinnipeds and sharks. Daily totals and seasonal means of immature elephant seals were estimated from linear interpolation of weekly counts (Huber *et al.*, 1985; PRBO, unpublished data). Tidal heights and currents (incoming or outgoing) were determined using the Tides program (Wallner, 1990), adapted for SFI. We examined several tidal calculations and found that the proportion of observation time (arcsine square root transformed) when tidal height was >1.3725 m (4.5 ft) explained the greatest amount of variation in shark behavior; hereafter, the variable "tide" refers to this calculation. We excluded portions of observation periods before sunrise and after sunset. Otherwise, we did not adjust for relative times of observation periods, because white shark activity at SFI varies little with time of day during daylight hours (Klimley *et al.*, 1992).

We examined both daily and interannual effects of environmental variables on white shark attacks and sightings. Because colinearity is a problem in analyses involving daily weather and oceanographic data, the results of both single-variable (including date adjustment) and multivariate analyses are presented to differentiate actual from confounded effects (see Richardson, 1978). To estimate the effects of environmental variables, we used linear multiple regression (Seber, 1977) in both single-variable and multivariate analyses, with the STATA computer statistical package (Computing Resource Center, 1992). To reduce skew and to adjust for time duration of daily observation periods, we used log [(events + 0.001) per 100 hours of observation] as the dependent variable. Analyses of covariance were used to compare trends.

The purpose of the interannual analyses was to determine whether season-long environmental conditions (e.g., El Niño events) (Philander, 1989) affected white shark abundance or behavior in a given year. For these analyses, we used as the dependent variable events per 100 hours of observation over the period September 1–November 30. For 1987 and 1988, we used actual data within the shortened periods of those years (see above), extended before and after using predicted values from event models based on all years (see below). For interannual analyses, environmental variables were represented by mean values of daily observations during the period September 1–November 30 each year.

We had two objectives in our use of multiple regression analyses: (1) to explore the effects of each environmental variable controlling for other effects and (2) to develop single predictive models of white shark activity based on date and those environmental variables with significant effects. Linear and quadratic terms, the latter estimating curvilinear effects, were fitted in a forward stepwise manner by (1) examining the date-adjusted effects of all terms independently and simultaneously, (2) refitting the models after eliminating terms that had insignificant effects in both single-variable and multivariate analyses, and (3) reexamining the effects of each previously dropped variable, independently, within final models. Except for date terms, final models included only those environmental variables having significant linear or quadratic effects. In the sightings model, terms were included if they had significant effects on shark surfacings, breaches, or both. Linear effects of variables with significant quadratic terms were determined by refitting the models without their quadratic terms. Significance was assumed when $p < 0.05$, although it should be noted that many unreported tests were performed; therefore, probability statements should be interpreted with caution.

Results

A peak in attack frequency with date and increased frequency with higher tide and number of immature elephant seals present (Table I) were confirmed as reported elsewhere (Ainley et al., 1985; Anderson et al., Chapter 25), and adjustments for these variables were made in the attack model (Table II). Neither the tide nor the number of seals present significantly affected nonpredatory sightings of white sharks after adjustment for date (Tables I and III). Wind direction, air temperature, barometric pressure, swell direction, and sea-surface salinity had no significant effects on either attacks or sightings.

Attack frequency increased significantly with swell height (Fig. 2) and decreased with water clarity (Fig. 3), in both single-variable (Table I) and multivariate (Table II) analyses. The significant linear and curvilinear pattern of attacks with clarity was similar in analyses using both original clarity scores only (1990–1992) and predicted (1987–1989) plus original (1990–1992) scores (Fig. 3). This finding justified our use of the latter in the overall attack model. Neither swell height nor clarity significantly affected sightings (Table I).

Attack frequency was not significantly affected by upwelling that day (Table I); however, the frequency of both attacks (Table II) and sightings (Table III) increased with higher upwelling the previous day. This correlation was similar in single-variable analyses of both attacks ($t = 2.23$, $p = 0.026$) and sightings ($t = 2.42$, $p = 0.015$). In contrast, sightings of white sharks increased significantly with decreased upwelling on the same day (Tables I and III). Sightings also increased significantly with decreased wind speed, increased cloud cover, and increased sea-surface temperature as well as during periods of decreased moonlight (Tables I and III). The moonlight effect was not completely confounded by a tidal effect, as it was still significant ($t = -2.31$, $p = 0.021$) when tide was added to the model. The negative single-variable effect of visibility on sightings (Table I) was likely the result of confounding with other variables (probably cloud cover and wind speed), as the effect of visibility was insignificant when it was added to the sightings model ($t = -0.99$, $p = 0.322$). Besides upwelling, no other significant lagged effects were found.

Most environmental effects on daily attack frequency were similar when examined using attacks on elephant seals only (Table II); however, only clarity affected attacks on California sea lions (Table II). The fewer significant effects with sea lions could be an artifact of smaller sample size. For sightings, the effects of wind speed and moonlight were significant with surfacings but not breaches, whereas significant effects of sea-surface temperature and upwelling were present with breaches but not surfacings (Table III). Surfacings increased linearly with cloud cover, whereas insignificant linear but negative curvilinear

TABLE I Environmental Effects on the Frequency of Observed Attacks and Sightings

Variable[a]	Mean	SD	Attacks		Sightings	
			t	p	t	p
Elephant seals[b]	505.700	158.00	1.92	0.056	0.96	0.340
Tide	0.344	0.23	3.73	0.000	0.92	0.356
Wind direction[c]			1.18	0.316	1.50	0.215
Wind speed	5.050	2.17	0.99	0.325	−2.94	0.003
Visibility	21.960	11.40	1.63	0.104	−2.21	0.028
Air temperature	15.640	1.60	−0.24	0.810	0.80	0.425
Barometric pressure	1014.300	3.92	−0.80	0.424	0.32	0.748
Cloud cover	5.217	3.90	0.57	0.571	2.91	0.004
Swell direction	284.500	3.22	0.08	0.936	0.01	0.988
Swell height	1.584	0.55	3.36	0.001	1.14	0.255
Sea temperature	13.710	1.12	1.83	0.069	2.45	0.015
Sea salinity	33.510	0.17	−0.44	0.663	−0.82	0.412
Water clarity[d]	2.655	0.89	−6.28	0.000	0.00	0.998
Water clarity[e]	2.721	0.76	−5.77	0.000	0.16	0.872
Upwelling	40.520	63.80	1.61	0.108	−2.57	0.010
Moonlight	49.780	29.10	−1.65	0.099	−2.56	0.011

[a]Date-adjusted single variable; see text for units of measurement.
[b]The effects of immature elephant seals present, without date adjustment, were $t = 6.77$, $p < 0.0001$ for attacks and $t = 2.97$, $p = 0.003$ for sightings.
[c]p values for wind direction are based on F statistics rather than t values.
[d]Scores recorded in 1990–1992 ($N = 229$). For all other variables, $N = 434$.
[e]Actual plus predicted values in 1987–1992 ($N = 434$).

TABLE II Environmental Effects on the Attack Frequency on Northern Elephant Seals and California Sea Lions

Variable	Overall attacks[a]		Elephant seals[b]		California sea lions[c]	
	t	p	t	p	t	p
Date	−0.55	0.585	1.42	0.155	−0.32	0.753
Date²	−2.65	0.008	−2.14	0.033	−1.49	0.138
Elephant seals	2.46	0.014	2.33	0.020	−0.52	0.601
Elephant seals²	3.92	0.000	2.03	0.043	1.58	0.115
Tide	3.28	0.001	2.81	0.005	1.48	0.141
Swell height	3.47	0.001	2.52	0.012	1.20	0.231
Water clarity	−5.27	0.000	−2.40	0.017	−2.04	0.042[d]
Water clarity²	−2.14	0.033	0.36	0.719	−2.06	0.040[d]
Upwelling the day before	2.42	0.012	2.58	0.010	−0.09	0.927

[a]$F_{(9,424)} = 15.19$, $p < 0.000$, adjusted $r^2 = 0.2289$, $N = 248$ attacks.
[b]$F_{(9,424)} = 7.80$, $p < 0.000$, adjusted $r^2 = 0.1245$, $N = 87$ attacks.
[c]$F_{(9,424)} = 1.22$, $p < 0.125$, adjusted $r^2 = 0.0046$, $N = 13$ attacks.
[d]Actual plus predicted values (see Table I).

TABLE III Environmental Effects on the Frequency of White Shark Sightings, Surfacings, and Breaches

Variable	Overall sightings[a]		Surfacings[b]		Breaches[c]	
	t	p	t	p	t	p
Date	2.96	0.003	1.74	0.082	2.62	0.009
Date²	−1.47	0.142	−0.54	0.590	−1.83	0.067
Wind speed	−2.22	0.027	−2.68	0.008	−0.00	0.998
Wind speed²	1.88	0.061	3.00	0.003	−0.23	0.821
Cloud cover	1.53	0.128	2.71	0.007	0.65	0.514
Cloud cover²	−2.40	0.017	−0.75	0.452	−2.46	0.014
Sea temperature	2.28	0.023	1.14	0.255	2.45	0.015
Upwelling	−2.23	0.027	−1.30	0.195	−1.68	0.093
Upwelling²	2.17	0.006	0.24	0.807	2.38	0.018
Upwelling the day before	2.79	0.005	1.17	0.244	1.84	0.066
Moonlight	−2.33	0.020	−2.44	0.015	−1.43	0.154

[a] $F_{(11,422)} = 4.97$, $p < 0.000$, adjusted $r^2 = 0.0927$, N = 68 events.
[b] $F_{(11,422)} = 3.88$, $p < 0.000$, adjusted $r^2 = 0.0691$, N = 29 events.
[c] $F_{(11,422)} = 3.22$, $p < 0.000$, adjusted $r^2 = 0.0541$, N = 32 events.

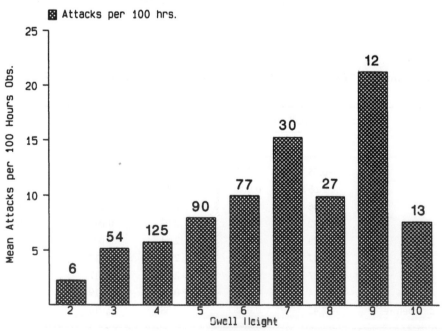

FIGURE 2 The mean number of white shark attacks observed per 100 hours relative to swell height, separated into increments of 0.305 m (1 ft); the sample sizes of the observation periods are shown above each bar.

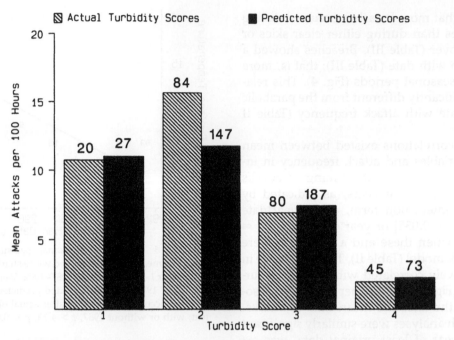

FIGURE 3 The mean number of white shark attacks observed per 100 hours relative to water clarity. Actual clarity scores (N = 229) were recorded during 1990–1992; "predicted scores" (N = 434) represent actual scores plus predicted values for 1987–1989, based on a model (see text). The sample sizes of the observation periods for each turbidity score category are shown above each bar.

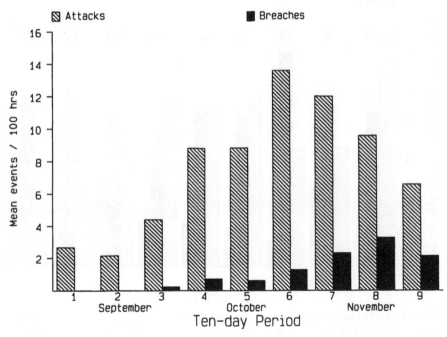

FIGURE 4 The mean number of white shark attacks and breaches observed per 100 hours by season, separated into 10-day periods, September 1–November 30 (period 5 contains 11 days). The curvilinear seasonal pattern of attacks (see Table II) differs from the linear seasonal pattern of breaches (Table III) [by 10-day period: analysis of covariance, $F_{(3,14)} = 5.17$, $p = 0.031$].

results indicated that more breaches occurred during partly cloudy skies than during either clear skies or complete cloud cover (Table III). Breaches showed a linear relationship with date (Table III); that is, more occurred later in seasonal periods (Fig. 4). This relationship was significantly different from the parabolic relationship of date with attack frequency (Table II and Fig. 4).

No significant correlations existed between mean environmental variables and attack frequency in interannual analyses; however, the timing of attacks varied significantly between years, as indicated by the effects of an interaction term, either year*date [$F_{(19,414)} = 4.47$, $p = 0.035$] or year*date^2 [$F_{(19,414)} = 6.64$, $p = 0.010$], when these and a year term were added to the attack model (Table II). This difference in timing was positively correlated with mean sea-surface temperature (Fig. 5); the warmer the mean seasonal temperature, the later the mean date of attacks. Differences in both analyses were similarly significant when 1987, the year of least original data, was excluded.

The location of attacks also varied significantly with date (Fig. 6). Early in the season, a higher proportion of attacks occurred to the southeast, whereas a higher proportion occurred later in the season to the

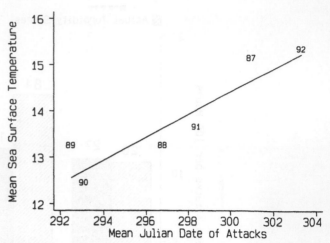

FIGURE 5 The mean seasonal sea-surface temperature by weighted mean date of observed white shark attacks per 100 hours, September 1–November 30, 1987–1992. Weighted mean date calculations for 1987 and 1988 included predicted observations for periods that occurred before and after actual observation periods (see text; with or without 1987, $t > 4.13$, $p < 0.026$).

north. Suspecting that this might be related to current patterns around SFI, we compared the locations of shark attacks and sightings with tidal current (in going versus outgoing) and found a similar signifi-

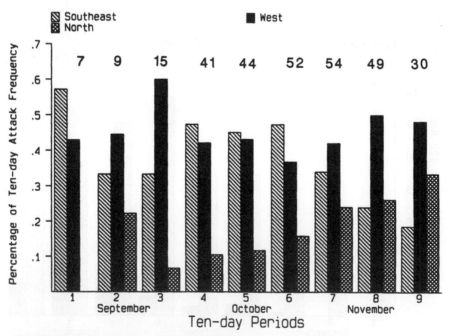

FIGURE 6 The proportion of mean white shark events recorded per 100 hours southeast, west, and north of SFI (see Fig. 1) during 10-day periods, September 1–November 30 (period 5 contains 11 days). Sample sizes during each 10-day period are indicated. The pattern of events by date between directional areas is significant (analysis of covariance): overall, $F_{(5,21)} = 8.68$, $p = 0.002$; between southeast and north, $F_{(3,14)} = 15.61$, $p = 0.001$; between west and north, $F_{(3,14)} = 6.16$, $p = 0.026$; but not between southeast and west $F_{(3,14)} = 3.59$, $p = 0.079$.

TABLE IV   Frequencies of White Shark
Attacks and Sightings Relative
to Location and Tidal Current

Direction	Tidal current	
	Rising	Falling
Southeast	58	54
West	62	73
North	14	40

cant difference (Table IV): more activity occurred southeast of the islands during incoming tides, whereas more activity occurred to the west and north during outgoing tides (G test, $G = 10.47$, df = 2, $p = 0.005$). A marginally significant interaction was also found between the locations of the attacks and the tide [analysis of covariance after adjustment with the attack model, $F_{(13,1279)} = 3.00$, $p = 0.050$]. More attacks occurred to the west compared to the southeast of SFI during higher tides. No significant interactions between other environmental effects and location of activity were found.

Discussion

SFI is located in one of five eastern boundary current systems known for their abundant marine biota (Ainley and Boekelheide, 1990) and in the midst of the largest upwelling center in the California Current system (Parrish *et al.*, 1982; Minerals Management Service, 1987). Three major oceanographic processes affect waters surrounding SFI in the fall: the southbound California Current, which is weakened at that time; the northbound Davidson Current, which strengthens and reaches the surface in early November; and coastal upwelling of nutrient-rich water, which is also weakened in fall (Bolin and Abbott, 1963; Hickey, 1979; Chelton, 1984). Weather patterns at SFI (Pyle *et al.*, 1993) are strongly influenced by these marine processes (Namais, 1969; Ainley and Boekelheide, 1990), and both the marine systems and, in turn, weather patterns in the area show substantial interannual variation (Norton *et al.*, 1985). Shark occurrence and behavior at SFI appear to be influenced by all of these climatological processes.

White shark attack frequency at SFI was not significantly affected by any nonoceanic weather variables, indicating that these variables influenced neither the behavior of prey nor our ability to detect attacks. The effect of wind speed on sightings (Table III), espe-

cially surfacings, however, probably does reflect our reduced ability to detect these events during rougher sea conditions. Our ability to detect sightings may have improved during greater cloud cover, which reduced solar glare. Alternatively, darker or less contrasting light conditions may have caused white sharks to occur closer to the water surface, explaining the increased number of surfacings during cloudier skies (Table III).

Both increased swell height and decreased water clarity were correlated with greater attack frequency, but neither of these variables influenced white shark sightings (Tables I–III). These findings suggest that both of these effects result, at least in part, from behavior of the prey. Increased swell height, like increased tidal height, reduces the haul-out area for elephant seals (see Chapter 25, by Anderson *et al.*), forcing them into the water and, in effect, increasing prey availability for white sharks (Fig. 2). We infer that decreased water clarity limited the ability of pinnipeds to detect stalking white sharks, thus increasing the chance of successful predation. The negative curvilinear correlation with clarity (Table II and Fig. 3) may result from a decreased ability by sharks to detect prey in the most turbid conditions. Although it is also possible that white shark activity generally increases with decreased clarity (see Myrberg, 1969; Cliff *et al.*, 1989), we might then expect to see a concomitant increase in the frequency of sightings, which we did not find (Table I). On the other hand, more white sharks may have moved into SFI waters during turbid periods, remaining at stalking depths during these improved conditions for predation.

The only correlation significant in analyses of both attacks and sightings was that of increased activity with increased upwelling the day before observation (Tables II and III). We infer from this that more white sharks are present around SFI during periods about 24 hours after bouts of coastal upwelling. This pattern was not directly related to water clarity, as clarity and upwelling the day before were not significantly related (date-adjusted analysis, 1990–1992; $t = -0.60$, $p = 0.549$). Although numerous studies have correlated fish movement in the California Current with upwelling on seasonal and interannual bases (e.g., Bakun and Parrish, 1980; Parrish *et al.*, 1981), few have reported on daily response patterns. Briggs *et al.* (1988; see also Minerals Management Service, 1987) found that Cassin's auklets *Ptychoramphus aleuticus* and their prey—euphausiids and juvenile fish—concentrated along frontal boundaries that appeared near SFI after intensified upwelling. Carey *et al.* (1982) found that a telemetered white shark preferred moving along a thermocline, perhaps for orientation or to enable sampling of two water masses. It is possible that white sharks concentrate along frontal boundaries created by upwelling near SFI for these reasons. Alternatively, the white shark may be a member of a pelagic fish assemblage showing movement patterns responding to upwelling-related changes in water mass, as has been documented in the Gulf of California with hammerhead sharks *Sphyrna lewini* (Klimley and Butler, 1988). Telemetry studies on tagged white sharks in the area would greatly elucidate the responses of sharks to local upwelling.

Although significant only with sightings and surfacings, an increase in shark presence at SFI during dark-moon periods was indicated in all analyses (Tables I and III; $t = -1.36$, $p = 0.175$, when moonlight was added to the attack model: Table II). It is possible that this weak effect might be caused by more nocturnal movement by (and predation on) pinnipeds around SFI during full-moon periods (see Trillmich and Mohren, 1981; Watts, 1993). Although an illumination threshold has yet to be quantified, evidence from retinal examinations (Gruber and Cohen, 1985), attack patterns versus time of day around SFI (Klimley *et al.*, 1992), and underwater movement patterns (Strong *et al.*, 1992) indicate that the white shark is mainly a diurnal predator. Thus, it is also possible that white sharks have an endogenous or direct response to ocean variables influenced by lunar cycles, as has been documented in other fish (reviewed by Gibson, 1978). Effects on white shark occurrence and predatory activity of periodic variables such as lunar phase, swell height, water clarity, and upwelling might explain the periodicity and occasional 7- to 10-day hiatuses of observed attacks at SFI (Klimley *et al.*, 1992).

The effects of environmental factors and date on white shark breaches (Table III and Fig. 4) indicate that this activity may be social rather than predatory in nature. Most breaches occurred late in fall, in a pattern significantly different from that of attack frequency, and when white sharks could be shifting from a fall predatory to a spring reproductive mode. Up to three repeated breaches were recorded on two occasions from SFI. The frequency of breaches was not significantly affected by prey abundance, tide, water clarity, or swell height, which might be expected if they represented failed predatory attempts. Higher sea-surface temperatures, the lack of immediate upwelling, and partly cloudy skies may possibly simulate environmental conditions on breeding grounds (probably in subtropical neritic waters south of Point Conception) (Klimley, 1985b), resulting in more breaches by white sharks at SFI during these conditions. We infer that the splash caused by breaches may repre-

sent social signaling (see Chapter 22, by Klimley *et al.*), perhaps by male white sharks defending a territory, attempting to attract a female, or both. Similar communication through breaching has been inferred for whales (Herman and Tavolga, 1980).

Strong *et al.* (1992) found that white sharks often circled downstream of islands and bait sources, apparently as part of a searching pattern. Our spatial results also reflect this phenomenon. The shift in attack frequency from southeast to north of the island as the season progressed (Fig. 6) is temporally consistent with the replacement at the surface of the southbound California Current by the northbound Davidson Current (Hickey, 1979; Chelton, 1984). Spatial results of tidal current and height further support this conclusion, although interactions between ocean and tidal currents (Noble and Gelfenbaum, 1990) likely complicate movement patterns by white sharks around SFI.

Although seasonal patterns of white shark occurrence differ between SFI and the adjacent coast (Ainley *et al.*, 1985), where many attacks on humans occur (Miller and Collier, 1981; Lea and Miller, 1985), some of our results might apply to the coast, and thus may be useful to humans interested in avoiding large white sharks. Those effects that we attribute primarily to pinniped behavior (i.e., increased tide and swell height and decreased water clarity) could be unrelated to attacks on humans; indeed, more attacks on humans seem to occur in clearer water along the coast (Miller and Collier, 1981; see also Limbaugh, 1963). We suggest, however, that effects of upwelling, moonlight, current patterns, and interannual patterns of water temperature be considered in relation to human attack frequency. Analyses of these and other environmental effects on white shark–human interactions, similar to those presented here, could elucidate these potential influences.

Summary

We examined the daily and seasonal effects of 13 weather, oceanographic, and lunar variables on the predatory and nonpredatory behavior of white sharks *C. carcharias* at SFI, California, during autumn 1987–1992. Effects were assessed using both single-variable and multivariate analyses. Daily frequency of observed white shark attacks on pinnipeds increased with swell height and upwelling the day before observation, and decreased with water clarity. Daily frequency of nonpredatory sightings (including surfacing and breaches) increased with cloud cover, sea-surface temperature, and upwelling the day before observation, and decreased with wind speed, upwelling the day of observation, and lunar stage. Spatial patterns of activity indicate that white sharks stalk prey "downstream" in oceanic and tidal currents. No significant interannual patterns were apparent relative to environmental factors; however, the temporal pattern of attacks shifted later in the season during years of warmer sea-surface temperature. White shark breaches showed a different temporal pattern than attacks, suggesting that they may represent social signaling rather than failed predatory attempts. On the basis of all results, we infer that sea-surface temperature, upwelling, and lunar illumination likely affected shark behavior; swell height and water clarity likely affected prey behavior relative to predation; and wind speed and possibly cloud cover likely affected our ability to detect shark activity. Some of our findings can possibly be used to help reduce the incidence of white shark attacks on humans.

Acknowledgments

The research program at the South Farallon Islands, part of the Farallon National Wildlife Refuge, is conducted through the cooperation and support of the U.S. Fish and Wildlife Service, the U.S. Coast Guard, and the Farallon Patrol. The white shark program was funded in part by the National Geographic Society, the Gulf of the Farallones Marine Sanctuary, and PRBO, and it depended heavily on the PRBO interns acknowledged in the work of Klimley *et al.* (1992) and Anderson *et al.* (Chapter 25). We especially thank L. Gilbert for preparing the computer files used in all analyses presented here. M. Silkey assisted with the preparation of the lunar data. P. Adams (National Marine Fisheries Service, Tiburón) and P. Walker (Scripps Institution of Oceanography, La Jolla) provided the recent oceanographic information; M. Elliot (U.S. Geologic Survey, Menlo Park), and E. Ueber and J. Roletto [Gulf of the Farallones National Marine Sanctuary (GFNMS), San Francisco] provided us the theodolites for determination of attack locations. GFNMS also permitted employment of our water turbidity buoy. D. G. Ainley, D. Evans, and W. J. Sydeman provided helpful comments on the manuscript. We especially thank D. G. Ainley for frequent feedback and unconditional support throughout the course of our white shark research program. This is PRBO contribution 583.

27

White Shark Predation and Scavenging on Cetaceans in the Eastern North Pacific Ocean

DOUGLAS J. LONG
Department of Ichthyology
California Academy of Sciences
Golden Gate Park
San Francisco, California

ROBERT E. JONES
Museum of Vertebrate Zoology
University of California
Berkeley, California

Introduction

Cetacean remains are frequently found in the stomachs of sharks (Table I). Thus, sharks are often considered to be predators of dolphins and porpoises. To better understand the extent of predation on cetaceans, it is essential to understand how a cetacean might end up in the stomach of a shark. A fundamental distinction must be made between scavenging a dead cetacean and the act of predation—attacking and/or killing a live cetacean. Of over 350 species of sharks, only about 6, including the white shark *Carcharodon carcharias*, are known to prey on small odontocetes (Table II).

Bites, punctures, and tooth rakes have been used with some success to identify species of sharks responsible for attacks on humans (Gudger, 1950; Martini and Welch, 1981; Egaña and McCosker, 1984; Lea and Miller, 1985; Collier, 1992; Nakaya, 1993), since tooth shape, size, spacing, and variability differ among genera and species (see Chapter 3, by Hubbell, and Chapter 8, by Purdy). This method has been used in studies of shark predation on pinnipeds (Brodie and Beck, 1983; Hiruki *et al.*, 1993) and on small odontocetes (Leatherwood *et al.*, 1972; Corkeron *et al.*, 1987; Cockcroft *et al.*, 1989) and has been especially important in understanding white shark predatory dynamics along the California coast (Ames and Morejohn, 1990; Le Boeuf *et al.*, 1982; Ainley *et al.*, 1985; McCosker, 1985; Long, 1991a; Long *et al.*, Chapter 24; Ames *et al.*, Chapter 28).

Along the Atlantic seaboard of North America, the white shark sometimes preys on harbor porpoises *Phocoena phocoena* (Day and Fisher, 1954; Arnold, 1972). Off South Africa, white sharks prey on bottle-nosed dolphins *Tursiops truncatus*, common dolphins *Delphinus delphis*, and Indo-Pacific humpback dolphins *Sousa plumbea* (Cockcroft *et al.*, 1989; Cliff *et al.*, 1989). In Australian waters, white sharks are known to prey on bottle-nosed dolphins (Corkeron *et al.*, 1987; Bruce, 1992). In the Mediterranean, white sharks may also feed on cetaceans, including the Risso's dolphin *Grampus griseus* and other species (see Chapter 30, by Fergusson). In New Zealand waters, white sharks apparently attack dusky dolphins *Lagenorynchus obscurus*, based on observations of individuals with healed bite scars and shark-damaged dorsal fins (M. Webber, personal communication) (Fig. 1D).

In contrast to the above records, shark predation on cetaceans in the eastern North Pacific Ocean is poorly documented. Norris (1967) said, " . . . apparently shark predation is of importance in some areas. We almost never see this in California; at least I can-

TABLE I Shark Species Whose Diets Are Known to Include Cetaceans

Common name	Species	Reference
Pigeye shark	*Carcharhinus amboinensis*	Cliff and Dudley (1991b)
Bronze whaler shark	*Carcharhinus brachyurus*	Cliff and Dudley (1992a)
Galápagos shark	*Carcharhinus galapagensis*	Compagno (1984a)
Bull shark	*Carcharhinus leucas*	Baughman and Springer (1950); Bass *et al.* (1975); Cliff and Dudley (1991b)
Blacktip shark	*Carcharhinus limbatus*	Dudley and Cliff (1993b)
Oceanic whitetip shark	*Carcharhinus longimanus*	Compagno (1984a); Stevens (1984)
Dusky shark	*Carcharhinus obscurus*	Compagno (1984a)
Sandbar shark	*Carcharhinus plumbeus*	Cliff *et al.* (1988a); Stillwell and Kohler (1993)
White shark	*Carcharodon carcharias*	Templeman (1963); Randall (1973); Ellis (1975); Carey *et al.* (1982); Pratt *et al.* (1982); Stevens (1984); McCosker (1985); Cliff *et al.* (1989); Bruce (1992)
Portuguese shark	*Centroscymnus coelolepis*	Clark and Merrett (1972); Ebert *et al.* (1992)
Tiger shark	*Galeocerdo cuvieri*	Bell and Nichols (1921); Baughman and Springer (1950); Compagno (1984a); Stevens (1984); Stevens and McLoughlin (1991); Randall (1992); Simpfendorfer (1992)
Sixgill shark	*Hexanchus griseus*	Ebert (1986, 1994)
Shortfin mako shark	*Isurus oxyrinchus*	Stillwell and Kohler (1982); Stevens (1984); Cliff *et al.* (1990)
Sevengill shark	*Notorynchus cepedianus*	Ebert (1991); D. J. Long (unpublished observation)
Blue shark	*Prionace glauca*	Strasburg (1958); Stevens (1973, 1984)
Sleeper shark	*Somniosus microcephalus*	Williamson (1963); Beck and Mansfield (1969)
Pacific sleeper shark	*Somniosus pacificus*	Crovetto *et al.* (1992)
Hammerhead shark	*Sphyrna* sp.	Leatherwood *et al.* (1972)

not think of a case." Leatherwood *et al.* (1972) and Ridgway and Dailey (1972) observed several instances of shark-wounded dolphins and of sharks attacking living dolphins in waters off southern California and western Mexico, but such observations were scarce, and none implicated the white shark.

Stroud and Roffe (1979), from Oregon, reported a white-sided dolphin *Lagenorynchus obliquidens* in which shark attack was the primary cause of death, and Long (1991a) reported a white shark attack on a live pygmy sperm whale *Kogia breviceps* on the basis of a white shark bite on the caudal peduncle. Slipp and Wilke (1953) and Sullivan and Houck (1979) re-

ported carcasses of Baird's beaked whale *Berardius bairdi*, and Roest (1970) reported a Cuvier's beaked whale *Ziphius cavirostris*, all having large shark bites, but they did not discuss the possibility of postmortem scavenging. Minasian *et al.* (1984) illustrated a live stranded Cuvier's beaked whale having a partially healed scar attributed to a shark, but upon examination of their photograph, we conclude that a shark was not the cause of the scar.

Examinations of white shark stomach contents collected along the West Coast of the United States from 1935 to 1984, reported by Klimley (1985b), failed to find cetacean remains. The first records of white shark

TABLE II Shark Species Identified as Predators on Small Odontocetes

Common name	Species	Reference
Bull shark	*Carcharhinus leucas*	Caldwell *et al.* (1965); Cockroft *et al.* (1989)
Oceanic whitetip shark	*Carcharhinus longimanus*	Leatherwood *et al.* (1972); D. J. Long (unpublished data)
Dusky shark	*Carcharhinus obscurus*	Cockroft *et al.* (1989)
White shark	*Carcharodon carcharias*	Day and Fisher (1954); Arnold (1972); Cockroft *et al.* (1989); Long (1991b)
Tiger shark	*Galeocerdo cuvieri*	Cockroft *et al.* (1989)
Shortfin mako shark	*Isurus oxyrinchus*	Leatherwood *et al.* (1972); Ridgway and Dailey (1972)

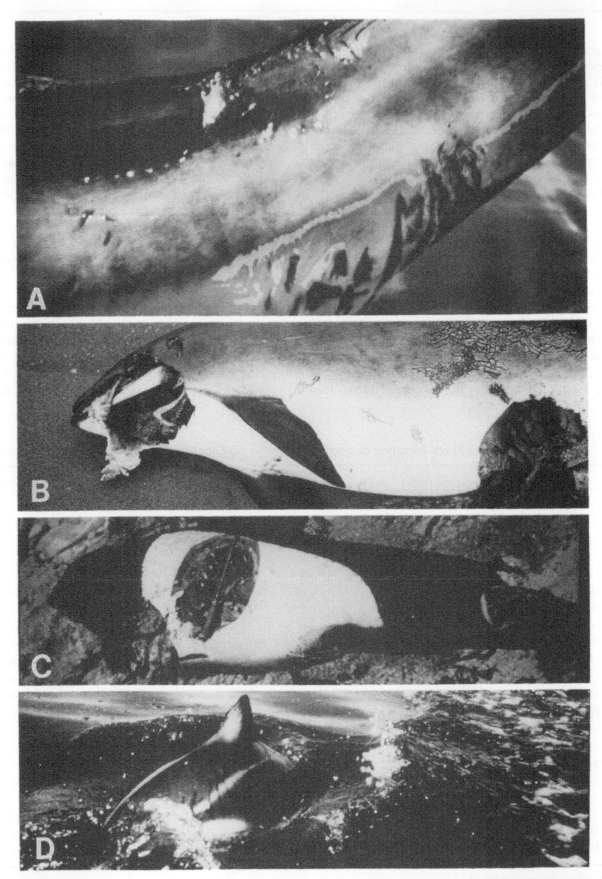

FIGURE 1 Small odontocetes wounded by white sharks. (A) The caudal peduncle of an immature pygmy sperm whale (case 15). (Photo by D. J. Long.) (B) A harbor porpoise, showing a fatal bite in the abdomen (34 cm wide) and a second bite on the lower jaw (case 19). (Photo by C. Keiper.) (C) A Dall's porpoise, showing a large (36-cm-wide) bite on the left flank and a smaller bite on the chest (case 1). (Photo by R. E. Jones.) (D) A dusky dolphin, showing healed scars from a white shark bite (southern New Zealand). (Photo by M. Webber.)

predation on cetaceans in this area were from wounded carcasses reported by Stroud and Roffe (1979) and Long (1991a). However, another possible record of predation on odontocetes, based on accounts by local fishermen, was a white shark attack on a harbor porpoise off Pacifica, San Mateo County, California on September 1, 1983, but some details are lacking (Szczepaniak, 1990). There are no records of white sharks preying on larger odontocetes or on adult mysticetes.

While predation by sharks on cetaceans is relatively rare, many sharks scavenge dead cetaceans (Long, 1991b) (Table III). Carcasses of mysticete or large odontocete whales may float for several weeks, providing a large amount of food for many species of sharks and, possibly, supply an important portion of the diet for white sharks during certain times of the year. White sharks are frequently observed feeding on carcasses of whales off Australia and the eastern United States (Randall, 1973; Ellis, 1975; Carey *et al.*, 1982; Pratt *et al.*, 1982; McCosker, 1985; Casey and Pratt, 1985), and some authors suggested that dead cetaceans may be a primary food source for white sharks in certain areas (Carey *et al.*, 1982; Castro, 1983). Scavenging by white sharks on the west coast of North America has not been well documented, but white sharks have been seen scavenging carcasses of basking sharks *Cetorhinus maximus* (Follett, 1966). Reviewed here are records collected from the coast of California of white sharks attacking and feeding on live cetaceans as well as scavenging cetacean carcasses; these accounts demonstrate the importance of cetaceans in the diet of white sharks in the eastern North Pacific.

Methods

Information about shark predation and scavenging on cetaceans in the eastern North Pacific from 1972 to 1992, spanning the entire coast from Washington state to southern California, was gathered through records from the National Marine Mammal Stranding Network (NMMSN), several museums, and scientific

TABLE III White Shark Predation on Small Odontocetes from the West Coast of the United States

No.	Common name	Species	Sex	Date	Locality[a]	Bite location[b]	Condition[c]
1	Dall's porpoise	*Phocoenoides dalli*	F	August 7, 1973	Bolinas Beach, California	2, 3, 6	Recently killed
2	Harbor porpoise	*Phocoena phocoena*	M	July 28, 1978	South Beach, California	1, 3	Alive
3	White-sided dolphin	*Lagenorhynchus obliquidens*	M	May 12, 1981	Fort Ord, California	1	Alive
4	Harbor porpoise	*Phocoena phocoena*	F	June 4, 1982	Pomponio Beach, California	4	Recently killed
5	Harbor porpoise	*Phocoena phocoena*	F	June 14, 1982	Stinson Beach, California	1	Recently killed
6	Harbor porpoise	*Phocoena phocoena*	F	June 23, 1982	Stinson Beach, California	1, 2	Recently killed
7	Harbor porpoise	*Phocoena phocoena*	M	September 2, 1983	Linda Mar Beach, California	1, 4	Recently killed
8	Harbor porpoise	*Phocoena phocoena*	M	October 1, 1983	Point Bonita, California	1, 3	Recently killed
9	Harbor porpoise	*Phocoena phocoena*		October 2, 1983	Fort Cronkhite, California	1, 2	Recently killed
10	Harbor porpoise	*Phocoena phocoena*	F	July 18, 1984	Dillon Beach, California	4	Recently killed
11	Harbor porpoise	*Phocoena phocoena*		December 4, 1985	Long Beach, Washington	1, 3, 4	Recently killed
12	Harbor porpoise	*Phocoena phocoena*		August 5, 1986	Franklin Point, California	2, 3	Recently killed
13	Dwarf sperm whale	*Kogia simus*	M	January 12, 1987	Stinson Beach, California	1	Alive
14	Risso's dolphin	*Grampus griseus*	F	July 8, 1989	Trinidad Beach, California	1, 2	Recently killed
15	Pygmy sperm whale	*Kogia breviceps*	M	August 31, 1989	Parjaro Dunes, California	1	Alive
16	Stejneger's beaked whale	*Mesoplodon stejnegeri*	M	May 21, 1990	Pacific Grove, California	1, 2	Alive
17	Cuvier's beaked whale	*Ziphia cavirostris*		June 19, 1990	South Beach, California	1	Recently killed
18	Harbor porpoise	*Phocoena phocoena*	F	April 4, 1991	Drake's Beach, California	3, 5	Recently killed
19	Harbor porpoise	*Phocoena phocoena*	F	February 15, 1993	Kehoe Beach, California	3, 5	Recently killed

[a]Location collected (see Fig. 2).
[b]Location of bite on the body (see Fig. 8).
[c]Condition of the animal at the time of first examination.

institutions (see Acknowledgments). All marine mammals are federally protected, and all live or dead stranded cetaceans that are found are examined. Information on species, sex, size, age, location, date, and overall condition is collected, and when possible, a gross necropsy is performed to determine the cause of death. Although the NMMSN spans the entire coastline of the western United States, and the area for California is well monitored, potential biases exist in the stranding data (see Chapter 24, by Long *et al.*).

Records of all cetaceans suspected to have been bitten by sharks were carefully reviewed, and white shark predation was determined by three criteria: (1) direct observation of an attack in which both predator and prey were positively identified (only one such encounter is known from this area), (2) observations of living cetaceans showing fresh bite wounds or healed bite scars that could be positively attributed to a white shark, or (3) necropsies on dead cetaceans indicating that white shark attack was the primary cause of death, or that bites were inflicted prior to death. Criteria included evidence of recent exsanguination from a bite area, trauma, subdermal muscular or osseous hemorrhage near a bite, or signs of blood clotting or healing around a bite area.

Jaws and tooth sets of predatory sharks from California were examined to determine the species responsible for inflicting wounds seen on odontocete and mysticete carcasses and to interpret feeding behavior. Jaws of white sharks (N = 27), shortfin mako sharks *Isurus oxyrinchus* (N = 28), blue sharks *Prionace glauca* (N = 61), sevengill sharks *Notorynchus cepedianus* (N = 15), sixgill sharks *Hexanchus griseus* (N = 8), tiger sharks *Galeocerdo cuvieri* (N = 11), bull sharks *Carcharhinus leucas* (N = 13), and dusky sharks *Carcharhinus obscurus* (N = 10) were studied (for data, see Long, 1994). Complete bite widths of white sharks (Table IV) were measured across the upper jaw behind the last posterior tooth, and behind the fifth tooth for partial bites.

Evidence of scavenging was based on direct observation of the sharks actually feeding on the carcass, or on bites and/or tooth marks left on a carcass that can be attributed to a white shark. Bites, punctures, and rake marks were measured and photographed when possible. We assessed seasonal and annual trends of white shark scavenging of large cetaceans stranded along the California coast from Del Norte County south to Monterey County, 1972–1992. The large cetacean category included all mysticete and large odontocete whales, such as sperm whales *Physeter catodon*, killer whales *Orcinus orca*, pilot whales *Globicephala macrorhynchus*, and beaked whales *Z. cavirostris* and *Mesoplodon* spp.

TABLE IV White Shark Lengths and Bite Gapes

Total Length (m)	Bite Width (cm)[a]	Incomplete Width (cm)[b]
1.63	11.7	10.2
1.96	17.7	13.8
2.64	23.4	19.4
2.69	27.0	22.4
3.68	32.0	30.1
5.10	48.0	39.5

[a]Measured across the upper jaw from behind the last tooth of the row.

[b]Measured across the upper jaw from behind the fifth tooth.

Results

Predation

We inspected 19 small odontocetes that showed signs of predation by white sharks (Table III). These specimens included 11 harbor porpoises, 2 Dall's porpoises *Phocoenoides dalli*, and 1 each of the white-sided dolphin, Risso's dolphin, pygmy sperm whale, dwarf sperm whale *Kogia simus*, Cuvier's beaked whale, and Stejneger's beaked whale *Mesoplodon stejnegeri*.

Bites were usually located on several areas of the body: 68% (N = 13) on the caudal peduncle, 31% (N = 6) in the urogenital region, 36% (N = 7) on the abdomen, 21% (N = 4) on the dorsum, 10% (N = 2) on the head, and 5% (N = 1) on the flanks. Of the 19 specimens, 1 was recovered from Washington state, and the others were collected from central and northern California (Humboldt, Marin, San Mateo, and Monterey counties; Fig. 2). Attacks on small odontocetes occurred year-round, but most were recorded during summer (52%), with fewer attacks reported in fall, winter, and spring (15% each; Fig. 3).

These records are the only known examples of white shark predation on odontocetes in the eastern North Pacific Ocean. For the Dall's porpoise, the dwarf sperm whale, and the Cuvier's and Stejneger's beaked whales, this report lists the first known instances of white shark predation on these species. Reported here is the first record of shark predation on the Risso's dolphin in North American waters, but Ian Fergusson (Chapter 30) believes that white sharks may prey on this species in the Mediterranean.

Scavenging

Records of shark-scavenged carcasses extended from Bandon, Oregon, and Crescent City, California,

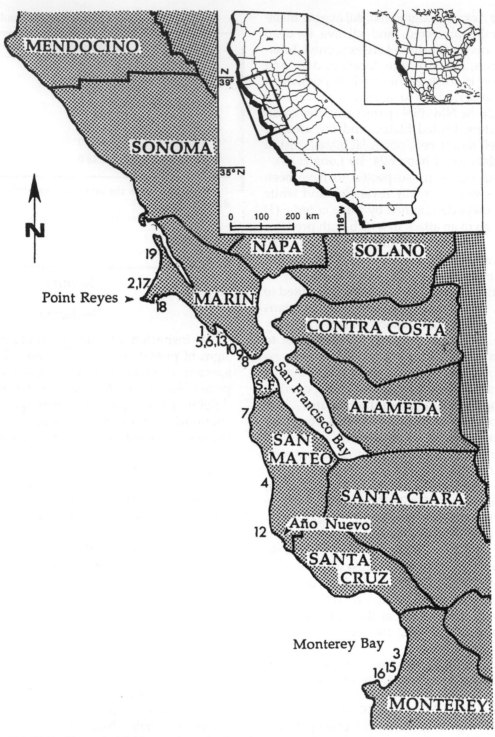

FIGURE 2 The central California coast, showing the stranding localities of small odontocetes listed in Table III; cases 11 and 14, both north of the area covered, are not shown.

south to San Diego, California. At least half of the strandings occurred off the central California coast, and these were the ones considered in this study. Three to 12 (average, 6.2) large cetaceans have stranded along the central California coast annually (Fig. 4), but many more probably sank or floated out to sea. Cetacean carcasses were available all year, but the abundance of cetacean carcasses increased mark-

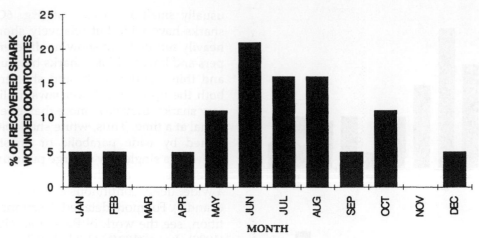

FIGURE 3 The seasonal distribution (by month) of odontocetes wounded by white sharks (N = 19).

edly during spring (Fig. 5). About 70% of carcasses were from gray whales *Eschrichtius robustus*.

It was difficult to estimate the percentage of cetacean carcasses that were scavenged by white sharks, because earlier stranding reports sometimes did not identify the species of shark that inflicted the bites. However, of the 37 large carcasses we examined, 56% (N = 21) were scavenged by white sharks; 21% (N = 8), by blue sharks; and 2% (N = 1), by sevengill sharks. Only 8% (N = 3) showed evidence of scavenging by both white and blue sharks.

Discussion

Proper Identification and Practical Uses of Shark Bites

Without firsthand observation of sharks feeding on marine mammals, it is difficult to correctly identify what species of shark may be involved when examining bites on a live or dead mammal. However, most species of large predatory sharks have a distinct dental morphology that, if properly examined, can lead

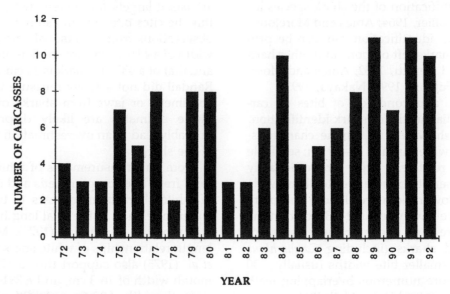

FIGURE 4 The annual numbers of large cetacean carcasses stranded along central California, 1972–1992.

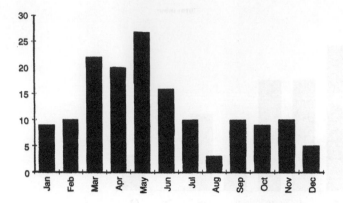

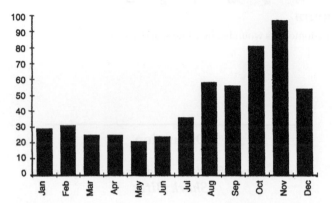

FIGURE 5   (A) The seasonal (by month) distribution of large cetacean carcasses stranded along central California, and (B) the seasonal occurrence of stranded pinnipeds showing white shark bites (after Chapter 24, by Long et al.).

to accurate specific or generic identification (see Chapter 3, by Hubbell). In special cases, tooth fragments recovered from a body or carcass provide an unquestionable identification of the shark species involved (Orr, 1959; Collier, 1964; Ames and Morejohn, 1980; Nakaya, 1993); identification can also be provided by serration marks left on bone and other hard objects (Deméré and Cerutti, 1982; Ames and Morejohn, 1980; Cigala-Fulgosi, 1990; Nakaya, 1993).

The size, shape, and condition of bites on carcasses can also be diagnostic for shark identification. White shark bites show several unique characteristics: the bite area is wide (usually 20–60 cm; see Table IV); the teeth are relatively large, proportionately long, and coarsely serrated; the upper teeth are wide and thin, and the lower teeth are slender and more conical; they have relatively few (12–13) teeth in the tooth row; and there is no overlap between adjacent teeth. Carcharhinid sharks (including *Carcharhinus* and *Prionace*) have smaller bite widths (usually <30 cm), smaller and more numerous overlapping teeth having finer serrations, and lower teeth that are very narrow and lack serrations; bites on cetaceans are

usually small and clean-cut (Fig. 6C and D). Tiger sharks have wide but relatively short teeth that are heavily serrated and show no difference between uppers and lowers. Mako sharks have unserrated, long, and thin anterior teeth, and shorter lateral teeth in both the upper and lower jaws; unlike other predatory sharks, they have more than one tooth row functional at a time. Thus, white shark bites can be determined by wide parabolic or arc-shaped wounds, showing a single row of large punctures that exhibit a difference between the upper and lower jaws, or else a wide, deep, continuous cut or circular bite (Figs. 1, 6, and 7). For more detailed descriptions of shark dentition, see the work of Bass *et al.* (1975), Compagno (1988), Long (1994), Hubbell (Chapter 3), Applegate and Espinosa-Arrubarrena (Chapter 4), and Purdy (Chapter 8).

Although analysis of bites can provide insight into the feeding activities of sharks, some caution is warranted. Several workers have measured the sizes of bites on carcasses as a way to estimate the relative size of the shark. Theoretically, the size of a white shark's gape increases with the total length of the shark, and the size of the bite is about equal to the size of the gape. Randall (1973) and Castro (1983) suggested that some white sharks may reach a maximum size of 750 cm, even though the largest reliably recorded white shark was about 720 cm (see Chapter 10, by Mollet *et al.*).

Size estimates based on bite sizes can be misleading due to several sources of error, detailed here.

Original Estimates of Bite Size and Total Length

Estimates of shark bite–body length relationships are based largely on the review by Randall (1973). In this, he cites bite–length estimates based on general observations from a personal communication: a bite width of a 441- to 457-cm shark is about 30.5 cm wide, and that of a 487-cm shark is 33 cm wide. Apparently, Randall did not actually measure the gapes of fresh specimens or jaws from sharks of known lengths. These estimates are likely disproportionate, and probably lead to an overestimation in the total length of the shark.

From our measurements of white shark jaws and from fresh bites on pinnipeds and cetaceans, the size of white shark bites appears to be proportionately larger in relation to the total length of the shark than was reported by Randall (1973). Mouth width–total length relationships of white sharks reported by Bass *et al.* (1975) also support this: a 170-cm shark had a mouth width of 16.3 cm, and a 391-cm specimen had a mouth width of 37.6 cm. Additionally, we measured a bite-sized piece of blubber, 40 cm across, that was

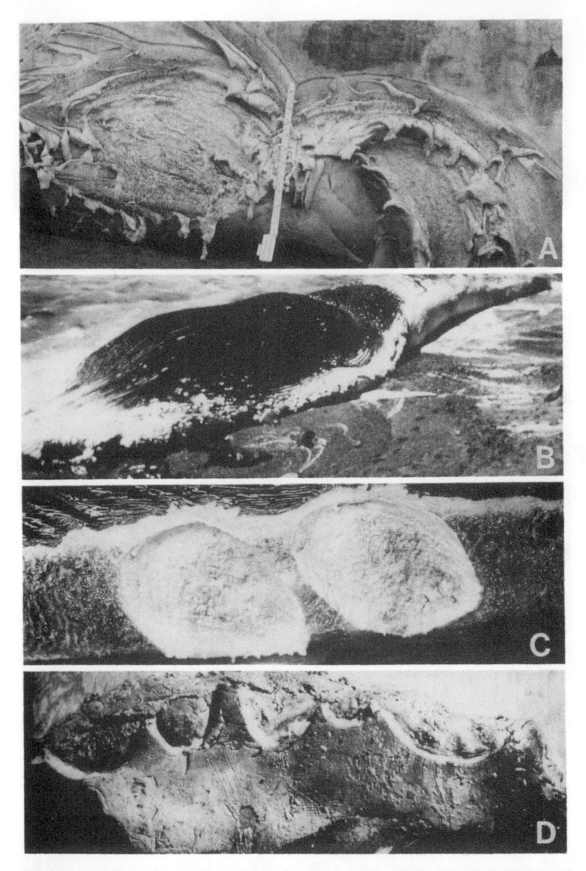

FIGURE 6 Shark bites on cetacean carcasses. (A) Two large bites on a mysticete carcass, showing the large size and often ragged edges made by white sharks (the measuring tape shows a 30-cm section for scale). (Photo by D. J. Long.) (B) A blue whale carcass, showing white shark scavenging on the entire lateral margin of the blubber layer. (Photo by R. E. Jones.) (C) Two small (16- to 18-cm-wide) blue shark bites on a mysticete carcass, showing the characteristic small circumference and clean bite edges. (Photo by K. Beckmann.) (D) A harbor porpoise carcass scavenged by a blue shark, showing clean-edged bites and selective feeding on the blubber layer. (Photo by D. J. Long.)

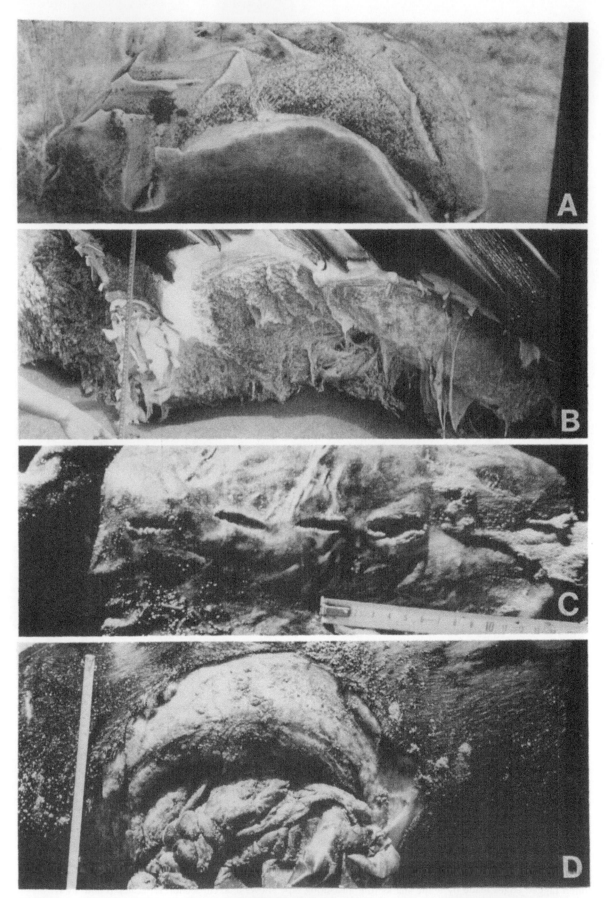

FIGURE 7 White shark bites on marine mammals, showing possible sources of interpretation error. (A) An incomplete bite on a mysticete carcass (the 23-cm-wide bite encompasses only the first half of the gape). (B) Three overlapping bites on a mysticete carcass, creating the illusion of a single bite >110 cm wide. (C) Tooth punctures on the hindflipper of an elephant seal that was curved when bitten (the flipper returned to its normal shape, but the punctures are linear and do not show the normal arc shape). (D) A large bite on a California sea lion *Zalophus californianus*, showing distortion and stretching of the bite area of the epidermis and a smaller bite arc in the blubber layer. (Photos by D. J. Long.)

regurgitated by a shark estimated to be about 450 cm. Similarly, we measured a fresh bite on an adult northern elephant seal *Mirounga angustirostris* that was 58 cm wide, even though the size of the attacking shark was <540 cm (for gape measurements, see Table IV).

Completeness of Single Bites

The width of the gape depends on where the measurements are taken. The gape is widest at the posterior of the jaw, and the gape becomes relatively smaller anterior to this. Accordingly, the size of the bite corresponds to the areal proportion of the bite, or simply, the amount of flesh that is within the gape (Table IV). If a white shark bites a carcass using only the anterior half of the gape, then the shark's total length would appear to be small (Fig. 7A). For an accurate estimate, the bite must encompass the whole of the gape, and data gathered should include a count of tooth punctures involved in the bite.

Angle and Force of Attack

Even parabolic bites were usually seen in cases in which the bite direction was relatively perpendicular to the body of the mammal. Single large uneven bites, however, were sometimes seen in some marine mammals that indicated a different direction in bite angle and/or force of bite impact against the body. The orientation of these abnormally large bites suggested that they could be inflicted if the attack angle was more oblique, or if the force of impact caused a "sliding" of the bite along the body. Tooth rakes, long cuts, and uncharacteristically wide bite parabolas were evidence of this phenomenon.

Overlapping Multiple Bites

We frequently encountered shark-scavenged cetacean carcasses that had seemingly huge single bites, but actually had two or three smaller bites that overlapped. This created the illusion of a single bite from a very large shark (Fig. 7B). Often, these bites overlapped in such a way that they created an even-looking parabola, a shape characteristic of a large single bite. More commonly, the large "bite" created by overlapping bites usually had a wide unnatural parabolic shape.

Increase of Bite Size by Scavengers

On several occasions, we have seen that the size of a single white shark bite had been increased by scavengers picking around the inner margin of the original bite, often feeding neatly without leaving evidence of their activity. On land, gulls *Larus* spp. and turkey vultures *Cathartes aura* often eat around a bite area, and in the water, blue sharks can do the same.

As with overlapping bites, this creates the illusion of a single large bite.

Postmortem Gassing and Expansion

When a carcass begins to decompose, gasses fill and expand the body cavity, blubber, and dermal tissues. This results is stretching and bloating of the outer tissues; a bite on the outer wall of the body can stretch as much as 25% larger than its original size in this way (Fig. 7D). On the basis of such bites, total length estimations of sharks would be exaggerated.

Decomposition around the Bite

As marine mammal carcasses decompose, tissue is sloughed off around exposed margins, including those of the bite. Advanced decomposition increases the apparent size of the bite area. Thus, any size estimates would be greater than the original.

Drying

In cases in which stranded carcasses are exposed to heat or sunlight for extended periods, portions may dry at different rates, causing shrinkage, stretching, and distortion. The epidermis, flippers, and fins usually dry first, and in some cases, different levels of tissue dry and distort at different rates (Fig. 7D). Therefore, sizes and shapes of bites on the body may be altered due to differential drying of tissues.

Flexion and Distortion

In some cases of predation on live animals or on moving carcasses, the shark may inflict a bite on a portion of the body that is flexed into a concave position. When the animal or carcass returns to a more linear shape, the typical parabolic shape of the bite stretches much wider (Figs. 1A and 7C). If the body or carcass had been arched in a convex direction, the size of the bite arc would decrease when the body returned to its natural shape.

To use shark bites as a valid estimate of shark size, bites and scars should (1) be on fresh carcasses that show little sign of gassing, bloating, decomposition, or scavenging; (2) be an even parabolic shape, to eliminate errors associated with overlapping bites and flexion and distortion; and (3) include individual tooth marks, to judge the areal proportion of the gape actually used in the bite.

Size Relationships of White Sharks and Cetacean Prey

General estimations of shark size based on measurements of bites from the carcasses, and comparisons of bites with gape measurements of known-length white sharks, show that large, but not small (<350-cm), white sharks prey on small odontocetes. This observation is supported by Arnold (1972), Stevens (1984), Cliff *et al.* (1989), and Bruce (1992), all of whom found cetacean remains to be more common in the stomachs of adult white sharks. This also agrees with Klimley's (1985b) and others' suggestion of a size-related ontogenetic shift in prey type, with small young white sharks feeding on fish and large adult white sharks preying more frequently on pinnipeds and cetaceans (Fig. 9).

Also, there seems to be a relationship between the size of the predator and the size of the prey. The largest shark-wounded animals in this study were juvenile Cuvier's and Stejneger's beaked whales, about 300 cm long, and the size of the attacking shark, estimated from the size and width of tooth punctures to be about 500 cm long. In all other specimens examined, the size of the shark was considerably larger than that of the prey, and in no cases have we seen premortem bites on a cetacean that was larger than the attacking shark. This implies an upper size limit to the prey of white sharks and may explain why larger cetaceans are not attacked.

Geographic Distribution of Predation

The majority of white shark-bitten odontocetes were recorded along the central California coast (Fig. 2); this is a pattern similar to that of shark-wounded pinnipeds (see Chapter 24, by Long *et al.*). Only one new record was from north of this area, that of a harbor porpoise from Long Beach, Pacific County, Washington, which is the only area north of California where white sharks have been recorded in significant numbers (Klimley, 1985b). Few reliable records of shark-bitten odontocetes have been confirmed south of central California, one being an adult female common dolphin collected at La Jolla, San Diego County, California, on May 27, 1989. This specimen had shark bites on the tail and caudal peduncle, but the species of shark was not determined.

It is possible that predation on odontocetes is more frequent than records indicate. White sharks feed heavily on pinnipeds, but pinnipeds can haul out onto land to heal from a shark injury without risk of another immediate attack (Klimley *et al.*, 1992; Long *et al.*, Chapter 24). On the other hand, cetaceans cannot retreat to shark-free areas, so they are more vulnerable to subsequent attacks. Most would be consumed, and few would be recorded by investigators. White sharks are known to hunt in nearshore waters (Klimley, 1985b; Klimley *et al.*, 1992), but most of the

species of cetaceans discussed here are found further offshore (Leatherwood *et al.*, 1982; Dohl *et al.*, 1983). The exception is the harbor porpoise, which is the most common inshore odontocete (Huber *et al.*, 1980; Dohl *et al.*, 1983; Szczepaniak, 1990). This may account for the relatively high incidence of predation on this species in comparison to the other small odontocete species found in the area, but offshore data are required to resolve this matter.

Another reason for the low number of shark attacks on cetaceans also relates to the availability and abundance of different species of pinnipeds, the primary prey items of white sharks in this area. In areas where white shark predation on odontocetes is more frequent (e.g., South Africa, Australia, and the Mediterranean), the abundance and diversity of pinnipeds are lower (Corkeron *et al.*, 1987; Cliff *et al.*, 1989; Cockcroft *et al.*, 1989; Bruce, 1992; Fergusson, Chapter 30). Last, the incidence of shark attack in eastern North Pacific waters may be low because only one species of shark is involved. In other areas, where predation levels on odontocetes are higher (as much as 30% or more), white sharks, as well as tiger sharks, bull sharks, and dusky sharks, may be involved (Wood *et al.*, 1970; Corkeron *et al.*, 1987; Cockcroft *et al.*, 1989). None of these other species live in the cooler waters off the central and northern California coast (Compagno, 1984a; Long, 1994; Seigel *et al.*, 1996). Overall, shark predation on small cetaceans along the North American west coast appears to be low.

Seasonality of Attacks

White shark attacks on cetaceans were documented year-round, but most wounded odontocetes were recorded during summer and autumn (Fig. 3). The timing of these attacks corresponded to the seasonal peak of shark-wounded pinnipeds along the central California mainland (see Chapter 24, by Long *et al.*). This indicates that during summer and early fall, a greater number of sharks may be feeding along this coast. The majority of records were from harbor porpoises, but this species is most abundant during autumn, and least abundant during summer (Szczepaniak, 1990). Additionally, harbor porpoises are most abundant in California waters north of Point Reyes (Dohl *et al.*, 1983), but all of the California records of white shark-bitten harbor porpoises were south of there. The same geographic patterns are exhibited among white shark-wounded pinnipeds (see Chapter 24, by Long *et al.*). This evidence indicates that seasonal trends of attacks are related to predator abundance, rather than prey abundance.

Method of Attack

White sharks may attack small odontocetes in relatively the same way that they attack pinnipeds and humans: by a swift initial bite that surprises and kills or disables the animal (Tricas and McCosker, 1984; Klimley, 1994). The usual attack pattern may vary when hunting dolphins and porpoises. Odontocetes have an anteriorly directed sonar and a lateral visual field, so a surprise attack must be in the "blind area" either from above, below, or behind to avoid visual detection or detection from the sonar (Fig. 8). To avoid detection, white sharks seem to focus their bites on particular areas of the cetacean body. On the basis of bite orientation on live and dead shark-wounded cetaceans, it appears that sharks attack four major areas of the body: (1) the caudal peduncle, (2) the urogenital region, (3) the abdominal area, and (4) the dorsum; bites on the head and the flanks are less common (Table III and Fig. 8). An ineffective initial attack, or "bite-and-spit" behavior (Tricas and McCosker, 1984; Klimley, 1994), would allow some animals to survive and escape with minor wounds, but this attack behavior has since been discounted (see Chapter 22, by Klimley *et al.*).

The caudal peduncle is a vulnerable area, because a single well-placed bite can sever swimming muscles, the spinal column, and major blood vessels (Burne, 1952), thus effectively immobilizing the cetacean. Previous accounts of shark predation on cetaceans confirm that this area is frequently attacked. Arnold (1972) reported that three harbor porpoises, in the stomach of a single white shark, each had the flukes severed from the tail stock. Previous studies on Hawaiian spinner dolphins *Stenella longirostris* (Norris and Dohl, 1980), bottle-nosed dolphins (Cockcroft *et al.*, 1989), and the pygmy sperm whale (Long, 1991a) also noted a high prevalence of bites in this area. Results of our investigation confirm the vulnerability of the caudal peduncle in odontocetes, 68% of the animals in this study showing bite wounds there.

Bites in the urogenital and abdominal regions can also be serious since these areas have dense networks of nerves, blood vessels, and vital organs (Burne, 1952). Many of the shark-wounded animals inspected in this study had bites in these areas. The dorsal areas have a much thicker blubber and muscle mass, and attacks in these areas would less likely cause death. This may be why healed shark wounds are more frequently seen on the backs of living dolphins and porpoises (Norris and Dohl, 1980; Corkeron *et al.*, 1987).

Scavenging

Whale carcasses are available off northern California throughout the year, but are particularly abun-

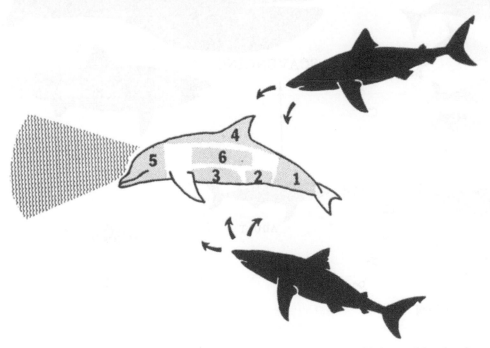

FIGURE 8 The major areas where white sharks bite small odontocetes: (1) the caudal peduncle, (2) the urogenital region, (3) the abdominal region, (4) the dorsal region, (5) the head, and (6) the flanks. Also illustrated is how white sharks probably attack from below, above, or behind, out of the range of the sonar field (wavy lines).

dant during spring (Fig. 5). Nearly 70% of carcasses are of the gray whale, which experience their peak natural mortality during their northward migration in spring (Poole, 1981). A smaller mortality peak among other large cetaceans occurs during fall and winter. Although the periodicity and locality of whale carcasses are unpredictable, they nonetheless are a regular food source for sharks in the area. The timing in abundance of cetacean carcasses corresponds to the time of the year that white shark predation on pinnipeds is low (see Chapter 24, by Long *et al.*) (Fig. 6). It is possible that white sharks shift from hunting pinnipeds in late summer and fall to scavenging large cetaceans in spring and early summer (Fig. 9).

A single large cetacean is a huge food source. A carcass may float for several weeks and exude a continuous slick of blood and oil that can attract sharks from long distances. Pratt *et al.* (1982) observed four to nine white sharks feeding on the carcass of a dead fin whale *Baleanoptera physalus* off New York, but they saw only one shark feeding at a time. One of us (D.J.L.) had the chance to observe a similar situation when the carcass of a blue whale *Baleanoptera musculus* was floating off San Francisco, California, in Au-

gust 1988. About five different white sharks fed on the carcass, but only one individual fed at a time. Even though blue sharks were in the area, none fed when the white sharks were feeding. Accordingly, we have found few carcasses on which bites from both white and blue sharks were evident (8% of 37 carcasses examined). All of this indicates, as noted by Pratt *et al.* (1982), that a feeding hierarchy exists among white sharks, and between white sharks and other species of sharks.

On the basis of relative sizes of bites on the whale carcasses we examined, it appears that only large (>350-cm) white sharks scavenge carcasses. Pratt *et al.* (1982) and Casey and Pratt (1985) observed only adult white sharks feeding on whale carcasses off the eastern United States, and Klimley (1985b) noted that only large white sharks scavenge basking shark carcasses off California. As with predation on small odontocetes, scavenging large carcasses may also be a form of ontogenetic dietary change (Fig. 9).

The stomach contents of one of the feeding sharks in the study by Pratt *et al.* (1982) weighed 28 kg, but our observations and estimates from examination and measurements of carcasses, based on the amount of

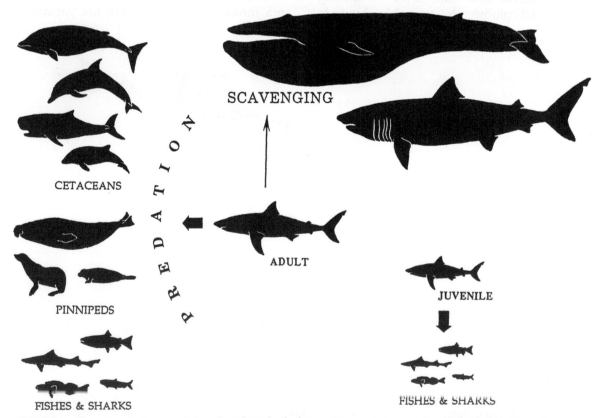

FIGURE 9 The ontogenetic prey shift in the white shark, from a diet consisting mainly of fishes among juveniles to one of mammals and larger fish among adults.

flesh removed by single white shark bites, show that large white sharks can easily consume at least 20 kg of flesh in a single bite. Pratt *et al.* (1982) also noted that the same individuals were seen around the carcass for at least 1 week. Both white and blue sharks seem to feed selectively on the blubber layer of a carcass (see Klimley, 1994); they move along the carcass and strip off and eat the outer layer (Fig. 6B and D). Carey *et al.* (1982) estimated that 30 kg of blubber may provide enough food to satisfy the energy needs of an average-sized white shark for 1.5 months. It seems, then, that the sharks may maximize their energy intake by preferentially feeding on cetacean blubber.

Summary

Nineteen specimens of odontocetes found stranded alive or recently dead on beaches along the West Coast of the United States, mostly from central California, showed bite wounds attributable to white sharks *C. carcharias*. Bites inflicted prior to death confirm that the white shark occasionally preys on small odontocetes in the eastern North Pacific Ocean. These prey include the harbor porpoise *P. phocoena*, Dall's porpoise *P. dalli*, Pacific white-sided dolphin *L. obliquidens*, Risso's dolphin *G. griseus*, pygmy sperm whale *K. breviceps*, dwarf sperm whale *K. simus*, Cuvier's beaked whale *Z. cavirostris*, and Stejneger's beaked whale *M. stejnegeri*. On the basis of stranding records, white shark predation on cetaceans in the region seems uncommon. Most wounded odontocetes were recovered during summer and early fall, and seasonal trends were likely due to white shark, not cetacean, abundance. Usually, large (>350-cm) white sharks attack cetaceans, and most bites on recovered carcasses are on the caudal peduncle or the abdomen. White sharks also commonly scavenge on carcasses of large cetaceans. Up to a dozen such carcasses are beached along the central California coast each year, and many show evidence of scavenging by sharks. In the eastern North Pacific, the natural mortality of large cetaceans offers a potential food source for white sharks during spring and early summer.

Acknowledgments

Thanks go to the Museum of Vertebrate Zoology, University of California, Berkeley; R. Bandar, I. D. Szczepaniak, L. Thomsen, D. Catania, B. Cutler, and T. Iwamoto of the California Academy of Sciences; K. D. Hanni, K. Beckmann, and M. Webber of the Marine Mammal Center, Sausalito; J. A. Seigel and J. Heyning of the Los Angeles County Museum of Natural History; H. J. Walker, Jr., of the Scripps Institution of Oceanography, La Jolla; and C. Keiper, J. Cordaro, and field volunteers of the NMMSN, National Marine Fisheries Service. The comments by anonymous reviewers and the editors certainly improved this chapter.

28

White Shark-Inflicted Wounds of Sea Otters in California, 1968–1992

JACK A. AMES
*California Department of
Fish and Game
Monterey, California*

JOHN J. GEIBEL
*California Department of
Fish and Game
Menlo Park, California*

FREDERICH E. WENDELL *and*
CHRISTINE A. PATTISON
*California Department of Fish and Game
Morro Bay, California*

Introduction

California sea otters *Enhydra lutris* occupy a relatively narrow band of nearshore habitat. They are easy to count, as they spend much time at the sea surface within view from land. Except for those drowned and held for hours or longer at depth [as with net-drowned otters; California Department of Fish and Game (CDFG), unpublished data], we believe that dead sea otters almost always float. Given the area's prevalent onshore winds, we believe that a sufficiently high percentage of carcasses reach shore to provide a reasonable sample of total mortality.

Despite biases that have occurred in the rate of recovery through time and throughout the sea otter's range (Jameson, 1986; Bodkin and Jameson, 1991; Gerrodette, 1983), available data collected provide insight into the relative contribution of various causes of death. Data collected in 1968–1992 were assessed to approximate the annual sea otter mortality caused by white sharks *Carcharodon carcharias*. This chapter provides an update of the work of Ames and Morejohn (1980) through 1992, using the same criteria to diagnose white shark bites. Of the sharks that inhabit the area, only white sharks are known to bite sea otters (Ames and Morejohn, 1980).

California sea otters doubled in numbers and geographical range from 1968 to 1992 (Fig. 1). In 1968, about 1200 sea otters (Wild and Ames, 1974; CDFG,

1976) occupied approximately 200 km of nearshore habitat between Cambria and Monterey, as measured along a smoothed 5-fm isobath (CDFG, unpublished data). In 1992, about 2400 sea otters (2101 counted; R. J. Jameson, personal communication) occupied approximately 400 km of range, from the Santa Maria River to Año Nuevo Island (CDFG, unpublished data).

To look for population effects on sea otters, we examined geographical subsets of the mortality data where both shark bites were most frequent and where detailed sea otter census data were available. We selected Año Nuevo Island and Point Joe for more detailed analysis (Fig. 1).

The sea otter range reached the vicinity of Point Joe in 1958, and the species occupied the area completely by 1962 (CDFG, 1976). The Point Joe area included the 10 km from Cypress Point to Point Pinos (Fig. 1). Therefore, the area had been part of the sea otter range for the 6 years before, and throughout, the 25 years of this study.

In 1982, the otter population began to expand into the Año Nuevo Island area, defined as the 32-km stretch of sea otter habitat between Sand Hill Bluff and Pigeon Point (Fig. 1). By 1993, the northern portion of this area was still beyond what was generally recognized as the northern extremity of the range, and we assume the area was not fully occupied (CDFG, unpublished data). Prior to 1982, only a few

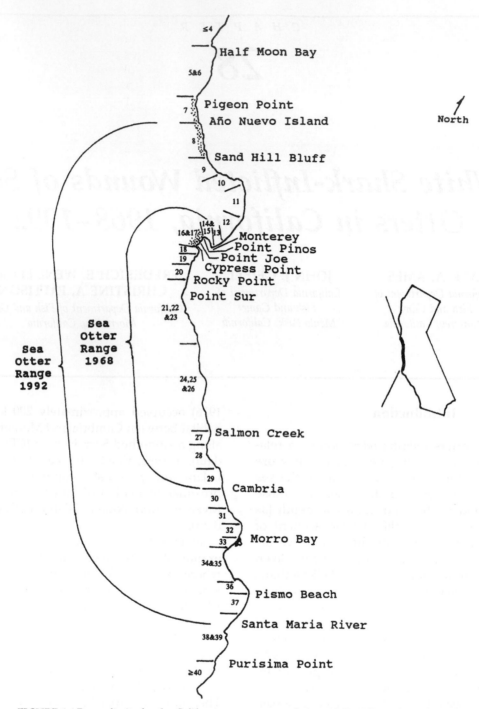

FIGURE 1 Range limits for the California sea otter population, 1968–1992, and area boundaries in which dead sea otters were recovered. Stippled areas are considered in detail in the text.

individuals and occasional small groups of sea otters had been observed in the area (CDFG, unpublished data).

Methods

Efforts to obtain data from all dead otters reported in California began in 1968. Most of the dead otters reported in 1968–1992 were examined, and many received a detailed laboratory necropsy. Some of these data have been reported previously (Mattison and Hubbard, 1969; Wild and Ames, 1974; Morejohn et al., 1975; Ames and Morejohn, 1980; Ames et al., 1983; Bodkin and Jameson, 1991; Riedman and Estes, 1990).

As a result of postmortem examinations, dead otters were assigned to one of 16 categories (Ames *et al.*, 1983). Three of these categories were characterized by major cuts on the carcass. Shark bites listed as certain or near certain were diagnosed by (1) white shark tooth enamel fragments present, (2) scratch patterns on bone or cartilage that match the serrate edge of white shark teeth, or (3) on various parts of otter carcasses, multiple cuts that are "stab-like" in appearance (Ames and Morejohn, 1980). Less certain, but in our view still likely, shark bite victims are those with major cuts but lacking any of the above, more diagnostic wounds. This is mainly because we have

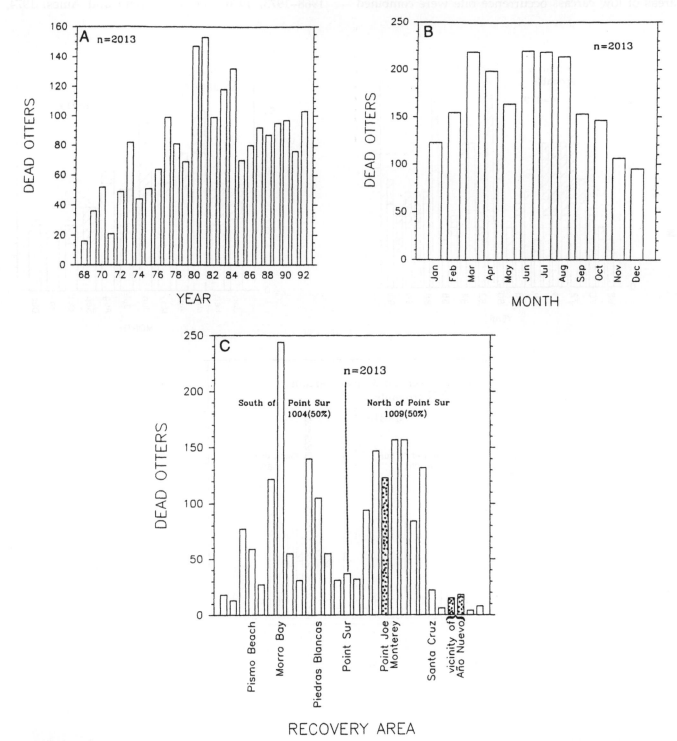

FIGURE 2 Total recorded sea otter mortality in California by (a) year, (b) cumulative month, and (c) area, 1968–1992. Stippled areas are considered in detail in the text.

occasionally seen dead otters having definitive shark bite wounds on one part of the body and cuts without pattern on other parts.

Each dead otter was assigned to one of 37 recovery areas (4–40, Fig. 1) of unequal size but defined by convenient landmarks (Morejohn *et al.*, 1975; Ames *et al.*, 1983; CDFG, unpublished data). Furthermore, areas of low carcass occurrence rate were combined

into the 28 larger areas used in this analysis (Figs. 1, 2c, 3c, and 4).

To assess the sea otter population size in the vicinity of Point Joe, we used data from 31 ground counts in 1980–1992 (Table I). We assumed the area supported a similar population size from 1968 to 1980 based on a variety of air and/or ground counts in 1968–1973, 1976, and 1979 (Wild and Ames, 1974;

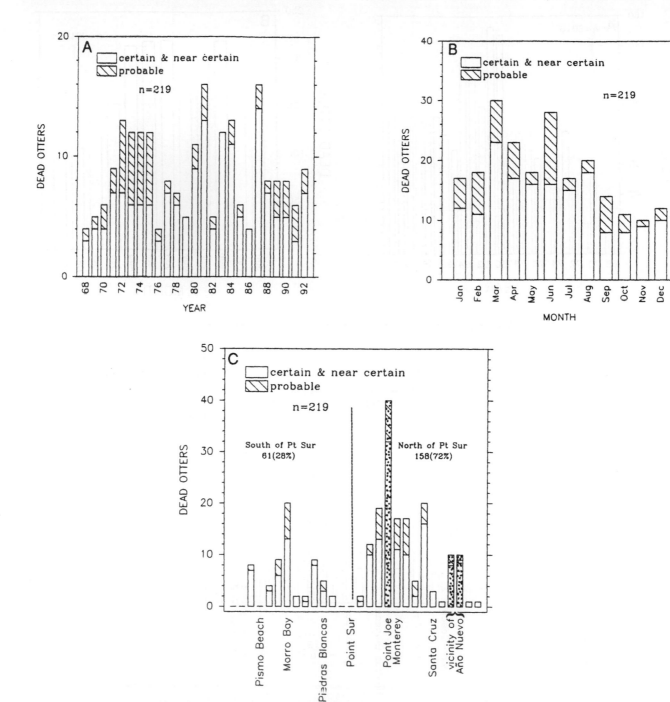

FIGURE 3 Shark-bitten dead sea otters in California by (a) year, (b) cumulative month, and (c) area, 1968–1992. Stippled areas are considered in detail in the text.

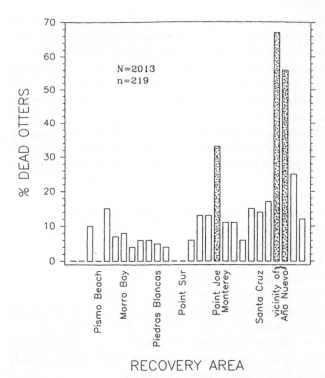

FIGURE 4 Shark-bitten dead California sea otters as a percentage of the total recorded mortality by area, 1968–1992. Stippled areas are considered in detail in the text.

CDFG, 1976, unpublished data). Previous studies also suggest that sea otter densities in the established range tend to be stable (Wendell *et al.*, 1986). Based partly on the 95% estimated counting accuracy of ground counts in California (Estes and Jameson, 1988), we judged that 85% of the sea otters would have been counted. The result was adjusted upward to account for the low viewing angle and great distances from viewers to otters (up to 1.4 km) in the estimates of local population size. For each year, 1980–1992, we estimated an average population size based on an expansion of the annual average count.

To develop a rough estimate of the local sea otter population size in the Año Nuevo area, we used aerial count data from 49 overflights conducted between February 1982 and November 1992 (Table II) (CDFG, unpublished data). Counts were averaged for each year and doubled based on the assumption that only about half of the otters were seen due to the usual prevalence of wind chop and dispersed animals (Wild and Ames, 1974; CDFG, 1976, unpublished data).

Results

White shark-bitten otters have occurred each year in all months (but not in each month each year) and

TABLE I Sea Otter Mortality and Population Size in the Point Joe Area[a] for 1968–1992

Year	Dead otters (all causes)	Shark-bitten otters[b]	Local otter population[c]	Shark bites as % of local otter population
1968	1	1	75	1
1969	1	1	75	1
1970	3	1	75	1
1971	6	4	75	5
1972	5	5	75	7
1973	6	1	75	1
1974	6	3	75	4
1975	4	3	75	4
1976	2	0	75	0
1977	1	1	75	1
1978	7	2	75	3
1979	3	0	75	0
1980	8	0	90	0
1981	11	2	65	3
1982	6	1	50	2
1983	9	2	50	4
1984	9	5	50	10
1985	4	1	75	1
1986	5	1	110	1
1987	7	2	75	3
1988	4	0	100	0
1989	4	2	75	3
1990	3	0	75	0
1991	2	0	85	0
1992	6	2	100	2
Total	123	40		2

[a]Defined here as a 10-km habitat measured along the 5-fm isobath extending from Cypress Point to Point Pinos.

[b]Includes all dead sea otters with cuts, based on the assumption that shark bites are the most likely source of mortality.

[c]Derived from 31 ground counts conducted between 1980 and 1992. The assumption was that 85% of the otters present were counted. Our population estimates were adjusted based on this assumption. We used 75 otters, the average of later estimates as a population estimate during 1968–1979, as a "ball park" population size, because the area had been fully reoccupied since 1962.

through a large part of the range. Curiously, no shark-bitten carcasses have been detected among the 68 recorded in the 100 km central portion of the range from Rocky Point to Salmon Creek (Figs. 1 and 3a).

Of the 2013 dead sea otters recorded, 219 (11%) were white shark bitten. If white sharks actually consume sea otters, or if they kill otters without cutting them, the actual mortality level in the population from this source would be higher. However, if sharks

TABLE II Sea Otter Mortality and Population Size in the Año Nuevo Island Area[a] for 1968–1992

Year	Dead otters (all causes)	Shark-bitten otters[b]	Local otter population[c]	Shark bites as % of local otter population
1968	0		0	
1969	0		0	
1970	0		0	
1971	0		0	
1972	0		0	
1973	0		0	
1974	0		0	
1975	0		0	
1976	1	0	d	
1977	1	0	d	
1978	1	0	d	
1979	0	1	d	
1980	1	1	d	
1981	0		d	
1982	0		2	
1983	2	2	24	17
1984	4	2	10	20
1985	0		14	
1986	2	0	20	0
1987	5	5	44	11
1988	1	1	26	4
1989	2	0	18	
1990	7	5	30	17
1991	4	2	28	7
1992	2	2	38	5
Total	33	20		8

[a]Defined here as the 32-km habitat measured along the 5-fm isobath extending from Sand Hill Bluff to Pigeon Point.

[b]Includes all dead otters with cuts, based on the assumption that shark bites are the most likely source of mortality.

[c]Derived from 49 census flights conducted during 1982–1992. The mean annual count was doubled, based on the assumption that only about half of the otters present were seen from the plane in this area.

[d]Unknown, but thought to be very small.

bite dead floating otters (a possibility that can often, but not always, be ruled out), an opposite bias would occur. There is no direct evidence that white sharks consume sea otters.

We recorded 1004 (50%) otter carcasses to the south and 1009 (50%) to the north of Point Sur (Fig. 2c); 61 (28%) to the south and 158 (72%) to the north were shark bitten (Fig. 3c). If we assume a binomial distri-

bution with equal probability of shark-bitten otters being recovered north and south of Point Sur, the 99% confidence bounds range from 78 to 133 occurrences in either direction. Consequently, rejecting the null hypothesis, we conclude that otter mortality due to shark bites occurred most frequently to the north of Point Sur. Areas north of Point Sur with the highest frequency of occurrence of shark-bitten carcasses were the Point Joe and Año Nuevo Island areas.

It appeared to us that shark-bitten otters were clustered in time. To test this hypothesis, we compared the time distribution of shark-bitten carcasses to those expected from a Poisson distribution having an interval of 41.5 days—the mean time between shark-bitten carcasses—using Pearson's test for goodness of fit (Bhattacharyya and Johnson, 1977, pp. 424–428) (Fig. 5); intervals with four, five, and six carcasses were combined because the χ^2 test is unstable if cells have low expected numbers. On this basis, we rejected the null hypothesis (χ^2 at α 0.01 = 13.28 < 14.50, df = 4, N = 220 intervals) and accepted the alternative hypothesis that there were several periods during which shark-bitten otters were clustered.

It also appeared to us that the occurrence of shark-bitten otters remained fairly constant during the study period. To test this, we divided the study period into four equal 6-year parts (eliminating 1968 to create equal-size groupings) and compared expected versus observed numbers of shark-bitten carcasses (Fig. 6). As a result, we accepted the null hypothesis (χ^2 at α 0.1 = 6.25 > 1.56, df = 3). So, despite an increasing otter population and an overall increase in carcass recoveries, no concomitant increase in shark-bitten carcasses recoveries occurred.

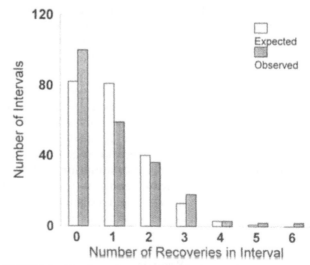

FIGURE 5 Number of expected and observed shark-bitten sea otter carcasses recovered in 220 time intervals of 41.5 days each, 1968–1992.

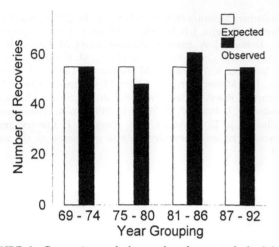

FIGURE 6 Comparison of observed and expected shark-bitten sea otter carcasses recovered during four time intervals of 6 years each, 1969–1992.

The population estimates in the vicinity of Point Joe varied between 50 and 110 sea otters from 1980 to 1992, and averaged 75 (Table I). Prior to 1980, we used the area's average estimated population size of 75 otters. Forty of the 123 dead otters (33%) recovered from the vicinity of Point Joe were classified as shark bitten (Figs. 2c and 3c and Table I). Annual recorded shark bite mortality varied from 0% to 10% of the estimated local otter population through the period, with a 25-year average of 2%.

Twenty of 33 dead sea otters (61%) recovered from the Año Nuevo Island area were classified as shark bitten (Figs. 2c and 3c). That otter population varied in size from 2 to 44, generally increasing through time (Table II). Annual recorded shark bite mortality, as a percentage of population size, ranged from 0% to 20%, and averaged 8% for the 11-year period 1982–1992.

Discussion

The effect of shark-induced mortality on sea otter population growth is unknown, but probably is not significant. On the basis of sea otter mortality data and published population estimates for the late 1960s and 1970s, and range wide counts for the 1980s and early 1990s (Wild and Ames, 1974; CDFG, 1976; Geibel and Miller, 1984; R. J. Jameson, personal communication), we estimated the average annual mortality due to white shark bites to be about 0.5%. However, the fact that a considerable proportion of this mortality occurred at the northern periphery of the range indicates that white sharks could have a range-limiting effect.

The lack of an increase in shark-bitten otters over the study period is remarkable, because otters doubled their numbers and range and the northern edge of the range expanded into areas of high white shark activity. Possible reasons for this situation are (1) a reduction in white shark numbers in the sea otter's range, (2) greater population growth of other prey species (e.g., elephant seal *Mirounga angustirostris*, California sea lion *Zalophus californianus*, and harbor seal *Phoca vitulina*) within the sea otter's range (Lowry *et al.*, 1992; Barlow *et al.*, 1993; Hanan, 1993), and (3) a preference by white sharks for these other prey (e.g., pinnipeds) over sea otters (Klimley, 1994).

Sea otters occupied the Point Joe area for the entire period of our study, and the population there was considered to have been stable. The effect of a localized 2% annual mortality on the local or range wide population is unknown.

Sea otters began the continuous occupation of the Año Nuevo Island area in the early 1980s, but the effect of the 8% white shark-induced mortality estimated for this area is not apparent. The fact that the vicinity is occupied primarily by male otters (CDFG, unpublished data) certainly complicates any assessment of the significance of shark-induced mortality on otter population expansion. The annual removal of 8% of the local sea otter population would have an effect on the otter prey base, tending to lengthen the time of expansion over a unit portion of range.

Summary

Lethal wounds on sea otters *E. lutris* caused by white sharks *C. carcharias* in California have been confirmed by (1) shark tooth enamel fragments remaining in otters' wounds, (2) scratch patterns on otter bone or cartilage that match the serrate edge of white shark teeth, and (3) multiple cuts on various aspects of otter carcasses, some of which may be "stab-like" in appearance. Conclusive evidence that white sharks eat sea otters is not available. It is possible that they merely bite and release otters, often killing them in the process.

Approximately 8% (163 of 2013) of California sea otter carcasses inspected from 1968 to 1992 exhibited very strong evidence of having been wounded by white sharks; including all dead otters with cuts (N = 219), the prevalence reaches 11%. Lethally bitten otters occurred in all months of the year and throughout most of the otter's California range, but seasonal and spatial concentrations were also apparent. The number of shark-bitten otter carcasses recovered annually, rangewide, varied little during the study period, despite a doubling of the sea otter population.

We estimate that the annual sea otter mortality caused by white sharks averaged 0.5% rangewide. However, in the vicinity of Año Nuevo Island, white shark-caused mortality reached 20% (averaging 8%) of that subpopulation in some years.

Acknowledgments

Several institutions and many people contributed to this chapter. It would be impossible to name them all, but notable are the Santa Barbara Museum of Natural History, the U.S. Fish and Wildlife Service, C. Benz, J. L. Bodkin, R. Burge, L. J. V. Compagno, D. Costa, E. Ebert, J. A. Estes, E. Faurot-Daniels, M. Flippo, W. Follett, A. Giles, B. Green-Ross, R. A. Hardy, M. Harris, B. Hatfield, G. L. Jameson, R. J. Jameson, M. Kenner, R. Lea, D. B. Lewis, J. Mason, J. A. Mattison, Jr., D. J. Miller, M. Odemar, S. Owen, M. L. Riedman, A. Roest, G. Sanders, S. Schultz, N. Siepel, M. Staedler, J. Vandevere, L. Wade, P. W. Wild, T. Williams, K. Wilson, and C. Woodhouse, Jr. D. G. Ainley, J. A. Estes, A. P. Klimley, and an anonymous reviewer provided helpful comments on the manuscript.

29

Records of White Shark-Bitten Leatherback Sea Turtles along the Central California Coast

DOUGLAS J. LONG

Department of Ichthyology
California Academy of Sciences
Golden Gate Park
San Francisco, California

Introduction

Adult white sharks *Carcharodon carcharias* feed on fishes and marine mammals, and are suspected of preying on sea turtles. Bigelow and Schroeder (1948) and Compagno (1984a) listed sea turtles as dietary items of white sharks, but did not identify the species of turtles eaten, nor did they cite specific examples or give further information. In addition, analyses of gut contents of white sharks (Bass *et al.*, 1975; Stevens, 1984; Klimley, 1985b; Cliff *et al.*, 1989; Bruce, 1992; Fergusson, Chapter 30) did not list sea turtles as prey items. The only published accounts of a white shark feeding on a sea turtle is Coles' (1919) observation of a large (approximately 550-cm) white shark attacking a live adult loggerhead sea turtle *Caretta caretta* off the coast of North Carolina, but additional published records of white sharks feeding on sea turtles do not exist. In this chapter, new evidence of white sharks feeding on sea turtles is presented.

Materials and Methods

Sea turtles in U.S. waters are federally protected; thus, all live and dead turtles that strand along the West Coast are investigated and documented (see Seagars and Jozwiak, 1991; Long *et al.*, Chapter 24).

Between January 1991 and December 1994, seven stranded leatherback sea turtles *Dermochelys coriacea* were reported from the coast of central California, including two sea turtles showing evidence of feeding by sharks. Wounds were examined and evaluated, following the criteria used by Long *et al.* (Chapter 24) and Long and Jones (Chapter 27) to determine the source of the wounds.

Results

On February 16, 1992, an adult female leatherback sea turtle (158 cm total carapace length) was collected on the beach at Half Moon Bay, San Mateo County, California (37°28′ N, 122°27′ W) (Fig. 1-1). Both the left front and right hind flippers were bitten off near the base. Examination of the wounds showed punctures, cuts, and serration marks attributable to an adult white shark (Fig. 2A). Portions of this specimen were retrieved and are curated at the California Academy of Sciences.

On August 11, 1993, another adult female (162 cm estimated total carapace length) washed ashore on Brighton Beach, Bolinas, Marin County, California (37°54′ N, 122°43′ W) (Fig. 1-2). The turtle had the posterior end of the carapace and the left hind flipper bitten off, and a large bite (about 48 cm wide) had

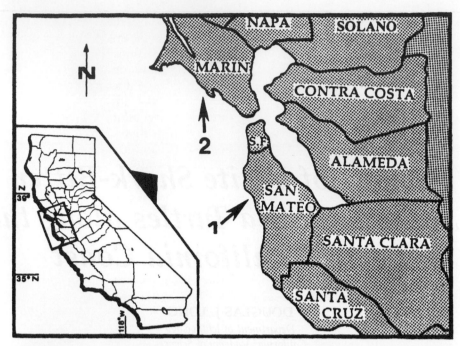

FIGURE 1 The central California coast, showing locations where the two white shark-bitten leatherback sea turtles were recovered. Numbers correspond to the specimen case numbers used in the text.

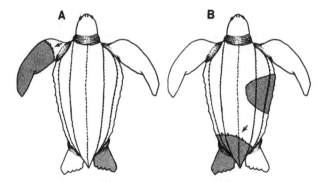

FIGURE 2 Illustrations of the two leatherback sea turtles, showing location, position, and shape of the shark bites (shaded areas). Arrows point to individual tooth punctures and cuts: (A) an adult female from Half Moon Bay (specimen 1); (B) an adult female from Bolinas (specimen 2).

removed the right central section of the carapace (Fig. 2B). Based on jaw measurements taken by Long and Jones (Chapter 27), the size of the white shark was estimated to be that of an adult 500–550 cm in total length. Portions of this turtle were collected and are curated at the University of California, Berkeley, Museum of Vertebrate Zoology.

Discussion

On the basis of the large deep bites and wide tooth punctures observed on the leatherback turtle carcasses, I conclude that the wounds were unmistakably inflicted by white sharks. Unfortunately, both turtles had been dead for several days prior to examination, so it was difficult to confirm whether white shark predation was the direct cause of death for the turtles, or whether the bites were inflicted after death. Postmortem examinations of the carcasses, however, did not reveal death from more common causes such as disease, drowning from net entanglement, trauma from boat collision, or complications from ingestion of plastics. In any case, the presence of large bites and removal of large portions of tissue from the carcasses are direct evidence of feeding (either predation or scavenging) by white sharks.

These records are important in several respects. First, the large bites taken out of the sea turtle shells attest to the strength and bite force of the white shark. Unlike most sea turtles, whose shells are covered with large bony plates, the leatherback turtle has a thick shell composed of small bony tile-like plates covered with a thick layer of cartilage and skin (Marquez, 1990). While not as solid as the shells of other sea turtles, the leatherback's shell is still very tough. The bites that penetrated both the carapace and the plastron in the second turtle specimen showed that a great amount of force was needed. The tooth morphology of tiger sharks *Galeocerdo curvieri* is believed to be adapted for preying on sea turtles (Witzell, 1987), and they are a common food item of this shark species (Randall, 1992; Simpendorfer, 1992; Stevens and McLoughlin, 1991). The morphology and microstructure of white shark teeth are also designed to penetrate and cut hard objects with application of heavy forces (Preuschoft *et al.*, 1974; Frazzetta, 1988), including mammal bones and turtle shells. Therefore, predation would not be prevented by the turtle's carapace.

Second, turtles previously have not been reported as dietary items for white sharks in the eastern North Pacific (Klimley, 1985b). It is not surprising that these new feeding records came from central California, as this area has a high incidence of white shark predation on marine mammals (Le Boeuf *et al.*, 1982; Klimley, 1985b; Long *et al.*, Chapter 24; Jones and Long, Chapter 27). Additionally, leatherbacks are the most common sea turtle along central California and are most frequently seen offshore or recovered as dead stranded individuals in the summer and fall (Starbird *et al.*, 1993). It is interesting that while one of the carcasses was collected in August, the other was collected in February, during a season when both leatherbacks are rare and white shark predation on pinnipeds is low (Klimley *et al.*, 1992; Starbird *et al.*, 1993; Long *et al.*, Chapter 24).

Last, these are the first records of the white shark as a possible predator on the leatherback sea turtle. Off southern Australia, Cropp (1979) described what he believed to be a leatherback's defensive behavior in the presence of a white shark, a behavior that included erratic diving, rolling on the surface of the water, and violent flailing of the turtle's flippers as it floated on its back. The shark, however, attempted no attack. Killer whales *Orcinus orca* are the leatherback's only positively confirmed predator (Caldwell and Caldwell, 1969), but tiger sharks are also suspected as predators (Keinath and Musick, 1993). From the evidence presented here, white sharks are also likely predators on the leatherback sea turtle.

Few records document shark predation on sea turtles, and thus, little can be said regarding trends in predatory dynamics. Sea turtles are probably uncommon dietary items for white sharks, because sea turtles are not abundant in temperate seas frequented by white sharks (Compagno, 1984a; Marquez, 1990). Alternatively, the seeming rarity of sea turtles in the diet of white sharks may be due to the turtles' possible low food value when compared to that of marine mammals. Klimley (1994) and Long and Jones (Chapter 27) discuss the shark's preference for prey with high fat content; it is possible that the sea turtles in this study were attacked and rejected after several bites. Only additional records and further observations of white shark–sea turtle interactions can lead to more concrete conclusions on the nature of their predatory interactions.

Summary

White sharks *C. carcharias* are alleged to feed on sea turtles, but little documented evidence exists. Presented here are two cases of white shark-bitten leatherback se turtles *D. coriacea*, collected along the central California coast. Direct predation or postmortem scavenging could not be discerned. These records are among the few cases documenting white sharks feeding on sea turtles, including the first records from the Pacific.

Acknowledgments

Data on the leatherback sea turtles were collected by me and other volunteer personnel from the National Marine Mammal Stranding Network under the auspices of the National Marine Fisheries Service and the National Oceanographic and Atmospheric Administration. I thank R. E. Jones (Museum of Vertebrate Zoology, University of California, Berkeley), S. Webb and K. Hansen (Point Reyes Bird Observatory), and R. Bandar (California Academy of Sciences, San Francisco) for assistance. D. G. Ainley, A. P. Klimley, P. Pyle, and an anonymous reviewer provided helpful comments on the manuscript.

30

Distribution and Autecology of the White Shark in the Eastern North Atlantic Ocean and the Mediterranean Sea

IAN K. FERGUSSON

European Shark Research Bureau
Welwyn Garden City, Hertfordshire, England

Introduction

The white shark *Carcharodon carcharias* ranges throughout temperate oceans, with large individuals known to penetrate tropical waters (Compagno, 1984a; see also Chapter 31, by Gadig and Rosa). The species has been studied in the field most regularly at only the few localities where it is relatively abundant, such as pinniped colonies along the central California coast (Ainley *et al.*, 1981, 1985; Le Boeuf *et al.*, 1982; Klimley, 1985b; Klimley *et al.*, 1992, Chapter 16; Pyle *et al.*, Chapter 26; Long, 1994; Long *et al.*, Chapter 24; Goldman, Chapter 11), the shelf waters of the mid-Atlantic Bight (Carey *et al.*, 1982; Pratt *et al.*, 1982; Casey and Pratt, 1985; Pratt, Chapter 13), South Australia (Tricas and McCosker, 1984; McCosker, 1987; Bruce, 1992; Strong *et al.*, 1992), and the Cape and KwaZulu–Natal provinces, South Africa (Ferreira and Ferreira, Chapter 35, and Cliff *et al.*, 1989, Chapter 36). A propensity for this species to occasionally attack humans has inspired much media attention and has served as the focus for a number of scientific studies (Bolin, 1954; Bini, 1960; Collier, 1964, 1993; Follett, 1966; Miller and Collier, 1981; Engaña and McCosker, 1984; Lea and Miller, 1985; McCosker and Lea, Chapter 39;

Levine, Chapter 40; West, Chapter 41; Burgess and Callahan, Chapter 42). Scientific attention has also been afforded to this species in those regions where it is less commonly encountered. Taylor (1985) reviewed occurrences at the Hawaiian Islands, and, more recently, Nakano and Nakaya (1987) and Uchida *et al.* (Chapter 14) gave contemporary data on Japanese specimens. Francis (Chapter 15) described a pregnant female and free-swimming neonates in New Zealand waters, and Ellis and McCosker (1991) have discussed Azorean records. Data on South American occurrences have been presented by Siccardi *et al.* (1981) and Gadig and Rosa (Chapter 31). Opportunity for study at all of these locales comes principally via the fortuitous capture of specimens taken as accidental fisheries by-catch.

This chapter presents a contemporary review and analysis of white shark occurrences within the eastern North Atlantic Ocean and the Mediterranean Sea. The literature is scarce citing the white shark (under numerous synonyms) as an inhabitant of the region, and few reliable data can be gleaned from what does exist. Few white shark captures within the Mediterranean have been reported to the scientific community, rendering problematic the collation of available data. With this study, I hope to inspire the future collation of workable material from the area and draw preliminary conclusions about the biology of white sharks in the region (principally, the Mediterranean Sea).

Materials and Methods

Over a 3-year period, I searched for all available historical and contemporary citations of white shark occurrences in the Mediterranean Sea and the eastern North Atlantic Ocean, using the Zoological Record and extensive facilities of the Library and Fish Section of the British Museum of Natural History and the Marine Biological Association, Plymouth, England. A number of largely unknown published works were made available to me by researchers based in other European Community nations. In 1990, I created a captures and sightings database, serving to computerize and standardize records. I prepared a questionnaire that I forwarded to researchers working in the study region. Pre-1987 specimens were identified from a variety of published and unpublished information; the majority of post-1987 specimens were accompanied by photographic evidence that confirmed the species' identity. In addition, stated values of total length (TL) were assessed (where not measured) by comparison of submitted photographs with a selection of morphometric and photographic white shark records housed at the California Academy of Sciences [ranging from 218 cm (specimen CAS 26366) to 510 cm (CAS 26245)]. Records of sharks held on database by the European Shark Research Bureau are quoted in this chapter with the catalog numbers commencing with M.

Commonly, the carcasses of white sharks taken in the region are dismembered, sold, or dumped prior to scientific examination. Consequently, a large proportion of the existing literature is anecdotal and, often, of doubtful accuracy. Many literature works in which, in my opinion, specimens were misidentified were thus excluded from consideration. The basking shark *Cetorhinus maximus*, the shortfin mako *Isurus oxyrinchus*, and the porbeagle *Lamna nasus* inhabit the region and are occasionally confused with *C. carcharias* (Casey and Pratt, 1985). In particular, among Mediterranean fishermen, the frequently encountered *I. oxyrinchus* is often confused with juvenile white sharks (L. del Cerro and F. Cigala-Fulgosi, personal communication).

All measurements are given as TL, as defined by Compagno (1984a, as TOT). Historical records were generally quoted in metric values and for the purposes of this study, it is assumed that these expressed TL, measured in a manner similar to that of the current methodology. The lengths of specimens >530 cm TL, when accompanied by mass that appeared erroneous (on the basis of the method of Mollet and Cailliet, Chapter 9), were estimated, when possible, by inspecting photographic evidence (see, e.g., an allegedly 700 cm TL female illustrated in Tortonese, 1965). Other morphometric values, when cited, follow the definitions and abbreviations of Compagno (1984a).

The majority of specimens were not examined to assess reproductive maturity. For most analyses, I sorted records into four TL classes: pups (<185 cm), juveniles (185–250 cm), subadults (251–450 cm) and adults (>450 cm). Essentially, a TL value of 457 cm was chosen to differentiate immature and mature sharks of either sex, following the method of Casey and Pratt (1985), although white sharks mature at size ranges that differ geographically (see Chapter 13, by Pratt, and Chapter 15, by Francis). The discrimination between mature and immature sharks in this study thus remains tentative, lying between 350 and 500 cm TL.

To investigate the effects of sea-surface temperature (SST) on occurrence, I cross-referenced early capture records with oceanographic charts from the Koninklijk Nederlands Meteorologisch Institut; records from the 1980s and 1990s were compared against charts prepared from satellite imagery (SATMER charts).

Results

Areal Distribution

Eastern North Atlantic Ocean

Carcharodon carcharias is rare in the eastern North Atlantic (J.-C. Quéro, cited in Whitehead *et al.*, 1984), but the limits of its distribution there are poorly known (Fig. 1). Springer (1979) suggested that the species is a "casual visitor" to the region, contrary to Mediterranean records, which indicate that seasonal occurrence at certain locales is consistent. However, the data are variable and, in keeping with Springer's contention, indicate that few (if any) areas of this region hosted either sizable or persistent populations. Most captures were from oceanic islands, particularly the Azores, but the origin of these sharks, likely not resident, remains unknown. A large number of Azorean specimens were mature, judged on the basis of size; most recently, an approximately 520 cm TL male, confirmed by good photographs, was taken off Ponta Garca, San Miguel, February 10, 1990 (G. Wood, personal communication). White sharks were reputedly common at this archipelago when

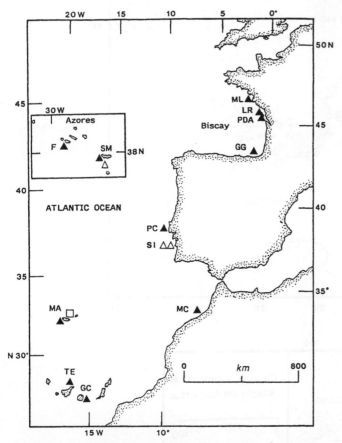

FIGURE 1 Distributional records of white sharks in the eastern North Atlantic Ocean, pooled from captures (solid triangles), attacks on boats (open triangles) and a human swimmer (open square). GC, San Cristobal, Gran Canaria; TE, Tenerife; MA, Madeira; MC, Moroccan coast; SI, Sines, Portugal; PC, Portuguese coast; GG, Gulf of Gascogne; PDA, Pertuis d'Antioche; LR, La Rochelle; ML, mouth of the Loire; F, Faial, Azores; SM, San Miguel, Azores.

commercial hunting for sperm whales (*Physeter macrocephalus*) was a viable industry, but few reliable accounts are available to verify this. In recent decades, white sharks of alleged record dimension have been taken (see Ellis and McCosker, 1991), but these records can be refuted upon closer scrutiny of the photographs (Randall, 1987). The continued occasional presence of this species at the islands is enigmatic and may indicate that certain individuals undertake transoceanic migrations of considerable distance (as noted by Compagno, 1984a). Rare midocean records are known from the Pacific, at the Hawaiian (Taylor, 1985), Marshall (Compagno, 1984a), and Easter islands (see Chapter 42, by Burgess and Callahan).

Other insular occurrences have been recorded at the Madeira, Canary, and Cape Verde archipelagos. Bellon and Mateau (1932) described a specimen

stranded at San Cristobal, Gran Canaria, and R. Payne (personal communication) reported an approximately 400 cm TL individual landed at Tenerife, during 1987. At the Cape Verdes, Cousteau and Dumas's (1953) report of a subsurface encounter with a white shark is consistent with confirmed captures at St. Vincent Island (Cadenat and Blanche, 1981).

Continental records in the region are confined to the coast between Senegambia and latitude 45° N. In European waters, the species has been reported as far north as La Rochelle (Charente-Maritime, France). A summary of historical records from the Biscay coast has been given by Quéro *et al.* (1978), along with a description of a 210 cm TL specimen taken in the Pertuis d'Antioche (offshore from La Rochelle, 46°03′ N, 01°17′ W) on May 24, 1977. I found no evidence to convincingly demonstrate the presence of white sharks (even as vagrants) in waters of northern Europe or the United Kingdom. I searched photographic records of lamnids taken by rod and line off the southwestern coast of the British Isles, from the early 1950s to 1990 (housed at the Shark Angling Club of Great Britain, Looe, Cornwall), confirming that all were either *L. nasus* or, especially in more recent years, *I. oxyrinchus*. Hamilton (1843) noted that U.K. records of white sharks were based on unverified reports; it is quite likely that both *L. nasus* and *I. oxyrinchus* have long been confused with white sharks here.

Records from continental waters of the eastern North Atlantic included specimens from Golfe de Gascogne (see comments in Quéro *et al.*, 1978), Portugal, Morocco, Western Sahara, and the Senegambia (captures near Dakar, at M'Bodiene village, in summer 1945, and l'Île de Gorée, in summer 1957; Cadenat and Blanche, 1981). The infrequency of recent records indicates that white sharks are probably highly transient along these coasts, but the paucity of capture data precludes further analysis of spatiotemporal trends.

Mediterranean Sea

A total of 123 reliable captures or sightings were available among historical and contemporary records (Fig. 2A and B and Appendix 1–3). On the basis of dividing the region into a western and eastern zone at longitude 15° E, 50 (61%) of the sites were in the west, as summarized below.

Western Basin. White sharks were reported from all national coasts of this region. The majority of Spanish records were from the Catalonian Sea in the area of Valencia [Perez-Arcas (1878) describes early records] and at offshore islands (Islas Columbretes

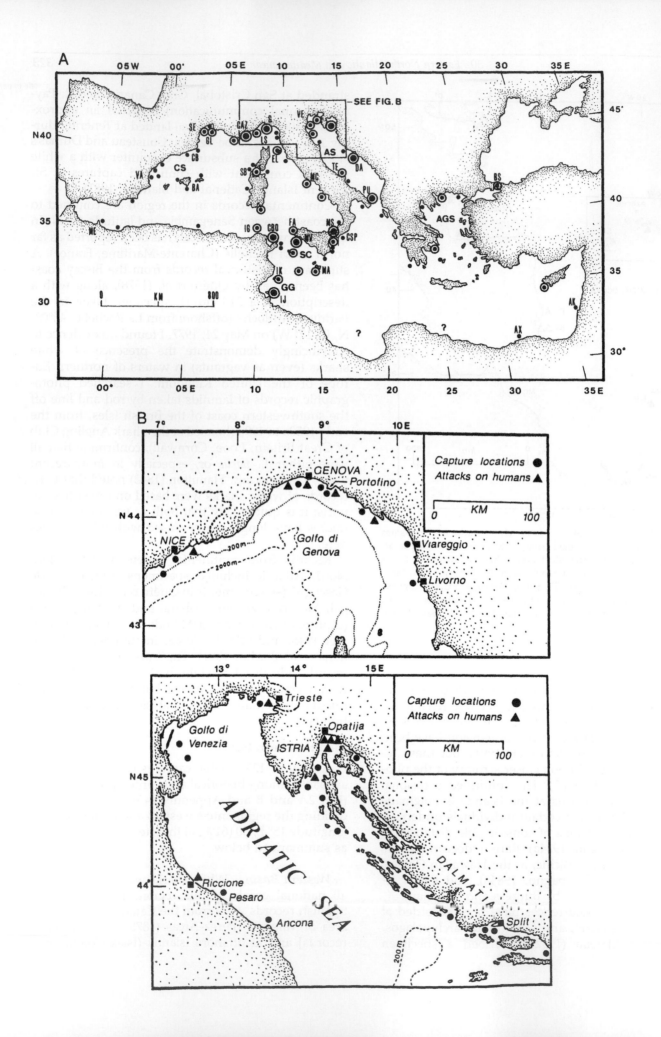

FIGURE 3 A female white shark, 202 cm total length, captured in August 1981, 45 km southwest of Pantelleria Island, Sicilian Channel. (Photo by F. Cigala-Fulgosi, at Mazara del Vallo, Sicily; MPA-0281J.)

and Balearics; Lozano Rey, 1928). Recently (November 17, 1992), an adult male (MSP-0492MM, 475 cm TL) was taken on the Costa Brava, north of Barcelona (my own unpublished data). A number of captures, including juveniles and adults, were reported further north, at Sète and nearby locales in the Gulf of Lions (France; 43°24′ N, 03°41′ E), as well as eastward, in the rapidly deepening waters from the Côte d'Azur to the Gulf of Genova (Fig. 2B). Juvenile and adult white sharks were reported on a number of occasions from tuna traps set in or near the Strait of Bonifacio, between Corsica and Sardinia. The species ranged along the entire western Italian coast and was implicated in attacks on humans at a number of locations (my own unpublished data).

The most consistent reports of *Carcharodon* in the entire study region originated from the Sicilian Channel and nearby waters (Sicily, Tunisia, and Malta).

Regionally rare elasmobranchs were occasionally reported from the intensive commercial fisheries that function within the Channel (principally for scombrids and swordfish *Xiphias gladius*), for example, *Carcharhinus obscurus* (Capapé *et al.*, 1979) and *Carcharhinus brachyurus* (Cigala-Fulgosi, 1983a). *Carcharodon carcharias*, however, was quite frequently recorded from these fisheries, including neonates and juveniles, especially by the fishery at Mazara del Vallo (Trapani, Sicily), between Sicily and Tunisia (F. Cigala-Fulgosi, personal communication) (Fig. 3). White sharks were reported regularly from Tunisian waters, primarily from the northern Gulf of Tunis and Cape Bon, but also from the southern regions of Sfax and the Gulf of Gabes (Postel, 1958; Capapé *et al.*, 1976; C. Capapé and J. Zaouali, personal communication). These specimens included mature adults (both sexes) and, more frequently, juveniles <300 cm TL, all taken

FIGURE 2 (A) Coastal records for white sharks in the Mediterranean Sea. Uncircled dots denote individual specimens; a number of specimens from the same location are denoted by small circled dots (1–5 specimens) and large dots (>5 specimens). Question marks in Libyan waters indicate a lack of data. AGS, Aegean Sea; AK, Acre, Israel; AS, Adriatic Sea; AX, Alexandria; BA, Balearic Islands; BS, Bosporus; CAZ, Côte d'Azur; CB, Costa Brava; CBO, Cape Bon; CS, Catalonian Sea; CSP, Cabo Spartivento; DA, Dalmatian coast; EL, Elba; FA, Favignana (the Egadi Islands); G, Genova; GG, Gulf of Gabès; GL, Gulf of Lions; IG, La Galite Island; IJ, Jerba and Zarzis; IK, Kerkenna Island; LS, Ligurian Sea; MA, Malta; MC, Monte Circeo; ME, Melilla; MS, Straits of Messina; MV, Mazara del Vallo; PU, Puglia; SB, Strait of Bonafacio; SC, Sicilian Channel; SE, Sète; TE, Termoli; VA, Valencia; VE, Venice. (B) Enlarged maps of the Ligurian Sea and the northern Adriatic.

on surface longlines set for swordfish. Off western Sicily (the Egadi Islands), adults were recorded either as tuna fishery by-catch or as free-swimming individuals observed in small aggregations by divers and fishermen working near shore (Appendix 1). White sharks were well known to the fishermen of Isole Pelagie (southwest of Malta), particularly among surface longliners operating over offshore shoals. I validated four captures from Malta between 1964 (an adult male; Fig. 4) and 1987; the last of these, an alleged 713 cm TL female taken April 1987 (Figs. 5 and 6), is discussed by Mollet *et al.* (Chapter 10).

Eastern Basin and the Adriatic Sea. A northerly skew of captures in the Eastern basin was apparent among the records, with only nominal and infrequent records from the warmer, more saline waters of the extreme southeastern Mediterranean. White sharks were rare in the Aegean, although some have entered

FIGURE 5 A female white shark, allegedly 713 cm total length and 2730 kg, captured off Filfla, Malta, in April 1987 by Alfredo Cutajar, on a 5-mm line set for swordfish. (Photo by J. Abela; MMA-0187MF.)

FIGURE 4 A male white shark, approximately 500 cm total length, captured off south Malta in 1964. Note the axillar blotch. (Photo by J. Abela; MMA-0164MM.)

FIGURE 6 Excised and dried jaws of the shark shown in Fig. 4, displayed here by John Abela. The upper jaw perimeter measured 1220 mm when fresh, but 1120 mm when remeasured by A. D. Testi in 1991.

the Dardenelles and the Sea of Mamara as far as the Bosporus, Turkey (see the photograph in Baldridge, 1974a). However, as yet, they have not been recorded from the Black Sea (Bini, 1967; my own unpublished data). The species is not known among the Dodecanese Islands (M. Corsini, personal communication), but it may be responsible for some of the more severe damage inflicted by sharks to commercial fishing gear deployed in the straits separating Kos and Kalymnos from Turkey. A handful of reports by British Armed Forces personnel, requiring confirmation, describe a large free-swimming white shark patrolling off Paphos, Cyprus, during 1993. The species is occasionally taken accidentally in Cypriot fisheries (Fischer *et al.*, 1987).

Historically, white sharks appear to have frequented the Adriatic Sea with some regularity, and they continue to be reported there. The species was noted there by Brunnich (1768), Stossich (1880), Doderlein (1881), Faber (1883), Graeffe (1886), and Carus (1889–1893). More contemporary reference was made by Tortonese (1956), Riedl (1963), Bini (1967), and Soljan (1975). Faber (1883) described two captures, totaling 880 kg, made during summer 1878. He also cited a 530 cm TL specimen, caught during September 1879, at Ustrine, as "one of the largest caught in Adriatic waters." He inferred that the species frequented the region during summer, a statement later seconded by Graeffe (1886), who noted its occurrence near Trieste, in pursuit of tunas during June and July. Quoting an economic value for white shark flesh, he stated that six to eight specimens were captured each year, mostly from Dalmatia and the Kvarner Gulf (Istrian Riviera). Tortonese (1956) examined a "large example" from Trieste, but gave no other details. A 500 cm white shark was captured at Sibenik, Croatia, in August 1993, shortly after a number of sightings by fishermen of a large shark some kilometers further north at Losinj Island (G. Notarbartolo di Sciara, personal communication).

Juveniles were reported from the Italian Adriatic coast at Ancona (43°07' N, 13°30' E), at Termoli (42° N, 15°01' E), and off Puglia. An adult female (>500 cm TL) was photographed swimming at the surface while disrupting a tuna-fishing trip off Rimini in September 1989 (photograph by G. Bartoletti, courtesy of A. D. Testi). Seemingly, the same shark frequented the region between Rimini and Pesaro from 1986 to 1989, and was sighted often enough that local fishermen dubbed it "Willie." Farther north, another large (approximately 480 cm TL) white shark was photographed during July 1977 by oceanographic researchers on a platform situated 12 km off the entrance to Venice Lagoon over a mud and sand bottom 16 m deep (L. Cavaleri, photographs and personal

communication). This shark was first observed cruising at the surface and was enticed closer, to within a few meters of the platform, by a large piece of meat. Earlier evidence of white sharks along this coast was afforded by a nonfatal attack on a spear fisherman off Riccione (International Shark Attack File No. 1220) during 1963. In the southern Adriatic, attacks attributable to this species have been reported from Corfu and Ippocampo (Manfredonia), Italy (August 22, 1988).

Seasonality

Few accurate capture dates are available in the early literature (70 contemporary records), however, they have allowed an assessment of seasonal occurrence patterns. Records were viewed especially in respect to the location of the 15°C and 26°C isotherms, on the basis of the work of Casey and Pratt (1985), who noted that the 15°C value is an important threshold for white shark movements in the western North Atlantic, and Cliff *et al.* (1989), who thought that the 26°C isotherm was at the upper range of tolerance for this species.

Catch data indicated that *Carcharodon* in the Mediterranean tolerated sea surface temperatures (SSTs) ranging from 7.5°C to 25°C, but few records were reported in waters above 23°C. The capture sites of sharks, mapped as a function of season (Fig. 7), indicated a northerly movement from southern areas (Tunisia and the Sicilian Channel), after the late spring–early summer, into cooler northern areas, such as the Ligurian Sea and the central and northern Adriatic Sea. However, there were a number of outliers in the data set. In particular, juvenile sharks were taken off Tunisia and Sicily during warmer summer months (C. Capapé and F. Cigala-Fulgosi, personal communication) but, nevertheless, close to the 25°C isotherm. As the Mediterranean is essentially isothermal (at 13°C), with only a thin layer of warmer surface waters above the thermocline, sharks may conceivably be able to range within areas of high surface SST by cruising at depth. Summer captures of neonates and juveniles from Mazara del Vallo (Sicily) were most likely made by offshore trawls, at depths >200 m (F. Cigala-Fulgosi, personal communication), and well below the thermocline. The data were insufficient to establish any significant pattern with respect to the 15°C isotherm, either within the Mediterranean or in Atlantic waters near the Straits of Gibraltar. The coldest SSTs in the data were for two February records from the Bosporus (where the monthly-mean SST was 7.5°C); for one of these, however (a 391 cm TL female), the shark was found dead on the beach.

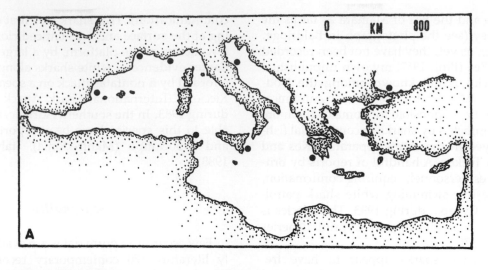

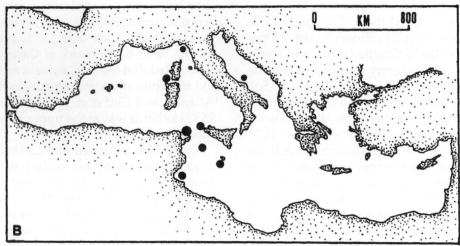

FIGURE 7 The distribution of white sharks within the Mediterranean, by season: (A) winter (mid-December to early March), (B) spring (March to mid-June), (C) summer (mid-June to mid-September), and (D) fall (mid-September to mid-December). Single specimens or sightings are denoted by small circles; 2–5 records, by midsized circles; and >5 records, by large circles.

Site Affinity

Among the various reports of captures, sightings, and interactions with humans, several sharks were resighted at a single location over prolonged periods. In some cases, sharks were resident at headland sites for an entire year (such as off Monte Circeo, Italy; Bini, 1960). Other examples were of sharks frequently resighted over a number of years (e.g., a mature female, dubbed "Willie," well known to the fishermen of Rimini, Italy, 1986–1989). Interestingly, the site at Monte Circeo was a focus for repeated white shark encounters during the 7-year period 1957–1964. Three divers were attacked (in September 1956, summer 1960, and September 1962), all within an area of about 10 km². The 1957 attack on scuba diver G. Lom-

bardo concluded with the capture of a 420 cm TL female white shark, taken by the victim with a baited line upon his return to the attack site 1 week later (Bini, 1960). Fishermen claimed that this shark had been resident at the locale during the previous year, scavenging from nets. At this site, another nonfatal attack occurred in 1960, followed by a fatal attack in 1962 (scuba diver Maurizio Sarra); then, in November 1964, diver Gino Felicioni was harassed by a large female and a smaller male white shark cruising together near the bottom. In a sequence of events not unlike those experienced by Lombardo in 1957, Felicioni revisited the site shortly afterward and landed a 300 cm TL, 380 kg male white shark (identification confirmed by news media photographs).

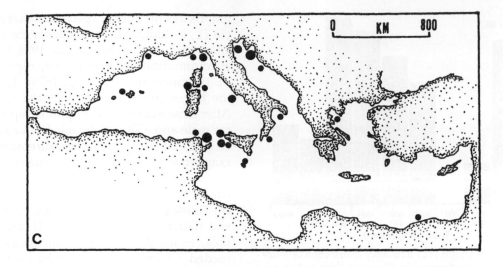

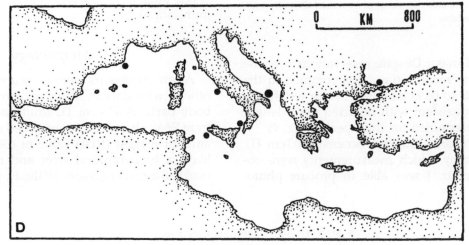

FIGURE 7 (Continued)

Size Distribution

The TL of 86 specimens (sexes combined) ranged from 135 to an alleged 713 cm (Fig. 8), and averaged 415.6 cm (SD = 151.4). Among records in which gender was recorded (N = 40), the TL of females (N = 27) ranged from 142 to 713 cm (X = 448.4 ± 140.6 cm), with 44% >500 cm. The TL of males (N = 13) ranged from 135 to 620 cm (X = 432.5 ± 114.7 cm), with 45% >450 cm. On the basis of Casey and Pratt's (1985) supposition that white sharks ≥457 cm TL are mature (sexes combined), and the arbitrary delineation that sharks >200 cm are subadults, the size class frequency of the 86 records was adults 42%, subadults 44%, and juveniles 14%.

Only 20 (23%) of these records included values for

Similar incidents were reported from the island of Elba, during the early and late 1980s (principally unprovoked attacks on small fishing boats). In the case of the Monte Circeo records, it is clear that the site was frequented by at least four different sharks (assuming that the first-reported individual was removed by Lombardo), importantly, of both genders. While segregation by sex has been reported elsewhere for this species (Strong *et al.*, 1992), at some sites the bias of males to females is in continual flux and combinations of both genders are frequently seen in close feeding aggregations (my own unpublished observation at Dyer Island, South Africa; L. J. V. Compagno, personal communication). Data on the sex of Mediterranean sharks were insufficient to deduce the existence or extent of sexual segregation.

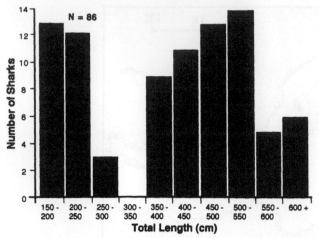

FIGURE 8 The total length (TL) frequency distribution of white sharks captured in the Mediterranean. The upper limit of the range is uncertain, as reliable measurements of 11 sharks allegedly >600 cm TL were not available.

graphs of a number of specimens that measured >500 cm TL, including three females reported as 700 cm TL (Tortonese, 1965; Anonymous, 1974; J. Abela, personal communication) and a male, reputedly of 620 cm TL (Piccinno and Piccinno, 1979). These are among the largest reported sizes for the species (Table I). After close examination of photographs (e.g., the 620-cm male and a 700-cm female from Favignana), I concluded that many of the measurements were greatly exaggerated. The largest specimens for which I considered the TL measurements to be valid were a 535 cm TL female, Favignana, Sicily, May 8, 1987; a 475 cm TL male, Tossa de Mar, Spain, November 17, 1992; and (fair reliability) a 520 cm TL male, Capo Granitola, Sicily, August 1990 (this shark may have exceeded 520 cm; photographs by Giovanni Rivas and Sergio Ragonese, courtesy of M. Zuffa, personal communication).

both length and mass. Despite reservations over the accuracy of some records (i.e., it is common in the region for sharks to be eviscerated before weighing), the data were consistent with reported mass–length relationships determined for this species (Fig. 9).

A variety of records listed specimens >600 cm TL, but the manner by which measurements were obtained was unclear. I was able to procure photo-

Morphology

A few specimens were measured more fully than others, with data available on girth, fins, and other body parts. A 475 cm TL adult male (MSP-0492MM) from Spanish waters was recorded with a low pectoral anterior margin (P1A) value of 630 mm, which was likely a typographical error and may actually have read 930 mm (Fergusson, 1994a; H. F. Mollet, personal

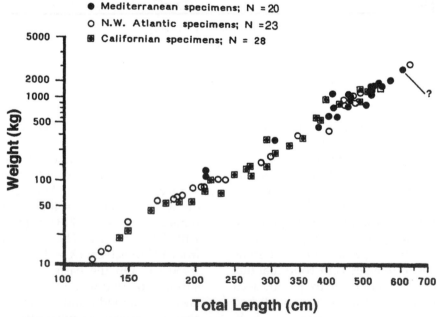

FIGURE 9 Mass at-length scatter diagram for *Carcharodon carcharias*, comparing data from the Mediterranean to those from the eastern North Pacific (Klimley, 1985b) and the western North Atlantic (Casey and Pratt, 1985b). The alleged 713-cm Malta female placed according to approximate mass, rather than total length.

TABLE I   Records of White Sharks >530 cm Total Length From the Mediterranean Sea

Capture location	Date	TL (cm)[a]	Mass (kg)	Sex	Remarks	Reference
Favignana Island, Sicily	May 8, 1987	535	2000	F	Taken in a net for *Thunnus thynnus*; left P1[b] posterior margin excised, photographs available	Author's data
Capo Spartivento, Italy	Mid-June 1989	550			Taken in a drift net for *Xiphias gladius*; stomach contained two dolphins	Anonymous (1990)
Islas Columbretes, Spain	January 1878	600			Taken in a net enclosure	Rey (1928)
Rijeka, Croatia (Yugoslavia)	September 9, 1934	600				Giudici and Fino (1989)
Monterosso, northern Italy		600	600		Eviscerated mass	Tortonese (1956)
Cape Bon, Tunisia	June 1953	600	1500		Mass low for TL (see Chapter 9, by Mollet and Cailliet; may be eviscerated mass;	Postel (1955)
Ganzirri, Sicily	March 9, 1965	620			alleged D1H.(1.6 m) suggests that TL is also exaggerated	Giudici and Fino (1989); Donato Nicola (personal communication)
10 km west of Gallipoli, Italy	September 18, 1979	620	1700	M	Caught by P. Alessandrelli in a dragnet, 72 m depth; mass approximated; may actually have been female; P1A = 98 cm, UCL = 120 cm	Piccinno and Piccinno (1979)
Ghar el Melh, Tunisia	January 4, 1989	670	2032	F	No data to confirm reported TL	G. Wood (personal communication); Giudici and Fino (1989)
Kraljevica, Croatia	September 1934	700			No source or details given; locality given as "Porto Re" (former name)	Giudici and Fino (1989)
Camogli, Italy	May 16, 1954	700	1400	F	Photograph published giving TL; this and others' photos indicate TL of 520–550 cm	Tortonese (1965)
Formica Island, Sicily	May 1974	700	1500	F	Nearer 520–530 cm; uncertainty about sex	Anonymous (1974)
Filfla Islet, Malta	April 17, 1987	713	2730	F	Captured by A. Cutajar; see Chapter 10, by Mollet *et al.*, for analysis of the validity of J. Abela's measurements	Abela (personal communication)

[a]Lengths are those given in the references. TL, total length.
[b]Morphometric measurements and abbreviations are given in Appendix 1 of Chapter 10, by Mollet *et al.*

communication). A seemingly low first dorsal fin height (D1H) value for an alleged 713 cm TL female from Malta (MMA-0187MF) is best explained through processes of negative allometry (see Chapter 10, by Mollet *et al.*). At other extremes, a specimen taken in 1965 at Ganzirri, Sicily, reputedly 620 cm TL, was reported to possess an orca-like D1H of 160 cm (again, surely a typographical error or gross exaggeration).

Mediterranean sharks fitted well the generic descriptions given by Bigelow and Schroeder (1948), Bass *et al.* (1975), and Compagno (1984a). On the basis of inspecting good color photographs of freshly captured specimens, I judged that Mediterranean white sharks were uniformly a dark, slate gray on the dorsum; two adult females (MMA-0187MF: 713 cm, Malta; and MSI-0287MF: 535 cm, Sicily) exhibited numerous small irregular dark blotches on the flanks posterior to the fifth gill slit. The majority of specimens had dark-colored ventral caudal lobes, with very little white or buff blotching, as is often seen in Australian and South African examples. The dorsal surfaces of the pelvic fins were always marked by a fairly extensive area of dark pigmentation. All specimens had black tips at the ventral apexes of the pectoral fins, but some examples of all size classes apparently lacked the black axillar blotch that commonly characterizes this species in other regions (Bigelow and Schroeder, 1948; Compagno, 1984a). Dark gray (not black) axillar blotches were present, however, on a few of the small juveniles, 135–200 cm TL, examined from Tunisia (C. Capapé, personal communication), but not on any from Sicily (F. Cigala-Fulgosi, personal communication), indicating that this feature, on occasion, may be either absent or rapidly lost as the shark matures. However, other large adults did display the blotch [e.g., MMA-0164MM: Malta (Fig. 4); MSI-0174M: Favignana; and MSA-0176MM: Sardinia]. Clearly, the presence of this feature is highly variable.

On the basis of photographic evidence, adults did not appear to display lacerations, scratches, and associated marks around the head and the anterior flanks, as is commonly observed in California, Australia, and South Africa. The lack of these features may be indicative of localized abundance. Assuming that such injuries are primarily (but not exclusively) inflicted by conspecifics as a consequence of intraspecific aggression or competition, then adult *Carcharodon* in the Mediterranean may rarely engage in close-contact feeding or other competitive aggregations. One female, 535 cm TL, taken at Favignana Island (Sicily) in May 1987, exhibited a conspicuous semicircular excision to the posterior margin of the left pectoral fin, bordered by smaller cuts measuring approximately 360 mm across. Such marks may be inflicted during courtship (see Chapter 15, by Francis), and it is instructive that this specimen was trapped in the same net with a an approximately 450 cm TL male (which escaped).

Reproduction

A pregnant female white shark was reported from the Mediterranean by Norman and Fraser (1937), and was later discussed by Bigelow and Schroeder (1948), Tricas and McCosker (1984), and Francis (Chapter 15). The specimen was taken at Agamy Beach, Alexandria, Egypt, during summer 1934 and measured 425 cm TL. This often-quoted report remains unverified, principally as the stated masses of the embryos are surely erroneous (Tricas and McCosker,

1984) and the identity of the species cannot be ascertained (I could not trace the photographs, cited by Norman and Fraser, and likely housed at the British Museum). Another, likely more valid record of a pregnant white shark was given by Sanzo (1912) and is discussed by Francis (Chapter 15). An embryo, examined and described by Sanzo (now in the Museum of Zoology, Florence), originated from a "pescecane" (Sicilian colloquial for *white shark*) taken by fishermen in the Straits of Messina some years prior (1903?) to his publication (the exact date of capture neither is mentioned nor can be ascertained). Sanzo's stated mass for the female shark (around 500 kg) is likely misleading. Because white sharks commonly have been eviscerated upon capture, the mass determined upon landing was much lower than would be expected for a mature female. Further analyses are currently under way to determine the identity of this embryo (A. D. Testi, personal communication), which may be *I. oxyrinchus*.

In September(?) 1992, a large (>500 cm TL) female white shark was landed by commercial fishermen at Sidi Daoud, on the northwest tip of Cape Bon, Tunisia (Fig. 1). The specimen had been caught in a net enclosure set for tuna (*Thunnus* spp.); this fishery has regularly taken white sharks as by-catch, principally during spring (see Postel, 1958). Upon landing at the dock, the shark was found to contain two fetuses, believed to be full-term. Seemingly unaware of the significance of the capture, the fishermen either butchered or dumped the shark and her embryos. Jeanne Zaouali (personal communication) was able to confirm the capture from fishermen, but no tangible evidence (i.e., photographs) have been traced. The shark was probably towed either moribund or dead, and most of a larger litter of pups may have been aborted and lost. The lack of data on this specimen is not particularly surprising; other large female white sharks (e.g., an approximately 530-cm specimen from June 1979; photographs courtesy of J. Zaouali) have been captured at the site, but only later have been discovered by researchers from preserved jaws and photographs.

Spatiotemporal Occurrences of Pregnant Females, Mature Pairs, and Small Juveniles

I combined records of pregnant females and sightings of free-swimming male and female pairs of large sharks >450 cm TL (Fig. 10A). The occurrence of male–female pairs is a possible indicator of mating activity (Casey and Pratt, 1985). These records were compared with the spatiotemporal distribution of

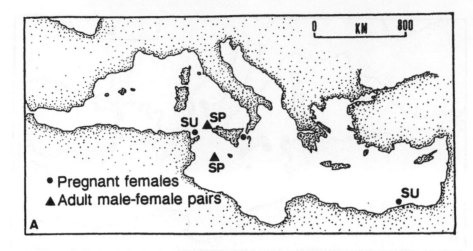

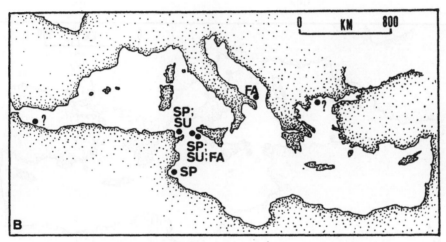

FIGURE 10 Spatiotemporal distributions of (A) pregnant females (circles) and adult male-
-female pairs (triangles) and (B) small juvenile (<185 cm total length) white sharks in the
Mediterranean Sea. Capture seasons designated as SP, spring; SU, summer; and FA, fall.

small juveniles (Fig. 10B). The size of neonate white sharks appears as equally variable as size at maturity in adults (see Chapter 15, by Francis); in this study, I defined pups as white sharks <185 cm TL. This being the case, pups were reported from neritic waters along the coasts of Tunisia, Sicily, Algeria, Melilla (Morocco), the Italian Adriatic Sea, and the northern Aegean Sea. In Tunisian waters, they were reported from the Gulf of Tunis, taken on longlines, from April to August (C. Capapé, personal communication). Neonates were also reported from the opposing (north) side of the Sicilian Channel, landed at Mazara del Vallo, southwestern Sicily (F. Cigala-Fulgosi, personal communication). Records of adult pairs and pregnant females were restricted to spring and summer; small juveniles were reported in similar areas

within the same seasons, with additional captures made during autumn within the southern Adriatic.

The distribution of small juveniles (Fig. 11A), centered in central regions of the Mediterranean, was similar to that of larger juveniles (Fig. 11B). Increasingly larger sharks were found in an increasingly wider range (Fig. 11C and D); however, sharks 250–450 cm TL, and probably approaching maturity, were largely confined to the western and central Mediterranean. Conceivably, then, this apparent east-to-west movement occurs with increasing maturity.

Diet

Stomach contents data from Mediterranean sharks were confined to a small sample of larger (>390 cm

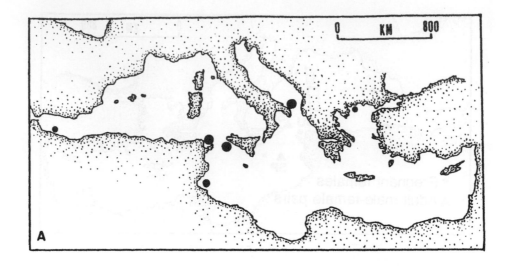

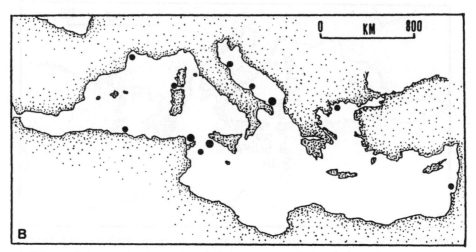

FIGURE 11 The areal distribution of white sharks, plotted as a function of total length (TL).
An index of the number of sharks recorded is given by symbol size (see Fig. 2A). The four size
classes (both sexes pooled) are as follows: (A) neonates and small juveniles, <185 cm TL); (B)
large juveniles, 185–250 cm; (C) subadults, 251–450 cm; and (D) adults, >450 cm.

TL) specimens (Table II). Pinnipeds were absent in the diet. The rare Mediterranean monk seal *Monachus*, having an endangered and widely scattered population, seems unlikely to be among sharks' regular prey. The remains of small cetaceans, however, were reported in many of the sharks examined, although the significance of white shark predation on small odontocetes within the region is poorly studied. Dolphin remains in shark stomachs have been found largely intact or partially dismembered, with tail stocks severed. While these prey may have been taken free-swimming, an earlier report (Condorelli and Perrando, 1909) described an attack by a 450 cm TL female white shark on a dolphin entrapped in a

net off Sicily. Lengthy field studies of social groups of bottle-nosed dolphins *Tursiops truncatus* at one locale frequented by adult *Carcharodon* (the northeastern Adriatic, off Croatia), did not reveal any shark-inflicted scarring, nor did mother–calf pairs elicit behaviors indicating avoidance of potential predators (G. Notarbartolo di Sciara, personal communication).

Predation on teleosts was confined to swordfish and tuna, including reports of both scavenging from commercial fisheries and pursuit of free-swimming prey. Healed shark bites are not uncommon on the flanks of regionally captured *Thunnus* spp. and *X. gladius* (my own unpublished observation) but, based on the size and shape of individual tooth punc-

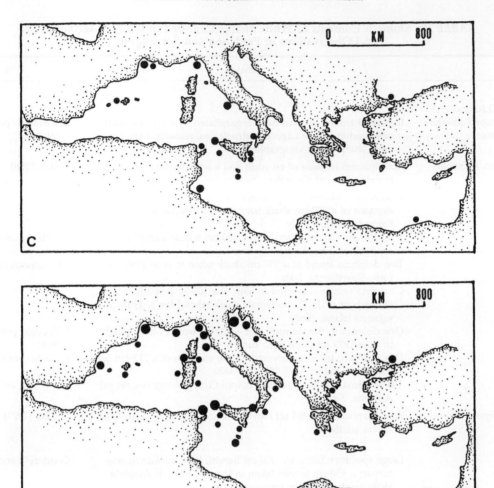

FIGURE 11 (*Continued*)

tures, largely result from interactions with *I. oxy-rinchus*. On the other hand, this could indicate that few of these fish survive the predatory attack of a large white shark. White sharks also prey on *I. oxy-rinchus* and *Prionace glauca*, either as scavenged by-catch or taken free-swimming; both species are frequently taken as incidental catch in Mediterranean fisheries and, in particular, may be readily scavenged from longlines. In the opinion of the captors of the 713-cm female white shark taken off Malta in 1987, the intact blue shark found in its stomach had not been scavenged from a set line. White shark predation on these shark species has also been recorded off South Africa (L. V. J. Compagno and S. Cook, personal communication). These sharks spend much time at the surface, and large *Carcharodon* quite conceivably

could capture them successfully by maneuvering into a position directly below, to then make a vertical rush at them (Fergusson, 1994b).

Human remains have been found rarely in white shark stomachs. To my knowledge, the three cadavers found within a 450 cm TL female, taken off Augusta, Sicily (Condorelli and Perrando, 1909), represent the highest number of individual humans recorded from a single white shark. Injurious attacks on humans by white sharks are also infrequent in the Mediterranean, totaling 28 reliably reported encounters since 1900, the majority from Italian and Croatian coastal waters (I. K. Fergusson and A. D. Testi, unpublished data). Antishark nets remain in place off beaches in the Kvarner Gulf (Croatia), where five shark fatalities (all attributable to white sharks) since

TABLE II   **Stomach Contents of White Sharks Captured in the Mediterranean Sea and Observations of Predation and Scavanging**

Prey	Remarks	Reference
Mammals		
Toothed whales (Odontocetes)		
Atlantic bottle-nosed dolphin *Tursiops truncatus*	Identified from a severed head regurgitated by a 480-cm shark observed at Venice Lagoon, Adriatic Sea; remains were recovered; shark was photographed	L. Cavaleri (personal communication)
Common dolphin *Delphinus delphinus*	Skeletonized remains of an adult and a fetus found in a 520-cm female taken off Cape Bon, Tunisia	Postel (1958)
Unidentified small odontocetes	Caudal musculature and peduncle of a dolphin found in the stomach of a 390-cm shark harpooned in December 1991 off Bagnara Calabra, Italy	
	Two dolphins in the stomach of a 450-cm female caught in January 1991 off Sète, France	B. Séret (personal communication)
	Two dolphins found in a 550-cm shark taken in June 1989 off Capo Spartivento, Italy	Anonymous (1990)
	Complete remains, severed in half, of a 100-kg dolphin in the stomach of a 535 TL cm female shark taken May 8, 1987, at Favignana Island, Sicily	
	One dolphin in the stomach of a ~520 cm female shark caught in June 1979 off Cape Bon, Tunisia	J. Zaouali (personal communication)
	One dolphin, tail stock severed, in the stomach of a 713-cm female captured in April 1987 off Malta	J. Abela (personal communication)
	450-cm female seen to attack a dolphin trapped in a net set off Augusta, Sicily	Condorelli and Perrando (1909)
Sperm whale *Physeter macrocephalus*	Shark captured in 1964 off Italy while scavenging on a juvenile sperm whale	Budker (1971)
Other mammals		
Humans	Large specimen taken in 1900 off Baratti, Italy, contained one corpse; a 450-cm female taken in January 1909 off Augusta, Sicily, contained three corpses	Condorelli and Perrando (1909)
Domestic goat *Capra hircus*	Horns found in the stomach of a 530-cm shark taken in spring 1974 off Favignana Island, Sicily	Anonymous (1974)
Reptiles		
Marine turtles (Chonidae) Green turtle *Chelonia mydas*	Carapace fragments found in May 1956 in a 520-cm female shark from Cape Bon, Tunisia	Postel (1958)
Loggerhead turtle *Caretta caretta*	60-cm-diameter carapace found in a 713 cm TL female caught in April 1987 off Malta	Albeta (personal communication)
Unidentified marine turtles	Remains in the stomach of a 470-cm, 1100-kg male white shark taken in August 1976 at Santa Caterina di Pittinurri, Sardinia	M. Zuffa (personal communication)
Cartilaginous fishes		
Mackerel sharks (Lamnidae) Shortfin mako *Isurus oxyrinchus*	100 cm TL juvenile found in a 520 cm TL female white shark taken in May 1956 off Cape Bon, Tunisia	Postel (1958)
Blue shark *Prionace glauca*	220 cm TL specimen found in April 1987 in a 713-cm TL female white shark caught in Malta	J. Abela (personal communication)
Bony fishes		
Swordfishes (Xiphiidae) Swordfish *Xiphias gladius*	Individual found in a 450-cm female taken in January 1991 off Sète, France	B. Séret (personal communication)
	Longlined specimen attacked in June 1987 by a large shark off Scilla, Italy	
	Vertebral column and skeletonized head found in the stomach of a 520-cm white shark taken off Favignana Island, Sicily, April 24, 1980	Bruno (1980)

(continues)

TABLE II (Continued)

Prey	Remarks	Reference
Tunas (Scombridae) Bluefin tuna *Thunnus thynnus* Albacore *Thunnus alalunga*	White sharks are often seen and captured in close association to tunas; various historical and contemporary reports describe predation on these species, yet few records of stomach contents	Authors data
Unidentified fish remains	Various teleost remains in the stomach of a male specimen, 470 cm and 1100 kg, caught in August 1976 off Santa Caterina di Pittinurri, Sardinia	M. Zuffa (personal communication)
Inanimate items Pair of shoes	Found in a 620-cm male taken off Gallipoli in September 1979. Authors suggest that the items represent a "tragic" meal, yet the shoes may have been ingested as garbage	Piccinno and Piccinno (1979)
Plastic liner containing garbage	Found in the stomach of a 713-cm female from Malta in April 1987; originated at an offshore drilling platform, located near the capture site	J. Abela (personal communication)
Plastic bottle and bag	Found in the stomach of a ~530-cm white shark taken in May 1974 off Formica Island, Sicily	Anonymous (1974)

TL, Total length.

1934 occurred at the closely spaced resorts of Opatija, Rijeka, Susak, and Ika (the last incident was reported in 1971).

Discussion

The paucity of eastern North Atlantic Ocean records of *C. carcharias*, examined in a historical context, infers that the species has never been common in the region. No specimens have been recorded from the Bay of Biscay since 1977 (J.-C. Quéro, personal communication). The rates of interchange among sharks from various portions of the species' range are not known, but are likely to be minimal with respect of the study region. The capture of adult specimens at the Azores Islands demonstrates that some degree of transoceanic movement may occur, possibly as a (largely) west-to-east nomadic journey within the Gulf Stream from North America. Equally, the possibility of interchange between the northern and southern portions of the eastern North Atlantic, along the African coast, is unclear. The species is very rare (and possibly absent) from Sierra Leone southward to the equator. Conceivably, transient sharks may undertake such a route through equatorial waters by cruising in cooler waters at depth.

As yet, no data confirm or refute the existence of Atlantic–Mediterranean transiency among white sharks. On the basis of all capture records pooled for the Mediterranean, neritic waters of the region are both a pupping and nursery ground. The sighting of adult pairs, and bite marks on adult females, indicate that mating itself occurs within the region. After analysis of the seasonal distribution of pregnant females and small juveniles, I hypothesize that parturition occurs late spring to late summer in the central Mediterranean, particularly from the Sicilian Channel east to the southern Adriatic. Given an apparent paucity of sharks 200–300 cm TL, it is possible that juveniles depart the Mediterranean into Atlantic waters before reaching maturity (e.g., the 210-cm specimen from Biscay; Quéro *et al.*, 1978). Unfortunately, few size frequency data were available from the eastern North Atlantic coasts to further demonstrate whether such a movement is likely. No capture records were available from the Straits of Gibraltar, the site where an Atlantic–Mediterranean interchange would be most obvious.

White sharks as predators of cetaceans and large teleosts within the Mediterranean are a subject that has been little studied (Fergusson, 1994b). The seasonal and geographic patterns in the abundance of *Carcharodon* and other lamnids in the region may be a response to the movements of certain teleost prey. Boero and Carli (1979) noted that a decline in catches of bullet tuna *Auxis rochei* within the Gulf of Genova coincided with an increasing absence of lamnids, but mention only a single white shark in their data (a 1400-kg female taken off Camogli, May 16, 1954). Published regional lists of elasmobranchs in the fisheries by-catch are few, and therefore, a full investigation of correlations between white shark abundance and availability of various teleost prey is problematic. White sharks clearly take both cetaceans and fish by means of opportunistic scavenging, but the bias between the importance of this activity against predation on free-swimming animals is unclear. Both blue fin tuna *Thunnus thynnus* and swordfish have declined within the Mediterranean due to overfishing, but both species remain available to natural predators to some extent. Odontocetes are also an accessible, perhaps primary, prey group, at least to adult white sharks in the Mediterranean, first, as free-swimming prey and, second, as a resource for scavenging. For example, striped dolphins *Stenella coeruleoalba* are abundant, numbering about 200,000 in the western Mediterranean (Forcada *et al.*, 1992). Moreover, their vulnerability to predation and availability to scavengers has likely increased as a result of debilitation by a major morbillivirus epizootic (Aguilar and Raga, 1990). As fisheries discards, thousands of dolphins are dumped at sea, having been incidentally captured in drift nets (Notarbartolo di Sciara, 1990).

In conclusion, I provide my opinions in regard to the relative abundance of white sharks in the Mediterranean Sea. Contrary to the comments of Ellis and McCosker (1991, p. 55), I suggest that it is inaccurate to describe *Carcharodon carcharias* as "exceedingly rare" in the Mediterranean. During the course of this research, I located no fewer than 38 white sharks captured since 1979, averaging 2.7 specimens per year but increasing to 5 sharks annually through 1991–1993 as reporting efficiency increased. Moreover, many specimens were, and continue to be, unreported (partially through a process of misidentification), and hence, the true rate of capture cannot be ascertained. Annual rates of capture may even be three to four times greater than is shown in the data presented here, and certain key localities deserve more diligent monitoring. For example, studies at two Sicilian Channel fisheries, 1970–1980, at the Gulf of Tunis, Tunisia (C. Capapé, personal communication), and Mazara del Vallo, southwest Sicily (F. Cigala-Fulgosi, personal communication), indicated that as many as 2 neonate or juvenile white sharks were taken per month in the by-catch during successive summers (biennially, based on existing data), but were generally ignored by fishermen (hence, most went unreported). Similar rates of capture existed for other species that, until that time (1970–1980), also had been regarded as nominal in the region (such as *Alopias superciliosus* and *C. brachyurus*; F. Cigala-Fulgosi, personal communication). Thus, the incidental capture of small white sharks is not an infrequent

phenomenon, at least in the Sicilian Channel, but only large specimens attract media attention, later to appear in the literature. Clearly, the paucity of existing records is both erroneous and misleading. Literature records of the 19th century are equally incomplete, and thus preclude any attempt to investigate longer-term temporal trends in rates of capture and sighting. It is prudent to classify the abundance of white sharks within the region according to the criteria of Potts and Swaby (1992), whereby the term *indeterminate* encompasses those species for which insufficient data exist to more accurately categorize abundance. In clarification, on the basis of capture rates of certain other large elasmobranchs found in the region (e.g., *C. maximus*, *Odontaspis ferox*, *Carcharias taurus*, and a number of other carcharhinids), *C. carcharias* appears to be "uncommon," rather than "scarce." Although *Carcharodon* is essentially rare wherever it occurs, the Mediterranean Sea, from a comparative viewpoint, should be classified as one of the global centers of reproduction and abundance for the species.

Summary

The areal distribution of the white shark *C. carcharias* in the eastern North Atlantic Ocean and the Mediterranean Sea has been described from historical and contemporary data. The species was more frequently reported from the Mediterranean Sea, with 85 capture and sighting sites, and principally within the Western Basin and the Adriatic Sea. Sharks were recorded throughout the year here, where SSTs ranged from 7.5°C to 25°C. Rising temperatures in lower latitudes during summer apparently displaced the population northward. The TLs among 86 Mediterranean specimens ranged from 135 cm to an alleged 713 cm, and averaged 415.6 cm.

The degree of Atlantic–Mediterranean transiency, recruitment rates, and geographic origins of Mediterranean white sharks are unknown. Available data indicated that reproductive activity was focused regionally in the south–central Mediterranean, and a recent record of a pregnant female from Tunisian waters has been described. On the basis of the spatiotemporal distribution of juveniles <185 cm TL, I suggest that parturition likely occurs in late summer and early fall, with nursery zones existing in Sicilian and Tunisian neritic waters. Data on feeding from 15 sharks >390 cm TL revealed feeding on odontocetes, pelagic sharks, chelonians, swordfish, and scombrid teleosts. Two records this century, one verifiable, included the discovery of identifiable human remains in the stomachs of white sharks taken in Italian waters. Fishermen reported white sharks scavenging on hooked teleosts and accidentally netted odontocetes.

The species appears to be uncommon, as opposed to scarce, in the Mediterranean. True capture rates are impossible to quantify on an annual basis, because the majority of specimens are not reported, or their identification is confused with that of other lamnids. A minimum of 38 specimens have been recorded from the region since 1979, including 15 sharks during 1991–1993. I suggest, given the number of captures and sightings and, equally, considering the cosmopolitan population structure of white sharks in the region, that the Mediterranean is a center of reproduction and abundance for this species.

Acknowledgments

My sincere gratitude goes to the following, who provided considerable advice and data throughout the course of this research:

L. J. V. Compagno, A. P. Klimley, M. P. Francis, H. L. Pratt, Jr., J. G. Casey, H. F. Mollet, G. Notarbartolo di Sciara, I. Bianchi, B. Séret, L. del Cerro, F. Cigala Fulgosi, C. Capapé, A.D. Testi, M. Zuffa, L. Cavaleri, J. Piza, J. Zaouali, J.-C. Quéro, G. H. Burgess, M. Callahan, and O. Crimmen. J. Abela kindly made available his original photographs and brought scientific attention to the Maltese specimens. Additional thanks to the following, for useful conversations in respect to this chapter and white sharks in general: T. Casey, J. E. McCosker, K. J. Goldman, D. J. Long, R. S. Collier, J. West, M. Cappo, P. Pyle, S. D. Anderson, W. R. Strong, Jr., Paul and Grace Niska Atkins, and C. Ferreira. Q. Bone, R. Earll, P. Batterbury, and anonymous reviewers provided useful comments on the manuscript. My thanks especially go to the editors for their constructive comments and suggested improvements to the final draft. B. Zahuranec (Office of Naval Research) kindly arranged the travel funding for my participation at the symposium; my gratitude also goes to American Institute of Biological Sciences for their assistance.

Appendix 1 **Location and Capture Details for White Sharks in the Western and Central Mediterranean Sea to Sicily**

No.	Location	Date	TL (cm)	BM (kg)	Sex	Catalog no.	Remarks	Reference
1	Melilla, North African coast	1920s	150				Examined at Melilla fish market	Rey (1928)
2	Golfo de Chafarinas, Morocco							Buen (1926)
3	Algerian coast		223				No precise locality	Dumeril (1865)
4	La Galite Jalta Island, Tunisia	1953–1976					A few specimens taken	Capapé *et al.* (1976)
5	Bizerte, Tunisia	1970s					A few specimens taken	Capapé *et al.* (1976)
6	Zembra Island, Tunisia	1970s					A few specimens taken	Capapé *et al.* (1976)
7	Ghar el Melh, Tunisia	April 1, 1989	570[a]	2032		MTU-0389MF	Original TL reported as 670 cm, but may be 570 cm based on reported BM	Giudici and Fino (1989)
8	Northwest Cape Bon, Tunisia	June 1953	600[a]	1500		MTU-0153M	TL likely at error based on low BM	Postel (1958)
9	Northwest Cape Bon	May 1954					Two specimens	
10	Northwest Cape Bon	May 16, 1956	520	1800	F	MTU-0256MF	Taken in a tuna net	Postel (1958)
11	Northwest Cape Bon	May 22, 1956	410	1300	M	MTU-0356MS	Taken in a tuna net	Postel (1958)
12	Northwest Cape Bon	May 1979	220		F	MTU-0279J		C. Capapé (personal communication)
13	Northwest Cape Bon	June 1979	500+		F	MTU-0379MF		J. Zaouali (personal communication)
14	Northwest Cape Bon	August 1981	185		M	MTU-0181J		C. Capapé (personal communication)

(continues)

Appendix 1 (*Continued*)

No.	Location	Date	TL (cm)	BM (kg)	Sex	Catalog no.	Remarks	Reference
15	Northwest Cape Bon	June 1983	175		F	MTU-0183J		C. Capapé (personal communication)
16	Northwest Cape Bon	April 1985	190		F	MTU-0185J		C. Capapé (personal communication)
17	Northwest Cape Bon	September 1992	500+		F	MTU-0592MF	Gravid female carrying two embryos	C. Capapé (personal communication)
18	Kelibia, Tunisia	1953–1976						Capapé *et al.* (1976)
19	Golfe de Gabes, Tunisia	1953–1976						Capapé *et al.* (1976)
20	Zarzis	March 21, 1972	400			MTU-0172MS		Quignard and Capapé (1972)
21	Zarzis	March 21, 1972	400			MTU-0272MS		Quignard and Capapé (1972)
22	Karzis	April 21, 1972	185			MTU-0372MS		Quignard and Capapé (1972)
23	Pantelleria Island, Italy	August 1981	202		F	MPA-0281J		F. Cigala-Fulgosi (personal communication)
24	Pantelleria Island	Spring 1992	530[a]		F	MPA-0292M	Photographed free-swimming near the surface	F. Andreoli (personal communication)
25	Banco di Mezziogorno, near Lampedusa Island, Italy	1992–1993					Aggregations encountered by fishermen	A. D. Testi (personal communication)
26	Maltese Islands	1964	500[a]		M	MMA-0164MM		J. Abela (personal communication)
27	Maltese Islands	March 1973	530		F	MMA-0173MF		J. Abela (personal communication)
28	Maltese Islands	May 1979	400		F	MMA-0179MS		J. Abela (personal communication)
29	Maltese Islands	April 17, 1987	713[a]	2730[a]	F	MMA-0187MF		Mollet *et al.* (Chapter 10)
30	Maltese Islands	August 18, 1993	500+		M	MMA-0393M	Shark made three passes at a 4.5-m boat; final pass swimming "on its side"	I. Maxwell (personal communication)
31	Egadi Islands, Sicily	May 1974	520[a]	1500	F[a]	MSI-0174M	Captured in a tuna net	Bruno (1980)
32	Egadi Islands	April 24, 1980	520[a]	1600	F[a]	MSI-0180M	Captured in a tuna net	Bruno (1980)
33	Egadi Islands	April 25, 1980	400			MSI-0280MS	Captured in a tuna net	Bruno (1980)
34	Egadi Islands	May 8, 1987	535	2000	F	MSI-0287MF	Captured in a tuna net at the harbor entrance	Bruno (1980)
35	Egadi Islands	May 8, 1987	450[a]		M		Captured in a tuna net at the harbor entrance	Bruno (1980)
36	Egadi Islands	1989					Captured in the harbor	

(*continues*)

Appendix 1 (*Continued*)

No.	Location	Date	TL (cm)	BM (kg)	Sex	Catalog no.	Remarks	Reference
37	Egadi Islands	August 13, 1989	300[a]			MSI-0289	Three white sharks observed by diver G. Bertin, Isola Galera	
38	Capo Granitola, Trapani, Sicily	August 25, 1990	520+		M	MSI-0190MM	Taken in a tuna net; photographs taken by G. Rivas and S. Ragonese	M. Zuffa (personal communication)
39	Mazara del Vallo, Sicily	August 1977	135		M	MSI-0377	Mostly taken in off-shore trawls, Sicilian Channel	F. Cigala-Fulgosi (personal communication)
40	Mazara del Vallo	August 1978	220			MSI-0278		F. Cigala-Fulgosi (personal communication)
41	Mazara del Vallo	June 25, 1985	220			MSI-0283		F. Cigala-Fulgosi (personal communication)
42	Mazara del Vallo	August 10, 1983	155		M	MSI-0383		F. Cigala-Fulgosi (personal communication)
43	Mazara del Vallo	August 11, 1983	142		F	MSI-0285J		F. Cigala-Fulgosi (personal communication)
44	Mazara del Vallo	June 1991	400			MSI-0591MS	Confirmed by dentition	F. Cigala-Fulgosi (personal communication)
45	Mazara del Vallo	November 1993	135			MSI-0493JN	Confirmed by dentition	F. Cigala-Fulgosi (personal communication)
46	Siracusa, Sicily	1852						Perez-Arcas (1878)
47	Siracusa	1970s	400					Touret (1992)
48	Cabo San Croce, Augusta, Sicily	January 26, 1909	450	800	F		Stomach contained remains of three humans	Condorelli and Perrando (1909)
49	Catania, Sicily	1880s	300					Doderlein (1881)
50	Messina, Sicily	1850–1992	500		F		Several captures and sightings, both historical and recent	Facciolà (1894); Riggio (1894)
51	Ganzirri, Sicily	March 9, 1965	620				TL dubious	Giudici and Fino (1989)
52	Capo Spartivento, Calabria, Italy	June 1989	550			MII-0189M	Taken in a swordfish drift net; stomach contained two dolphins	Anonymous (1990)
53	Gallipoli, Italy	September 18, 1979	620		M	MII-0479M		Piccinno and Piccinno (1979); Giudici and Fino (1989)
54	Monte Circeo, Terracina, Italy	September 1957	420	600	F	MIT-0157FS	Captured after an attack on a diver	Bini (1960)
55	Monte Circeo	November 1964	300	380	M	MII-0264MS		Felicioni and Polidori (personal communication)

(*continues*)

Appendix 1 (*Continued*)

No.	Location	Date	TL (cm)	BM (kg)	Sex	Catalog no.	Remarks	Reference
56	Elba Island, Italy	August 1938	510+		F[a]	MIT-0138M	Taken in a tuna net	
57	Livorno, Italy						General locality only	
58	Viareggio, Italy		315					Giglioli (1880)
59	Monterosso, Italy		600					Tortonese (1956)
60	Santa Margherita Ligure, Italy			640				Sassi (1846)
61	Portofino, Italy	1876	446		M			Tortonese (1938)
62	Camogli, Italy	May 16, 1954	520[a]	1400	F	MIL-0154MF	Published photographs claiming 700 cm TL, but TL is probably less, although >500 cm	Tortonese (1965)
63	Capo Testa, Sardinia	June 1977	500+			MSA-0177M	Caught in a tuna trap; photographs of preserved jaws	M. Zuffa (personal communication)
64	Capo Testa	April 1978	200			MSA-0178J	Caught in a tuna trap 100 m offshore	M. Zuffa (personal communication)
65	Capo Falcone, Sardinia	April 1971	400+		F[a]	MSA-0171	Taken in a tuna trap 100 m offshore	M. Zuffa (personal communication)
66	Santa Caterina di Pittinurri, Sardinia	August 1976	470[a]	1100	M	MSA-0176MM	Taken in a tuna trap	M. Zuffa (personal communication)
67	Ilot des Moines, Corsica	September 1984	500	850		MCO-0184	Taken in a commercial net	Touret (1992)
68	Nice, France						Several specimens reported	Risso (1810); Moreau (1881); Carus (1889–1893)
69	Antibes, France	January 1991						Touret (1992)
70	Toulon, France							Carus (1889–1893)
71	Toulon	1950s					Specimen responsible for an attack on fishermen	I. K. Fergusson and A. D. Testi (unpublished data)
72	Marseilles, France	1989	400+			MFR-0489	Photographed pursuing a school of tuna	Moreau (1881); Touret (1992)
73	Martigues, France						Jaws preserved at Monaco	Roule (1912)
74	Grau-de-Roi, France	1943, 1946, 1956					One specimen described as 8 m TL exhibited during WWII throughout Provence region	Granier (1964)
75	Palavas, France	1943						Granier (1964)
76	Sète, France	August 1875	400				Girth 223 cm	Moreau (1881)
77	Sète	1876	242				Girth 150 cm	Moreau (1881)
78	Sète	1976[a]	450[a]					B. Serét (personal communication)
79	Sète	January 1991	450		F	MFR-0191MF	Taken in a trawl	B. Serét (personal communication)

(*continues*)

Appendix 1 (*Continued*)

No.	Location	Date	TL (cm)	BM (kg)	Sex	Catalog no.	Remarks	Reference
80	Tossa de Mar, Spain	November 17, 1992	475	1000	M	MSP-0492MM	Washed ashore moribund	Piza (personal communication); Fergusson (1994a)
81	Tarragona, Spain							De Buen (1926)
82	Vinaroz, Spain						Jaws examined	Rey (1928)
83	Islas Columbretes, Spain	January 1878	600				Caught in a net enclosure	Rey (1928)
84	Castellon de la Plana, Spain						"Large" specimen	Perez-Arcas (1878)
85	Valencia, Spain		500					Perez-Arcas (1878)
86	Puerto de Mazarrón, Spain							Perez-Arcas (1878)
87	Cabo Salinas, Majorca Island						Taken in a net	Rey (1928)
88	Andraitx, Majorca Island	July 1992	500			MBA-0392	Two specimens taken by a commercial fisherman	Piza (personal communication)
89	Menorca Island							Barcelo and Combis (1868)
90	Spanish waters	1984					Taken as by-catch in a swordfish fishery	Rey *et al.* (1986)

TL, Total length; BM, body mass.
[a] Some uncertainty exists as to the accuracy of this information.

Appendix 2 **Location and Capture Details for White Sharks in the Adriatic Sea**

No.	Location	Date	Total length (cm)	Body mass (kg)	Sex	Catalog no.	Remarks	Reference
1	Golfo di Venezia, Italy		490					Doderlein (1881); Carus (1889–1893)
2	Golfo di Venezia	July 1977	480			MIA-0277M	Photographed near a research platform in Venice Lagoon	L. Cavaleri (personal communication)
3	Trieste, Italy	1873	460		M			Doderlein (1881); Graeffe (1886)
4	Riccione, Italy	1963	450[a]			ISAF 1220	Responsible for an attack on a human	
5	Pesaro, Italy	September 1989	500+			MIA-0188M	Photographed	A. D. Testi (personal communication)
6	Ancona, Italy	December 17, 1991	210	180		MIA-0291J		A. D. Testi and M. Zuffa (personal communication)
7	Termoli, Italy	March 1992	230			MIA-0192J	Four or five juveniles caught during fall near shore	A. D. Testi and Bianchi (personal communication)

(*continues*)

Appendix 2 (Continued)

No.	Location	Date	Total length (cm)	Body mass (kg)	Sex	Catalog no.	Remarks	Reference
9	Kvarner Gulf, Croatia	June 1879						Graeffe (1886); Tortonese (1956)
10	Rijeka, Croatia	August 1934	600					Giduici and Fino (1989)
11	Cres Island, Croatia	June 17, 1878	371					Carus (1889–1893)
12	Losinj, Croatia	August 1993				MCA-0193	Repeated sightings by fishermen	G. Notarbartolo di Sciara (personal communication)
13	Sibenik, Croatia	August 1993	500			MCA-0293M		G. Notarbartolo di Sciara (personal communication)
14	Kraljevica, Croatia	September 1934	700					Giudici and Fino (1989)
15	Ustrine, Croatia	September 1879	530					Faber (1883)
16	Moschiena, Croatia	September 1934	500					Giudici and Fino (1989)
17	Dalmatian Coast	1877–1879					A number of captures reported	Doderlein (1881); Graeffe (1886)
18	Split, Croatia	July 23, 1879	445					Carus (1889–1893)
19	Gradac, Croatia	November 5, 1879	250					Carus (1889–1893)

[a]Some uncertainty exists as to the accuracy of this information.

Appendix 3 Location and Capture Details for White Sharks in the Eastern Mediterranean, the Aegean Sea, and the Bosporus

No.	Location	Date	Total length (cm)	Body mass (kg)	Sex	Catalog no.	Remarks	Reference
1	Greek waters	Various					General locality in Doderlein (1881)	Thompson (1947); Ondrias (1971); Economidis (1973); Doderlein (1881)
2	Corfu Island	July 1951				ISAF 277	White shark responsible for a fatal attack on a swimmer	
3	Aspra Spitia, Greece	December 1984	700[a]			MGR-0284M	White shark, estimated at 700 cm TL, killed a diver	
4	Makryalos, Thermaïkos Gulf, Greece	September 1972	460	1300	M	MAG-0472MM	Caught 600 m offshore	Economidis and Bauchot (1976)
5	Kavállah, North Aegean Sea, Greece		230					Konsuloff and Drenski (1943)
6	Thásos, North Aegean Sea, Greece		180					Konsuloff and Drenski (1943)

(continues)

Appendix 3 (*Continued*)

No.	Location	Date	Total length (cm)	Body mass (kg)	Sex	Catalog no.	Remarks	Reference
7	Alexandroúpolis, North Aegean Sea, Greece							Konsuloff and Drenski (1943)
8	Síros, Cyclades, Greece		500				Listed as *Lamna cornubica*	Heldreich (1878)
9	Bosporus, Turkey	February 1881	391				Washed ashore at Seraglio Point, Istanbul; girth 335 cm; stranded at Beglerbey	
10	Bosporus	November 17, 1881	470	1500	F			
11	Bosporus	February 1962	500+	3750	F	MTK-0162M	Mass surely in error; photograph available from the Popufalo Picture Agency	G. Wood (personal communication)
12	Paphos, Cyprus	1993	500+		F[a]	MCY-0593M	Sighted subsurface by divers on a few occasions	Fisher *et al.* (1987)
13	Acre, Israel		200				Taken by rod and line	Ben-Tuvia (1971)
14	Agamy Beach, Alexandria, Egypt	1934	425		F		Carrying nine embryos; identity doubted by various authors, and stated TL is low for a mature female; original photographs not located at the BMNH	Norman and Fraser (1937)

[a]Some uncertainty exists as to the accuracy of this information.

31

Occurrence of the White Shark along the Brazilian Coast

OTTO B. F. GADIG *and* RICARDO S. ROSA
Universidade Federal da Paraíba, CCEN
Departamento de Sistematica e Ecologia,
Campus Universitario
João Pessoa, Paraíba, Brazil

Introduction

The white shark *Carcharodon carcharias* is rare along the Brazilian coast, where its occurrence to date is known from only five published records (Ribeiro, 1923; Rocha, 1948; Ruschi, 1965; Ellis and McCosker, 1991; Arfelli and Amorim, 1993). Archaeological records, however, show that the white shark was one of the most common species in eastern and southern Brazil about 4000–6000 years ago. Many teeth of this species have been collected at sites in Rio de Janeiro, São Paulo, Santa Catarina, and Rio Grande do Sul (Duarte, 1968; Richter, 1986; Franco and Barbosa, 1991).

We report here eight additional recent records (including an attack on a human), and comment on the species' distribution in the western South Atlantic.

Materials and Methods

Most records lack precise measurements or major information on biological features and, for the most part, were based on examination of photographs and/or jaws. To estimate the total length (TL) of photographed specimens, we compared the animal's length with those of adjacent persons or inanimate objects of known size in the photographs. For jaw measurements, we used the methodology proposed by Randall (1973), measuring the perimeter of the palatoquadrate by laying a string over its curvature, at the base of the teeth, as well as measuring the height of the enamel of the largest upper tooth. Data were compared with those of Randall (1973) to estimate TL and mass.

Whenever possible, persons involved with the captures were contacted to obtain additional information.

Results

The occurrences of white sharks recorded to date for the Brazilian coast are as follows, in chronological order.

1. An individual was captured in waters of the Ceará coast, northeast Brazil, at the end of the last century; its jaws were deposited in a local collection (Rocha, 1948).

2. A specimen, estimated to be 6 m TL, was displayed at the Praça XV public market, Rio de Janeiro, in 1907; its jaws were preserved (Ribeiro, 1923).

3. A shark, estimated to be 5.2 m TL and 1200 kg in mass, was captured in a gill net set close to the shore at Barra de Guaratiba, Rio de Janeiro, in 1931. The shark was photographed and then cut into small pieces at the Praça XV public market (Fig. 1). This shark was first identified as a basking shark *Ce-*

FIGURE 1 A white shark caught at Barra de Guaratiba, Rio de Janeiro, in 1931, previously identified as a basking shark (record 3).

torhinus maximus (Tomas and Gomes, 1989). However, the conical snout, gill slits that did not reach the dorsal outline, relatively large eyes, black marks on the pectoral fin axil, and the relatively small mouth, are clear characteristics of *C. carcharias*. Moreover, the shark in question had a stingray's tail in its stomach, a finding not consistent with the strictly planktivorous feeding habit of *Cetorhinus*.

4. An individual, estimated to be 4.7 m TL and >1000 kg in mass, was captured in a gill net at Barra do Jucu, Espírito Santo, during the 1950s. The jaw had a perimeter of 98 cm, and the biggest upper tooth was 4.1 cm long.

5. A specimen, captured off Espírito Santo in the early 1960s, was preserved at the Professor Mello Leitao Museum (Ruschi, 1965); we did not examine the material.

6. A specimen, >4 m TL and about 900 kg in mass, was captured by a spear diver at Angra dos Reis, Rio de Janeiro, in 1968; a photograph is published in the book by Ellis and McCosker (1991).

7. An individual was captured by hook and line at Atafona, Rio de Janeiro, during summer in the 1970s. The jaw perimeter measured 102 cm, and the height of the largest upper tooth was 5 cm. On the basis of these measurements, we estimate 5.4 m TL and about 1300 kg mass (Fig. 2).

8. A shark, about 4 m TL (estimated from a photograph) and about 800 kg in mass, was caught in a gill net set 30 m from shore at Saquarema Beach, Rio de Janeiro, in January 1974.

9. A second individual from Atafona, Rio de Jan-

FIGURE 2 Jaw set of a white shark, about 5.4 m total length, caught during the 1970s at Atafona, Rio de Janeiro (record 7).

eiro, was captured by hook and line during summer in the early 1980s. Only some teeth were preserved, and we estimate the size at 4 m TL.

10. A nonfatal white shark attack occurred at Cabo Frio, Rio de Janeiro, in February 1981. A diver was attacked while spear fishing. The bite marks on his left arm (Fig. 3) show the broad teeth and intertooth gaps, similar to the pattern described by Ames and Morejohn (1980) and Collier (1992) as characteristic of the white shark.

11. A specimen, measuring 5 m TL and about 1100 kg in mass, was caught by longline at Acaraú, Ceará, in the early 1990s. The palatoquadrate perimeter was 98 cm, and the height of the largest upper tooth was 4.7 cm.

12. A second individual was captured at Acaraú, Ceará, in June 1991. It was estimated to be 5.5 m TL and about 1400 kg in mass. The palatoquadrate perimeter measured 104 cm, and the height of the largest upper tooth was 5.1 cm.

13. A female white shark, about 5.3 m TL and 2500 kg in mass, was captured off Cananéia, São Paulo, on December 8, 1992 (Arfelli and Amorim, 1993). Although not reported by these authors, this specimen had bite marks in the left pectoral fin, possibly inflicted by a male white shark during copulatory behavior.

Discussion

The specimens reported here ranged from 4 to 6 m TL, assuming that the Ribeiro (1923) estimate for the largest one is correct. TLs of four sharks, estimated from measurements of the jaws (Fig. 4A and B), ranged from 4.7 to 5.5 m. Thus, all records of *C. carcharias* reported for Brazil were of large individuals (>4 m). The fishermen of Atafona, Rio de Janeiro, where this species is called *cação-boto* (dolphin shark), reported the capture of an individual measuring about 1.5 m, but this information could not be confirmed.

Mass was estimated for seven individuals (Fig. 4C). Not included were the records of Ribeiro (1923),

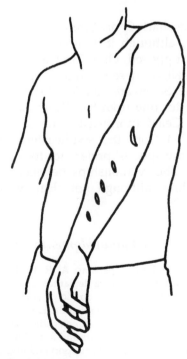

FIGURE 3 Wounds likely inflicted by a white shark during an unprovoked nonfatal attack on a diver at Cabo Frio, Rio de Janeiro, in 1981 (record 10). Note the pattern of teeth marks and the gaps between them.

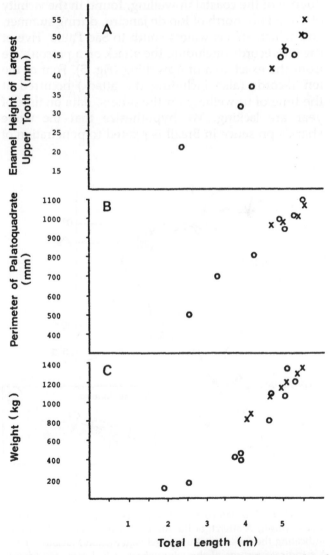

FIGURE 4 Relationships between total length and (A) height of enamel of the largest tooth, (B) perimeter of palatoquadrate, and (C) estimation of mass. X, The present study; O, Randall (1973).

Rocha (1948), and Ruschi (1965), as well as records 9 (teeth only) and 10 (attack on a human), due to lack of precise TL measurements. Our estimates indicated a mass ranging from 800 to 1300 kg. The only directly weighed specimen (record 13) was that from Cananeia, São Paulo (Arfelli and Amorim, 1993). The individual captured in Barra de Guaratiba, Rio de Janeiro (record 3) (Fig. 1), according to the local fishermen, weighed about 1200 kg. Plotting this information in the TL–mass graphic resulted in an estimate of 5.2 m TL, consistent with the estimate made from examination of the photograph.

The concentration of records in the area between Rio de Janeiro and Espírito Santo (10 of 13, or 77%, including the likely attack on a person) reflects the species' preference for cooler temperate waters (Compagno, 1984a). In this region, which otherwise is tropical and subtropical, these records reveal the influence of the coastal upwelling, found in the vicinity of Cabo Frio, north of Rio de Janeiro, during summer, and which affects waters south to São Paulo. Five of the ten records (including the attack on a person), are from the exact area of upwelling (Fig. 5). Four of the ten records (also including the attack) occurred at the time of upwelling; for the others, data on time of year are lacking. We hypothesize that the white shark's presence in Brazil is related to penetration of

cold water, as a function of seasonal upwelling at Cabo Frio, Rio de Janeiro. There, the continental shelf of eastern Brazil is narrow, promoting the coastal penetration of cold waters (Matsuura, 1986).

In the western South Atlantic, white sharks have also been reported along the coast of Argentina (Siccardi *et al.*, 1981). These authors raised the possibility that a trans-Atlantic migration of the white shark occurs between South Africa and South America. This hypothesis cannot be evaluated at present. In Uruguay, white sharks have been reported by Ximenes (1962) and O. Mora (personal communication). Ellis and McCosker (1991) pointed out that a number of pinniped colonies along the Argentina coast could well attract a resident population of white sharks, as is true in Australia, South Africa, and California. Colonies of the pinnipeds *Otaria flavescens* and *Arctocephalus australis* occur along the Rio Grande do Sul coast, in southern Brazil, and in adjacent Uruguay and Argentina (Pinedo *et al.*, 1992). However, observations to link the white shark with these potential prey species have not been conducted (Rosas, 1989).

The three occurrences in Ceará, in equatorial waters of Brazil, cannot be explained at present due to the scarcity of data.

Summary

The white shark *C. carcharias* is rare along the Brazilian coast, although it is more frequent than was previously supposed. Only 13 records of this species exist for Brazil; of these records, we present eight for the first time. Seven are from Rio de Janeiro, two from Espírito Santo, one from São Paulo (southeast), and three from Ceará (northeast). The concentration of records (77%) from the southeastern coast is explained by the species' preference for cooler temperate waters. The relationships between *C. carcharias* and upwelling, although not clear, was discussed here.

Acknowledgments

We thank the following persons: A. R. G. Tomas and L. A. Zavala-Camin (Instituto de Pesca de Santos–São Paulo) and U. L. Gomes (Universidade do Estado do Rio de Janeiro) for comments on the manuscript, L. Capistrano (Fundacao Brasileira de Conservacao da Natureza) and R. Otoch (Universidade Federal do Ceará) for specimen data, M. R. de Oliveira and R. S. Britto (Universidade Catolica de Santos–São Paulo) for support and inspiration, and A. P. Klimley (Bodega Marine Laboratory) and D. G. Ainley (Point Reyes Bird Observatory) for invaluable support and the invitation to participate in the white shark symposium.

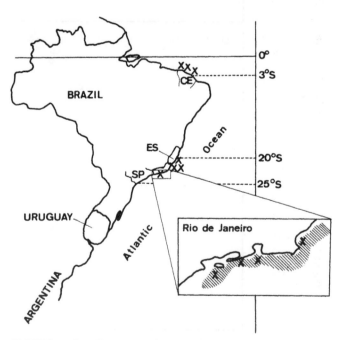

FIGURE 5 Brazilian coast, showing the occurrence of the white shark. (inset) A stretch of the Rio de Janeiro coast, with shading indicating the presence of upwelled water (from Matsuura, 1986); X's indicate records of the white shark. CE, Ceará; ES, Espírito Santo; SP, São Paulo.

Catches of White Sharks in KwaZulu–Natal, South Africa and Environmental Influences

GEREMY CLIFF
Natal Sharks Board
Umhlanga Rocks, South Africa

SHELDON F. J. DUDLEY
Natal Sharks Board
Umhlanga Rocks, South Africa

MARK R. JURY
Oceanography Department
University of Cape Town
Rondebosch, South Africa

Introduction

Catches of white sharks *Carcharodon carcharias* in KwaZulu–Natal, formerly known as Natal, are largely confined to the shark nets that protect swimmers against shark attack at more than 40 beaches along the coast (Cliff *et al.*, 1989). The first nets were laid in Durban in 1952, followed by the widespread installation of nets in the mid-1960s. In 1994, a total of 40 km of nets was permanently installed and maintained by the Natal Sharks Board (NSB). Between 1978 and 1993, these nets caught an average of 1354 sharks annually, including 39 white sharks.

White sharks are infrequently caught by recreational anglers in KwaZulu–Natal. These catches are poorly documented, but the most publicized have been those made by anglers from the piers demarcating the entrance to Durban Harbour. Using whale meat as bait, these anglers targeted the large sharks that followed the whale carcasses in tow to the processing factory. The largest white shark, landed in 1953, weighed 754 kg, and the last white shark on record, weighing 234 kg, was landed in 1968 (Mara, 1985). Catches of white sharks by these anglers declined as a result of the cessation of whaling from Durban Harbour in 1976. Consequently, this study is confined to catches of white sharks in the shark nets.

In April 1991, the South African government declared the white shark a protected species, making it illegal to catch or kill any white shark without a permit issued by the Director-General of Environment Affairs (Compagno, 1991). This was preemptive legislation to allow the scientific community time to assess the stocks of white sharks along the coast and the impact of fishing-induced mortality. At present, the shark net catches provide the only long-term index of changes in white shark stocks in southern Africa. This chapter examines trends in the catch rates of white sharks in the shark nets from 1966 to 1993.

A relationship between white shark catch rates and environmental parameters, including El Niño–Southern Oscillation (ENSO), is investigated. In its warm phase, or El Niño, ENSO leads to decreased summer rainfall and increased westerly wind flow in southern Africa (Lindesay, 1988). Increased easterly winds and summer rainfall and reduced coastal sea-surface temperatures (SSTs) are associated with the cold phase, or La Niña (Schumann *et al.*, 1996).

Cliff *et al.* (1989) described the general biology of 429 white sharks caught in the nets between 1978 and 1988. The length–mass and length–length equations published by these authors are improved in this chap-

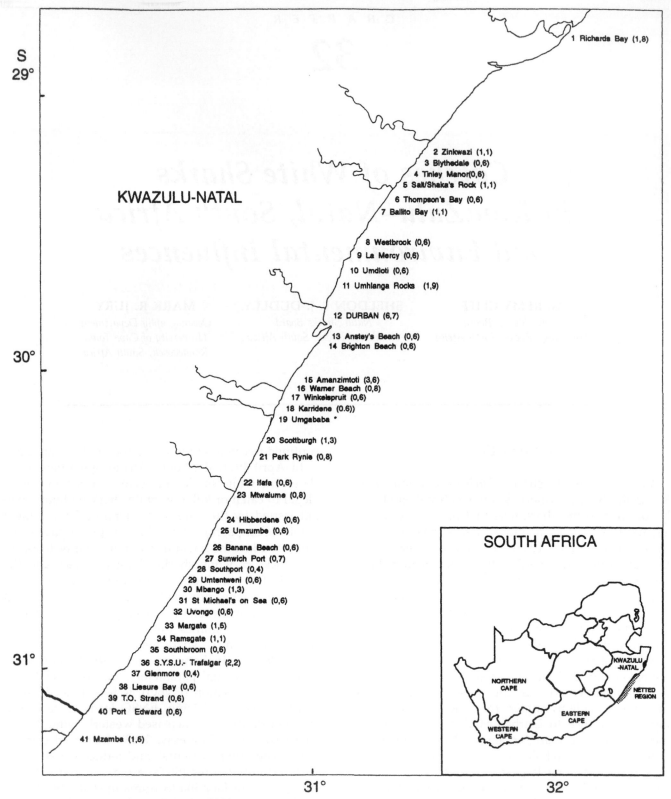

FIGURE 1 Netted beaches on the KwaZulu–Natal coast and, in parentheses, the length of nets (in kilometers) in January 1993. *, Nets were removed from Umgababa in 1990.

ter, which also examines the stomach contents of white sharks caught from 1989 to 1993. This study details the capture of the first mature male white sharks in the nets and the capture of possible young-of-the-year.

Methods

The KwaZulu–Natal shark nets are distributed along a 325-km stretch of coast, with a single net installation, Mzamba, in Eastern Cape (Fig. 1). A description of net specifications and deployment has been given elsewhere (Cliff *et al.*, 1988a).

Although nets were first installed in 1952, records kept by the NSB date back only to 1966. The early data were not as accurate or complete as those collected after 1977, for reasons discussed by Dudley and Cliff (1993a). To provide a comparable sequence of catch rates from 1966 to 1993, catches at several beaches that lacked continuous data were excluded. These locations included Durban, Anstey's, and Brighton beaches (Fig. 1). Catches at Richards Bay, where nets were installed in 1980, have been treated separately. Because of its geographic isolation—84 km from the nearest net installation at Zinkwazi—Richards Bay may have fished a virgin shark stock, depending on residency and movement patterns of the sharks.

White shark catches (excluding catches from Richards Bay and Durban, Anstey's, and Brighton beaches) were analyzed relative to ENSO, which was quantified using the dimensionless Southern Oscillation Index (SOI). The SOI measures the surface pressure anomaly difference between Darwin and Tahiti and is supplied by the U.S. Climate Analysis Center. As ENSO influences summer rainfall and coastal sea surface temperature (SST), these parameters were formulated into indices for comparison with white shark catches.

Analysis was approached in two ways. First, a generalized linear model was fitted by means of GLIM, a component of GENSTAT, which treats the response variable (catch) as Poisson-distributed frequency data. Independent variables considered in addition to SOI were monthly fishing effort and, to account for seasonal effects, various dimensionless annual and biannual oscillatory functions (Table I).

Second, two physical parameters, average rainfall and SST, were incorporated. Local rainfall data were obtained for the inland coastal plains from Durban southward; the rainfall in northern KwaZulu–Natal was excluded, due to the random influence of tropical cyclones. SST was taken at the Amanzimtoti shark

TABLE I Relationship between Quarterly White Shark Catches and Southern Oscillation Index, January 1978 to December 1993

Independent variable[a]	Estimate	SE	*t*
Time[b]	−0.0270	0.0076	−3.56
Time^2	0.00027	0.0001	2.85
Time^3	−0.0000008	0.0000003	−2.42
sin [(month − $\frac{1}{2}$) × π × $\frac{1}{12}$]	−0.4779	0.0644	−7.42
SOI	0.0435	0.0183	2.38
Constant	2.780	0.148	18.77
r^2	40.28%		
Degrees of freedom	63		

SE, Standard error; SOI, Southern Oscillation Index.
[a]The dependent variable is log catch.
[b]Time taken as the middle of each quarter.

nets (Fig. 1) beginning in 1981. For comparison, the time series were normalized by subtracting the historical mean for each month and dividing by the standard deviation. Some of the longer time series exhibited bias in the period 1981–1993, so minor adjustments were made to "zero" the departure means and create records with similar variance.

Statistical lag cross-correlation analysis was carried out on the unsmoothed normalized departures using Pearson's product moment method of linear regression. As pairwise cross-correlations using the monthly means were poor, and due to the winter peak in shark catches and the summer peak in SOI, SST, and rainfall, correlations were computed using these seasonal data.

Shark lengths used in the text are precaudal measurements (PCLs), measured as a straight line, parallel to the body, from the tip of the snout to the precaudal notch. Geometric regression relationships between PCL and fork length (FL), total length (TL), upper caudal length (UCL), and mass (M) were determined using STATGRAPHICS 6.0. This package identified outlier data points, which had previously been included by Cliff *et al.* (1989). Points with a standardized residual >3 were omitted, as the chance that such points were not outliers is <1%. These points made a large difference to the regressions and are assumed to have been the result of unacceptably large errors in measurement.

FL was measured as a straight line, parallel to the body, from the tip of the snout to the fork in the tail. Casey and Pratt (1985) defined TL as "a straight line measurement along the body axis from tip of snout and intersecting a perpendicular line dropped from the tip of the upper caudal lobe." Compagno (1984a) measured TL by depressing the upper caudal lobe to lie parallel to the body axis.

Results and Discussion

Shark Net Catches and Catch Rates

Two features are evident in the catch rates of white sharks (Fig. 2). One is the sharp decline that lasted until 1972. This decline was evident among other shark species, species groups, and all species combined (Dudley and Cliff, 1993a). The drop was interpreted as the removal of the resident community, with catches sustained by the steady influx of sharks from adjacent waters (Wallett, 1983; Cliff *et al.*, 1988b). Such a catch pattern is understandable in more sedentary or resident shark species such as the bull shark *Carcharhinus leucas* (Cliff and Dudley, 1991a), but it seems unusual for the white shark, thought to be a wide-ranging species. On the other hand, a tagged 150-cm white shark was recaptured exactly 1 year later at the same location (see Chapter 36, by Cliff *et al.*), suggesting that white shark movement may not be random and that individuals may return to fixed localities at regular intervals. This pattern has also been shown by the recapture of tagged sharks in Western Cape (see Chapter 36, by Cliff *et al.*) and by research in California (Ainley *et al.*, 1985; Klimley and Anderson, Chapter 33).

The second feature of the catch rates was the marked cyclical trend, with peaks every 4–6 years (1968, 1973, 1978, 1984, and 1989), presumably reflecting natural variation in abundance within the netted region. The effect of the initial "fishing out of residents" phase is superimposed on the natural cycle, and in interpreting catch rate trends, the influence of this initial phase should be isolated. Based on the results shown in Fig. 2, it is tempting to conclude that the "fishing-out" phase ended in 1972; however, it may have ended earlier, with 1971 and 1972 being years of low natural abundance.

The catch rate of white sharks from 1973 to 1993 declined significantly ($r = -0.527$, $p = 0.014$; two-tailed t test). From 1974 on, the trend was still significant ($r = -0.459$, $p = 0.041$), but in 1978–1993, the period for which the data were the most accurate, there was no significant decline in the catch rate ($r = -0.046$, $p = 0.076$). At Richards Bay, where 45 white sharks were caught in 14 years of netting, the catch rate displayed a steep initial decline (Fig. 2). Marked fluctuations followed, and there was no significant trend in the catch rate.

Cliff *et al.* (1989) reported no significant decline in the catch rate of white sharks along the entire coast from 1974 to 1988. The exclusion of certain beaches and the inclusion of an additional 5 years of data produced a decline that was statistically significant. Twenty white sharks were caught in the first 7.5 months of 1994. Given that catches are highest in the

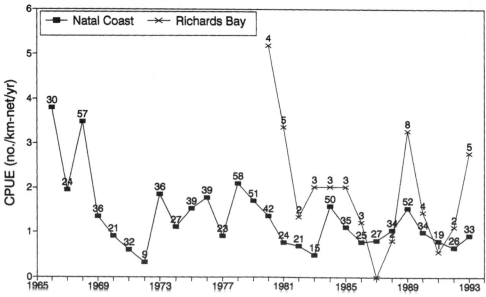

FIGURE 2 The annual catch of white sharks per unit of effort (CPUE) in all shark nets, except those at Durban, Anstey's, and Brighton beaches and Richards Bay, 1966–1993 (Natal coast) and for Richards Bay, 1980–1993. Annual catches are given next to each datum point.

second half of the year (Cliff *et al.*, 1989), it is possible that the catch for 1994 may constitute one of the 4- to 6-year peaks—the previous peak being in 1989. If so, there may be no significant trend in catch rates. Given the interannual variability in catches, the significant decline reported may be an artifact.

Despite the uncertainty surrounding the trend in catch rates, being an apex predator, the white shark is unlikely to be abundant. The sum of mortalities in the shark nets and catches of white sharks in Western Cape, which numbered at least 50 per year prior to the fishing ban (T. Ferreira, personal communication), may have had a significant impact on the stocks.

The median annual length of the 247 females, caught and accurately measured, declined between 1978 and 1993 ($r = -0.607$, $p < 0.01$, df = 14); no such trend was evident among the 185 males (Fig. 3). There is no reason that only the females should show such a trend. Theo Ferreira (personal communication) noted a decline in the number of very large (>4-m) white sharks observed at western Cape congregating sites.

Mark–recapture studies provide another insight into the population dynamics of the white shark in southern Africa. In the 5-year period 1989–1993, 27 (14.4%) of the white sharks caught were found alive in the nets. Until the mid-1980s, live white sharks were killed, but by 1989 all were released. Of the 22 tagged, to date three have been recaptured, and the results have been used *inter alia* to provide an initial population estimate of 1279 white sharks (coefficient of variation, 24%) for the south and east coasts of South Africa (see Chapter 36, by Cliff *et al.*). Contin-

ued tagging and the recapture of sharks already tagged will improve this estimate.

White sharks have been caught in the nets throughout the year. Between 1989 and 1993, catches were highest from August to January, with a peak in August (Fig. 4). This trend was similar to that observed between 1978 and 1988, when catches were highest in July (Cliff *et al.*, 1989).

Correlation between White Shark Catches and El Niño

The monthly SOI values, summed to give an annual ENSO index, were plotted against annual catch rates from 1966 to 1993 (Fig. 5). A close relationship was apparent between these two parameters, except during the first 7 years, when catch rates declined sharply.

Both catch rates (Fig. 4) and SOI were highly seasonal. Thus, the seasonal effects were removed and the detrended monthly catch was plotted together with SOI (Fig. 6). It was apparent that SOI led catches by about 6 months in a phase of increasing SOI, while the reverse occurred in declining SOI, which lagged catches by about 6 months. A model relating quarterly catches to SOI was obtained (Table I) and used to compare quarterly catches against catches predicted by the model (Fig. 7). Catches in 1981–1983 and 1991–1992 were far lower than predicted. All factors in the regression were significant, but an r^2 of 40.3% (df = 63) indicates that some variability is unexplained. The

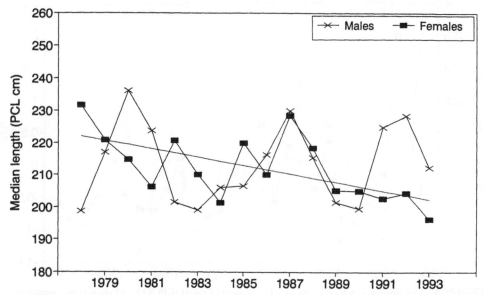

FIGURE 3 The median precaudal length (PCL) of male and female white sharks from the shark nets, 1978–1993, and the regression line fitted to the annual values for females.

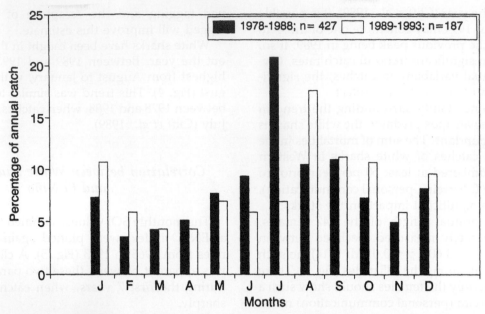

FIGURE 4 The seasonality of white sharks caught in the shark nets.

regression procedure did not take into account autocorrelation within variables.

Pairwise cross-correlation analysis using the seasonal data revealed several associations. Summer (January–March) SOI was associated with summer rainfall ($r = 0.82$, $p = 0.001$, df = 10), as well as rainfall in the preceding spring (October–December). (Fig. 8). Very dry conditions prevailed during El Niño 1983 and 1992, whereas in other summers of high SOI, such as in 1989, the relationship was weaker. There was a correlation between the winter (June–September) shark catch rate and SOI ($r = 0.52$, $p = 0.07$, df = 11) and SST ($r = -0.48$, $p = 0.09$, df = 11) of the preceding summer (January–March). Other correlations between seasonal statistics at various lags were not significant.

A multiple linear regression model using all three environmental parameters provided the best fit:

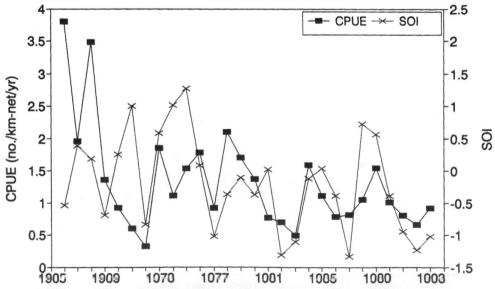

FIGURE 5 The annual catch of white sharks per unit of effort (CPUE) in all shark nets, except those at Durban, Anstey's, and Brighton beaches and Richards Bay, plotted with the annual Southern Oscillation Index (SOI), 1966–1993.

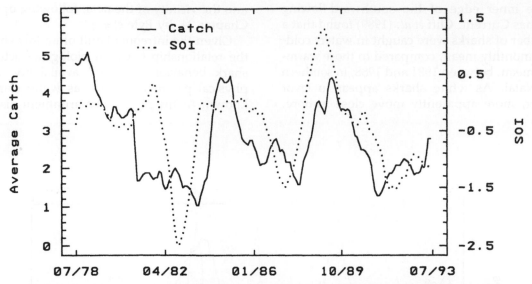

FIGURE 6 The monthly values for white shark catches per unit of effort (CPUE) and the Southern Oscillation Index (SOI), both of which were detrended by the removal of seasonal effects.

winter shark CPUE = (0.495 × summer rain) − (0.4497 × summer SST) − 0.031 (r^2 = 0.448, p = 0.02).

Thus, an increased (decreased) winter shark CPUE correlated with a high (low) rainfall and a low (high) SST during the preceding summer.

The lag-zero correlations between winter shark catch and each of the three physical parameters were lower than those for the previous summer. Rainfall is highest in summer, and water turbidity inshore is lowest at this time. It is not certain whether the decreased turbidity persists into winter following high summer rainfall. Other SOI signals include increased easterly winds and coastal upwelling that lead to low SST in Eastern Cape during summer in the cold phase of ENSO (La Niña) (Schumann *et al.*, 1996). These processes also affect KwaZulu–Natal as borne out by the negative relationship between shark catch and SST. It seems that easterly winds enhance SST gradi-

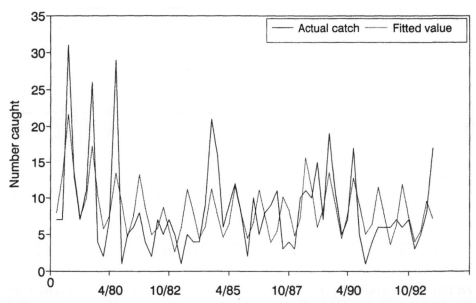

FIGURE 7 Quarterly catches of white sharks, 1978–1993, plotted with the values predicted by the model presented in Table I.

ents on the inner edge of the southward-flowing warm Agulhas Current. Cliff *et al.* (1989) found that a higher number of sharks were caught in waters colder than the monthly mean, compared to those warmer than the mean, between 1981 and 1988, in southern KwaZulu–Natal. As white sharks appear to favor cooler water, more apparently move close inshore,

into the vicinity of the nets, following upwelling (see Chapter 34, by Pyle *et al.*).

Given the infrequent and often low shark catches, the relationship between biological factors, such as shark behavior and food availability, and several physical parameters, such as water clarity, are unknown. As the physical environment displays a lead

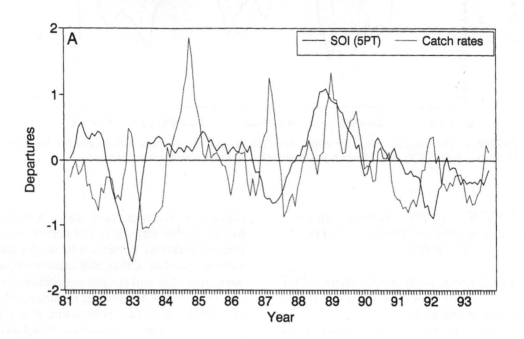

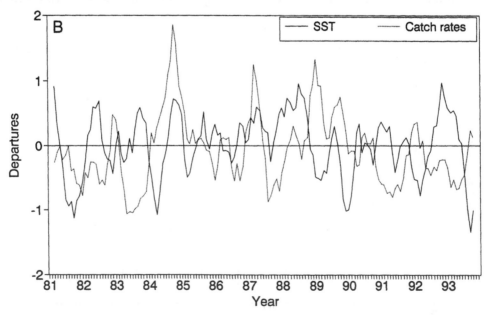

FIGURE 8 Normalized departures in monthly values for white shark catch rates and the following physical parameters: (A) Southern Oscillation Index (SOI), (B) sea-surface temperature (SST), and (C) rainfall. A 5-point running mean was applied to the SOI.

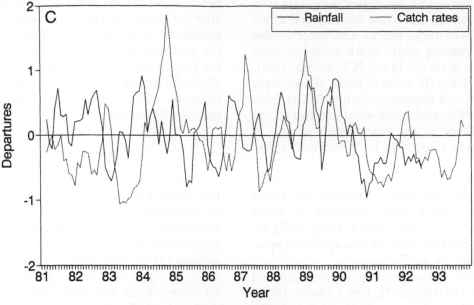

FIGURE 8 (*Continued*)

time of approximately 6 months over shark catches, long-range forecasts may assist in predicting periods of high shark abundance in the nearshore zone.

Mature Males

The first confirmed record of a mature male (293 cm PCL, 371 kg) from the shark nets was made in 1989. Two other mature males were caught in 1992 (Table II), the largest, 373 cm (892 kg), being the longest and heaviest white shark examined since reliable data became available. Two males, both heavier and longer than the smallest mature shark, had soft claspers and were immature (Table II). The siphon sacs of four of the five sharks were measured and found to be about 10% of the PCL. In carcharhinid sharks, the siphon sac of mature males is well developed, extending to the pectoral girdle and constitut-

ing about 40% of the PCL (G. Cliff, unpublished data).

Three large male white sharks, estimated to be 370, 320, and 320 cm, were seen feeding on a whale carcass 1 km off Durban in August 1993. These sightings, together with the capture in the shark nets of four males (Table II) and two adolescent females >3 m PCL (342 and 320 cm PCL), between 1989 and 1993, indicates that large sharks do enter the study region. Nevertheless, shark nets may be inadequate sampling gear for very large white sharks (Cliff *et al.*, 1989).

Small Sharks and Pupping Grounds

Although the nets select sharks >150 cm PCL, 12 white sharks <150 cm PCL were examined between 1976 and 1993. An additional shark was caught at the

TABLE II Largest Male White Sharks Examined from Shark Nets, 1989–1993

| Date of capture | Locality | Reproductive state | Measurement | | | | | | Hepatosomic index | Inner clasper as % of PCL | Siphon as % of PCL |
			M (kg)	PCL (cm)	FL (cm)	TL (cm)	TOT (cm)	UCL (cm)			
June 4, 1992	Trafalgar	Mature	892	373	415	452	474	98	13.6	17.2	9.7
April 29, 1992	T.O. Strand	Mature	550	317	359	384	402	79	16.0	19.7	13.2
June 1, 1992	Mzamba	Immature	442	306	345	387	390	78	12.7	19.6	—
December 27, 1991	Ifafa	Immature	420	295	324	374	376	85	13.8	20.6	10.2
December 6, 1889	Mzamba	Mature	371	293	313	350	369	77	15.1	21.1	10.2

M, Body mass; PCL, precaudal length; FL, fork length; TL, total length; TOT, total length, depressed caudal; UCL, upper caudal length.

surface 15 km offshore by an angler, who reported that "3–4 sharks of this size were seen several times that day." Details of these sharks and those of the smallest free-swimming white shark recorded from southern Africa, a male of 114 cm PCL (Smith, 1951), are presented in Table III. Most of these sharks were caught in August and September, when SST is coldest (Cliff *et al.*, 1989), and at the southern end of the netted region. Three of the sharks were caught at Mzamba, in Eastern Cape (Fig. 1). "Umbilical" scars, were noted on four sharks, but this does not necessarily imply that they were absent on the others, which include the specimen examined by Smith (1951). Such scars have been observed on other young free-swimming white sharks (Ellis, 1975), although the reproductive mode of this species is aplacental (see Chapter 15, by Francis).

The 10 smallest free-swimming sharks recorded worldwide were 140–160 cm TL (see Chapter 15, by Francis), the smallest being reported by Smith (1951). These lengths equate to 114–129 cm PCL. It is not known how long the "umbilical" scars are visible after birth, but many of the sharks caught in the shark nets (Table III) may be young-of-the-year. All other southern African records of white sharks <160 cm TL were from eastern Cape (L. J. V. Compagno, unpublished data). Very little sampling of white sharks has been undertaken in many parts of Eastern Cape,

but it would appear that Eastern Cape is the pupping area for white sharks. This would account for the occasional incursion of newborn white sharks into the southern waters of KwaZulu–Natal during winter. Large white sharks appear to be most common in Western Cape (L. J. V. Compagno, personal communication), which would reduce the possibility of cannibalism by adults.

Stomach Contents

In the period 1989–1993, 14 of 105 sharks examined had everted their stomachs. Among the remainder, 33 stomachs (36%) contained food (Table IV). The most common prey, by frequency of occurrence, were elasmobranchs (51%), marine mammals (30%), and teleosts (21%). One shark ate 477 sardines *Sardinops sagax*, a clupeid fish that migrates into KwaZulu–Natal waters from the south during midwinter. Both dusky shark *Carcharhinus obscurus* and pinniped remains were found in 18.2% of stomachs. The latter were probably remains of the South African fur seal *Arctocephalus pusillus*, as it is the only locally resident pinniped. One stomach contained both pinniped remains and 24 small well-worn pebbles that weighed 50 g. An intact white gumboot and a sheet of cardboard were also found in one of the stomachs that contained seal remains.

TABLE III **Smallest White Sharks Examined from the Shark Nets, 1976–1993**

Date of capture	Locality	Sex	"Umbilical" scar	M (kg)	PCL (cm)	FL (cm)	TL (cm)	TOT (cm)	UCL (cm)	Hepatosomic index
December 1950[a]	Algoa Bay	M		20	114	125	140			17.0
August 23, 1990	Mzamba	F		45	131	146	167	168	38	15.1
August 30, 1985	Mzamba	F	Present	42	131				38	13.8
August 19, 1976	Umdloti	M		38	133				38	
September 5, 1990	Mzamba	F		50	135	152	175	177	40	13.2
August 8, 1978	Leisure Bay	M	Present	42	138				40	
August 9, 1990	Scottburgh	F		48	140	161	184	186	44	10.2
August 15, 1988	Margate	M		46	142				44	13.3
October 25, 1989	Mzamba	M		56	142	162			44	13.8
September 19, 1976	Leisure Bay	F	Present		143				42	
September 13, 1983	Brighton	M	Present	46	144				43	11.5
September 15, 1974	Umzumbe	F	Present		145				42	
September 28, 1989	T.O. Strand	F		60	145	161			41	13.0
September 10, 1976	Marina Beach	F			145				39	

M, Body mass; PCL, precaudal length; FL, fork length; TL, total length; TOT, total length, depressed caudal; UCL, upper caudal length.
[a]The smallest white shark recorded in South Africa (Smith, 1951). The Algoa Bay shark was not caught in the shark nets.

TABLE IV Stomach Contents of White Sharks Caught 1989–1993

| Common name | Scientific name | Measurement[a] | | |
		% F	% N	% M
Sharks/Rays (elasmobranchs)		51.1	3.6	37.0
Whale shark	*Rhincodon typus*	3.0	0.2	0.1
Dusky shark	*Carcharhinus obscurus*	18.2	1.2	14.1
Milk shark	*Rhizoprionodon acutus*	3.0	0.2	1.9
Hammerhead sharks	*Sphyrna* spp.	3.0	0.2	9.1
Unidentified small shark		12.1	0.8	3.4
Unidentified large shark		6.1	0.4	6.2
Unidentified shark		3.0	0.2	2.0
Unidentified shark/ray		9.1	0.6	0.2
Bony fishes (teleosts)		21.2	93.4	39.1
South African sardine	*Sardinops sagax*	3.0	91.7	33.4
Spotted grunter	*Pomadasys commersonni*	6.1	1.0	5.5
Slinger	*Chrysoblephus puniceus*	3.0	0.2	0.1
Unidentified bony fishes		9.1	0.6	0.0
Mammals		30.3	1.9	23.7
Common dolphin	*Delphinus delphis*	3.0	0.2	1.2
Unidentified dolphin		6.1	0.4	3.2
Unidentified pinniped		18.2	1.2	19.2
Unidentified mammal		3.0	0.2	0.1
Invertebrates		12.1	0.8	0.1
Squids (Teuthoidea)		3.0	0.2	0.0
Cuttlefishes (Sepiidae)		6.1	0.4	0.1
Other invertebrates		3.0	0.2	0.1
Totals		33	520	938.2

[a]Percentage total of stomachs (F), number of prey items (N), and mass (M, in kilograms), respectively.

Because of their high resistance to digestion, teleost otoliths and cephalopod beaks, when found without accompanying soft tissue, were excluded from the analysis. Otoliths from white seacatfish *Galeichthys feliceps* were found in three stomachs, and the beaks from nine cephalopods, seven of which were *Loligo* sp., were found in four stomachs.

These results generally agree with those of Cliff *et al.* (1989), the most noticeable difference being a higher incidence of pinniped remains in this study (18.2% versus 7.3% for the earlier period). It is surprising that 9 of 10 sharks that had ingested mammals were <250 cm, as Cliff *et al.* (1989) found that sharks >250 cm have a far higher incidence of marine mammals in their stomachs than smaller sharks.

Morphometric Relationships

PCL, FL, and TL have all been used to measure the size of white sharks. Of these, PCL is the least prone to error in sharks with a precaudal notch. TL is widely used, despite its being measured in at least three ways (Dudley and Cliff, 1993b). The geometric regression equations below enable the conversion of various lengths to PCL.

Fork length:

FL = 1.100 PCL + 3.554 (N = 142; range, 131–373 cm PCL; 95% confidence limits on slope, 1.086 and 1.114; r^2 = 0.9943)

Total length (Casey and Pratt, 1985):

TL = 1.251 PCL + 5.207 (N = 36; range, 131–307 cm PCL; 95% confidence limits on slope, 1.233 and 1.268; r^2 = 0.9984)

Total length (TOT in Compagno, 1984a):

TL = 1.245 PCL + 7.975 (N = 40; range, 131–373 cm PCL; 95% confidence limits on slope, 1.231 and 1.260; r^2 = 0.9987)

Where TL is defined as the sum of PCL and a proportion of the UCL, the relationship between PCL and UCL is

$UCL = 0.265\ PCL + 4.105$ (N = 360; range, 131–373 cm PCL; 95% confidence limits on slope, 0.258 and 0.272; $r^2 = 0.9318$)

The following length–mass equation was also generated:

$M = 2.14 \times 10^{-5}\ PCL^{2.944}$ (N = 383; range, 131–373 cm PCL; $r^2 = 0.9534$)

Summary

Annual catch rates of white sharks *C. carcharias* in shark nets, set along the KwaZulu–Natal coast, varied considerably from 1966 to 1993. A cyclical trend peaked at 4- to 6-year intervals, but a significant decline in sharks caught, the biological significance of which is uncertain, occurred between 1973 and 1993. Low catches followed El Niño, and significant, but certainly not causative, relationships existed between catch rates during winter and the rainfall and SST of the preceding summer. The catch included several juveniles, which may have been young-of-the-year. The first of three mature males ever taken in the shark nets was caught in 1989. Improved regression equations were given for the relationships between PCL and several other parameters in white sharks.

Acknowledgments

We are indebted to the field and laboratory staff of the Natal Sharks Board, who provided the white sharks and the labor to dissect them. S. E. Piper (Psychology Department, University of Natal, Durban) fitted the catch model.

Population Biology

CHAPTER

33

Residency Patterns of White Sharks at the South Farallon Islands, California

A. PETER KLIMLEY
Bodega Marine Laboratory
University of California, Davis
Bodega Bay, California

SCOT D. ANDERSON
Inverness, California

Introduction

Our present understanding of white shark *Carcharodon carcharias* occurrence patterns at pinniped colonies is largely based on baiting studies. In one study, four sharks were attracted to a bait source over a 5-week period at the South Farallon Islands, California, a locality inhabited by several pinniped species (Klimley, 1987a). Although one shark was observed seven times during 4 weeks, the other three were observed one or two times. In a second study, 20 white sharks were attracted during a 9-day period to a bait source near Dangerous Reef, inhabited by a breeding colony of Australian sea lions *Neophoca cinerea*, in Spencer Gulf, South Australia (Strong *et al.*, 1992). Sixteen sharks were also identified at a bait source near the Neptune Islands, inhabited by a breeding colony of New Zealand fur seals *Arctocephalus forsteri*, during a month-long period. Six individuals were sighted repeatedly during 1- to 2-week periods, three of these sharks were again observed 5 months later, and one was seen 580 days later. However, of 58 tagged sharks, only 22% returned to the bait source on the second day, and 19% on the third. Strong *et al.* suggested that the sharks had a high degree of attachment to sites having pinniped colonies. In a third study, 40 white sharks were tagged when attracted to a bait source near Dyer Is-

land, South Africa. Forty-five percent of these sharks were not seen again after the tagging day, despite a 6-month observation period, and only 20% were recorded during the following week. However, two individuals were observed after a period of 2 months. L. J. V. Compagno (personal communication) concluded that sharks tend to disappear from a particular site after less than 1 week's stay, often in less than 3 days.

There are two potential shortcomings in the use of baiting as a sampling technique to estimate the size of the shark populations at a site. First, the estimate of population size depends on the extent of area over which the chemical attractants disperse. If currents were to transport attractants away from the site, sharks would be drawn from the surrounding ocean rather than the area adjacent to the pinniped colony. Sharks hunting close to shore (within the high-risk zone described by Klimley *et al.*, 1992) might not pass through the bait corridor, and therefore could not follow the olfactory stimulus to its source. Second, the period of residency of sharks at the colony might be influenced by the degree of baiting. Sharks might stay longer than usual at the colony to feed on the continuous supply of bait, or might leave prematurely after becoming satiated from feeding on bait.

Ainley *et al.* (1985) suggested that a small number

of white sharks fed in a particular year on pinnipeds at the South Farallon Islands, 1972–1982. The small number of sharks recorded, as the pinniped population was recovering from anthropogenic decimation, appears to be lower than the 13–20 sharks attracted to baits at the Neptune Islands and Dangerous Reef over shorter periods (Strong *et al.*, 1992). In contrast to the Australian study, that at the South Farallones did not involve baiting. The small population estimate at the Farallones was confirmed by a marked reduction in the number of sharks seen after four were captured (see Chapter 34, by Pyle *et al.*).

This study adds to earlier studies of occurrence patterns at the South Farallon Islands (Ainley *et al.*, 1981, 1985; Klimley *et al.*, 1992). Incidental reports prior to 1987 indicated that most predation occurred during autumn, coinciding with the arrival of immature northern elephant seals *Mirounga angustirostris*, a frequent prey of white sharks. First, by discriminating among sharks on the basis of unique markings and differing sizes, we estimate the minimum number of sharks that visited the islands each autumn, 1988–1992. Second, we describe the temporal and spatial patterns of visitation by individual sharks.

Materials and Methods

White sharks were identified at the Farallon Islands from 1988 to 1992 on the basis of unique markings and differing sizes. Distinguishing marks were seen in direct observations, photographs, and video records taken when sharks fed on pinnipeds. Eighty-one percent of the predatory attacks occurred 25–449 m from shore (at most, 1.8 km) (Klimley *et al.*, 1992). Most feeds were witnessed from the 102-m-high peak of Lighthouse Hill (see Fig. 1 in Klimley *et al.*, 1992). A few feeds were seen from other locations on the island. Sharks were also identified when they investigated inanimate objects that were tethered at East Landing (see the map in Klimley *et al.*, 1992). Sharks either surfaced and slowly swam by these objects or attacked them as they drifted away (see Chapter 20, by Anderson *et al.*). Objects used in this study included air-filled mattresses, styrofoam blocks, and surfboards. Hours of observer effort were recorded (Fig. 1). A few sharks were sexed by noting the presence or absence of claspers.

We identified 18 white sharks on the basis of their fin and body markings. For example, Cut Caudal was missing the upper lobe of her caudal fin (see CC in Fig. 2). Cut Dorsal was lacking the tip of his first dorsal fin and had a small, rounded, backward projection on the upper posterior edge of the fin (see CD).

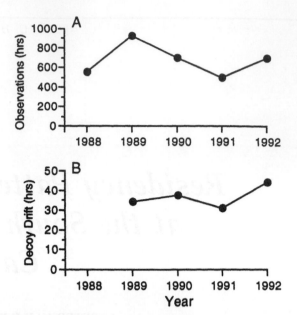

FIGURE 1 (A) Hours per year, 1988–1992, spent on Lighthouse Hill searching for white shark feeding bouts. (B) Time per year, 1989–1992, that decoys were deployed off East Landing to attract white sharks to the surface for identification.

White Slash had a long oval scar extending from above the gill slits down its side toward the mouth (see WS). White Peduncle had a oval white scar at the base of her caudal fin (see WP).

Knowledge of a shark's size was useful, particularly when unique marks were unavailable for identification. Shark lengths were determined in the following manner. The distance between the first dorsal and caudal fins was measured in pixels on video frames converted to digital format. Fin separations were measured on three to six frames when the shark appeared perpendicular to the axis of the camera. The maximum value was used for the shark's size. The pixel measurement was multiplied by 1.8, the proportional relationship between the above dimension and the shark's total length (based on the diagram in Bigelow and Schroeder, 1948).

The length (in pixels) and the distance to the shark from the camera [determined with a theodolite (Klimley *et al.*, 1992)] were entered into a polynomial equation that calculated the shark's length (in centimeters). The equation was based on measurements (in pixels) of an 8-m PVC pipe inscribed with alternating 25-cm black-and-white sections. Pixel measurements were made with the pole positioned perpendicular to the camera at 200-m intervals from 0.28 to 2.45 km from Lighthouse Hill. The polynomial described the curve that best fitted the decreasing number of pixels measured per meter at increasing distances from the camera.

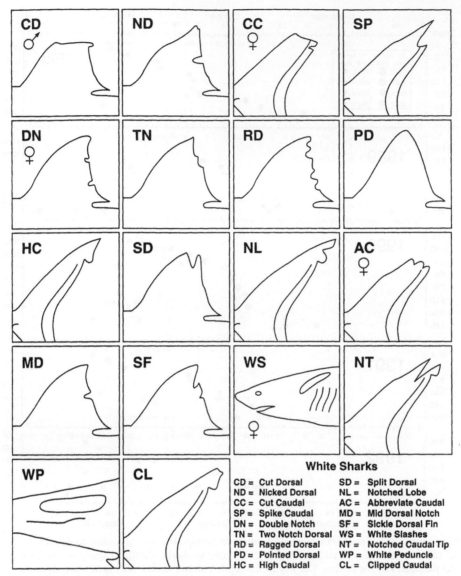

White Sharks

CD =	Cut Dorsal	SD =	Split Dorsal
ND =	Nicked Dorsal	NL =	Notched Lobe
CC =	Cut Caudal	AC =	Abbreviate Caudal
SP =	Spike Caudal	MD =	Mid Dorsal Notch
DN =	Double Notch	SF =	Sickle Dorsal Fin
TN =	Two Notch Dorsal	WS =	White Slashes
RD =	Ragged Dorsal	NT =	Notched Caudal Tip
PD =	Pointed Dorsal	WP =	White Peduncle
HC =	High Caudal	CL =	Clipped Caudal

FIGURE 2 Fin and body characteristics used to identify 18 white sharks that visited the South Farallon Islands from 1988 to 1992. The names of the sharks (based on their individual markings) are given, together with corresponding alphabetical codes.

We used different polynomials for fin separation measurements made at the center or sides of the monitor screen, to minimize error from spherical aberration in the camera's lens. Use of two focal length settings of the zoom lens (50 or 88 mm) permitted accurate measurements over a wide range of distances. Between-fin measurements (in pixels) were entered into a spreadsheet, where they were converted into lengths.

We assumed two unmarked sharks to be the same shark when their length measurements differed by ≤±5 cm within the same year. We believe that failures to discriminate between different sharks were uncommon because our length criterion was small relative to the 234-cm separation between the smallest observed shark (346 cm, Middle Dorsal Notch) and the largest (586 cm, unidentifiable shark). We considered unmarked sharks different when their sizes varied by >±5 cm. Separate measurements during a given attack of individually identified sharks were usually ≤+5 cm apart.

Results

A record of the attendance of white sharks at the South Farallon Islands is shown here for the period from late August to early December over a 5-year

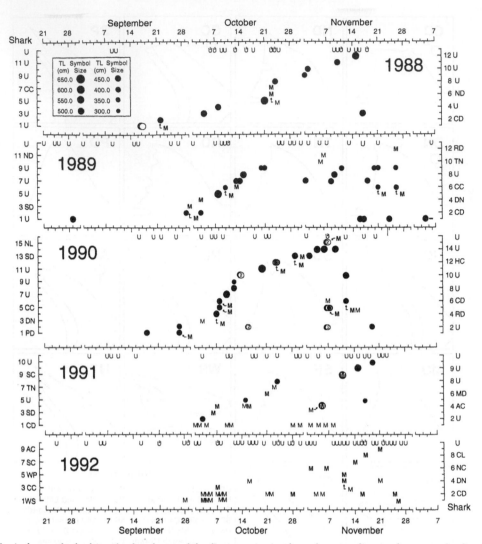

FIGURE 3 Chronological record of white sharks observed feeding on pinnipeds and responding to decoys at the South Farallon Islands, 1988–1992. Alphabetical codes refer to sharks pictured in Fig. 2; U's refer to sharks identified solely on the basis of size. Relative sizes are indicated by the varying diameters of the circles. The key in the top left shows circles with increasing diameters corresponding to 50-cm increments in shark total length (TL). Boldface M's are used for sharks recognized during feeding bouts on pinnipeds; lightface M's indicate responses to decoys. U's along the top (with the number of sharks indicated) show predatory attacks in which the predator was not recognized from either markings or size.

period, 1988–1992 (Fig. 3). Twelve different white sharks were observed during 1988, 12 in 1989, and 15 during 1990. We identified only 10 and 9 sharks during 1991 and 1992, respectively. The decrease in the number of individuals identified during these years did not necessarily indicate that the population of white sharks was shrinking. In fact, less observational effort was spent at Lighthouse Hill due to poor weather conditions during 1991 (Fig. 1A) and less effort was expended deploying decoys from East Landing (Fig. 1B). Sharks were not separated by sizes during 1992, because video records were not taken with the calibrated video recorder. Predatory attacks were actually more frequent during 1992, yet during many

of these attacks the identity of the shark could not be determined from the video records (see many U's in Fig. 3).

White sharks were observed feeding at the island for periods of ≤3 days during early autumn, and some fed for similar periods during late autumn. This pattern of visitation was particularly evident in 1989 and 1990 (Fig. 3A). Only one or two feeding bouts per shark were observed during the first 3 weeks of October (see single and paired circles sloping upward in Fig. 3). Temporary visits, with one or two feeding bouts per shark, were also observed from the second week of October to the second week of November 1990. The sharks may have left, because either they were satiated after feeding or they were forcibly displaced by newly arriving sharks. However, there was not a procession of successively larger sharks that visited the island during these periods (see variability to the circle diameters). Some white sharks that visited during October returned later in November (see the scatter of circles during November).

Although sharks usually appeared to make brief visits, this was not always the case. Cut Dorsal and White Slash were exceptions to this trend. For example, Cut Dorsal investigated the decoy drifting off East Landing on 5 and 6 different days over 10-day periods in 1992 and on 6 days during 1 week in 1991 (Fig. 3B). White Slash investigated decoys on 5 days during 1 week in 1992.

Individual sharks often fed at the same locations on similar dates on successive years. Cut Caudal (Fig. 4A), Double Notch, and Cut Dorsal (Fig. 4B) showed this pattern of behavior at the islands. Cut Caudal fed on an elephant seal on October 10, 1989, near Maintop Reef off West End Island (solid square 2 in Fig. 5). The shark returned on October 8, 1990, to feed near Indian Head on West End Island (solid diamond 6). On October 7, 1992, Cut Caudal was first seen feeding >150 m away from the 1990 location off Maintop Reef (solid inverted triangle 9). We saw Double Notch feed on a seal on October 3, 1989 (solid circle in Fig. 3A). Although never observed feeding during 1990, she was recognized while attacking a decoy on October 3, 1990 (lightface M in Fig. 3). Finally, Cut Dorsal arrived at the island during the same 9-day period on each of 4 successive years (1989–1992) (open symbols 2, 4, 7, and 19 in the chronological record in Fig. 5). He exhibited fidelity to a particular site, East Landing, where, after successfully feeding in 1988, he often investigated decoys and occasionally fed during 1989 and 1990. Although not feeding at this location until late in 1992, he repeatedly attempted to do so, evidenced by repeated visits to a decoy floating there. Of interest was the absence of Cut Dorsal from his "familiar" spot after October 10, 1992 (clear triangle 10), and his appearance in an "unfamiliar" location off Maintop Reef (clear triangle 11) after the arrival of a large female white shark at East Landing.

Discussion

The number of sharks identified at the South Farallon Islands during autumn ranged from 9 in 1992 to 15 in 1989. These totals should be considered minimum estimates of the total number that may have been present. The actual numbers might be larger, because we considered sharks with <±10 cm of size

FIGURE 4 Two sharks that repeatedly returned to particular locations at the South Farallon Islands at the same time each year. (A) Cut Caudal was missing the upper lobe of her caudal fin, and (B) Cut Dorsal was lacking the tip of his dorsal fin and had a small, rounded, backward projection on the upper posterior edge of the fin. (Photographs by S. D. Anderson.)

difference and no distinguishable markings to be the same shark, but mainly because we did not detect sharks unless they fed on pinnipeds or investigated decoys. In addition, poor weather conditions prevented us from searching for attacks some of the time. The average of 11.4 sharks per 3-month observation period (or 3.8 sharks per month) during the 5 years of the study was consistent with the total of 4 sharks attracted to a bait source over a 1-month period during 1985 (Fig. 6). Unlike the studies at Dangerous Reef, baiting sessions were limited to <4 hours. For this reason, the mixture of macerated fish and blood may not have dispersed far from the island, attracting a greater number of sharks to the island from surrounding waters.

We observed roughly twice the number of sharks at the South Farallon Islands over a 4-month period during each of the years from 1987 to 1992 than was estimated by Ainley *et al.* (1985) in 1982. Our search effort was certainly much greater, but the shark popu-

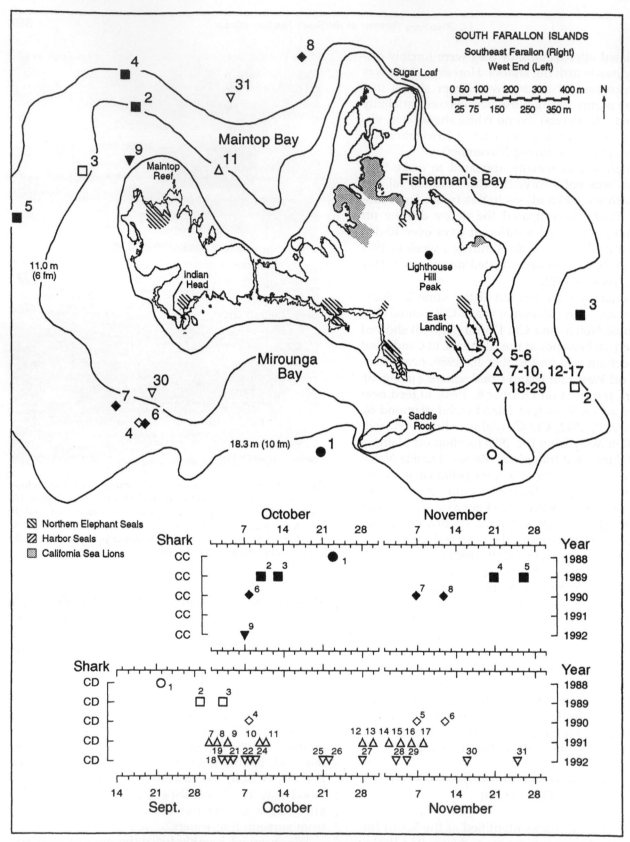

FIGURE 5 Chronological record of visits by Cut Caudal (CC) and Cut Dorsal (CD) to the South Farallon Islands, 1988–1992. CC sightings are indicated by solid symbols and corresponding numbers on the top graph; CD is denoted by clear symbols in the bottom graph. Locations of the sightings are shown on the map; the symbols and numbers on the map and on the chronological records are the same. Pinniped colonies are indicated by hatching and stippling. Note that CC was observed off West End less than 3 days apart during 1989 (solid square 2), 1990 (solid diamond 6), and 1992 (solid inverted triangle 9).

lation probably has grown as well, attracted to the growing elephant seal population. The numbers of immature northern elephant seals visiting the island during the years prior to 1982 were generally lower than the numbers present during our study (see Chapter 34, by Pyle *et al.*). The increase in the number of observed sharks may have partly been due to greater observational effort. Since 1987, continuous daytime watches for predatory attacks have been carried out from Lighthouse Hill during autumn (Klimley *et al.*, 1992).

On the other hand, we observed considerably fewer sharks than were identified during considerably shorter periods at pinniped colonies on Dangerous Reef and at the Neptune Islands in Spencer Gulf, South Australia. Strong *et al.* (1992) attracted 20 white sharks to a bait source at Dangerous Reef over one 9-day period, February 16–24, 1991, and 16 sharks were identified at a bait station at the Neptune Islands over a 31-day period, August 2–September 1, 1990. The sharks observed in the study by Strong *et al.* may not necessarily have been drawn from close to

pinniped colonies, but may have been attracted from the surrounding waters due to the dispersal of chemical attractants by ocean currents. A dye-tracer experiment was undertaken to attempt to estimate the extent and rate of dispersal of chemical attractants at Dangerous Reef (B. Bruce, personal communication). The experiment led to the tentative conclusion that attractants dispersed up to 5 km from the source of attractants. However, this was based on various assumptions on the behavior of the dye and the minimum detection concentration for sharks to respond. Sharks may have been attracted while moving between islands (i.e., during the inter-island cruising pattern described by Strong *et al.*, 1992). Several other islands exist in the vicinity of Dangerous Reef, the nearest being 16 km to the north. Small seal colonies exist on two of these islands (B. Bruce, personal communication). Further supporting this possibility are the observations of a member of our group (A.P.K.) and David Short (South Australian Research and Development Institute, Adelaide) at Dangerous Reef, 1992. A mixture of tuna oil, blood, and macerated fish

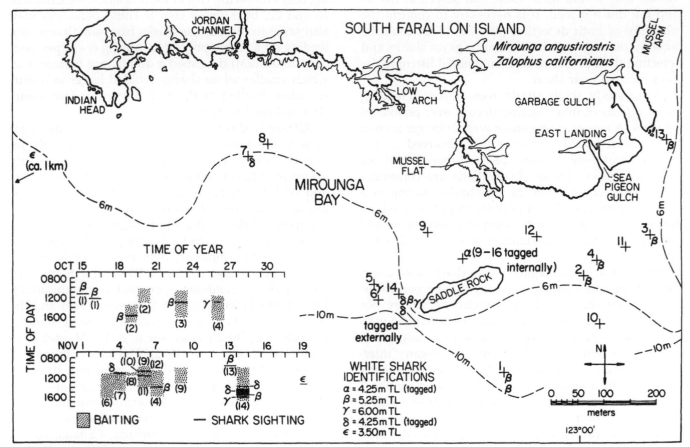

FIGURE 6 Sharks attracted to bait sources in Mirounga Bay of the South Farallon Islands during 1985 (from Klimley, 1987a). The bait sources are indicated by crosses with numbers that refer to observation sessions indicated in the chronological graph. Baiting periods are denoted by stippling; the presence of sharks is indicated by solid lines.

was dispensed for a period of 3 days. A surface slick of oil was observed leading away from the reef, presumably accompanied by dissolved chemical attractants, blood, and particles of fish. On the third day (March 31, 1992), a 4.3 m total length female arrived at the odor source and was tagged with a coded ultrasonic beacon. However, this shark was not detected by an automated recording device moored close to Dangerous Reef over the next 7 days. The shark was attracted to the site after half a day of baiting after the automated recorder was interrogated. The odor corridor also extended offshore at this time.

Our ideas about white shark movement patterns are dependent on the reliability of our sampling methods—observations of sharks during feeding bouts on pinnipeds and investigatory strikes on decoys. We found that our inability to observe a particular shark feeding was not always proof that the shark was absent from the island. For example, Double Notch was observed not to feed, but rather to investigate a decoy, on October 3, 1991, exactly 1 year after being observed preying on a pinniped (see DN in Fig. 3). Both methods detected only hunger-motivated sharks, and would miss those that stayed at the island but did not feed. It is necessary to confirm the reliability of both detection methods. This could be done by placing coded ultrasonic tags on sharks and detecting their presence with automated listening devices moored near shore.

In our study, white sharks were observed feeding on pinnipeds or investigating decoys over periods of ≤3 days. As soon as one shark was no longer seen at the Farallon Islands, another was observed over a similar short period. Some of the same individuals were seen later. These observations are consistent with a nomadic lifestyle of the sharks moving in a somewhat regular circuit and staying at sites where pinnipeds are abundant. Telemetry studies would provide additional information to evaluate this hypothesis.

Strong et al. (1992) tagged white sharks with color-coded tags and acoustic tags. Eighty-one percent of the 32 individuals with color-coded tags were not sighted after 2 days. Similarly, the sharks tagged with acoustic transmitters either remained for only 1 day, "island patrolling," or they left and began "interisland cruising." The shark that A. P. Klimley tagged in 1993 with a sonic beacon was not seen for the 7 days following tagging, suggesting that she was a temporary visitor to the reef.

We do not know why some sharks leave after brief visits to the Farallon Islands. The sharks may have become satiated. Carey et al. (1982) argued that one meal of whale flesh was sufficient to satisfy the metabolic requirement of a white shark for 3 months. However, satiation does not explain why some feeding sharks appear to depart upon the arrival of other sharks. Cut Dorsal, a longtime resident of East Landing, moved to Maintop Bay to feed and was absent from the former site for 2 weeks following the arrival of a larger shark (CD in Fig. 5). Cut Dorsal had fed at East Landing for the previous 3 years, continued to feed at the site later that year, and did so the following year. If sharks were displaced from areas around the island by other sharks, the dominant individuals were not always larger (see the variation in the diameters of the circles, indicating shark sizes in October 1989 and 1990, in Fig. 3).

Investigators have suggested that white sharks exclude each other from food resources. Pratt et al. (1982) reported that only a single white shark fed on a whale at a time, despite the fact that many sharks visited the whale in succession. Klimley et al. (Chapter 22) described a "tail slap" display, which appears to discourage other sharks from stealing pinnipeds. Sharks performed other agonistic behaviors, such as "breach," "side-by-side," and "tilting," during feeding bouts involving two or more sharks (see Chapters 16 and 22, by Klimley et al.). These behaviors may also serve to discourage sharks from simultaneously feeding on a single prey item. Strong et al. (personal communication) described a dominance hierarchy in which smaller white sharks avoided larger individuals when feeding at the bait sources at Dangerous Reef and the Neptune Islands.

Although short visits by sharks were most common at the Farallon Islands, some occasionally remained for longer periods. Cut Dorsal was repeatedly observed at East Landing over 10-day periods in 1991 and 1992. White Slash was also seen repeatedly during a 10-day period during 1992. Goldman et al. (Chapter 11) detected a shark, tagged with an ultrasonic tag, on 5 days over a 2-week period in 1992, and two sharks intermittently over periods of 10 and 13 days, respectively, during 1993 at the South Farallones (K. J. Goldman, personal communication). Five of 58 sharks tagged at Dangerous Reef and the Neptune Islands were observed at the same sites for periods of up to 2 weeks (see 6, 8, 11, 24, and 36 in Table 2 of Strong et al., 1992). Our observations and those of Strong et al. (1992) are consistent with the idea that most sharks at pinniped colonies are "transient," while a few are longer-term "residents." Killer whales Orcinus orca that prey on pinnipeds at colonies at the Oak Bay Islands off Victoria, Canada, have a similar dual social system (Baird and Stacey, 1988).

We found that two sharks, Cut Caudal and Cut Dorsal, fed on successive years at the same locations.

It is tempting to believe that sharks regulate the timing of their movements so that they arrive at familiar locations at those times of year when they successfully fed in the past. Such timing is remarkable in the context of the species' propensity at times to make extensive movements. Carey *et al.*, (1982) tracked a white shark, tagged with an ultrasonic transmitter, 190 km over a period of 2.5 days. Bruce (1992) reported that sharks were captured 18–220 km from release points, with times at liberty ranging from 30 to 78 days. Cliff *et al.* (Chapter 36) recorded a mean distance of 365 km (range, 0–1409 km) traveled by six sharks during an average 275 days since tagging. White sharks would have to possess an accurate internal "clock" for such precise timing of their arrivals at localities used in successive years. Given the existence of such clocks in other vertebrates, it would be surprising if white sharks did not possess them also.

Summary

We have described here occurrence patterns of white sharks *C. carcharias* at the South Farallon Islands, 1988–1992. During this period, 18 individuals were identified on the basis of total length and fin and body scarring. Minimum estimates of the number of sharks visiting the study area were 12 each during 1988 and 1989, 14 during 1990, 10 during 1991, and 9 during 1992. The decrease in 1991–1992 was likely due to decreased observation effort (1991) and a change in our method of observation (1992). During 1989 and 1990, individuals were observed feeding on pinnipeds for periods ≤3 days before other sharks were seen feeding for similar periods. Two exceptions to this rule were sharks that remained for 10-day periods during 1991 and 1992. Most white sharks were transient visitors, but a few were short-term residents. Three individuals returned at the same times in successive years. These sharks may regulate the timing of their movements so that they arrive at a particular site at that time of year when they had successfully fed there before.

Acknowledgments

We thank M. Elliot of the U.S. Geological Survey for lending two Wild T-2 theodolites for use in the study. Members of the Farallon Patrol transported investigators and equipment back and forth to the South Farallon Islands. D. G. Ainley provided much council and administrative commitment to our effort at the South Farallon Islands and read a draft of the manuscript. This is contribution 596 of the Point Reyes Bird Observatory.

34

Trends in White Shark Predation at the South Farallon Islands, 1968–1993

PETER PYLE
Point Reyes Bird Observatory
Stinson Beach, California

SCOT D. ANDERSON
Inverness, California

DAVID G. AINLEY[1]
Point Reyes Bird Observatory
Stinson Beach, California

Introduction

During 1968–1986, observations of white shark *Carcharodon carcharias* activity at the South Farallon Islands (SFI), 48 km west of San Francisco, California, were conducted consistently, but at a low to moderate effort level (Ainley *et al.*, 1981, 1985). An increase in shark predation on pinnipeds from 1970 to 1982 was correlated with an increase in the number of their preferred prey, immature (1- to 3-year-old) northern elephant seals *Mirounga angustirostris*. On October 2, 1982, a fisherman removed four adult white sharks from SFI waters, an event that negatively affected the frequency of white shark sightings from the islands in 1983–1984, despite continued prey abundance (Ainley *et al.*, 1985). In 1983, the population of elephant seals at SFI stabilized at 350–400 breeding animals (Stewart *et al.*, 1994), and in 1987, we standardized our heretofore opportunistic observation effort by maintaining a directed continuous watch for shark activity during all daylight hours in autumn (Klimley *et al.*, 1992). In this chapter, we assess population trends of adult white sharks at SFI on the basis of observations both prior to and subsequent to (1) the removal of four adults in 1982, (2) the stabilization of the seal population, and (3) the standardization of our observation program.

[1]Present address: H. T. Harvey & Associates, Alviso, CA.

Methods

Because of the changed protocol as of 1987, we separately examined trends during the periods 1968–1986 and 1987–1992. We standardized observational data as much as possible by confining analyses to attacks observed during September 1–November 30, when >95% of shark activity at SFI occurs (Ainley *et al.*, 1985). We assessed attacks rather than nonpredatory sightings because the former were far more obvious events (see Chapter 26, by Pyle *et al.*); hence, frequency of detection was not strongly influenced by observer biases.

To assess trends in attack frequency, we used simple and multiple linear regression (Stata Corporation, 1993, pp. 110–125) on log-transformed values (see Chapter 26, by Pyle *et al.*). For 1968–1986, we simply used the number of attacks observed during each autumn. The dependent variable for 1987–1992 was attacks recorded per 100 hours of observation from the lighthouse, arcsine square root transformed to equalize the variances (Sokal and Rohlf, 1981, pp. 380–387). No observation from the lighthouse was performed in September 1987, affecting certain analyses. For all analyses, we reexamined trends in attack frequency after statistically adjusting for the mean number of elephant seals present each autumn, as determined from weekly counts [Huber *et al.*, 1985; Point Reyes Bird Observatory (PRBO), unpublished data].

Results

Abundance of immature elephant seals increased dramatically during the period 1968–1983 ($t = 23.64$, $p < 0.001$) (Fig. 1), but leveled off after 1982 ($t = 0.48$, $p = 0.630$; 1983–1993). The number of observed shark attacks also increased from 1968 to 1982 ($t = 4.57$, $p = 0.001$) (Fig. 1), but after adjustment for the number of seals, no correlation was evident ($t = 1.20$, $p = 0.254$). Conversely, a significant correlation between attacks and number of seals ($t = 3.81$, $p = 0.002$) became insignificant after adjusting for year ($t = -0.45$, $p = 0.664$), indicating colinearity and making it difficult to ascertain which effect (seals or year) caused the increase in shark observations. The removal of four large sharks in 1982 resulted in a decline in attacks observed in the years 1983–1985, significantly below what would be expected given the increasing trend in 1968–1982 (Fig. 2). Attacks observed rebounded between 1983 and 1986 (Fig. 1) ($t = 14.86$, $p = 0.005$ unadjusted; $t = 13.51$, $p = 0.047$ adjusted for seal abundance).

The mean number of attacks per 100 hours of observation increased between 1987 and 1992 (Fig. 3), both before ($t = 2.30$, $p = 0.020$) and after ($t = 2.28$, $p = 0.023$) adjustment for the number of seals. When the unadjusted indices were separated by month for analyses (Fig. 4), no trends were found in September ($t = -0.476$, $p = 0.635$; 1988–1993) or October ($t = 0.47$, $p = 0.639$), but an increase in attack frequency was evident in November ($t = 2.25$, $p = 0.026$). Levels of significance were the same for trends in all three months, after adjustment for the mean number of seals present.

Discussion

Our results support the observations of Ainley *et al.* (1985) that the increase in white shark attack frequency at SFI through the mid-1980s was correlated with the abundance of immature elephant seals, and that the removal of four adult sharks in 1982 impacted the shark population in island waters. Observed increases in attack frequency became insignificant when adjusted for seal abundance; however, colinearity between attack frequency and prey abundance made it difficult to infer whether or not the shark population was increasing. The trends for this period can be explained by (1) a static shark population with an increasing individual predation rate (due to an increase in prey), (2) an increase in sharks with a

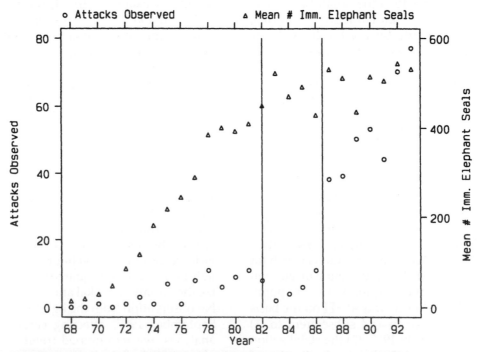

FIGURE 1 The number of white shark attacks observed and the mean number of immature northern elephant seals at the South Farallon Islands during September 1–November 30, 1968–1993. Vertical lines indicate 1982, when four adult white sharks were removed from the population, as well as the break between 1986 and 1987, when our observation effort was intensified.

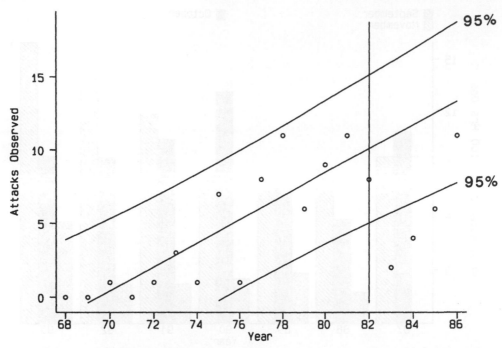

FIGURE 2 The trend in the number of white shark attacks observed at the South Farallon Islands from 1968 to 1986 (central regression line) and 95% confidence intervals calculated from the sum of the standard error in the prediction (slope) plus the residual error (Stata Corporation, 1993, pp. 79–80).

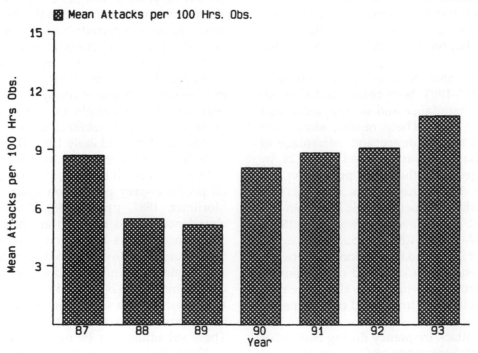

FIGURE 3 The frequency of white shark predation observed during a standardized effort at the South Farallon Islands from 1987 to 1993. The higher attack frequency in 1987 may partially reflect the lack of observation during September of that year, when frequency is typically lower (see Fig. 4).

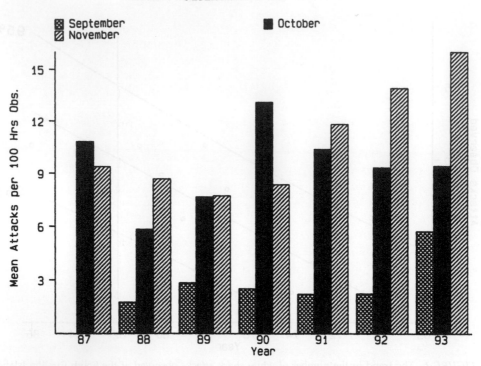

FIGURE 4 The frequency of white shark predation observed during September, October, and November 1987–1993.

constant individual predation rate, or, perhaps most likely, (3) increases in both of these factors. The increase in attacks during 1983–1986, after the removal of four adults and the stabilization of the seal population, indicates an increase in shark abundance during this period (see Chapter 33, by Klimley and Anderson).

Attack frequency also increased significantly during the period 1987–1993, both before and after adjustment for seal abundance and despite a constant population size of seals. These results, along with increases in both the incidence and percentage of shark-bitten elephant seals at SFI (see Chapter 24, by Long *et al.*), suggest that the shark population had increased during this period. It is unlikely that changes in sea-surface temperature (SST) explain this trend. Although SST at SFI averaged warmer in 1991–1993 than in 1987–1988 (PRBO, unpublished data), Pyle *et al.* (Chapter 26) found no relationship between SST and attack frequency. Continued observation of attack frequency from SFI will help to clarify the relationship between observed trends and environmental effects.

An increase in attack frequency during November 1987–1993, but not September or October, indicates either that shark occurrence patterns had shifted later each autumn or that shark abundance increased dur-

ing November for other reasons. Pyle *et al.* (Chapter 26) showed that attacks are more frequent later in the fall during years of warmer mean SST. This seasonal effect may have contributed to the trend, although we may then have expected attack frequency to decrease in September, an effect not observed (see Fig. 4).

We suggest that the inferred increase in the SFI shark population from 1983 to 1993 may be related to increased recruitment of younger white sharks, 10–15 years of age. This might be expected given (1) the rapid increase and stabilization in the prey resource during the 1970s and early 1980s (Fig. 1), (2) the slow growth rate of white sharks, which take an estimated 9–12 years to reach maturity (Cailliet *et al.*, 1985), and (3) predator–prey population dynamics (Begon and Mortimer, 1981, pp. 89–138). An increase in prey abundance should result in an increase in predator abundance, with a lag period roughly equivalent to the maturation rate of the predator. On November 15, 1993, Pyle and Anderson documented an unprecedented number (12–15) of different sharks at one attack, most of them smaller animals, 3–4.5 m in length (corresponding to subadults) (Cailliet *et al.*, 1985). These subadults may be more abundant around SFI in November, after the larger adult sharks have fed and/or departed. Again, continued observation of activity and predation frequency of white sharks at SFI is needed to elucidate the population dynamics of this apex predator.

Summary

Trends in predation frequency by white sharks *C. carcharias* at SFI were assessed. The increased frequency of observed attacks from 1968 to 1983 could be attributed to increases in preferred prey, immature northern elephant seals *M. angustirostris* and/or an increase in sharks, whereas an increase subsequently, from 1987 to 1993, likely reflected an increase in the shark population at SFI. Significantly fewer attacks than expected were observed in 1983, 1984, and 1985, after the removal of four adults from SFI waters in 1982. This indicates the possible vulnerability of the shark population at SFI. Recent increases in shark predation in November may reflect recruitment into the population of younger white sharks, an expected occurrence given (1) increases in pinniped prey during the 1970s, (2) the slow growth rate of white sharks, and (3) predator–prey dynamics.

Acknowledgments

We thank the U.S. Fish and Wildlife Service and the Farallon Patrol for logistical support. Our studies on white sharks at SFI have been supported by the National Geographic Society, the Gulf of the Farallones National Marine Sanctuary, PRBO, and especially by A.P. Klimley. Feedback and comments from A. P. Klimley and two reviewers greatly improved the manuscript. This is PRBO contribution 586.

Population Dynamics of White Sharks in South Africa

CRAIG A. FERREIRA *and* THEO P. FERREIRA

South African White Shark Research Institute
V & A Waterfront
Cape Town, South Africa

Introduction

White sharks *Carcharodon carcharias* occur along the entire southern African coastline, from Swartkopmund, Namibia, to Natal. The center of its distribution appears to be the southern and western Cape (Bass *et al.*, 1975). Human settlement and development along this coastline, with emphasis on the area from Saldanha Bay, southwestern Cape, to East London, eastern Cape, has resulted in heavy human use of these waters and increased encounters between white sharks and humans. A range of injurious and fatal encounters, with negative and sensationalized media coverage, has brought reprisals toward the white shark and other cartilaginous fishes.

Until 1990, *in situ* research on white sharks in southern Africa had been limited to the Natal anti-shark netting program (see, e.g., Cliff *et al.*, 1989, Chapter 32). Over the past 5 years, interest in the natural history of the white shark has increased. The White Shark Research Project [currently the White Shark Research Institute (WSRI/P)] led the way in South Africa. The South African coastline is vast, rugged, and often hostile. The WSRI/P commenced with searching for white sharks at a number of sites. Dyer Island, Quoin Point, and Struis, False, and Mossel bays were explored and identified as white shark sites. Algoa Bay, although not yet explored by the WSRI, is well known for its white sharks among fishermen, divers, and researchers from the Port Elizabeth aquarium. Numerous white shark-inflicted injuries on jackass penguins *Spheniscus demersus* and Cape fur seals *Arctocephalus pusillus* indicate the white shark's presence at the islands of Algoa Bay.

This chapter is a preliminary report on the population dynamics of the white shark in coastal waters of southern Africa. We hope to inspire cooperation within the scientific community in South Africa, so that a population estimate and management program for the white shark will result.

Materials and Methods

During the period October 1990 to August 1994, the WSRI conducted research trips to Dyer Island and False, Struis, and Mossel bays. Most of the work was conducted at Dyer Island and Struis Bay. Once the WSRI boat was suitably anchored at a research site, a number of environmental recordings were made.

Sharks were attracted to the boat by means of odor corridors. Chum consisted of a mixture of fresh sardines, unrefined fish oil, and seawater. The mixture was dispensed into the water at a low volume, the concentration of which was determined on the basis of sea conditions: with stronger current and more adverse seas, more mixture was dispensed. Chum volume did not exceed 1 liter/min. Two baits, tied to a natural fiber rope, drifted 10 m astern of the boat, on small white floats. Baits were always natural meat, such as seal, whale, dolphin, shark, or fish, and were 5–10 kg in mass. If the sharks were active and readily took baits, we attached larger (10- to 20-kg) baits to the floats.

Once a shark arrived and showed interest in the baits, we attempted to lure the shark into tagging distance by gradually pulling the bait closer to the boat. It was often possible to pull a shark to the side of the boat and lift its head from the water. This often afforded the opportunity to sex the shark. A floating 5-m-long PVC pipe, color divided into 1-m segments, was used for estimating total length (TL). Once a white shark had a plaque tag attached to it, a photograph was taken. In a clear photograph, it was then possible to confirm length estimates by using the plaque as a scale.

Sharks were free-tagged with Oceanographic Research Institute Hallprint streamer dart tags, incorporating the WSRI plaque tag. The plaque tag was 70 × 40 × 1.5 mm, conspicuous yellow in color, with large black three-letter codes. Tags were placed as close to the base of the dorsal fin as possible, using a 4-m-long aluminum pole.

Results

White Shark Sites

White sharks frequent a number of localities in South Africa (Fig. 1), including seal and penguin colonies, large reefs and banks, and seemingly desolate stretches of coastline. These areas were inspected by the WSRI.

Dyer Island (Fig. 2A)

Dyer Island consists of a small group of islands lying east of Danger Point near Gaans Bay, about 4 km from the mainland shore. There are two main islands; the remainder of the group consists of shallow reefs.

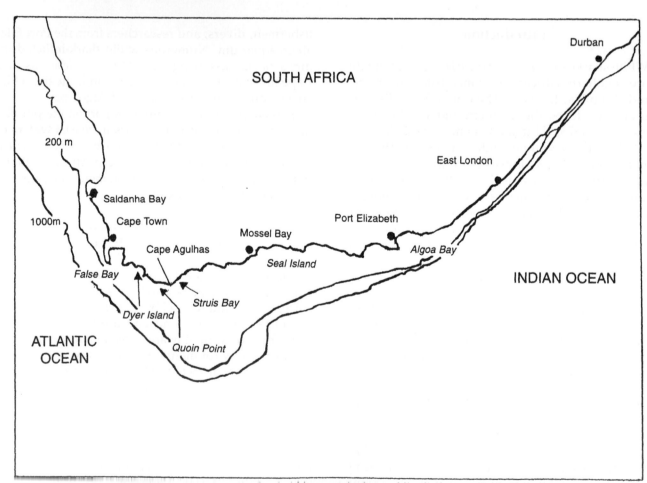

FIGURE 1 Sites along the South African coast where white sharks are reliably found.

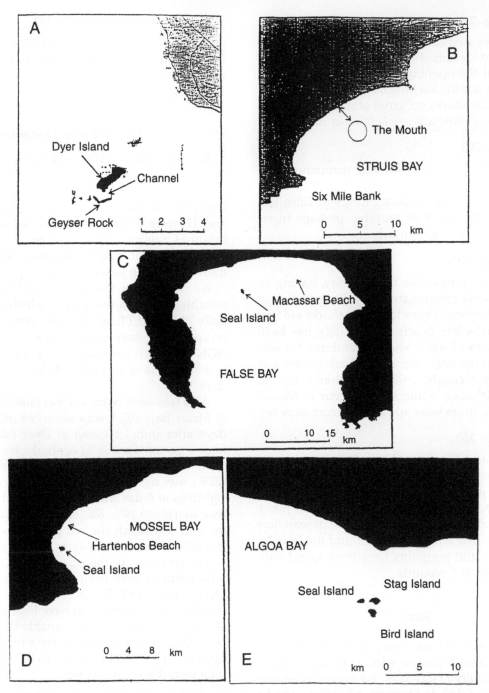

FIGURE 2 White shark sites: (A) Dyer Island, (B) Struis Bay, (C) False Bay, (D) Mossel Bay, and (E) Algoa Bay.

The most prominent of the islands is Dyer, a low-profile island about 1.5 km long and 0.5 km wide. The island is home to jackass penguins and a host of other marine birds. Adjacent to Dyer Island is Geyser Rock, about 0.5 km long and 150 to 180 m wide and home to a colony of Cape fur seals.

Separating the two islands is a 150-m-wide channel ranging from 3 to 6 m in depth. The eastern opening of the channel is unobstructed and 250 m wide. The western entrance is cluttered by shallow reefs and huge kelp beds. The western end, although obstructed, is navigable by small craft in fair weather, and on one occasion, we observed a humpback whale *Megaptera novaeangliae* passing through the channel. When the sharks are present at Dyer Island, they can be seen patrolling the channel.

Struis Bay (Fig. 2B)

Struis Bay lies east of Cape Agulhas. Several white shark sites were identified within this area. Die Mond (The Mouth), at the opening of the Kars River, is one site. Other sites are the Six Mile Bank and the wreck of the *Pioneer*. White sharks occurred at these sites, especially when the yellowtail *Seriola lalandei* moved in.

False Bay (Fig. 2C)

White sharks were encountered throughout the bay, but occurred mainly at two sites: Seal Island and Macassar Beach. White shark sightings in False Bay increased over the summer months, perhaps in response to an influx of forage fish.

Mossel Bay (Fig. 2D)

Mossel Bay is comparable to False Bay, having an island where seals congregate and a long stretch of coastline. The two main sites for white sharks are Seal Island and Hartenbos Beach. Mossel Bay has been known as an area of white shark abundance for several decades. In the past, one white shark hunter was reported to have caught >60 white sharks around Seal Island. Although white sharks occur in Mossel Bay year-round, more were apparent during summer.

Algoa Bay (Fig. 2E)

The Bird Island Group in Algoa Bay is an area of white shark abundance. Geographically and environmentally, this group is comparable to Dyer Island. The Port Elizabeth Museum and University for many years have studied sea birds here. The researchers have often reported white sharks around the islands, and many seals and penguins have been found with white shark-inflicted wounds.

Sizes

We divided our data into two periods, because our sea time was limited during winter, in order to exclude any possibility of size bias. We tagged 92 sharks in Struis Bay and 182 at Dyer Island. White sharks ranged from 150 to 500 cm TL. Average length varied considerably within and between sites and indicated no distinct size segregation (Fig. 3). There was, however, a marked preponderance of sharks 260–300 cm TL. The greatest prevalence of sharks in this size class was apparent at Struis Bay and Dyer Island. The sizes of females at Dyer Island and Struis Bay averaged 350 and 340 cm TL, respectively. The mean length for males differed greatly: 330 cm TL for Dyer Island and 270 cm TL for Struis Bay.

Sex Ratio

A degree of sex segregation occurred at both Dyer Island and Struis Bay, with females outnumbering males (females, 73% and 85%, to males, 27% and 14%, respectively).

Tagging and Resighting

Between January 1992 and August 1994, 102 white sharks were tagged at Dyer Island, 43 in Struis Bay, and 2 in Mossel Bay (Table I). A total of 255 individual white sharks have been recorded by the WSRI; sex and TL were estimated for 147, which constituted a balanced sample (Table I) (See end of chapter for Tables I–II.). Conclusions discussed below were drawn mostly from the latter and the tagged individuals.

All but two of the sharks, AFD and ACA, were resighted at the same area in which they were originally observed (Table II). One shark, AGT, was observed on 11 days over a period spanning 65 days, AGN was seen on 5 days over a period of 57 days, and AFD was observed on 5 days over a span of 100 days.

AFD was seen twice at Dyer Island and three times at Struis Bay. ACA was observed at Struis Bay, 545 days after initial tagging at Dyer Island. A notably higher percentage (36%) of sharks was resighted over a 1-day interval (Fig. 4). A sharp decrease in sightings (22%) was noted for 2-day intervals and longer. Resightings at 4-day intervals decreased to 2% and then rose sharply to 14%, 8%, and 14% over the 8- to 32-day interval. Although the sample was not substantial, a possible cycle of short-term mobility was apparent.

We received numerous reports from fishermen who claim to have seen WSRI tags on sharks in the Dyer Island and Struis Bay areas. Unfortunately, these fishermen prefer to have little to do with white sharks, and thus, we were unable to get tag numbers from them. On the basis of these reports, however, it appears that if sea time were to be increased, resightings would increase as well.

Seasonality

The abundance of white sharks at Dyer Island and Struis Bay, varied little year-round, except for fluctuations over short periods. A peak in sightings was observed, however, over winter (August) at Dyer Island; the peak for Struis Bay (23%) occurred in July (Fig. 5).

Fluctuations in water temperature apparently did not play a substantial role in affecting patterns in the

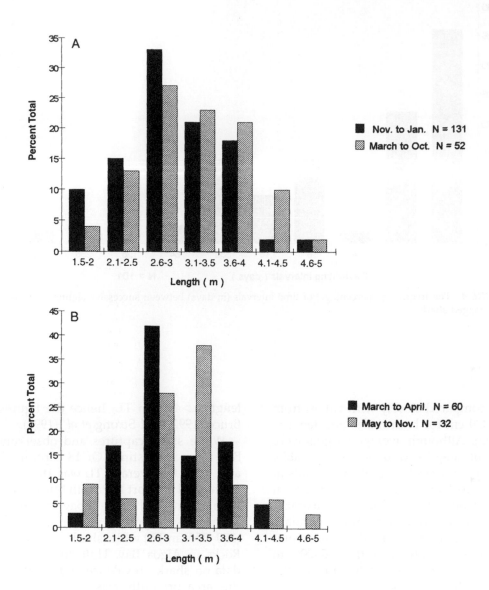

FIGURE 3 Sizes of white sharks at (A) Dyer Island and (B) Struis Bay.

number of sharks sighted (Fig. 6). Sharks were sighted through all temperature ranges. Shark sightings, however, did peak in the higher temperatures, and at Dyer Island, sightings decreased notably when temperatures reached 11–12°C.

Discussion

Size Distribution

The mean TLs of male and female sharks observed by us at Dyer Island and Struis Bay differed consid-erably from a sample of 495 sharks taken from the antishark nets in KwaZulu–Natal, 1960–1988. The average length of sharks at Dyer Island and Struis Bay was 320 cm TL, compared to 250 cm TL at Natal (Wallett, 1973; Cliff *et al.*, 1989).

White sharks >450 cm TL at Dyer Island and Struis Bay constituted only 2% and 3%, respectively, of the total number of individuals sighted. These figures are far lower than the 24% reported by Bruce (1992) and 12% documented by Strong *et al.* (1992) for sharks >450 cm TL off South Australia. The mean TLs for male and female sharks at Dyer Island (330 and 350

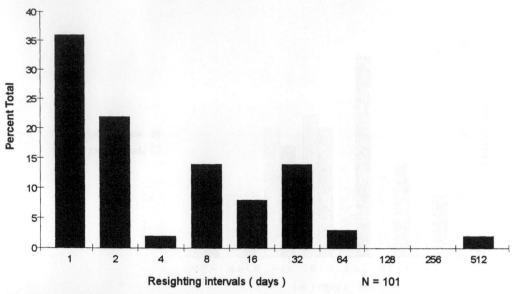

FIGURE 4 The frequency (percentage) of time intervals (in days) between successive sightings of the same tagged shark.

cm, respectively) compared closely with values from South Australia: 370 cm for both males and females (Strong *et al.*, 1992). Although average lengths were similar, the mix of lengths differed considerably. Sharks 260–300 cm comprised 33% of individuals at Dyer Island and Struis Bay, compared to only 17% in South Australia (Strong *et al.*, 1992). In Australia, sharks in the 360 cm TL group were most abundant (33%); at Dyer Island and Struis Bay, this group comprised 17% of the size mix. Sharks in the 150–200 cm TL group were absent from the Australian sample, but comprised 7% of the WSRI sample.

The relatively few white sharks <200 cm and >400 cm TL may be the result of white shark hunting over the past three decades along the South African coastline. Traditionally, white shark fishermen have always targeted the largest animals, and in this way, could have depleted the stock of potential breeders. In Australia, hunting of white sharks has occurred for a much longer period than in South Africa, and the hunting efforts have been far more concerted. The absence there of very small and very large individuals could also have been the result of fishing for large white sharks. White shark hunting in Australia has, in recent years, been sharply criticized, leading to a decrease in hunting efforts there. Thus, it is possible that with less fishing for large sharks during recent years, sharks have had the opportunity to grow to

lengths >400 cm TL, hence the figures reported by Bruce (1992) and Strong *et al.* (1992).

White shark captures and observations in Algoa Bay were interesting. Of 18 captures between 1979 and 1991, the average TL was 180 cm (L. V. J. Compagno and the Port Elizabeth Museum, personal communication). L. J. V. Compagno (personal communication) also reported numerous sightings of large (300–550 cm TL) white sharks at Bird Island and Black Rocks in Algoa Bay. Unfortunately, no quantitative data on shark prevalence and length estimates from this area presently exist, but the small sharks here could indicate the presence of a pupping ground.

Sex Ratios

The relatively high degree of segregation at Dyer Island and Struis Bay was quite different from catches off KwaZulu–Natal. The sex ratio of males to females for Dyer Island and Struis Bay, combined, was 1:3.5. The ratio for KwaZulu–Natal, from a sample of 591 white sharks, 1974–1988, was 1:1.4 (Cliff *et al.*, 1989).

Although sexual segregation occurred at specific sites in South Australia (Strong *et al.*, 1992), overall segregation was not evident; with sites combined, the ratio was almost 1:1, which is similar to the ratio at KwaZulu–Natal, but completely different from those at Dyer Island and Struis Bay.

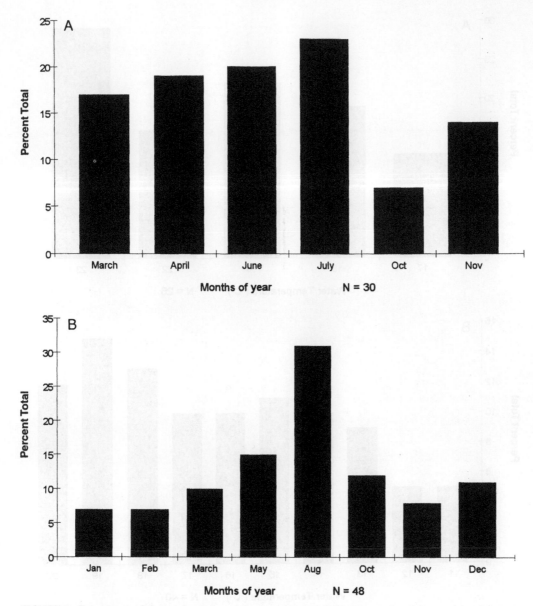

FIGURE 5 Percentage of the total number of sharks sighted in different months of the year at (A) Struis Bay and (B) Dyer Island.

Seasonality

In South Africa, Australia, and the west coast of North America, white sharks occur year-round, but with peaks of sightings over specific periods. The peak periods differ between the three seasons: winter (this study), summer (Bruce, 1992), and autumn (Ainley *et al.*, 1981; Klimley, 1985b), respectively. Along the east coast of North America, sightings are greater during summer (Casey and Pratt, 1985). At Dyer Island and Struis Bay, winter brings rough weather, but as in Australia, this did not have a negative effect on white shark abundance (Strong *et al.*, 1992).

In regard to water temperatures, the pattern appears to correspond with that observed by Bass (1978), who concluded that water temperature did not play an important role in South Africa, as large (>270 cm TL) sharks were caught off Natal from Feb-

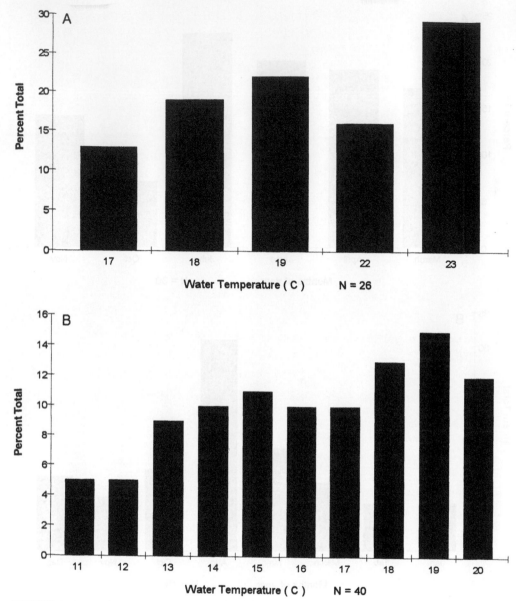

FIGURE 6 Percentage of sightings in relation to water temperature at (A) Struis Bay and (B) Dyer Island.

ruary to June, when water temperatures were low, but were absent from September to January as well, when temperatures were similar. It is likely that the movement of prey has much more of an effect than does water temperature on the sharks' movements.

Summary

Population dynamics of white sharks *C. carcharias* were studied along the South African coast from Oc-
tober 1991 through August 1994. Study areas included Dyer Island and Struis, False, and Mossel bays. During the study, 255 white sharks were recorded, 147 of which were tagged. Of the latter, 30 individuals were resighted 59 times; one shark (AGT) was resighted 10 times. Resighting intervals of individual sharks ranged from 1 to 545 days. Although gender separation was not absolute, a notably high percentage of females occurred at both Struis Bay and Dyer Island. Sharks observed ranged from 150 to 500 cm TL. No notable size segregation occurred, al-

though a high percentage of sharks 260–300 cm TL was observed. White shark abundance overall showed no substantial change during the study. Abundance, however, fluctuated at sites over short periods.

Acknowledgments

We thank the entire WSRI staff, both permanent and otherwise, for their enthusiastic support and work. Special thanks go to T. and J. Ferreira. We thank C. Fallows, M. Hughes, M. Boon, and N. Du Plooy for the many hours spent in the field. Many thanks go as well to A. P. Klimley and D. G. Ainley for much-needed advice and guidance. We wish to thank Mitchell Du Plessis and Associates and the V & A Waterfront management for their unselfish logistical support, the Department of Nature Conservation for their logistical support, and the fishermen and businessmen of Gaans Bay and Struis Bay for their advice and support. The Mazda Wildlife Fund, Marine Car Hire, and Two Oceans Marine sponsored the work, and the Oceanographic Research Institute in Durban supplied the dart tags. Thanks go to C. Ferreira for her hours of voluntary work and N. Ferreira for her logistical and financial support.

TABLE I White Sharks Tagged by the White Shark Research Institute, January 1992 to August 1994

No.	Code	Date	Locality	Sex	TL (m)	No.	Code	Date	Locality	Sex	TL (m)
1	ACA	January 13, 1992	DI	M	4.0	30	ABM	November 18, 1992	DI	F	3.0
1a	ADB	January 13, 1992	DI	F	4.0	31	ABJ	November 18, 1992	DI	F	3.5
2	ADC	January 14, 1992	DI	M	2.0	32	ABV	November 18, 1992	DI	F	3.3
3	ADE	January 14, 1992	DI		2.5	33	AAP	November 18, 1992	DI	F	3.0
4	ADB	January 14, 1992	DI	F	4.0	34	ABE	November 18, 1992	DI	F	3.5
5	ADG	January 16, 1992	DI		3.5	35	B11351	November 18, 1992	DI		3.5
6	ADD	March 16, 1992	DI	F	3.8	36	B11354	November 18, 1992	DI		2.9
7	ABD	January 22, 1992	DI		3.5	37	B11352	November 18, 1992	DI	F	4.0
8	ABY	March 16, 1992	DI		2.5	38	AFI	December 1, 1992	DI	F	3.8
9	AAX	March 17, 1992	DI	M	2.5	39	AFA	December 1, 1992	DI		3.8
10	AAO	March 17, 1992	DI	F	2.0	40	AFH	December 1, 1992	DI		3.2
11	AAP	March 17, 1992	DI	M	2.5	41	AFC	December 1, 1992	DI		3.
12	AAR	March 17, 1992	DI		3.5	42	AFO	December 2, 1992	DI		2.8
13	AAZ	May 10, 1992	DI		2.5	43	AFP	December 2, 1992	DI	F	2.6
14	AAV	May 10, 1992	DI		2.5	44	AFE	December 2, 1992	DI	F	2.5
15	AAW	May 10, 1992	DI	F	4.5	45	AFF	December 2, 1992	DI	M	3.5
16	AAM	May 10, 1992	DI	F	4.5	46	AFG	December 2, 1992	DI	F	3.2
17	ABI	October 27, 1992	DI		2.7	47	AFM	December 2, 1992	DI		3.0
18	ABK	October 27, 1992	DI		3.5	48	AFJ	December 2, 1992	DI		3.0
19	ABH	October 27, 1992	DI	M	3.7	49	AFK	December 2, 1992	DI		3.0
20	ABD	October 27, 1992	DI		3.5	50	AFD	December 3, 1992	DI	F	3.7
21	ABQ	October 27, 1992	DI	M	4.5	51	AFL	December 3, 1992	DI		2.8
22	ABR	October 27, 1992	DI	F	3.5	52	AAF	December 3, 1992	DI		3.0
23	ABS	October 27, 1992	DI	F	3.2	53	AFB	December 3, 1992	DI		2.5
24	ABC	October 27, 1992	DI		3.7	54	B113883	December 1992	DI	F	3.4
25	ABU	October 27, 1992	DI		4.0	55	AFU	December 11, 1992	DI		3.8
26	ABP	November 1, 1992	DI	F	2.8	56	ACS	December 11, 1992	DI	M	3.2
27	ABX	November 1, 1992	DI	F	3.8	57	ACZ	December 26, 1992	DI		1.8
28	ABW	November 1, 1992	DI	F	3.4	58	AET	December 26, 1992	DI	F	3.0
29	ABU	November 18, 1992	DI		2.6	59	AFR	December 26, 1992	DI		3.0

(continues)

TABLE I (*Continued*)

No.	Code	Date	Locality	Sex	TL (m)	No.	Code	Date	Locality	Sex	TL (m)
60	AFY	December 26, 1992	DI		3.0	105	ADM	June 1, 1993	SB	F	4.5
61	ACO	December 26, 1992	DI		1.8	106	AGO	October 11, 1993	SB	M	2.7
62	AEA	December 26, 1992	DI		3.2	107	AGW	October 11, 1993	SB	M	3.5
63		December 1992	DI		3.0	108	AGX	October 12, 1993	SB		3.0
64	AEV	December 26, 1992	DI	F	2.7	109	AHB	October 15, 1993	SB		3.2
65	AFT	December 26, 1992	DI	M	3.5	110	AHD	October 15, 1993	SB	M	3.3
66	AEO	December 26, 1992	DI			111	AHA	October 15, 1993	SB	F	3.5
67	AEV	December 26, 1992	DI			112	AHC	October 16, 1993	MB	F	3.3
68	ACV	December 26, 1992	DI		2.5	113	AGZ	November 2, 1993	SB		2.9
69	AEH	January 5, 1993	DI	M	2.8	114	AHE	November 2, 1993	SB	M	1.9
70	AEI	January 5, 1993	DI		1.7	115	AHG	November 2, 1993	SB		2.3
71	AEP	January 5, 1993	DI	M	3.5	116	AGC	November 2, 1993	SB		2.7
72	AEE	January 5, 1993	DI		2.5	117	AGR	November 2, 1993	SB		2.6
73	AFQ	January 5, 1993	DI		3.3	118	AGS	November 3, 1993	SB	F	4.5
74	ACA	January 6, 1993	DI		3.6	119	AHH	November 13, 1993	DI	F	4.0
75	AEF	January 6, 1993	DI		3.5	120	AGT	November 13, 1993	DI		2.9
76	AEN	March 4, 1993	SB	F	2.8	121	AGV	November 13, 1993	DI		3.3
77	ADD	March 4, 1993	SB	F	2.8	122	AGI	December 5, 1993	DI		2.8
78	ADC	March 4, 1993	SB	F	2.0	122	aAGN	December 21, 1993	DI	F	2.8
79	AEB	March 4, 1993	SB	F		123	AHS	Decmeber 28, 1993	DI		2.6
80	ACP	March 4, 1993	SB	F	3.8	124	AHT	December 28, 1993	DI	F	2.5
81	AFW	March 4, 1993	SB	F	2.8	125	AHP	January 15, 1994	DI		2.3
82	ADA	March 4, 1993	SB	M	2.2	126	AA	January 15, 1994	DI		2.5
83	AEK	March 4, 1993	SB	F	3.3	127	AHV	August 21, 1994	DI	F	3.5
84	ADK	March 7, 1993	SB		2.6	128	AIC	August 21, 1994	DI	F	3.6
85	AD	March 17, 1993	SB		3.0	129	AIP	August 21, 1994	DI	F	3.0
86	ACK	March 22, 1993	SB		2.2	130	AIG	August 21, 1994	DI		3.0
87	ADJ	April 15, 1993	SB	F	4.0	131	AHR	August 21, 1994	DI		2.8
88	AEC	April 15, 1993	SB	F	3.9	132	AIR	August 22, 1994	DI	F	3.7
89	AED	April 15, 1993	SB	F	3.5	133	AIE	August 22, 1994	DI	F	3.5
90	ADV	April 24, 1993	SB		3.0	134	AHI	August 22, 1994	DI		3.0
91	ADP	April 24, 1993	SB	F	4.5	135	AIF	August 22, 1994	DI	F	2.6
92	ADD	April 24, 1993	SB		2.0	136	AIV	August 22, 1994	DI	F	4.1
93	ADX	April 25, 1993	SB	F	2.9	137	AIY	August 22, 1994	DI	F	4.0
94	ADB	April 24, 1993	SB	F	3.4	138	AIB	August 22, 1994	DI	M	3.2
95	ADS	April 25, 1993	SB	F	3.5	139	AID	August 22, 1994	DI		3.3
96	ADZ	April 25, 1993	SB	F	3.0	140	AIT	August 22, 1994	DI	M	3.0
97	ADV	April 25, 1993	SB		2.8	141	AIS	August 22, 1994	DI	F	4.2
98	B12815	April 25, 1993	SB		2.5	142	AHN	August 22, 1994		M	4.0
99	AND	May 23, 1993	SB	F	2.6	143	AHO	August 22, 1994	DI	F	4.7
100	ADW	May 23, 1993	SB	F	3.5	144	AHX	August 22, 1994	DI	F	4.0
101	ADY	June 1, 1993	SB	F	3.9	145	ACR		MB		3.0
102	ADT	June 1, 1993	SB	F	3.2	146	ACV		MB		3.5
103	ADR	June 1, 1993	SB		3.8	147	AER	March 4, 1993	SB	F	
104	AEQ	June 1, 1993	SB	F	3.9						

TL, Total length; DI, Dyer Island; SB, Struis Bay; MB, Mossel Bay.

TABLE II　**Sightings of White Sharks at Dyer Island and Struis Bay, January 1992 to August 1994**

Code	Locality	Date	Interval between sightings (days)[a]	Code	Locality	Date	Interval between sightings (days)[a]
ADB	DI	January 13, 1992	1, 8	AGK	DI	December 24, 1993	3
ADE	DI	January 14, 1992	2	AHJ	DI	December 27, 1993	1, 1, 2
ABY	DI	March 16, 1992	1	AHS	DI	December 28, 1993	1
AFA	DI	December 1, 1992	2	AHT	DI	December 28, 1993	17
AFK	DI	December 2, 1992	8	AET	DI	December 26, 1992	30
AFJ	DI	December 2, 1992	36	AFW	SB	March 4, 1993	1, 2, 1
ACS	DI	December 11, 1992	25, 1, 1	AEN	SB	March 4, 1993	1, 3, 1
ACO	DI	December 26, 1992	10	ADA	SB	March 4, 1993	1, 2
AFD	DI	December 3, 1992	33, 59, 57, 2	ADI	SB	March 7, 1993	5
AEA	DI	December 26, 1992	10, 2	AER	SB	March 4, 1993	42
ACZ	DI	December 26, 1992	10	AEK	SB	March 4, 1993	42
AEH	DI	January 5, 1993	1, 1	SCAR	SB	April 15, 1993	1
ACA	DI	January 13, 1993	545	AEC	SB	April 15, 1993	9
AGT	DI	November 13, 1993	17, 12, 10, 3, 3, 1, 1, 1, 15, 16	ADX	SB	March 8, 1993	48
				ADV	SB	April 25, 1993	37
AGN	DI	December 21, 1993	3, 3, 1, 19				

DI, Dyer Island; SB, Struis Bay.

[a]Only sightings over 1 day or more after the initial sighting are listed.

36

First Estimates of Mortality and Population Size of White Sharks on the South African Coast

GEREMY CLIFF
Natal Sharks Board
Umhlanga Rocks, South Africa

RUDY P. VAN DER ELST *and*
ANESH GOVENDER
Oceanographic Research Institute
Durban, South Africa

TRAIL K. WITTHUHN
Struis Bay, South Africa

ELINOR M. BULLEN
Oceanographic Research Institute
Durban, South Africa

Introduction

The white shark *Carcharodon carcharias* occurs along the entire South African coast, but the center of its distribution is the temperate waters of the Western Cape (Bass *et al.*, 1975). The species' range extends into Namibia and possibly southern Mozambique (Compagno *et al.*, 1989).

This species has bitten humans in Cape waters and also in KwaZulu–Natal, where a gill-netting program, introduced in 1952, has reduced the incidence of shark attack (Davies, 1964; Wallett, 1983; Cliff, 1991). At present, 40 km of nets, maintained by the Natal Sharks Board (NSB), catches an average of 1354 sharks per year, including 39 white sharks (see Chapter 32, by Cliff *et al.*). Initially, all potentially dangerous sharks found alive in the nets were killed. Since 1988, however, most live sharks, including dangerous ones, have been released and, whenever possible, tagged (Cliff and Dudley, 1992b).

Other sources of fishing mortality of white sharks are big-game angling, spear fishing, and incidental catches in commercial fisheries in Western Cape waters. None of these mortalities has been quantified, although one angler captured 18 white sharks in False Bay over a 4-year period (Wallett, 1983). Some spear fishermen carry a "powerhead," that is, a rifle bullet modified to fit over the end of the spear, which is fired at any threatening shark. The considerable commercial value of the jaws and teeth may lead to increased targeting and possible overexploitation (Compagno, 1991). In South Australia, fishermen and divers claim that the species has undergone a serious decline in recent years (Bruce, 1992).

In South Africa, research has been carried out on white sharks caught in the shark nets (Bass *et al.*, 1975; Cliff *et al.*, 1989, Chapter 32). In 1990, the Shark Research Center of the South African Museum initiated research into the movements, habitat use, behavior, abundance, and population structure of the species (Compagno, 1991).

In the absence of information about the status of the stock, government protection was granted to *C. carcharias* in April 1991, making it illegal to "catch or kill any white shark except on the authority of a permit issued by the Director-General of Environment Affairs" (Compagno, 1991).

Clearly, there is an urgent need to investigate the population dynamics of white sharks to assess the validity and effectiveness of this preemptive legislation. The tagging of white sharks under the auspices of a national marine fish tagging program, initiated by the Oceanographic Research Institute (ORI) in 1984, has provided this opportunity. The mark-and-recapture study described in this chapter is a first attempt at quantifying the stock of white sharks along a large section of the South African coast.

Materials and Methods

Tagging and Shark Capture

Each dart tag had a stainless steel head and a plastic streamer, at first red (Floy Tag), but lately yellow (Hallprint) (van der Elst, 1990). The latter tag also was used by Bruce (1992), and a tag of similar design has been used extensively by the U.S. National Marine Fisheries Service (Casey, 1985). Shark lengths used in the text are precaudal measurements (PCLs). Details about the white sharks tagged and recaptured are given in Table I.

The shark nets are distributed along 325 km of KwaZulu–Natal coast (Fig. 1), Richards Bay being the northernmost location. Mean monthly sea-surface temperature (SST) ranges from 19.9°C in August to 25.3°C in February (Cliff and Dudley, 1991a). Specifications of the nets and the manner in which they are deployed and serviced have been given by Wallett (1983) and Cliff et al. (1988a).

In 1990, one of the authors (T.K.W.), a commercial line fisherman, started tagging white sharks in the vicinity of Struis Bay (Fig. 1), 6 km east of Cape Agulhas, the southernmost point on the African continent. The mean monthly SST there ranges from 15°C in July to 21°C in January (Greenwood and Taunton-Clark, 1992). The sharks were caught on rod and 40-kg line, using small live sharks as bait. SST was measured (by T.K.W.) when most white sharks were tagged.

Estimating Population Size

A modified Petersen estimate (Seber, 1982) was used to calculate the population size, N, at the end of time interval i as

$$N_i = \frac{(M_i + 1)(n_i + 1)}{(m_i + 1)} - 1 \qquad (1)$$

where m_i is the number of marked animals recaptured in the total catch, n_i, of white sharks, and M_i is the number of marked animals alive at the end of interval i.

In this study, the tagging and recapture of white sharks were undertaken concurrently over a 5-year period. This period was split into five 1-year periods, and a Petersen estimate was obtained for each year. The total population size of white sharks, $\bar{N}$, was estimated as

$$\bar{N} = \frac{\sum N_i}{(s - 1)} \qquad (2)$$

where s is the number of samples.

In applying the Petersen estimate, it was assumed that all tagging took place at the beginning of time interval i and that all recaptures occurred at the end. Allowance was made for the rerelease of tagged sharks upon recapture and the loss of a constant fraction of tagged animals, $1 - \alpha$, either through immediate tag shedding or death from capture stress. Assuming that m_0 animals are marked at the start of a time interval, $i = 0$, then the number effectively tagged, M_0, is

$$M_0 = \alpha m_0 \qquad (3)$$

The number surviving to the next time interval, $i = 1$, is equal to

$$M_1 = \alpha m_0 e^{(-z)} \qquad (4)$$

TABLE I Details of Tagged and Recaptured White Sharks

Locality[a] tagged	Sex[b]	Precaudal length (cm)	Recapture locality	Time free (days)	Distance travelled (km)	Growth (cm)
Ballito	M	180	Algoa Bay	27	774	
Marina Beach	F	180	Trafalgar	1	4	0
Struis Bay	F	250	Struis Bay	373	0	
Struis Bay	F	250	Struis Bay	357	0	
Struis Bay	F	300	Durban	527	1409	
Glenmore	F	150	Glenmore	366	0	15

[a]Localities shown in Figure 1.
[b]Abbreviations: F, female; M, male.

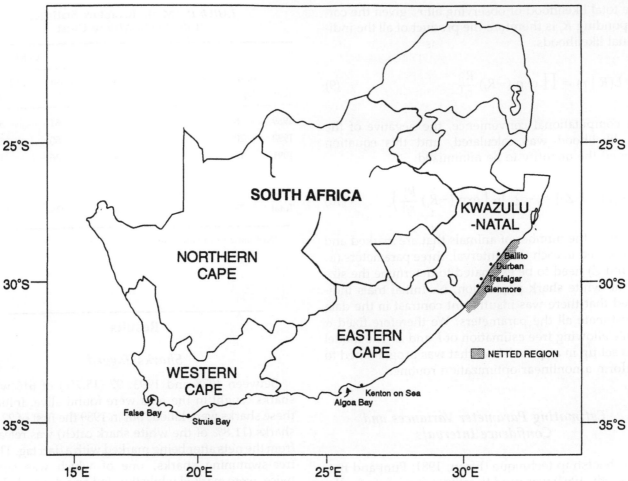

FIGURE 1 The southern African distribution of the white shark (after Compagno *et al.*, 1989). The shark-netted region in KwaZulu–Natal and the localities at which white sharks were tagged or recaptured are indicated.

where Z is the instantaneous rate of total mortality, taken to be constant. If, at this time, a further m_1 animals are tagged and released or if recaptured animals are rereleased, then the number surviving to the next time interval, M_1, is

$$M_1 = \alpha(me^{-Z} + m_1 + W_1) \qquad (5)$$

where W_1 is the number of marked animals rereleased. In general,

$$M_{i+1} = M_i + \alpha(m_{i+1} + W_{i+1}) \qquad (6)$$

In order to apply the Petersen estimate, we need to know the total annual catch, n_i, which comprises a known component, K_i (the sum of the catch in the shark nets and that of T.K.W.), and an unknown component, U_i (the sum of trawl net catches and those made by spear fishermen and other big-game anglers), such that $n_i = K_i + U_i$.

Estimating Mortality

Using the Baranov catch equation (Ricker, 1975), the estimated number of tag recaptures, R_i, that are reported, assuming that all such recaptures are reported, is

$$R_i = M_i \frac{F}{Z} (1 - e^{-Z}) \qquad (7)$$

where F is the instantaneous fishing mortality rate, taken to be constant throughout the experiment. Hilborn (1990) has shown that the sampling distribution of the tag recoveries can be approximated by the Poisson distribution. The likelihood of the expected number of recoveries, R_i, given the observed number of tag recoveries, r_i, is

$$L(R_i \mid r_i) = \exp(-R_i) \frac{R_i^{r_i}}{r_i!} \qquad (8)$$

The total likelihood of observing all r_i, given the corresponding R_i is therefore the product of all the individual likelihoods:

$$L(R \mid r) = \prod \exp\left(-R_i\right) \frac{R_i^{r_i}}{r_i!} \qquad (9)$$

For computational convenience, the negative of the log likelihood was calculated, and this equation formed the quantity to be minimized:

$$-\ln\left[L(\alpha, F, Z)\right] = -\sum \ln\left(\exp\left(-R_i\right) \frac{R_i^{r_i}}{r_i!}\right) \qquad (10)$$

Given the number of animals that are marked and recaptured in each time interval, three parameters (α, F, and Z) need to be estimated to determine the size of the white shark population, $\tilde{N}$. Initial trials indicated that there was insufficient contrast in the data to estimate all the parameters. We therefore fixed α while allowing free estimation of F and Z. The model was set up in a spreadsheet that was programmed to perform a nonlinear optimization routine.

Estimating Parameter Variances and Confidence Intervals

A bootstrap technique (Efron, 1981; Punt and Butterworth, 1993) was used to estimate variances for the parameters F, Z, N_i, and $\tilde{N}$. A large number of artificially generated recapture data sets were randomly drawn from a Poisson distribution using the procedure described in the book by Press *et al.* (1986). The procedure requires, as input, the expected mean number of recoveries in each interval (observed mean, 1) (Table II). The number of rereleases of white sharks in each time interval, i, was calculated as $0.6(G_i)$, where G_i is the random deviate from a Poisson distribution of mean value = 1, and 0.6 represents the average number of recaptured white sharks that have been rereleased (Table II). To each pseudodata set a new set of parameters and derived quantities (such as the N_i's) are estimated. The standard error of a parameter or derived quantity, Q, is then obtained from

$$SE(Q) = \sqrt{\sum_1^n \frac{(Q^n - \bar{Q})^2}{n-1}} \qquad (11)$$

where Q^n is the value of Q from the nth data set and $\bar{Q}$ is the mean of the Q^n's. The 95% confidence intervals were calculated using the percentile method.

TABLE II   **Mark–Recapture Statistics off the South African Coast**

Year	Number tagged	Recaptures, m_i	Rereleases, W_i	Catch Known, K_i	Catch Unknown, U_i
1989	6	1[a]	1[a]	61	50
1990	20	1	0	50	50
1991	16	1	1	36	10
1992	13	3	2	38	10
1993	18	0	0	53	10
Total	73	6	4	238	130

[a]Not used in the analyses.

Results

Sharks Tagged

Between 1978 and 1993, 97 (15.7%) of 616 white sharks caught in the nets were found alive. Initially, these sharks were killed, but in 1989 the first of 22 live sharks (11.8% of the white shark catch) was released from the nets after being marked with a dart tag. Three free-swimming sharks, one of which was tagged twice, were marked while they fed on a dead whale off Durban. Of the 25 sharks tagged, 13 were females and 11 were males. The males ranged from 130 to 370 cm PCL, with a mode of 200 cm; the two largest males, 320 and 370 cm, were both feeding on the whale carcass. The females ranged from 150 to 265 cm PCL, with a mode of 180–200 cm (Fig. 2).

In the Struis Bay area, 46 were tagged (by T.K.W.). They included 32 females (range, 150–450 cm; mode, 300 cm) and 8 males (range, 250–350 cm; mode, 350 cm) (Fig. 2). These sharks were tagged in water where the average SST was 18.6°C (range, 16.2–21.8°C; N = 42).

A 180-cm unsexed specimen was tagged by a commercial angler off Kenton-on-Sea, and a 158-cm female was tagged and released from a beach seine net in False Bay. In total, between 1989 and 1993, 73 white sharks were tagged along the south and east coasts of South Africa (excluding sharks tagged by members of the White Shark Research Institute, Cape Town).

Recaptures

Six sharks (8.2% of those tagged) were recaptured (Table I). The mean distance traveled while at liberty

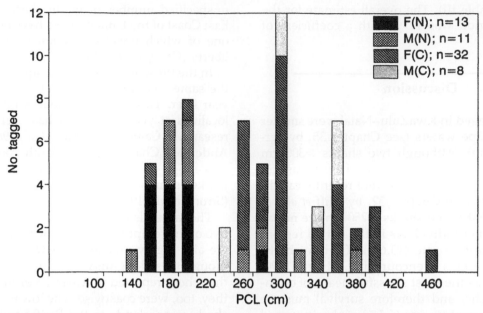

FIGURE 2 Size–frequency distribution of female (F) and male (M) white sharks tagged in KwaZulu–Natal (N) and in the western Cape (C). Sharks of unrecorded sex are excluded. PCL, Precaudal length.

was 365 km (SE = 244; range, 0–1409 km), and the mean time at liberty was 275 days (SE = 86; range, 1–527 days). Three of the 22 sharks released from the shark nets were recaptured. One traveled 4 km in 1 day and was rereleased. Another, of 150 cm, was recaptured in the same shark net installation at Glenmore 366 days later; it had grown by 15 cm. It was also released. A third shark, of 180 cm, traveled 774 km from Ballito to Algoa Bay in 27 days, a rate of at least 28.7 km/day.

Three sharks tagged in the western Cape (by T.K.W.) were recaptured, two at the same locality (Struis Bay) after 357 and 373 days, respectively; both sharks were released. The third shark was recaptured in the Durban shark nets, having traveled 1409 km in 527 days.

Estimate of Mortality and Population Size

The number of sharks tagged, the number recaptured, the number of recaptures rereleased, and the known and assumed unknown catches are shown annually for the 5-year period 1989–1993 (Table III). The shark that was recaptured 1 day after release was excluded from the analysis, because the time at liberty was too short to assume complete mixing of marked and unmarked white sharks. Given a survival factor, $\alpha = 0.9$ (see Discussion), the computed val-

TABLE III White Shark Population Estimates off the South African Coast, 1989–1993

Year	Survivors,[a] M_i	Estimated Total Catch, n_i	Estimated Population,[b] N_i	CV (%)
1989	5.4	111	716 (147–591)[c]	24
1990	21.2	100	1119 (388–1685)	43
1991	27.8	46	676 (350–1582)	63
1992	29.9	48	377 (439–1701)	123
1993	33.8	63	2227 (718–2772)	34
Mean			1279 (839–1843)	24

CV, Coefficient of variation.
[a]See Eq. 8 in the text.
[b]See Eq. 1 in the text.
[c]95% confidence limits of the population estimate are given in parentheses.

ues for instantaneous rates of mortality were $Z = 0.53$ and $F = 0.055$ year^{-1}. Of these two parameters, Z was a better estimate than F, as shown by the lower coefficient of variation (Table IV). Assuming that the average annual unknown catch was 50 for the years 1989 and 1990, and 10 in the years 1991–1993 (T. Ferreira, personal communication), then the annual population estimate ranged from 377 (1992) to 2227 sharks (1993) (Table III). The overall estimate for the 5-year study period was 1279, with a coefficient of variation of 24%.

Discussion

The sharks tagged in KwaZulu–Natal were smaller than those in Cape waters (see Chapter 35, by Ferreira and Ferreira). Although two sharks >300 cm PCL were tagged while feeding on a whale carcass off Durban and four mature males were recently caught in the shark nets (see Chapter 32, by Cliff et al.), it would appear that white sharks >250 cm are not as common as smaller individuals in the netted region.

The high recapture rate (13.6%) of white sharks released from the nets is encouraging. Murru (1990) regards gill nets as the most stressful means of catching elasmobranchs, and therefore survival must be low. The lower recapture rate (6.5%) of sharks tagged by T.K.W. may be due to the ban on angling for this species in 1991.

The overall white shark recapture rate of 8.2% was similar to or slightly higher than that of other large coastal sharks marked in the ORI tagging program. van der Elst and Bullen (1992) reported the following recaptures: 7.3% of 96 tiger sharks Galeocerdo cuvier, 6.8% of 88 bull sharks Carcharhinus leucas, 3.5% of 1122 raggedtooth sharks Carcharias taurus, and 1.6% of 252 broadnose sevengill cowsharks Notorynchus cepedianus.

Our white shark recapture rate (8.2%) was lower than that in South Australia, where 13.6% of 22 white sharks tagged were recaptured, with traveling distances of 18–220 km and times at liberty of 30–78 days (Bruce, 1992). The higher recapture rate and lower mean time at liberty (60 days) suggest that South Australian white sharks may incur a higher fishing mortality, which is understandable, as no legislation prohibits angling for this species. Only two of a small but

unspecified number of white sharks tagged on the East Coast of the United States have been recaptured, one of which traveled 614 km during 1.3 years at liberty (Casey et al., 1991).

In the present study, four recaptures were made in the same area; three of the four occurred close to 1 year later. This pattern of sharks returning to fixed localities at yearly intervals has been suggested by research in California (Ainley et al., 1985; Klimley and Anderson, Chapter 33). In South Australia, there is also a high degree of site fidelity, with 36% of the 58 marked sharks resighted, all at their original locations (Strong et al., 1992).

The large distances (774 and 1409 km) covered by two of the recaptured sharks indicate that the animals are also highly mobile in South African waters. Although these long-distance movements were farther than those reported in Australia or the United States, they, too, were coastwise. The low incidence of white sharks near islands in the Pacific and Indian oceans indicates that there may be limited transoceanic movement, and the warm tropical surface waters may act as a barrier to regular transequatorial movement of white sharks. Although the population of white sharks along the southern African coast may not be geographically isolated, there may be little recruitment from other centers of abundance, highlighting the need for local legislation to prevent possible overexploitation. Genetic studies are being conducted to compare white sharks from different regions (see Chapter 6, by Martin).

Several assumptions, some of which were discussed by Ricker (1975), were made in using the Petersen estimate to determine the size of the white shark population.

1. Instantaneous mortality rates, Z and F, are constant. The assumption of constant natural mortality may not be unreasonable, given that the duration of the study is relatively short compared to the life span of white sharks. Furthermore, the sharks tagged comprised few very small animals, which may be more prone to predation, and no very large sharks, which were approaching the end of their life span. On the other hand, fishing mortality would be affected by any change in fishing effort during the study period. The largest components of this effort are the shark nets, which remained constant, and big-game angling, whose effort was heavily curtailed by the protective legislation introduced in 1991. There were no reports of any change in effort in the trawl fishery and by spear fishermen. The extent of the reduction in F is unknown.

2. There is a 10% instantaneous tag shedding and tag-induced mortality. Tags inserted into sharks

TABLE IV   Parameters in White Shark Mortality Estimates off the South African Coast

Parameter	Value	Statistic		
		CV (%)	L (95%)	R (95%)
α	0.9 (fixed)			
F	0.055	46	0.015	0.100
Z	0.530	12	0.420	0.660

CV, Coefficient of variation; L, left (lower) 95% confidence limit of estimate; R, right (upper) 95% confidence limit of estimate.

caught in the nets were checked to ensure that they were firmly embedded before the animals were released. Due to the stress of capture on a baited line or in a gill net, some mortality is likely. Telemetry may provide an assessment of capture mortality. In this study, we assumed that 90% of tagged animals survived the stress of capture, hence $\alpha = 0.9$.

3. There is no long-term tag shedding. The good condition of the tag in a shark recaptured after 527 days indicates good tag retention. Tag shedding is therefore thought to be low. A large bull shark retained its dart tag for 11 years in captivity before being released. Double-tagging experiments on white sharks may provide more information on long-term tag shedding rates.

4. All recaptures of tagged sharks are reported. All recaptures in the shark nets should be reported; however, following the introduction of protective legislation, recaptures by big-game anglers may be unreported for fear of prosecution. Some spear fishermen are known to shoot sharks that threaten them. Any tagged sharks killed in this way may pass unnoticed by the diver, and hence may not be reported.

5. The distribution of tagged fish or the fishing effort is random. Despite the high degree of site fidelity discussed earlier, the average time to recapture of 0.754 year is ample time to allow for the mixing of tagged and untagged sharks. The recapture of a shark 1 day after release was excluded from the population analyses, because there was insufficient time for mixing. Fishing effort is not random, as the nets are permanently installed at fixed localities, while the efforts of T.K.W. are concentrated in the Struis Bay area.

6. Recruitment or emigration is negligible. Recruitment of newborn white sharks to the fishery may be balanced by a combination of natural mortality and inaccessibility of particularly large specimens. Tagging occurred between Richards Bay, KwaZulu–Natal, and Struis Bay, Western Cape, or only part of the shark's southern African range. There will be considerable movement of sharks into and out of the tagging region, resulting in a gradual reduction in the proportion of tagged animals in the study area.

The Petersen estimate of 1279 sharks (Table III) applies only to the region between Richards Bay and Struis Bay and excludes all the northern and much of the western Cape coasts, where the many colonies of South African fur seal *Arctocephalus pusillus pusillus* (Oosthuizen and David, 1988) may attract a large number of white sharks. The estimate is fairly insensitive to decreased values of the unknown component of the annual catch, which, when halved, results in a 14% decline in the population, that is, to 1098

sharks. The coefficient of variability (24%) of the estimate is low, considering the small sample sizes.

In this study, $F = 0.055$, $Z = 0.53$, and $Z - F = 0.48$ year^{-1}, which represents the sum of M and U, where M is the instantaneous rate of natural mortality and U is the sum of the instantaneous rate of emigration and long-term tag shedding. M is unknown, but is likely to be low for this apex predator, with its apparent slow growth and low fecundity. In the porbeagle shark *Lamna nasus*, another member of the family Lamnidae, $M = 0.18$ year^{-1} (Aasen, 1963). M is unlikely to exceed this value in white sharks of the size range tagged in this study. This results in $U = 0.3$ year^{-1}. As mentioned above, long-term tag loss is likely to be low, and emigration, whereby tagged white sharks move out of the study region, is the major component of U.

A possible yardstick in assessing the validity of protective legislation is to ensure that fishing mortality, now mainly that due to shark nets, does not exceed natural mortality. In this study, F is considerably lower than the sum of M and U. Many of the sharks caught in this study were released alive; consequently, the real F may be lower than 0.055 year^{-1} and, in our opinion, does not represent overfishing of white shark stocks. Improved estimates of mortality and emigration are needed, however, before relaxation of the current protective legislation can be considered.

Summary

The ORI Tagging Programme tagged 73 white sharks *C. carcharias* in South African waters between January 1989 and December 1993. Anglers in temperate Cape waters tagged 48 (66%) of the sharks; the remainder were tagged by the NSB. Cape specimens were larger than those from KwaZulu–Natal; most of the sharks were 150–400 cm PCL. Six of the sharks (8.2%) were recaptured within a mean of 275 days (range, 1–527 days) and a mean distance traveled of 365 km (range, 0–1409 km). A modified Petersen estimate was used to determine the size of the white shark population for each of the 5 years of the study. Allowance was made for capture-induced mortality and the rerelease of tagged sharks that were recaptured. Fishing mortality was assumed to be constant, despite the introduction of protective legislation in 1991. The overall estimate was 1279 sharks (95% confidence limits, 839–1843) for the region Richards Bay in KwaZulu–Natal to Struis Bay in Western Cape. Mortality rates were estimated as $F = 0.055$ year^{-1} (95% confidence limits, 0.015–0.10) and $Z = 0.53$

year^{-1} (95% confidence limits, 0.42–0.66). Improved estimates of mortality are needed before any relaxation of the protective legislation can be considered.

Acknowledgments

We are indebted to the members of the ORI Tagging Programme, particularly the field staff of the NSB, who have tagged

white sharks. The financial support of Stellenbosch Farmers Winery and the Southern Africa Nature Foundation, sponsors of the tagging program, is gratefully acknowledged. A. E. Punt, L. Beckley, S. F. J. Dudley, and V. Peddemors commented on the manuscript. The senior author (G.C.) thanks the National Audubon Society and the Steinhart Aquarium, California Academy of Sciences, for financial assistance in attending the symposium.

37

Population Dynamics of White Sharks in Spencer Gulf, South Australia

WESLEY R. STRONG, JR.
The Cousteau Society
Chesapeake, Virginia, and
Department of Biological Sciences
University of California
Santa Barbara, California

BARRY D. BRUCE
Division of Fisheries Research
Commonwealth Scientific and Industrial
Research Organization
Hobart, Tasmania, Australia

DONALD R. NELSON
Department of Biological Sciences
California State University
Long Beach, California

RICHARD D. MURPHY
The Cousteau Society
Chesapeake, Virginia

Introduction

Only recently has the problem of unregulated exploitation of many of the world's shark stocks gained widespread attention (see Compagno, 1990c; Heneman and Glazer, Chapter 45). Rightfully, most of this attention has been aimed at species subject to heavy commercial fishing. The white shark *Carcharodon carcharias* has also been heavily exploited, but is unique in having been sought almost solely for sport or for economic gain derived primarily on its jaws and teeth (Compagno, 1990c).

It has been difficult to assess the impact of exploitation on white shark populations. Due to a variety of challenges, most aspects of the species' biology and behavior are poorly understood, including growth rates, longevity, age at maturity, fecundity, gestation period, population abundance, age composition, distribution, and migratory patterns (Cailliet *et al.*, 1985; see also many other chapters in this volume).

In spite of these difficulties, the species has recently been afforded varying degrees of protection by the governments of South Africa and the state of California (see Chapter 45, by Heneman and Glazer). The justification for protection, in both cases, has been based almost entirely on the rarity of the species, as indicated by anecdotes and incidental evidence of declining local stocks. While these actions are prudent and highly commendable, there is clearly a need to develop a consistent means of assessing white shark population size. The Australian Society for Fish Biology stressed this when, in 1989, it included *C. carcharias* in the "uncertain status" category of the Australian Threatened Fishes List. The white shark remains under consideration for protection pending "acquisition of more reliable data regarding its abundance and distribution in South Australian waters."

This chapter is a direct result of the need for more information on white sharks in South Australia: in 1989, in collaboration with the South Australian Department of Fisheries, the Cousteau Society committed its vessel *Alcyone* to a 2.5-year study of white shark biology and behavior in South Australian waters. Here, we assess white shark abundance in the lower Spencer Gulf, using two techniques: a relative index of population density based on sightings per unit of effort and estimates of population size based on resightings of tagged individuals. We discuss spatial and temporal dynamics of the population as well.

Methods

Five expeditions were conducted over a 2.5-year period—April 19–May 22, 1989; January 26–March 16, 1990; August 2–September 1, 1990; January 19–February 25, 1991; and September 13–24, 1991—in waters adjacent to eight small islands lying to the south and east of Port Lincoln in the mouth of Spencer Gulf (Fig. 1). Effort was concentrated around Dan-

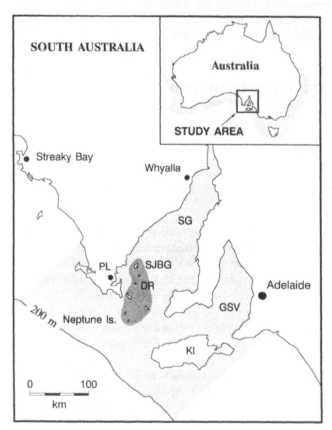

FIGURE 1 A map of the study area, showing the portion in which abundance was indexed (darkly stippled) and the surrounding region where tagged sharks have been resighted or recaptured (lightly stippled). (Inset) The study area relative to the species' normal range, extending throughout the southern portion of the continent (stippled). SG, Spencer Gulf; SJBG, Sir Joseph Banks Group; DR, Dangerous Reef; PL, Port Lincoln; GSV, Gulf St. Vincent; KI, Kangaroo Island.

gerous Reef (34°49′ S, 136°13′ E) and the Neptune Islands (35°14′ S, 136°03′ E), where previous records indicated a high abundance of white sharks. Each island was classified as either "inshore" or "offshore" on the basis of proximity to the coast: inshore sites were <37 km offshore, in waters <30 m deep; offshore sites were >37 km from the mainland, in waters >30 m deep. The sites were distributed within an area of about 260 km² (darkly stippled region in Fig. 1).

Standardized Baiting Procedure

Sharks were attracted via a three-dimensional odor corridor consisting of known amounts of mammal blood (in liquid and dried form), tuna oil, and frozen minced tuna. The beef, sheep, and pig blood was obtained from a local slaughterhouse; relative percentages of the amounts of each type used varied between orders, but most was beef blood. Whole undiluted blood was collected from the animals, and sodium citrate anticoagulant was added. Some of the blood was refrigerated, but most was dehydrated and then pulverized into a fine powder.

Whole bait fish, minced meat, and oil were derived from skipjack tuna *Katsuwonus pelamis*. Minced meat consisted primarily of chopped heads and entrails and occasionally of whole bodies. Relatively pure tuna oil was rendered during the canning process. Except for the dried blood, all bait was stored on deck in large iceboxes.

Two methods were used to create a robust odor corridor. First, a dried blood solution, mixed in a 1.33-m³ container filled with approximately 1200 liters of seawater and 1.4 kg of powderized blood, was continuously siphoned from the tank into the water at an average rate of 6.15 liters/min via a 13-mm-diameter hose. The fluid level was not allowed to fall below approximately 250 liters and was replenished at about 2.5-hour intervals. Second, a viscous chum, prepared by mixing 10 liters of whole mammal blood, 7 liters of minced tuna, 250 ml of tuna oil, and 100 liters of seawater, was delivered overboard at a rate of approximately 18 liters/hour. Initially, this was done by hand, but during the fourth and fifth expeditions, we used an automated system incorporating a pump based on the Archimedes' screw principle. The pump was fed from a 30-liter pan, refilled as necessary from a large container of chum stock. The amounts of the aforementioned ingredients were selected based on availability, mixing properties, and intuition.

As needed, whole tuna and large chunks of horse meat were thawed, cut into pieces weighing usually 1 kg, and suspended from two bait lines. Near the end of each line, a single toy balloon was tied so as to buoy the bait at the surface. Crewmen manipulated the lines to allow only moderate consumption ($\bar{X}$ = 1.0 ± 1.1 kg/hour for a subset of 14 sharks over 114 hours).

The entire baiting process was continuous, round the clock, as the ship lay anchored about 600 m offshore the island study sites. In addition to carefully monitoring bait output, we also recorded environmental parameters, hourly, that affect bait dispersal: wind and current speed and direction as well as sea-surface temperature.

Identification of Individuals

Whenever possible, sharks were marked using conventional fisheries dart tags (Hallprint), color-coded to facilitate recognition of individuals by adding short (10- to 20-cm) sections of polyvinyl "spaghetti" tubing. Tags were applied using rubber-sling-powered pole spears, both topside and by divers in antishark cages; 22% of the sharks received a second, backup tag. Other sharks were recognized by natural body markings. The total length (TL) of each shark was estimated and rough measurements were obtained in 10 cases by comparing the shark's length to that of a free-floating graduated line, tied near the bait so as to stream out along the shark's body as it swam with the bait in its mouth.

Short-Term Movement Patterns

Acoustic tracking was conducted using two types of crystal-controlled ultrasonic transmitters. Standard 40-kHz pingers, 30 mm diameter × 110 mm long and containing updated versions of the hybrid circuits described by Nelson and McKibben (1981), were used in seven trackings (Ultrasonic Telemetry Systems). Depth-sensing 38-kHz pingers (Vemco), 39 mm diameter × 184 mm long, were applied to two sharks. Both types of transmitters produced signals detectable at distances >1.5 km. Telemetered depths were accurate to within 1 m during field tests conducted on site prior to tracking. Dukane N15A235A and Vemco VR-60 ultrasonic receivers were used to monitor and decode incoming signals.

Transmitters either were concealed in baits and fed to the sharks (n = 5) or were applied with a pole spear in a manner similar to that used for conventional tags (n = 4). Neither attachment technique induced visible behavioral changes. Following deployment of a transmitter, all baiting was stopped. Invariably, the sharks soon left the vicinity. A tracking team in an

aluminum skiff (4–5.4 m long) followed, maintaining a mean estimated distance of 185 ± 130 m (n = 272; estimated based on signal amplitudes obtained during range tests). At 15-minute intervals, we recorded the shark's bearing, range (estimated from signal amplitude), and depth, if applicable. The 31-m-long research vessel followed, plotting the tracking boat's course at the designated intervals, while maintaining a distance of 1–3 km. *Alcyone*, being an optionally wind-powered vessel, was silent during the majority of the tracking.

Fluorometry

We used fluorometric dye tracing in an attempt to characterize the nature of bait dispersal at Dangerous Reef. At 0815 hours on January 21, 1991, 10 liters of rhodamine B (AFirm) was poured from the stern of *Alcyone*, forming a bright red cloud. We monitored its subsequent dispersal from a skiff equipped with a calibrated Sequoia–Turner 111 fluorometer (Unipath) and a high-sensitivity flow-through door. Samples were drawn continuously from a depth of 1 m as we conducted transects across the dissipating cloud. Vertical profiles were taken intermittently as well. See the papers by Fischer (1968) and Wilson (1968) for extensive treatments of water-tracing methods.

Abundance Index

We compared shark abundances by site, season, and sex, using 1 hour of standardized baiting as the basic unit of sampling effort. Only sharks sighted for the first time during a given expedition were included. Further, since remnants of the odor corridor may have lingered for several hours after the cessation of baiting (see the discussion on "downstream circling"), no records were used from areas where baiting had been conducted within the preceding 48 hours by our vessel or any other (see Discussion).

Population Estimates

Estimates were calculated using the Jolly–Seber method for multiple mark–recapture sampling from an open population (i.e., changing size due to births, deaths, and movements). Jolly's (1965) equation is

$$N_t = \frac{M_t}{a_t}$$

where N_t is the estimated size of the population just before sample time t, M_t is the estimated size of the marked population available for recapture just before sample time t, and a_t is the estimated proportion of animals marked and includes a correction for small samples.

Additional parameters used to solve N_t are given in the footnotes to Table IV. Formulas and detailed discussions about the model can be found in the work of Seber (1982) and Krebs (1989) and in the Discussion. Throughout this chapter, averages are expressed ±1 standard deviation.

Results

Population Composition

We identified 67 sharks and tagged 40 using conventional tags. Shark size varied considerably, ranging from 220 to 550 cm TL ($\bar{X}$ = 360 ± 70 cm) (Table I). The population did not appear to be segregated by size (Table II): TL averaged 360 ± 60 cm in males (n = 30) and 370 ± 70 cm in females (n = 29).

Twenty-three sharks (34%) were resighted by our team, always at their original locations (Tables I and III). Eighteen (27%) were seen during a single expedition only, on as many as 10 different days. Five (8%) were sighted during each of two subsequent expeditions. One of these, a 290 cm TL female, was seen on 23 different days over a 580-day period spanning three expeditions. Two other females (430 and 360 cm TL) were observed 15 times each over respective periods of 179 and 207 days, spanning two expeditions.

Several of the tagged sharks were resighted between our sampling periods by fishermen and sport divers at other islands in the Sir Joseph Banks Group. None, however, were positively identified. Since our final expedition, other tagged sharks have been resighted by a number of observers at Dangerous Reef, the Sir Joseph Banks Group, the Neptunes, and as far away as Whyalla and Coffin Bay (see Fig. 1). Unfortunately, the sharks could not be individually identified, because the tag color codes had become obscured by algal growths. Three tagged sharks had been killed by fishermen within 30, 73, and 78 days of tagging (Fig. 2) (see Bruce, 1992). A fourth tagged shark was killed at Little English Island, 18 km from Dangerous Reef, but its tag was not saved. Thus, a minimum of 6% of the sharks we observed have been killed to date.

Marked spatiotemporal separation existed between males and females (Table II). Females were most abundant at Dangerous Reef and other "inshore" islands (34 females to 9 males), especially during winter (note: the total for females includes resightings of individuals on subsequent expeditions),

but males predominated at the Neptunes and other "offshore" islands (1 female to 20 males), especially during summer. Juvenile sharks, as small as 220 cm TL, also exhibited this pattern. Of 9 males seen inshore at Dangerous Reef, eight were present during late winter (Table II). Thus, with rare exception, winter was the only period when mixing of male and female white sharks occurred.

Horizontal Movement Patterns

Six individuals were tracked in the vicinity of Dangerous Reef. The TLs of subjects ranged from 300 to 500 cm ($\bar{X}$ = 370 cm; 2 males and 4 females). Continuous contact was maintained for periods ranging from 8.5 to 27.5 hours ($\bar{X}$ = 17 hours; n = 7). Telemetry contact was occasionally reestablished with several of the sharks as they returned intermittently to Dangerous Reef over periods of up to 28 days.

Continuous tracking revealed three basic patterns of horizontal movement. The first, appearing during the initial phase of every tracking session, was referred to as "downstream circling." Each of the sharks crisscrossed several kilometers downstream of our baiting station for periods of up to 12 hours after transmitter attachment and the concomitant cessation of chumming (Figs. 3 and 4). Minimum and maximum distances downstream averaged 2.3 ± 3.1 and 6.3 ± 2.9 km (n = 7), respectively.

Four sharks exhibited an "island-patrolling" pattern, ranging more freely around Dangerous Reef, but remaining in the vicinity after having spent several hours in downstream circling. One 360 cm TL female, tracked on January 31, 1991, had patrolled an area covering 50 km² along Dangerous Reef for at least 27.5 hours, after which the track was terminated (Fig. 3). None of the sharks considered to be patrolling swam more than 8.6 km from the reef during the tracking period.

The third pattern, "interisland cruising," was characterized by a more linear path, with occasional sorties toward islands encountered along the shark's route (Fig. 4). A 430 cm TL female, tracked for 26 hours, crisscrossed the area north of Dangerous Reef for a period of approximately 8.5 hours (downstream circling), then departed on a winding northerly course covering 35 km (Fig. 4A). The shark approached to within 200 m of one of the islands. Another female took a similar route, passing close to several islands of the Sir Joseph Banks Group, and was eventually lost in bad weather (Fig. 4B). See the paper by Strong et al. (1992) for more detail about individual tracks.

TABLE I White Sharks Observed in Lower Spencer Gulf,
South Australia, between April 1989 and September 1991

Shark no.	Est. total length (m)	Sex	Initial observation date	Site no.[a]	Shark no.	Est. total length (m)	Sex	Initial observation date	Site no.[a]
1	3.4		April 20, 1989	1	35	3.7	F	September 1, 1990	1
2	4.0		April 20, 1989	1	36	3.7	F	September 1, 1990	1
3	3.0	M	April 26, 1989	2	37	3.7	F	January 21, 1991	1
4	3.4	F	May 20, 1989	3	38	4.9	F	January 25, 1991	1
5	4.3		January 29, 1989	1	39	4.6	M	February 16, 1991	2
6	4.6	F	January 28, 1990	1	40	3.2	M	February 16, 1991	2
7	3.6	F	February 5, 1990	1	41	3.2	M	February 18, 1991	2
8	2.9	F	February 5, 1990	1	42	3.0	M	February 2, 1991	2
9	3.7	M	February 9, 1990	4	43	4.0	M	February 21, 1991	2
10	2.2	F	February 16, 1990	3	44	2.3	M	February 21, 1991	2
11	2.9	F	February 17, 1990	1	45	4.3	M	February 21, 1991	2
12	3.4	F	February 22, 1990	1	46	4.3	M	February 21, 1991	2
13	4.0		February 23, 1990	1	47	4.0	M	February 21, 1991	2
14	4.0	F	March 1, 1990	1	48	4.3	M	February 22, 1991	2
15	3.4	F	March 3, 1990	1	49	3.6	M	February 22, 1991	2
16	3.7	F	March 5, 1990	1	50	3.0	M	February 22, 1991	2
17	4.3	F	March 6, 1990	1	51	3.7	M	February 22, 1991	2
18	3.4	M	March 9, 1990	5	52	4.3	M	February 22, 1991	2
19	3.4		March 11, 1990	1	53	4.6	M	February 22, 1991	2
20	3.0	M	March 14, 1990	1	54	4.6	M	February 23, 1991	2
21	3.0	F	June 9, 1990	1	55	4.0		February 23, 1991	2
22	3.0	M	August 2, 1990	1	56	2.4	M	February 24, 1991	2
23	4.3	F	August 9, 1990	1	57	4.0	M	February 24, 1991	2
24	3.4	F	August 9, 1990	1	58	4.6	F	February 24, 1991	2
25	3.0	F	August 13, 1990	1	59	4.0	M	September 15, 1991	1
26	2.2		August 16, 1990	1	60	3.5	F	September 15, 1991	1
27	3.7	M	August 16, 1990	1	61	2.6	M	September 16, 1991	1
28	3.4	M	August 16, 1990	1	62	3.0	F	September 16, 1991	1
29	4.0	F	August 16, 1990	1	63	3.0		September 18, 1991	1
30	3.0	F	August 17, 1990	1	64	4.0	M	September 19, 1991	1
31	3.7	F	August 17, 1990	1	65	3.0	M	September 19, 1991	1
32	3.8	F	August 30, 1990	1	66	3.5	M	September 23, 1991	1
33	4.0	F	August 31, 1990	1	67	4.6	F	September 23, 1991	1
34	5.5	F	September 1, 1990						

[a]1, Dangerous Reef[b]; 2, South Neptune Island[c]; 3, Sibsey Island[b]; 4, North Neptune Island[c]; 5, Buffalo Reef.[c]
[b]Inshore sites: <37 km off the coast, water depth <30 m.
[c]Offshore sites: >37 km off the coast, water depth >30 m.

Rate of Movement

Rate of movement, based on point-to-point estimates within the horizontal paths of four sharks, averaged 3.2 km/hour (N = 145 fifteen-minute segments). Expressed in body lengths per second (bl), rates of movement ranged from 0.21 to 0.25 bl/sec.

Vertical Movement Patterns

Telemetry from two individuals revealed a strong tendency to swim either near the surface (1–3 m) or along the bottom, which consistently averaged about 20 m throughout the study area. The sharks spent very little time in midwater. When traveling between

TABLE II Relative Abundances of White Sharks at Sites
in Lower Spencer Gulf, South Australia

Site	Summer[a] No. and sex	Effort (hours)	Observation rate (n/hour)	Fall[a] No. and sex	Effort (hours)	Observation rate (n/hour)	Winter[a] No. and sex	Effort (hours)	Observation rate (n/hour)
Inshore sites[b]									
Dangerous reef	1M:8F:3U	724.4	0.017	1F	11.0	0.091	8M:23F:2U	580.0	0.057
Sibsey Island	1F	51.5	0.019	1F	22.2	0.045			
Langton Island	0	41.3	N/A	0	23.2	N/A			
Pearson Island				0	89.7	N/A			
Greenly Island				0	48.0	N/A			
English Island				0	21.7	N/A			
Stickney Island				0	9.9	N/A			
Spalding Island				0	10.5	N/A			
Fossil Point				0	37.0	N/A			
Subtotal	1M:9F:3U	818.2	0.016	2F	273.2	0.007	8M:23F:2U	580.0	0.057
Offshore sites[c]									
North Neptune Island	1M	43.8	0.023				0	38.7	N/A
South Neptune Island	14M	64.0	0.219						
Buffalo Reef	1M	20.4	0.049						
Subtotal	16M	128.2	0.125				0	38.7	N/A
Overall total							25M:34F:5U	1838.3	0.035

[a]Seasons: summer, December 22–March 19; fall, March 20–June 20; winter, June 21–September 22.
[b]Sites <37 km off the coast, water depth <30 m.
[c]Sites >37 km off the coast, water depth >30 m.

the surface and the bottom, however, they ascended or descended slowly and steadily, usually taking 4–6 minutes to make the 20-m depth transition. Further, the sharks almost never reversed direction in midwater (e.g., upon descending 5–10 m, the animal would continue until it reached the bottom). A t-test ($P >$ 0.001) showed that daytime cruising depths averaged 11.5 ± 8.5 m (n = 135) and were significantly shallower than those recorded at night ($\bar{X}$ = 16.5 ± 8.0 m; n = 92).

Fluorometry

When 10 liters of rhodamine B was poured into the sea at 0815 hours, the resultant opaque cloud began traveling with the current, 125° SE. By 0945 hours, it had moved 2.3 km and covered a visible area approximately 200 m in diameter. Wind direction was consistently 45° and an oil slick, originating at the bait source, could be seen traveling away in that direction, 80° apart from the direction that the dye cloud was moving. At 1015 hours, rhodamine concentrations reached 0.5 µl/liter at the center of the cloud, 2.7 km from the source. At 1045 hours, the tidal current (and

the dye) shifted abruptly to 80° E. At 1245 hours, the dye was no longer visible on the surface, although vertical profiles indicated that irregularly distributed concentrations of up to 0.025 µl/liter were still resident in the water column, mostly in the upper 6 m. Detectable (i.e., above background) levels of dye were present over an area about 1.1 km in diameter, the leading edge of which was approximately 4.1 km downstream of the source. Thus, central dye concentrations decreased more than an order of magnitude during the 2.5 hours, as the cloud traveled about 1 km. Note that a sustained 45% reduction in current velocity was associated with the changing tide (dropping from 35.8 to 19.7 cm/sec). Wind velocity and sea-surface temperature averaged 31 km/hour and 20.3°, respectively.

Fluorometry substantiated our observation that bait corridors separate into several fractions during transport downcurrent and downwind. Components such as lightweight oils, blood and other dissolved matter, and large negatively buoyant particles "behaved" differently and may result in more than one distinct, useable gradient (Fig. 5). Therefore, we could not determine their relative importance in

TABLE III　**Resightings of White Sharks Tagged between April 1989 and September 1991 in Lower Spencer Gulf, South Australia**

Shark no.[a]	Days between sightings[b]
6	10, 10, 4, 1, 1, 5, 1
7	4
8	2, 10, 1, 3, 1, 6, 2, 2, 1, 1, 5, 3, 1, **169**
9	1
11	1, 1, 3, 2, 5, 1, 1, 1, 1, 1, 6, 1, **144**, 4, 2, 8, 2, 5, 1, 7, **382**, 1
14	1, 1
15	1, 1, 1, 6, **322**, 8, 6
16	6, **161**, 5, 5, 1, 1
17	6, 13
24	4, 5, 1, 3, 2, 1, 4, 1, 1, 1, **142**, 8, 5, 1
27	1, 1
28	1, 8
29	1, 1
32	1
33	1
35	1, **379**
36	10, 2, 1, 1, 1, 3, 4, 1
40	1
53	1
55	1
56	1
58	4
59	1
62	7, 8

[a]Identities of white sharks are given in Table I. All resightings occurred at the original tagging locations.

[b]Boldface numbers denote sightings on subsequent expeditions.

shark attraction or the ranges at which each is effective. If oceanographic conditions differed markedly and consistently between sites, odorant dispersal and subsequent attraction would have been biased. However, although a more detailed analysis of environmental data is warranted, current and wind velocities from Dangerous Reef (inshore) and the Neptune Islands (offshore) did not differ markedly. For Dangerous Reef and the Neptunes, respectively, current velocities averaged 14.9 ± 8.7 cm/sec (range, 0.4–51; N = 316) and 13.2 ± 7.8 cm/sec (range, 1.3–36.5, N = 126), while wind velocities averaged 25.5 ± 15 km/hour (range, 0–70.9; N = 357) and 27.8 ± 13.5 km/hour (range, 2–59.6; N = 129). Thus, the main factors affecting the physical distribution of odorants were similar between sites.

Abundance Index

The relative abundance of white sharks was measured using 1 hour of standardized baiting as the basic unit of sampling effort, an approach similar to the widely used catch per unit of effort (CPUE) (see, e.g., Ricker, 1975) (see Table II). Sixty-four valid sightings (i.e., from independent baiting sessions) were documented during a total of 1838.3 hours of standardized baiting, an average rate of 0.035 arrivals per hour (i.e., each new sighting occurring at average intervals of 28.6 hours). This rate varied considerably within and between sites. The highest rate of arrival, observed at South Neptune Island during summer, was more than six times greater than the mean rate (Table II).

Jolly–Seber Population Estimates

Four mark–resight sampling periods yielded two estimates of the white shark population in waters immediately surrounding Dangerous Reef (Table IV): 191.7 [95% confidence interval (CI), 36.5–1612.2] and 18 (95% CI, 3.9–157.6) individuals during the second and third census periods, respectively. The probability of surviving from the second to third sample periods, including losses from emigration and natural mortality, was 0.2.

Discussion

Population Composition

The size distributions of males and females were similar, but the sexes were spatiotemporally separated (Table II). Sexual segregation has been reported previously among sharks, even white sharks (e.g., Pratt, 1979; Klimley, 1985b; Stevens, 1987; Ferreira and Ferreira, Chapter 35), but we know of no other records of a large coastal species being segregated over such a small geographic area.

It appeared to us that major concentrations of both sexes were positioned on opposite sides of an expansive physical and biological gradient. Oceanographic features, such as fronts and thermoclines, are well known to produce discontinuities in a variety of marine populations (Brandt and Wardley, 1981; Kiorboe et al., 1988; Longhurst, 1981; Owen, 1981). Inshore and offshore islands in our study area are separated by a seasonally dominant frontal zone at the mouth of Spencer Gulf. This boundary is most prominent in summer and fall and results in a significant reduction in exchange between Gulf and shelf waters (Sher-

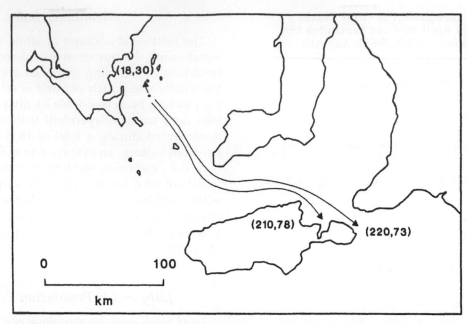

FIGURE 2 Recaptures of white sharks tagged at Dangerous Reef. The numbers in parentheses are the distance (in kilometers) from the point of tagging and the time until recapture (in days), respectively. The lines represent minimum point-to-point paths between tagging and capture locations, not necessarily the actual paths taken. (From Bruce B. D. Preliminary Observations on the biology of the white shark. *Aust. J. Mar. Freshwater Res.* 1992, **43**, 1–11, with permission.)

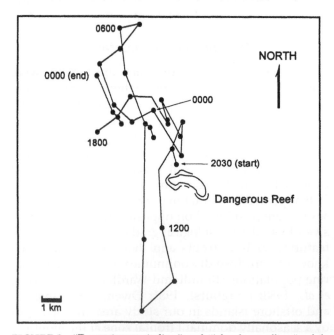

FIGURE 3 "Downstream circling" and "island patrolling" exhibited by a female white shark (360 cm total length) tracked at Dangerous Reef, South Australia, January 31–February 2, 1991. Positions are shown at hourly intervals. (From Strong *et al.*, 1992.)

ringham, 1990). Marked discontinuities in physical [e.g., temperature, salinity, and nutrient levels (Lennon *et al.*, 1987; Bruce and Short, 1992)] as well as biological properties [e.g., larval fishes (Bruce and Short, 1992)] occur across this zone, and it is conceivable that they act to limit the movements of white sharks in the area (see Chapter 34, by Pyle *et al.*).

A reversal of wind stress, combined with seasonal cooling of Gulf waters during autumn and winter, resulted in the breakdown of the frontal zone and resumption of exchange between Gulf and shelf waters (Lennon *et al.*, 1987). Interestingly, with rare exceptions, winter was the only period when mixing of male and female white sharks occurred (Table II). This seasonal influx of large males may correspond to the mating period. The species is certainly capable of moving across environmental gradients, such as those encountered between either side of the frontal zone. A 460 cm TL white shark, tracked by Carey *et al.* (1982) during open-ocean swimming, showed a "clear preference for the thermocline" but not for any absolute temperature.

Within the last 20 years, this pattern of segregation

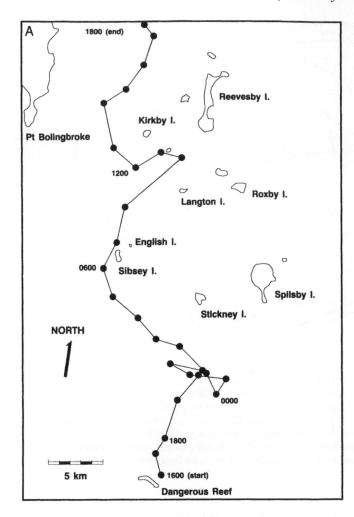

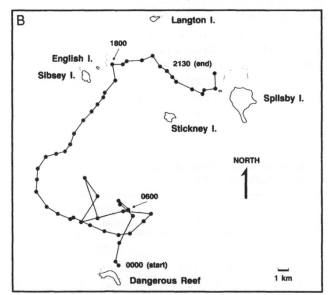

FIGURE 4 Routes taken by two female white sharks tracked at Dangerous Reef, South Australia. (A) A 490 cm TL shark, January 25–26, 1991 (positions are shown at hourly intervals), and (B) a 430 cm TL shark, August 9, 1990, positions shown are hourly to 1200 hours and at 15-minute intervals thereafter.

has fluctuated spatially and, perhaps, temporally: for many years, only male white sharks were seen at Dangerous Reef during summer (R. Fox, R. Taylor, V. Taylor, J. Johnston, and E. Palmer, personal communication). While anecdotal, this point is supported by a considerable amount of photographic evidence. No information is available concerning winter trends, because few dive trips have been conducted during winter.

Demographics of white shark populations on the east and west coasts of North America have been described by Casey and Pratt (1985) and Klimley (1985b), respectively. Although the pattern is more exaggerated on the west coast, size (approximate age) stratification occurs in both populations with juveniles concentrated south of the main adult populations. Large males are generally limited to the northern portions of the populations whereas mature females (>400 cm TL) are captured throughout, apparently because they migrate southward to pup in warmer distant waters (Klimley, 1985b). Size segregation is also present off southern Africa (see Chapter 36, by Cliff *et al.*) and in the Mediterranean (see Chapter 30, by Fergusson).

South Australian population structure contrasts with this longitudinal pattern in that juvenile white sharks are found in and immediately adjacent to areas shared by subadult and adult sharks (Bruce, 1992; Strong *et al.*, 1992). This contrast is probably the result of geographic orientations of the respective coastlines and subsequent marked environmental differences. While the North American coastlines cross many latitudes and faunal boundaries, the expansive southern coast of Australia is comparatively homogeneous. As stated, however, South Australian gulf waters are segregated by a seasonal frontal zone. These zones provide a notably different habitat, rich in species preyed on by juvenile white sharks, and require only short migrations to and from the main populations. This probably explains why the gulfs, rather than the distant east or west coasts, appear to be the primary natal areas for white sharks in southern Australia. Small white sharks, however, do occur along other areas of the coast (Bruce, 1992).

Horizontal Movement Patterns

Continuous tracking of six individuals revealed three basic horizontal movement patterns: downstream circling, island patrolling, and interisland cruising. Downstream circling was almost certainly a searching pattern in response to remnants of the odor corridor. Bait particles drifting down and concentrating along the bottom over a period of days may have

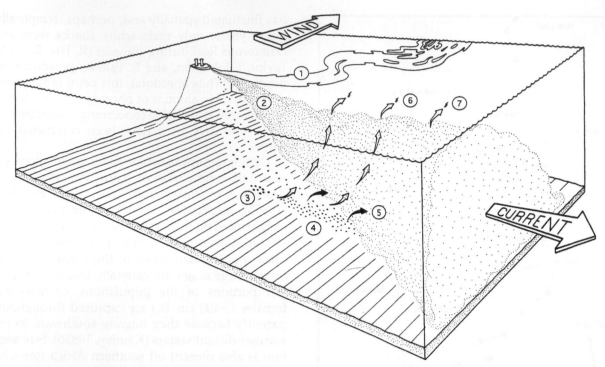

FIGURE 5 Hypothetical representation of the complex nature of odorant dispersal with the stern of the boat as the point of origin. The odorants consist of (1) the primary oil slick, with its direction mainly influenced by wind and surface currents; (2) the primary corridor, consisting of blood, other organic materials, and small suspended particles moving downcurrent; (3) large pieces of bait sinking to the bottom near the point of origin; (4) increasingly smaller pieces of bait settling out farther away from the point of origin, (5) settled pieces that decay or are broken apart by scavengers, (6) surrendering dissolved organics into the water column; (7) nonmiscible hydrocarbons returning to the surface, sometimes far away from the primary slick; and (8) floating hydrocarbons released as suspended particles break up.

created a "secondary bait source" that each shark investigated after we had stopped baiting (Figs. 3–5). An alternative explanation is that the sharks continued searching well beyond the time at which any trace of the corridor could be detected. Regardless, this pattern represents a consistent approach to the problem of a diminishing or lost odor source.

Island patrolling, when continued over several days, could be a home-ranging pattern around islands regularly visited and supporting colonies of pinnipeds. However, island patrolling could be a response to the long history of baiting in the study area. A more absolute determination would require tracking for many days after baiting had ceased. Goldman *et al.* (Chapter 11) obtained evidence of island patrolling by a white shark tracked in the vicinity of the

TABLE IV Jolly–Seber Population Estimates from Tag–Resightings
of White Sharks at Dangerous Reef, South Australia

Sample, t	Proportion marked, a_t	Size of marked population, M_t	Population estimate, N_t	Confidence limits (95%)	Probability of survival, O_t
1	a	0.0	a	a	3.125[b]
2	0.13	25.0	191.7	36.5–1612.2	0.200
3	0.50	9.0	18.0	3.9–157.6	a
4	0.25		a	a	a

Sampling was conducted during Cousteau expeditions 2–5 (for dates, see Methods).
[a]No estimate was possible with the available data.
[b]Survivorship is >1 due to the artifact of starting with the estimated marked population (O_t) = 0.

Farallon Islands, California, where little baiting has been conducted, and none during the year of their study.

Interisland cruising was a linear pattern characteristic of nomadic behavior. These sharks were no longer under the influence of our odor corridor, as they continued on northerly paths (Fig. 4). Interestingly, many of the large-scale directional changes observed during our tracking sessions appeared to have been oriented swims toward individual islands encountered along the shark's route (Fig. 4A and B).

Rate of Movement

The rate of movement averaged 3.2 km/hour (n = 145 fifteen-minute segments). This is the same speed reported for a large white shark tracked off the northeast coast of the United States (Carey *et al.*, 1982) and off the Farallon Islands (Goldman *et al.*, Chapter 11).

Abundance Index

Based on a simple standardized baiting protocol, the abundance index was a convenient means for grossly assessing white shark population size. However, the method is wrought with many spatial and temporal biases, the standardization of which would be dependent on the resources and logistics of researchers. Thus, results from such an effort must be interpreted with caution.

To exclude many biases within our study, only sharks sighted for the first time during a given expedition were included in the analysis. Further, since remnants of the odor corridor may have lingered for several hours after the cessation of baiting, as discussed above, no records were used from areas where baiting had been conducted within the preceding 48 hours by our vessel or any other.

There is clearly a need for a standardized nonlethal means of assessing shark populations. Acoustic attraction (see, e.g., Nelson and Johnson, 1972) has the important advantage of being evenly distributed, but is subject to background noise, has relatively limited range, and, based on preliminary trials, may not be highly attractive to white sharks. Visual lures (see Chapter 20, by Anderson *et al.*, and Chapter 21, by Strong), monitored directly or remotely, have promising potential for white sharks, but as discussed by Strong (Chapter 21), may not be useful with other shark species. While conventional baiting is subject to many biases, it is the most widely used and available option. Thus, additional work on odorant transport, separation, and decay is worthwhile, perhaps by direct sampling of labeled organic material.

Jolly–Seber Method Assumptions

Before considering the population estimates, we discuss here the assumptions of the method.

Every Individual, Marked or Unmarked, Has the Same Probability (a_t) of Being Caught in the *t*th Sample, Given That It Is Alive and in the Population When the Sample Is Taken

We conducted a zero-truncated Poisson test of equal catchability, a powerful test for use with populations experiencing negligible mortality (Caughley, 1977). The resultant χ^2 value (2.47, df = 1) did not exceed the critical value for a = 0.05 ($\chi^2/0.05 = 3.84$). The null hypothesis of equal catchability was therefore accepted. Some sources of unequal catchability may not be revealed by such tests, however, and we must rely on knowledge and intuition to recognize them (Krebs, 1989). While most mark–recapture surveys involve species that can be readily caught and handled, this study necessitated visual identification of large dangerous animals attracted to bait. The validity of any such assessments depends heavily on details of the baiting technique; as observed of many shark species, strong relationships exist between the amounts (and types) of bait used and arrival frequencies. We attempted to negate baiting biases by using a standardized technique. The primary concern, then, becomes differential attraction or avoidance of the sampling area by individual sharks as a function of season or other factors. This question merits further investigation.

We must also consider the possible influences of social dominance on the probability of sighting an individual once it arrives at the bait source area. W. R. Strong, Jr., *et al.* (unpublished data) observed linear size-based feeding hierarchies in this population of white sharks; individuals deferred to larger conspecifics during direct interactions and appeared to limit their use of the feeding area as necessary in order to avoid close proximity (see Chapter 23, by Barlow, and Chapter 27, by Long and Jones). However, subordinate individuals were not completely excluded from the area.

Finally, and most importantly, we must evaluate the question of whether we were sampling from a single population. Since the sexes mixed only during winter at inshore sites (Table II), males and females did not have equal opportunity to be sampled, depending on season and location. Therefore, although sampling was conducted throughout the study area, we only computed an estimate for the Dangerous Reef site, where the majority of the tagging and recoveries occurred.

Every Marked Individual Has the Same Probability, O_t, of Surviving from the tth to the $(t + 1)$st Sample

Mortality from sport and commercial fishing fluctuates and, as discussed below, is considerable at times. The most likely effect from fishing is that unreported killings of untagged sharks may reduce the proportion of untagged to tagged individuals observed during subsequent samples, thereby reducing the estimate. The severity of this effect, in turn, depends on the spatial and temporal proximity of the mortality event to a given sample.

While concrete data on longevity are not available, vertebral banding patterns crudely indicate a maximum age of 25–30 years (Cailliet et al., 1985). Other evidence indicates a high annual survivorship for such long-lived sharks with virtually no natural enemies. We therefore disregarded effects of natural mortality during the study period; other than fishing, emigration, therefore, must be primarily responsible for "mortality."

Each Newly Encountered Animal That Is Not Already Recorded in the Current Sampling Period Has the Same Probability, v_t, of Being Tagged

Ideally, baiting effort, and hence, tagging, would have been spread as uniformly as possible throughout the region, but as Siniff et al. (1977) pointed out concerning a similar study of Weddell seals *Leptonychotyes weddelli*, the economic and logistical constraints were highly prohibitive. The focus on Dangerous Reef was a trade-off between sampling design and maximizing the number of sharks encountered and thus available for other purposes. Although this limits the usefulness of the estimate, this assumption should not be violated.

Individuals Do Not Lose Their Marks, and Marks Are Not Overlooked during Subsequent Sightings

While there are methods for estimating the tag loss rate, this problem applies more to studies of smaller animals, especially teleosts, in which tags must be of lightweight construction and the tissues into which they are embedded may tear easily. To use the unmodified Jolly–Seber estimates, one needs only to ensure that the loss rate is low enough that bias is negligible (Arnason and Mills, 1981). As mentioned earlier, 22% of the sharks received two tags. No double- (or single-) tagged shark was later found to have lost a tag. The standard dart tags used to mark fish in this study have an outstanding reliability record, having been recovered after periods of up to 22.6 years (J. G. Casey, personal communication).

Since the sighting history of each individual must be known in order to use the Jolly–Seber and other open population models, we maintained a detailed logbook of ancillary descriptions based on extended underwater observations as well as motion and still photography. Records of natural diagnostic features gained importance when, after periods of 1 year or more, tag color codes became obscured by algal fouling. We dealt with this, however, by attaching a replacement tag, then pulling the obscured tag free as a bait handler lured the shark into position. The tag could then be cleaned and read. While errors may have occurred on rare occasions, our identification of individual white sharks was successful.

Sampling Time Is Negligible in Relation to Intervals between Samples

The ideal tagging period is "instantaneous" (i.e., occurring over a very brief period), the intersample period being at least several times longer. Sampling design is often influenced by the biology and behavior of the species under study, however, and a variety of interpretations of this assumption have been used. Dadswell (1979), for example, partitioned several years of relatively continuous sampling into 100-day periods of equal effort in order to estimate the number of sturgeon *Acipenser* sp. in the St. John River, and probably violated the requirement that all tagged individuals receive ample time to become well mixed in the population.

Our four census periods averaged 33 days in length. Animals tagged during a given census were commonly resighted, often weeks apart (Table III), but none of these resightings was used to compute our estimate. Only those resightings made during subsequent expeditions, after intervals of at least 138 days ($\bar{X} = 181.8$ days), were considered valid. Given the species' demonstrated movement patterns, individuals had sufficient time to become well mixed.

Jolly–Seber Population Estimates

The population estimates, 191.7 and 18, are reasonable, considering that (1) sampling intervals were long; (2) estimates were made during winter and summer, respectively (females being more abundant during winter); and (3) Dangerous Reef is the area of greatest known white shark abundance in South Australia. The confidence intervals for the first estimate are very broad due to the low sample size.

Extrapolating observed arrival rates from the abundance index for the winter and summer periods at inshore sites, mainly Dangerous Reef, enables a crude comparison between the index and the Jolly–

Seber estimates. By multiplying the winter arrival rate (0.057 sharks per hour) and the sampling interval over which the population estimate was derived (33 days), we can expect to have seen 45.1 different individuals (as compared to the 191.7 predicted by the Jolly–Seber model). By the same method, the summer arrival rate (0.016 sharks per hour) yields 12.7 individuals (as compared to 18 predicted by the model).

Recent trips conducted for filming purposes during 1994 and 1995 have consistently failed to bait-in white sharks at rates approaching those recorded during this study (R. Fox, C. Roessler, and N. Wu, personal communication). Anecdotal comparisons of sightings made by tour operators suggest that over the past decade, the highest white shark abundances occurred during this study. Thus, interpreted conservatively, the Jolly–Seber estimates represent maximum values for a population that either was at the peak of a long cycle or was unusually large due to other unknown causes. Even if we accept the upper intervals as our estimates, therefore, the white shark population in the study area is relatively small, given that the first estimate was derived at a prime location during peak abundance.

General Discussion

White shark abundances are influenced by a wide variety of abiotic and biotic factors. Although direct evidence is limited, two main anthropogenic factors had much impact on white shark populations during recent centuries: (1) direct and incidental fishing pressure and (2) dramatic shifts in prey abundances as primarily influenced by human activities, especially harvesting. In South Australia, as elsewhere, small white sharks feed more on teleosts than do larger individuals (Bruce, 1992). Australian "snappers" *Pagrus auratus*, which grow to 130 cm TL and form large schools, are preyed on by small white sharks, but are also heavily sought by commercial and sport fishers. Nonetheless, it is extremely difficult to assess the extent of this competition.

Interactions with surface-dwelling mammals by both white sharks (see, e.g., Klimley *et al.*, 1992) and humans (see, e.g., Stirling, 1972) are easier to deduce. Pinniped abundances have fluctuated enormously over the last 100 years, experiencing a nearly global downward trend due to excessive harvesting near the end of the 19th century (see, e.g., Stirling, 1972). Around this time, sealing stopped because animals became commercially extinct or protected. The population trend was then reversed. Many pinniped populations still appear to be recovering, and some are growing rapidly; in South Australia, a recent sur-

vey indicated that during the 20-year period, 1970–1990, the number of fur seal *Arctocephalus forsteri* pups on South Neptune Island has doubled, increasing from 1000 to approximately 2000 (Shaugnessy, 1990). The effects that such sudden substantial increases in pinniped numbers are having on white shark populations cannot presently be determined, but by conducting follow-up estimates of white shark abundance in conjunction with pinniped surveys, quantitative indications of interactions can be obtained.

Although the collection of relevant data, until recently, has been opportunistic, sport and commercial fishing on white sharks continue to have the most discernible impact on population size (see Chapter 40, by Levine, and Chapter 41, by West). Bruce (1992) roughly estimated recreational captures for South Australia at <10 per year. Estimates of commercial captures (mostly incidental) are very crude, but it is highly probable that they outnumber the local sport catch (Bruce, 1992). That the study population commonly encounters commercial fishing gear was evidenced by the fact that 10% of the sharks we observed bore short remnants (<2 m) of longlines and gill nets.

Sport-fishing data from the east coast of North America and southeastern Australia indicate declines in the proportions of white sharks taken relative to other shark species caught over the last several decades. Casey and Pratt (1985) reported a drop from 1:67 (white sharks to all others captured) in 1965 to 1:210 in 1983 along the coast of the eastern United States, and Pepperell (1992) reported average ratios of 1:22 in the 1960s, 1:38 in the 1970s, and 1:651 in the 1980s for the coastal area near Sydney, Australia (see Chapter 41, by West). Similar trends are evident in South Africa (see Chapter 36, by Cliff *et al.*, and Chapter 40, by Levine). Capture data are subject to considerable error from differences in effort, reporting, etc. (see, e.g., Bruce, 1992) and anecdotes are generally unreliable. Nonetheless, it is difficult to explain these purported dramatic decreases in white shark abundance as anything other than the gradual disappearance of this specialized apex predator from temperate neritic communities.

At a minimum, we should attempt to validate or refute this trend in one or more white shark populations by monitoring total or relative population sizes, survival rates, and shifts in population structure. Much could be learned, for example, if we better understood the events and conditions surrounding the shift from male to female predominance that took place in the inshore portion of the study area over the last 20 years. Was this due to marked changes in hydrographic conditions or population structure? It may be that higher levels of inshore fishing pressure depleted the male segment in such a way as to directly alter its distribution or, alternately, that the reduced male population was competitively excluded by proportionally greater numbers of large female sharks. Additionally, greater attention must be given to obtaining regular reliable mortality data from fishermen and to increasing our understanding of prey population dynamics, especially of the delphinids that are important prey in the Spencer Gulf (Bruce, 1992). Generalizations about white shark abundances and behaviors have, historically, been based on observations conducted under widely variable baiting regimes. It is hoped that future research efforts will utilize and build on the standardized baiting procedure detailed here in order to more effectively compare white shark populations in South Australia and elsewhere.

The South Australian Department of Fisheries is continuing to tag white sharks throughout the region. Given the large number of sharks now tagged, we hope to identify the primary geographic boundaries of this population. Do individuals generally restrict their movements to the Spencer and St. Vincent gulfs and the adjacent coastline, or, over time, do they distribute more uniformly throughout southern Australia? Any such pattern will naturally influence the population's responses to local exploitation and will therefore affect management decisions. Acquisition of this knowledge may be expedited through the use of genetic analyses (e.g., of tissues collected via biopsy punches attached to pole spears) to compare allele frequencies of potential subpopulations.

The inshore islands of the Spencer Gulf also have historically sustained the majority of the mortality resulting from sport fishing, apparently because the sites are most accessible (see Fig. 1). Since this area is presently inhabited by the female segment of the population, females may have been disproportionately exploited. In 1989, legislation was passed to restrict chumming at Dangerous Reef. We recommend that chumming be restricted throughout the Sir Joseph Banks Group to provide limited protection for locally concentrated female white sharks. Further, we hope that any such restrictions will include special dispensations for educational endeavors, because documentaries and ecotourism contribute to the appreciation of white sharks as living resources. Dangerous Reef and the Neptune Islands have long been renowned for hosting the greatest numbers of white sharks in Australian waters. Even if no appreciable drops are indicated in future surveys at these sites, it must be recognized that high local densities might be maintained through density-dependent recruitment

into these areas, which act as major sinks for the population. Thus, future population assessments should include areas well outside the Spencer Gulf region.

Summary

Movements and population dynamics of white sharks C. carcharias were studied during five expeditions to lower Spencer Gulf, South Australia. We used a variety of observational techniques, including standardized baiting, conventional tagging, and ultrasonic telemetry. Sixty-seven individuals were attracted, ranging from 220 to 550 cm TL ($\bar{X} = 360 \pm 70$ cm). Among 40 individuals tagged, we resighted 23, the most frequent of which was seen on 23 different days over a 580-day period. Sharks were segregated by sex: females were most abundant at "inshore" islands during winter, whereas males predominated at "offshore" islands during summer. Tracked sharks demonstrated what we call downstream circling, island patrolling, and/or inter-island cruising; sharks swam either at the surface or along the bottom, but not in midwater. An index of abundance yielded a rate of 0.035 sharks arriving per hour of baiting. Using a Jolly–Seber analysis, the population size at Dangerous Reef was estimated to be 191.7 (95% CI, 36.5–1612.2) and 18 (95% CI, 3.9–157.6) individuals during our second and third expeditions, respectively. The probability of survival between samples, as predicted by the model, was low (0.20).

Acknowledgments

We are grateful to the members and employees of the Cousteau Society for financial and logistical support. We are especially indebted to J.-M. Cousteau and the dedicated crew of *Alcyone*: C. Jouet-Pastre, N. Durassoff, M. Deloire, C. Davis, P. Martin, S. Arrington, P. Allioux, M. Blessington, D. Brown, J. Cramer, B. Gicquel, P. Pique, L. Prezlin, M. Richards, A. Rousset, T. Stern, C. Van Alphen, M. Westgate, C. Wilcox, and Y. Zlotnicka. Superb logistical and technical support was provided by J. Brody, Y. Goldman, L. Sullivan, and the rest of the Cousteau team. We are likewise indebted to the South Australian Department of Fisheries for administrative support and for making *Ngerin* and her excellent crew available for our fifth expedition: N. Chigwiddin, K. Branden, B. Davies, C. Fooks, D. Kerr, D. Short, and N. Wigan as well as G. Dubas and I. Gordon. A. Ebeling, R. Fox, B. Talbot, R. Taylor, and V. Taylor contributed their unique talents and much inspiration to this effort. Sincere thanks go to the National Parks and Wildlife Service of South Australia for permission to conduct research in park waters and to the following for equipment loans and donations: New South Wales Fisheries Research Institute (tags), the California Department of Fish and Game (ultrasonic receiver), Ocean Technology Systems (underwater communications), J.B.L. Enterprises (tagging spears), and Pelican Products (waterproof cases). Finally, we deeply appreciate the help of G. Wright, L. Botsford, and the editors of this volume.

An Evaluation of Methodologies to Study the Population Biology of White Sharks

GREGOR M. CAILLIET[1]
Moss Landing Marine Laboratories
Moss Landing, California

Commentary

Methodology for Studying Populations

The eight speakers in this plenary session on population biology covered five general areas of the world's oceans and came up with at least eight methods of studying populations and several problems associated with these. I suggest that we need to use as many of these approaches as many times as possible in each place, and exchange these ideas so that the approaches that one team uses, such as the tagging and sighting of marked individuals or identification on the basis of dorsal fin shape, are used in as many places as possible.

The first approach, a qualitative and observational one, has been taken in Europe and the Mediterranean (see Chapter 30, by Fergusson). It is certainly a start, the research in other areas having already been through that phase.

The second method used is to look at catches over time. In many places, it is impossible to use fishery catch because these data are not reported accurately. It is certainly an approach worth following, however,

and, of course, has been used in the United States (Casey and Pratt, 1985; Klimley, 1985b).

The third approach is to use catch per unit of effort. This is better than just catch alone because it standardizes effort. This method has been used in South Africa (see Cliff *et al.*, 1989; see also Chapters 32 and 36, by Cliff *et al.*). A subset of that would be fishery observations, which, again, as Ian Fergusson has pointed out, are being used in Europe and the Mediterranean.

The fourth approach, which I am really happy to see being used, because I think it is more powerful than a lot of us think in spite of shaky assumptions, is tag and recapture. I distinguish between tagging and marking. With marking, we do not know who the individuals are, but with tagging we do. What I would stress here is that those of you who are using the technique in South Africa and Australia (see Chapter 35, by Ferreira and Ferreira; Chapter 36, by Cliff *et al.*; and Chapter 37, by Strong *et al.*) use as many of these estimates as possible. From my experience, mainly in exercises for population biology class, the one assumption that the the Jolly–Seber method does not have to worry about is emigration and immigration, because we take into account the first and last times that animals in the population are resighted. The big assumption is that the population is "open," not "closed." I would encourage those who use the Peterson estimate to try Jolly–Seber estimate,

---

[1]Rapporteur Cailliet's comments summarize results presented during talks on population biology given in a plenary session during the Symposium on the Biology of the White Shark, held March 4–7, 1993, in Bodega Bay, California.

and vice versa. Conforming to the assumptions and knowledge of the extent of the area one is covering are both essential.

The fifth approach is field identification on the basis of natural markings, which, in a way, is the same as the method of tag–capture. The problems, of course, are that (1) there may be more than one animal with the same kind of scar—apparently not a problem thus far—and (2) one may or may not always see the animals, even though they are present. That is a problem with any tag–capture methodology. I think this approach is certainly something to encourage and the one place where it has been done is here at the South Farallon Islands (see Chapter 33, by Klimley and Anderson).

The sixth technique, not yet used in the study of white shark populations, is the genetic approach. There is only one paper that I know of on the topic of the population genetics of sharks, and it was based on work in Australia (Lavery and Shaklee, 1989). On the other hand, I note that a start has been made in the study of white shark genetics by Martin *et al.* (Chapter 6) in this symposium. Those of you who do catch animals should get in contact with population geneticists. As such information builds up, it will be extremely interesting relative to the movements of individuals of the species, population mingling, and interindividual and interlocation differences in genetic composition.

The seventh approach is to study the frequency of shark attack, taking into account the relative abundance of prey. This technique has been used in regard to predation on pinnipeds at the South Farallon Islands (see Chapter 34, by Pyle *et al.*). No one has mentioned shark attacks on humans, and I am glad, because I do not think that the frequency of these attacks is an indication of anything, given that there is greater human use of waters now and there is not the same kind of increase in the rate of shark attacks on humans. Nevertheless, this approach is worth trying in certain places.

The eighth approach is a need to understand as much as we can about ontogenetic movements. We do not know nursery grounds, mating and courting grounds, and where birth occurs. For many bony fishes, we can assess abundance of eggs, larvae, and other stages of the life cycle, especially if they can be sampled randomly. We do not have any information of this type for white sharks.

Problems with Methodologies

There are some problems with these methods of studying white shark populations. First, the movements of the species are not well known, whether ontogenetic, periodic, or irregular. We just do not know the rates of emigration and immigration from areas, the paths from pupping areas to nursery grounds, or those from feeding areas to mating areas. This is a big problem. Second, we really do not have any way of estimating the rates of natural or fishing mortality. The more information we obtain on the size and age compositions of populations, the better we are going to be able to deal with this problem. Third is the need to obtain random samples. Can one ever randomly sample a white shark? I do not think so. But we have to do our best and not worry about this assumption—however, it is a problem. Finally, we have a necessity to learn what motivation leads white sharks to the areas where we see them. Without an understanding of whether it is mating, feeding, or courting, our observations will not be complete.

Interactions
with
Humans

White Shark Attacks in the Eastern Pacific Ocean: An Update and Analysis

JOHN E. McCOSKER
California Academy of Sciences
San Francisco, California

ROBERT N. LEA
Marine Resources Division
California Department of Fish and Game
Monterey, California, and
California Academy of Sciences
San Francisco, California

Introduction

Attacks by white sharks on humans in the eastern North Pacific Ocean have attracted considerable, if not undeserved, attention (McCosker, 1981; Lea, 1987; Klimley, 1994). In fact, in consideration of the large human population living between Mexico and Canada [38,320,000 in 1991 (Anonymous, 1992)], its proximity to the coast, and the increasing recreational use of nearshore waters made possible by modern sporting equipment, it is surprising that so few shark attacks on humans have occurred during the latter half of this century.

We recognize the difficulty in obtaining an accurate estimate of the time spent by humans in activities that would allow for shark–human interactions. However, we presume that few, if any, white shark attacks on humans in North America, other than those neither observed nor recovered, during the latter half of this century have been overlooked. The attention paid by the news media to shark attacks would guarantee reporting. The first recorded shark attack in California occurred in 1926, and the next incident was not reported until 1950. Since then, attacks have occurred on a nearly annual basis. It is also likely that Pacific coastal Native Americans, such as the

Chumash, Ohlone, Salinan, Pomo, Esselen, and coastal Miwoks, encountered sharks (Loeb, 1926; Gifford, 1939). Their fishing craft were not significantly different in configuration from many of today's kayaks. Loeb referred to a prayer said by the Pomo of central California prior to swimming to the offshore rocks when hunting seals. A similar litany is repeated today by many surfers upon entering the "shark triangle," which ranges from Año Nuevo Island to Bodega Head and out to the Farallon Islands.

In California, the majority of unprovoked shark attacks on humans involve *Carcharodon carcharias*, particularly north of Point Conception. All known shark attacks in California from 1926 to 1979 were reviewed by Miller and Collier (1981). Subsequently, Lea and Miller (1985) added records for California and Oregon, 1980–1984, and Collier (1992, 1993) provided additional descriptions of attacks. Ellis and McCosker (1991) listed eastern North Pacific white shark attacks between 1926 and 1991 and described several in detail (see Fig. 1).

In the eastern North Pacific since Lea and Miller's (1985) review, we have learned of three reports of attacks by sharks other than *C. carcharias*. One appeared to have been an unprovoked attack, the second was provoked, and there is inadequate evidence

WASHINGTON

OREGON

CALIFORNIA

PT. CONCEPTION

BAJA

GUADALUPE

FIGURE 1 Locations of confirmed unprovoked attacks by white sharks in the eastern North Pacific, 1926–1993. Those attacks in which the white shark is implicated but not demonstrated are identified by a question mark.

onace glauca. Obviously, this was a provoked attack and has no significance to this study. Finally, we discount the reported attack of a "mako shark" on John Mark Regan (Haldane, 1992) at San Onofre, Orange County, on November 29, 1992. The purported event lacks authenticity, and all attempts to verify the account have found nothing.

Methods

In this chapter, we (1) review recent attacks by white sharks, primarily in the eastern North Pacific, based on unpublished reports, newspaper articles, and interviews with victims and witnesses; (2) provide expanded discussions of disputed cases of attacks by white sharks; and (3) more extensively describe those incidents that are particularly instructive or that have been misrepresented in media reports. Enumeration of incidents prior to 1984 follows the Shark Attack File (SAF) of Schultz and Malin (1963), updated by the International Shark Attack File [ISAF of the American Elasmobranch Society (G. H. Burgess, personal communication)], and the report of Lea and Miller (1985). The majority of the eastern North Pacific attacks presented here can be ascribed to *C. carcharias.* Several may have been caused by *C. carcharias*, but, as we note, confirming evidence is inadequate (California incidents 2 and 55) and we have so designated those with a question mark. In two cases (California incidents 28/29 and 54/55), it appears that two victims were successively attacked by the same shark; we treat each pair as single incidents. Finally, two California incidents (cases 1 and 8), described here, are not included in our analysis due to our uncertainty of the involvement of *C. carcharias.* Measurements are presented in metric units, but nearly all victims described their experience in terms of English feet/pounds. We either have not converted those measurements when quoting victims and witnesses or have done so and rounded them so as to avoid creating an incorrect impression of precision.

Results

Attacks in Chile, Panama, and El Salvador

Reported attacks by white sharks in the eastern Pacific, other than those in waters of North America, are few, and the evidence is often unreliable. Attacks on three divers in Chile, reliably attributable to *C. carcharias* were reported by Egaña and McCosker

to believe that the third actually occurred. We shall treat these briefly. Robert Rodriguez was attacked on May 26, 1985, while surfing at Huntington Beach, Orange County, California. He sustained minor injuries caused by an undetermined species of shark. Larry Stroup, an amateur scuba diver, was attacked at Santa Barbara Island, California, on September 3, 1989, while hand-feeding squid to blue sharks *Pri-*

(1984). A fourth Chilean diver was attacked and killed by a white shark near Valparaíso in 1988 (Eduardo Reyes, cited in Ellis and McCosker, 1991). An attack also occurred on March 23, 1994, on two swimmers alongside a vessel in the open sea 590 km east of Easter Island. Photographs clearly identify the attacking shark as *C. carcharias* (G. H. Burgess, unpublished data).

The 1943 death of a swimmer, resulting from a shark attack in the Gulf of Panama, was reported by Kean (1944). On the basis of the tips of two teeth removed from the victim, J. T. Nichols of the American Museum of Natural History identified them as from "a small so-called maneater shark *C. carcharias*, and from a small individual of this species probably not more than 7 feet or so long." Unfortunately, attempts by Jim Atz and Norma Feinberg to discover the existence of those teeth were unsuccessful. We seriously doubt that *C. carcharias* was involved, for several reasons. First, the Gulf of Panama is an unlikely locality for *C. carcharias*. Despite Meek and Hildebrand's (1923) inclusion in their review of marine fishes in Panama, they admit that it "is not recorded from either coast of Panama." Second, we are unaware of extant specimens or valid records of *C. carcharias* from Pacific Panama. Finally, due to ontogenic changes in diet, the length of the attacker is much too short to be among the white sharks involved in attacks on humans or other mammalian prey (Tricas and McCosker, 1984; McCosker, 1985). We find it more likely that the attacker was a species of *Carcharhinus*.

During August and September 1993, newspapers reported attacks by "great white sharks" on oyster divers and surfers in the vicinity of the Bay of Fonseca, El Salvador. Extensive conversations by one of us (J.E.M.) with Jaime Ruiz, the manager of the fishing cooperative, disclosed that the attacks occurred in estuaries and at river mouths, a habitat more typical of bull sharks *Carcharhinus leucas* than of white sharks. Lacking additional evidence, we doubt the involvement of white sharks.

Attacks at Guadalupe Island, Mexico

White sharks are known from the waters near the offshore Mexican islands of Cedros, San Benito, and Guadalupe, and are occasionally seen and/or captured along the outer coast of the Baja California peninsula. *Carcharodon carcharias* is also known, although rare, in the vicinity of the Gulf of California (Kato, 1965; Galvan-Magana *et al.*, 1989). Although it is possible that attacks on humans, particularly swimmers and sport and commercial fishermen and divers, may

have occurred and were not widely reported, we are aware of only two documented attacks along the west coast of Mexico, both at Guadalupe Island (Collier, 1992).

Isla de Guadalupe is 300 km southwest of San Diego, California. The island has long been a refuge for populations of three species of pinnipeds: the northern elephant seal *Mirounga angustirostris*, the California sea lion *Zalophus californianus*, and the Guadalupe fur seal *Arctocephalus townsendi*. White shark sightings, as well as pinnipeds bearing shark attack scars, are not uncommon at Guadalupe. Anecdotal reports exist of white shark encounters with commercial abalone divers and sport fishermen and divers, particularly along the southwest end of the island (see Chapter 19, by Collier *et al.*).

California Attacks, 1985–1993

The following records, summarized in Table I, update the reports of Miller and Collier (1981), Lea and Miller (1985), and Collier (1964, 1992, 1993). In addition, on the basis of new evidence, we reexamine several shark attacks that occurred before 1984.

1?. Norman Peixotto (SAF 215), swimmer; July 8, 1926; Alameda County, San Francisco Bay. Miller and Collier (1981, p. 101) reported that "[a] swimmer and his dog were swimming off Bay Farm Island [in Alameda] when a 5 ft shark first attacked the dog, then the swimmer. Injuries were serious and the victim was rescued with help from a boat launched from shore." We are unable to identify this shark with certainty. It is possible that the attacker was a sevengill shark *Notorynchus cepedianus*, but the unprovoked nature of the attack makes this unlikely. Herald (1968) mentioned the aggressiveness of *Notorynchus*, but recognized that attacks were in response to provocation. The purported length seems too short and the location of the attack (well within San Francisco Bay) seems unlikely for *C. carcharias*. We cannot reasonably assign such an attack to a shark other than *N. cepedianus* or *C. carcharias*. The incompleteness of the report and the considerable difference (e.g., in ship traffic, water quality, and faunal composition) in the environment of San Francisco Bay at that time makes the attack by either species a possibility.

2?. R. Campbell (SAF 733), swimmer; October 8, 1950; San Diego County, Imperial Beach. Miller and Collier (1981, p. 101) reported that the "[v]ictim was one of four swimmers (body surfing) treading water in 12 ft of water. Shark bit victim on leg then attacked a second time from in front. Victim kicked at shark on second attack and was apparently bitten on thigh;

TABLE I Confirmed and Possible Attacks on Humans by *Carcharodon carcharias* in the Eastern Pacific

No.	Date	Name	Sex	Location	Activity	Distance to shore (m)[a]	Water depth (m)	Attack depth (m)[a]	Time (hours)	Kayak/surfboard Color	Kayak/surfboard Length (cm)	Injury	ISAF no.	Reference no.[b]
Washington														
1	April 12, 1989	Robert Harms	M	Pacific Beach	SU	100	3	S	1045			Minor bite on arm	2490	1
Oregon														
1	August 24, 1976	Mike Shook	M	Winchester Bay	SU		2.5	S				Only to board	1883	2
2	November 27, 1979	Kenny Doudt	M	Cannon Beach	SU		2.5	S	1030			Major	1903	2
3	October 27, 1980	Christopher Cowan	M	Off Umpqua River	SU	150	3	S	P.M.	Orange	220	Minor	1918	3
4	October 20, 1983	Randy Weldon	M	Cape Kiwanda	SU	75	2.5–5	S	1000	Off-white[c]	183	Only to board	1968	3
5	September 30, 1984	Robert Rice	M	Cape Kiwanda	SU	15–25	3–5	S	1530	Yellow–green	193	Only to board	1983	3
6	October 23, 1988	Wyndham Kapan	M	Cannon Beach	SU	100	5	S	1730	Yellow	203	Bitten on leg	1997	1
7	February 24, 1991	Tony Franciscone	M	Neskowin	SU		3	S				Only to board	2155	1
8	March 8, 1992	Mike Allman	M	Winchester Bay	SU		3	S	1030			Shoulder bitten, board broken	2107	1
9	September 13, 1992	Jerad Brittain	M	Gold Beach	SU		3	S	1700			Uninjured	2438	1
10	January 2, 1993	Bill Weaver	M	Bastendorff Beach	SU		2.5	S				Uninjured	2470	1
California														
1	July 8, 1926	Norman Peixotto	M	San Francisco Bay	SW		3	S				Serious	215	2
2	October 8, 1950	R. Campbell	M	Imperial Beach	SW		4	S	1200			Bitten on leg and thigh	733	2
3	December 7, 1952	Barry Wilson	M	Pacific Grove	SW		6–9	S	1400			Fatal	236	4
4	February 6, 1955	James Jacobs	M	Pacific Grove	SK		6	S	0900			Minor	240	5
5	August 14, 1956	Douglas Clarke	M	Pismo Beach	SW		1	S	1630			Bitten on thigh, sides, and hand	246	2
6	April 28, 1959	Peter Savino	M	Morro Bay	SW		3	S				Body not recovered	249	2
7	May 7, 1959	Albert Kogler, Jr.	M	Bakers Beach	SW		5	S	1745			Fatal	372	6
8	June 14, 1959	Robert Pamperin	M	La Jolla Cove	SK	100	10	S	1710			Body not recovered	37	1,2

9	October 4, 1959	James Hay	M	Bodega Bay	SK	5	S	1500	Bitten on swim fin	554	2
10	April 24, 1960	Frank Gilbert	M	Tomales Point	SK	6	S	1400	Major	683	7
11	May 19, 1960	Suzanne Theriot	F	Aptos	SW	2.5	S	1300	Bitten on leg	686	2
12	May 21, 1961	Rodney Orr	M	Tomales Point	SK	8	NB	0930	Uninjured, wet suit bitten	2092	1, 2
13	August 20, 1961	David Vogensen	M	Salmon Creek	SW	5	S	1530	Bitten on foot and groin	917	2
14	January 14, 1962	Floyd Pair, Jr.	M	Farallon Islands	SC/SP	10	S	1030	Major leg wound	1001	2
15	November 11, 1962	Leroy French	M	Farallon Islands	SC/SP	30	S	1345	Major bites to thigh and leg	1115	2
16	January 11, 1964	Jack Rochette	M	Farallon Islands	SC/SP	15	S	1200	Major bites to thigh and leg	1247	9
17	January 22, 1966	Donald Barthman	M	Cypress Point	SK	18–20	S	1000	Bitten on arm and thigh	1398	2
18	July 27, 1968	Frank Logan	M	Bodega Rock	SK	6	NB	1100	Major bites to midbody	1569	2
19	July 20, 1969	R. Colby	M	Pigeon Point	SK	9	S	1300	Bitten on foot	2042	2
20	September 6, 1969	Donald Joslin	M	Tomales Point	SK	8	NS	1120	Major bites on leg	1647	2
21	October 2, 1971	Calvin Ward	M	Sea Ranch	SC	5	5	1200	Bitten on legs	2093	2
22	May 28, 1972	Helmut Himmrich	M	Tomales Point	SK	3–4	S	1330	Bitten on legs and buttock	1474	7
23	July 19, 1972	Kenneth Gray	M	Point Purisima	H	6	B		Major	1778	2
24	September 9, 1972	Hans Kretschmer	M	Point Sur	SU	10	S	1000	Bitten on legs and board	2043	2
25	May 26, 1974	Leroy Hancock	M	Tomales Point	SK	3	S	1130	Bitten on leg	1780	2
26	July 26, 1974	Robert Kehl	M	Albion	SK	12–16	6–10		Minor bite on foot	1725	2
27	August 5, 1974	R. Sanders	M	San Gregorio Beach	SU	5	S		Surfboard and hand bitten	2094	2
28	September 2, 1974	Dale Webster	M	Franklin Point	SC	10–13	S	1730	Minor bite on foot	1726	2
29	September 2, 1974	Jack Greenlaw	M	Franklin Point	SC	10–13	S	1730	Minor injury to hand	1726	2
30	September 14, 1974	Jon Holcomb	M	Farallon Islands	H	7–10	NB	1335	Major	1727	2
31	September 28, 1974	Kirk Johnston	M	North of Point Sur	SU	3	S	0720	Major injuries to thigh, lower abdomen, and surfboard	1784	2

(continues)

TABLE I (Continued)

No.	Date	Name	Sex	Location	Activity	Distance to shore (m)[a]	Water depth (m)	Attack depth (m)[a]	Time (hours)	Kayak/surfboard			ISAF no.	Reference no.[b]
										Color	Length (cm)	Injury		
32	July 19, 1975	Gary Johnson	M	Point Conception	H		7–9	B	1330			Swim fin bitten	2095	2, 10
33	July 23, 1975	Robert Rebstock	M	Point Conception	SC		9	S	1430			Major	1739	2, 10
34	August 9, 1975	Gilbert Brown	M	Bear Harbor	SK		7–9	S	1330			Major bite on arm	1786	2
35	December 6, 1975	Robin Buckley	M	Farallon Islands	SC/SP	100	12	5	1200			Bitten on leg	1745	9
36	October 18, 1976	William Kennedy	M	Moonstone Beach	SU		2.5–3	S	1230			Bitten on leg	1787	2
37	December 18, 1976	Jack Worrell	M	San Miguel Island	H	100	11–13	5–6	0900			Bitten on buttocks	1788	9
38	August 14, 1977	Glen Friedman	M	Tomales Point	SK		11	NS	1300			Major injuries to lower body	1790	2
39	May 11, 1979	Calvin Sloan	M	Año Nuevo Island	SC		10	6	1000			Bitten on swim fin	2097	2
40	October 17, 1980	Curt Vikan	M	Moonstone Beach	SU	100	2–3	S	0930			Uninjured	2203	3
41	December 19, 1981	Lewis Boren	M	Spanish Bay	SU			S	P.M.	Yellow	163	Fatal	1864	3
42	February 7, 1982	D. Harvey Smith	M	Stillwater Cove	SC	400	125	13	1100			Bitten on calf	1950	3
43	July 24, 1982	Casimir Pulaski	M	Point Buchon	P	1600	45	S	1100	Yellow	357	Uninjured, board was bitten	2204	3
44	August 29, 1982	John Buchanon	M	Morro Bay	SU		2–3	S	0930	Red	185	Uninjured, board was bitten	2233	3
45	September 19, 1982	Michael Herder	M	Bear Harbor	SK	5	4–5	NB	1430			Bitten on thigh	1960	3
46	September 14, 1984	Omar Conger	M	Pigeon Point	SK	150	5	S	0830			Fatal	1978	3
47	September 30, 1984	Paul Parsons	M	Tomales Point	SK		9	NS	1000			Major bites on legs and buttocks	1979	3
48	February 18, 1985	Chris Massahos	M	San Miguel Island	SC	14		S	1245			Minor bruises	2615	9
49	September 28, 1985	Rolf Ridge	M	Elephant Rock	SK		5	4				Uninjured	2616	1

50	December 6, 1986	Frank Gallo	M	Monastery Beach	SK	100	21	NS				Major wounds to neck, shoulder, and forearm	2032	1
51	August 15, 1987	Craig Rogers	M	Tunitas Beach	SU		3	S		0730		Hand and surfboard	2031	1
52	April 24, 1988	Mark Rudy	M	North of Morro Rock	SU		3	S				Uninjured	2335	1
53	August 11, 1988	Carl Lafazio	M	Crescent City	SU		3	S				Bitten on thigh	2007	1
54	January 26, 1989	Tamara McAllister	F	North of Malibu	K			S		See text		Fatal	1999	1
55	January 26, 1989	Roy Jeffrey Stoddard	M	North of Malibu	K			S		See text		Body not recovered	1999	1
56	September 9, 1989	Mark Tisserand	M	Farallon Islands	H	70	10	8–9	1230			Wounds to leg	2048	1, 9
57	January 12, 1990	Sean Sullivan	M	Montara Beach	SU	250	3	S			255	Uninjured	2337	1
58	August 28, 1990	Rodney Swan	M	North of Trinidad Head	SU	50	3	S	1650	White	200	Minor	2293	1
59	September 5, 1990	Matt Hinton	M	Trinidad State Beach	K			S	1700		245	Uninjured	2294	1
60	September 8, 1990	Rodney Orr	M	Jenner	SK	150		S	1300			Minor bites	1824	1
61	November 3, 1990	Eloise Tavares	F	Monastery Beach	SC	150	19	S	1500			Major bite on leg	2154	1
62	July 1, 1991	Eric Larsen	M	Davenport	SU	150	3–5	S	0830	White	213	Bitten on legs and arms	2106	1
63	October 5, 1991	John Ferreira	M	Davenport	SU	150	3	S	0815			Bitten on arms and back	2105	1
64	December 4, 1991	David Abernathy	M	Shelter Cove	H			S				Minor	2617	1
65	October 29, 1992	Andy Schupe	M	San Miguel Island	H		20	20	1100			Bitten on foot	2618	1
66	November 14, 1992	Ken Kelton	M	Año Nuevo Island	K	150	5	S	1250	Red	350	Kayak bitten	2441	1
67	March 12, 1993	Don Berry	M	Pedro Point	SK	125	10	5–6	1430			Bitten on swim fin	2504	1
68	August 11, 1993	David R. Miles	M	Westport Union Landing	SK	70	8	D	1440			Major bites to head and shoulders	2533	1
69	October 10, 1993	Rosemary Johnson	F	Jenner	K	1500		S	1400	Dark blue	275	Kayak bitten	2537	1

(continues)

TABLE I (*Continued*)

No.	Date	Name	Sex	Location	Activity	Distance to shore (m)[a]	Water depth (m)	Attack depth (m)[a]	Time (hours)	Kayak/surfboard Color	Length (cm)	Injury	ISAF no.	Reference no.[b]
Guadalupe Island, Baja California, Mexico														
1	September 9, 1973	Al Schneppershoff	M	Caleta Melpomene	SK/SP	135	12	S	1645			Fatal Major bite to leg	2044	10
2	September 11, 1984	Harry Ingram	M	Caleta Melpomene	SK/SP	300		S	1750			Minor, diver was not bitten	2205	10
Chile														
1	September 29, 1963	Crisologo Urizar G.	M	El Panul	SK/SP	50		S	1100			Fatal, consumption assumed	2330	11
2	January 5, 1980	José Larenas-Miranda	M	Pichidangui	H	20		U	1100			Fatal, largely consumed	2331	11
3	March 4, 1981	Carlos Veraga M.	M	Bahia Totoralillo	SK/SP	50		1.5	1030			Minor, bitten on foot	2332	11
4	December 15, 1988	Juan Tapia-Avalos	M	Valparaiso	SK							Fatal	2333	12

Activity: H, hookah diver; K, kayaker; P, paddleboarder; SC, scuba diver; SK, skin diver; SP, spear fisherman; SU, surfer; SW, swimmer; attack depth: B, bottom; D, descending; NB, near bottom; NS, near surface; S, surface; U, underwater. ISAF, International Shark Attack File.

[a] We caution readers with regard to the accuracy of the "distances from shore" and "depths of water."

[b] References: (1) This study; (2) Miller and Collier, 1981; (3) Lea and Miller, 1985; (4) Bolin, 1954; (5) Fast, 1955; (6) Gilbert *et al.*, 1960; (7) Follett, 1974; (8) Collier, 1964; (9) Collier, 1993; (10) Collier, 1992; (11) Egaña and McCosker, 1984; (12) Ellis and McCosker, 1991.

[c] With black and purple stripes.

wounds were major. Aid to victim was immediate; a nearby lifeguard placed the victim on his board and brought him ashore. The shark was reported to be 'large.'" The description of the attack seems much like that of *C. carcharias*. Lacking additional evidence, such as a tooth fragment, and considering the possibility that it could have involved another, possibly tropical shark species, we are cautious in identifying the attacker.

8?. Robert Pamperin (SAF 376), skin diver; June 14, 1959; San Diego County, La Jolla Cove. This incident, presumably resulting in fatality, is shrouded in mystery. The date most often cited, January 14, is incorrect. The lack of a corpse has caused several investigators to question whether a shark attack had actually occurred. Presuming that it was a shark attack, the attacking species remains uncertain. Miller and Collier (1981, p. 101) related that "[t]he abalone freediver was swimming at the surface in 30 ft of water when a large shark, most probably a tiger shark, grabbed the victim from behind or lower side. The initial (probably only) attack raised the victim partially out of the water; the shark then submerged and was seen with victim held between the jaws. The body was not recovered." The incident and/or its aftermath was witnessed from the beach by a number of bystanders, several of whom we have interviewed. The size of the shark ("20 feet"), as estimated by Pamperin's dive partner, Gerald Lehrer, allows for only two possible attacking species: the white shark or the tiger shark *Galeocerdo cuvier*. White sharks are more likely to occur within California, whereas tiger sharks are known from only a few records off this coast. However, evidence for *G. cuvier* as the attacker includes interviews conducted by the late Conrad Limbaugh (R. H. Rosenblatt, personal communication) with Lehrer, who described the long tail and small teeth of the shark. In addition, the attack occurred during the end of the 1957–1959 El Niño (Chelton *et al.*, 1982). The marked intensity of that event resulted in dramatic shifts in the temperature and direction of the otherwise southward-flowing California Current, as well as the northward occurrence of numerous southern species (Radovich, 1959, 1961). Lacking additional evidence, we suspect that the identity of the attacker may never be determined.

California White Shark Attacks Subsequent to the Review by Lea and Miller (1985)

48. Chris Massahos, scuba diver at the surface; February 18, 1985; Santa Barbara County, San Miguel Island, between Point Bennett and Castle Rock. An extensive report has been provided by Collier (1993).

49. Rolf Ridge, free diving and intending to spear fish in waters >5 m deep; September 28, 1985; Marin County, Elephant Rock, near Tomales Point. Water visibility was described as "15 feet." As Ridge, an adult white male, descended, he was bumped on his left hip and turned to see a "12–15 foot long" shark pass by his side. He thrust his meter-long speargun at the shark's mouth and swam backward to shore. The shark swam slowly past the diver and did not contact him again.

50. Frank Gallo, free diver; December 6, 1986; Monterey County, Carmel Bay, Monastery Beach. Gallo, an adult white male, had been scuba diving in the giant kelp bed about 100 m off the north end of Monastery Beach. After completing his dive, he removed his tank and attached it and his surf mat to the kelp canopy. He then free dove with his speargun in pursuit of fish. He was struck in the upper right side while nearing the surface upon returning from a descent, and suffered massive wounds to his neck, shoulder, forearm, and chest. He was taken by his dive partner to shore, where he received medical attention. Gallo, a paramedic, was in extremely good physical condition and, in part, orchestrated his own rescue and emergency medical treatment, factors probably contributing to his survival. He received emergency medical treatment by the Carmel Highlands Fire Department at the beach within 10 minutes of the attack and was transported to the Community Hospital of the Monterey Peninsula, arriving <1 hour after the attack. After recovery, Gallo continued to dive and compete in spear-fishing competitions for many years thereafter.

51. Craig Rogers, surfer; August 15, 1987; San Mateo County, Tunitas Beach. Rogers, an adult white male, was sitting on a surfboard at 0730 hours; a shark bit the board and cut his hand, requiring six stitches to two fingers. Rogers surfed to shore.

52. Mark Rudy, surfer; April 24, 1988; San Luis Obispo County, just north of Morro Rock. Rudy was uninjured following his attack.

53. Carl Lafazio, surfer; August 11, 1988; Del Norte County, Crescent City. Lafazio, an adult white male, was surfing near the mouth of the Klamath River when attacked by a shark described as "8–10 ft long." "Shark bit once, swam away, and apparently was coming back for a second bite just as Lafazio made it to shore with the help of his friends." His right thigh required 27 stitches.

54 and 55?. Tamara McAllister and Roy Jeffrey Stoddard, kayakers; January 26, 1989; Los Angeles County, north of Malibu. McAllister, an adult white

female, and Stoddard, an adult white male, were kayaking from Paradise Cove, Malibu, and left at 0700 hours on January 26. They had planned an outing of approximately 5 km total distance, which would have required about 3 hours' duration; we therefore assumed that they were attacked in the morning, between 0700 and 1000 hours. Their two ocean kayaks (one white, the other yellow, each about 15 feet 4 in. in length) were discovered the following day, tied together, about 80 km northwest of Malibu. The underside of one of the kayaks bore major abrasions and three large holes, suggesting the impact of a biting *C. carcharias*. McAllister's body was found floating at the surface on January 28, roughly 8 km from shore and 84 km northwest of their entry site. She had several bruises on her hands and several shark bites, including a massive 33-cm gouge removed from her left thigh. Her lungs were without water, suggesting that she had not drowned before she was bitten, and the damage to her hands would indicate that she had fought with her attacker. Stoddard was not recovered. We consider McAllister's death to be the result of exsanguination following attack by *C. carcharias*. Although we are unable to positively explain the cause of Stoddard's death, we assume it to be directly related to the attack on McAllister. For the purposes of this study, we consider this incident to be a single event.

56. Mark Tisserand, hookah diver; September 9, 1989; Southeast Farallon Island. An extensive report has been provided by Collier (1993).

57. Sean Sullivan, surfer; January 12, 1990; San Mateo County, Montara Beach. Sullivan, an adult white male, was surfing approximately 250 m offshore along a sandy beach, due north of a rocky headland. The day was cold and rainy, and the wind was offshore. Sullivan was lying on the surfboard (255-cm triple-fin white fiberglass), facing the ocean, beyond the offshore sandbar and the surf impact zone. A "large" shark, estimated (but not seen before the attack) by Sullivan to have been 3–4 m in length, struck and bit the board, knocking Sullivan into the water. Uninjured, he got back onto his board and paddled in, without seeing the shark again. The board was examined by the beach lifeguard, who noted tooth penetrations, but found neither teeth nor tooth fragments. From the description, it is most likely that the attacker was *C. carcharias*.

58. Rodney Swan, surfer; August 28, 1990; Humboldt County, north of Trinidad Head. An adult white male, Swan was 50 m offshore on a 200-cm white surfboard at 1650 hours, wearing a full black wet suit. The water was said to be exceptionally warm. The board was hit from beneath and somewhat left of center, then the shark pivoted around the board and took two quick bites at the right rear, slightly wounding Swan. The board was then briefly pulled underwater and released.

59. Matt Hinton, kayaker; September 5, 1990; Humboldt County, Trinidad State Beach. Alone in a 245-cm kayak off Trinidad State Beach at about 1700 hours, Hinton was attacked from below and overturned. He saw an "8–10 ft white shark, hit it with [his] paddle, and it swam away."

60. Rodney Orr, spear fisherman on a surfboard; September 8, 1990; Sonoma County, Jenner. An adult white male, Orr was free diving and spear fishing at Russian Gulch at 1300 hours. He dove to spear a lingcod *Ophiodon elongatus*, when he was bitten on the head and neck. He hit the shark with his hands and the speargun, was carried by the shark, and then was released. He then paddled ashore. Serious injuries had resulted to his face and neck.

61. Eloise Tavares, scuba diver surface swimming; November 3, 1990; Monterey County, Carmel Bay, Monastery Beach. Tavares, an adult white female, at 1500 hours was swimming on the sea surface to the dive site as part of a scuba-diving class. Her dive buddy was 1 m away, and they were 150 m off the beach. She felt a "bang" to her leg but thought it was her dive buddy. Upon a second hit, she turned, expecting to see her buddy, but saw nothing. She then felt pressure on her leg and yelled, "I'm stuck!" She then felt that her leg was in a "trap" and became concerned about freeing herself, not knowing how she was "stuck." Turning to grab her leg, she realized that "the trap I thought my leg was in came out of the water—it was a shark and was about a foot away from my face. . . . I think the shark still had my foot in its mouth and I think it either bit deeper or was releasing my leg because I felt more pain and not just pressure." She screamed and yelled, among other things, "Shark!" Her buddy, now 10 m away, did not hear her but did see the shark. At this point, Tavares apparently blacked out and is not sure of what happened next. She and her buddy tried signaling for help and then started swimming to the nearest kelp bed. She was missing her left swim fin (from the bitten leg). Several divers came to Tavares' aid and swam her through the surf. She was treated on the beach by paramedics and then taken by ambulance to Community Hospital of the Monterey Peninsula, where she was held overnight and treated for major injuries, including nerve damage. The white shark was estimated at "12–13 ft."

62. Eric Larsen, surfer; July 1, 1991; Santa Cruz County, Davenport Landing (about 12 km south of Año Nuevo Island). An adult white male, Larsen was

sitting on a white board at 0830 hours, roughly 150 m offshore. He was bitten on his left leg and left and right arms. The shark then released him, and he paddled to shore. His injuries were major, requiring 400 stitches.

63. John Ferreira, surfer; October 5, 1991; Santa Cruz County, Horseshoe Reef off Davenport. Ferreira, an adult white male, was among 12 surfers, at 0815 hours, "150 yds" offshore from Scott Creek. He was bitten twice by a "12 ft shark" in the arm and back. A tourniquet was applied by other surfers, whereupon he was taken to shore and received further treatment. Injuries were major.

64. David Abernathy, hookah urchin diver; December 4, 1991; Mendocino County, Shelter Cove, north of Fort Bragg. Abernathy was at the surface when a shark became tangled in his hose; it dragged and then released him. He was uninjured, although his swim fin and the diving hose were bitten.

65. Andy Schupe, hookah diver; October 29, 1992; San Miguel Island, Castle Rock. Hookah diving for urchins, Schupe, an adult white male, was on a jumbled rock and sparse kelp stalk bottom, 20 m deep, at 1100 hours. The sky was clear but water visibility was murky, at about 6 m visibility. Down about 6 minutes, he felt a tug on his foot and thought it was caused by a seal. He then realized he had been bitten and observed a shark surge beneath him. He recalled it, swimming away, to be "7–8 ft long with a really dark gray back." He estimated the shark to have had a "18–24 inch girth." The bite, through his neoprene bootie, along the top and side of his left foot, required 26 stitches. His Scubapro Jetfin received a cut 7 cm long with a triangular tooth mark on the underside.

66. Ken Kelton, kayaker; November 14, 1992; San Mateo County, Año Nuevo Island. Kelton, an adult white male, was alone in a 350-cm red Dancer river kayak. His paddle blades were white plastic. He approached Steel Reef, north of Año Nuevo Island and about 150 m offshore over a 5-m-deep channel. The water was clear (estimated 2.5 m visibility), the air was warm, the sky had a high overcast, the surf was 0.5–1 m, and the tide was extremely high. At 1250 hours, Kelton felt but did not see the initial impact as a shark bit the hindportion and then thrust the kayak out of the water. After 5–10 seconds of thrashing, the bow was lifted up, and the shark then released it. The event, after the initial attack, was observed by Kelton's partner, kayaker Michael Chin, approximately 8 m away. Chin's description of the size and coloration of the shark is befitting of *C. carcharias*, and Chin stated that although not observing the initial contact, he did observe the shark to place "his jaws on the boat and then lunge horizontally out of the water

with the boat clamped in his jaws, his belly skimming the water surface. I saw a dark dorsal fin, and the profile of the shark's immense bulk, dark on top turning to silver on the sides and fading to white on the belly. From the top of the fin straight down to the belly spanned about 3'" (M. Chin, personal communication with J. E. McCosker, December 8, 1992). Kelton, uninjured, then paddled the damaged kayak to shore, and the shark was not seen again. J. E. McCosker and Ken Goldman examined the kayak and interviewed Kelton on November 20, 1992. The widest bite measurement was 37 cm, indicating that the kayak was taken well into the shark's mouth. The dental pattern and the tooth penetration were appropriate for *C. carcharias*, but no teeth or fragments remained. It was difficult to estimate the size of the shark from the bite pattern.

67. Don Berry, free-diving spear fisherman; March 12, 1993; San Mateo County, Pedro Point, Linda Mar Beach. Berry, an adult white male, was spear fishing without scuba, wearing a black neoprene wet suit with blue sleeves, and free diving from a 3.3-m paddleboard about 125 m from the beach. The board was anchored in water 10 m deep. The day was cloudy but warm, the water was typically murky, the current was strong, and the tide was incoming. Berry had been in the water for nearly 1 hour and had speared a cabezon *Scorpaenichthys marmoratus* and placed it on his board. At 1430 hours, he dived to the bottom and was returning to the surface. At 5–6 m beneath the surface, he felt something grab his left swim fin (black Dacor) from behind and then push him forward a short distance through the water. With his speargun in his right hand, he pushed the shark with his left hand and was then released. He was able to observe a large shark that he described as "a cream color, not white, which then changed over into a gun-metal grey." Uninjured, he swam to his board, paddled farther from shore to warn others, and then returned to shore. The punctures and scratches of the fin were examined by Ken Goldman and J. E. McCosker, and Berry was interviewed by them on March 15, 1993. The evidence was not conclusive, but indicated that the damage was most likely the result of a *C. carcharias* bite.

68. David R. Miles, skin diver; August 11, 1993; Mendocino County, Westport Union Landing, 30 km north of Fort Bragg. Miles, an adult white male, was wearing a black neoprene suit and free diving alone for abalone at 1440 hours, about 300 m offshore above an 8-m rocky bottom. Sky and water conditions were clear. While descending, he was bitten on the head and shoulders, then released by the shark. He then swam to an offshore rock. He was aided by other

divers to shore, where he was airlifted to a hospital and treated for major injuries.

69. Rosemary Johnson, kayaker; October 10, 1993; Sonoma County, Jenner. Johnson, an adult white female, was kayaking near Arched Rock on a warm overcast day. Sea conditions were smooth. She and three men paddled approximately 1.5 km offshore and 1.5 km south of the river inlet at Russian Gulch. The others were in an 335-cm-long white Scrambler, a 275-cm turquoise Kiwi, and a 335-cm white gondola-shaped plastic ocean kayak. Johnson was in a 275-cm-long dark blue Frenzy kayak, on the oceanside of the arch. At about 1400 hours, two of the others were about 200 m away and saw a "huge, white shark" lift her kayak into the air. Their reference to *white* concerns its ventral surface coloration. Johnson thought she had hit a rock, was confused, and fell out. Two of the others rapidly paddled toward her and tried to return her to her kayak; it had a large hole and she could not stay in it. She boarded the largest kayak and was paddled in. No one saw the shark again. One of the men thought the shark to be "15–18 ft" in length; another thought it was larger. Johnson suffered no injuries.

Oregon Attacks

Attacks by white sharks in Oregon between 1976 and 1984 have been described by Miller and Collier (1981), Lea and Miller (1985), Griffith (1993), and S. Cook (unpublished manuscript). Lacking adequate data, we do not include the attack in 1951 on a swimmer ("OSU student" in Cook's unpublished manuscript) near Florence among probable *C. carcharias* victims. Attacks on Oregon surfers are summarized in Table I (Nos. 1–5).

6. Wyndham Kapan, surfer; October 23, 1988; Clatsop County, Cannon Beach. An adult white male, Kapan was bitten on the leg, requiring 21 stitches.

7. Tony Franciscone, surfer; February 24, 1991; Tillamook County, Neskowin. As Franciscone paddled for a wave, a shark came from the side and bit the rear of the board, catching the surf leash and pulled Franciscone into the water. No injury occurred.

8. Mike Allman, surfer; March 8, 1992; Douglas County, Winchester Bay, south of the jetty off the Umpqua River. Allman, an adult white male, was surfing at 1030 hours when a 3.5-m shark struck him and his board, breaking the board. He was bitten on the shoulder and pulled underwater, then released from the shark's mouth. Upon surfacing, he recalls that "the shark was sitting on the top of the water in front of me." Still tethered to his surfboard, he paddled to his surfing partner, who helped him to the beach. He was taken by ambulance to Coos Bay, where he was treated for cuts to his left shoulder and a 5-cm gash on his left side. Allman estimated the shark to be 4–5 m in length. Injuries were major.

9. Jerad Brittain, surfer; September 13, 1992; Curry County, Gold Beach, mouth of the Rogue River, 20 m south of the south jetty. At about 1700 hours, a large shark attempted to bite the rear portion of Brittain's surfboard and became tangled in the leash. Pulled from his board and into the water, he tried to remain above water and shove the shark away as the shark twisted with the leash in its mouth. The leash broke. Brittain's brother, on a surfboard, came to his aid and paddled ashore with Brittain hanging onto the side of the board. Brittain was not injured.

10. Bill Weaver, surfer; January 2, 1993; Coos County, Bastendorff Beach, at Coos Bay, between Winchester and Simpson's reefs. A shark bit the nose of Weaver's surfboard; he slid off the back without injury.

Washington Attacks

1. Robert Harms, surfer; April 12, 1989; Grays Harbor County, Pacific Beach. Harms was paddling on his purple surfboard at 1045 hours, about 100 m from shore, when he felt a sharp pain to his left arm. He resisted and pulled his arm away, and then saw a swirl next to his board and what looked like a "large, grayish fish." After several seconds of disorientation, he caught a wave to shore. On shore, he removed his wet suit and observed blood covering his arm and a number of lacerations and superficial puncture wounds extending from the base of his thumb to about 10 cm up his arm. He and his surfing partner felt that the wounds were not serious and treated them with antiseptic and adhesive bandages. During this period, Harms was noted as pale and probably suffering from minor shock. No additional medical treatment was received.

Discussion

We begin by recalling Baldridge's (1974b) careful clarification of the obvious—that the location and timing of shark attacks on humans are strongly and directly correlated with human behavior.

Unprovoked attacks on humans by *C. carcharias* in the eastern North Pacific appear to be similar to those in Chilean waters (Egaña and McCosker, 1984), South Africa (Cliff, 1991; Levine, Chapter 40), Australia (Hughes, 1987; West, Chapter 41), and elsewhere

(Ellis and McCosker, 1991). Recent reports from Japan (Nakaya, 1993) and Italy (Testi, 1993; Fergusson, Chapter 30) have also implicated *C. carcharias*. Whereas white sharks are found worldwide in temperate coastal waters, attacks are typically centered where pinnipeds are abundant, such as nearshore rocky outcrops and islands and often near the entrances of rivers. Adult *C. carcharias* appear to inhabit those locations where pinnipeds are abundant and available as prey, and recreational and commercial water users are thereby at greater risk at those locations.

Although the localities in the eastern North Pacific appear to be no more dangerous than elsewhere in the world, more than half of all known attacks by *C. carcharias* have occurred in California and Oregon. As stated elsewhere in this chapter, we attribute this to a higher number of recreational and commercial water users rather than to a higher number of potential attackers. Burgess and Callahan (Chapter 42) report on 138 unprovoked attacks by *C. carcharias* and found that (subsequent attacks and minor differences in our records modify these percentages only slightly): 51% occurred on the U.S. Pacific Coast, 21% in Australia, and 11.3% in South Africa; 28.5% resulted in fatality. West (Chapter 41) reports on 34 attacks in Australia from 1791 to 1993 and found that 11 were swimmers, 9 were surfers, 4 were spear fishermen, and 3 were skin divers; 21 of these attacks resulted in fatality. Levine (Chapter 40) describes 58 attacks in South Africa between 1922 and 1991 and found that 17 were surfers and bodyboarders, 15 were spear fishermen, and 20 were swimmers; 14 of these attacks resulted in fatality. By comparison, the 9% of attacks that were fatal in the eastern North Pacific is small, compared to 80% in Chile, 62% in Australia, and 24% in South Africa. We attribute these differences to the nearly universal usage in the United States of the "buddy system," whereby a victim is aided soon after attack and taken ashore, where rapid transportation and expert medical attention have been generally available.

We are familiar with 27 unprovoked attacks on humans by *C. carcharias* since the review by Lea and Miller (1985). Summarizing these subsequent attacks, they involved 24 males and 3 females; there were 13 surfers, 4 kayakers, 3 hookah divers, 2 scuba divers, and 5 free divers. One attack was in southern California, with 20 in northern California, 5 in Oregon, and 1 in Washington; 25 (93%) were at the surface and 2 (7%) were beneath the surface; 1 (4%) was fatal, 11 (41%) resulted in significant but not fatal injury, and 15 (56%) resulted in minor wounds, if any. Only 1 (4%) individual saw the shark before the attack; nearly all victims (except kayakers) wore black wet suits; all victims were adults (as opposed to children); all attacks occurred during daylight; all attacks in Oregon and Washington were on surfers; and attacks occurred in all months except June.

We conclude the following in regard to eastern North Pacific attacks since 1950. Attack is usually at the surface (Tables I and II). Two individuals were attacked while ascending or descending, and 5 of 24 breath-hold divers were attacked beneath the surface or near the bottom. Six of 7 hookah divers were at the bottom or descending; 1 was at the surface. Eight of 12 scuba divers were at the surface, exhibiting behavior similar to that of a free diver when attacked; the other 4 were near the bottom or at middepths.

Attacks by white sharks in the eastern North Pacific have occurred during all months except June (with the possible exception of R. Pamperin) (Table III). We found no correlation between timing and latitude or

TABLE II Activities of Humans When Attacked by White Sharks in the Eastern North Pacific

	Swimming	Surface diving[a]	Surfing	Hookah diving	Scuba diving	Kayaking
Washington			1			
Oregon			10			
Central and northern California[b]	5	29	15	6	4	3
Southern California						1
Guadalupe Island		2				
Total (N = 76)[c]	5	31	26	6	4	4
Fatalities (N = 7)	3	2	1			1

[a]Breath-hold, hookah, and scuba divers who were attacked while at the surface.
[b]Includes attacks at San Miguel Island.
[c]Incidents 28/29 and 54/55 are each counted as 1.

TABLE III Monthly Attacks by White Sharks in the Eastern North Pacific

	California and Baja	Oregon and Washington	Total[a]
January	5	1	6
February	3	1	4
March	2	1	3
April	3	1	4
May	5	0	5
June	0	0	0
July	8	0	8
August	10	2	12
September	14	2	16
October	7	2	9
November	3	1	4
December	6	0	6

[a]Incidents 28/29 and 54/55 were each counted as 1.

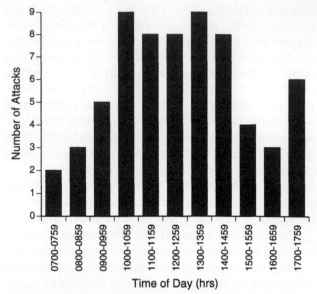

FIGURE 2 The known time of attack (Pacific Standard Time corrected for daylight saving in North America) by white sharks in the eastern Pacific Ocean (N = 69).

other obvious factors, and concluded that many factors associated with human behavior could skew any such relationship. For example, the timing of the sport abalone season (variable within a state and in the past), commercial urchin fishery season, school vacations, and sea and air conditions (generally favoring warmer temperatures) that enhance surfing and swimming should be taken into consideration. We therefore find it curious that there are no confirmed recorded attacks during June.

The frequency of white shark attacks in the eastern Pacific (and elsewhere; cf. Ellis and McCosker, 1991) upon men and women is significantly different (82 males versus 4 females); however, lacking the necessary data concerned with comparable time spent at risk by the population of each sex, we are unable to draw any conclusions. Known white shark attacks have occurred only during daylight hours and are more prevalent at midday (Fig. 2). Again, we lack comparable data for the time of day spent at risk by the human population to allow a significant conclusion to be drawn.

The behavior of humans, when attacked, has changed somewhat during this century, as human recreational activity has shifted. The history of attacks in California before 1972 indicates that swimmers and surface-swimming abalone divers were the most common attack victims, but this is no longer the case. Since the attack on Hans Kretschmer, while surfing at Point Sur in 1972, surfers, paddleboarders, and bellyboarders have become the most common victims. [The first recorded attack by a white shark on a surfer occurred not in California, but at Makaha, Oahu, Hawaii, in 1969 (Balazs and Kam, 1981).] Tricas and McCosker (1984) described the history of shark attacks

upon surfers in California and suggested that "since the early 1970s, the trend in surfboard design has been toward an increase in flotation, reduction in board length, multiple posterior fixed rudders ('skegs'), and bifurcated 'v' tails. All of these modifications have enhanced the similarity between the silhouette of a surfer and that of a pinniped, and [they] suggest this may increase the possibility of attack." McCosker (1985) further hypothesized that the problem had been limited to the northern coast of California and Oregon. In part, this may be explained by the fact that surfing in southern California is confined mostly to long expanses of sandy beaches, well away from pinniped aggregations and rookeries. Surfing along the northern California and Oregon coastlines, and now along Washington state, British Columbia, and southeastern Alaska, is often conducted in the vicinity of rocky headlands and river mouths, where pinnipeds congregate.

The apparent discrepancy in the frequency of attack by C. carcharias in northern and southern California (Table IV and Fig. 1) may be explained by differences in geography, human behavior, and legislation. As mentioned above, surfing in southern California is confined mostly to sandy beaches, which are unlikely habitats for pinnipeds and the sharks that feed on them. In addition, sport diving for abalone in northern California is limited by legislation to breath-hold diving, and recreational abalone fishermen are allowed to use scuba only south of Yankee Point, Monterey County. Breath-hold divers spend the majority

TABLE IV Country, State, and County of Attacks by White Sharks in the Eastern North Pacific

Location	No.[a]	Locality	No.
United States		California (continued)	
Washington		San Francisco	1
Grays Harbor	1	Farallon Islands	6
		San Mateo	9
Oregon		Santa Cruz	3
Clastsop	2	Monterey	8
Tillamook	3	San Luis Obispo	5
Lincoln	0	Santa Barbara	3
Lane	0	San Miguel Island	3
Douglas	3	Ventura	0
Coos	1	Los Angeles	1
Curry	1	Orange	0
		San Diego	1 (?)
California			
Del Norte	1	**Mexico**	
Humboldt	4	Baja California	
Mendocino	5	Guadalupe Island	2
Sonoma	7		
Marin	8		

[a]Not included are the attacks on Peixotto in 1926 and Pamperin in 1959. Attack incidents 28/29 and 54/55 are each counted as 1.

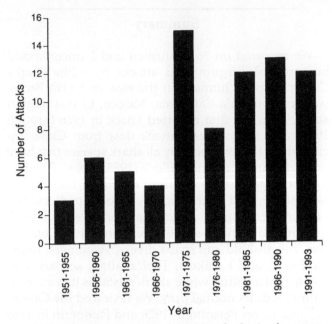

FIGURE 3 Confirmed attacks by white sharks in the eastern North Pacific, recorded in 5-year intervals (except 1991–1993).

of their dive time at the surface, and thereby increase the likelihood that they would be mistaken for pinnipeds by *C. carcharias*. Other than the two kayakers attacked somewhere off Malibu in 1989, confirmed attacks by white sharks on recreational water users have not yet occurred in southern California. There is a likelihood that surface-swimming divers and/or kayakers in the vicinity of the offshore islands (the Farallones, Año Nuevo Island, and the Channel Islands) are at higher risk of attack by white sharks.

An examination of the trend in attacks suggests that since the early 1970s, the frequency rose dramatically but then remained level for two decades (Fig. 3). It is likely that the number of attacks in the current decade will exceed previous levels. However, in consideration of the presumed increase in recreational water use in the Pacific Northwest, we are surprised that the attacks are so few and that the periodicity is relatively stable. Lacking accurate historical and recent data on human behavior patterns, we are unable to provide further quantitative analysis. We suggest, however, that numerous qualitative factors are involved, including, but not limited to, these: (1) humans have learned, with some exceptions, to avoid geographic locations where previous attacks have occurred, particularly the Farallon Islands, Tomales Point and Bird Rock (Marin County), as well as Año Nuevo Island; (2) educational programs have changed diver behavior, for example, avoidance of offshore exposed reefs and swimming in the vicinity

of pinniped rookeries; and (3) surface spear fishermen now are less likely to attach their catch to themselves.

It has been suggested that more attacks occur during the warm waters of El Niño. El Niño 1957–1959 included attacks by white sharks on three humans in northern California and a possible attack by a tiger shark in southern California. During El Niño 1982–1984, another period of especially warm water, four California attacks occurred, all in 1982. Although those numbers are higher than the annual average, we are unable to observe any significant correlation. Moreover, the most attacks in California during any single year occurred in 1974, a cold-water period.

Despite the accumulation of considerable data on attacks by white sharks in the eastern North Pacific, we find ourselves unable to explain with certainty many of the factors involved. Debate continues as to several issues: Why do white sharks attack humans? Is there a periodicity or repeatability to those attacks? Also, why are so few humans consumed after attack? With the recent passage of California legislation (Assembly Bill 522) to protect white sharks from targeted capture, an improved opportunity exists to observe their natural behavior as well as their interactions with humans. The potential for attack on humans by white sharks remains extremely low, and it is our hope that, by better understanding the behavior of *Carcharodon carcharias*, we can reduce this likelihood even further.

Summary

We reported on 76 confirmed and 2 unconfirmed but probable unprovoked attacks by white sharks *C. carcharias* on humans in the eastern North Pacific (waters from Baja California, Mexico, to Washington state), from the first reported attack in 1926 through 1993, and summarized attack data from Chile. Included are the 29 attacks by all shark species that have occurred since Lea and Miller's (1985) report. The majority of attacks have occurred at or near the surface, near shore, and in the vicinity of pinniped colonies and/or river mouths. Attacks have occurred during all months, with the possible exception of June, on surfers, breath-hold and scuba divers, swimmers, hookah divers, and kayakers. Typical attack scenarios indicate that an adult white shark mistakes its victim for a pinniped, its normal prey. We reviewed the California attacks on Peixotto in 1926 and Pamperin in 1959 and question the involvement of *C. carcharias*. We also reviewed and commented on attack data from elsewhere in the eastern Pacific, although these data are scarce and often unreliable, and discounted the involvement of *C. carcharias* in attacks on humans in Panama and El Salvador.

Acknowledgments

Many individuals contributed to our study. In particular, we thank the victims and witnesses who provided us with invaluable information and advice. We thank many other individuals, including, but certainly not limited to, the following, for their assistance with this project: D. G. Ainley, S. D. Anderson, J. Atz, G. Balazs, D. H. Baldridge, K. Barsky, S. Barsky, G. H. Burgess, G. M. Cailliet, G. Cliff, R. S. Collier, S. Cook, H. Davidson, L. Dearborn, A. C. Egaña, R. E. Ellis, N. Feinberg, the late W. I. Follett, A. Giddings, K. J. Goldman, J. Griffith, J. Grissim, J. Hughes, A. P. Klimley, D. J. Miller, R. Mooi, W. A. Palsson, D. Perlman, A. Pont, P. Pyle, J. E. Randall, E. Reyes, A. I. Roest, R. H. Rosenblatt, J. Ruiz, J. Seigel, L. Taylor, D. Thompson, T. C. Tricas, R. Warner, and J. West.

40

Unprovoked Attacks by White Sharks Off the South African Coast

MARIE LEVINE

Shark Research Institute
Princeton, New Jersey

Introduction

White sharks *Carcharodon carcharias* range along the entire coast of southern Africa, from Namibia to KwaZulu–Natal and southern Mozambique, but the center of the population appears to be the coastal waters of the western Cape Province of South Africa (Compagno, 1984a).

White sharks have been implicated in attacks on humans in South Africa (Davies, 1961, 1964; Davies and D'Aubrey, 1961a,b; Wallett, 1973, 1983; Compagno, 1984a, 1987; Cliff, 1991; Levine, 1996), and elsewhere (e.g., Baldridge, 1973); including the northwestern Pacific (Collier, 1964, 1992, 1993; McCosker and Lea, Chapter 39; Miller and Collier, 1981), Australia (West, Chapter 41), and the Mediterranean (Fergusson, Chapter 30).

I attempted to assemble accurate information about shark attacks in southern Africa by collecting data on 297 unprovoked attacks and 116 incidents (provoked attacks, attacks on boats, and posthumous scavenging). Various species of sharks were involved, and information on unprovoked attacks was uncovered as far back as 1852. Attacks that had never been recorded in the scientific literature were investigated, and many additional details were revealed in previously documented attacks. Approximately 1200 people were interviewed; these included victims, witnesses, lifesavers, paramedics, physicians, medical examiners, commercial fishermen, and government officials. Information was supplied by the victim or eyewitnesses in all but 20 of the unprovoked attacks (all

species) in South Africa since 1944, and in 10 of the cases before 1944. However, the quality and amount of information regarding attacks prior to 1922 varied considerably; although a few of the attacks had been examined in detail by contemporary researchers, most were not. To prevent distortion by data of variable quality all pre-1922 attacks were excluded from this study. Of the 297 unprovoked attacks, 225 took place from 1922 to July 1994, a period of 72 years; 63 of these cases involved white sharks. This study is limited to these 63 cases.

Methods

A *shark attack* was defined as any incident in which a shark initiated aggressive behavior toward a human and in which physical contact occurred. In some cases divers were struck with force by the sharks, but their wetsuits protected them from abrasions; the divers used their spearguns as billys and/or the shark bit their fins, speared fish, or scuba tank, leaving the diver unharmed. Contacts with riders' boards have been included even when the persons were not bitten; contacts with boats were excluded. When the victims' behavior elicited aggression or the sharks were stressed as the result of shark-fishing activity or capture, the events were considered "provoked"; such incidents were not included.

In addition to the cases discussed here, another 40 attacks may have involved white sharks, but the reliability of eyewitnesses was questionable and/or other factors made white shark involvement uncertain. In the 63 valid cases, white shark involvement was determined through (A) recovery of white shark tooth fragments, (B) definitive bite patterns in the victim's body or equipment, (C) reliable eyewitness and/or victim's identification of species, and (D) recovery of body parts and/or bloodstained swimsuit of the victim from the shark's gut. Shark sizes are expressed as total length (TL).

Results

On the basis of the four criteria for inclusion of incidents in this study, sample sizes were as follows: A, 2 cases; B, 15; AB, 2; C, 17; AC, 4; BC, 19; CD, 2; and ABC, 2.

Locations

No unprovoked attacks by white sharks were recorded in the frigid waters of the northern Cape Province. Thirty-one (49.2%) of the 63 attacks took place in the Western Cape; 17 (27%), in the Eastern Cape (1, in the northern transitional zone formerly known as Transkei); and 15 (23.8%) in the subtropical seas of KwaZulu–Natal (Fig. 1).

Activity of the Victims

The 63 cases include 22 attacks on board riders (18 surfers, 3 bodyboarders, and 1 paddleskier), 21 on divers (16 spear fishermen, 3 skin divers, and 2 scuba divers), and 20 on swimmers (Fig. 2).

Time of Year and Water Temperature

Davies (1963) hypothesized that shark attacks were unlikely in water temperatures <21–22°C, because few people swam in chilly water, and those who did so remained in the sea for only a short period. In KwaZulu–Natal, the sea is coldest in winter: the inshore (<1 nautical mile from shore) sea-surface temperature (SST) averages 19°C. The summer SST averages 24°C, thus encouraging more people to enter the water and for longer periods than in winter. The number of shark attacks, Davies believed, was directly related to the number of swimmers in the sea. Although attacks have taken place in every month,

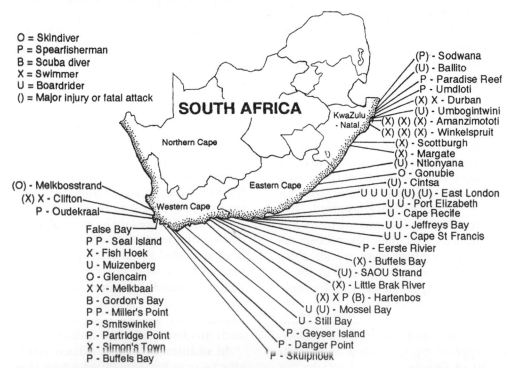

Locations of unprovoked white shark attacks on humans in South Africa

FIGURE 1 The South African coast, showing fatalities, sites, and activities of people bitten by white sharks in unprovoked attacks.

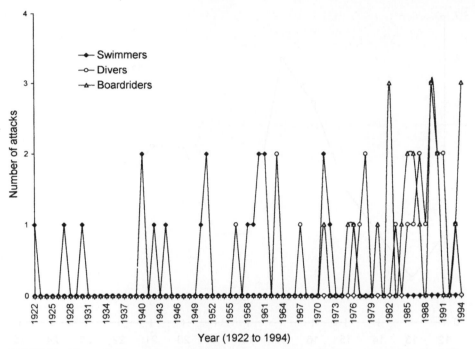

FIGURE 2 The activities of people attacked by white sharks along the coast of South Africa, 1922–1994. Note the shift from swimmers to board riders and divers.

the number of attacks seems to peak in midsummer, and of 27 attacks during midwinter (June–October), only one involved a swimmer (Fig. 3).

Wet suits, however, now permit divers and board riders to linger in water considerably cooler than 21°C. In the 41 instances in which the SST was measured or estimated, it ranged from 12°C to 26.1°C; 26 (65%) of the attacks took place in water <21°C (Fig. 4).

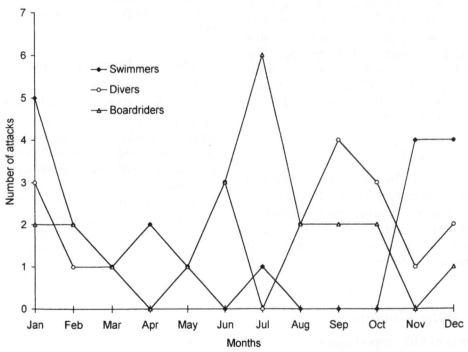

FIGURE 3 The months in which white shark attacks occurred in this study.

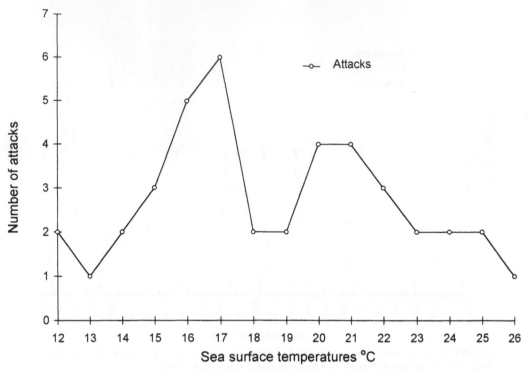

FIGURE 4   Sea-surface temperatures in 41 white shark attacks, when it was measured or estimated; 75% took place in water <21°C.

Time of Day

Anatomical studies indicate that white sharks are capable of color vision (Gruber and Cohen, 1985). Color vision has limited value to nocturnal species, but would be an asset to a species active by day, such as the white shark. Attacks have occurred from 0650 to 1935 hours; with one exception, all took place during daylight hours (see Chapter 21, by Strong). The attacks correspond to recreational use of the sea, with a peak in late morning, 1100–1200 hours. Attacks are spread throughout the afternoon, peaking in the late afternoon, 1530–1659 hours (Fig. 5).

Color

Of the 21 divers attacked by white sharks, 16 (76%) were wearing black wet suits. Eleven (61%) of the 18 surfers were using white boards; three surfboards and the paddleski were blue. However, most surfboards manufactured in South Africa are white (with blue a close second), and the majority of divers wear black wet suits.

Water Visibility

Water visibility is of little importance to swimmers or board riders. However, because good visibility is a requisite for divers and their face masks permit good vision underwater, divers are more likely to be aware of sharks in the area and are better equipped to take evasive or defensive actions when threatened: 86% of the divers sustained no injuries or minor injuries, compared with board riders (63%) and swimmers (35%).

Distance From Shore and Depth of Water

Among attacks on swimmers, 63% took place <50 m from shore (Fig. 6), and 43% occurred in water <2 m deep (Fig. 7). Among board riders, 60% clashed with the sharks >50 m from shore, and 90% of the attacks occurred in waters 2–5 m deep. Seventy-six percent of the divers encountered the sharks ≥100 m from shore. Of significance, however, is that only one incident involving divers (case 400) took place below the surface. Although their activity was described as "diving" when attacked, 20 (95.2%) of the 21 individuals in this category could be considered "swimmers" wearing face masks, although 7 (35%) of the 20 were carrying or towing bleeding fish.

Environmental Factors

In most cases, environmental factors that may have contributed to the attack were present. In 59

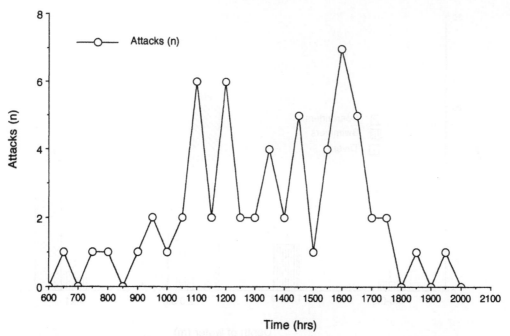

FIGURE 5 The time of day when attacks attributed to white sharks have occurred in this study.

cases, the following environmental factors were present: local rivers were in flood or sewage and/or effluent was present in at least 14; 14 attacks took place close to kelp beds, on a reef, adjacent to an estuary, or near an upwelling of cold water; 15 took place in the vicinity of a pinniped rookery or haul-out; and in 14, marine mammals were observed in the immediate vicinity close to the time of the attack. In 19 cases, shoals of fish were in the area or there was some fishing activity. The above are elements of habitats favorable to white sharks; some reflect a favorable transitory condition, but in the absence of control

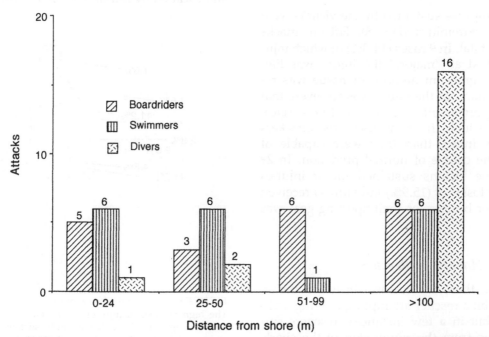

FIGURE 6 The distance from shore in 58 white shark attacks: 63% of attacks on swimmers ≤50 m from shore; 60% on boardriders ≥50 m from shore; 84% on divers ≥100 m from shore.

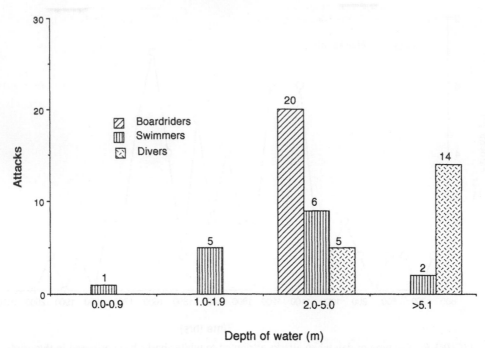

FIGURE 7 The depth of water at attack sites when known: 43% of attacks on swimmers occurred in water <2 m deep.

data (days in which the same conditions exist and no attack takes place), it is not possible to establish the degree of risk these factors represent.

Extent of Injuries

Most of the injuries sustained by the victims were on their body extremities (Fig. 8). Fifteen attacks (23.8%) proved fatal. In 9 cases (14.3%) in which injuries are described as "major," the injury was life-threatening, a significant amount of tissue was removed by the shark, or the injury was so severe that it resulted in permanent disability and/or surgical amputation of a limb. In most cases, the attackers caused far less injury than they were capable of delivering in the course of normal predation. In 29 cases (46%), the victims sustained minor injuries (with no tissue loss); 10 (15.9%) individuals received abrasions and/or bruises, or their sporting gear was bitten.

Hazard to Rescuers

In this study, there are no cases in which the attacker actually bit a rescuer attempting to bring a victim to shore, but in a few instances, rescuers sustained abrasions from the rough skin of the shark. Like many other predators, white sharks appear to

concentrate solely on the selected prey item. However, there was one instance in this study in which a white shark, deprived of its initial victim, attacked a second victim with increased aggression. The first victim (case 410), a surfer, managed to repel the shark and escaped by remounting his board and catching a

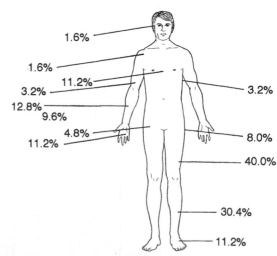

FIGURE 8 Areas of the body injured during unprovoked attacks. The high number of injuries on the hands, forearm, lower legs, and feet include defensive wounds sustained in repelling the sharks. Note that the total exceeds 100% because in some cases more than one body part was bitten.

wave inshore; seconds later, the shark attacked another surfer (case 411) and severed his leg. The same behavior was noted in an attack on March 23, 1994, off Easter Island in the Pacific Ocean; moments after a white shark was repelled by the first victim, it attacked a second victim and severed her leg (see Chapter 39, by McCosker and Lea). This behavior is not confined to white sharks. On February 13, 1974, in KwaZulu–Natal, a carcharinid shark lacerated the shin of a swimmer and was kicked away. A moment later, it bit another swimmer and removed a large amount of tissue, which led to surgical amputation of his lower leg (Wallett, 1983).

Characteristics of White Shark Predatory Attacks

In 43 (74%) of 58 cases in which the direction of approach is known, the shark approached from behind and/or below the victim (Fig. 9). In some cases in which the shark approached from behind or below, the victims suffered tissue loss. This suggests that these attacks may have been motivated by hunger, rather than curiosity.

In 23 cases (36.5%), it appeared that the shark may have intended to feed on the victim. In 5 of the cases, the shark submerged the victim and the body was not recovered. It was assumed that the victims drowned or were exsanguinated (Klimley, 1994) and were at least partly consumed by the shark (Table I).

TABLE I Attacks in Which the Body Was Not Recovered

Case no.	Disoriented by shark	Disabled by shark	Submerged by shark	No. of bites	Type of injury	TL (m)
067	Yes	Yes	Yes	2	Fatal	4.5
079	No	Yes	Yes	>2	Fatal	
097	Yes	Yes	Yes	2	Fatal	3.6
300	Unknown	Yes	Yes	>3	Fatal	2.4
378	Unknown	Yes	Yes	>1	Fatal	

TL, Total length.

In another 10 cases, the shark removed a considerable amount of tissue from the victim. All of these victims sustained a forceful initial bite that reduced their ability to escape, a strategy also recorded by Tricas and McCosker (1984) and McCosker (1985). All but 1 of the victims were either catapulted above the water and/or submerged by the shark. Five of the victims sustained multiple bites, and 6 died from their injuries (Table II).

There are 8 additional cases in which the behavior of the shark suggests an aborted feeding attempt; in all of these cases, the victims sustained an initial disabling bite. Six of the 8 victims were carried underwater by the shark, and 3 were struck with such force they were temporarily rendered incapable of resistance (Table III).

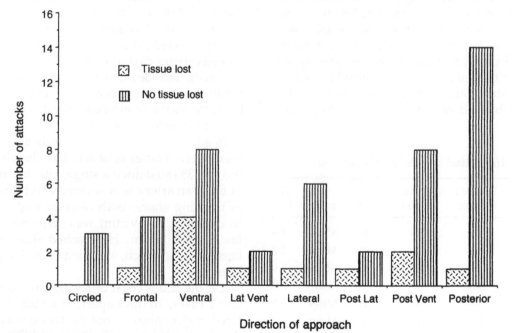

FIGURE 9 The direction of approach by the shark. In 42 cases, the shark approached from behind and/or below the victim. Lat Vent, Lateral ventral; Post Lat, posterior lateral; Post Vent, posterior ventral.

TABLE II **Attacks Resulting in Tissue Loss**

Case no.	Disoriented by shark	Disabled by shark	Submerged by shark	No. of bites	Type of injury	TL (m)
049	Yes	Yes	No	>1	Fatal	3.0
106	No	Yes	No	1	Major	2.8
122	Yes	Yes	No	3	Fatal	
137	Yes	Yes	Yes	3	Major	2.2
140	Yes	Yes	No	>1	Fatal	
208	No	Yes	Yes	>3	Fatal	
285	Yes	Yes	No	1	Major	
333	Yes	Yes	Yes	1	Major	
386	No	Yes	Yes	1	Fatal	4.5
411	Yes	Yes	No	1	Fatal	4.0

TL, Total length.

Discussion

Nature of Predatory Attacks

In 11 of the 23 cases listed in Tables I–III, the shark used its mass to ram and stun the victim. Despite its large size, a shark may weigh little when submerged, due to the buoyancy afforded by displacement of seawater and the oil held in its liver. McCosker (1985) observed that once a prey is sighted, the white shark rapidly ascends and at close range (i.e., less then one-half body length) begins one of several modal action feeding patterns. A transmitter placed on a white shark indicated that it had an average cruising speed of 3.2 km/hour (Carey *et al.*, 1982), but white sharks are capable of short bursts of speed in excess of this (my own unpublished observation; S. D. Anderson,

TABLE III **Attacks With No Tissue Lost**

Case no.	Disoriented by shark	Disabled by shark	Submerged by shark	No. of bites	Type of injury	TL (m)
043	No	Yes	Yes	1	Minor	3.6
210	No	Yes	Yes	1	Minor	2.9
254	No	Yes	Yes	1	Minor	4.0
272	Yes	Yes	Yes	1	Fatal	2.7
338	Yes	Yes	No	2	Major	2.5
373	No	Yes	Yes	1	Major	>3.0
381	No	Yes	Yes	1	Major	5.5
410	Yes	Yes	No	1	Minor	4.0

TL, Total length.

personal communication). The mass and speed provide momentum, which permit the shark to apply considerable force to prey, stunning or immobilizing it.

The amount of kinetic energy in a moving object is $(\frac{1}{2})MV^2$, where M represents mass and V indicates velocity). A shark of a mass M_s moving at a velocity V_s impacts the prey mass, M_p, and gives it a velocity V_p. In the simplest scenario, a shark strikes the ventral section of prey resting on the surface of the water. Kinetic energy is transferred to the prey. The prey has a velocity upward V_p, which is counteracted by the force of gravity ($g = 9.8$ m/sec), and the prey reaches a maximum height H above the water surface: $V_p = V_s \times (M_s/M_p)^{\frac{1}{2}}$.

Conservation of kinetic energy gives $M_p \times V_p{}^2 = M_s \times M_p{}^2$. For example, a shark with a mass of 700 kg, traveling at a speed of 3.2 kph, impacts prey that has a mass of 70 kg. When we know V_s, we can calculate V_p, as $V_p{}^2 = 10\ V_s{}^2$. Next, we can determine H, as $H = V_p{}^2/2g = 0.334V_s{}^2$ (Fig. 10). When the charge of the shark is halted by impact, its kinetic energy is transferred to the prey. And, when the prey is on the surface, the prey may be flung well above the surface of the water.

In at least 11 cases, the shark indeed moved upward at such velocity that, upon impact, the victim was flung out of the water. When this happens, the energy of impact is absorbed, to some degree, by the victim's body. The effect of such contact on human tissue varies considerably. The skin is pliable, strongly resistant to traction forces, and is unlikely to be damaged if the victim is wearing a wetsuit (the rough hide of the shark would, however, cause an abrasion on unprotected skin, as it did in case 140). The subcutaneous tissue cushions the effect of the impact and the elastic muscles usually escape damage, but air in the lungs may be violently compressed. At the very least, the victim is momentarily disoriented and incapable of effective resistance.

In addition to the 12 fatalities listed in Tables I–III, there were 3 other fatal attacks. The victims in cases 063 and 353 sustained a single bite on a flexed leg, but in each, an artery was severed and there was a delay in reaching shore; both died en route to a hospital. In case 062, the victim was swimming alone when his leg was bitten. He reached shore unaided, collapsed on the beach, and died in the hospital 23 hours later.

On the south coast of KwaZulu–Natal, there were also 2 cases, 16 days apart, in which the victims suffered major injuries, but no tissue was removed by the sharks. In case 143, Davis and D'Aubrey (1961b) reported that the shark made a series of "tentative" bites. The shark had ample opportunity to remove tissue from the 1.47-m victim, but it did not. Unfor-

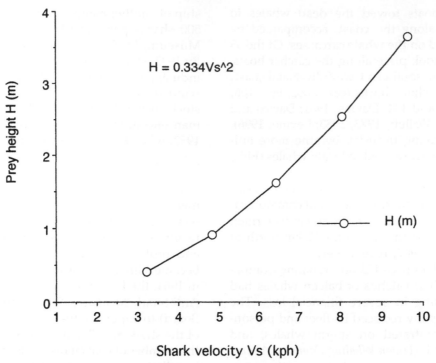

$$H = 0.334Vs^2$$

FIGURE 10 The height that a 700-kg shark may fling a 70-kg prey floating stationary, at first, on the surface.

tunately, the victim's foot was virtually severed by the shark, and subsequently amputated. In case 144, the shark made a single bite on the victim's flexed leg. Three days later, after an arterial graft failed, the leg was surgically amputated. In both of these cases, the sea was turbid and rivers were in flood; case 144 took place near a channel, and in case 143, the victim was standing on a rock on the rim of a channel. The attacks occurred 27 and 33 km south of Durban in 1961, and it may be worth noting that in 1961 sei whale abundance off Durban, as measured by both catch and sightings data, was at its highest in 14 years (*Union Whaling Company Annual Report*, 1962).

When examining white shark attacks in KwaZulu–Natal it is helpful to be aware of factors that impact on the nearshore marine environment: the annual sardine run, shore-based whaling (1908–1975), the shark-fishing industry (1928–1932), and shark nets (1952–1994). These factors are reviewed below.

Sardine Run

Each summer, vast shoals of sardines *Sardinops ocellata* appear in the northernmost sector of the eastern Cape, and they follow the cold inshore countercurrent northward. The sardines, in turn, provide fodder for hordes of predatory fish: *Pomatomus saltatrix, Scomber japonicus, Auxis thazard, Atractoscion aequidens, Thryssa vitrirostris, Scomberomorus plurilineatus, Trachurus capensis,* and *Lichia amia;* dolphins: *Tursiops ad-uncus, Delphinus delphis,* and *Sousa plumbea;* birds: *Sula capensis;* and sharks: *C. carcharias, Isurus oxyrinchus, Galeocerdo cuvier, Carcharias taurus, Carcharhinus leucas, C. obscurus, C. plumbeus, C. brachyurus, C. limbatus, C. brevipinna, Sphyrna mokarran, S. zygaena, S. lewini,* and *Rhizoprionodon acutus* (Wilson, 1985). The sardine shoals move into southern KwaZulu–Natal coastal waters around June or July. Often, pockets of 1 million fish or more run close inshore, where wind and current drive them into the surf zone. The Natal Sharks Board attempts to remove their nets before the shoals move close to the beach, but during the 1971 run, over 1000 sharks were removed from the nets in 10 days (Wallett, 1983). One of the earliest recorded shark attacks in KwaZulu–Natal occurred during the sardine run of 1897, when a young boy wading after stray fish was taken by a shark, species unknown.

Shore-Based Whaling

There were shore-based whaling stations in Kwa-Zulu–Natal from 1908 to 1975. In that 67 year period, a whale-processing plant was located 60 km south of Durban at Park Rynie, and as many as six whale-processing plants were operating at the same time on Durban's "Bluff," a large sand spit at the harbor entrance (Ellis, 1991). Of the 30 shark attacks (Davies, 1964) at Durban beaches between 1908 and 1975, 2 (cases 097 and 107) involved white sharks at the beach adjacent to the harbor entrance.

Whale-catcher boats towed the dead whales to processing plants along the coast, accompanied by large sharks that fed on the whale carcasses. Of the 33 shark attacks that took place along the catcher boats' inshore route up the southern KwaZulu–Natal coast, at least 8 involved white sharks (cases 062, 063, 106, 122, 137, 140, 143, and 144; Davies, 1961; Davies and D'Aubrey, 1961a,b; Wallett, 1973, 1983; Levine, 1996).

In 1954, the whaling industry became more efficient; spotter planes were used to locate whales (Ellis, 1991). In addition to the southern route, catcher boats were towing dead whales to Durban down the northern coast as well (K. Pinkerton, personal communication). Two white shark attacks on spear fishermen (cases 165 and 195) occurred 24 and 32 km north of Durban in 1963 and 1967, respectively.

In 1967, the Durban-based Union Whaling Company reported that their catches of baleen whales had plummeted, resulting in severe financial loss. The next year, the company reduced its fleet and personnel by 50%, concentrated on sperm whales, and turned a small profit. (*Union Whaling Company Annual Report*, 1968). Stocks of sperm whales remained unchanged, and whaling was profitable until 1975. As company policy, Union Whaling Company had decided that whaling would be abandoned before sustained financial losses occurred. The rise in fuel prices in the mid-1970s resulted in a decision to cease whaling in 1976.

In 1967, when the whalers recorded the decline in baleen whales, the shark anglers of Durban also noted a sudden decline in large white sharks. From 1947 to 1975, the Bowman Trophy was awarded annually for the heaviest shark landed by a shore-based fisherman in South Africa (Mara, 1985). Without exception, the trophy shark was caught with whale meat bait and landed 2 km from the whale-processing plant at Durban's South Pier. For 19 years, 1947–1966, all but one of the Bowman Trophy winners was a white shark (Fig. 11). During the next 8 years, 1967–1975, only one was a white shark, and when whaling ceased in 1975, the shark anglers caught no more trophy sharks off South Pier.

Shark Fishing Industry

Shark attacks were recorded in KwaZulu–Natal prior to establishment of the shore-based whaling industry in Durban in 1908; in 1907, the city erected a large bathing enclosure to protect bathers from sharks (Davies, 1963). By the time the enclosure was demolished in 1928, an intensive shark-fishing industry had developed alongside the whaling factories on Durban's Bluff. One company processed 6681 sharks in the first 10 months of operation (*Natal Fisheries Department Annual Report*, 1931). The floating factory

ship of another company was capable of dealing with 500 sharks per day (Archives of the Local History Museum, Durban; *Saturday Magazine*, July 15, 1939). By 1932, stocks of resident species were depleted, the industry collapsed, and the shark-fishing fleet was scuttled (*Natal Advertiser*, November 11, 1932). Shark stocks began to recover in the 1940s, and shark–human encounters increased off Durban beaches until 1952, when barrier nets were installed (Davies, 1964).

Shark Nets

During the 1957–1958 holiday season, no swimmers were attacked by sharks at Durban, but there were 8 shark attacks at beaches south of the city. As a result, a number of coastal resorts installed gill nets and protective barriers. In time, their maintenance became a financial burden to the local authorities, and in 1964, the Provincial Government created the Natal Anti-Shark Measures Board (now the Natal Sharks Board) to supervise the installation and maintenance of the shark nets (Wallett, 1973; Compagno, 1987). As the number of coastal resorts grew, the number of net installations increased, but the catch per unit of effort of white sharks declined between 1966 and 1990 (Cliff, 1991). From 1974 to 1988, annual catches of white sharks ranged from 22 to 61, or 2.7% of the total species caught in the nets (Cliff *et al.*, 1989). To date, however, there has been only one white shark attack (case 285) at a netted beach (Wallett, 1983). Although it is not possible to measure the degree to which shore-based whaling and the annual sardine run have contributed to shark attacks in KwaZulu–Natal, nor the degree that shark fisheries and shark nets have contributed to the reduction of shark attacks, it is probable that all have had significant roles.

In KwaZulu–Natal, the earliest attack in which white shark involvement could be confirmed took place in 1940. Between 1940 and 1975, there were at least 22 white shark attacks in South African waters: 12 (55%) were in KwaZulu–Natal and 10 (45%) in the Cape provinces. When whaling ceased in 1975, shark nets had already been installed at 39 beaches in KwaZulu–Natal (Wallett, 1983). With the cessation of shore-based whaling, statistics changed dramatically; from 1975 to 1994, there were 39 white shark attacks—only 3 (8%) were in KwaZulu–Natal, and the remainder (92%) were in the Cape provinces.

Attacks on Swimmers

Following the installation of gill nets and the cessation of whaling, white shark attacks on swimmers ceased in KwaZulu–Natal, although there were 7 unprovoked attacks by other shark species. The last white shark attack on a swimmer in Cape waters took

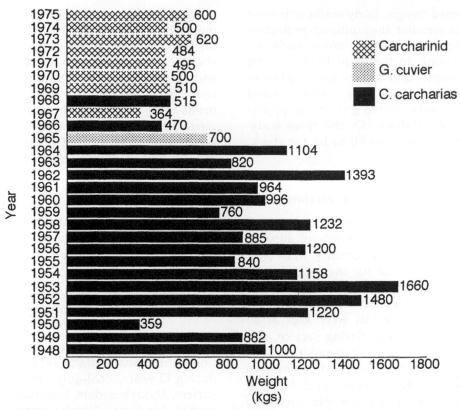

FIGURE 11 The size and species of the largest shark caught in the Bowman Trophy shark derby, by year. G. cuvier, *Galeocerdo cuvier*; C. carcharias, *Carcharodon carcharias*. All of these sharks were caught at Durban.

place in 1976, but since that time, there have been 12 attacks on swimmers by other species (Levine, 1996).

Attacks on Divers

Despite the increasing popularity of diving in the subtropical waters of KwaZulu–Natal, there have been no white shark attacks on divers since 1978. This suggests that there are fewer white sharks in the areas where diving activity has been taking place. In KwaZulu–Natal, diving takes place seaward of the shark nets, but there are no pinniped rookeries or haul-outs in the province. There have been 3 white shark attacks on divers in Natal; 2 occurred prior to 1975, and 1 took place in 1978. By comparison, there were 2 white shark attacks on divers in Cape waters prior to 1975, and 17 attacks from 1976 to 1994.

Compared to spear fishing, scuba diving was slow to develop as a sport in South Africa. By the mid-1960s, the SAUU was training divers, and by the late 1970s, international certification agencies (NAUI, PADI, and CMAS) were active in the country. Although there are no official statistics, Tim Condon (personal communication), publisher of South Africa's dive magazine *Underwater*, estimates that there are 100,000 active free divers and spear fishermen,

and 20,000–25,000 scuba divers. In this study, the ratio of attacks on free divers and spear fishermen to scuba divers is 19:2. However, all of the free divers and spear fishermen were on the surface when attacked. This was true, too, of one scuba diver, who suffered severe blood and tissue loss during the attack, and died soon afterward. The other scuba diver (case 400) was submerged and was not injured when the shark mouthed his tank. In 8 (38%) of the cases involving divers, the victim described the behavior of the shark as "investigatory." These include cases 355 and 370, in which a shark grabbed a diver's hand/forearm, towed him along the surface for a short distance, and then departed; both incidents involved a 2.5-m white shark, but they took place 856 days and 440 km apart.

Attacks on Board Riders

In the late 1960s, the surfing beaches of South Africa gained international fame, and as the numbers of surfers, bodyboarders, and windsurfers increased, so did shark attacks involving board riders. In part, the increase may be due to insulated clothing, which permits board riders (and divers) to remain in the water for extended periods and, in part, also due to the evolution of surfboard design. Early surfboards were long and cumbersome, but they offered protection from sharks because the surfer's entire body remained atop the board. In the early 1970s, shorter boards came into vogue, and surfers sustained more severe injuries from sharks. Body boarders fared even worse; their boards offered little or no protection whatsoever from a shark. Of the three bodyboarders in this study, one was killed by the shark, and two sustained major injuries.

Increased Survival Rates of Victims

Due to improved trauma care protocols, today's shark attack victims have a higher rate of survival than those attacked earlier in this century. In December 1957, a special trauma kit for the treatment of shark attack victims came into use in South Africa (Wallett, 1983). Known as the Feinberg Pack, after the doctor who devised it, these kits were kept at all beaches served by the Surf Life Saving Service, and lifeguards were trained in their use. The kit contains equipment to stop arterial bleeding, and intravenous fluids to prevent shock. Oxygen is also available at all Surf Life Saving stations. Prior to the introduction of the Feinberg Pack, 54.6% of the victims attacked by white sharks died of their injuries. From 1958 to 1994, only 17.3% of the attacks proved fatal.

Future Efforts

Of the 225 shark attack victims previously mentioned, 75% did not suffer any tissue loss; and in the 63 cases involving white sharks, 62% of victims sustained no such loss. This supports Baldridge's (1988a) hypothesis that in some cases hunger is not the causal factor in an attack. In virtually all cases, however, the victim immediately ceased his or her activity. The victims in this study were spearfishing, diving, surfing, swimming, floating, treading water, wading, or standing in the water. However, *we* have defined these activities, not the shark. We need to decipher what the shark perceives. If born blind, we cannot truly understand color; born deaf, we cannot experience a symphony; and born without an electric sense, we are ill equipped to comprehend the integral sensory information provided to the shark. However, we can grasp some general principles (Levine, 1994).

Shark attacks may have a devastating financial impact on communities that rely heavily on tourism. Because attacks are rare, it has been difficult to assemble a useful database to counter public hysteria (Gifford, 1993). A multidisciplinary approach to the subject is needed; attacks must be actively investigated and input must be assembled from forensic scientists,

medical practitioners, marine biologists, shark taxonomists, shark ethologists, and commercial fishermen. Victims and witnesses must be interviewed, environmental data gathered and assessed, the sequence and extent of injuries require interpretation by a forensic scientist, and the species of attacking shark needs to be identified. Through interdisciplinary cooperation, it may be possible to identify factors that predispose attack or that trigger attacks. Achieving insights into shark attacks is by no means the same as discovering effective means of preventing attacks, but it is one of the requisite conditions. However, we have learned some ways of lessening the risk of an attack: high-risk areas and seasons can be identified, and recreational use of the sea can be restricted when the risk of shark attack is highest.

Summary

White sharks *C. carcharias* were involved in 63 of 225 unprovoked shark attacks off the South African coast during 72 years, 1922–July 1994. Victims included 18 surfers, 3 bodyboarders, 1 paddleskier, 16 spear fishermen, 3 skin divers, 2 scuba divers, and 20 swimmers. Fifteen of the attacks were fatal, and 9 other victims suffered major injuries. In 29 cases, the victims were bitten, but sustained no tissue loss. In another 10 cases, the individuals sustained bruises and abrasions, or their sporting gear was damaged by the shark. By geographic area, most attacks took place in the temperate waters of the Western Cape (N = 31); others occurred off the Eastern Cape (N = 17) and in the subtropical seas of KwaZulu–Natal (N = 15).

Acknowledgments

I thank A. Gifford, L. J. V. Compagno, G. Charter, and A. Bowmaker for their support. I am also grateful for the assistance of D. G. Ainley, A. P. Klimley, G. Cliff, C. Martinez del Rio, M. Marks, M. Coutts, B. Levine, M. B. McMahan, and two anonymous reviewers. Much is owed to other investigators; without their efforts, much information would have been lost. After the name of each individual are listed cases to which they have contributed: M. Anderson-Reade, 355; G. Askew, 377; J. Bass, 210, 218; D. Cawston, 063; G. Charter, 246, 254, 258, 270, 272, 285, 333; G. Cliff, 316, 328, 338, 353, 367, 377, 378, 386, 387, 400, 401; T. Condon, 377; D. J. D'Aubrey, 143, 195; D. H. Davies, 137, 140, 143, 144, 165; B. Davis, 333; S. F. J. Dudley, 371, 372; A. Gifford, 378, 386, 405, 406, 408, 410, 411; A. Heydorn, 213; R. Horn, 301, 303; R. Joseph, 371, 372; P. Lansberg, 359; A. Munro, 373; Oceanographic Research Institute, 169; W. Pople, 270; H. Robson, 062; M. Smale, 353, 379; J. Stone, 381; P. van der Walt, 386; J. Wallace, 208, 258; T. S. Wallett, 137, 140, 143, 144, 208, 258, 272, 285; K. Watt, 338; and G. Wilson, 043. Most of all, I thank the shark attack victims and their families for their unanimous and continuing support and encouragement in this research.

APPENDIX 1 White Shark Attacks along the Coast of South Africa

Case no.	Name of victim	Sex	Age	Activity[a]	Date of attack	Location of attack	Time of attack (hrs)	SST[a] (°C)	Water visibility (m)	Injury	Location of bite	Method of species ident.[a]	Size of shark (m)
043	Pells	M	18	SW	May 6, 1922	Simon's Town, False Bay	A.M.	?	Clear	Minor	Abdomen, thigh	T, W	3.6
049	Heyns	M	17	SW	Dec. 28, 1927	Little Brak River	1100	Warm	Clear	Fatal	Leg	W, A	4.4
052	LeRoux	M	23	SW	Jan. 16, 1930	Melkbaai, False Bay	1730	?	?	Minor	Torso, arm	W, B	
062	Lund	M	17	SW	Feb. 22, 1940	Amanzimtoti	1630	23.8	Clear	Fatal	Thigh, knee	B	
063	Lees	M	25	SW	March 31, 1940	Winkelspruit	1100	?	Clear	Fatal	Calf	B	
067	Bergh	M	18	SW	Nov. 1, 1942	Clifton	1230	Cold	Clear	Fatal	Body not recovered	W	4.5
079	Schmidt	M	17	SW	Jul. 22, 1944	Hartenbos	1630	Warm	?	Fatal	Body not recovered	W	
097	Dumayne	M	14	SW	Feb. 11, 1950	Durban	1430	?	None	Fatal	Body not recovered	W	3.7
106	Plowman	M	24	SW	Nov. 28, 1951	Winkelspruit	1715	?	Clear	Major	Leg severed	W	2.4
107	Tait	M	23	SW	Nov. 29, 1951	Durban	1300	?	Turbid	Minor	Thigh abraded	W	1.8
114	Smith	M	29	SK	Oct. 27, 1956	Glencairn, False Bay	?	18*	Clear	Minor	Foot	W	
122	Prinsloo	M	42	SW	Jan. 9, 1958	Scottburgh	0730	?	None	Fatal	Multiple	W, A	**
129	Schreuder	M	18	SW	Jan. 15, 1959	Melkbaai, False Bay	1615	19*	Clear	Minor	Thigh	W	1.9
137	Hely	M	16	SW	April 30, 1960	Amanzimtoti	1535	23.0	<1.0	Major	Multiple	B	2.1
140	Sithole	M	25	SW	Dec. 24, 1960	Margate	1630	24.0	Poor	Fatal	Legs severed	T	3.0
143	Land	M	13	SW	Jan. 6, 1961	Winkelspruit	1935	24*	Turbid	Major	Legs	B	
144	Murphy	M	15	SW	Jan. 22, 1961	Amanzimtoti	1420	26.1	Turbid	Major	Thigh	B	
165	Passmore	M	24	SP	Jan. 6, 1963	Umdloti	1600	21.0	6.5	Minor	Arm	B	
169	Coetzee	M	?	SP	Dec. 20, 1963	Hartenbos	1200	?	6.0	Minor	Foot	W, B	1.7
195	Jones	M	32	SP	Mar. 19, 1967	Paradise Reef	0650	21*	7.0	Minor	Buttock/fish	W	3.0
208	Klein	M	>50	SW	April 11, 1971	Buffels Bay	1100	Cold	3.5	Fatal	Multiple	W, B	
210	Scheltema	M	21	SU	Jun. 30, 1971	Mossel Bay	1535	16.4	Clear	Minor	Leg/surfboard	T, W, B	2.9
213	Teague	F	16	SW	Dec. 16, 1971	Fish Hoek, False Bay	1445	19.5	Clear	Minor	Forearm	W, B	3.0
218	Brink	M	?	SW	Dec. 24, 1972	Hartenbos	0900	Cool	1.5	None	Swim fin	W, B	>3.0
246	Robertson	M	19	SU	Aug. 17, 1975	Cape St. Francis	1330	16*	Clear	Minor	Leg/surfboard	B	2.5
254	Flanagan	M	20	SU	Oct. 6, 1976	Cape St. Francis	1100	Cold	>3.0	Minor	Thigh/surfboard	T, B	3.5
258	Spence	M	19	SW	Nov. 27, 1976	Clifton	1605	17.0	30.0	Minor	Torso	W, B	3.0
263	Hartman	M	25	SP	Oct. 30, 1977	Partridge Point, False Bay	1600	?	5.0	None	Fish	W	5.0
270	Lombard	M	27	SP	Sep. 27, 1978	Miller's Point, False Bay	1600	Warm	6.0	None	Fish	W	3.7
272	Steenkamp	M	23	SP	Dec. 12, 1978	Sodwana	1645	cold	12.0	Fatal	Legs	T, W	2.3
285	Wright	M	24	B	Jan. 31, 1980	Ballito	1445	25.0	5.5	Major	Foot	B	2.8
300	Macun	M	27	SU	Jun. 29, 1982	Ntlonyana	0930	Warm	Clear	Fatal	Surfboard, No body	W, B	2.4
301	Plumb	M	38	P	July 24, 1982	Nahoon, East London	1600	16*	Turbid	None	Paddleski	T	2.4
303	Fawkes	M	16	SU	Sep. 30, 1982	Nahoon, East London	1450	17.0	3.0	None	Surfboard	B	2.8

(continues)

APPENDIX 1 (Continued)

Case no.	Name of victim	Sex	Age	Activity[a]	Date of attack	Location of attack	Time of attack (hrs)	SST[a] (°C)	Water visibility (m)	Injury	Location of bite	Method of species ident.[a]	Size of shark (m)
316	Leuw	M	29	SP	Aug. 20, 1983	Seal Island, False Bay	1000	12*	5.0	Minor	Thigh	W	5.0
328	Doran	M	27	SU	Jun. 15, 1984	Muizenberg, False Bay	1730	12*	Turbid	None	Surfboard	T, W	2.4
332	James	M	34	SP	Jan. 4, 1985	Buffels Bay, False Bay	1130	20*	10.0	Minor	Torso	W, B	3.5
333	Eldridge	M	18	SU	Jan. 17, 1985	Umbogintwini	1830	25*	1.0	Major	Thigh, calf	T, W	
338	Gee	M	22	B	Oct. 24, 1985	East London	0800	17.0	4.0	Major	Thigh, calf	T, B	2.5
344	Carcary	M	21	SU	Feb. 18, 1986	East London	1200	>22.0	6.0	None	Surfboard	W	>2.4
349	d-Bruyn	M	35	SP	Aug. 10, 1986	Geyser Island	1100	Cold	20.0	None	Fish	W	
353	Olls	M	21	B	Dec. 22, 1986	SAOU Strand	1700	21.0	6.0	Fatal	Thigh/calf	B	3.0
355	Botha	M	30	SP	Jan. 28, 1987	Eerste Rivier	1200	20*	8.0	Minor	Hand, fish	W, B	>2.5
356	McCallum	M	24	SU	Sep. 13, 1987	Still Bay	1100	21*	Clear	Minor	Torso	B	3.5
359	Smit	M	21	SP	Oct. 11, 1987	Seal Island, False Bay	1215	15*	7.0	Minor	Thigh, calf	W, B	
367	v. Rensburg	M	36	SP	Jun. 19, 1988	Skulphoek	0945	14*	9.0	Minor	Arm	W, B	2.5
370	Krouse	M	28	SK	Jun. 3, 1989	Gonubie	1530	17*	>2.0	Minor	Forearm	W, B	3.0
371	Jacobs	M	16	SU	July 5, 1989	Jeffreys Bay	1615	22.0	>1.0	None	Surfboard	W, B	2.5
372	Rezanno	M	34	SU	July 20, 1989	Jeffreys Bay	1035	14.0	Turbid	Minor	Torso	B	>3.4
373	v. Broembsen	M	21	SU	Aug. 22, 1989	Mossel Bay	1045	?	2.0	Major	Multiple	W, B	
377	v. Niekerk	M	29	SP	Sept. 17, 1989	Smitswinkel, False Bay	1430	17*	10.0	Minor	Torso	B	
378	v. Niekerk	M	29	SK	Nov. 18, 1989	Melkbosstrand	1205	?	Clear	Fatal	Leg recovered	B	
379	Botha	M	23	SU	March 14, 1990	Cape Recife	1400	20*	2.5	Minor	Lower leg	B	2.3
381	Forrester	M	22	SU	May 6, 1990	Cintsa	1200	22.0	10.0	Major	Thigh	T, W, B	5.5
386	Pace	F	21	SC	Jun. 24, 1990	Hartenbos	1545	18*	3.0	Fatal	Thigh	W, B	4.5
387	Gasant	M	26	SP	Sep. 15, 1990	Oudekraal	1320	13*	9.0	Minor	Hand	W	5.0
401	Hyman	M	20	SP	Feb. 12, 1991	Miller's Point, False Bay	1230	16*	6.0	Minor	Foot	W	5.5
400	Marais	M	32	SC	May 19, 1991	Gordon's Bay, False Bay	1330	15*	6.0	None	Scuba tank	W	3.5
405	LeRoux	M	28	SU	July 30, 1993	Port Elizabeth	1500	16.0	1.5	Minor	Torso, surfboard	W, B	4.0
406	Bouwer	M	36	SP	Sep. 26, 1993	Danger Point	1130	15*	4.0	Minor	Calf	W	4.0
408	Anderson	M	15	SU	Feb. 12, 1994	Port Elizabeth	1630	20*	1.0	Minor	Leg, surfboard	W, B	3.0
410	Carter	M	31	SU	July 9, 1994	Nahoon, East London	1335	17*	Turbid	Minor	Leg, surfboard	W, B	4.0
411	Corby	M	22	SU	July 9, 1994	Nahoon, East London	1336	17*	Turbid	Fatal	Leg severed	W, B	4.0

[a] Abbreviations: Activity, B = bodyboarder, P = paddleskier, SC = scuba diver, SK = skindiver, SP = spearfisherman, SU = surfer, SW = swimmer; temperature, SST = sea-surface temperature; method of species identification, A = artifacts from shark's gut, B = bite pattern, T = tooth fragments, W = witnesses.

*Estimated value.

**Body mass = 126 kg.

White Shark Attacks
in Australian Waters

JOHN WEST

Taronga Zoo
Mosman, Sydney, Australia

Introduction

White shark *Carcharodon carcharias* attacks on humans in Australia evoke extreme emotional reactions from some parts of the community and the media. This chapter reviews white shark attacks in Australian waters from 1791 through 1992, presenting data from the Australian Shark Attack File (ASAF), and compares the frequency of attacks over time to apparent trends in the population of white sharks, based on data from several sources. In this analysis, I also sought information on the behavioral patterns of both the white sharks and the people they attacked.

Attacks by white sharks have occurred in widely separated parts of the world, and in both warm and cool, shallow and deep waters (Baldridge, 1974a). This chapter analyzes data from cases in which a white shark was or was strongly suspected to be the attacking species. The 34 such cases represented only 7% of all the recorded shark encounters in the ASAF (Appendix 1). I also mention here cases in which the white shark *may* have been involved in an attack, but these records will not be included in the analysis because in New South Wales, Queensland, and Western Australia, tiger sharks *Galeocerdo cuvier* and large whalers *Carcharhinus* spp. could have been involved.

Materials and Methods

The ASAF was established in 1984 and is archived at Taronga Zoo, Sydney. The file is affiliated with the International Shark Attack File, coordinated by the American Elasmobranch Society.

Information on attacks in my analysis was compiled from a variety of sources, including the work of Baldridge (1969, 1974a,b, 1988a), Coppleson (1988), Gilbert (1963b), Green (1976), the International Shark Attack File, the ASAF, and G. Whitley (unpublished data). The ASAF data are continually updated. The criteria used to decide whether or not to include attacks in the ASAF are as follows: any interaction in which injury occurred to a human (alive), the equipment worn or held was damaged, or imminent contact was averted by diversionary action by the victim or others (no injury to the human occurred). Incidents involving kayaks and small canoes, but not boats, are included, but the information is held in a separate file. The files are described elsewhere (West, 1993).

Information in the ASAF, and for this analysis, was gathered on the basis of returned ASAF questionnaires; access to a national news-clipping service; radio, television, and printed news media reports; networks of personal and professional contacts; newspaper archives in libraries; and coroners' and police reports. As of January 1,1993, 484 shark attacks (all species) were recorded in Australian waters (within 200 nautical miles of Australia, the Torres Strait Islands, and Cocos Island); 180 of these attacks were fatal. The total number of recorded attacks includes provoked and unprovoked encounters, bites while removing sharks from nets, bites on kayaks and small canoes, and bites from sharks held in captivity.

In a number of cases, the white shark may have been involved, but information was insufficient to include them in this analysis. These "possible" cases include attacks recorded as fatal or having extensive injury, within the range of the white shark, in areas other than a river, and containing no evidence implicating other shark species. However, this issue is complicated by the fact that in New South Wales, Queensland, and Western Australia, tiger sharks and large whalers could have been involved. Therefore, "possible" attacks by white shark will not be included (Fig. 1).

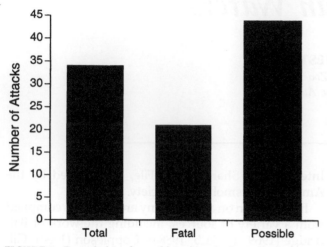

FIGURE 1 Recorded (and possible) white shark attacks in Australian waters, 1791–1992.

Results

The human population of Australia has increased steadily from 1,145,600 in 1851 to 3,765,300 in 1900, 8,307,500 in 1950, and 17,529,000 in 1992, averaging 15% more people per decade (S. Bull, New South Wales Bureau of Statistics, personal communication). Most of this increase has occurred near the coast. The Australian coastline extends some 36,000 km, and more and more, especially since the 1950s, human swimmers, divers, surfers, fishermen, and sailboarders have used the adjacent coastal waters. These activities are usually undertaken well away from the meshed beaches of the Sydney and Brisbane regions (Fig. 2). Even with the majority of the population living close to the coast and with increasing numbers enjoying the coastal resources, the number of recorded shark encounters and fatalities has declined (Fig. 3).

Attacks

In Australia, the white shark is known to occur in southern Queensland, New South Wales, Victoria, South Australia, southern Western Australia, and Tasmania (Compagno, 1984a). The majority (65%) of confirmed white shark attacks, however, have occurred in the cooler waters of South Australia, Victoria, and Tasmania (Table I).

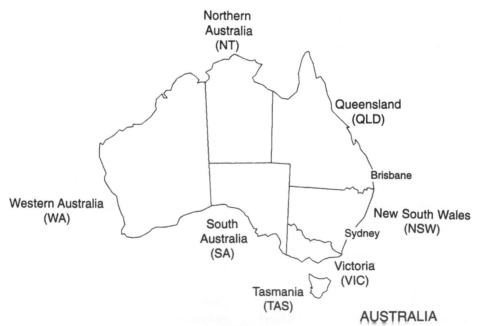

FIGURE 2 A map of Australia, showing localities mentioned in the text.

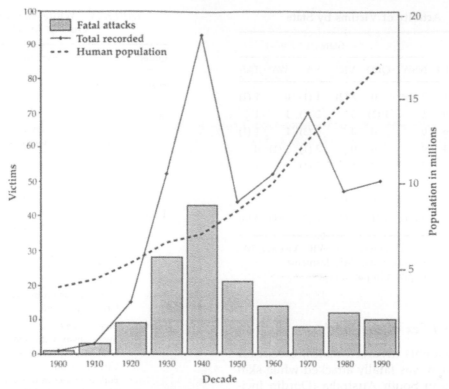

FIGURE 3 Reported shark attacks and fatalities in Australian waters (all species), 1900–1990.

White shark attacks totaled 34: 21 cases were fatal; in 9, the victim recovered from injuries; and in 4, no injury occurred ("possible" white shark attacks account for an additional 44 fatalities; Table I and Fig. 1). The majority of cases occurred before 1950. Continued investigation may produce additional evidence to positively identify the attacking species among the "possible" cases.

Activities of the Victims

The activities of victims at the time of the attack included 11 swimming, 9 surfing, 6 scuba diving, 4 spear fishing, 3 snorkeling or skin diving, and 1 diving into the water from a pier (Tables II and III).

TABLE I **South Australian White Shark Attacks on Humans by State**

State	Total No.	%	Fatal No.	%	No. possibly fatal
South Australia	11	33	7	64	8
New South Wales	8	23	6	75	14
Victoria	7	20	3	43	4
Tasmania	4	12	3	75	5
Western Australia	3	9	1	33	4
Queensland	1	3	1	100	9
Northern Territory	0	0	0	0	0
Total	34	100	21	62	44[a]

[a]All persons in this category are thought to have been swimming at the time of the attack.

TABLE II **Activities of Victims in White Shark Attacks**

Activity	Total	Fatal No.	%	Recovered	Uninjured
Swimming	11	10	91	1	0
Surfing	9	2	22	4	3
Scuba diving	6	4	67	1	1
Spear fishing	4	2	50	2	0
Skin diving and snorkling	3	2	67	1	0
Diving into water	1	1	100	0	0
Total	34	21	62	9	4

TABLE III **Activity of Victims by State**

Activity	Total	NSW	QLD	VIC	SA	WA	TAS
Swimming	11	7 (6)[a]	0	2 (2)	1 (1)	0	1 (1)
Surfing	9	1	1 (1)	3	2 (1)	1	1
Scuba diving	6	0	0	0	4 (3)	1	1 (1)
Spear fishing	4	0	0	0	3 (1)	1 (1)	0
Skin diving and snorkling	3	0	0	1	1 (1)	0	1 (1)
Diving into water	1	0	0	1 (1)	0	0	0
Total	34	8 (6)	1 (1)	7 (3)	11 (7)	3 (1)	4 (3)

NSW, New South Wales; QLD, Queensland; VIC, Victoria; SA, South Australia; WA, Western Australia; TAS, Tasmania.
[a]Numbers of fatalities are given in parentheses.

Age and Sex of the Victims

All recorded incidents involved males, except for 1, in which a woman was fatally attacked while skin diving for scallops in South Australia (Durdin incident, 1985). Witnesses observed the shark devouring her body at the scene of the attack.

The victims were 13–47 years old (N = 26); 10 persons were 10–20, 12 were 20–30, and 4 were >30 years old.

Behavior of the Shark

The behavior of the white shark was analyzed in regard to the sequence of the attack: before the attack, during initial contact, and after the attack (Table IV). In most cases, the victim was grabbed and quickly released. In only 5 cases was the shark seen before the attack.

White Shark Size

The total length (TL) of the sharks, estimated by the victims or witnesses, ranged from 2 to 6 m (N = 20). Only one of the estimates was confirmed, as in that case, the white shark was captured (Fig. 4).

The Victims' Injuries

Among the 34 white shark encounters, 4 (12%) resulted in no injury (1 scuba diver and 3 surfers), injuries (15%) are not known (not reported) in 5, and in 9 (26%) the bodies were not recovered.

In categorizing injuries, I used the following general categories: "legs" involved injuries to any part of

TABLE IV **White Shark Behavior During Attack**

Order	Description	No.
First	Circled victim	3
	Swam toward victim	2
Second	Grabbed victim by body	8
	Brabbed victim by leg	7
	Grabbed victim and board	3
	Bit surfboard	4
	Swam toward victim	1
	Grabbed victim	1
	Struck victim with nose	1
Third	Swam away	5
	Swam away with body in mouth	4
	Made second attempt to bite	2
	Circled victim	2
	Grabbed victim by leg	1
	Bit surfboard	1
	Grabbed victim by arm	1
	Grabbed victim	1
Fourth	Swam away	4
	Swam away with body in mouth	3
	Grabbed victim by leg	1
	Circled victim	1
	Followed victim to shore	1
	Bit victim several times	1
	Stayed around attack site	1

the foot, calf, knee, or thigh; "arms" involved anywhere from the hand to the upper arm; and "torso" involved the buttocks, stomach, chest, etc. Considering all victims, 7 were injured on the legs, 2 on the torso, and 2 on the arm; no injuries were reported to the head. In addition, there were 5 reports of multiple bites: torso/arms (1), torso/legs (1), torso/legs/arms (1), and legs/arms (2). Of the 9 persons whose bodies

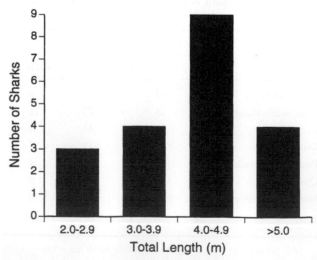

FIGURE 4 Total lengths of the white sharks implicated in attacks.

were not recovered (26%), 3 were scuba diving (2 recreational and 1 collecting abalone), 3 were swimming, 2 were skin diving or snorkeling, and 1 was diving into the water from a pier.

Time and Season of the Attacks

The time of the attack was recorded in 18 cases. The majority occurred from 12 pm to 3 pm. Of the 34 recorded attacks, 21 (62%) occurred during summer months December–March (Fig. 5).

Attack Site

Geographically, 11 attacks occurred in South Australia, 8 in New South Wales, 7 in Victoria, 4 in Tasmania, 3 in Western Australia, and 1 in Queensland (see Fig. 1 and Table V). In all cases, the white shark attacks took place in areas having direct access to the open ocean: 20 occurred at ocean beaches, 6 in bays open to the ocean, 3 next to a wharf (in large bays open to the ocean) over a sandy bottom, 3 in the open ocean with a rocky seabed, and 2 at a reef (temperate water) in the open ocean.

In 19 cases, the distance from shore was recorded: 2 were <50 m from shore, 2 were between 50 and 100 m, 11 were 100–500 m, 1 was between 500 and 1000 m, and 3 were >1000 m offshore (2 spear fishers at 1000 m and 1 scuba diver at 2000 m).

In 11 cases, the presence of other animals was noted: fish were present in 6 cases, 1 recorded fish and other sharks, 2 recorded seals (Australian fur seal *Arctocephalus pusillus* in 1 and Australian sea lion *Neophoca cinerea* in the other), and 2 recorded dolphins (unidentified, possibly bottle-nosed dolphins *Tursiops truncatus*).

Trends in White Shark Numbers: Catch Statistics

The numbers of white sharks (*C. carcharias*) captured off New South Whales by game fish anglers from 1961 to 1990 averaged 5.5 per year. By decade, however, the average number caught per year dropped from 7.6 in the 1960s and 9.1 in the 1970s to 0.4 in the 1980s (Pepperell, 1992). Pepperell also stated that white sharks have been the least common of the shark species caught in New South Wales; the ratio of white sharks to all species caught changed from 1:22 in the 1960s to 1:38 in the 1970s and 1:651 in the 1980s. The majority of white sharks (109, or 67.3%) weighed <150 kg, ranging from <10 to 550 kg; however, only 8 weighed >300 kg.

The number of white sharks captured by the Game Fishing Club of South Australia between 1938 and 1990 has also shown a marked decline. After 1951–1952, only 25 specimens have been caught (Pepperell, 1992). Bruce (1992) stated that

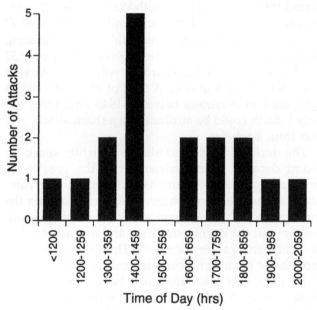

FIGURE 5 Times of day when the white shark attacks occurred.

TABLE V **Months of White Shark Attacks**

State	January	February	March	April	May	June	July	August	September	October	November	December
SA	2		3					1	2			3
NSW		5	2	1								
VIC		2	1	1						1	2	
TAS	1		1				1			1		
WA			1						2			
QLD										1		
Total	3	7	8	2	0	0	1	1	4	3	2	3

SA, South Australia; NSW, New South Wales; VIC, Victoria; TAS, Tasmania; WA, Western Australia; QLD, Queensland.

Capture rates over the last 10 years by the Game Fishing Club of South Australia have averaged 1.4 sharks per year, and total captures by recreational fishers in South Australia are probably less than 10 per year. The rate of capture of white sharks by commercial fishers is difficult to assess because captures are usually not recorded, and in many cases, captured sharks are either not landed intact or not landed at all. Local estimates of commercial captures range from less than 10 per year to greater than 100 per year, but is more than likely that such captures exceed recreational captures.

The number of white sharks meshed off New South Wales, including the Sydney region (25 beaches), Central Coast (12 beaches), Newcastle (11 beaches), and Woolongong (5 beaches), 1951–1990, totaled 474 and averaged 11.8 per year (J. G. Pepperell, personal communication). During the period 1972–1990, only 185 white sharks were meshed with the average for this period, being 10.5 per year. The sex ratio was 59:41 (58.9% female); 50.8% were alive in the nets. The majority of specimens captured were 200–300 cm TL, ranging from 100 to 550 cm. The majority were captured between July and December (Reid and Krogh, 1992).

There has been a marked decline in the average number of white sharks captured in mesh nets, from 38.3 per year in 1950–1952 to 3.6 per year by 1987–1990 (Collins, 1972; Anonymous, 1989). Reid and Krogh (1992), stated,

C. carcharias are rarely caught in the Woolongong region, with only six being recorded throughout the full study period [and 3 from the Central Coast region from 1978–90]. In the Sydney region, there were protracted declines in the catch of C. carcharias from 1953–54 to 1971–72 and from 1972–73 to 1989–90. In the Newcastle region, catches of C. carcharias were fairly stable from 1950–51 to 1977–78 and then declined until 1989–1990.

Discussion

Characteristics of the Attacks

All white shark attacks occurred in, or had direct access to, the open ocean. The only areas where more than one white shark attack has been recorded are Aldinga Beach and the Carrackalinga Heads area, Adelaide, South Australia, where 4 attacks occurred on skin divers who were spear fishing in 1961, 1962, 1963, and 1991 (2 fatal).

The high percentage (26%) of bodies consumed or taken away by white sharks does not conform, generally, to the "bite-and-spit" behavior of white sharks when feeding on mammalian prey, as hypothesized by Ellis and McCosker (1991). The large size of the sharks may be a contributing factor, as 65% of the attacks involved sharks >400 cm TL.

No attacks by white sharks were recorded in the months of May or June (winter), but 62% occurred between December and April (summer to fall). This pattern may be more a function of humans' avoiding the water when temperatures are cold than of white shark behavior or numbers. Surfboard riders, scuba divers, and spear fishermen using wet suits have been the predominate victims of white shark attacks during the cooler months, April–November. It was not until 1964 that a fatality occurred to a victim wearing a wet suit during winter (spear fishing in August). Since 1961, all but one of the victims were wearing a wet suit and were either scuba diving, surfing, or skin diving.

Trends in Attacks

Compared to the number of deaths from any other form of water-related activity, those due to shark attack (all species) are very low in Australia. To put the issue of shark attacks into perspective, during the period 1968–1991 in New South Wales alone, 121 fishermen were swept off the rocks and drowned and 37 surfers drowned (S. Bull, personal communication). During the same period in New South Wales, only 32 shark encounters were recorded, with only 1 fatality (not a white shark attack). A total of 292 scuba divers have died in Australia between 1945 and 1986, but only 1 death could be attributed to a shark attack, and that from a whaler.

The decline in reported attacks by white sharks in recent decades is inconsistent with the greatly increased population of humans and their much greater use of inshore marine waters of Australia during the same period. The decline in attacks, however, is consistent with the decline in fished and meshed catches of white sharks (Pepperell, 1992; Bruce, 1992), although the situation is complicated; both of these authors note that less effort has been conducted by game fishers in inshore waters in recent years. Nevertheless, Pepperell (1992) believes that the decline in white sharks warrants an urgent investigation into the status of the species off southeastern Australia. Even more so than earlier, humans using Australian inshore waters are subject to little danger from white sharks compared to other factors (West, 1993).

Summary

Reviewed here were data from the ASAF, including all recorded incidents of attacks on humans by white sharks C. carcharias in Australian waters, 1791–1992. Among the 34 white shark attacks, 21 were fatal (9 bodies not recovered), 9 persons recovered from their injuries, and 4 persons were not injured in the encounter. Activities of the victims included 11 swim-

ming, 9 surfing, 6 scuba diving, 4 spear fishing, 3 skin
diving, and 1 diving into the water. The majority of
the attacks (62%) occurred during summer, December
–March. The TLs of the sharks, recorded for 20 cases,
ranged from 2 to 6 m. The human population, and its
use of coastal waters, in southern Australia has been
increasing dramatically during recent decades, but
the number of white shark attacks has been declining. This trend is due to an apparent decline in the
numbers of white sharks during the same period.

Acknowledgments

I thank the Zoological Parks Board of New South Wales and the
Association of Zoo Friends (New South Wales) for their support of
the ASAF, and the Director and Chief Executive of Taronga Zoo, J.
Kelly, and the Executive Director, G. Smith, for their encouragement. D. H. Baldridge allowed the use of data from the Analytical
Data on Shark Attacks (1988), the Australian Museum allowed access to unpublished data from the Gilbert Whitley archives, S. Bull
provided data from the New South Wales Bureau of Statistics, and
A. P. Klimley and D. G. Ainley provided many helpful comments
to improve the manuscript.

APPENDIX I White Shark Attacks in Australian Waters

ISAF no.	ASAF no.	Name	Date	Location	State	Activity
471	14	Peter Rooney	February 6, 1876	Albert Park/St. Kilda Pier	VIC	SW
6	58	Milton Coughlan	February 4, 1922	Coogee Beach, Sydney	NSW	SW
95	88	Norman Clark	February 15, 1930	Middle Brighton	VIC	D
31	118	Herbert Mcfarlane	March 2, 1935	North Narrabeen, Sydney	NSW	SW
32	119	Ernest Mcdonald	March 9, 1935	Maroubra Beach, Sydney	NSW	SW
93	127	Raymond Bennett	January 22, 1936	West Beach, Adelaide	SA	SW
33	128	David Patton	February 4, 1936	South Styne, Manly	NSW	SW
35	137	John Welsh	February 13, 1937	Bar Beach, Newcastle	NSW	SW
1100	212	Theo Brown	March 1, 1955	Woodmans Point	SW	SC
55	213	Noel Langford	March 9, 1955	Wamberal, Gosford	NSW	SW
97	217	John Wishart	April 4, 1956	Portsea Beach	VIC	SW
383	226	Brian Derry	January 17, 1959	Safety Cove, Port Arthur	TAS	SW
842	247	Brian Rogers	March 12, 1961	Aldinga Beach	SA	SP
1122	262	Jeoff Corner	December 9, 1962	Carrickalinga Head	SA	SP
1235	267	Rodney Fox	December 8, 1963	Aldinga Beach	SA	SP
1344	269	Heneri Bource	November 26, 1964	Lady Julia Percy Island	VIC	SN
1406	271	Raymond Short	February 27, 1966	Coledale Beach	NSW	SW
1463	280	Robert Bartle	August 19, 1967	Jurien Bay	WA	SP
2317	302	Terry Manuel	January 9, 1974	Streaky Bay	SA	C, SC
1756	310	Robert Slack	July 1, 1975	Adventure Bay, Bruny Island	TAS	C, SC
1887	403	Philip Horley	August 1, 1977	Cactus Beach	WA	SU
2357	320	Gert Taleen	March 1, 1982	South Cape Bay	TAS	SK
2318	344	Neil Williams	December 1, 1983	South Neptune Island	SA	SC
1986	325	Shirley Durdin	March 3, 1985	Wiseman's Beach, Port Lincoln	SA	SK
2188	342	Terrance Gibson	September 18, 1987	Merino Rocks, Adelaide	SA	SC
2005	432	Mathew Foale	March 9, 1989	Waitpinga Beach, Encounter Bay	SA	SU
2443	441	STeven Jillett	October 22, 1989	Launceston	TAS	SU
2444	445	Garry White	November 24, 1989	Kilcunda	VIC	SU
2445	446	John Benson	October 1, 1989	Surfers Point, Phillip Island	VIC	SU
2446	449	Tony Patton	April 6, 1990	Fingal, North Coast of NSW	NSW	SU
2122	472	Jonathon Lee	September 8, 1991	Aldinga Beach	SA	SC
2447	477	Mark Jepson	March 10, 1992	Point Lonsdale	VIC	SU
2448	481	Jason Bates	August 25, 1992	Lipson Cove, Tumby Bay	SA	SU
2442	482	Michael Docherty	September 4, 1992	Morton Island, Brisbane	QLD	SU

ISAF, International Shark Attack File; ASAF, Australian Shark Attack File; VIC, Victoria; NSW, New South Wales; SA, South Australia; SA, Western Australia; TAS, Tasmania; QLD, Queensland; SW, swimming; D, diving off a pier; SC, scuba diving; SP, spear fishing; SN, snorkeling; C, collecting abalone; SU, surfing; SK, skin diving.

Worldwide Patterns of White Shark Attacks on Humans

GEORGE H. BURGESS *and* MATTHEW CALLAHAN

International Shark Attack File
Florida Museum of Natural History
University of Florida
Gainesville, Florida

Introduction

The white shark *Carcharodon carcharias* is the largest species of predatory shark and is a well-known threat to humans. Baldridge (1974a), in his seminal treatment of shark attack, noted that the white shark was the species most often cited in attacks on humans, with 32 recorded attacks. Although a series of reports have reviewed individual and regional attacks by white sharks on humans (e.g., Bolin, 1954; Fast, 1955; Smith, 1963; Whitley, 1963; Collier, 1964; Follett, 1974; Miller and Collier, 1981; Egaña and McCosker, 1984; Lea and Miller, 1985; Collier, 1992; Nakaya, 1993; Levine, Chapter 40; West, Chapter 41), and others have discussed shark attack in general (Coppleson, 1963; Schultz, 1963; Baldridge, 1974a, 1988b), no definitive work has examined circumstances surrounding white shark attacks worldwide. In this study, 159 documented unprovoked attacks on humans by white sharks were analyzed for patterns involving geographical, historical, demographic, temporal, environmental, morphological, and behavioral parameters associated with both the attacking shark and its victim. Trends in white shark attacks on humans are compared to recent perspectives on the natural feeding behavior of the species (Tricas and McCosker, 1984; McCosker, 1985; Klimley *et al.*, 1992, Chapter 16).

Data were generated from investigations archived at the International Shark Attack File (ISAF), a compendium of documented shark attacks on humans (Schultz, 1963; Burgess, 1991). Founded in 1958, the ISAF is curated at the Florida Museum of Natural History, University of Florida, where it is operated under the aegis of the American Elasmobranch Society. New and historic shark attacks are added to the ISAF through the efforts of a worldwide network of cooperating investigators. Data derived from original investigative reports, medical records, autopsies, and print/video media are coded and computerized for in-house and cooperative analyses.

Methods

We studied unprovoked attacks on live humans located within the water column and at the sea surface when associated with surfboards, kayaks, rafts, and inner tubes. Not included were (1) provoked attacks and attacks on boats; (2) incidents involving divers in protective cages and baited sharks (considered provoked); (3) "scavenge" cases, in which the victim was dead or presumed dead due to drowning or other causes prior to the attack; and (4) attacks in which victims disappeared without witnesses' observing an attack, but shark-damaged gear and/or bodies subsequently were recovered, and circumstantial evidence indicated provocation. Inclusion of the latter was, of course, difficult to decide.

Similarly, attributing an attack specifically to the white shark was frequently difficult. In some cases, knowledge of the zoogeography of large shark species facilitated assignment of the attack to the white shark; in others, tooth remains or bite patterns provided clues. Most problematic were attacks in which more than one large shark species occurs and supporting bite documentation was lacking.

We chose to exercise a conservative approach toward validating cases for inclusion. Because validation was often subjective, one might anticipate differences in the absolute numbers for a given geographical locality between this and regional analyses (see, e.g., Chapter 40, by Levine, and Chapter 41, by West). In addition, ISAF coverage of some geographical areas is more complete than for others. For example, ISAF coverage of the Pacific coast of North America, Japan, New Zealand, and Australia is largely complete. John West, director of the Australian Shark Attack File, and Robert N. Lea, who maintains a large eastern Pacific database, have long shared shark data with the ISAF. Thus, essentially, the same white shark data sets used by those investigators in their regional analyses (see Chapter 41, by West, and Chapter 39, by McCosker and Lea) were used here. Other areas, most notably South Africa (see Chapter 40, by Levine) and the Mediterranean Sea (see Chapter 30, by Fergusson), are not as completely covered in the ISAF.

Our data categories mirror those used by Baldridge (1974a), with few exceptions. We formulated the gear/clothing color protocol used in that subsection. Since its inception, ISAF measurements have been recorded in English units; we have converted English measures to metric (to one decimal place) throughout the chapter. Temperature measurements similarly have been converted from Fahrenheit to Celsius. Individual cases are cited using the ISAF case number.

Results

The completeness of data coverage of the 159 white shark attack reports in the ISAF varies from case to case. Therefore, certain data categories are more robust than others. In our analyses, some categories, for example, geographical region (N = 158) and year of attack (N = 159), were nearly or completely covered, while others were represented by less complete data. In the following subsections, the number of data points used in the analysis of each category is indicated in brackets at the end of the

respective subsection; percentages were calculated using the N for that category. In some data categories, multiple responses were possible; therefore, total percentages within those categories may exceed 100%.

Geographical Region

Greater than one half (50.6%) of all white shark attacks occurred along the Pacific coast of the United States (Table I). California, with 67 attacks, accounted for 42.4% of the world total; 11 attacks (7%) occurred in Oregon, and 2 (1.3%) in Washington. Other areas with historically high numbers of attacks include Australia (32, 20.2%), South Africa (16, 10.1%), and New Zealand (8, 5.1%). Fewer attacks have been reported at scattered localities in North and South America, Asia, and Europe (Table I). The northernmost attack was from Puget Sound, Washington (approximate latitude 47°50′ N; ISAF 2592) and the southernmost from Campbell Island, New Zealand (latitude 52°32′ S; ISAF 2236). [N = 158]

Historical Number of Attacks

The earliest documented white shark attack on a human occurred in 1876 in Victoria, Australia. There has been a gradual rise in the number of attacks

TABLE I Geographic Areas of White Shark Attacks on Humans

Geographic area	Attacks	
	No.	%
United States		
Pacific coast	80	50.6
California	67	42.4
Oregon	11	7.0
Washington	2	1.3
Atlantic coast	6	3.8
Hawaii	1	0.6
Australia	32	20.2
South Africa	16	10.1
New Zealand	8	5.1
Japan	4	2.5
Italy	3	1.9
Chile	2	1.3
Mexico (Pacific)	2	1.3
Easter Island	2	1.3
Argentina	1	0.6
Papua New Guinea	1	0.6
Total	158	100.0

throughout the 20th century (Fig. 1), with the greatest increase coming since 1960. Attacks were somewhat fewer than expected during the 1940s relative to the overall trend. On the basis of the first 4.5 years of the 1990s (January 1, 1990–June 30, 1994), the current decade promises to continue this upward trend, with a projected increase to 80–85. [N = 159]

Outcome

Although injuries often are severe, white shark attacks on humans resulted in fatalities only 26.1% of the time. Prior to 1950, 91.7% of reported attacks resulted in death, and during the 1950s the fatality rate was 50%. Recent fatality rates have declined to only 15.2% since 1970, reflecting the pronounced pattern of reduction observed in historical data (Fig. 2). [N = 157]

Time of Day

Humans were attacked by white sharks almost exclusively during daylight, with only one attack (ISAF 1100) occurring between 2000 and 0559 hours (Fig. 3). The period 0800–1759 hours accounted for 91.1% of all attacks; more than two thirds (67.7%) occurred between 1000 and 1559 hours. Markedly few attacks occurred from 1200 to 1359 hours. [N = 124]

Water Clarity

In the absence of technical testing equipment, determination of water clarity is subjective. Despite dif-

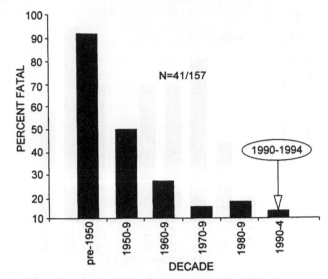

FIGURE 2 Temporal trends in the fatality rates of victims of white shark attack; 1990s figures are through June 1994.

ficulties in qualifying this parameter, 55.3% of white shark attacks were reported to occur in "clear" water and 44.7% were described as occurring in "turbid," "murky," or "muddy" waters. [N = 76]

Water Temperature

White shark attacks occurred over a wide range of seawater temperatures, from 8°C (New Zealand; ISAF 2236) to 24°C (4 cases), with 89.5% recorded from 12°C to 24°C. Nearly two thirds (60.5%) of attacks occurred at temperatures ≤20°C. A weakly de-

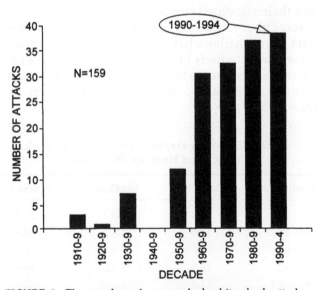

FIGURE 1 The number of unprovoked white shark attacks on humans by decade. The single attack occurring before the 1910s (1876) is not depicted; 1990s figures are through June 1994.

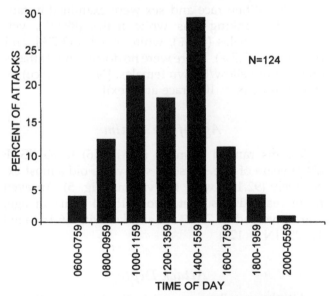

FIGURE 3 The time of day when white shark attacks on humans occurred; the nighttime period, 2000–0559 hours, is compressed.

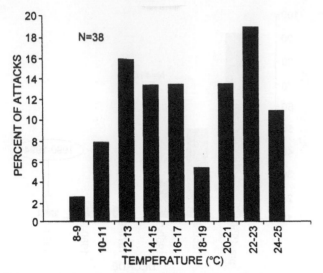

FIGURE 4 White shark attacks on humans, as a function of water temperature.

fined bimodal pattern of attacks relative to water temperature is apparent (Fig. 4). [N = 35]

Race and Sex of the Victim

Victims of most races and both sexes were attacked. Following the criteria of Baldridge (1974a), racial groupings were made according to skin pigmentation. Victims were overwhelmingly white (92.7%), with fewer numbers of black (2.7%) and yellow/ brown pigments (4.6%). Males (96.6%) were attacked predominantly. Attacks on females have increased in recent years, with 8.6% of 1990s attacks involving females, compared to 6.7% in the 1980s and 1.2% pre-1980. When race and sex were examined in tandem, the ranking was white males (89.0%), yellow/brown males (4.6%), white females (3.7%), and black males (2.7%). There were no documented attacks on black or yellow/brown females. [N = 109 (race), N = 149 (sex), N = 109 (race and sex)]

Age of the Victim

Victims ranged from 10 (ISAF 246) to 55 (ISAF 2504) years of age. Persons <40 years old almost exclusively (92.1%) were the victims (Fig. 5). An even more restrictive subset, those 15–34 years of age, were involved in 79.5% of attacks, with 66.1% aged 15–29. [N = 127]

Attack Depth

Attacks usually occurred in the upper portion of the water column (83.3% from 0 to 3.2 m) (Table II).

TABLE II Depths at which White Sharks Attacked Humans (N = 150)

Depth (m)	Attacks (%)	Depth (m)	Attacks (%)
0–1.7	78.7	0–3.2	83.3
1.7–3.2	4.7	0–6.2	90.0
3.2–6.2	6.6	0–9.3	94.0
6.3–9.3	4.0		
9.3–24.5	6.0		

Typically, attacks took place directly at the surface (78.7% from 0 to 1.7 m). The deepest recorded attack was at 25 m, in waters off Japan in 1992 (ISAF 2320; Nakaya, 1993). [N = 150]

Water Depth

White shark attack was predominantly a shallow-water phenomenon (Table III), with 84.9% occurring at depths of <12.3 m. Approximately one third (31.1%) occurred at depths of 0–3.2 m, and over half (53.8%) from 0 to 6.2 m. The deepest water where an attack occurred was 3000 m off Easter Island in 1994 (ISAF 2599 and ISAF 2625); numerous cases are documented from depths of <1.5 m. [N = 106]

Attack Habitat

Nearshore waters, including beaches, harbors, bays, lagoons, tidal river/creek mouths and their upstream portions, and areas immediately adjacent to land, hosted most attacks (65%) (Table IV). Beaches and their associated surf zones were cited in 20.3% of cases. Other frequently recorded inshore attack habitats were harbors/bays (9.8%) and breakwaters/ jetties/docks/wharfs (4.9%). Fewer attacks (35%) occurred in offshore waters, with 24.5% taking place at offshore reefs/banks, and 10.5% in the open sea. [N = 143]

TABLE III Water Depth at Site of White Shark Attacks on Humans (N = 106)

Depth (m)	Attacks (%)	Depth (m)	Attacks (%)
<1.7	11.3	0–3.2	31.1
1.7–3.2	19.8	0–6.2	53.8
3.2–6.2	22.6	0–9.3	72.6
6.3–9.3	18.9	0–12.3	84.9
9.3–12.3	12.3		
≥12.4	15.1		

TABLE IV Habitats in which White Sharks Attacked Humans (N = 143)

Habitat	Attacks (%)
Nearshore waters	65.0
Nearshore waters, beaches (no specifics)	21.7
Inside breaker, surf line	14.0
Harbor, bay	9.8
Breakwater, jetty, dock, wharf	4.9
Just beyond breaker, surf line	4.9
Mouth of tidal river, creek	4.2
Tidal river, creek other than mouth	2.1
Waters between sandbar and shore or other bar	1.4
Lagoon	1.4
Inside meshed area	0.7
Offshore waters	35.0
Offshore reef, bar, bank	24.5
Open sea	10.5

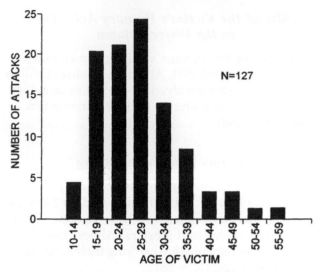

FIGURE 5 The ages of the victims attacked by white sharks.

Activity of the Victim

Victims were engaged in a variety of aquatic activities at the time of attack (Table V). White shark attacks occurred almost equally in the three major types of activities: surfing/kayaking/rafting (32.2%), diving/snorkeling (33.5%), and swimming/floating/wading (31.6%). Attacks on humans entering or exiting the water were minimal (2.5%). Surfing produced the most attacks (30.4%) for any single activity. Swimmers (12.7%) and free divers (12%) were also frequently attacked. [N = 158]

Activity of Others in the Area

Nonvictimized humans in the attack area were engaged in aquatic activities similar to those of victims, but in somewhat different proportions (Table VI). Some type of swimming/floating/wading occurred in half (50%) of those cases that noted other humans in the water near the victim. Bathing/swimming was most often cited (25.9%). Diving/snorkeling and surfing were reported in 33% and 32.1% of the cases, respectively. [N = 85]

TABLE V Activity of Victims at the Time of Attack (N = 158)

Activity	% of Total
Diving, snorkeling	33.5
Free diving (mask, fins, with or without snorkel)	12.0
Scuba diving	7.6
Snorkeling (mask, fins, with or without snorkel)	3.8
Other diving, subsurface activity	3.8
Subsurface, diving activity, (no details)	3.8
Hard-hat diving	2.5
Surfing, kayaking, rafting	32.2
Riding surfboard	30.4
Aboard kayak, float, raft, inner tube	1.9
Swimming, floating, wading	31.6
Swimming	12.7
Treading water	4.4
Other surface activity	4.4
Floating, little or no motion	3.2
Surface activity (no specifics)	2.5
Wading	2.5
Shallow water activity (no specifics)	1.3
Other shallow water activity	0.6
Entry, exit	2.5
Entering water (jump, fall, dive)	1.9
Leaving water (ladder, side of boat)	0.6

TABLE VI Activites of Others in the Area at the Time of Attack (N = 112)

Activity	Attack (%)[a]
Diving, snorkeling	33.0
Surfing, kayaking, rafting	32.1
Surfing	32.1
Swimming, floating, wading	50.0
Bathing, swimming	25.9
Other activities	9.8
Splashing, horseplay	4.5
Thrashing, flailing	3.6
Loud voices, noises	3.6
Wading	2.7

[a]Since multiple activities often occurred, total percentages exceed 100%.

Site of the Victim's Primary Activity in the Water Column

Activities of the victims were largely centered at the water surface (65.2%). About one quarter (29.7%) of the victims were involved in subsurface activities, and 5.1% were standing in shallow water when attacked. [N = 158]

Victims Who Were Fishing

Victims were fishing in one third (34.8%) of cases. Of these, 49% were spear fishing and 58.2% were gathering marine animals, such as abalones, pearl oysters, pen shells, and lobsters. Almost two thirds (60%) of those fishing were carrying their catch at the time of the attack. [N = 55]

Other Humans Fishing in the Attack Area

Nonvictimized humans were engaged in fishing activities in the attack area 34.8% of the time. Their activities were similar to those of the victims: spear fishing (53.1%) and harvesting marine animals (53.1%). Somewhat fewer were carrying their catch (43.8%) than were victims (60%). [N = 32]

Gear and Clothing Worn or Carried by the Victim (Exclusive of Surfboards or Kayaks)

Wet suits were worn by fewer than one half (40.8%) of the victims (Table VII). Other commonly worn or carried items included face masks (30.6%), swim fins (29.3%), weapons (e.g., spear guns or knives; 23.1%), and swimsuits (19%). Gear less frequently worn included scuba equipment, dive bags/stringers, life jackets, hard-hat dive suits, and nonaquatic clothing. [N = 147]

Color of the Victim's Gear and Clothing

The primary (dominant), secondary, and tertiary colors of victims' gear were assigned to one of nine major color groups. The black/gray/slate group was predominant (Table VIII), the case in 75.9% of attacks and representing the primary color in 74.1%. Four color groups, white/silver/talc, purple/violet/mauve, green/olive/teal/lime/avocado, and orange/peach/apricot, were not involved as a primary color. Secondary colors were split among seven groups, with yellow/gold/khaki/lemon (27.3%) most commonly cited. Tertiary colors also were spread among seven color groups (Table VIII). When primary and secondary color groups were viewed in combination (Table IX), black/yellow (21.9%), black/white (15.6%), black/orange (12.5%), and black/blue (12.5%) were the most frequent. [N = 54 (primary), N = 33 (secondary), N = 13 (tertiary), N = 32 (primary–secondary–tertiary combination)]

Other Humans Near the Victim at the Time of Attack

The presence of other humans near the victim at the time of attack was examined at three ascending ranges: 0–3 m, 3–15.2 m, and "in general area." No other humans were present within 3 m of the victim in the vast majority (84.8%) of cases; 1 and "2 or more" humans were nearby 10.1% and 5.1% of the time, respectively. No humans were within 3–15.2 m in half (51%) of the attacks. One to 5 were present in 37.9%, 6–10 in 6.1%, and "several" in 5.1% of the cases. Other humans were predominantly (91.3%) found within the "general area." [N = 99 (0–3 m), N = 98 (3–15.2 m), N = 127 ("in general area")]

Size of the Shark

Estimating the length of an attacking shark is very difficult; the data used here were generated from victims' and/or witnesses' estimates or from estimates based on bite sizes. Attacking sharks were mostly large (Table X), with 88% 260 cm or more in length and 80% in the 260-to 500-cm range. The largest attacking shark (ISAF 1247) was reportedly 610–760 cm, larger than the confirmed maximum size of the spe-

TABLE VII　Clothing and Gear Worn or Carried by the Victim at the Time of Attack (N = 147)

Clothing and gear	Attacks (%)
Wet suit/dry suit	40.8
Face mask with or without snorkel	30.6
Swim fins	29.3
Weapons (e.g., knife, speargun, or powerhead)	23.1
Swimsuit	19.0
Scuba	15.0
Other gear	12.2
Dive bag/stringer	8.8
Life jacket or other float	5.4
Hard-hat suit	2.7
Non-swim clothing	1.4

TABLE VIII Colors of White Shark Attack Victims' Gear and Clothing

Color	Order by area of color group (%)[a]			Total (%)[b]
	Primary	Secondary	Tertiary	
Black/gray/slate	74.1	3.0		75.9
Blue/aqua/turquoise	14.8	9.1	7.7	22.2
Yellow/gold/khaki/lemon	5.6	27.3	23.1	27.8
Red/pink/rose/magenta/coral/maroon	3.7	12.1	7.7	13.0
Brown/tan/buff/rust/sienna/cocoa/beige	1.8		7.7	3.7
Green/olive/teal/lime/avacado		18.2	15.4	14.8
White/silver/talc		15.2	30.8	16.7
Orange/peach/apricot		15.2	7.7	11.1
Purple/violet/mauve				0

[a]Number of observations: primary (N = 54), secondary (N = 33), tertiary (N = 13), and total (N = 54).
[b]Represents the percentage of cases in which a color group was present to some degree on a victim's gear or clothing.

TABLE IX Color Combinations of Victims' Gear in White Shark Attacks on Humans (N = 32)

Color	Attacks (%)
Black and yellow	21.9
Black and white	15.6
Black and blue	12.5
Black and orange	12.5
Black and green	9.4
Black and red	6.2
Yellow and red	6.2
Blue and black	3.1
Blue and orange	3.1
Blue and yellow	3.1
Brown and green	3.1
Yellow and green	3.1

cies; the smallest was 150–180 cm (ISAF 1741). [N = 100]

Close Passes by the Shark

The attacking shark was not seen before the attack in 64.4% of the cases. When seen, sharks frequently (80.9%) made one or more close passes near, but not touching, the victim during the attack sequence (Table XI). A single close pass was noted in 45.2% of attacks, and two or more passes were cited 35.7% of the time. The highest precise number of close passes recorded was 5, although one witness described them as "too many to specify." [N = 42]

TABLE X Lengths of White Sharks Attacking Humans (N = 100)

Length (m)	Attacks (%)
1.4–2.0	3.0
2.0–2.6	9.0
2.6–3.2	17.0
3.2–3.8	17.0
3.8–4.4	20.0
4.4–5.0	26.0
5.0–5.6	4.0
5.6–6.2	2.0
6.3+	2.0

TABLE XI Number of Close Passes by Attacking White Sharks (N = 42)

Passes	Attacks (%)
None	19.1
One	45.2
More than one	35.7
Two	2.4
Three	7.1
Five	2.4
"Few"	2.4
"Several"	11.9
"Many"	7.1
"Too many to specify"	2.4

Direction of the Attack

The direction from which the victim was attacked by the white shark was categorized into five classes. Attacks occurred primarily from below (31.3%) and from the side (32.5%). Attacks from behind were reported in 25.3% of the cases. Frontal attacks (8.4%) and attacks from above (2.4%) were rare. [N = 83]

Shark Behavior

Shark behavior was examined at the time of the first strike, during subsequent strikes, and after the final strike. The first strike nearly always was characterized as sudden and violent (92.4% versus minimum of turmoil, 7.6%). A single discrete bite occurred in half (56.8%) of the attacks (Table XII), and multiple bites were noted in about one third (32.8%). In 10.4% of the cases, the shark did not bite the victim or associated gear. Three discrete bites was the highest precise number recorded, but reports of "many" and "too many to specify" suggest that repetitive bites may exceed this number. In those encounters involving repetitive bites, subsequent attacks most often (68.3%) included multiple deliberate strikes or releasing of a bite hold, then quickly biting again (19.5%); frenzied behavior (12.2%) was less often encountered. After the final strike, the shark was seen leaving the area or was not seen again in 69.8% of the attacks (Table XIII). [N = 132 (first strike), N = 125 (discrete strikes), N = 41 (subsequent strikes), N = 96 (final strike)]

Discussion

Shark Attack in Perspective

Interpretation of shark attack statistics must be undertaken acknowledging that scientific protocols,

TABLE XII   Number of Bites by the Attacking White Sharks on Its Victim and/or Associated Gear (N = 125)

Bites	Attacks (%)
None	10.4
One	56.8
More than one	32.8
Two	16.8
Three	6.4
"Several"	3.2
"Many"	4.0
"Too many to specify"	2.4

TABLE XIII   Behavior of the Attacking White Shark after Final Strike on Its Victim (N = 96)

Behavior	Attacks (%)
Presumably left area	69.8
Not seen after final strike	37.5
Seen leaving area	32.3
Remained in area	30.2
Remained in immediate area	18.8
Followed victim/rescuers to shore	6.3
Remained attached to victim, had to be forcibly removed	3.1
Remained attached to victim, released without force	2.1

considered standard in experimental design, cannot be provided. This report, and any such analysis of shark attack data, is and will always be based on *statistics*, not *statistical treatment* of experimentally derived data. Shark attack data are gathered in a systematic yet fortuitous manner; despite efforts to develop a worldwide network of cooperating investigators, reporting of attacks varies greatly from region to region and case to case. In many Third World areas, attacks are reported slowly, if at all, while in some developed countries reports of shark attacks are intentionally suppressed or irrationally amplified.

The inability to provide controls for virtually all data categories is another problem. Consider, for example, an attempt to gather control data on the color of victims' gear and clothing. One would have to record the number of person-hours spent in all types of aquatic activities by every color/color combination concurrently in use throughout the world, at all major areas, since gear color/clothing choices vary regionally. Even if this monumental task could be accomplished, data from only a single point in time would be generated, as fashion trends change rapidly. How could an investigator recreate last year's color use pattern? Or that of 10, 20, or 50 years ago? There is simply no way to backtrack.

Control data on use of the marine environment is also difficult to obtain. Armed with actual person-hours in the water, we could calculate attack rates. But any such effort to inventory person-hours would have to examine use on a worldwide, seasonal, and diurnal basis, for all types of aquatic activity. Once again, the result would produce only a one-time snapshot, since human use patterns have certainly changed historically.

Therefore, we must be cautious when interpreting shark attack statistics. At the risk of stating the obvious, it is important to remember that both shark and human abundance and behavior patterns affect the chance of an interaction. Because the densities of both interactive parties seldom are equal, the abundance of either the shark or the human is more likely to influence the odds of an interaction taking place. While it is true that there are areas of high white shark abundance [e.g., the Farallon Islands (Klimley *et al.*, 1992)], as one examines certain statistical categories, it becomes obvious that human densities and activity patterns frequently are the driving forces behind many such interactions.

A number of statistics addressed in this study reflect trends in human demography. The number of attacks in a particular region surely is affected by human use patterns; it is not surprising to see that the U.S. Pacific coast, particularly California (Table I), with its large and aquatically active population, leads the world in white shark attacks. Similarly, the continuous rise in the number of attacks throughout this century (Fig. 1) does not indicate that water-loving humans are under siege by sharks, but simply mirrors increases in worldwide population growth and aquatic activity. The number of attacks during the 1940s, a period dominated by World War II, decreased presumably due to reduced recreation in the white shark's preferred habitats. Young white males, the group that represents the white shark's statistical targets (see below), were the major demographic group recruited as combatants during the conflict. Since little military action occurred in areas of white shark abundance during this period, white shark attacks decreased.

Human demographic and behavioral patterns also explain the racial, sexual, and age structures of attack victims and the timing of attacks. Although control data are lacking, it seems most likely that characterization of the typical white shark attack victim as young, white, and male, again, is strongly influenced by the historical heavy-use pattern demonstrated by this group. Increased attacks on females in the 1980–1990s probably are related to increased participation in traditionally male-dominated aquatic activities, such as surfing and diving. The data related to the timing of attacks (Fig. 3) demonstrate especially well how human behavior influences attack statistics. The concentration of attacks between 1000 and 1759 hours matches the time distribution curve of 1018 bathers monitored at Myrtle Beach, South Carolina (Baldridge, 1974a). The decrease in white shark attacks in the 1200–1359 hours period mirrors a trend shown by Baldridge (1974a), using data combined for shark species. While both these midday reductions could reflect real shifts in shark behavior, it seems more logical to attribute them to human feeding activity!

The observed reduction in fatality rate through time (Fig. 2) also is explainable on the basis of human activities. Improved medical care and, more importantly, increases in the number of emergency treatment facilities in the proximity of attack sites, in major attack areas, has greatly reduced the mortality rate. In addition, the high pre-1960 rates probably are also attributable, in part, to biases in reporting. Most attack data from this period were derived from the print media, rather than personal investigations by ISAF cooperators (the ISAF was founded in late 1958). We believe that press coverage of shark attacks during this period often was limited to human deaths and that nonfatal attacks frequently did not warrant media coverage.

Lack of control data hinders interpretation of our data on human activity at the attack site. Attack statistics (Table V) indicate an even distribution of victims' activities. If statistics addressing the types of activity of nearby humans (Table VI) are an accurate approximation of relative effort, then the rates of attacks on each user group are proportionately similar. Somewhat fewer victims were engaged in swimming than were nearby nonvictims. Conversely, the aquatic activity of nearby persons does not measure person-hours in the water, only the presence or absence of a particular type of activity.

Without control data, it is also difficult to analyze results related to the fishing activities of the victim and those nearby. We cannot ascertain whether or not divers engaged in spear fishing, or other gathering of marine animals, are more prone to attack than nonfishing divers. Attack victims, and nonvictims swimming nearby, were equally engaged in some type of fishing (34.8% each) and expended similar efforts in spear fishing (49% and 53.1%, respectively) and other types of fishing (58.2% and 53.1%, respectively), indicating a certain degree of homogeneity. Thus, the somewhat higher percentage of attack victims (60%) noted carrying their catches, compared to nonvictims (43.8%) swimming nearby, may be of some significance.

Attacks on Humans Compared to Natural White Shark Biology and Behavior

How do the patterns observed in white shark attacks on humans relate to natural biological and behavioral patterns of white sharks? It is instructive to compare white shark attack patterns with our current perceptions of the predatory strategies, behavior, and ecology of white sharks.

1. Large white sharks (>300 cm; most, 350–500 cm) predominantly feed on solitary pinnipeds and cetaceans (Le Boeuf *et al.*, 1982; Tricas and McCosker,

1984; Ainley *et al.*, 1985; McCosker, 1985; Long, 1991a; Long *et al.*, Chapter 24). Evidence from our analyses indicates that humans present an image similar enough to that of pinnipeds and cetaceans, in terms of size, shape, color, and behavior, that predatory behavior is triggered in large white sharks (see Chapter 20, by Anderson *et al.*, and Chapter 21, by Strong).

Statistics indicate that attacking sharks were large (88% >260 cm and 63% in the 320- to 500-cm range; Table X). Attack victims were primarily adults and young adults (95.3% aged 15 or older; Fig. 5), and therefore of adult human size, >150 cm and >45.4 kg or heavier. These sizes compare favorably with those of the white shark's preferred prey of several pinniped/cetacean species. The shape of a human in the water or on a surfboard or kayak certainly approximates that of marine mammals (see the compelling photograph in Tricas and McCosker, 1984; see also Chapter 20, by Anderson *et al.*, yet see Chapter 19, by Collier *et al.*, for attacks on dissimilar objects).

Gear/clothing worn by victims was overwhelmingly dark (88.9%: black, 74.1%; blue, 14.8%; Table VIII), similar to the dark coloration exhibited by many marine mammals. These dark colors also enhance the development of a distinct near-surface silhouette when viewed from below (see Chapter 21, by Strong). Our data indicate that black was selected over other colors. One might argue that the high percentage of victims wearing black simply is a reflection of the number of victims wearing wet suits. Although wet suits are generally colored black, in recent years other color schemes have become popular. Significantly, wet/dry suits reportedly were worn in only 40.8% of reported attacks (Table VII), leaving a large number of attacks that occurred on non–wet-suited, but still black-appareled, victims.

Large white sharks normally attack solitary marine mammals. Isolated humans were more frequently attacked (no humans within 3.1 m of the victim in 84.8% of the attacks, none within 3.1–15.2 m in 51%) than those clumped in distribution.

While white sharks are known to perform "investigational strikes" at objects of varying sizes, shapes, and colors (see Chapter 19, by Collier *et al.*, and Chapter 20, by Anderson *et al.*), it seems reasonable that they would most frequently strike an object similar in shape and color to traditional prey items. This is borne out by Strong's (Chapter 21) observation that white sharks moved toward a seal-shaped plywood sheet more often than to a square-shaped sheet. Darkly attired adult-sized humans swimming alone near the water's surface probably produce a visually acceptable facsimile of a white shark's pinniped/small cetacean search image.

2. Large white sharks normally take their large prey items at or near the surface (Ainley *et al.*, 1981; Tricas and McCosker, 1984; McCosker, 1985; Klimley *et al.*, Chapter 16). Most (78.7%) attacks on humans occurred in the upper 1.7 m of the water column (Table II), confirming this observation. However, since 70.3% of human activity was spent at or near the surface or standing in shallow water, this statistic is not unequivocally useful in this regard.

3. Large white sharks attack violently and by surprise (Tricas and McCosker, 1984; McCosker, 1985; Klimley *et al.*, 1992, Chapter 16). "Sudden and violent" attacks were reported in 92.4% of attacks on humans. By contrast, in Florida shark attacks, in which white sharks have not been implicated, over half of initial attacks are cited as occurring "with minimum turmoil" (G. H. Burgess, unpublished data).

4. Large white sharks attack from beneath and behind (Tricas and McCosker, 1984; Ainley *et al.*, 1985; McCosker, 1985; Klimley *et al.*, Chapter 16) versus from above (Klimley *et al.*, Chapter 16). Human posture in the water varies from vertical to horizontal, depending on the activity (e.g., surfing), and a victim's sense of the direction of attack is based on his or her sight perspective during the initial interaction. Most (89.1%) white shark attacks on humans occurred from "below," "behind," or "from the side." Since human activity was focused primarily at or near the surface (70.3%), one would expect attacks "from above" to be physically impossible much of the time, and they proved to be virtually nonexistent (2.4%). The low incidence of frontal attacks (8.4%) and the high percentage (64.4%) of victims who never saw the attacking shark prior to contact indicate that approach from below and behind is a preferred strategy on a prey item located near the surface. The observation by Klimley *et al.* (Chapter 16) of so many beheaded seals may indicate an alternative white shark feeding behavior predicated on attacking diving, rather than surfacing, seals.

5. Large white sharks use a "bite and spit" mode of prey capture (Tricas and McCosker, 1984; McCosker, 1985) versus an "exsanguination"/rejection feeding strategy (Klimley *et al.*, Chapter 16). "Bite and spit," in which the prey item is attacked, released, and allowed to bleed to death or lapse into shock prior to a second attack, has been proposed by Tricas and McCosker (1984) as the white shark's natural feeding mode. Klimley *et al.* (Chapter 16) advanced an alternative, "exsanguination" hypothesis, which is actually a variation on the same theme. They suggest that a white shark carries an acceptable prey item underwater for an extended period until it is exsanguinated, releases it, and then returns to resume

feeding. Unacceptable prey, such as penguins, pelicans, sea otters, humans, and decomposing pinnipeds, are rejected after an initial attack and not subjected to repeat attacks. They further suggest that high-fat species constitute the most acceptable prey, based on palatability.

Our statistics confirm neither hypothesis. Over half (56.8%) of human victims received only a single bite (Table XII), and in 69.8% of the attacks, the shark did not return after its last (or only) bite (Table XIII). In some cases, the victim was carried, occasionally for a prolonged distance, and in others there was initial bite and release. If the "bite and spit" hypothesis were in effect, one might expect more frequent multiple bites from the attacking shark and more persistent attempts at consumption. This may be explained, though, on the basis of humans' being more capable of escape after an initial attack due to the intervention of rescuers (Baldridge, 1988a; see also Chapter 44, by Baldridge) or by the shark's being frightened by a sudden loud sound such as a boat motor (Myrberg *et al.*, 1978; Klimley and Myrberg, 1979; Klimley *et al.*, Chapter 16). Alternatively, humans may be rejected by white sharks after an initial attack, as has been suggested also for attacks on sea otters (Ames and Morejohn, 1980; Klimley *et al.*, Chapter 16). However, the relatively high percentage (32.8%; Table XII) of cases in which multiple bites occurred, including many instances in which victims were wholly consumed, argues that humans are not uniformly unpalatable. Perhaps humans are treated somewhat differently than natural food items, because they are novelties in the domain of the white shark; as such, they may elicit more cautious secondary attack behavior by the shark. This allows for escape using routes and methods (e.g., boarding a boat or receiving aid from other humans) unavailable to natural prey species. Another plausible explanation, which is also applicable to both hypotheses of predatory strategy, is that we mainly are observing failed or poorly executed attacks on humans (or natural prey) when we see those that are released; those not released (representing 100% mortality) may represent successful attacks.

6. Large white sharks are persistent after the first attack (Tricas and McCosker, 1984). White shark statistics are somewhat ambivalent on this matter. Two or more bites were recorded in only 32.8% of the cases (Table XII), yet in 80.9%, one or more close passes were noted (Table XI). In those attacks in which repetitive bites were recorded, 68.3% were described as "deliberate." Once again, it is possible that intervening humans may have interfered with the white shark's full predatory sequence on a human victim.

Also, a victim's perception of whether or not a shark is "deliberate" in attack is certainly a biased one.

7. White sharks attack in nearshore (25–450 m from land) shallow waters (4–12 m) (Ainley *et al.*, 1980; Klimley *et al.*, Chapter 16; Pyle *et al.*, 26). Nearshore waters were described as the sites of white shark attacks in 65% of the cases. Attacks on humans occurred at the same depths (Table III) as shark predation on natural prey in the Farallon Islands: 84.9% in waters ≤12.3 m and 53.8% in water 3.2–12.3 m deep. We have no control data on the choice of distance from shore or depth zones by humans, but it is safe to assume that shallow nearshore waters are more highly used than deeper offshore waters. Thus, we cannot rule out the possibility that the predominance of attacks in shallow nearshore waters simply is a reflection of human use patterns; parenthetically, we are equally unaware of the extent of white shark natural predation in deeper offshore waters.

8. Large white shark attacks occur most often in turbid water (see Chapter 26, by Pyle *et al.*). Descriptions of water clarity are highly subjective. Humans were attacked less frequently in turbid (44.7%) than in clear (55.3%) water. This is in contrast to trends observed in Florida's non-white shark attacks (G. H. Burgess, unpublished data), with waters usually turbid, but approximates Baldridge's (1974a) worldwide (all shark species) figure of 48.5% under turbid conditions.

9. White sharks are most common in waters warmer than 12–14°C (Tricas and McCosker, 1984). Although white shark attacks have been documented in waters as low as 8°C, 89.5% were recorded in seawater temperatures ≥12°C. Most attacks (60.5%) occurred in temperatures ≤20°C. This would appear to represent a truly white shark-influenced trend, since Baldridge (1974a) contended that the worldwide predominance of multispecies attacks at 21–29°C was largely based on human comfort requirements rather than shark preferences.

Advice to Those Who Share Waters with White Sharks

White shark attacks on humans are relatively rare events, yet the potential consequences are so great that the problem cannot be dismissed. How can humans engaged in aquatic activities within the natural range of white sharks reduce the chances of encountering one? We discuss possible courses of action below, but also advise consideration of the points made by Baldridge (Chapter 44).

Since white shark abundance is seasonal in certain areas (Ainley *et al.*, 1981, 1985; Klimley and Anderson, Chapter 33; Pyle *et al.*, Chapter 26), humans can limit their time in the water in areas and during periods of peak shark abundance. If a white shark is sighted or an incident occurs at a particular site, continued human use of the site should be discouraged, inasmuch as the species exhibits at least some site specificity (Goldman *et al.*, Chapter 11; Klimley and Anderson, Chapter 33; Strong *et al.*, Chapter 37). Caution should be taken to avoid regions of high pinniped and cetacean abundance, especially pinniped haul-out areas. Activity at the water surface should be limited whenever possible; this is most applicable to divers, but is obviously not feasible for surfers. Scuba divers can minimize their surface and midwater time by making underwater travel swims just off the bottom (as practiced by elephant seals; see Chapter 17, by Le Boeuf and Crocker) and reducing the number of trips to surface vessels. Kelp forests may offer some protection (see Chapter 28, by Ames *et al.*). The color black appears to encourage white shark interest. Thus, we would advise the use of alternatively colored swimwear (cf. Chapter 44, by Baldridge). Carrying one's catch should be avoided. Perhaps the most important suggestion to those engaged in aquatic activities is to stay in groups; white sharks clearly favor solitary individuals. Surfers, by the nature of their sport, almost always violate this suggestion.

Steady worldwide increases in human recreational activity in nearshore waters, accompanied by gradually rebounding marine mammal populations in some geographical regions, almost ensure that the number of white shark attacks will continue to rise. When humans enter the marine realm, they are visiting a foreign environment, one that includes risks as well as the potential for personal gratification. There are acknowledged risks involved in all aquatic activities, not the least of which is drowning. Swimmers, surfers, and divers encounter adverse physical forces such as strong currents, undertow, and breaking surf. Surfers also risk potential injuries such as spinal damage, cuts, and contusions, and divers always must consider embolisms. Thus, a conscious decision is made every time a human enters the sea; risks are weighed and taken. White sharks are a risk that humans must consider when entering their domain; using proper caution, that risk can be significantly minimized, but never completely avoided.

Summary

Attacks by white sharks *C. carcharias* on humans draw international attention, but they have not been critically examined on a worldwide scale. In this study, 159 documented unprovoked attacks by white sharks on live humans were analyzed for patterns involving geographical, historical, demographic, temporal, environmental, morphological, and behavioral parameters associated with both the attacking shark and its victim. One half (50.6%) of the attacks occurred along the Pacific coast of the United States, with smaller numbers primarily occurring in Australia (20.2%) and South Africa (10.3%). Absolute numbers of attacks have steadily increased since the 1950s, mirroring trends in human recreational use of nearshore waters. Attacks were largely nonfatal (73.9%), with the fatality rate dropping from 91.7%, before 1950, to 15.2% for the years since 1970. Attacks occurred almost exclusively during daylight, especially in the 1000–1600 hours period. Attacks took place most often in clear water and at water temperatures ranging from 12°C to 20°C. Victims were primarily young (15- to 34-year-old) white males. Most attacks occurred at or near the surface, with over half in water depths ≤6.2 m. Primary attack sites were nearshore waters, especially beaches and the associated surf zone, and offshore reefs/banks. Victim activity was almost equally divided among surfing/rafting, diving/snorkeling, and swimming. Of those divers reported to be engaged in fishing (including shellfish gathering), half were holding a catch at the time of attack. Less than one half (40.8%) of victims were wearing wet suits. The primary color of a victim's gear/clothing was black/gray (74.1%), with approximately equal secondary occurrences of a spectrum of colors. Attacks took place on solitary individuals; 84.8% of attacks occurred when no other humans were within 3 m, and half with no others within 3–15.2 m. The size of the attacking shark was estimated to be 260–500 cm total length. Most victims noted one or more close passes without contact prior to or during the attack sequence. Attacks largely occurred from below and the side of the victim. The initial strike was almost uniformly sudden and violent. Only a single discrete bite occurred in one half of the attacks; multiple bites occurred one third of the time. Following the last strike, the attacking shark was said to leave the area two thirds of the time. These patterns were discussed in relation to white shark feeding patterns on human prey.

Acknowledgments

We thank the many ISAF cooperators, past and present, who made this study possible through their unselfish investigative efforts; in particular, we are indebted to P. Gilbert for his role in the development and maintenance of the ISAF and to H. D. Baldridge for providing his reduced data (Baldridge, 1988b). We especially thank G. Cliff, C. Duffy, R. N. Lea, J. E. McCosker, K. Nakaya, and J. West for sharing information on white shark attacks in their respective regions. A. P. Klimley and D. G. Ainley freely shared observations of white shark behavior and offered stimulating comments in review of the manuscript. E. Watts provided valuable assistance during the data reduction, and S. J. Walsh graciously assisted with the figures. We gratefully acknowledge the financial assistance provided by R. C. Dorion through the Stewart Springer Fund.

Chemical Repellent Tests on White Sharks, with Comments on Repellent Delivery Methods

DONALD R. NELSON
Department of Biological Sciences
California State University
Long Beach, California

WESLEY R. STRONG, JR.
Department of Biology
University of California, Santa Barbara
Santa Barbara, California, and
The Cousteau Society
Chesapeake, Virginia

Introduction

Certain surfactant chemicals have been shown to be effective in repelling sharks if delivered in a directed manner in sufficient quantity and concentration. Based on this, we have been researching the feasibility of a practical directed-application repellent device. Previous field tests have been primarily on bait-attracted blue sharks, *Prionace glauca*, while laboratory studies have used various small sharks due to the constraints of the apparatus. Prior to the results presented here, no trials had been accomplished on sharks as large and as dangerous as adult white sharks.

The identification of surfactants as shark repellents originated with the discovery by Clark (1974, 1983) that a secretion of the Moses sole, *Pardachirus marmoratus*, was aversive to sharks and the observation by Zlotkin and Barenholz (1983) that pardaxin (the active component of the secretion) possessed surfactant (detergent) properties damaging to cell membranes. Subsequent tests of several commercial surfactants showed that one compound—sodium dodecyl sulfate [SDS; also known as sodium lauryl sulfate (SLS), especially in the relatively impure commercial form]—re-pelled small captive lemon sharks, *Negaprion brevirostris*, better than the dried/redissolved Moses sole secretion (Gruber and Zlotkin, 1982; Gruber *et al.*, 1984). Field tests on blue sharks showed that moderate doses of SDS (e.g., 100 ml of 15% solution or 250 ml of 5–10% solution), delivered to the mouth cavity, nearly always resulted in mouth gaping, head shaking, and permanent departure at accelerated speed (Nelson, 1991; Gruber *et al.*, 1984). Control doses of seawater had no effect. Thresholds of repellency for SDS and other chemicals were determined in laboratory studies by Smith (1991) and Sisneros (1993).

The opportunity for repellent tests on large white sharks occurred during a Cousteau Society expedition to South Australia (Cousteau and Richards, 1992). Using the Society's *R/V Alcyone* and the South Australian Fisheries' *R/V Ngerin*, five multiweek visits were made to islands in the Spencer Gulf region for the purpose of filming and studying white sharks. The primary research objectives were acoustic tracking of shark movements (Strong *et al.*, 1992, Chapter 37), observations of behavior at bait sources (Strong, Chapter 21), and population assessments (Bruce, 1992; Strong *et al.*, Chapter 37). The chemical repellent tests were conducted near the termination of the

FIGURE 1 (A) W. R. Strong delivers a measured dose of chemical repellent to a bait-attracted white shark from the stern platform of the *R/V Alcyone*. (Courtesy of the Cousteau Society.) (B) The air-powered syringe gun used in the repellent trials. The tapered fiberglass barrel is made from a heavy-duty fishing rod blank. The acrylic cylinder contains 250 ml of test solution (to the right of the piston) and compressed air (60 psi; to the left). The device is discharged by opening the thumb-operated valve, which allows the compressed air to drive the piston to the right, pushing the contents out the barrel in approximately 1 second. A one-way poppet valve at the tip of the barrel prevents leakage of the chemical prior to discharge. Test nozzles provide either a broad conical spray pattern for in-mouth delivery or a narrower pattern for outside-mouth delivery. The length of the gun is 122 cm.

last two of these visits, because driving away the sharks would interfere with other objectives.

Methods

The tests were conducted at Dangerous Reef, South Australia. In each trial, 250 ml of a 10% seawater solution of SDS was delivered into the mouth of the shark by an air-powered syringe gun (Fig. 1). These trials were conducted by one of us (W.R.S.) from the swimstep of the vessel on sharks brought within reach by a large piece of bait on a line. Individuals were recognized by a combination of size, sex, body markings, and, in some cases, color-coded dart tags. At the moment of the chemical shot, each shark had its head out of the water to approximately the first gill slit, and three had the piece of bait in the mouth. Views of behaviors during the underwater departures were poor, because all observers were topside. Attempts to bring sharks close enough for trials from the underwater shark cages were unsuccessful.

Results

Four directed-application trials were accomplished as described below from our field notes (Table I). Upon receiving the chemical, all sharks responded vigorously (latency of approximately 2 seconds), rejecting the baits and accelerating away underwater. Three of the four trials resulted in permanent departures, with the shark not being seen again for the remainder of the baiting session.

Trial 1 (Shark 17)

The shark received only a partial dose, as it pulled its head away at the moment of the shot. It then passed under the stern, vigorously shook its head and body, and rapidly accelerated away near the surface. Two minutes later, a shark (possibly No. 17) was seen on the surface 25 m away. At 12 minutes after the shot, shark 17 began passing through the baited area more regularly, but did not feed as readily.

Trial 2 (Shark 11)

A full dose was injected into the mouth as the shark went for the bait. With quick beats of its tail, the shark departed rapidly almost straight down. Two minutes later, a shark (identified as No. 11) swam slowly past relatively deep, but did not show interest in the baits. Shark 11 was not seen again for the remainder of the trip (4 days), but was seen on the next trip (144 days later).

Trial 3 (Shark 15)

This shark had fed vigorously and was attacking baits more actively than any other shark on this trip. A successful shot was delivered at full dose. After a latency of about 2 seconds, the shark made several quick tail beats and accelerated downward at an angle of about 70°. Shark 15 was not seen again for remainder of the trip (approximately 2.5 days), but was seen two trips (322 days) later.

Trial 4 (Shark 62)

The shark had been seen intermittently since the day before, but a repellent trial was not conducted until after dark (2030 hours). A full dose of SDS was delivered. The shark paused very briefly, then slammed its head back into the water and accelerated downward and away. Baiting and observing were continued for 2 more hours prior to the vessel's departure for port. No sharks were seen.

TABLE I **Directed-Application Repellent Trials on White Sharks at Dangerous Reef, South Australia**

Date	Shark no.	Total length (m)	Sex	Time signted (hours)	Time of trial (hours)	Amount of chemical (m)[a]	Next sighting near bait
March 6, 1990	17	4.3	F	1615	1819	125	12 minutes later
March 12, 1990	11	2.9	F	1530	1552	250	144 days later
March 12, 1990	15	3.4	F	1305	1611	250	322 days later
September 24, 1991	62	3.0	F	0100	2030	250	—

[a]10% seawater solution of sodium dodecyl sulfate (SDS).

Discussion

Although we accomplished just four trials, our results confirm that large white sharks can be effectively repelled by the directed application of a reasonable quantity of surfactant solution. Results also show that the dosage used is not lethal to the sharks, as two of the individuals repelled were seen on later trips. One remaining question is whether large white sharks would require a larger dose (volume or concentration) of repellent than would smaller species such as the blue shark. For both blue and white sharks, 250-ml shots of 10% SDS were sufficient for a strong reaction, if a substantial part of it reached the mouth cavity and remained there long enough to stimulate the relevant receptors. There is no reason to suspect that the white shark's irritation receptors are less sensitive than those of other sharks. However, the larger volume of the white shark's mouth cavity might lead to a greater dilution of the chemical before it reaches the receptors.

From laboratory studies (see below), we know that if the entire 250-ml shot of 10% SDS were to remain within the mouth cavity for several seconds, the dose would be several orders of magnitude above the threshold for repellency. In the field situation, however, this does not happen. In tests by divers on free-swimming blue sharks, the placement of the syringe gun nozzle tip varied from trial to trial, resulting in some shots delivering most of the chemical in front of or out the side of the mouth rather than inside it. The movement of the shark's head immediately after the shot also affected the net dose received. If the head, with mouth open, was swung to one side, most of the chemical would be washed clear of the mouth in a fraction of 1 second. Rapid flushing also happens when the mouth-open shark accelerates forward. A relatively large dose is needed, therefore, to insure that, in spite of these losses, enough residual chemical will remain in the mouth long enough to effect the response.

In questioning the practicality of surfactants as shark repellents, one must consider the method by which the chemical is delivered. In the directed-application method, as described above, a relatively concentrated dose of chemical is delivered to the shark's sensitive area. It requires the user to aim and discharge the delivery device at close range, and, as such, is not suitable in some applications, for example, when the person is unconscious or otherwise unable or unwilling to face the shark. A repellent device of this type would be useful for divers, life raft occupants, etc., as a combined shark billy/discharge gun. It could initially be used as a billy (physical prod), but would have the capability of chemical discharge, if the shark did become discouraged by physical means alone.

The enveloping-cloud (surrounding-cloud) repellent method depends on the creation and maintenance around the person of a cloud of dilute chemical potent enough to turn back a shark that enters it. This was the desired mode of action of the original "Shark Chaser" (i.e., the dispersable packet developed during World War II by the U.S. military), since withdrawn from service because of its ineffectiveness (Nelson, 1983). However, an "effective life jacket shark repellent packet" remains an "unsatisfied requirement" for the military (Baldridge, 1990) for passively protecting survivors of an air crash or ship sinking. The idea of a small chemical packet attached to survival gear is attractive, but a major problem exists: no chemical has been found that is potent enough to create the necessary enveloping cloud.

Based on an analysis of attack situations and the kinetics of chemical dispersion at sea, Johnson and Baldridge (1985) established a set of requirements for a practical life jacket repellent packet for the military. One requirement was that the chemical be "essentially instantaneously effective" at a concentration of 0.1 ppm, or effective with an exposure integral of 0.1 $sec/g/m^3$. Tests of surfactants in a laboratory situation simulating enveloping-cloud repellency were conducted by Smith (1991) and Sisneros (1993). In special roundabout tanks, swimming sharks were directed to pass through sections of water containing the chemical in uniformly dispersed form. Thresholds ($EC_{50}s$) for useful repellency of horn sharks, *Heterodontus francisci*; swell sharks, *Cephaloscyllium ventriosum*; and leopard sharks, *Triakis semifasciata* were in the vicinity of 100 ppm for SDS, the most potent of the several surfactants tested. In these studies, thresholds for "strong" repellent responses ranged from 75 to 175 ppm, while those for "minimum noticeable" responses ranged from 40 to 95 ppm. It is obvious from these results that surfactants such as SDS, while useful as repellents in the directed-application mode, are not nearly potent enough (by three orders of magnitude) to be practical in the enveloping-cloud mode.

What is the prognosis for eventually finding or developing a repellent chemical potent enough to meet the requirements set by Johnson and Baldridge (1985)? First, the human user must be able to remain in the central, most concentrated part of the cloud for a prolonged period. The chemical, therefore, must be very innocuous to the person but rapidly very effective against the shark. This is a more serious problem for the enveloping-cloud method than for the directed-application method, in which the shark, not the person, gets the most concentrated dose. Furthermore, the surfactant chemicals used in the above directed-application tests act primarily as irritants, most likely sensed by irritation receptors, not olfactory receptors. A squirt of SDS to a shark is comparable to a spray of Mace or pepper gas to a person. Irritants, while eliciting immediate strong responses, normally do not operate at very small concentrations (Nelson, 1983). Odorants, however, are sensed by olfactory receptors and are commonly detectable by animals at concentrations much smaller than the 0.1-ppm criterion for a life jacket repellent. Baldridge (1990) was pessimistic about the chances of finding a suitable chemical among those eliciting "direct physiological changes" (i.e., toxicants and irritants) and felt that if one is ever found, it will be among those that convey information, that is, semiochemicals, such as pheromones.

It is interesting, as Baldridge pointed out, that the original impetus for one of the components of Shark Chaser (the acetate ion) was that it was found in quantity (as ammonium acetate) in decaying shark flesh. Shark fishermen of the time believed that shark carcasses left in the water tended to drive other sharks out of the area, resulting in poor subsequent catches. If this is true, then the responsible chemical (probably not ammonium acetate) is acting at very small concentrations, undoubtedly via the olfactory sense, and probably as a semiochemical warning of a danger to sharks in that area. However, even if such a repellent semiochemical is identified, there is no guarantee that its mode of action would be rapid enough to produce the immediate strong withdrawal response needed for a personal shark repellent.

In regard to the particular situation in which white sharks attack humans, how effective would the enveloping-cloud or directed-application method of repellency be? Most attacks have been on persons relatively near shore and actively engaged in water sports, such as swimming, diving, and surfing (Miller and Collier, 1981; Tricas and McCosker, 1984; see also other chapters in this volume). The attack is typically a direct strike from below and behind, with the victim rarely, if ever, seeing the shark before the attack. This is not an appropriate situation for any enveloping-cloud repellent, a concept intended for air/sea disaster victims passively awaiting rescue on the surface, usually far at sea. Probably no chemical method could prevent a first-strike attack by an unseen white shark on an actively moving person. After the person is seized, however, a device capable of directed delivery of a strong repellent would give the victim a significantly better chance to fight off the shark and counter any return attacks.

Summary

The opportunity to test a chemical repellent on large white sharks occurred during a Cousteau Society filming/research expedition in South Australia. Four directed-application trials were accomplished on sharks bait-attracted to the water surface near the boat. The surfactant SDS was delivered to the mouth cavity of the sharks by an air-powered syringe gun. The full dosage (250 ml of 10% solution) was sufficient to cause accelerated, permanent departure by the sharks. The practicality of repelling sharks with chemicals was discussed in terms of directed-application and enveloping-cloud methods of delivery. The latter method is not yet feasible with presently known chemicals, especially in attack situations involving white sharks.

Acknowledgments

We are grateful to the Cousteau Society, especially R. Murphy and J.-M. Cousteau, for providing the opportunity to conduct these tests on white sharks. We also acknowledge the cooperation of B. Bruce and the South Australian Department of Fisheries for additional logistical support. The U.S. Office of Naval Research provided initial support for the program on shark repellent research, of which this study is a part.

44

Comments on Means for Avoidance or Deterrence of White Shark Attacks on Humans

H. DAVID BALDRIDGE, JR.

Sarasota, Florida

Introduction

White sharks *Carcharodon carcharias* have attacked humans engaged in the full range of marine recreational and occupational activities. Bathers have been attacked at beaches; surfers and kayakers, farther offshore; and all forms of divers and spear fishermen, both on the surface and at depth.

Of all species, white sharks and tiger sharks *Galeocerdo cuvieri* are the ones most often implicated in attacks. The median length of all attackers as a group is estimated at about 2.1 m. Tiger sharks reach sizes twice that, and white sharks grow to total lengths perhaps three times the median for attackers in general.

Consequently, means for avoidance or deterrence of attack by a white shark, of necessity, must incorporate all those measures historically proposed for mollifying attacks in general, while at the same time taking into account the special considerations occasioned by a predator that is very much larger than its infrequent human prey.

Consideration of attacks on humans by very large white sharks, apart from aggressions by all species and sizes of sharks, is analogous, considering automobile accidents in general, to those patently one-sided high-speed engagements between very large trucks and subcompact cars. Clearly, a vastly different set of statistics will result. In both cases, avoidance would appear to be far more workable than deterrence.

Discussion

Time to React

One important survival strategy exploits a frequently observed attack behavior of large white sharks, termed bite-and-release, which has afforded more than a few victims the time after an initial strike to escape or be rescued. Having been suddenly taken in a powerful grasp and dragged or propelled to about the limit of breath-holding, the victim is just as suddenly released. Unfortunately, for some victims, the time factors are not always so favorable. Nevertheless, bite-and-release behavior has been observed often enough to become a recognized factor in surviving an attack by a large white shark.

Some researchers believe bite-and-release to be an attack tactic used by white sharks whereby prey is mortally wounded in the initial onslaught, after which the shark has only to disengage and await the victim's demise (Tricas and McCosker, 1984). Others have proposed that wounded prey might be released because the shark sensed something unpalatable in its physical or chemical makeup (see Chapter 16, by Klimley *et al.*; Chapter 28, by Ames *et al.*). Let me now propose a couple of reasonable alternatives that would not require cognition by the shark.

First, at the same time that the shark is drowning its ensnared victim, human or marine animal, the victim is also drowning the shark. Being a mouth-breather, the shark must continually take water into its mouth in order to oxygenate its gill tissues at a linear flow rate of about 1 m/sec (estimated minimum swimming speed) (see Chapter 11, by Goldman *et al.*; Chapter 37, by Strong *et al.*). With the shark's mouth jammed full of prey, such flow ceases, and the only way to reinstate it is for the shark to release its victim. Relaxing the bite might simply be a voluntary or involuntary response by the shark to a sensory signal indicating inadequate flow of water over the gills.

Second, there is a real possibility that a shark is incapable of bringing its full biting force to bear except from a wide-opened jaw starting position. While a large white shark often severs whatever it takes in its initial bite, it might not be physically able to execute that same force from a closed-mouth condition of muscle contraction. Having failed for whatever reason to remove an ingestible portion in the initial bite, and not able to swallow the prey whole in its present orientation, the shark would have no alternative but to release its victim, at least momentarily.

Regardless of the reason for bite-and-release, this behavioral pattern happens regularly enough to make it an important consideration in surviving an attack, when the victim is close enough to shore to escape or companions in a boat are near enough to effect a quick rescue.

Good Common Sense

The watchwords for minimizing the hazard of attack by a white shark, large or small, are the same as for attacks in general: *good common sense.* For people who are determined to venture well away from heavily populated beaches (e.g., divers, spear fishermen, kayakers, or surfers), good common sense would certainly include becoming familiar with intended areas of operation—with particular emphasis on evaluating the up-to-date but ever-changing biological profile. One might envision a broad communication network involving dive shops, clubs, governmental agencies, etc., from which interested parties might receive regularly revised information on shark sightings, unusual activity among prey species, and so forth. Nevertheless, the final acceptance of risk would always remain with the individual, for, particularly in our present litigious society, no such network, or any component thereof, should ever offer any guarantee of minimal chance of attack in any area at any time.

Good common sense requires that careful consideration be given to visual presentations made to white sharks by potential victims, especially to the heavy predominance of black or gray as the primary color of gear reportedly used by known victims (see Chapter 40, by Levine; Chapter 42, by Burgess and Callahan). On the other hand, how important is color, when attacks are so often directed at silhouetted victims (see Klimley, 1994, Chapter 20, by Anderson *et al.*; Chapter 21, by Strong)?

Other Considerations

The U.S. Navy currently provides, as life jacket survival equipment, a small (1-in. × 6-in. cylinder) battery-operated strobe light (TEKNA-MINI-STROBE). Would a white shark be as likely to strike such brightly flashing prey (body, board, or boat) either clearly discernible or in confusing silhouette, intentionally or mistakenly?

The use of any type of weapon to deter a large white shark would be far more successful under conditions in which the human is the aggressor, that is, a chemically charged squirt gun used to discourage and/or drive away an uncommitted potential attacker (see Chapter 43, by Nelson and Strong). Good common sense in any such scenario would take on a new level of meaning.

Once in the sudden grasp of a large white shark, some victims have survived by intentionally doing absolutely nothing, and others, by doing everything possible in the way of probing and/or striking every part of the shark within reach. What course of action should be taken, if "best" cannot be concluded with conviction? The numbers are statistically too small, and no control data are available on those who tried the same tactics unsuccessfully. It is certainly reasonable that action or inaction by survivors had little, if anything, to do with either the fact or timing of their release by the shark before they were either more seriously injured or drowned. In the behavior of bite-and-release, it appears to be the sole prerogative of a large white shark to release its victim as much as it was to attack in the first place.

As long as humans are attracted to the sea for play or profit, total avoidance of attack by a white shark is simply not realistic. The sea is a risky place, at best, and to use it in relative safety is to accept that risk on an enlightened basis, with the clear understanding that the highly improbable will never become the impossible. I made a series of recommendations on how to avoid shark attacks in 1974, and these are given in Appendix 1. Most of the advice is just as valid today as it was then, when much less was known about white sharks.

APPENDIX 1[a]

Advice to Bathers and Swimmers

Always swim with a companion, and do not wander away from a cohesive group of other bathers and thereby isolate yourself as a prime target for attack.

Do not swim in water known to be frequented by dangerous sharks. Leave the water if sharks have been recently sighted or are thought to be in the area.

Although not conclusively proven, human blood is highly suspect as an attractant and excitant for sharks. Keep out of the water if possessed of open wounds or sores. Women should avoid swimming in the sea during menstrual periods.

It is not always convenient, but very murky or turbid water of limited underwater visibility should be avoided if possible. In any event, a particularly watchful eye should be maintained for shadows and movements in the water. If there is any doubt, get out at once.

Refrain from swimming far from shore, where encountering a shark becomes more probable.

Avoid swimming alongside channels or drop-offs to deeper water, which provide ready access for a shark.

Leave the water if fish are noticed in unusual numbers or behaving in an erratic manner.

Take no comfort in the sighting of porpoises, for this does not at all mean that sharks are not about.

Avoid uneven tanning of the skin prior to bathing in the sea, for sharks apparently respond to such discontinuities of shading.

Use discretion in putting human waste into the water.

Avoid swimming with an animal, such as a dog or a horse.

Take time to look around carefully before jumping or diving into the sea from a boat.

Particularly at low tide, take notice of a nearby offshore sandbar or reef that might have entrapped a shark.

Avoid swimming at dusk or at night, when many species of sharks are known to be searching for food.

It might be a good idea to select other than extremely bright colors for swimwear.

Never, in any form or fashion, molest a shark no matter how small it is or how harmless it might appear.

Keep a wary eye out toward the open sea for anything suggestive of an approaching shark.

Advice to Divers

NEVER DIVE ALONE. Not only might the very presence of your diving buddy deter the shark, but together you have a far better chance of becoming aware of a nearby shark in time to take effective countermeasures. Furthermore, if something did happen to you, at least there would be assistance close at hand.

Do not in any way provoke even a small shark—not by spearing, riding, hanging onto its tail, or anything else that might seem like a good idea at the time. Even a very small shark can inflict serious, possibly fatal, injury to a human.

Do not keep captured fish, dead or alive, about your person or tethered to you on a stringer or similar device. Remove all speared or otherwise wounded fish from the water immediately.

Do not spear fish in the same waters for such extended periods that curious sharks may be drawn to the area by either your prolonged quick movements or an accumulation of body juices from numbers of wounded fish.

Leave the water as soon as possible after sighting a shark of reasonable size, even if it appears to be minding its own business.

Submerged divers, as opposed to surface swimmers, have a better chance of seeing a shark making investigatory passes prior to being committed to attack. Use smooth swimming strokes, making no undue commotion, in reaching the safety of a boat or the shore. To the greatest extent possible, remain submerged, where chances are greater for watching the shark and countering its charge if attack occurs. Do not count on the shark's either circling or passing close at hand without contact before it makes a direct run.

Use discretion in the choice of wetsuit colors in terms of conditions and sea life prevalent in the waters of intended operations. Do not take a chance on being mistaken for the area shark's natural prey of choice.

Carry a shark billy or plan to use the butt of a speargun for this purpose, if necessary. Such devices have been shown to be very effective in holding an aggressive shark at bay until its ardor cools.

Take full advantage of your submerged position and limits of visibility to be aware always of nearby movements and presences. Shark attack case histories indicate that such vigilance has played a major role in lowering injuries and mortality rates among diver-victims.

Do not maneuver a shark into a trapped position between yourself and any obstacle such as the beach, reef, sandbar, or possibly even a boat.

As with swimmers, do not wander away from an established group of other divers and thereby possibly give an appearance of fair game. Avoid diving at dusk and at night.

Advice to Victims

Try to remain calm and take full advantage of weapons available to you.

Use any object at hand to fend off the shark while at the same time not intentionally provoking it further.

Keep fully in mind the limitations of such devices as powerheads, gas guns, and spearguns, and do not expect them to accomplish the impossible. Such weapons, if used improperly, may serve only to further agitate the shark.

Use available spears and knives first to fend off the shark, and attempt to wound the fish only as a last resort. Sharks often seem to react with increased vigor to efforts at sticking it with pointed objects.

Discretion should be used in making aggressive movements toward a shark. One that had not yet committed itself to attack might be "turned on" by such movements if interpreted by it as a threat. On the other hand, quick movements toward a shark close at hand might produce a desirable startle response.

Once contact has been made or is imminent, fight the shark as best you can. Hit it with your bare hands only as a last resort. Probing the shark's eyes especially, and perhaps also its gills, has often been effective. Startle responses, which at least buy valuable time, have been produced occasionally by such actions as shouting underwater or blowing bubbles. Do anything that comes to mind, because the seconds or minutes of time during which the shark might withdraw as a result could be sufficient to effect your rescue.

Most shark attacks produce wounds that are readily survivable. Bleeding should be controlled as quickly as possible—even before the victim has been brought ashore. Treatment by a physician is indicated even where wounds are relatively minor.

---

[a]From Baldridge, H. D. (1974). Shark attack: A program of data reduction and analysis. *Contrib. Mote Mar. Lab.* **1**(2).

More Rare Than Dangerous: A Case Study of White Shark Conservation in California

BURR HENEMAN
Living Resources
Bolinas, California

MARCI GLAZER
Center for Marine Conservation
San Francisco, California

Introduction

In the first week of October 1993, those of us who had been working to win protection for white sharks *Carcharodon carcharias* in California could only wait for the final act to play out. Assembly Bill (AB) 522, which the authors of this chapter had sponsored and shepherded through the California legislature, was on the governor's desk, awaiting its fate as the first attempt at white shark protection in the United States. The governor's signature would mean that, with a few exceptions, no white sharks could be taken in California for at least 5 years. A veto would provide a discouraging precedent for any similar effort in the future. Moreover, it would mean a disappointing end to a year's hard work by an unusual alliance that included legislators and legislative staff members, commercial and sport fishermen, surfers, commercial and sport divers, sea kayakers, scientists, and environmentalists.

As an avid sport diver, AB 522's author, Assemblyman Dan Hauser, had needed some convincing to carry a bill to protect white sharks. Once persuaded, he and his staff then became superb champions of the measure. With their collaboration, we managed to get it through four committees, the Assembly, and the Senate without any votes cast against it, in a year when bitter partisan fights were the rule. The joke passed around was that, with politics so brutal, pro-

tection for white sharks became the easiest consensus issue.

Dan Hauser is a Democrat, and his office happened to face Republican Governor Pete Wilson's offices across a small courtyard in the capitol building. As the deadline for the governor's action approached, our anxiety grew as the list of vetoed bills lengthened. To break the tension, one of Hauser's staff put a hand-drawn poster in one of their courtyard windows: a shark fin protruding from waves, with the message "Please sign AB 522." On October 11, the last day for Wilson to act on bills, a large poster appeared in one of the governor's windows: a giant shark leaping clear of the water, mouth open and wearing hot-pink sunglasses, and a marker-pen message in the governor's handwriting, "Dan—Cordially, Pete Wilson."

When the legislation went into effect on January 1, 1994, California became the second jurisdiction in the world (after South Africa) to protect "the supreme hunter–killer; the largest game fish in the world; the most dangerous of the sharks; the *ne plus ultra* of predators . . . one of the few animals on Earth that we fear can—or worse, *will*—eat us . . . the stuff of which legends are made," a creature that is, at the same time, more legend than real and "more rare than they are dangerous" (quotations from Ellis and McCosker, 1991).

This chapter is a case study of AB 522 that describes its genesis, the considerations that went into deciding what to include in the bill, the strategy for winning approval by the legislature and the governor, and the process of assembling a remarkably disparate alliance to support it. Our narrative emphasizes two themes that we believe were critical to the ultimate success of the bill. First, white shark research and the close collaboration of white shark researchers with conservationists were both crucial in developing the unprecedented state policies embodied in AB 522. Second, the attitudes of people in the marine user groups that most frequently encounter white sharks were extraordinarily positive. Both of these elements are likely to be similarly important in attempts to protect white sharks anywhere else in the world.

Our reconstruction of the political process is fairly detailed in the hope that this case study will be helpful in white shark conservation efforts elsewhere in this species' range, as well as future efforts in California. AB 522 adds support to the truism that conservation victories are temporary. The California legislature routinely places an automatic expiration date on new regulations; the assumption is that people will make the effort to renew the regulations if there is enough support for them. AB 522 has such a "sunset" clause: its provisions will end after 5 years—on January 1, 1999—unless the legislature removes or extends that date in the meantime.

Some Prehistory of AB 522

Nine years ago . . . I witnessed for the first time, off the California coast, a predatory attack by a great white shark. The attack took place close to me, less than 25 meters from shore. The first signs were dramatic. The blue water suddenly turned a vivid crimson, stained by the blood of a struggling northern elephant seal. The huge shark, much of its body out of the water in clear view, swam vigorously back and forth. It appeared to be enjoying its meal. (Klimley, 1994)

In 1982, Peter Klimley, David Ainley, then in charge of the Point Reyes Bird Observatory (PRBO)'s Farallon Islands research, and one of the authors (B.H.), then executive director of PRBO, first discussed collaborating on white shark research at the Farallones. Beginning in 1970, PRBO Farallones biologists and volunteers had been documenting all observed white shark attacks on pinnipeds, but Klimley's experience in shark research suggested a range of other opportunities. Farallon Island white shark research escalated in 1985 and 1987 and eventually involved Klimley, Ainley, Peter Pyle, and Scot Anderson of PRBO, and John McCosker and Ken Goldman of the California Academy of Sciences. We refer to the ele-

ments of that research relevant to this chapter, but the full descriptions are covered in several other chapters in this volume. B. Heneman left PRBO at the end of 1984 but continued to follow the Farallon white shark research as it matured to produce additional interesting contributions to the white shark literature (e.g., Ainley et al., 1985; Klimley et al., 1992). Of particular importance to white shark conservation was support for the picture of a small population with low reproductive potential vulnerable to even modest levels of mortality (see Chapter 34, by Pyle et al.).

In 1992, the authors, both on the staff of the Center for Marine Conservation (CMC), were responsible for developing marine conservation initiatives on the Pacific coast. Given the Farallon Islands research and CMC's involvement in shark conservation along the Atlantic and Gulf coasts, white shark conservation seemed to be a likely candidate. Consultation with Ainley and Pyle inspired development of a campaign to protect white sharks in California late in 1992, timed to take advantage of the 1993 legislative session in Sacramento.

Our broad political strategy was based on several assumptions. We assumed that the proper forum was the state rather than the federal government. The potential take of white sharks that we were most concerned about was in state waters, within 3 miles of the mainland or any islands. White sharks had been taken in California waters from 1980 to 1992 (Table I). Furthermore, management of any shark fisheries off California was a state responsibility, because there was no federal fishery management plan for sharks. Even if a case could be made for federal management of shark fisheries, achieving it would be a several-year process, with the outcome even less certain than what we were attempting at the state level.

After reviewing the state fish and game code, we concluded that white shark protection could not be achieved under existing legislation; new legislation would be required. In other words, we would be initiating a highly political process in which even a modest misstep could kill the legislation and stall the effort for at least 1 year.

We assumed that passing a law to protect an animal best known in the public's mind for killing and eating people would require broad support and negligible opposition. To develop that support and minimize any potential opposition, we needed to do two things: (1) marshal a wide variety of arguments to appeal to different perspectives and (2) from the beginning, consult with and involve everyone we could think of who might support—or oppose—the protection of white sharks. Our political strategy remained based on these assumptions throughout the cam-

TABLE I White Shark Landings in California by Port

Year	Total Landings[a] (kg)[b]	Eureka (kg)	San Francisco (kg)	Monterey (kg)	Santa Barbara (kg)	Los Angeles (kg)	San Diego (kg)
1980	753.0		752.1		0.4		
1981	19.1				19.1		
1982	3652.3		3617.0		35.4		
1983	287.6		3208.7		35.4		
1984	2767.8		2495.2		186.9		85.7
1985	1234.7		103.4	407.8	103.5		20.0
1986	1202.5		181.0	31.8	262.2	68.0	59.4
1987	608.7				187.8	259.0	162.4
1988	996.1				18.1	259.0	162.4
1989	595.1			303.9	284.0	259.0	2.7
1990	120.2				87.1	11.8	21.3
1991	54.9				13.2	29.9	11.8
1992	434.1				135.2	248.6	50.3
1980–1992	12,726.4		8647.3	734.4	2081.1	841.0	413.7

[a]Data from the California Department of Fish and Game landing receipts reported in *Commercial Fish Landings, 1980–1992*. Landing receipt masses are normally for dressed mass (head and gutted). Extremely low values may represent shark parts or could be the result of data entry error.

[b]Catch data were converted from pounds to kilograms.

paign. Nothing that happened over the next several months caused us to question their correctness and fundamental importance.

If our political strategy was correct from the beginning, we fell well short of the mark initially in choosing our policy objectives; there may be a lesson here for white shark conservation efforts elsewhere. Given the white shark's reputation, aggravated by the still potent imagery of the 1975 movie *Jaws*, we were concerned that a proposal to protect this species might be met with widespread skepticism and hostility from some quarters. We assumed that we had to set our sights low for this initial effort. Although we thought the evidence argued for coastwide protection, we doubted we could find enough support for such an ambitious plan. After consulting with the Farallones researchers, we decided the minimum objective would be seasonal protection (July—February?) within 1 mile of the Farallon Islands, an absolutely defensible and enforceable proposal. Year-round protection at the Farallones would be one step better, and our most ambitious plan envisioned similar seasonal or year-round protection at Año Nuevo Island, another location on the northern California coast where very high seasonal concentrations of pinnipeds attracted large white sharks.

Launching the Proposal

(b) The white shark, as the principal apex predator off the California coast, plays a vital role in maintaining the overall health and stability of California's marine environment and is also a species of great interest to the public. (c) A healthy white shark population off the California coast is critical to maintaining a balance in our marine ecosystems. (d) The white shark has been, and continues to be, uncommon, and its population off the California coast is sensitive to any increased level of sport or commercial taking, as has been demonstrated in the recent past. (policy of the state of California, excerpted from the findings written into AB 522)

In January 1993, we faced two tactical decisions as the deadline for introducing new bills in the legislature approached. First, we needed a legislator who would carry our bill. Assemblyman Dan Hauser, chairman of the legislature's Joint Committee on Fisheries and Aquaculture, was the obvious choice. Hauser represented the north coast, the center of California's fishing industry, and he authored most of the fisheries-related legislation.

Second, we needed to expand the circle of people we were talking to beyond the "family" of CMC, PRBO, and the California Academy of Sciences. We decided to try out the idea on Zeke Grader, Executive Director of the Pacific Coast Federation of Fishermen's

Associations (PCFFA), the main fishing industry association in California. PCFFA and Grader, a friend and colleague from various issues, had a solid history of working closely with the environmental and scientific communities on such issues as offshore oil and gas development, the impact of logging practices and water diversions on salmon streams, San Francisco Bay dredge spoils disposal, and conflicts between gill nets and wildlife. Grader also worked closely with Assemblyman Hauser and Mary Morgan, the outstanding staff person for Hauser's committee.

We thought that the fishing industry reaction might be mixed. On our side was the role of white sharks in controlling seal and sea lion populations and the fact that there was no existing commercial fishery directed at white sharks. Against us was the possibility that industry might resist putting any potential target species out of bounds. PCFFA and the fishing industry were the most organized marine user group in Sacramento; given Grader's conservation sympathies, we knew our proposal would be in trouble if it got an unfavorable response from him. If he was enthusiastic, we would have an invaluable ally. The authors decided we would present the "ambitious" year-round Farallones and Año Nuevo plan when we met with Grader in early January 1993.

Grader was, indeed, enthusiastic and thought industry would generally be supportive. Moreover, he had a surprising suggestion—that we go for white shark protection on the entire California coast. He also made a welcome offer—that he include white shark protection on PCFFA's shortlist of new fisheries bills for Assemblyman Dan Hauser and Mary Morgan. In the space of a few minutes, the fortunes of white shark conservation in California underwent breathtaking change. When, with Grader's help, we persuaded Hauser to carry the bill, our campaign had found the best possible advocate within the legislature. On the other hand, instead of trying to protect white sharks near a couple of specks along the coast, we found ourselves committed to a coastwide measure.

Gearing Up for the Legislative Season

The white shark may be the last of Earth's primary predators capable of invoking irrational fear and response in humans. Although landbound predators such as lions, tigers, and wolves once held this authority, they have all been subdued, nearly eradicated, or restricted to wildlife preserves where we point at them from the safety of our automobiles. But *Homo sapiens* remains out of its element in the ocean, and it is from this perspective that the white shark still lurks in the primal quarters of the human mind. By revealing the white shark's natural story, we hope to supplant fear

and vindictiveness with respect and understanding for this beleaguered citizen of the sea. (Pyle, 1992)

The vastly expanded geographic scope of the proposed legislation added significant challenges to our lobbying campaign. We would need to enlist the help of more organizations in more parts of the state. After Hauser agreed to carry a bill, the first step in broadening our base of support was to invite PRBO and PCFFA to be the principal cosponsors. Both organizations agreed, creating a strong trio of supporters: PRBO, whose Farallon Islands base was home to the most extensive white shark research in the United States; PCFFA, the principal commercial fishing industry organization in the state; and CMC, a national environmental organization that was already a leader in shark conservation.

We also initiated an AB 522 contact group, compiling a lengthy list of organizations that might be interested in the bill. We wanted feedback on the bill's provisions, and we wanted to make sure that user groups first heard about the bill from us, not a newspaper story. To meet the deadline for introducing new bills, Assemblyman Hauser had introduced a preliminary version of AB 522 in January. We knew we would have to refine the bill to address concerns that had not occurred to us. We wanted to learn what those concerns were as early in the legislative process as possible.

We had witnessed what could happen to a bill when unanticipated concerns were raised during hearings. All too often, legislators, under time pressure and operating without sufficient information, yield to the temptation to actually write the laws they make, drafting and adopting new language in the space of minutes. On such occasions, the bill's proponents usually can do little but watch—and hope the wreckage will be limited. Committee hearings are not the place to write legislation.

On March 2, 1993, CMC and PRBO invited a shortlist of groups to discuss the bill at a "white shark conservation scoping meeting." It was our first gathering of organizations whose views we did not know in advance. We met at the offices of the Gulf of the Farallones National Marine Sanctuary, a jurisdiction that was as supportive of our efforts as a federal agency could be on a state issue. The California Department of Fish and Game, which would be commenting on our bill, sent observers at our invitation. Interest groups represented included surfers, sea kayakers, commercial sea urchin divers, and those advocating marine mammal conservation and local shark conservation. Organizations representing gillnetters and commercial abalone divers did not attend, but wrote to us with their concerns.

Those who wrote and attended the meeting were generally supportive, but they raised three issues. The local (Monterey Bay) shark conservation group proposed including protection for basking sharks *Cetorhinus maximus* in the bill. As sponsors of the bill, we decided against adding another species for which there was much less reliable information available and which was the target of a small directed fishery. We thought that the result might be protection for neither. In writing, the California Gillnetters Association asked us to exempt their white shark by-catch, and the California Abalone Association (CAA) expressed concern about divers being prosecuted if they killed or injured white sharks in self-defense. (We describe how we addressed those concerns below, under Navigating Sacramento.)

The meeting produced two important outcomes. Before the legislative hearings began, we had identified significant additional support, and, as it turned out later, we had learned of all but one negotiable issue that would be raised about AB 522.

The next day, we gave white shark protection its first airing in the state capitol at the annual Fisheries Forum. Hosted by Assemblyman Hauser's Joint Committee on Fisheries and Aquaculture, the forum provided an opportunity for legislators to hear what commercial and sport-fishing interests considered to be important issues in the coming year. As the largest regular meeting of California fisheries interests, it was an excellent opportunity to promote our bill in both the hallways and the hearing. Again, the reception received by our plan was encouraging, with a few people suggesting ways to strengthen the bill. Even commercial fisherman Mike McHenry, notorious for catching four large white sharks at the Farallon Islands in 1 day in 1982 (see Chapter 34, by Pyle *et al.*), said he favored protection for them.

During the public comments, a representative of the party boat industry expressed the frustration much of the fishing industry felt about the escalating California sea lion population and the general concern for dwindling salmon numbers; he proposed a slogan that was well received: "Save a salmon, kill a sea lion." White shark protection was next on the agenda, with comments from one of us (B.H.). Believing that the best tactic might be to leave them laughing, he picked up on the sea lion issue, suggesting an alternative, more legal slogan: "Save a salmon, save a great white shark." When the line got a warm response, we knew we were off to a running start in Sacramento.

The killing of the four sharks at the islands possibly nullified our study, but their capture and the subsequent reduction in shark activity did indicate some interesting points. First, it became apparent that only a few, and perhaps no more than six different, large white sharks were responsible for the level of activity we observed in recent years. . . . Second, the fact that capture of the four sharks significantly reduced shark sightings and shark/pinniped interactions indicated that the individual sharks were somewhat "resident" during the fall. . . . A week after capturing the four sharks, the same commercial vessel returned, and fished for but caught no sharks. . . . This indicates that it may be relatively easy to "overfish" a local concentration of large white sharks. (Ainley *et al.*, 1985)

We were familiar with the Farallon Islands research and were confident in the arguments for white shark protection for that area and Año Nuevo Island, which was fundamentally similar. The expansive geographic scope of our bill meant that we needed to acquaint ourselves with much more information on white sharks to respond to potential objections to coastwide protection. The crash course we got in the most up-to-date research at the Symposium on the Biology of the White Shark at Bodega Marine Laboratory, March 4–7, 1993, was an enormous help. We also were able to try out ideas for the legislation on white shark specialists from Australia, New Zealand, South Africa, the Mediterranean, and southern California. The information we gleaned from the symposium enabled us to further shape the bill and refine our arguments. The reinforcement received on our decision to seek coastwide protection was an additional benefit.

The Arguments for AB 522

$1,000 REWARD for 1st Great White Shark in 1993 on the *California Dawn* (San Diego party boat advertisement, *Western Outdoor News*, February 26, 1993)

I am writing to express my support for AB 522 to protect white sharks in California. As you may know, I was mauled by a shark thought to be a Great White on July 1, 1991 while surfing near Davenport, CA. My experience with the shark convinced me that sharks are an important part of the natural order of things. Any creature which is as well-adapted to its environment as the shark deserves a lot of respect. (Eric Larsen, letter to Assemblyman Hauser, April 17, 1993)

These two quotes embody the fundamental arguments for white shark conservation: the threat of an outmoded attitude we can no longer afford, and the promise of a new outlook that we in the scientific and conservation communities would do well to both encourage and learn from.

In order not to miss an opportunity for persuasion, we and our collaborators in the scientific community assembled a more complex suite of arguments. Its basic form was that these magnificent and uncommon animals are an important component of our coastal marine ecosystems, and that their population—typical of top predators—was naturally low and vulnerable to even a modest level of sport or commercial take. We buttressed that statement with a variety of supporting evidence and arguments, summarized below.

The Scientific Holistic Argument

This cluster of arguments has to do with the general value of biological diversity for its own sake, the more specific importance of apex predators to ecosystems, and the direct application of those attitudes to white sharks and the California coastal marine environment.

For some, this argument was simply about trying to maintain natural balances between sharks, mammals, fish, and other organisms in the marine environment. Others, particularly in the commercial and sport-fishing communities, saw a system that was terribly out of balance and in need of correction. As the Sportfishing Association of California (the organization of southern California party boat operators) put it in supporting the bill, "We only hope that an expanded population of white sharks helps to begin to control the runaway population of California sea lions that have overrun our near shore areas and the sport fishing fleets' fishing grounds." Indeed, large white sharks may be the main check on the rapid population increase of pinniped species that many fishermen consider to be serious competitors with humans for certain fish.

The role of white sharks in controlling pinniped populations was a surprising point of consensus, mentioned in more letters of support than any other argument. Organizations using this argument included some of the strongest marine mammal protection organizations in the state, including the Marine Mammal Center, Friends of the Sea Otter, and the American Cetacean Society. The California Urchin Divers Association probably analyzed the situation correctly in their letter of support: "Most people have no desire to control burgeoning marine mammal populations. White sharks may be one of the few lethal controls on marine mammals that are acceptable to the general public."

The Biological Argument

This argument emphasized what was known about white shark biology and populations and what might be inferred from other sharks. The conclusion to be drawn from the evidence was that the white shark population, at least that of adults, was low along the California coast, that circumstantial evidence indicated the population was not growing, and that white sharks are slow-growing animals with low fecundity, and therefore highly vulnerable to overfishing.

The very general conclusions about population size relied heavily on the research of known individuals at the Farallon Islands and the reduction in observed shark attacks on pinnipeds at the Farallones after four large sharks were caught in 1982.

Three pieces of evidence helped support the argument that the California population has not increased in the previous two decades. First, shark attacks on seals and sea lions at the Farallon Islands had not increased at a more rapid rate than the pinniped populations (see Chapter 34, by Pyle *et al.*). Second, although the southern sea otter population had doubled since 1968, evidence of white shark attacks on them had actually decreased since 1975 (see Chapter 28, by Ames *et al.*). Finally, there had been no increase in white shark attacks on humans in California, based on comparing the 1973–1982 and 1983–1992 decades (Fig. 1) (see Chapter 39, by McCosker and Lea).

Research by Cailliet *et al.* (1985) suggested that white shark females are at least 9–10 years old when they first breed, and they have fewer than 10 pups per litter. If their breeding behavior is like that of

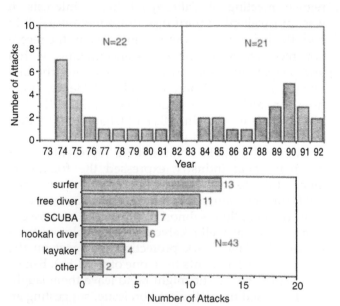

FIGURE 1 Fact sheet on white shark attacks on humans by decade (1973–1982 and 1983–1992) and by water activity of the victims. (Prepared by the Center for Marine Conservation from data provided by J. E. McCosker and R. N. Lea.)

some other large sharks, they may breed only every 2–3 years. All of this information supports the slow-growth/low-fecundity/vulnerability to overfishing argument.

The Threat of Endangerment Argument

This argument summarized the evidence that trophy or thrill fishermen might well be an imminent threat to the white shark status quo. We knew of ads such as the one above, for example, and an Australian white shark fisherman was widely quoted as saying he was going to fish white sharks in California next. The financial incentive certainly existed to take large white sharks. We had seen a medium-sized set of white shark jaws for sale for $8000 in a Miami store, and we heard numerous accounts from knowledgeable people of large jaws fetching $8000–$10,000 in Australia and Japan. Furthermore, the combination of rapidly escalating world demand for sharks for food and the institution of a restrictive federal fishery management plan for sharks on the Atlantic coast could result in unpredictable fishing pressures on various West Coast species, including white sharks.

Although there was no existing fishery directed at white sharks (an argument for action in itself, since no fishery would be shut down), we argued that the state had an excellent opportunity to take the virtually unprecedented step of protecting a vulnerable species before it got into trouble. Moreover, here was a chance to avoid another Endangered Species Act crisis.

The Precedent Argument

South Africa's total protection of white sharks, instituted in 1991, provided a precedent (Compagno, 1991). South African waters are famous for being shark infested; if that government had protected white sharks, perhaps it really was reasonable for California to do the same. The South African law may also have been an implicit challenge: Did California have the guts to follow suit? We also were able to say that South Australia was considering similar regulations, a statement that is still true as of this writing (B. Bruce, personal communication).

The Spiritual Argument

This argument was closely related to the scientific holistic argument, and many of the bill's supporters wove the two together. As with the Eric Larsen letter quoted above, we most often heard *rational* fear bal-

anced by respect, awe, and acceptance of the white shark as a creature with its own right to exist. We and others contended that these are magnificent and rare animals that should not be condemned as evil or malevolent. As one white shark biologist put it, "This is not the indiscriminate killer that the media portrays [sic], but simply one of the ocean's many organisms, tending to its livelihood" (Pyle, 1992). In other words, it was time to get past the "*Jaws* mentality," just as, on land, we no longer believed in killing the bears and wolves to make it safe to walk in the woods.

The sponsors and the other supporters of AB 522 blended these five basic arguments in a variety of ways. Here, we have tried to tease them apart in order to recognize and characterize the different components.

Dealing with Our Client's Negatives

Great White Shark Devours Newlywed Diving Off Australia (*San Francisco Chronicle* headline, June 10, 1993)

What shark scientists find remarkable about attacks on humans by the most terrifying animal in the sea is not that there are so many of them, but that there are so few. And that so very few white shark attacks are fatal (*Sacramento Bee*, March 8, 1993)

In developing arguments and printed material for AB 522, the authors had to decide how to confront one issue that was impossible to ignore: white shark attacks on humans. One reality that we could do nothing about was that the legislative session would be reaching a climax just as the risk for white shark attacks was peaking along the California coast. We settled on a strategy of presenting factual information and assuming people would conclude that this was a nonissue (Fig. 1).

The shark attack records provided to us (J. E. McCosker, personal communication) show that the number of shark attacks in California is low (about two per year, on average, since 1973), that the rate of attacks has not increased in the past 20 years, and that only about 10% of the attacks are fatal. It also seemed reassuring to us and to others to point out that many more people are killed by lightning, by bee stings, by grizzly bears in Glacier National Park, or by pigs in Iowa than by white sharks.

Furthermore, we felt that it was important to help disseminate the growing evidence that people can reduce the risk of shark attacks by being sensible about where and when they are in the water (see Chapter 42, by Burgess and Callahan, and Chapter 44, by Baldridge).

The Coalition

Suppose you're a 4,000-pound ocean predator with a mouthful of dagger-like teeth and a nasty habit of occasionally snacking on humans. Where do you look for friends? Where else but California, amigo? (San Diego Union Tribune, March 12, 1991)

When we step off the Pacific Rim into the Pacific we have to understand whose domain we're in. The ocean is a wild and wonderful place and it won't be nearly as wonderful if we eliminate the wild. We will always be in more danger on the highway than we will ever be while in the ocean. And just as we wouldn't want to bump into a tiger in the jungle, or a bear in the woods, the world wouldn't be as great a place without them. (Caughlan, 1987)

In our efforts to develop support for the white shark conservation campaign, one of our earliest contacts was Rob Caughlan, who could count among his many accomplishments being president for several years of Surfrider Foundation, the largest and most politically active organization of surfers in California. Caughlan, despite a few scary experiences of his own, had written several pro-white shark articles for surfing publications. He put us in touch with other key surfers, an important marine constituency where we had few contacts. These early names were the nucleus of a contact list that expanded to over 75 individuals, most of them representatives of marine user groups up and down the coast of California.

With Pyle and McCosker, we put together fact sheets that explained AB 522 and the importance of white shark protection, described the Farallon Islands research, and provided reliable information on shark attacks on humans in California (e.g., Fig. 1). Our outreach effort consumed many days, as we communicated with people on our contact list through mailings, by phone, and at meetings. We devoted similar effort to a media campaign. The single most important element of the political process was the large investment of time in explaining, answering questions, reassuring, and soliciting letters of support to the legislature and, eventually, the governor.

The result of these efforts: sympathetic newspaper and television stories seen by legislators throughout the state, a stack of letters from an impressive list of supporters, and absolutely no opposition. Supporters included three key recreational marine user groups: the Central California Council of Diving Clubs, Surfrider Foundation, and Bay Area Sea Kayakers. [Interestingly, the strongest support for AB 522 among sport divers was in northern California, where more than 90% of white shark attacks on humans in California have occurred (see Chapter 39, by McCosker and Lea). Divers in southern California expressed the most reluctance.] The Marine Mammal Center, Friends of the Sea Otter, and the American Cetacean Society had large California memberships and were organizations well known in Sacramento. From the fishing industry, PCFFA was joined by the Sportfishing Association of California and the Golden Gate Fishermen's Association (the two organizations that represent all the party fishing boat operators) as well as the California Urchin Divers Association. The California Academy of Sciences, which rarely enters a political fray, had all the more impact when it weighed in, thanks to John McCosker's efforts within that respected institution. The American Elasmobranch Society and PRBO were other scientific organizations supporting the bill. National environmental groups included the Natural Resources Defense Council, Earth Island Institute, Defenders of Wildlife, and CMC.

Navigating Sacramento

(a) It is unlawful to take any white shark (Carcharodon carcharias) . . . except under permits used pursuant to Section 1002 for scientific or educational purposes or pursuant to subdivision (b) for scientific or live display purposes.

(b) Notwithstanding subdivision (a), white sharks may be taken incidentally by commercial fishing operations using set gill nets, drift gill nets, or roundhaul nets. White sharks taken pursuant to this subdivision shall not have the pelvic fin severed from the carcass until after the shark is brought ashore. White shark taken pursuant to this subdivision, if landed alive, may be sold for scientific or live display purposes.

(c) Any white shark killed or injured by any person in self-defense may not be landed.

The department shall cooperate, to the extent that it determines feasible, with appropriate scientific institutions to facilitate data collection on white sharks taken incidentally by commercial fishing operations.

This article shall remain in effect only until January 1, 1999, and as of that date is repealed, unless a later enacted statute, which is enacted before January 1, 1999, deletes or extends that date. (from AB 522)

The actual legislative hearings were an anticlimax in the AB 522 campaign. Luck, preparation, and the enlightened attitude of supporters meant the bill sailed through four hearings without questions and with no votes cast against it. What drama there was took place away from the hearing rooms as the authors and Mary Morgan of Assemblyman Hauser's staff negotiated three amendments to meet the concerns of commercial net fishermen, commercial abalone divers, and a large sea aquarium company.

Two commercial fishing interests early in the process requested modifications to the bill as originally drafted, and we were able to work out changes acceptable both to them and to us as sponsors. Mike

Kitihara, representing CAA, asked us to exempt from prosecution anyone who killed or injured a white shark in self-defense. Since neither we nor the legal advisors that we consulted could imagine anyone being prosecuted for self-defense, we were happy to oblige. When CAA proposed as language "Any white shark killed or injured by any person in self-defense may not be landed," we, of course, accepted it.

Tony West, of the California Gillnetters Association, presented us with a thornier problem. According to West, although gillnetters did not target white sharks, they occasionally caught them incidentally (Fig. 2 and Tables II and III). These generally were not the large white sharks, which usually just left a hole

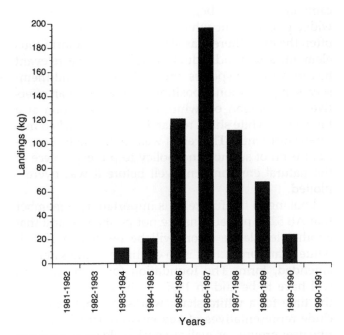

FIGURE 2 Drift gill net white shark landings (in kilograms), from all ports. The data are a subset of those presented in Table I and refer to fishing season rather than calendar year.

TABLE III   Number of White Sharks Caught in Southern California Net Fisheries

Fishery	1991 Los Angeles	1991 San Diego	1992 Los Angeles	1992 San Diego	Total
Set gill net	1	2	1	1	5
Drift net	0	0	0	0	0

National Marine Fisheries Service (NMFS) observed 10–12% of the gill net and 10–13% of drift net effort. Data were provided by D. Holts, NMFS, Southwest Fisheries Science Center.

in the net. Even the smaller white sharks that were caught caused considerable damage to nets. Allowing gillnetters to land and sell these incidentally caught sharks would mean the fisherman could recover some of the cost of repairing the net.

We did not want the California Gillnetters Association to oppose AB 522, nor did we want to create a loophole through which fishermen might be able to catch significant numbers of white sharks. In order to get accurate information on the extent of white shark landings, we analyzed the available landing and observer data collected by the California Department of Fish and Game and the National Marine Fisheries Service (Tables I and II). We learned that there were three types of net fisheries that caught white sharks incidentally: roundhaul net fisheries for squid and mackerel; nearshore set gill net fisheries for halibut, white sea bass, white croaker, angel shark, and a few other species; and the drift net fishery for thresher shark and swordfish.

The available data and conversations with several fishermen indicated that the white shark catch was low, that none of the gear types would be efficient for targeting white sharks, and that nearly all of the

TABLE II   Number of White Sharks Caught by Gill Net Fishing Set near Shore

Common name	Scientific name	Caught	Sold	Retained	Returned alive	Returned dead
California halibut	*Hippoglossus stenolepis*	4	2	1	1	
White sea bass	*Atrastoscion nobilis*	2	2			
Soupfin shark	*Galeorhinus zyopterus*	2	2			
White croaker	*Genyonemus lineatus*	0				
Angel shark	*Squatina californica*	0				

The California Department of Fish and Game monitored 2–5% of the nets set for California halibut, white sea bass, and white croaker and 1–2% of the effort for soupfin and angel sharks. (From *A Summary of 1983–1989 Southern California Gill Net Observation Data With an Overview on the Effects of Gill Nets on Recreational Catches.*)

white shark catch was in the nearshore set net fisheries. Moreover, restrictions on nearshore set nets that went into effect in 1987 had greatly reduced the take of white sharks in northern and central California (Table I). Finally, a total prohibition on gill nets in state waters off southern California, approved by the state's voters in 1990, would eliminate virtually all white shark take in that area if, as appeared likely, it survived a legal challenge. (The gill net prohibition was upheld by the courts, and it went into effect in 1994.) Confident that the exemption requested by West would not subvert white shark protection, the sponsors of the bill agreed to the language in section (b) (see above).

A few days before AB 522's first legislative hearing (April 13), the Sacramento lobbyist for a large sea aquarium corporation submitted a letter of opposition to Assemblyman Hauser, and then withdrew it, while making it clear that he still had objections to the bill. We were taken by surprise by concerns raised so late in the process by an organization that we thought would likely be an ally. Our efforts to resolve the problem were frustrated for weeks by difficulties in communicating with the lobbyist and by the corporation's policy of not allowing anyone knowledgeable on the staff to talk directly with us about anything to do with legislation. Adding to the confusion, the lobbyist had commented on an out-of-date version of the bill. He finally suggested a minor change in the language about live display; we readily agreed, since it did not change the meaning of the bill. The incident, a reminder that no support can be taken for granted, was a strange end to the evolution of AB 522.

Shark swallows man, then spits him out (*San Francisco Chronicle* headline, August 13, 1993)

As the legislative season dragged on into the higher-risk white shark season along the northern California coast, we kept a wary eye on the newspapers for stories that might be relevant to AB 522's prospects. On August 12, a white shark made a classic nonfatal attack on a recreational abalone diver. It took the man in its mouth headfirst, then spit him out. The victim was able to swim 30 m to shore, where friends helped him out of the water. The attack was off the Mendocino coast, in Assemblyman Hauser's district. Hauser never wavered.

In September, as if to underscore the need for protective legislation, a large white shark caught off Ventura, California, was sold for $10,000 to someone who wanted to display the dead animal. At the end of the month, the legislature took final action, and the bill went to the governor.

Discussion

In retrospect, although AB 522 was surprisingly noncontroversial, protection for white sharks in California was far from inevitable. The most important element in this successful campaign was the attitude that had evolved among those most likely to encounter white sharks—divers, surfers, kayakers, fishermen—as well as conservationists, that mirrored the change in popular feeling in recent decades about large terrestrial predators. Although the AB 522 process did not create that positive attitude, it did cause many people in marine user groups to examine their opinions about white sharks and decide how this species could best be protected. In addition, the AB 522 campaign certainly began the process of educating a wider public about white shark conservation. As is often the case, there was also an important fortuitous element: a few individuals who had some relevant history with the species and each other found themselves in professional positions that allowed an effective collaboration on white shark conservation at a time when white shark research in California had matured sufficiently. The result was an unusually beneficial fusion of science and policy to protect a piece of the natural environment well before it was overexploited.

Looking to the future, it is important to remember that AB 522's protections are not permanent and that another legislative effort will be needed in 1998 if white shark protection is to continue in the same, or amended, form. The foundation for such an effort will have to be laid in 1997. Perhaps that would be the time for a white shark workshop that could include representatives of key environmental and marine user groups, as well as shark biologists, and have white shark conservation, in general, and the continuation of white shark protection in California as agenda topics.

We foresee several topics as likely to arise during an effort to extend AB 522's provisions. One question may be whether white sharks should be protected at all. It would be useful, though probably not essential, to have additional information on white shark populations or sensitivity to take. One element of the law that may warrant refinement is the permit process for scientific, educational, or live display capture of white sharks. Finally, chumming to attract white sharks for recreational divers is a controversy that arose in the Monterey Bay National Marine Sanctuary after passage of AB 522; a workshop should at least consider how that practice can either further or interfere with white shark research.

When the new legislative effort is launched in California, this case study may be useful to whoever undertakes it. Moreover, the authors hope it can be a blueprint adaptable to similar efforts on the Atlantic coast of the United States or in other parts of the white shark's range.

Summary

We provided an account of successful efforts to protect white sharks *C. carcharias* in California waters, including the political process of developing arguments for the species' protection and building of the coalition that eventually supported the legislation. Relevant information on white shark attacks on humans, as well as on white shark landings and observer data from California fisheries, was reviewed. The legal protections for white sharks, as currently written, expire after 5 years, on January 1, 1999. We suggested ways to approach the extension of white shark protection in California beyond 1998.

Acknowledgments

D. Holts and T. West provided us with new information relevant to a history of AB 522. We greatly appreciate the suggestions provided by J. Visick and the editors in reviewing this chapter.

Bibliography

Aasen, O. (1963). Length and growth of porbeagle (*Lamna nasus* [Bonnaterre]) in North West Atlantic. *Norw. Fish. Mar. Invest.* **13**:20–37.

Abela, J. (1989). Lo squalo bianco più grande del mondo. *AQUA* **January:**20–21.

Agassiz, L. (1833–1844). "Recherches sur les Poissons Fossiles," Vols. 1–5. Impimérie de Petitpierre, Neuchâtel, Switzerland.

Aguilar, A., and Raga, J. A. (1990). Mortandad de delfins en el Mediterraneo. *Politic Cient.* **25**:51–54.

Ainley, D. G., and Boekelheide, R. J. (1990). "Seabirds of the Farallon Islands: Ecology, Dynamics, and Structure of an Upwelling-System Community." Stanford University Press, Stanford, California.

Ainley, D. G., Strong, C. S., Huber, H. R., Lewis, T. J., and Morrell, S. H. (1981). Predation by sharks on pinnipeds at the Farallon Islands. *Fish. Bull.* **78**:941–945.

Ainley, D. G., Henderson, R. P., Huber, H. R., Boekelheide, R. J., Allen, S. G., and McElroy, T. L. (1985). Dynamics of white shark/pinniped interactions in the Gulf of the Farallones. *South. Calif. Acad. Sci., Mem.* **9**:109–122.

Alcock, J. (1989). "Animal Behavior: An Evolutionary Approach." Sinauer, Sunderland, Massachusetts.

Allen, S. G., Huber, H. R., Ribic, C. A., and Ainley, D. G. (1989). Population dynamics of harbor seals in the Gulf of the Farallones, California. *Calif. Fish Game* **75**:224–232.

Ames, J. A., and Morejohn, G. V. (1980). Evidence of white shark, *Carcharodon carcharias*, attacks on sea otters, *Enhydra lutris. Calif. Fish Game* **66**:196–209.

Ames, J. A., Hardy, R. A., Wendell, F. E., and Geibel, J. J. (1983). "Sea Otter Mortality in California," unpublished report. Marine Resources Branch, California Department of Fish and Game, Monterey.

Anonymous (1974). Lo squalo di Favignana. *Subaqueo* **August–September:**41.

Anonymous (1989). "NSW Agriculture and Fisheries Annual Report 1988/89." Government Printers, Sydney, Australia.

Anonymous (1990). Cetacei spiaggiati lungo le coste italiane. 4. Rendiconto 1989. *Atti. Soc. Ital. Sci. Nat. Mus. Civ. Stor. Nat. Milano* **131**(27):413–432.

Antunes, M. T. (1970). Presence de *Alopias superciliosus* (Lowe) dans les mers du Portugal remarques sur les *Alopias* (Selachii) recents et fossiles. *Arq. Mus. Bocage, Ser. 2* **2**:363–378.

Applegate, S. P. (1965). Tooth terminology and variation in sharks with special reference to the sand shark, *Carcharias taurus* Rafinesque. *Los Angeles County Mus. Nat. Hist., Contrib. Sci.* **86**:3–18.

Applegate, S. P. (1966). The mystical fascination of the shark. *Mus. Alliance Q.* **5**(2):4–10.

Applegate, S. P. (1977). A new record-size bonito shark, *Isurus oxyrinchus* Rafinesque, from southern California. *Calif. Fish Game* **63**:126–129.

Applegate, S. P. (1991a). A paleontologist looks at the modern lamniform sharks. Paper presented at the Seventh Annual Meeting of the American Elasmobranch Society, New York (abstract).

Applegate, S. P. (1991b). A status report on the genus *Carcharodon*, the great white shark, and its fossil record. *J. Vertebr. Paleontol., Suppl.* **11**(3):14A–15A.

Arambourg, C. (1925). Révision des poissons fossiles de Licata (Sicilie). *Ann. Palaeontol. (Paris)* **14**:1–96.

Arambourg, C. (1927). Les poissons fossiles d'Oran. *Mater. Carte Geol. Alger., Ser. 1: Palaeontol.* **6**:1–298.

Arambourg, C., and Signeux, J. (1952). Les vertebres fossiles des gisements de phosphates (Maroc–Algerie–Tunisie). Service des Mines et de la Carte Geologique. *Rabat Notes Mem.* **92**:1–372.

Archer, J. (1976). The organization of aggression and fear in vertebrates. *In* "Perspectives in Ethology" (P. P. G. Bateson and P. H. Klopfer, eds.), pp. 231–298. Plenum, New York.

Arfelli, C. A., and Amorim, A. F. (1993). Notes on the white shark (*Carcharodon carcharias*) caught off Cananéia, São Paulo–Brazil. Paper presented at the Annual Meeting of the American Elasmobranch Society, University of Texas at Austin, May 27–June 2, 1993 (abstract).

Armstrong, E. A. (1947). "Courtship and Display Amongst Birds." Lindsay Drummond, London.

Arnason, A. N., and Mills, K. H. (1981). Bias and loss of precision due to tag loss in Jolly–Seber estimates for mark–recapture experiments. *Can. J. Fish. Aquat. Sci.* **38**:1077–1095.

Arnold, P. W. (1972). Predation on harbour porpoise, *Phocoena phocoena*, by a white shark, *Carcharodon carcharias*. *J. Fish. Res. Board Can.* **29**:1213–1214.

Avise, J. C. (1994). "Molecular Markers, Natural History, and Evolution." Chapman & Hall, New York.

Baird, R. W., and Stacey, P. J. (1988). Foraging and feeding behavior of transient killer whales. *Whalewatcher* **22**:11–15.

Bakun, A. (1973). Coastal upwelling indices, west coast of North America, 1946–1971. *NOAA Tech. Rep., NMFS* **NMFS SSRF-671.**

Bakun, A., and Parrish, R. H. (1980). Environmental inputs to fishery population models for eastern boundary current regions. *Intergov. Ocean. Comm. Rep.* **28**:67–104.

Balazs, G. H., and Kam, A. K. H. (1981). A review of shark attacks in the Hawaiian Islands. *Elepaio* **41**:97–105.

Baldridge, H. D. (1969). "International Shark Attack File Data Assimilation Program," final report. Mote Marine Laboratory, Sarasota, Florida.

Baldridge, H. D. (1973). "Shark Attack Against Man." U.S. Office of Naval Research, Washington, D.C.

Baldridge, H. D. (1974a). "Shark Attack." Droke House/Hallus, Anderson, South Carolina.

Baldridge, H. D. (1974b). Shark attack: A program of data reduction and analysis. *Mote Mar. Lab., Contrib.* **1**:1–98.

Baldridge, H. D. (1988a). Shark aggression against man: Beginnings of an understanding. *Calif. Fish Game* **74**:208–217.

Baldridge, H. D. (1988b). "Analytical Data on Shark Attacks." Privately printed, Sarasota, Florida.

Baldridge, H. D. (1990). Shark repellent: Not yet, maybe never. *Mil. Med.* **155**:358–361.

Baldridge, H. D., and Williams, J. (1969). Shark attack: Feeding or fighting? *Mil. Med.* **134**:130–133.

Barceló y Combis, D. F. (1868). Catálogo metódico de los peces que habitan o frecuentan las costas de las Islas Baleares. *Rev. Prog. Cien. Ex., Fisc. Nat.* **18**(3–4):1–46.

Barlow, G. W. (1968). Ethological units of behavior. *In* "The Central Nervous System and Fish Behavior" (D. Ingle, ed.), pp. 217–232. University of Chicago Press, Chicago.

Barlow, G. W. (1977). Modal action patterns. *In* "How Animals Communicate" (T. A. Sebeok, ed.), pp. 98–134. Indiana University Press, Bloomington.

Barlow, J., Boveng, P., Lowry, M. S., Stewart, B. S., Le Boeuf, B. J., Sydeman, W. S., Jameson, R. J., Allen, S. G., and Oliver, C. W. (1993). Status on the northern elephant seal population along the U.S. west coast. *U.S., NMFS, Southwest Fish. Sci. Cent., Admin. Rep.* **LJ-93-01.**

Bass, A. J. (1973). Analysis and description of variation in the proportional dimensions of scyliorhinid, carcharhinid and sphyrnid sharks. *S. Afr. Assoc. Mar. Biol. Res., Oceanogr. Res. Inst., Invest. Rep.* **32**.1–20.

Bass, A. J. (1978). Problems in studies of sharks in the southwest Indian Ocean. *In* "Sensory Biology of Sharks, Skates and Rays" (E. S. Hodgson and R. F. Mathewson, eds.), pp. 545–594. U.S. Office of Naval Research, Arlington, Virginia.

Bass, A. J., D'Aubrey, J. D., and Kistnasamy, N. (1975). Sharks of the east coast of Southern Africa. 4. The families Odontaspididae, Scapanorhynchidae, Isuridae, Cetorhinidae, Alopiidae, Orectolobidae and Rhiniodontidae. *Invest. Rep. Oceanogr. Res. Inst., Durban, S. Afr.* **39**:1–102.

Bastock, M., Morris, D., and Moynihan, M. (1953). Some comments on conflict and thwarting in animals. *Behaviour* **6**:66–84.

Bateman, A. J. (1948). Intra-sexual selection in *Drosophila*. *Heredity* **2**:349–368.

Bauchot, R., Platel, R., and Ridet, J.-M. (1976). Brain–body weight relationships in Selachii. *Copeia* **1976**:305–309.

Baughman, J. L., and Springer, S. (1950). Biological and economic notes on the sharks of the Gulf of Mexico, with special reference to those of Texas, and with a key for their identification. *Am. Midl. Nat.* **44**:96–152.

Beck, B., and Mansfield, A. W. (1969). Observations on the Greenland shark, *Somniosus microcephalus*, in northern Baffin Island. *J. Fish. Res. Board Can.* **26**:143–145.

Begon, M., and Mortimer, M. (1981). "Population Ecology: A Unified Study of Animals and Plants." Blackwell, Oxford.

Bell, J. C., and Nichols, J. T. (1921). Notes on the food of Carolina sharks. *Copeia* **92**:17–20.

Bellon, G., and Mateau, L. (1932). *Carcharodon carcharias. Notas Inst. Esp. Oceanogr.* **2**(53):11.

Bendix-Almgreen, S. E. (1983). *Carcharodon megalodon* from the Upper Miocene of Denmark, with comments on elasmobranch tooth enamel: Coronoin. *Bull. Geol. Soc. Den.* **32**(1–2):1–32.

Ben-Tuvia, A. (1971). Revised list of the Mediterranean fishes of Israel. *Isr. J. Zool.* **20**:1–39.

Berggren, W. A., and Hollister, C. D. (1974). Paleogeography, paleobiogeography, and the history of circulation in the Atlantic Ocean. *Spec. Publ.—Soc. Econ. Paleontol. Mineral.* **20**:126–186.

Berry, T. M., and Hutchins, J. B. (1990). A specimen of megamouth shark, *Megachasma pelagios* (Megachasmidae) from western Australia. *Rec. West. Aust. Mus.* **14**:651–656.

Bhattacharyya, G. K., and Johnson, R. A. (1977). "Statistical Concepts and Methods." Wiley, New York.

Bigelow, H. B., and Schroeder, W. C. (1948). "Fishes of the Western North Atlantic," Part 1. Yale University, New Haven, Connecticut.

Bigelow, H. B., and Schroeder, W. C. (1953). Fishes of the Gulf of Maine. *Fish. Bull.* **53.**

Bigelow, H. B., and Schroeder, W. C. (1958). A large white shark, *Carcharodon carcharias*, taken in Massachusetts Bay. *Copeia* **1958**:54–55.

Bini, G. (1960). Attacco documentato di pescecane (*Carcharodon carcharias*). *Bol. Pesca, Piscicolt. Idribiol. Ann.* 36 16:100 109.

Bini, G. (1967). "Atlante dei Pesci delle Coste Italiane. 1.

Leptocardi–Ciclostomi–Selaci." Mondo Sommerso Editrice.

Block, B. A., and Carey, F. G. (1985). Warm brain and eye temperatures in sharks. *J. Comp. Physiol. B* **156**:229–236.

Bodkin, J. L., and Jameson, R. J. (1991). Patterns of seabird and marine mammal carcass deposition along the central California coast, 1980–1986. *Can. J. Zool.* **69**:1149–1155.

Boero, F., and Carli, A. (1979). Catture di elasmobranchi nella tonnarella di Camogli (Genova) dal 1950 al 1974. *Mus. Ist. Biol. Univ. Genova, Boll.* **47**:27–34.

Bolin, R. L. (1954). Report on a fatal attack by a shark. *Pac. Sci.* **8**:105–108.

Bolin, R. L., and Abbott, D. P. (1963). Studies on the marine climate and phytoplankton of the central coastal area of California, 1954–60. *Calif. Coop. Ocean. Fish. Invest. Rep.* **9**:23–45.

Bonham, K. (1942). Records of three sharks on the Washington coast. *Copeia* **1942**:264–266.

Bonnot, P. (1928). Report on the seals and sea lions of California. *Calif. Dep. Fish Game, Fish Bull.* **14**:1–62.

Bowen, E. S. (1930). The role of the sense organs in aggregations of *Ameiurus melas. Ecol. Monogr.* **1.**

Brandt, S. B., and Wardley, V. A. (1981). Thermal fronts as ecotones and zoogeographic barriers in marine and freshwater ecosystems. *Proc. Ecol. Soc. Aust.* **11**:13–26.

Branstetter, S. (1981). Biological notes on the sharks of the north central Gulf of Mexico. *Contrib. Mar. Sci.* **24**:13–34.

Branstetter, S., and McEachran, J. E. (1986). First record of *Odontaspis noronhai* (Lamniformes: Odontaspididae) for the western North Atlantic, with notes on two uncommon sharks from the Gulf of Mexico. *North. Gulf Sci.* **8**:153–160.

Briggs, K. T., Ainley, D. G., Spear, L. B., Adams, P. B., and Smith, S. E. (1988). Distribution and diet of Cassin's auklet and common murre in relation to central California upwellings. *Acta Congr. Int. Ornithol., 19th* **1**:982–990.

Brodie, P., and Beck, B. (1983). Predation by sharks on the grey seal (*Halichoerus grypus*) in eastern Canada. *Can. J. Fish. Aquat. Sci.* **40**:267–271.

Brooks, D. R., and McLennan, D. A. (1991). "Phylogeny, Ecology, and Behavior: A Research Program in Comparative Biology." University of Chicago Press, Chicago.

Bruce, B. D. (1992). Preliminary observations on the biology of the white shark, *Carcharodon carcharias*, in South Australian waters. *Aust. J. Mar. Freshwater Res.* **43**:1–11.

Bruce, B. D., and Short, D. A. (1992). Observations on the distribution of larval fish in relation to a frontal zone at the mouth of Spencer Gulf, South Australia. *Bur. Rural Resour. Proc.* **15**:124–137.

Brunnich, M. T. (1768). Squalus carcharias. *In* "Icthyologia Massiliensis, Sistens Piscium Descriptiones," pp. 5–6. Rome.

Bruno, C. (1980). "Morte Bianca" a Favignana. Mondo Sommerso, July **1980**:124–125.

Budker, P. (1971). "The Life of Sharks." Weidenfeld & Nicolson, London.

Bullock, D. A. (1975). The general water circulation of Spencer Gulf, South Australia, in the period February to May. *Trans. R. Soc. S. Aust.* **99**:43–53.

Burgess, G. H. (1991). Shark attack and the International Shark Attack File. *In* "Discovering Sharks" (S. H. Gruber, ed.), pp. 101–105. Amer. Littoral Soc., Spec. Publ. 14.

Burghardt, G. M. (1970). Defining "communication." *In* "Communication by Chemical Signals" (J. W. Johnston, Jr., D. G. Moulton, and A. Turk, eds.), pp. 5–18. New Appleton-Century-Crofts, New York.

Burne, R. H. (1923). Some peculiarities of the blood vascular system of the porbeagle shark, *Lamna cornubica. Philos. Trans. R. Soc. London* **212**:209–257.

Burne, R. H. (1952). "Handbook of Cetacean Dissection." British Museum (Natural History), London.

Cadenat, J., and Blanche, J. (1981). Requins de Méditerranée et d'Atlantique. *Faune Trop.* **21**:1–330.

Cailliet, G. M., Kusher, D., Wolf, P., and Welden, B. A. (1983a). Techniques for enhancing vertebral bands in age estimation of California elasmobranchs. *NOAA Tech. Rep., NMFS* **8**:157–165.

Cailliet, G. M., Martin, L., Harvey, J., Kusher, D., and Welden, B. (1983b). Preliminary studies on the age and growth of blue (*Pronace glauca*), common thresher (*Alopias vulpinus*), and shortfin mako (*Isurus oxyrinchus*) sharks from California waters. *NOAA Tech. Rep., NMFS* **8**:179–188.

Cailliet, G. M., Natanson, L., Welden, B., and Ebert, D. (1985). Preliminary studies on the age and growth of the white shark, *Carcharodon carcharias*, using vertebral bands. *South. Calif. Acad. Sci., Mem.* **9**:49–60.

Cailliet, G. M., Love, M. S., and Ebeling, A. W. (1986). "Fishes: A Field and Laboratory Manual on Their Structure, Identification, and Natural History." Wadsworth, Belmont, California.

Caldwell, D. K., and Caldwell, M. C. (1969). Addition of the leatherback sea turtle to the known prey of the killer whale, *Orcinus orca. J. Mammal.* **50**:636.

Caldwell, M. C., Caldwell, D. K., and Siebenaler, J. B. (1965). Observations on captive and wild Atlantic bottlenosed dolphins, *Tursiops truncatus*, in the northeastern Gulf of Mexico. *Los Angeles County Mus. Nat. Hist., Contrib. Sci.* **91**:1–10.

California Department of Fish and Game (CDFG) (1976). "A Proposal for Sea Otter Protection and Research, and Request for Return of Management to the State of California, January 1976," Vol. I, unpublished report. California Department of Fish and Game, Sacramento.

Campagna, C., Le Boeuf, B. J., Blackwell, S. B., Crocker, D. E., and Quintana, F. (1995). Diving behavior and foraging location of female southern elephant seals from Patagonia. *J. Zool.* **236**:55–71.

Cantaluppi, G., and Mori, D. (1977). Studio morfostrutturale di denti tertologici di *Carcharodon megalodon* AG. del messiniano di Messina. *Natura* **15-VI 1977**:3–15.

Capapé, C., Chaldi, A., and Prieto, R. (1976). Les Sélaciens dangereux des côtes tunisiennes. *Arch. Inst. Pasteur Tunis* **53**(1–2):61–106.

Capapé, C., Zaouali, J., and Desoutter, M. (1979). Note sur la présence en Tunisie de *Carcharhinus obscurus* (Lesueur, 1818) (Pisces, Pluerotremata) avec clé de détermination des Carcharhinidae des côtes tunisiennes. *Bull. Off. Nation. Peches Tunisie, Tunis* **3**(2):171–182.

Cappetta, H. (1987). Handbook of Paleoichthyology. Chondrichthys II: Mesozoic and Cenozoic Elasmobranchii. *Handb. Paleoichthyol.* **3B**:1–193.

Cappo, M. (1988). Size and age of the white pointer shark *Carcharodon carcharias* Linnaeus: Was Peter Riseley's white pointer a world record? *Safish (Adelaide, South Australia)* **13**:11–13.

Caras, R. A. (1964). "Dangerous to Man." Chilton Books, New York.

Carey, F. G., and Teal, J. M. (1969). Mako and porbeagle: Warm bodied sharks. *Comp. Biochem. Physiol.* **28**:199–204.

Carey, F. G., Teal, J. M., Kanwisher, J. W., Lawson, K. D., and Beckett, J. S. (1971). Warm bodied fish. *Am. Zool.* **11**:137–145.

Carey, F. G., Teal, J. M., and Kanwisher, J. W. (1981). The visceral temperatures of mackerel sharks (Lamnidae). *Physiol. Zool.* **54**:334–344.

Carey, F. G., Kanwisher, J. W., Brazier, O., Gabrielsen, G., Casey, J. G., and Pratt, H. L. (1982). Temperature and activities of a white shark, *Carcharodon carcharias. Copeia* **1982**:254–260.

Carey, F. G., Kanwisher, J. W., and Stevens, E. D. (1984). Bluefin tuna warm their viscera during digestion. *J. Exp. Biol.* **109**:1–20.

Carey, F. G., Casey, J. G., Pratt, H. L., Urquhart, D., and McCosker, J. E. (1985). Temperature, heat production, and heat exchange in lamnid sharks. *South. Calif. Acad. Sci., Mem.* **9**:92–108.

Carrick, R., and Ingham, S. E. (1962). Studies on the southern elephant seal, *Mirounga leonina* (L.). V. Population dynamics and utilization. *CSIRO Wildl. Res.* **7**:198–206.

Carus, J. V. (1889–1893). "Prodromus Faunae Mediterraneae. 2. Plagiostomi–Selachoidei." Stuttgart.

Case, G. R., and Cappetta, H. (1990). The Eocene selachian fauna from the Fayum depression in Egypt. *Palaeontogr., Abt. A* **212**:1–30.

Casey, J. G. (1985). Transatlantic migrations of the blue shark; a case history of cooperative shark tagging. *In* "World Angling Resources and Challenges. Proceedings of the First World Angling Conference" (R. H. Stroud, ed.), pp. 253–268. International Game Fishing Association, Florida.

Casey, J. G., and Pratt, H. L., Jr. (1985). Distribution of the white shark, *Carcharodon carcharias*, in the western North Atlantic. *South. Calif. Acad. Sci., Mem.* **9**:2–14.

Casey, J., Pratt, H., Kohler, N., and Stillwell, C. (1991). "The Shark Tagger 1990 Summary." Cooperative Shark Tagging Program, U.S. National Marine Fisheries Service, Narragansett, Rhode Island.

Casier, E. (1947a). Constitution et evolution de la racine dentaire des euselachii. I. Note preliminaire. *Bull. Mus. R. Hist. Nat. Belg.* **23**:7–15.

Casier, E. (1947b). Constitution et evolution de la racine dentaire des euselachii. II. Etude comparative des types. *Bull. Mus. R. Hist. Nat. Belg.* **23**:9–14.

Casier, E. (1947c). Constitution et evolution de la racine dentaire des euselachii. III. Evolution des principaux

characteres morphologiques et conclusions. *Bull. Mus. R. Hist. Nat. Belg.* **23**:1–45.

Casier, E. (1950). Contributions a l'étude des poissons fossiles de la Belgique IX. La faune des formations dites "paniseliennes." *Bull. Inst. Sci. Nat. Belg. Bruxelles* **26**(42):1–52.

Casier, E. (1960a). Note sur la collection des poissons Paleocenes et Eocenes de L'Enclaver de Cabinda (Congo). *Ann. Mus. R. Congo Belge, Bruxelles* (A. 3) 1, 2:1–48.

Casier, E. (1960b). Transformation des systemes de fixation et de vascularisation dentaires dans l'evolution des selaciens du sous-ordre des Squaliformes. *Inst. R. Sci. Nat. Belg., Mem.* **2**:1–60.

Castro, J. (1983). "The Sharks of North American Waters." Texas A&M University Press, College Station.

Caughlan, R. (1987). "Livin' in the Food Chain. Making Waves." Surfrider Foundation.

Caughley, G. (1977). "Analysis of Vertebrate Populations." Wiley, London.

Chelton, D. B. (1984). Seasonal variability of alongshore geostrophic velocity off central California. *J. Geophys. Res.* **89**:3473–3486.

Chelton, D. B., Bernal, P. A., and McGowan, J. A. (1982). Large-scale interannual physical and biological interaction in the California Current. *J. Mar. Res.* **40**:1095–1125.

Cifelli, R. L. (1979). The role of circulation in the parcelling and dispersal of North Atlantic planktonic foraminifera. *Proc. Ann. Biol. Colloq. Sel. Pap.* 37:417–425.

Cigala-Fulgosi, F. (1983a). Confirmation of the presence of *Carcharhinus brachyurus* (Günther, 1870) in the Mediterranean. *Doriana* **5**(249):1–5.

Cigala-Fulgosi, F. (1983b). First record of *Alopias superciliosus* (Lowe, 1840) in the Mediterranean, with notes on some fossil species of the genus *Alopias. Estratto Dagli Ann. Mus. Civ. Stor. Nat. Genova* **84**:211–229.

Cigala-Fulgosi, F. (1990). Predation (or possible scavenging) by a great white shark on an extinct species of bottlenosed dolphin in the Italian Pliocene. *Tert. Res.* **12**(1):17–36.

Cigala-Fulgosi, F. (1992). Addition to the fish fauna of the Italian Miocene. The occurrence of *Pseudocarcharias* (Chondrichthyes, Pseudocarchariidae) in the lower Serravallian of Parma Province, northern Apennines. *Tert. Res.* **14**:51–60.

Cigas, J., and Klimley, A. P. (1987). A microcomputer interface for decoding telemetry data and displaying them numerically and graphically in real time. *Behav. Res. Methods Instrum. Comput.* **19**:19–25.

Cione, A. L., and Reguero, M. (1994). New records of the sharks *Isurus* and *Hexanchus* from the Eocene of Seymour Island, Antarctica. *Proc. Geol. Assoc.* **1994**:1–14.

Clark, E. (1974). The Red Sea's sharkproof fish. *Natl. Geogr. Mag.* **145**:718–727.

Clark, E. (1983). Shark repellent effect of the Red Sea Moses sole. *AAAS Sel. Symp.* **83**:135–150.

Clark, E., and von Schmidt, K. (1965). Sharks of the central gulf coast of Florida. *Bull. Mar. Sci.* 15:10 00.

Clarke, M. R., and Merrett, N. (1972). The significance of

squid, whale and other remains from the stomachs of bottom-living deep-sea fish. *J. Mar. Biol. Assoc. U.K.* **52**:599–603.

Cliff, G. (1991). Shark attacks on the South African coast between 1960–1990. *S. Afr. J. Sci.* **87**:513–518.

Cliff, G., and Dudley, S. F. J. (1991a). Sharks caught in the protective gill nets off Natal, South Africa. 4. The bull shark *Carcharhinus leucas* Valenciennes. *S. Afr. J. Mar. Sci.* **10**:253–270.

Cliff, G., and Dudley, S. F. J. (1991b). Sharks caught in the protective gill nets off Natal, South Africa. 5. The Java shark *Carcharhinus amboinensis* (Mueller & Henle). *S. Afr. J. Mar. Sci.* **11**:443–453.

Cliff, G., and Dudley, S. F. J. (1992a). Sharks caught in the protective gill nets off Natal, South Africa. 6. The copper shark *Carcharhinus brachyurus* (Gunther). *S. Afr. J. Mar. Sci.* **12**:663–674.

Cliff, G., and Dudley, S. F. J. (1992b). Protection against shark attack in South Africa, 1952 to 1990. *Aust. J. Mar. Freshwater Res.* **43**:263–272.

Cliff, G., and Dudley, S. F. J. (1993). Sharks caught in the protective gill nets off Natal, South Africa. 4. The black-tip shark *Carcharhinus limbatus* Valenciennes. *S. Afr. J. Mar. Sci.* **13**:237–254.

Cliff, G., Dudley, S. F. J., and Davis, B. (1988a). Sharks caught in the protective gill nets off Natal, South Africa. 1. The sandbar shark *Carcharhinus plumbeus* (Nardo). *S. Afr. J. Mar. Sci.* **7**:255–265.

Cliff, G., Dudley, S. F. J., and Davis, B. (1988b). An overview of shark catches in Natal's shark nets, 1966 to 1986. *S. Afr. Natl. Sci. Prog., Rep.* **157**:84–90.

Cliff, G., Dudley, S. F. J., and Davis, B. (1989). Sharks caught in the protective gill nets off Natal, South Africa. 2. The great white shark *Carcharodon carcharias* (Linnaeus). *S. Afr. J. Mar. Sci.* **8**:131–144.

Cliff, G., Dudley, S. F. J., and Davis, B. (1990). Sharks caught in the protective gill nets off Natal, South Africa. 3. The shortfin mako shark *Isurus oxyrinchus* (Rafinesque). *S. Afr. J. Mar. Sci.* **9**:115–126.

Cockcroft, V. G., Cliff, G., and Ross, G. J. B. (1989). Shark predation on Indian Ocean bottlenose dolphins *Tursiops truncatus* off Natal, South Africa. *S. Afr. J. Zool.* **24**:305–310.

Cohen, D. H., Duff, T. A., and Ebbesson, S. O. E. (1973). Electrophysiological identification of a visual area in shark telencephalon. *Science* **182**:492–494.

Coles, R. J. (1919). The large sharks of Cape Lookout, North Carolina. The white shark or maneater, tiger shark and hammerhead. *Copeia* **69**:34–43.

Collier, R. S. (1964). Report on a recent shark attack off San Francisco, California. *Calif. Fish Game* **50**:261–264.

Collier, R. S. (1992). Recurring attacks by white sharks on divers at two Pacific sites off Mexico and California. *Environ. Biol. Fishes* **33**:319–325.

Collier, R. S. (1993). Shark attacks off the California Islands: Review and update. *In* "Third California Islands Symposium: Recent Advances in Research on the California Islands" (F. G. Hochberg, ed.), pp. 453–462. Santa Barbara Museum of Natural History, Santa Barbara.

Collins, A. J. (1972). Shark meshing off New South Wales surfing beaches. *Fisherman* **September**:11–19.

Compagno, L. J. V. (1967). Tooth pattern reversal in three species of sharks. *Copeia* **1967**:242–243.

Compagno, L. J. V. (1970). Systematics of the genus *Hemitriakis* (Selachii: Carcharhinidae), and related genera. *Proc. Calif. Acad. Sci., Ser. 4* **38**:63–98.

Compagno, L. J. V. (1973). Interrelationships of living elasmobranchs. *J. Linn. Soc. London, Zool.* **53**(suppl. 1):15–61.

Compagno, L. J. V. (1977). Phyletic relationships of living shark and rays. *Am. Zool.* **17**:303–322.

Compagno, L. J. V. (1984a). FAO species catalogue. Vol. 4. Sharks of the world. An annotated and illustrated catalogue of shark species known to date. Part 1: Hexanchiformes to Lamniformes. *FAO Fish. Synop.* **125**:1–249.

Compagno, L. J. V. (1984b). FAO species catalogue. Vol. 4. Sharks of the world. An annotated and illustrated catalogue of shark species known to date. Part 2: Carcharhiniformes. *FAO Fish. Synop.* **125**:250–655.

Compagno, L. J. V. (1987). Shark attack in South Africa. *In* "Sharks: An Illustrated Encyclopedic Survey by International Experts" (J. D. Stevens, ed.), pp. 134–147. Struik, Cape Town, South Africa.

Compagno, L. J. V. (1988). "Sharks of the Order Carcharhiniformes." Princeton University Press, Princeton, New Jersey.

Compagno, L. J. V. (1990a). Alternative life-history styles of cartilaginous fishes in time and space. *Environ. Biol. Fishes* **28**:33–75.

Compagno, L. J. V. (1990b). Relationships of the megamouth shark, *Megachasma pelagios* (Lamniformes: Megachasmidae), with comments on its feeding habits. *NOAA Tech. Rep., NMFS* **NMFS 90**:357–379.

Compagno, L. J. V. (1990c). Shark exploitation and conservation. *NOAA Tech. Rep., NMFS* **NMFS 90**:391–414.

Compagno, L. J. V. (1991). Government protection for the great white shark (*Carcharodon carcharias*) in South Africa. *S. Afr. J. Sci.* **87**:284–285.

Compagno, L. J. V., Ebert, D. A., and Smale, M. J. (1989). "Guide to the Sharks and Rays of Southern Africa." Struik, Cape Town, South Africa.

Computing Resource Center (1992). *Stata Reference Manual: Release 3*, 5th ed. Computing Resource Center, Santa Monica, California.

Condorelli, M., and Perrando, G. G. (1909). Notizie sul *Carcharodon carcharias* L., catturato nelle acque di Augusta e considerazioni medico-legali su resti umani trovati nel suo tubo digerente. *Soc. Zool. Ital., Boll.* **1909**:164–183.

Condy, P. R. (1978). The distribution and abundance of southern elephant seals, *Mirounga leonina* (Linn.), on the Prince Edwards Islands. *S. Afr. J. Antarct. Res.* **8**:42–48.

Condy, P. R., van Aarde, R. J., and Bester, M. N. (1978). The seasonal occurrence and behaviour of killer whales *Orcinus orca* at Marion Island. *J. Zool.* **184**:449–464.

Conover, W. J. (1971). "Practical Nonparametric Statistics." Wiley, New York.

Cook, S. F., and Brzycki, S. J. (1986). Recent records of white shark distribution and feeding behavior in cold waters of the North Pacific Ocean and Bering Sea. *Ann. Meet., Am. Elasmobranch Soc.* (abstracts).

Cooper, J. (1974). The predators of the jackass penguin. *Bull. Br. Ornithol. Club* **94**:21–24.

Coppleson, V. M. (1963). Patterns of shark attacks for the world. *In* "Sharks and Survival" (P. W. Gilbert, ed.), pp. 389–421. Heath, Boston.

Coppleson, V. (1988). "Shark Attack," rev. ed. Angus & Robertson, North Ryde, New South Wales, Australia.

Corkeron, P. J., Morris, R. J., and Bryden, M. M. (1987). Interactions between bottlenose dolphins and sharks in Moreton Bay, Queensland. *Aquat. Mammal.* **13**:109–113.

Cortes, E., and Gruber, S. H. (1990). Diet, feeding habits and estimates of daily ration of young lemon sharks, *Negaprion brevirostris* (Poey). *Copeia* **1990**:204–218.

Cousteau, J.-M., and Richards, M. (1992). "Cousteau's Great White Shark." Abrams, New York.

Cousteau, J.-Y., and Dumas, F. (1953). "The Silent World." Harper & Row, New York.

Crile, G., and Quiring, D. P. (1940). A record of body weight and certain organ and gland weights in 3690 animals. *Ohio J. Sci.* **40**:219–259.

Crocker, D. E. (1993). "Reproductive Effort and Age in Female Northern Elephant Seals," unpublished M.S. thesis. University of California, Santa Cruz.

Crocker, D. E., Le Boeuf, B. J., and Costa, D. P. (1996). Drift diving in female northern elephant seals: Implications for food processing. *Can. J. Zool.* in press.

Cropp, B. (1979). Where ocean giants meet. *Oceans* **12**:43–46.

Crovetto, A., Lamilla, J., and Pequeno, G. (1992). *Lissodelphis peronii*, Lacepede 1804 (Delphinidae, Cetacea) within the stomach contents of a sleeping shark, *Somniosus* cf. *pacificus*, Bigelow & Schroeder 1944, in Chilean waters. *Mar. Mammal Sci.* **8**:312–314.

Dadswell, M. J. (1979). Biology and population characteristics of the shortnose sturgeon, *Acipenser brevirostrum* LeSeur 1818 (Osteichthyes:Acipenseridae), in the Saint John River Estuary, New Brunswick, Canada.

Daniel, J. F. (1934). "The Elasmobranch Fishes." University of California Press, Berkeley.

D'Aubrey, J. D. (1964). A carcharhiniid shark new to South African waters. *S. Afr. Assoc. Mar. Biol. Res., Invest. Rep.* **9**:3–16.

Daugherty, A. E. (1964). The sand shark, *Carcharias ferox* (Risso) in California. *Calif. Fish Game* **50**:4–10.

Davidson, H. (1992). Sharks: Why the recent surge in attacks? *Surfer* **33**(3):56–61.

Davies, D. H. (1961). Shark attack off the east coast of South Africa, 22nd January 1961. *Oceanogr. Res. Inst., Invest. Rep.* **4**.

Davies, D. H. (1963). Shark Attack and its relationship to temperature, beach patronage and the seasonal abundance of dangerous sharks. *Oceanogr. Res. Inst., Invest. Rep.* **5**.

Davies, D. H. (1964). "About Sharks and Shark Attack." Shuter & Shooter, Pietermaritzburg, Natal, South Africa.

Davies, D. H., and D'Aubrey, J. D. (1961a). Shark attack off the east coast of South Africa 24 December 1960, with

notes on the species of shark responsible for the attack. *Oceanogr. Res. Inst., Invest. Rep.* **2**.

Davies, D. H., and D'Aubrey, J. D. (1961b). Shark attack off the east coast of South Africa, 6 January 1961. *Oceanogr. Res. Inst., Invest. Rep.* **3**.

Davis, R. W., Le Boeuf, B. J., Marshall, G., Crocker, D., and Williams, J. (1993). Observing the underwater behavior of elephant seals at sea by attaching a small video camera to their backs. *Bienn. Conf. Biol. Mar. Mammals, 10th, Galveston, Tex., Nov. 11–15, 1993 (abstract).*

Dawkins, R. (1976). "The Selfish Gene." Oxford University Press, New York.

Day, L. R., and Fisher, H. D. (1954). Notes on the great white shark, *Carcharodon carcharias*, in Canadian waters. *Copeia* **1954**:295–296.

De Beaumont, G. (1959). Recherches sur la denture et la cavité orale d'*Alopias vulpinus* Bonat. (Selachii). *Rev. Suisse Zool.* **66**:387–410.

de Blainville, H. M. (1818). Sur les ichthyolites ou les poissons fossiles. *In* "Nouveau Dictionnaire d'Histoire Naturelle Appliquee aux Arts, a l'Agriculture, a l'Economie Rurale et Domestique, a la Médicine," pp. 310–395. Paris.

de Buen, F. (1926). Catálogo ictiológico del Mediterráneo espanol y de Marruecos. *Res. Campanas Int. Inst. Espanol Oceanogr.* **2**:153–161.

DeGange, A. R., and Vacca, M. M. (1989). Sea otter mortality at Kodiak Island, Alaska, during summer 1987. *J. Mammal.* **70**:836–838.

DeLong, R. L., and Stewart, B. S. (1991). Diving patterns of northern elephant seal bulls. *Mar. Mammal Sci.* **7**:369–384.

DeLong, R. L., Stewart, B. S., and Hill, R. D. (1992). Documenting migrations of northern elephant seals using day length. *Mar. Mammal Sci.* **2**:155–159.

Demèrè, T. A., and Cerutti, R. A. (1982). A Pliocene shark attack on a cetotheriid whale. *J. Paleontol.* **56**:1480–1482.

Demski, L. S. (1991a). Neural substrates for photic control of elasmobranch sexual development and behavior. *J. Exp. Zool., Suppl.* **5**:121–129.

Demski, L. S. (1991b). Elasmobranch reproductive biology: Implications for captive breeding. *J. Aquar. Aquat. Sci.* **5**:84–95.

Demski, L. S. (1993). The terminal nerve. *Acta Anat.* **148**:81–95.

DeMuizon, C., and DeVries, T. J. (1985). Geology and paleontology of the late Cenozoic marine deposits in the Sacaco area (Peru). *Geol. Rundsch.* **74**:547–563.

Desbrosses, P. (1930). Presence du squale feroce: *Odontaspis ferox* Agassiz dans le Golfe de Gascogne. *Soc. Fr., Bull.* **55**:232–235.

Deutsch, C. J., Haley, M. P., and Le Boeuf, B. J. (1990). Reproductive effort of male northern elephant seals: Estimates from mass loss. *Can. J. Zool.* **68**:2580–2593.

Deutsch, C. J., Crocker, D. E., Costa, D. P., and Le Boeuf, B. J. (1994). Variation in reproductive effort of the northern elephant seal in relation to age and sex. *In* "Elephant Seals: Population Ecology, Behavior and Physiology"

(B. J. Le Boeuf and R. M. Laws, eds.), pp. 169–210. University of California Press, Berkeley.

Doderlein, P. (1881). "Manuale Ittiologico del Mediterraneo," Parts 1–2:1–120. Palermo, Italy.

Dodrill, J. W., and Gilmore, R. G. (1979). First North American continental record of the longfin mako (*Isurus paucus* Guitart Manday). *Fla. Sci.* **42**:52–58.

Dohl, T. P., Guess, R. C., Duman, M. L., and Helm, R. C. (1983). Cetaceans of central and northern California, 1980–1983: Status, abundance, and distribution. *U.S. Minerals Manage. Serv., Contrib.* **14-12-0001-29090**.

Dollard, J., Doob, L. W., Miller, N. E., Mowrer, O. H., and Sears, R. R. (1939). "Frustration and Aggression." Yale University Press, New Haven, Connecticut.

Doroff, A. M., and DeGange, A. R. (1992). "Experiments to Determine Drift Patterns and Rates of Recovery of Sea Otter Carcasses Following the Exxon *Valdez* Oil Spill," NRDA Mar. Mammal Study 6, final report. U.S. Fish and Wildlife Service, Office of the Oil Spill, Anchorage.

Doudt, K. (1992). "Surfing With the Great White Shark." Shark Bite Publications, Hawaii.

Duarte, P. (1968). "O Sambaqui Visto Atraves de Alguns Sambaquis." Institute of Pre-history, University of São Paulo, São Paulo.

Dudley, S. F. J., and Cliff, G. (1993a). Some effects of shark nets in the Natal nearshore environment. *Environ. Biol. Fishes* **36**:243–255.

Dudley, S. F. J., and Cliff, G. (1993b). Sharks caught in the protective gill nets off Natal, South Africa. 7. The blacktip shark Carcharhinus limbatus (Valenciennes). *S. Afr. J. Mar. Sci.* **13**:237–254.

Duméril, A. (1865). "Histoire Naturelle des Poissons ou Ichthyologie Générale," Vol. 1. Librairie Encyclopédique de Roret, Paris.

Duncan, I. J. H., and Wood-Gush, D. G. M. (1971). Frustration and aggression in the domestic fowl. *Anim. Behav.* **19**:500–504.

Duncan, I. J. H., and Wood-Gush, D. G. M. (1972). Thwarting of feeding behavior in the domestic fowl. *Anim. Behav.* **20**:444–451.

Ebert, D. A. (1986). Biological aspects of the sixgill shark, Hexanchus griseus. *Copeia* **1986**:131–135.

Ebert, D. A. (1989). Life history of the sevengill shark, Notorynchus cepedianus (Peron), in two northern California bays. *Calif. Fish Game* **75**:102–112.

Ebert, D. A. (1991). Observations on the predatory behavior of the sevengill shark Notorynchus cepedianus. *S. Afr. J. Mar. Sci.* **11**:455–465.

Ebert, D. A. (1994). Diet of the sixgill shark Hexanchus griseus off Southern Africa. *S. Afr. J. Mar. Sci.* **14**:213–218.

Ebert, D. A., Compagno, L. J. V., and Cowley, P. D. (1992). A preliminary investigation of the feeding ecology of squaloid sharks off the west coast of southern Africa. *S. Afr. J. Mar. Sci.* **12**:601–609.

Economidis, P. S. (1973). Catalogue de poisons de la Grece. *Hell. Oceanol. Limnol.* **11**:421–598.

Economidis, P. S., and Bauchot, M.-L. (1976). Sur une collection de poissons des mers hélleniques (mers Égée et Ionienne) déposée au Muséum national d'histoire naturelle. *Bull. Mus. Nat. Hist. Nat., Paris, Ser. 3, No. 392, Zool.* **274**:871–903.

Edmunds, M. (1974). "Defense in Animals." Longman, London.

Efron, B. (1981). Nonparametric estimates of standard error: The jackknife, the bootstrap and other methods. *Biometrika* **68**:589–599.

Egaña, A. C., and McCosker, J. E. (1984). Attacks on divers by white sharks in Chile. *Calif. Fish Game* **70**:173–179.

Eibl-Eibesfeldt, I. (1970). "Ethology." Holt, Rinehart, & Winston, New York.

Eibl-Eibesfeldt, I., and Hass, H. (1959). Erfahrungen mit haien. *Z. Tierpsychol.* **16**:733–746.

Ellis, R. (1975). "The Book of Sharks." Grosset & Dunlap, New York.

Ellis, R. (1991). "Men and Whales." Knopf, New York.

Ellis, R., and McCosker, J. E. (1991). "Great White Shark." HarperCollins, New York; Stanford University Press, Stanford, California.

Elton, C. (1927). "Animal Ecology." Sidgwick & Jackson, London.

Emery, S. H. (1985). Hematology and cardiac morphology in the great white shark, Carcharodon carcharias. *South. Calif. Acad. Sci., Mem.* **9**:73–80.

Endler, J. A. (1986). "Natural Selection in the Wild." Princeton University Press, Princeton, New Jersey.

Espinosa-Arrubarrena, L. (1987). "Neogene Species of the Genus *Isurus* (Elasmobranchii, Lamnidae) in Southern California, USA and Baja California Sur, Mexico," unpublished M.S. thesis. California State University, Long Beach.

Estes, J. A., and Jameson, R. J. (1988). A double survey estimate for sighting probability of sea otters in California. *J. Wildl. Manage.* **52**:70–76.

Ewert, J. P. (1984). Tectal mechanisms that underlie prey-catching and avoidance behaviors in toads. *In* "Comparative Neurology of the Optic Tectum" (H. Vanegas, ed.), pp. 247–416. Plenum, New York.

Faber, G. L. (1883). "Fisheries of the Adriatic and the Fish Thereof." Quarich, London.

Facciolà, L. (1894). Cattura di un Carcharodon rondeleti M. H. nel mare di Messina. *Nat. Sicil.* **13**:182.

Farris, J. S. (1989). The retention index and the rescaled consistency index. *Cladistics* **5**:417–419.

Fast, T. N. (1955). Second known shark attack on a swimmer in Monterey Bay. *Calif. Fish Game* **41**:348–351.

Feder, M. E., and Lauder, G. V. (1986). Commentary and conclusion. *In* "Predator–Prey Relationships" (M. E. Feder and G. V. Lauder, eds.), pp. 180–189. University of Chicago Press, Chicago.

Felsenstein, J. (1985). Confidence limits on phylogenies: An approach using the bootstrap. *Evolution* **39**:783–791.

Felsenstein, J. (1988). Phylogenies from molecular sequences: Inference and reliability. *Annu. Rev. Genet.* **22**:521–565.

Felsenstein, J., and Kishino, H. (1993). Is there something wrong with the bootstrap on phylogenies? A reply to Hillis & Bull. *Syst. Biol.* **42**:193–200.

Fergusson, I. K. (1994a). A review of white shark (*Carcharodon carcharias*) occurrence in the Catalonian Sea (Spain), with notes on a recent adult male specimen. *Chondros* 5(2).

Fergusson, I. K. (1994b). Preliminary notes on white shark (*Carcharodon carcharias*) predation upon odontocetes in the Mediterranean Sea. *In* "Proceedings 2nd European of the Shark and Ray Workshop, February 1994" (S. Fowler and R. C. Earll, eds.). J.N.C.C., Peterborough, England.

Fischer, H. B. (1968). "Methods for Predicting Dispersion Coefficients in Natural Streams With Applications to Lower Reaches of the Green and Duwamish Rivers, Washington," ·Prof. Paper 582-A. U.S. Geological Survey, Washington, D.C.

Fischer, W., Schneider, M., and Bauchot, M.-L. (1987). "Fiches FAO d'Identification des Especes pour les Besoins de la Peche. Méditerranée et Mer Noire, Zone de Peche 37," Vol. 2. Food and Agriculture Organization, Rome.

Fitch, J. E. (1949). The great white shark *Carcharodon carcharias* Linnaeus in California waters during 1948. *Calif. Fish Game* 35:135–138.

Fleury, P. (1991). Model II regression. *Sysnet, SYSTAT Network* 8(2).

Follett, W. I. (1966). Man-eater of the California coast. *Pac. Disc.* 19(1):18–22.

Follett, W. I. (1974). Attacks by the white shark, *Carcharodon carcharias* (Linnaeus), in northern California. *Calif. Fish Game* 60:192–198.

Forcada, J., Aguilar, A., Hammond, P., Pastor, X., and Aguilar, R. (1992). Population abundance of striped dolphins inhabiting the western Mediterranean Sea. *Eur. Res. Cetaceans* 6:105–107.

Franco, T. C. B., and Barbosa, D. R. (1991). Ocorrencia de dentes de *Carcharodon carcharias* (Linnaeus, 1758) (Elasmobranchii, Lamnidae) no contexto das populacoes prehistoricas. *Congr. Bras. Zool. Univ. Fed. Bahia, 18th* p. 554 (abstract).

Frazzetta, T. H. (1988). The mechanics of cutting and the form of shark teeth (Chondrichthyes, Elasmobranchii). *Zoomorphology* 108:93–107.

Fujita, K. (1981). Oviphagous embryos of the pseudocarchariid shark, *Pseudocarcharias kamoharai*, from the central Pacific. *Jpn. J. Ichthyol.* 28:31–44.

Galvan-Magana, F., Nienhuis, H. J., and Klimley, A. P. (1989). Seasonal abundance and feeding habits of sharks of the lower Gulf of California, Mexico. *Calif. Fish Game* 75:74–84.

Garman, S. (1913). The Plagiostoma. *Mus. Comp. Zool., Harv., Mem.* 36:1–515.

Garrick, J. A. F. (1960). Studies on New Zealand elasmobranchii. Part XII. The species of *Squalus* from New Zealand; and a general account and key to the New Zealand squaloidea. *Trans. R. Soc. N.Z.* 88(3):519–557.

Garrick, J. A. F. (1967). Revision of sharks of the genus *Isurus* with a description of a new species (Galeoidea, Lamnidae). *Proc. U.S. Natl. Mus. Nat. Hist.* 118:663–690.

Garrick, J. A. F. (1974). First record of an odontaspidid shark in New Zealand waters. *N.Z. J. Mar. Freshwater Res.* 8:621–630.

Gauld, J. A. (1989). Records of porbeagles landed in Scotland, with observations on the biology, distribution and exploitation of the species. *Scott. Fish. Res., Rep.* **45.**

Geibel, J. J., and Miller, D. J. (1984). Estimation of sea otter, *Enhydra lutris*, population, with confidence bounds, from air and ground counts. *Calif. Fish Game* 70:225–233.

Gentry, R. L., and Kooyman, G. L., eds. (1986). "Fur Seals: Maternal Strategies on Land and at Sea." Princeton University Press, Princeton, New Jersey.

Gentry, R. L., Kooyman, G. L., and Goebel, M. E. (1986). Feeding and diving behavior of northern fur seals. *In* "Fur Seals: Maternal Strategies on Land and at Sea" (R. L. Gentry and G. L. Kooyman, eds.), pp. 61–78. Princeton University Press, Princeton, New Jersey.

Gerrodette, T. (1983). "Review of the California Sea Otter Salvage Program," final report. U.S. Marine Mammal Commission, Washington, D.C.

Gibson, R. N. (1978). Lunar and tidal rhythms in fish. *In* "Rhythmic Activity of Fishes" (J. E. Thorpe, ed.), pp. 201–213. Academic Press, New York.

Gifford, A. (1993). Sharks: Fear and the fish. *Afr. Wildl.* 46(6):251–253.

Gifford, E. W. (1939). The coast Yuki. *Anthropos* 34:318–321.

Giglioli, E. H. (1880). *Carcharodon rondeleti. In* "Elenco dei Mammiferi, degli Uccelli e dei Rettilli Ittiofagi od Interessanti per la Pesca, Appartenenti alla Fauna Italiana," Vol. 2, pp. 52–112. Firenze.

Gilbert, P. W. (1963a). The visual apparatus of sharks. *In* "Sharks and Survival" (P. W. Gilbert, ed.), pp. 283–326. Heath, Boston.

Gilbert, P. W., ed. (1963b). "Sharks and Survival." Heath, Boston.

Gilbert, P. W., Schultz, L. P., and Springer, S. (1960). Shark attacks during 1959. *Science* 132:323–326.

Gillette, D. D. (1984). A marine ichthyofauna from the Miocene of Panama and the Tertiary Caribbean faunal province. *J. Vertebr. Paleontol.* 4(2):172–186.

Gilmore, R. G. (1983). Observation on the embryos of the longfin mako, *Isurus paucus*, and the bigeye thresher, *Alopias superciliosus. Copeia* 1983:375–382.

Gilmore, R. G. (1993). Reproductive biology of lamnoid sharks. *Environ. Biol. Fishes* 38:95–114.

Gilmore, R. G., Dodrill, J. W., and Linley, P. A. (1983). Reproduction and embryonic development of the sand tiger shark, *Odontaspis taurus* (Rafinesque). *Fish. Bull.* 81:201–225.

Gingerich, P. D. (1992). Marine mammals (Cetaea and Sirenia) from the Eocene of Gebel Mokattam and Fayum, Egypt: Stratigraphy, age, and paleoenvironments. *Univ. Mich. Pap. Paleontol.* 30:1–84.

Giudici, A., and Fino, F. (1989). "Squali del Mediterraneo." Edizioni Atlantis, Rome.

Gluckmann, L. S. (1964). "Sharks of the Paleogene and Their Stratigraphic Significance." Nauka, Moscow. (In Russian.)

Gottfried, M. D. (1995). Miocene basking sharks (Lamniformes: Cetorhinidae) from the Chesapeake Group of

Maryland and Virginia. *J. Vertebr. Paleontol.* **15**:443–447.

Gottfried, M. D., Compagno, L. J. V., and Bowman, S. C. (1992). Skeletal anatomy of *Carcharodon megalodon*: Inferences based on comparisons with the Recent species *Carcharodon carcharias. J. Vertebr. Paleontol., Suppl.* **12**(3):30A.

Gould, S. J. (1966). Allometry and size in ontogeny and phylogeny. *Biol. Rev. Cambridge Philos. Soc.* **41**:587–640.

Graeber, R. C. (1978). Behavioral studies correlated with central nervous system integration of vision in sharks. *In* "Sensory Biology of Sharks, Skates and Rays" (E. S. Hodgson and R. F. Mathewson, eds.), pp. 195–225. U.S. Office of Naval Research, Arlington, Virginia.

Graeber, R. C., Schroeder, D. M., Jane, J. A., and Ebbesson, S. O. E. (1978). Visual discrimination following partial telencephalic ablations in nurse sharks (*Ginglymostoma cirratum*). *J. Comp. Neurol.* **180**:325–344.

Graeffe, E. (1886). *Carcharodon rondeleti,* in Uebersicht der Seethierfauna des Golfes von Triest, etc. *Arb. Zool. Inst. Univ. Wien, Zool. St. Trieste* **7**:446.

Granier, J. (1964). Les Eusélaciens dans le golfe d'Aigues-Mortes. *Bull. Mus. Hist. Nat., MarseilleS* **25**:33–52.

Green, J. (1976). "Shark Attacks in Australian Waters." Green, Gosford, New South Wales, Australia.

Greenwood, C., and Taunton-Clark, J. (1992). "An 'Atlas' of Mean Monthly and Yearly Average Sea Surface Temperatures Around the Southern African Coast," Int. Rep. 124. Sea Fisheries Research Institite, Department of Environmental Affairs, Cape Town, South Africa.

Griffith, J. (1993). Great white shark: Predator par excellence. *Oregonian* **March 18**:D4–D5.

Gruber, S. H., ed. (1991). "Discovering Sharks," Spec. Publ. 14. American Littoral Society, Highlands, New Jersey.

Gruber, S. H., and Cohen, J. L. (1985). Visual system of the white shark, *Carcharodon carcharias,* with emphasis on retinal structure. *South. Calif. Acad. Sci., Mem.* **9**:61–72.

Gruber, S. H., and Compagno, L. J. V. (1982). Taxonomic status and biology of the bigeye thresher, *Alopias superciliosus. Fish. Bull.* **79**:617–640.

Gruber, S. H., and Zlotkin, E. (1982). Bioassay of surfactants as shark repellents. *Nav. Res. Rev.* **34**:18–27.

Gruber, S. H., Zlotkin, E., and Nelson, D. R. (1984). Shark repellents: Behavioral bioassays in laboratory and field. *In* "Toxins, Drugs, and Pollutants in Marine Animals" (L. Bolis and J. Zadunaisky, eds.), pp. 26–42. Springer-Verlag, Berlin.

Gubanov, E. P. (1974). The capture of a giant specimen of the mako shark (*Isurus glaucus*) in the Indian Ocean. *J. Ichthyol.* **14**:589–590.

Gubanov, E. P. (1985). Presence of the sharp tooth sand shark, *Odontaspis ferox* (Odontaspididae), in the open waters of the Indian Ocean. *J. Ichthyol.* **25**:156–158.

Gudger, E. W. (1937). Abnormal dentition in sharks, Selachii. *Am. Mus. Nat. Hist., Bull.* **73**(art. II):249–280.

Gudger, E. W. (1950). A boy attacked by a shark, July 25, 1936 in Buzzard's Bay, Massachusetts. *Am. Midl. Nat.* **44**:714–719.

Guitard, D., and Milera, J. F. (1974). El monstruo marino de Cojímar. *Mar Pesca* **104**:10–11.

Guitart Manday, D. (1966). Nuevo nombre para una especie de tiburon del genero *Isurus* (Elasmobranchii: Isuridae) de aguas cubanas. Popyana, ser. A:1–9.

Guitart Madany, D. (1975). Las pesquerias pelagico-oceanicas de corto radio de accion en la region noroccidental de Cuba. *Oceanogr. Inst. Acad. Sci., Havana, Ser. Oceanol.* **31**:1–41.

Haldane, D. (1992). Shark attack called mistake. *Los Angeles Times* **November 30.**

Hallacher, L. E. (1977). On the feeding behavior of the basking shark, *Cetorhinus maximus. Environ. Biol. Fishes* **2**:297–298.

Hamilton, R. (1843). Genus *Carcharias.* The white shark. *In* "A History of British Fishes," pp. 304–305. Hardwicke & Bogue, London.

Hamilton, W. D. (1971). Geometry for the selfish herd. *J. Theor. Biol.* **31**:295–311.

Hanan, D. (1993). "Status of the Pacific Harbor Seal Population on the Coast of California in 1992," final report. U.S. National Marine Fisheries Service, La Jolla, California.

Hanan, D., Scholl, J., and Diamond, S. (1989). Harbor seal, *Phoca vitulina richardsi,* census in California, May–June 1988. *U.S., NMFS, Southwest Fish. Sci. Cent., Admin. Rep.* **LJ-89-13.**

Harland, W. B., Armstrong, R., Cox, A., Craig, L., Smith, A., and Smith, D. (1990). "A Geologic Time Scale." Cambridge University Press, Cambridge.

Hayward, T. L. (1993). Preliminary observations of the 1991–92 El Niño in the California Current. *Calif. Coop. Ocean. Fish. Invest. Rep.* **34**:21–29.

Heldreich, Th. de. (1878). "La Faune de Grèce, Rapport sur les Travaux et Recherches Zoologigues faites en Grèce", p. 91. Imprimerie de la Philocalie, Athens.

Helfman, G. S. (1988). Patterns in the life history of anguillid eels. *Proc. Congr. N.Z.* **23.**

Herald, E. S. (1968). Size and aggressiveness of the sevengill shark (*Notorynchus maculatus*). *Copeia* **1968**:412–414.

Herman, J. (1979). Reflexions sur la systematique des galeoidei sur les affinities du genere *Cetorhinus* a l'occasion de la decouverte d'elements de la denture d'un exemplaire fossile dans les sables du Kattendijk a Kallo (Pliocene inferieur, Belgique). *Ann. Soc. Geol. Belg.* **102**:357–377.

Herman, L. M., and Tavolga, W. N. (1980). The communication systems of cetaceans. *In* "Cetacean Behavior: Mechanisms and Functions" (L. M. Herman, ed.), pp. 149–210. Wiley, New York.

Hickey, B. M. (1979). The California Current system—Hypothesis and facts. *Prog. Oceanogr.* **8**:191–279.

Hilborn, R. (1990). Determination of fish movement patterns from tag recoveries using maximum likelihood estimators. *Can. J. Fish. Aquat. Sci.* **5**:217–222.

Hill, R. D. (1994). Theory of geolocation by light levels. *In* "Elephant Seals: Population Ecology, Behavior and Physiology" (B. J. Le Boeuf and R. M. Laws, eds.), pp. 237–246. University of California Press, Berkeley.

Hillis, D. M., and Bull, J. J. (1993). An empirical test of bootstrapping as a method for assessing confidence in phylogenetic analysis. *Syst. Biol.* **42**:182–192.

Hillis, D. M., and Huelsenbeck, J. P. (1992). Signal, noise, and reliability in molecular phylogenetic analyses. *Heredity* **83:**189–195.

Hillis, D. M., Huelsenbeck, J. P., and Cunningham, C. W. (1994). Application and accuracy of molecular phylogenies. *Science* **264:**671–677.

Hinde, R. A. (1970). "Animal Behaviour: A Synthesis of Ethology and Comparative Psychology." McGraw-Hill, New York.

Hindell, M. A., Slip, D. J., and Burton, H. R. (1991). The diving behaviour of adult male and female southern elephant seals, *Mirounga leonina* (Pinnipedia: Phocidae). *Aust. J. Zool.* **39:**595–619.

Hiruki, L. M., Gilmartin, W. G., Becker, B. L., and Stirling, I. (1993). Wounding in Hawaiian monk seals (*Monachus schauinslandi*). *Can. J. Zool.* **71:**458–468.

Hobson, E. S. (1963). Feeding behavior in three species of sharks. *Pac. Sci.* **17:**171–194.

Hodgson, E. S., and Mathewson, R. F., eds. (1978). "Sensory Biology of Sharks, Skates and Rays." U.S. Office of Naval Research, Arlington, Virginia.

Hoenig, J. M., and Brown, C. A. (1988). A simple technique for staining growth bands in elasmobranch vertebrae. *Bull. Mar. Sci.* **42:**334–337.

Horning, D. S., Jr., and Fenwick, G. D. (1978). Leopard seals at the Snares Islands, New Zealand. *N.Z. J. Zool.* **5:**171–172.

Hubbell, G. (1990). Nuevas apreciaciones sobre los antepasados del gran tiburon blanco. *Bol. Lima* **68:**27–28.

Hubbs, C. L. (1954). Changes in the fish fauna of western North America correlated with changes in ocean temperature. *J. Mar. Res.* **7:**459–482.

Huber, H. R. (1991). Changes in the distribution of California sea lions north of the breeding rookeries during the 1982–83 El Niño. *In* "Pinnipeds and El Niño: Responses to Environmental Stress" (F. Trillmich and K. A. Ono, eds.), pp. 129–137. Springer-Verlag, Berlin.

Huber, H. R., Ainley, D. G., and Morrell, S. H. (1980). Sightings of cetaceans in the Gulf of the Farallones, California, 1971–1979. *Calif. Fish Game* **68:**183–190.

Huber, H. R., Beckham, C., Nisbet, J., Rovetta, A., and Nusbaum, J. (1985). Studies of marine mammals at the Farallon Islands 1982–1983. *U.S., NMFS, Southwest Fish. Sci. Cent., Admin. Rep.* **LJ-85-01C.**

Huelsenbeck, J. P. (1991). Tree-length distribution skewness: An indicator of phylogenetic information. *Syst. Zool.* **40:**257–270.

Hughes, R. (1987). Shark attack in Australian waters. *In* "Sharks" (J. D. Stevens, ed.), pp. 108–121. Golden Press, Sydney.

Humason, G. L. (1979). "Animal Tissue Techniques." Freeman, San Francisco.

Hussakoff, L. (1909). A new goblin shark, *Scapanorhynchus jordani*, from Japan. *Am. Mus. Nat. Hist., Bull.* **26:**257–263.

Imber, M. J. (1971). Seabirds found dead in New Zealand in 1969. *Notornis* **18:**305–309.

Irwin, D. M., Kocher, T. D., and Wilson, A. C. (1991). Evo-

lution of the cytochrome *b* gene of mammals. *J. Mol. Evol.* **32:**128–144.

Jaekel, O. (1895). Unter-Tertiare Selachier aus Sudrussland. *Mem. Com. Geol.* **9**(4):1–35.

Jameson, G. L. (1986). "Trial Systematic Salvage of Beach-Cast Sea Otter, *Enhydra lutris*, Carcasses in the Central and Southern Portion of the Sea Otter Range in California," final report. U.S. Marine Mammal Commission, Washington, D.C.

Jamieson, B. G. M. (1994). Phylogeny of the Brachyura with particular reference to the Podotremata: Evidence from a review of spermatozoal ultrastructure (Crustacea, Decapoda). *Philos. Trans. R. Soc. London, Ser. B* **345:**373–393.

Janvier, P., and Welcomme, J. L. (1969). Affinites et paleobiologie de l'espece *Carcharodon megalodon* AG. Squale Geant des Faluns de la Touraine et de l'Anjou. *Rev. Fed. Soc. Sci. Nat.,* 3 *Ser.* **8**(34):1–6.

Jefferson, T. A., Stacey, P. J., and Baird, R. W. (1991). A review of killer whale interactions with other marine mammals: Predation to co-existence. *Mammal Rev.* **21:**151–180.

Johnson, C. S., and Baldridge, H. D. (1985). "Analytic Indication of the Impracticability of Waterborne Chemicals for Repelling an Attacking Shark—A Second, Confirming Look," Tech. Doc. 843. Naval Ocean System Center, San Diego.

Johnson, R. H., and Nelson, D. R. (1973). Agonistic display in the gray reef shark, *Carcharhinus menisorrah*, and its relationship to attacks on man. *Copeia* **1973:**76–84.

Johnson, R. H., and Nelson, D. R. (1978). Copulation and possible olfaction-mediated pair formation in two species of carcharhinid sharks. *Copeia* **1978:**439–542.

Jolly, G. M. (1965). Explicit estimates from capture–recapture data with both death and immigration—Stochastic model. *Biometrika* **52:**225–247.

Jones, N., and Jones, R. C. (1982). The structure of the male genital system of the Port Jackson shark, *Heterodontus portusjacksoni* with particular reference to the genital ducts. *Aust. J. Zool.* **30:**523–541.

Jones, R. C., and Lin, M. (1992). Ultrastructure of the genital duct epithelium of the male Port Jackson shark, *Heterodontus portusjacksoni*. *Aust. J. Zool.* **40:**257–266.

Jordan, D. S. (1898). Description of a species of fish (*Mitsukurina owstoni*) from Japan, the type of a distinct family of lamnoid sharks. *Proc. Calif. Acad. Sci., Ser. 3* **1:**199–204.

Jordan, D. S. (1905). "A Guide to the Study of Fishes," Vol. 1. Holt, New York.

Jordan, D. S., and Hannibal, H. (1923). Fossil sharks and rays of the Pacific slope of North America. *South. Calif. Acad. Sci., Bull.* **23:**27–63.

Jury, K. (1987). Huge 'white pointer' encounter. As told to the editor K. Jury. *Safish (Adelaide, South Australia)* **11**(3):12–13.

Karinen, J. F., Wing, B. L., and Straty, R. R. (1985). Records and sightings of fish and invertebrates in the eastern Gulf of Alaska and oceanic phenomena related to the

1983 El Niño event. *In* "El Niño North: El Niño Effects in the Eastern Subarctic Pacific Ocean" (W. S. Wooster and D. L. Fluharty, eds.), pp. 253–267. University of Washington, Seattle.

Kato, S. (1965). White shark *Carcharodon carcharias* from the Gulf of California with a list of sharks in Mazatlan, Mexico, 1964. *Copeia* 1965:384.

Kean, B. H. (1944). Death following attack by shark, *Carcharodon carcharias. JAMA* 12:845–846.

Keinath, J. A., and Musick, J. A. (1993). Movements and diving behavior of a leatherback turtle, *Dermochelys coriacea. Copeia* 1993:1010–1017.

Kellogg, R., and Whitmore, F. C. (1957). Mammals. *Geol. Soc. Am., Mem.* 67:1021–1024.

Kemp, N. R. (1991). Chondrichthyans in the Cretaceaous and Tertiary of Australia. *In* "Vertebrate Paleontology of Australasia" (P. Vickers-Rich, J. M. Monaghan, R. F. Baird, and T. H. Rich, eds.), Ch. 15. Pioneer Design Studio, Lilydale, Victoria, Australia.

Kennedy, J. S. (1992). "The New Anthropomorphism." Cambridge University Press, Cambridge.

Kenney, R. D., and Winn, H. E. (1986). Cetacean high-use habitats of the northeast United States continental shelf. *Fish. Bull.* 84:345–357.

Kenyon, K. W. (1959). A 15-foot maneater from San Miguel Island, California. *Calif. Fish Game* 45:58–59.

Keyes, I. W. (1972). New records of the elasmobranch *C. megalodon* (Agassiz) and a review of the genus *Carcharodon* in the New Zealand Fossil Record. *N.Z. J. Geol. Geophys.* 15(2):229–242.

Kimura, M. (1980). A simple method for estimating evolutionary rate of base substitutions through comparative analysis of nucleotide sequences. *J. Mol. Evol.* 16:111–120.

King, J. E. (1985). "Seals of the World," 2nd ed. Comstock, Ithaca, New York.

Kiorboe, T., Munk, P., Richardson, K., Christiansen, V., and Paulsen, H. (1988). Plankton dynamics and larval herring growth, drift and survival in a frontal area. *Mar. Ecol. Prog. Ser.* 44:205–219.

Klimley, A. P. (1980). Observations of courtship and copulation in the nurse shark, *Ginglymostoma cirratum. Copeia* 1980:878–882.

Klimley, A. P. (1981). Grouping behavior in the scalloped hammerhead. *Oceanus* 24:65–71.

Klimley, A. P. (1985a). Schooling in *Sphyrna lewini*, a species with low risk of predation: A non-egalitarian state. *Z. Tierpsychol.* 70:297–319.

Klimley, A. P. (1985b). The areal distribution and autoecology of the white shark, *Carcharodon carcharias*, off the west coast of North America. *South. Calif. Acad. Sci., Mem.* 9:15–40.

Klimley, A. P. (1987a). Field studies of the white shark, *Carcharodon carcharias*, in the Gulf of the Farallones National Marine Sanctuary. *In* "Current Research Topics in the Marine Environment" (M. M. Croom, ed.), pp. 33–36. Gulf of the Farallones National Marine Sanctuary, San Francisco.

Klimley, A. P. (1987b). The determinants of sexual segregation in the scalloped hammerhead shark, *Sphyrna lewini. Environ. Biol. Fishes* 18:27–40.

Klimley, A. P. (1994). The predatory behavior of the white shark. *Am. Sci.* 52:122–133.

Klimley, A. P., and Butler, S. B. (1988). Immigration and emigration of a pelagic fish assemblage to seamounts in the Gulf of California related to water mass movements using satellite imagery. *Mar. Ecol. Prog. Ser.* 49:11–20.

Klimley, A. P., and Myrberg, A. A., Jr. (1979). Acoustic stimuli underlying withdrawal from a sound source by adult lemon sharks, Negaprion brevirostris (Poey). *Bull. Mar. Sci.* 29:447–458.

Klimley, A. P., Anderson, S. D., Henderson, R. P., and Pyle, P. (1989). A description of predatory attacks by white sharks on pinnipeds. *Am. Soc. Ichthyol. Herpetol./ Am. Elasmobranch Soc. Annu. Meet.* (abstract).

Klimley, A. P., Anderson, S. D., Pyle, P., and Henderson, R. P. (1992). Spatiotemporal patterns of white shark (*Carcharodon carcharias*) predation at the South Farallon Islands, California. *Copeia* 1992:680–690.

Knowlton, N., Weight, L. A., Solorzano, L. A., Mills, D. K., and Bermingham, E. (1993). Divergence in proteins, mitochondrial DNA, and reproductive compatibility across the Isthmus of Panama. *Science* 260:1629–1632.

Konsuloff, S., and Drenski, P. (1943). Die Fischfauna der Aegais. *Ann. Univ. Sofia Fac. Sci.* 39:293–308.

Kooyman, G. L., and Gentry, R. L. (1986). Diving behavior of South African fur seals. *In* "Fur Seals: Maternal Strategies on Land and at Sea" (R. L. Gentry and G. L. Kooyman, eds.), pp. 142–152. Princeton University Press, Princeton, New Jersey.

Kozuch, L., and Fitzgerald, C. (1989). A guide to identifying shark centra from southeastern archeological sites. *Southeast. Archaeol.* 2:146–157.

Krebs, C. J. (1989). "Ecological Methodology." Harper & Row, New York.

Kretzmann, M. B. (1990). "Maternal Investment and the Post-weaning Fast in Northern Elephant Seals: Evidence for Sexual Equality," unpublished M.S. thesis. University of California, Santa Cruz.

Kruska, D. C. T. (1988). The brain of the basking shark (*Cetorhinus maximus*). *Brain, Behav. Evol.* 32:353–363.

Kuhry, B., and Marcus, L. F. (1977). Bivariate linear models in biometry. *Syst. Zool.* 26:201–209.

Lavenberg, R. J., (1991). Megamania, the continuing saga of megamouth sharks. *Terra* 30:30–39.

Lavery, S., and Shaklee, J. B. (1989). Population genetics of two tropical sharks, *Carcharhinus tilstoni* and *C. sorrah*, in northern Australia. *Aust. J. Mar. Freshwater Res.* 40:541–547.

Laws, R. M. (1994). History and present status of southern elephant seal populations. *In* "Elephant Seals: Population Ecology, Behavior and Physiology" (B. J. Le Boeuf and R. M. Laws, eds.), pp. 49–65. University of California Press, Berkeley.

Lea, R. N. (1987). "Pacific Coast Shark Attacks: What Is the

Danger? Proceedings of the Conference: Sharks, an Inquiry into Biology, Behavior, Fisheries, and Use." Oregon State University Extension Service, Corvallis.

Lea, R. N., and Miller, D. J. (1985). Shark attacks off the California and Oregon coasts: An update, 1980–84. *South. Calif. Acad. Sci., Mem.* **9**:136–150.

Leatherwood, J. S., Perrin, W. F., Garvie, R. L., and La Grange, J. C. (1972). Observations of sharks attacking porpoises (*Stenella* spp. and *Delphinus* cf *D. delphis*). *Nav. Undersea Cent., San Diego* **TN 908**:1–7.

Leatherwood, S., Reeves, R. R., Perrin, W. F., and Evans, W. E. (1982). Whales, dolphins, and porpoises of the eastern North Pacific and adjacent Arctic waters. *NOAA Tech. Rep., NMFS* **NMFS 444.**

Le Boeuf, B. J. (1974). Male–male competition and reproductive success in elephant seals. *Am. Zool.* **14**:163–176.

Le Boeuf, B. J. (1994). Variation in the diving pattern of northern elephant seals with age, mass, sex and reproductive condition. *In* "Elephant Seals: Population Ecology, Behavior and Physiology" (B. J. Le Boeuf and R. M. Laws, eds.), pp. 237–252. University of California Press, Berkeley.

Le Boeuf, B. J., and Laws, R. M., eds. (1994). "Elephant Seals: Population Ecology, Behavior and Physiology." University of California Press, Berkeley.

Le Boeuf, B. J., and Reiter, J. (1988). Lifetime reproductive success in northern elephant seals. *In* "Reproductive Success: Studies of Individual Variation in Contrasting Breeding Systems" (T. H. Clutton-Brock, ed.), pp. 344–362. University of Chicago Press, Chicago.

Le Boeuf, B. J., Riedman, M., and Keyes, R. S. (1982). White shark predation on pinnipeds in California coastal waters. *Fish. Bull.* **80**:891–895.

Le Boeuf, B. J., Costa, D. P., Huntley, T., Kooyman, G. L., and Davis, R. (1986). Pattern and depth of dives in northern elephant seals. *J. Zool.* **208**:1–7.

Le Boeuf, B. J., Costa, D. P., Huntley, A. C., and Feldkamp, S. D. (1988). Continuous, deep diving in female northern elephant seals, *Mirounga angustirostris*. *Can. J. Zool.* **66**:446–458.

Le Boeuf, B. J., Naito, Y., Huntley, A. C., and Asaga, T. (1989). Prolonged, continuous deep diving by northern elephant seals. *Can. J. Zool.* **67**:2514–2519.

Le Boeuf, B. J., Naito, Y., Asaga, T., Crocker, D., and Costa, D. P. (1992). Swim speed in a female northern elephant seal: Metabolic and foraging implications. *Can. J. Zool.* **70**:786–795.

Le Boeuf, B. J., Crocker, D. E., Blackwell, S. B., Morris, P. A., and Thorson, P. H. (1993). Sex differences in diving and foraging behavior of northern elephant seals. *Symp. Zool. Soc. London* **66**:149–178.

Le Boeuf, B. J., Morris, P., and Reiter, J. (1994). Juvenile survivorship of northern elephant seals. *In* "Elephant Seals: Population Ecology, Behavior and Physiology" (B. J. Le Boeuf and R. M. Laws, eds.), pp. 121–136. University of California Press, Berkeley.

Lehner, P. N. (1979). "Handbook of Ethological Methods." Garland STPM, New York.

Leidy, J. (1877). Description of vertebrate remains, chiefly from the phosphate beds of South Carolina. *J. Acad. Nat. Sci. Philadelphia* **8**:209–261.

Lennon, G. W., Bowers, D. G., Nunes, R. A., Scott, B. D., Ali, M., Boyle, J., Wenju, C., Herzfeld, M., Johansson, G., Nield, S., Petrusevies, P., Stephanson, P., Suskin, A. A., and Wijffels, S. E. A. (1987). Gravity currents and the release of salt from an inverse estuary. *Nature (London)* **327**:695–697.

Leriche, M. (1910). Les poissons Oligocenes de la Belgique. *Mus. R. Hist. Nat. Belg., Mem.* **5**(2):229–363.

Leriche, M. (1926). Les poissons Neogenes de la Belgique. *Mus. R. Hist. Nat. Belg., Mem.* **32**:368–472.

Leriche, M. (1927). Les poissons de la Molasse suisse. *Mem. Soc. Paleontol. Suisse* **46–47**:1–120.

Leriche, M. (1936). Sur l'importance des squales fossiles dans l'establissement des synchronismes de formations a grandes distances et sur la repartition stratigraphique et geographique de quelques especes Tertiaires. *Mus. R. Hist. Nat. Belg., Mem.* (2)3:739–772.

Levine, M. (1994). "Sharks, Questions, and Answers." New Holland, London.

Levine, M. (1996). "Sharks and Shark Attacks of Southern Africa, 1852 to the Present." Struik, Cape Town, South Africa. In press.

Limbaugh, C. (1963). Field notes on sharks. *In* "Sharks and Survival" (P. W. Gilbert, ed.), pp. 63–94. Heath, Boston.

Lindesay, J. A. (1988). "The Southern Oscillation and Atmospheric Circulation Changes Over Southern Africa," unpublished Ph.D. dissertation. University of Witwatersrand, Johannesburg.

Ling, J. K., and Bryden, M. M. (1981). Southern elephant seal *Mirounga leonina* Linnaeus, 1785. *In* "Handbook of Marine Mammals" (S. H. Ridgway and R. J. Harrison, eds.), Vol. 2. Academic Press, New York.

Loeb, E. M. (1926). Pomo folkways. *Univ. Calif. Publ. Am. Archeol. Ethnol.* **19**:149–405.

Long, D. J. (1991a). Apparent predation by a white shark *Carcharodon carcharias* on a pygmy sperm whale *Kogia breviceps*. *Fish. Bull.* **89**:538–540.

Long, D. J. (1991b). A review of shark predation on cetaceans. *Bienn. Conf. Biol. Mar. Mammals, 9th, Shedd Aquarium and the Brookfield Zoo, Chicago* (abstract).

Long, D. J. (1992a). Paleoecology of Eocene Antarctic sharks. The Antarctic paleoenvironment: A perspective on global change. *Antarct. Res. Ser.* **56**:131–139.

Long, D. J. (1992b). Sharks from the La Meseta Formation (Eocene), Seymour Island, Antarctic Peninsula. *J. Vertebr. Paleontol.* **12**(1):11–32.

Long, D. J. (1994). "Historical Biogeography of Sharks From the Northeastern Pacific Ocean," unpublished Ph.D. dissertation. University of California, Berkeley.

Long, D. J., and Hanni, K. D. (1993). Dynamics of white shark (*Carcharodon carcharias*) predation on Steller sea lions (*Eumetopias jubatus*) in California. *Bienn. Conf. Biol. Mar. Mammals, 10th, Galveston, Texas* (abstract).

Long, D. J., and Spencer, C. L. (1995). Cow sharks (Hexanchidae) as predators on the harbor seal (*Phoca vitulina richardsi*) in San Francisco Bay, California. *Chondros* **6**(2):9.

Longhurst, A. R. (1981). Significance of spatial variability. *In* "Analysis of Marine Ecosystems" (A. R. Longhurst, ed.), pp. 415–441. Academic Press, New York.

Lopez, J. C., and Lopez, D. (1985). Killer whales (*Orcinus orca*) of Patagonia, and their behavior of intentional stranding while hunting nearshore. *J. Mammal.* **66:**181–183.

Loughlin, T. R., Perlov, A. S., and Vladimirov, V. A. (1992). Range-wide survey and estimation of total number of Steller sea lions in 1989. *Mar. Mammal Sci.* **8:**220–239.

Lowry, M. S., Boveng, P., DeLong, R. J., Oliver, C. W., Stewart, B. S., DeAnda, H., and Barlow, J. (1992). Status of the California sea lion (*Zalophus californianus*) population in 1992. *U.S., NMFS, Southwest Fish. Sci. Cent., Admin. Rep.* **LJ-92-32.**

Lozano Rey, L. (1928). Ictiología Ibérica (Fauna Ibérica). Peces (Generalidades, Ciclóstomos y Elasmobranquios). *Mus. Nac. Cienc. Nat., Madrid* **1:**1–692.

Luiten, P. G. M. (1981). Two visual pathways to the telencephalon in the nurse shark (*Ginglymostoma cirratum*). II. Ascending thalamo-telencephalic connections. *J. Comp. Neurol.* **196:**539–548.

Lynn, R. (1966). "Attention, Arousal, and the Orientation Reaction." Pergamon, Oxford.

Maddison, W. P., and Maddison, D. R. (1992). "MacClade: Analysis of Phylogeny and Character Evolution." Sinauer, Sunderland, Massachusetts.

Maisey, J. G. (1985). Relationships of the megamouth shark, *Megachasma*. *Copeia* **1985:**228–231.

Mara, J. (1985). "A Fisherman's Tale: Fifty Years of Angling Along the Natal Coast." Angler Publications & Promotions, Durban, Natal, South Africa.

Marquez, M. R. (1990). FAO species catalogue. Vol. 11. Sea turtles of the world. *FAO Fish. Synop.* **125:**1–81.

Martin, A. P. (1992). "Mitochondrial DNA Evolution in Elasmobranch Fishes," Ph.D. dissertation. University of Hawaii, Honolulu.

Martin, A. P. (1995). Mitochondrial DNA sequence evolution in sharks: Rates, patterns, and phylogenetic inferences. *Mol. Biol. Evol.* **12:**1114–1123.

Martin, A. P., and Palumbi, S. R. (1993). Protein evolution in different cellular environments: Cytochrome *b* in sharks and mammals. *Mol. Biol. Evol.* **10:**873–891.

Martin, A. P., Naylor, G. J. P., and Palumbi, S. R. (1992). Rates of mitochondrial DNA evolution in sharks are slow compared with mammals. *Nature (London)* **357:**153–155.

Martin, H. R., Kingsley, M. C. S., and Ramsay, M. A. (1994). Diving behaviour of narwhals (*Monodon monoceros*) on their summer grounds. *Can. J. Zool.* **72:**118–125.

Martinez, D. R., and Klinghammer, E. (1970). The behavior of the whale *Orcinus orca*: A review of the literature. *Z. Tierpsychol.* **27:**828–839.

Martini, F. H., and Welch, K. (1981). A report on a nonfatal shark attack in the Hawaiian Islands. *Pac. Sci.* **35:**237–240.

Mate, B. R., Nieukirk, S., Mesecar, R., and Martin, T. (1992). Application of remote sensing methods for tracking large cetaceans: North Atlantic right whales (*Eubalaena glacialis*). *U.S., Minerals Manage. Serv., Outer Continental Shelf Study* **MMS 91-0069.**

Matsuura, Y. (1986). Contribuicao ao estudo da estrutura oceanografica da regiao sudeste entre Cabo Frio (RJ) e Cabo de Santa Marta (SC). *Cienc. Cult.* **38**(8):1439–1450.

Matthews, L. H. (1950). Reproduction in the basking shark, *Cetorhinus maximus* (Gunner). *Philos. Trans. R. Soc. London, Ser. B* **234:**247–316.

Matthews, L. H., and Parker, H. W. (1951). Notes on the anatomy and biology of the basking shark (*Cetorhinus maximus* (Gunner)). *Proc. Zool. Soc. London* **120:**535–576.

Mattison, J. A., Jr., and Hubbard, R. C. (1969). Autopsy findings on thirteen sea otters (*Enhydra lutris nereis*) with correlations with captive animal feeding and behavior. *In* "Sixth Annual Conference on the Biology of Sonar and Diving Mammals," pp. 99–101. Stanford Research Institute, Menlo Park, California.

Maul, G. E. (1955). Five species of rare sharks new for Madeira including two new species. *Not. Nat. Acad. Nat. Sci. Philadelphia* **279:**1–13.

McArdle, B. H. (1988). The structural relationship: Regression in biology. *Can. J. Zool.* **66:**2329–2339.

McCleery, R. H. (1978). Optimal behavior sequences and decision making. *In* "Behavioral Ecology: An Evolutionary Approach" (J. R. Krebs and N. B. Davies, eds.), pp. 377–410. Sinauer, Sunderland, Massachusetts.

McCosker, J. E. (1981). Great white shark. *Science* **81:**42–51.

McCosker, J. E. (1985). White shark attack behavior: Observations of and speculations about predator and prey strategies. *South. Calif. Acad. Sci., Mem.* **9:**123–135.

McCosker, J. E. (1987). The white shark, *Carcharodon carcharias*, has a warm stomach. *Copeia* **1987:**195–197.

McFarland, D. J. (1966). On the causal and functional significance of displacement activities. *Z. Tierpsychol.* **23:**217–235.

McLaren, I. A., and Smith, T. G. (1985). Population ecology of seals; retrospective and prospective views. *Mar. Mammal Sci.* **1:**54–83.

McNeill, D., and Freiberger, P. (1993). "Fuzzy Logic." Simon & Schuster, New York.

Meek, S. E., and Hildebrand, S. F. (1923). The marine fishes of Panama. Part I. *Field Mus. Nat. Hist. Publ. Zool.* **15.**

Mesnick, S. L., and Le Boeuf, B. J. (1991). Sexual behavior of male northern elephant seals: II. Female response to potentially injurious encounters. *Behaviour* **117:**262–280.

Meyer, A., and Wilson, A. C. (1990). Origin of tetrapods inferred from their mitochondrial DNA affiliation to lungfish. *J. Mol. Evol.* **31:**359–364.

Michelotti, G. (1861). Description de quelques nouveaux fossiles du terrain Miocène de la colline de Turin. *Rev. Mag. Zool. Pure Appl., Paris* (2)13:353–355.

Milinkovitch, M. C., Meyer, A., and Powell, J. R. (1994). Phylogeny of all major groups of cetaceans based on DNA sequences from three mitochondrial genes. *Mol. Biol. Evol.* **11:**939–948.

Miller, D. M. (1984). Reducing transformation bias in curve fitting. *Am. Stat.* **38:**124–126.

Miller, D. J., and Collier, R. S. (1981). Shark attacks in California and Oregon, 1926–1979. *Calif. Fish Game* **67:** 76–104.

Minasian, S. M., Balcomb, K. C., and Foster, L. (1984). "The World's Whales." Smithsonian Books, New York.

Minerals Management Service (1987). California seabird ecology study. II. Satellite data analysis. *U.S., Minerals Manage. Serv., Outer Continental Shelf Study* **MMS 87-0056.**

Mitchell, E. (1989). A new cetacean from the late Eocene La Meseta Formation, Seymour Island, Antarctic Peninsula. *Can. J. Fish. Aquat. Sci.* **46:**2219–2235.

Moreau, E. (1881). "Histoire Naturelle des Poissons de la France," Libr. Acad. Med., Vol. 1. Masson, Paris.

Morejohn, G. V., Ames, J. A., and Lewis, D. B. (1975). Post mortem studies of sea otter, *Enhydra lutris* L., in California. *Calif. Dept. Fish Game, Mar. Resour. Tech. Rep.* **30.**

Moreno, J. A., and Morón, J. (1992a). Comparative study of the genus *Isurus* (Rafinesque, 1810), and description of a form ('Marrajo Criollo') apparently endemic to the Azores. *Aust. J. Mar. Freshwater Res.* **43:**109–122.

Moreno, J. A., and Morón, J. (1992b). Reproductive biology of the bigeye thresher shark, *Alopias vulpinus* (Bonnaterre, 1788) (Squaliformes: Alopiidae) en el Atlantico nor-oriental y Mediteraneo occidental. *Sci. Mar.* **53:** 37–46.

Moreno, J. A., Parajua, J. I., and Morón, J. (1989). Biologia reproductiva y fenologia de *Alopias vulpinus* (Bonnaterre, 1788) (Squaliformes: Alopiidae) en el Atlantico nor-oriental y Mediterraneo occidental. *Scientia Marina* **53:**37–46.

Moritz, C., Dowling, T. E., and Brown, W. M. (1987). Evolution of animal mitochondrial DNA: Relevance for population biology and systematics. *Annu. Rev. Ecol. Evol.* **18:**269–292.

Moss, S. (1967). Tooth replacement in the lemon shark, *Negaprion brevirostris. In* "Sharks, Skates and Rays" (P. W. Gilbert, R. F. Mathewson, and D. P. Ralls, eds.), pp. 319–329. Johns Hopkins University Press, Baltimore.

Moss, S. (1984). "Sharks, an Introduction for the Amateur Naturalist." Prentice-Hall, New York.

Müller, J., and Henle, F. G. J. (1838). On the generic characters of cartilaginous fishes, with descriptions of new genera. *Charlesworth Ann. Mag. Nat. Hist.* **2**(2):33–37, 88–91.

Muller-Schwarze, D., and Muller-Schwarze, C. (1975). Relations between leopard seals and Adélie penguins. *Rapp. P.-V. Reun., Cons. Int. Explor. Mer* **169:**394–404.

Murru, F. (1990). The care and maintenance of elasmobranchs in controlled environments. *NOAA Tech. Rep., NMFS* **NMFS 90:**203–209.

Myagkov, N. A. (1991). The brain sizes of living elasmobranchii as their organization level indicator. I. General analysis. *J. Hirnforsch.* **32:**553–561.

Myrberg, A. A., Jr. (1969). Shark attraction using a video-acoustic system. *Mar. Biol.* **2:**264–276.

Myrberg, A. A., Jr., and Gruber, S. H. (1974). The behavior of the bonnethead shark, *Sphyrna tiburo. Copeia* **1974:** 358–37.

Myrberg, A. A., Jr., and Nelson, D. R. (1991). The behavior of sharks: What have we learned? *In* "Discovering

Sharks" (S. H. Gruber, ed.), Spec. Publ. 14, pp. 92–100. American Littoral Society, Highlands, New Jersey.

Myrberg, A. A., Jr., Gordon, C. R., and Klimley, A. P. (1978). Rapid withdrawal from a sound source by open-ocean sharks. *J. Acoust. Soc. Am.* **64:**1289–1297.

Naito, Y., Le Boeuf, B. J., Asaga, T., and Huntley, A. C. (1989). Long term diving records of an adult female northern elephant seal. *Nankyoku Shiryo* **33:**1–9.

Nakano, H., and Nakaya, K. (1987). Records of the white shark *Carcharodon carcharias* from Hokkaido, Japan. *Jpn. J. Ichthyol.* **33:**414–416.

Nakaya, K. (1971). Descriptive notes on a porbeagle, *Lamna nasus,* from Argentine waters, compared with the north Pacific salmon shark, *Lamna ditropis. Bull. Fac. Fish., Hokkaido Univ.* **21:**269–279.

Nakaya, K. (1993). A fatal attack by a white shark in Japan and a review of shark attacks in Japanese waters. *Jpn. J. Ichthyol.* **40:**35–42.

Nakaya, K. (1994). Distribution of white shark in Japanese waters. *Fish. Sci.* **60:**515–518.

Namais, J. (1969). Seasonal interactions between the North Pacific Ocean and the atmosphere during the 1960's. *Mon. Weather Rev.* **97:**123–192.

National Marine Fisheries Service (1992). "Recovery Plan for the Steller Sea Lion (*Eumetopias jubatus*)." U.S. National Marine Fisheries Service, Silver Spring, Maryland.

Nelson, D. R. (1983). Shark attack and repellency research: An overview. *AAAS Sel. Symp.* **83:**11–74.

Nelson, D. R. (1991). Shark repellents: How effective, how needed? *In* "Discovering Sharks" (S. H. Gruber, ed.), Spec. Publ. 14, pp. 106–108. American Littoral Society, Highlands, New Jersey.

Nelson, D. R., and Johnson, R. H. (1972). Acoustic attraction of pacific reef sharks: Effect of pulse intermittency and variability. *Comp. Biochem. Physiol. A* **42A:**85–95.

Nelson, D. R., and McKibben, J. N. (1981). Timed-release, recoverable, ultrasonic/radio transmitters for tracking pelagic sharks. *In* "Proceedings of the Third International Conference on Wildlife Biotelemetry" (F. M. Long, ed.), pp. 90–104. University of Wyoming, Laramie.

Nelson, D. R., Johnson, R. R., McKibben, J. N., and Pittenger, G. G. (1986). Agonistic attacks on divers and submersibles by gray reef sharks, *Carcharhinus amblyrhynchos:* Antipredatory or competitive? *Bull. Mar. Sci.* **38:**68–88.

Neter, J., Wassermann, W., and Kutner, M. H. (1983). "Applied Linear Statistical Models." Irwin, Homewood, Illinois.

Noble, M., and Gelfenbaum, G. (1990). A pilot study of currents and suspended sediment in the Gulf of the Farallones. *U.S. Geol. Surv., Open-File Rep.* **90-476.**

Norman, J. R., and Fraser, F. C. (1937). "Giant Fishes, Whales and Dolphins." Putnam, London.

Norris, K. S. (1967). Aggressive behavior in cetacea. *UCLA Forum Med. Sci.* **7:**225–241.

Norris, K. S., and Dohl, T. P. (1980). Behavior of the Hawaiian spinner dolphin, *Stenella longirostris. Fish. Bull.* **77:**821–849.

Northcutt, R. G. (1977). Elasmobranch central nervous system organization and its possible evolutionary significance. *Am. Zool.* **17**:411–429.

Northcutt, R. G. (1978). Brain organization in cartilaginous fishes. *In* "Sensory Biology of Sharks, Skates and Rays" (E. S. Hodgson and R. F. Mathewson, eds.), pp. 117–193. U.S. Office of Naval Research, Arlington, Virginia.

Northcutt, R. G. (1989). Brain variation and phylogenetic trends in elasmobranch fishes. *J. Exp. Zool., Suppl.* **2**: 83–100.

Norton, J., McLain, D., Brainard, R., and Husby, D. (1985). The 1982–83 El Niño event off Baja and Alta California and its ocean climate context. *In* "El Niño North: Niño Effects in the Eastern Subarctic Pacific Ocean" (W. S. Wooster and D. L. Fluharty, eds.), pp. 44–72. University of Washington, Seattle.

Notarbartolo di Sciara, G. (1990). A note on the cetacean incidental catch in the Italian driftnet swordfish fishery, 1986–1988. *Rep. Int. Whaling Comm.* **40**:459–460.

Nyberg, D. W. (1971). Prey capture in the largemouth bass. *Am. Midl. Nat.* **86**:128–144.

O'Brien, W. J. (1979). The predator–prey interaction of planktivorous fish and zooplankton. *Am. Sci.* **67**:573–581.

Okada, Y., Aoki, M., Sato, Y., and Masai, H. (1969). The brain patterns of sharks in relation to habit. *J. Hirnforsch.* **11**:347–365.

Olson, A. M. (1954). The biology, migration, and growth rate of the school shark, *Galeorhinus australis* Macleay (carcharhinidae) in south-eastern Australian waters. *Aust. J. Mar. Freshwater Res.* **5**:353–410.

Ondrias, L. C. (1971). A list of the fresh and sea water fishes of Greece. *Hell. Oceanol. Limnol.* **10**:23–96.

Oosthuizen, W. H., and David, J. H. M. (1988). Non-breeding colonies of the South African fur seal *Arctocephalus pusillus pusillus* in southern Africa. *Invest. Rep. Sea Fish. Res. Inst. S. Afr.* **132**.

Oritsland, T. (1977). Food consumption of seals in the Antarctic pack ice. *In* "Adaptations Within Antarctic Ecosystems" (G. A. Llano, ed.), pp. 749–768. Smithsonian Institution, Washington, D.C.

Orr, R. T. (1959). Sharks as enemies of sea otters. *J. Mammal.* **40**:617.

Otake, T., and Mizue, K. (1981). Direct evidence for oophagy in thresher shark, *Alopias pelagicus*. *Jpn. J. Ichthyol.* **28**:171–172.

Overstrom, N. A. (1991). Estimated tooth replacement rate in captive sand tiger sharks (*Carcharias taurus* Rafinesque, 1910). *Copeia* **1991**:525–526.

Owen, R. W. (1981). Fronts and eddies in the sea: Mechanisms, interactions and biological effects. *In* "Analysis of Marine Ecosystems" (A. R. Longhurst, ed.), pp. 197–233. Academic Press, New York.

Parer, D., and Parer-Cook, E. (1991). Suddenly from the deep. *Int. Wildl.* **21**:34–37.

Parker, T. J. (1887). Notes on *Carcharodon rondeletii*. *Proc. Zool. Soc. London* **1**:27–40.

Parker, H. W., and Boseman, M. (1954). The basking shark, *Cetorhinus maximus*, in winter. *Proc. Zool. Soc. London* **124**:185–194.

Parr, A. E. (1956). On the original variates of taxonomy and their regressions upon size in fishes. *Am. Mus. Nat. Hist., Bull.* **110**:369–398.

Parrish, R. H., Nelson, C. S., and Bakun, A. (1981). Transport mechanisms and reproductive success of fishes in the California current. *Biol. Oceanogr.* **1**:175–203.

Parrish, R. H., Bakun, A., Husby, D. M., and Nelson, C. S. (1982). Comparative climatology of selected environmental processes in relation to eastern boundary current fish reproduction. *FAO Fish. Rep.* **291**:731–777.

Parsons, G. R. (1982). The reproductive biology of the Atlantic sharpnose shark, *Rhizoprionodon terraenovae*. *Fish. Bull.* **80**:61–73.

Parsons, G. R., and Grier, H. J. (1992). Seasonal changes in shark testicular structure and spermatogenesis. *J. Exp. Zool.* **261**:173–184.

Paterson, R. (1986). Shark prevention measures working well. *Aust. Fish.* **March**:12–18.

Paust, B., and Smith, R. (1986). Salmon shark manual. The development of a commercial salmon shark, *Lamna ditropis*, fishery in the North Pacific. *Alaska Sea Grant Rep.* **86-01**.

Pavlov, I. P. (1927). "Conditioned Reflexes." Oxford University Press, London.

Payne, P. M., Selzer, L. A., and Knowlton, A. R. (1984). "Distribution and Density of Cetaceans, Marine Turtles, and Seabirds in the Shelf Waters of the Northeastern United States, June 1980–December 1983, Based on Shipboard Observations," contract NA-81-FA-C0023, final report. U.S. National Marine Fisheries Service, Narragansett, Rhode Island.

Penny, R. L., and Lowry, G. (1967). Leopard seal predation on Adélie penguins. *Ecology* **48**:878–882.

Pepperell, J. G. (1992). Trends in the distribution, species composition and size of sharks caught by gamefish anglers off south-eastern Australia. *Aust. J. Mar. Freshwater Res.* **43**:213–225.

Perez-Arcas, L. (1878). Nota sobre los peces escuálidos. *Carcharodon carcharias* (*Carcharias lamia*). *Ann. Soc. Esp. Hist. Nat., Actas* **7**(2):9–19.

Peterson, R. S., and Bartholomew, G. A. (1967). "The Natural History and Behavior of the California Sea Lion," Spec. Publ. 1. American Society of Mammalogy.

Philander, S. G. (1989). "El Niño, La Niña, and the Southern Oscillation." Academic Press, San Diego.

Piccinno, A., and Piccinno, F. (1979). Cattura di un enorme *Carcharodon* al largo di Gallipoli (Puglia). *Thalassa Salentina* **December 30**:89–90.

Pinedo, M. C., Rosas, F. C. W., and Marmontel, M. (1992). "Cetaceos e Pinipedes do Brasil. Uma Revisao dos Registros e Guia para Identificacao." U.N. Environmental Program, Fund. University Amazonas.

Platt, C. J., Bullock, T. H., Czéh, G., Kovaevi, N., Konjevi, D. J., and Gojkovi, M. (1974). Comparison of electroreceptor, mechanoreceptor, and optic evoked potentials in the brain of some rays and sharks. *J. Comp. Physiol.* **95**:323–355.

Poole, M. M. (1981). Migration corridors of gray whales along the central California coast, 1980–1982. *In* "The

Gray Whale" (M. L. Jones, S. L. Swartz, and S. Leatherwood, eds.), pp. 389–407. Academic Press, New York.

Popenoe, P. (1985). Cenozoic depositional and structural history of the North Carolina margin from seismic–stratigraphic analyses. *In* "Geologic Evolution of the United States Atlantic Margin" (C. W. Poag, ed.), pp. 125–187. Van Nostrand–Reinhold, New York.

Postel, E. (1955). Sur quelques captures et échouages d'animaux rares en Tunisie. *Bull. Sta. Oceanogr. Salammbô* **52**:47–48.

Postel, E. (1958). Sur la présence de *Carcharodon carcharias* L., 1758 dans les eaux tunisiennes. *Mus. Paris, Bull., Ser. 2* **30**:342–344.

Potts, G. W. (1980). The predatory behavior of *Caranx melampygus* (Pisces) in the channel environment of Aldabra Atoll (Indian Ocean). *J. Zool.* **192**:323–350.

Potts, G. W., and Swaby, S. E. (1992). "Evaluation of the Conservation Requirements of Rarer British Marine Fishes and Appendices," contract HF3. 03. 417, final report. Nat. Conserv. Council, Peterborough, England.

Pratt, H. L., Jr. (1979). Reproduction in the blue shark, *Prionace glauca. Fish. Bull.* **76**:445–470.

Pratt, H. L., Jr. (1988). Elasmobranch gonad structure: A description and survey. *Copeia* **1988**:719–729.

Pratt, H. L. (1993). The storage of spematozoa in the oviductal glands of western North Atlantic sharks. *Environ. Biol. Fishes* **38**:139–149.

Pratt, H. L., and Casey, J. G. (1990). Shark reproductive strategies as a limiting factor in directed fisheries, with a review of Holden's method of estimating growth parameters. *NOAA Tech. Rep., NMFS* **NMFS 90**:97–109.

Pratt, H. L., and Castro, J. I. (1991). Shark reproduction: Parental investment and limited fisheries, an overview. *In* "Discovering Sharks" (S. H. Gruber, ed.), Spec. Publ. 14, pp. 56–60. American Littoral Society, Highlands, New Jersey.

Pratt, H. L., Jr., and Tanaka, S. (1994). Sperm storage in male elasmobranchs: A description and survey. *J. Morphol.* **219**:297–304.

Pratt, H. L., Casey, J. G., and Conklin, R. E. (1982). Observations on large white sharks, *Carcharodon carcharias*, off Long Island, New York. *Fish. Bull.* **80**:153–156.

Press, W. H., Flannery, B. P., Teukolsky, S. A., and Vetterling, W. T. (1986). "Numerical Recipes. The Art of Scientific Computing." Cambridge University Press, Cambridge.

Preuschoft, H., Reif, W. E., and Muller, W. H. (1974). Functional adaptations of shape and structure in shark teeth. *Z. Anat. Entwicklungsgesch.* **143**:315–344. (In German.)

Punt, A. E., and Butterworth, D. S. (1993). Variance estimates for fisheries assessment: Their importance and how best to evaluate them. *Can. Spec. Publ. Fish. Aquat. Sci.* **120**:145–162.

Purdy, R., McLellan, J. H., Schneider, V. P., Applegate, S. P., Meyer, R. L., and Slaughter, B. H. (1996). Preliminary study of the Neogene fish faunas from the Texasgult, Inc., Lee Creek Mine, North Carolina. *Smithson. Contrib. Paleobiol.* In press.

Pyle, P. (1992). Sympathy for a predator. *Point Reyes Bird Observ., Observer* **93**.

Pyle, P., Nur, N., Henderson, R. P., and DeSante, D. F. (1993). The effects of weather and lunar cycle on nocturnal migrant landbirds at Southeast Farallon Island, California. *Condor* **95**:343–361.

Quéro, J. C. (1984). Lamnidae. *In* "Fishes of the Northeastern Atlantic and the Mediterranean" (P. J. P. Whitehead, M. L. Bauchot, J. C. Hureau, J. Nielsen, and E. Tortonese, eds.), Vol. 1, pp. 33–38. U.N. Environmental Science Organization, Paris.

Quéro, J. C., Verron, R., and Cattin, Y. (1978). Observations icthyologiques effectuées en Charente-Maritime en 1977. *Ann. Soc. Sci. Nat. Charente-Marit.* **6**(5):428–439.

Quignard, J. P., and Capapé, C. (1972). Complément a la liste commentée des Sélaciens de Tunisie. *Bull. Inst. Oceanogr. Peche, Salammbo* **2**(3):445–447.

Quinn, W. H., and Neal, V. T. (1987). El Niño occurrences over the past four and a half centuries. *J. Geophys. Res.* **92**:14449–14461.

Rabalais, S. C., and Rabalais, N. N. (1980). The occurrence of sea turtles on the south Texas coast. *Contrib. Mar. Sci.* **23**:123–129.

Radovich, J. (1959). Redistribution of fishes in the eastern north Pacific Ocean in 1957 and 1958. *Calif. Coop. Ocean. Fish. Invest. Rep.* **7**:163–171.

Radovich, J. (1961). Relationships of some marine organisms of the northeast Pacific to water temperatures particularly during 1957 through 1959. *Calif. Dep. Fish Game, Fish Bull.* **112**.

Randall, J. E. (1973). Size of the great white shark (*Carcharodon*). *Science* **181**:169–170.

Randall, J. E. (1987). Refutation of lengths of 11.3, 9.0, and 6.4 meters attributed to the white shark, *Carcharodon carcharias. Calif. Fish Game* **73**:163–168.

Randall, J. E. (1992). Review of the biology of the tiger shark (*Galeocerdo cuvieri*). *Aust. J. Mar. Freshwater Res.* **43**:21–31.

Randall, B. M., Randall, R. M., and Compagno, L. J. V. (1988). Injuries to jackass penguin (*Spheniscus demersus*): Evidence for shark involvement. *J. Zool.* **213**:589–599.

Ray, C. E. (1976). Geography of phocid evolution. *Syst. Zool.* **25**:391–406.

Reid, D. D., and Krogh, M. (1992). Assessment of catches from protective shark meshing off New South Wales beaches between 1950 and 1990. *Aust. J. Mar. Freshwater Res.* **43**:283–296.

Reiter, J., Stinson, N. L., and Le Boeuf, B. J. (1978). Northern elephant seal development: The transition from weaning to nutritional independence. *Behav. Ecol. Sociobiol.* **3**:337–367.

Rey, J. C., Caminas, J. A., Alot, E., and Ramos, A. (1986). Captures de Requins associées a la pêcherie espagnole de palangre en Méditeranée occidentale, 1984, 1985. 1. Aspects halieutiques. *Rapp. P.-V. Reun., Cons. Int. Explor. Mer* **30**,5(8):240.

Ribeiro, A. M. (1923). Fauna Brasiliense. Peixes 1. *Arch. Mus. Nacl., Rio de Janeiro* **2**(1):1–50.

Richardson, W. J. (1978). Timing and amount of bird migration in relation to weather: A review. *Oikos* **30**:224–272.

Richter, M. (1986). Elasmobranquios e osteictes Quaternarios da Bacia de Pelotas, costa sul do Rio Grande do Sul. *Reun. Grupo Trabalho Pesca Pesq. Tubaroes Raiasno Bras., Univ. Fed. Maranhao*, 2nd (abstract).

Ricker, W. E. (1973). Linear regression in fishery research. *J. Fish. Res. Board Can.* **30:**409–434.

Ricker, W. E. (1975). Computation and interpretation of biological statistics of fish populations. *Bull., Fish. Res. Board Can.* **191:**1–382.

Ricker, W. E. (1979). Growth rates and models. *In* "Fish Physiology" (W. S. Hoar, D. J. Randall, and J. R. Brett, eds.), pp. 677–743. Academic Press, New York.

Ridet, J.-M., Bauchot, R., Delfini, C., Platel, R., and Thireau, M. (1973). L'encéphale de *Scyliorhinus canicula* (Linné) (Chondrichthyes, Selacii, Scyliorhinidae). Recherche d'une grandeur de référence pour des études quantitatives. *Cah. Biol. Mar. Roscoff* **14:**11–28.

Ridgway, S. H., and Dailey, M. D. (1972). Cerebral and cerebellar involvement of trematode parasites in dolphins and their possible role in stranding. *J. Wildl. Dis.* **8:**33–43.

Riedl, R. (1963). "Fauna und Flora der Adria." Verlag Paul Parey, Hamburg, Germany.

Riedman, M. L., and Estes, J. A. (1990). The sea otter (*Enhydra lutris*): Behavior, ecology, and natural history. *U.S. Fish Wildl. Serv., Biol. Rep.* **90**(14).

Riggio, C. (1894). Cattura di *Carcharodon rondeletii* nelle acque di Capo Gallo di Isola delle Femine. *Il Nat. Sicil.* **13:**130–133.

Riggs, S. R. (1984). Paleoceanographic model of Neogene phosphorite deposition, U.S. Atlantic continental margin. *Science* **223:**123–131.

Risso, A. (1810). "Icthyologie de Nice." Schoell, Paris.

Roberts, B. L. (1978). Mechanoreceptors and the behaviour of elasmobranch fishes with special reference to the acoustico-lateralis system. *In* "Sensory Biology of Sharks, Skates and Rays" (E. S. Hodgson and R. F. Mathewson, eds.), pp. 117–193. U.S. Office of Naval Research, Arlington, Virginia.

Rocha, D. (1948). Subsidio para o estudo da fauna cearense (catalogo das especies por mim coligidas e notadas). *Rev. Inst. Ceara* **62:**102–135.

Roedel, P. M., and Ripley, W. E. (1950). California sharks and rays. *Calif. Dep. Fish Game, Fish Bull.* **75:**1–45.

Roest, A. I. (1970). *Kogia simus* and other cetaceans from San Luis Obispo County, California. *J. Mammal.* **51:**410–417.

Rosas, F. W. C. (1989). "Aspectos da Dinamica Populacional e Interacao com a Pesca, do Leao-Marinho do Sul, *Otaria flavescens* (Shaw, 1800) (Pinnipedia, Otariidae), no Litoral Sul do Rio Grande do Sul, Brasil," unpublished M.S. thesis. Fundamental University of Rio Grande, Rio Grande do Sul, Brazil.

Roule, L. (1912). Notice sur les sélaciens conservés dans les collections du Musée Océanographique. *Bull. Inst. Oceanogr.* **243:**1–36.

Roux, C., and Geistodoerfer, P. (1988). Shark teeth and tympanic bullae of cetaceans: Nuclei or manganese nodules collected in the Indian Ocean. *Cybium* **12**(2):129–137.

Royce, W. F. (1963). First record of white shark *Carcharodon carcharias* from southeastern Alaska. *Copeia* **1963:**179.

Ruschi, A. (1965). Lista dos tubaroes, raias e peixes de agua doce e salgada do Estado do Espirito Santo e uma observacao sobre a introducao do dourado no Rio Doce. *Bol. Mus. Biol. Prof. Mello Leitao* **25A:**1–22.

Sadowsky, V. (1970). On the dentition of the sandshark, *Odontaspis taurus*, from the vicinity of Cananeria, Brazil. *Inst. Oceanogr., Univ. Sao Paulo, Bol.* **18:**37–44.

Sanders, A. E. (1980). Excavation of Oligocene marine fossil beds near Charleston, South Carolina. *Natl. Geogr. Soc., Res. Rep.* **12:**601–621.

Sanderson, M. J., and Donoghue, M. J. (1989). Patterns of variation in levels of homoplasy. *Evolution* **43:**1781–1795.

Sanzo, L. (1912). Embrione di *Carcharodon rondeletii* M. Hle (?), con particolare disposizione del sacco vitellino. *Reg. Com. Talassograf. Ital. Venez., Mem.* **11:**1–10.

Sassi, A. (1846). "Catalogo de Pesci di Liguria." Ferrando, Genoa, Italy.

Sato, Y., Takatsuji, K., and Masai, H. (1983). Brain organization of sharks, with special reference to archaic species. *J. Hirnforsch.* **24:**289–295.

Savin, S. M., Douglas, R. G., and Stehli, F. G. (1975). Tertiary marine paleotemperatures. *Geol. Soc. Am., Bull.* **86:**1499–1510.

Scattergood, L. W., Trefethen, P. S., and Coffin, G. W. (1951). Notes on Gulf of Maine fishes in 1949. *Copeia* **1951:**297–298.

Schefler, W. C. (1979). "Statistics for the Biological Sciences." Addison-Wesley, Reading, Massachusetts.

Scholl, J. P. (1983). Skull fragments of the California sea lion (*Zalophus californianus*) in stomach of a white shark (*Carcharodon carcharias*). *J. Mammal.* **64:**332.

Schroeder, W. C. (1939). Additional Gulf of Maine records of the white shark *Carcharodon carcharias* Linnaeus from the Gulf of Maine in 1937. *Copeia* **1939:**48.

Schroeder, D. M., and Ebbesson, S. O. E. (1974). Nonolfactory telencephalic afferents in the nurse shark (*Ginglymostoma cirratum*). *Brain, Behav. Evol.* **9:**121–155.

Schumann, E. H., Cohen, A. L., and Jury, M. R. (1996). Wind-driven coastal SST variability along the south coast of South Africa and relationships to regional and global climate. *J. Mar. Res.* **53:**231–248.

Schultz, L. P. (1963). Attacks by sharks as related to the activities of man. *In* "Sharks and Survival" (P. W. Gilbert, ed.), pp. 425–452. Heath, Boston.

Schultz, L. P., and Malin, M. H. (1963). A list of shark attacks for the world. *In* "Sharks and Survival" (P. W. Gilbert, ed.), pp. 509–567. Heath, Boston.

Scud, B. E. (1962). Measurements of a white shark, *Carcharodon carcharias*, taken in Maine waters. *Copeia* **1962:**659–661.

Seagers, D. J., and Jozwiak, E. A. (1991). The California Marine Mammal Stranding Network, 1972–1987: Implementation, status, recent events, and goals. *NOAA Tech. Rep., NMFS* **NMFS 98.**

Seagers, D. J., Masters, D. P., and DeLong, R. L. (1985). A survey of historic rookery sites for California and northern sea lions in the southern California bight.

U.S., NMFS, Southwest Fish. Sci. Cent., Admin. Rep. **LJ-85-13.**

Seagers, D. J., Lecky, J. H., Slawson, J. J., and Stone, H. S. (1986). Evaluation of the California Marine Mammal Stranding Network as a management tool based on records for 1983 and 1984. *U.S. Dept. Comm., Natl. Mar. Fish. Serv., Southwest Region Admin. Rep.* **86-5.**

Seber, G. A. F. (1965). A note on the multiple recapture census. *Biometrika* **52:**249–259.

Seber, G. A. F. (1977). "Linear Regression Analysis." Wiley, New York.

Seber, G. A. F. (1982). "The Estimation of Animal Abundance and Related Parameters," 2nd ed. Griffin, London.

Seigel, J. A., and Compagno, L. J. V. (1986). New records of the ragged-tooth shark, *Odontaspis ferox*, from California waters. *Calif. Fish Game* **72:**172–176.

Seigel, J. A., Long, D. J., and Hernandez, J. J. (1996). The tiger shark, *Galeocerdo cuvieri*, f in coastal southern California waters. *Calif. Fish Game* **81(4):** in press.

Seret, B. (1987). Decouverte d'une faune a *Procarcharodon megalodon* (Asassiz, 1835) en Nouvelle-Caledonie (Pices, Condrichthyes, Lamnidae). *Cybium* **11(4):**389–394.

Shaugnessy, P. D. (1990). First survey of fur seals and sea lions in Western Australia and South Australia. *Aust. Ranger Bull.* **5(4):**1990.

Sherringham, K. (1990). "A Frontal Zone and Its Role in Inverse Estuarine Dynamics," unpublished M.S. thesis. Flinders University School of Earth Sciences, Adelaide, South Australia.

Siccardi, E., Gosztony, A. E., and Menni, R. C. (1981). La presencia de *Carcharodon carcharias* e *Isurus oxyrinchus* en el mar Argentino Chondrichthyes, Lamniformes. *Physis (Buenos Aires), Sec. A* **39(97):**55–62.

Silber, G. K. (1986). The relationship of social vocalizations to surface behavior and aggression in the Hawaiian humpback whale (*Megaptera novaeangliae*). *Can. J. Zool.* **64:**2075–2080.

Simpfendorfer, C. (1992). Biology of tiger sharks (*Galeocerdo cuvieri*) caught by the Queensland shark meshing program off Townsville, Australia. *Aust. J. Mar. Freshwater Res.* **43:**33–43.

Simpson, G. G., Roe, A., and Lewontin, R. C. (1960). "Quantitative Zoology," rev. ed. Harcourt Brace, New York.

Siniff, D. B., DeMaster, D. P., Hofman, R. J., and Eberhardt, L. L. (1977). An analysis of the dynamics of a Weddell seal population. *Ecol. Monogr.* **47:**319–335.

Sintsov, I. F. (1899). Notizen uber die Jura-, Kreide-, und Neogen-Ablagerungender gouvernments Saratow, Simbirsk, Samara und Orenburg. *Odessa Univ., Zapiski* **77:**1–106.

Sisneros, J. A. (1993). "Effect of Molecular Structure on the Shark Repellent Potency of Anionic Surfactants," unpublished M.S. thesis. California State University, Long Beach.

Siverson, M. (1992). Biology, dental morphology and taxonomy of lamniform sharks from the Campanian of the Kristianstad Basin, Sweden. *Palaeontology* **35:**519–554.

Siverson, M. (1995). Revision of the Danian cow sharks, sand tiger sharks, and goblin sharks (Hexanchidae, Odontaspididae, and Mitsukurinidae) from southern Sweden. *J. Vertebr. Paleontol.* **15:**1–12.

Slipp, J. W., and Wilke, F. (1953). The beaked whale *Berardius* on the Washington coast. *J. Mammal.* **34:**105–113.

Smith, J. L. B. (1951). A juvenile of the man-eater, *Carcharodon carcharias* Linn. *Ann. Mag. Nat. Hist., Ser. 12* **4:**729–736.

Smith, J. L. B. (1963). Shark attacks in the South African seas. *In* "Sharks and Survival" (P. W. Gilbert, ed.), pp. 363–368. Heath, Boston.

Smith, L. J., Jr. (1991). The effectiveness of sodium lauryl sulphate as a shark repellent in a laboratory test situation. *J. Fish. Biol.* **38:**105–113.

Smith, R. L., and Rhodes, D. (1983). Body temperature of the salmon shark, *Lamna ditropis*. *J. Mar. Biol. Assoc. U.K.* **63:**243–244.

Sokal, R. R., and Rohlf, F. J. (1981). "Biometry," 2nd ed. Freeman, San Francisco.

Soljan, T. (1975). "I Pesci dell' Adriatico." Mondadori, Milan.

Springer, S. (1939). The great white shark, *Carcharodon carcharias* Linnaeus, in Florida waters. *Copeia* **1939:**114–115.

Springer, S. (1948). Oviphagous embryos of the sand shark, *Carcharias taurus*. *Copeia* **1948:**153–157.

Springer, S. (1960). Natural history of the sandbar shark. *Fish. Bull.* **61:**1–38.

Springer, S. (1967). Social organization of shark populations. *In* "Shark, Skates and Rays" (P. W. Gilbert, R. F. Mathewson, and D. P. Rall, eds.), pp. 149–174. Johns Hopkins University Press, Baltimore.

Springer, S. (1979). *Carcharodon carcharias* (Linnaeus). *In* "Checklist of the Fishes of the North-eastern Atlantic and of the Mediterranean" (J. C. Hureau and T. Monod, eds.), Vol. 1. CLOFNAM, UNESCO Press, Paris.

Springer, V. G. (1964). A revision of the carcharhinid shark genera Scoliodon, Loxodon, and Rhizoprionodon. *U.S. Natl. Mus. Nat. Hist.* **115(3493):**559–632.

Springer, S., and Gilbert, P. W. (1976). The basking shark, *Cetorhinus maximus*, from Florida and California, with comments on its biology and systematics. *Copeia* **1976:**47–54.

Springer, V., and Gold, J. (1989). "Sharks in Question." Smithsonian Institution Press, Washington, D.C.

Sprugel, D. G. (1983). Correcting for bias in log-transformed allometric equations. *Ecology* **64:**209–210.

Squire, J. L. (1967). Observations of basking sharks and great white sharks in Monterey Bay, 1948–1950. *Copeia* **1967:**247–250.

Stanley, H. P. (1963). Urogenital morphology in the chimaeroid fish *Hydrolagus collei* (Lay and Bennett). *J. Morphol.* **112:**99–127.

Starbird, C. H., Baldridge, A., and Harvey, J. T. (1993). Seasonal occurrence of leatherback sea turtles (*Dermochelys coriacea*) in the Monterey Bay region, with notes on other sea turtles, 1986–1991. *Calif. Fish Game* **79:**54–62.

Stata Corporation (1992). "Stata Reference Manual: Release 3.1," 6th ed. Stata Corporation, College Station, Texas.

Steffens, F. E., and D'Aubrey, J. D. (1967). Regression analysis as an aid to shark taxonomy. *S. Afr. Assoc. Mar. Biol. Res., Oceanogr. Res. Inst., Invest. Rep.* **18**:1–16.

Stenzel, L. (1988). PRBO'S beached bird survey, part I. Life and death in our coastal waters. *Point Reyes Bird Observ., Newsl.* **82**:1–4.

Stevens, J. D. (1973). Stomach contents of the blue shark (*Prionace glauca* L.) off south-west England. *J. Mar. Biol. Assoc. U.K.* **53**:357–361.

Stevens, J. D. (1974). The occurrence and significance of tooth cuts on the blue shark (*Prionace glauca* L.) from British waters. *J. Mar. Biol. Assoc. U.K.* **54**:373–378.

Stevens, J. D. (1983). Observations on reproduction in the shortfin mako, *Isurus oxyrinchus. Copeia* **1983**:126–130.

Stevens, J. D. (1984). Biological observations on sharks caught by sport fishermen off New South Wales. *Aust. J. Mar. Freshwater Res.* **35**:573–590.

Stevens, J. D. (1987). "Sharks." Facts on File, New York.

Stevens, E. D., and McLeese, J. M. (1984). Why bluefin tuna have warm tummies: Temperature effect on trypsin and chymotrypsin. *Am. J. Physiol.* **246**:487–494.

Stevens, J. D., and McLoughlin, K. J. (1991). Distribution, size and sex composition, reproductive biology and diet of sharks from northern Australia. *Aust. J. Mar. Freshwater Res.* **42**:151–199.

Stevens, J. D., and Paxton, J. R. (1985). A new record of the goblin shark, *Mitsukurina owstoni* (family Mitsukurinidae), from eastern Australia. *Proc. Linn. Soc. N.S.W.* **108**:37–45.

Stewart, B. S., and DeLong, R. L. (1993). Seasonal dispersion and habitat use of foraging northern elephant seals. *Symp. Zool. Soc. London* **66**.

Stewart, B. S., and DeLong, R. L. (1994). Postbreeding foraging migrations of northern elephant seals. *In* "Elephant Seals: Population Ecology, Behavior and Physiology" (B. J. Le Boeuf and R. M. Laws, eds.), pp. 290–309. University of California Press, Berkeley.

Stewart, B. S., and Yochem, P. K. (1985). Radio-tagged harbor seal, *Phoca vitulina richardsi*, eaten by a white shark, *Carcharodon carcharias*, in the Southern California Bight. *Calif. Fish Game* **71**:113–115.

Stewart, B. S., Le Boeuf, B. J., Yochem, P. K., Huber, H. R., DeLong, R. I., Sydeman, W. J., and Allen, S. G. (1994). History and present status of the northern elephant seal population. *In* "Elephant Seals: Population Ecology, Behavior and Physiology" (B. J. Le Boeuf and R. M. Laws, eds.), pp. 29–48. University of California Press, Berkeley.

Stillwell, C. E., and Kohler, N. E. (1982). Food, feeding habits, and estimates of daily ration of the shortfin mako (*Isurus oxyrinchus*) in the northwest Atlantic. *Can. J. Fish. Aquat. Sci.* **39**:407–414.

Stillwell, C. E., and Kohler, N. E. (1993). Food habits of the sandbar shark *Carcharhinus plumbeus* off the U.S. northeast coast, with estimates of daily ration. *Fish. Bull.* **91**:138–150.

Stirling, I. (1972). Observations on the Australian sea lion, *Neophoca cinerea* (Peron). *Aust. J. Zool.* **20**:271–279.

Storms, M. (1901). Sur un *Carcharodon* du Terrain Bruxellien. *Soc. Belg. Geol. Paleontol. Hydrol., Ser. 2, T* **5**:201–213.

Stossich, L. (1880). *Carcharodon rondeletti*, in Prospetto della Fauna del Mare Adriatico. *Boll. Soc. Adriat. Sci. Nat., Trieste* **5**:68.

Strasburg, D. (1958). Distribution, abundance, and habits of pelagic sharks in the central Pacific Ocean. *Fish. Bull.* **138**:335–361.

Strong, W. R., Jr., Murphy, R. C., Bruce, B. D., and Nelson, D. R. (1992). Movements and associated observations of bait-attracted white sharks, *Carcharodon carcharias*: A preliminary report. *Aust. J. Mar. Freshwater Res.* **43**:13–20.

Stroud, R. K., and Roffe, T. J. (1979). Causes of death in marine mammals stranded along the Oregon coast. *J. Wildl. Dis.* **15**:91–97.

Sullivan, R. M., and Houck, W. J. (1979). Sightings and strandings of cetaceans from northern California. *J. Mammal.* **60**:828–833.

Sutherland, N. S. (1960). Visual discrimination of shape by *Octopus*: Open and closed forms. *J. Comp. Physiol. Psychol.* **53**:104–112.

Swofford, D. L. (1990). "PAUP: Phylogenetic Analysis Using Parsimony," version 3.1.1. Illinois Natural History Survey, Champaign.

Szczepaniak, I. D. (1990). "Abundance, Distribution, and Natural History of the Harbor Porpoise (*Phocoena phocoena*) in the Gulf of the Farallones, California," unpublished M.A. thesis. California State University, San Francisco.

Taniuchi, T. (1970). Variation in the teeth of the sand shark, *Odontaspis taurus* (Rafinesque), taken from the east China Sea. *Jpn. J. Ichthyol.* **17**:37–44.

Taylor, L. (1985). White sharks in Hawaii: Historical and contemporary records. *South. Calif. Acad. Sci., Mem.* **9**:41–48.

Taylor, L. R., Compagno, L. J. V., and Struhsaker, P. J. (1983). Megamouth—A new species, genus and family of lamnoid shark (*Megachasma pelagios*, family Megachasmidae) from the Hawaiian Islands. *Proc. Calif. Acad. Sci.* **43**:87–110.

Templeman, W. (1963). Distribution of sharks in the Canadian Atlantic (with special reference to Newfoundland waters). *Bull., Fish. Res. Board Can.* **140**:1–77.

Templeton, A. R. (1983). Phylogenetic inference from restriction endonuclease cleavage site maps with particular reference to the evolution of humans and the apes. *Evolution* **37**:221–244.

Testi, A. D. (1993). A review of shark attacks in Italy: 1952–1993. *Chondros* **4**:1–3.

Thompson, A. W. (1947). "A Glossary of Greek Fishes." London.

Thorson, P. H., and Le Boeuf, B. J. (1994). Developmental aspects of diving in northern elephant seal pups. *In* "Elephant Seals: Population Ecology, Behavior and Physiology" (B. J. Le Boeuf and R. M. Laws, eds.), pp. 271–289. University of California Press, Berkeley.

Tinbergen, N. (1952). Derived activities: Their causation, biological significance, origin and emancipation during evolution. *Q. Rev. Biol.* **27**:1–32.

Tinbergen, N. (1953). "The Herring Gull's World." Collins, London.

Tinbergen, N., and van Iersel, J. J. A. (1947). "Displacement reactions" in the three-spined stickleback. *Behaviour* 1:56–63.

Tomas, A. R. G., and Gomes, U. L. (1989). Observacoes sobre a presenca e *Cetorhinus maximus* (Gunnerus, 1765) (Elasmobranchii, Cetorhinidae) no Sudeste e Sul do Brasil. *Bol. Inst. Pesca* 16(1):111–116.

Tortonese, E. (1938). Revisione degli squali del Museo Civico di Milano. P 292, *Carcharodon carcharias* (L.). *Atti Soc. Ital. Sci. Nat. Milano* 77:283–318.

Tortonese, E. (1956). Leptocardia, Celostomata, Selachii. *In* "Fauna d'Italia," Vol. 2. Calderini, Bologna, Italy.

Tortonese, E. (1965). "Pesci e Cetacei del Mar Ligure." Bozzi, Genova.

Touret, F. (1992). Requins, alerte rouge en Méditerranée. *Newslook* **1992**(108):68–70.

Tricas, T. C. (1985). Feeding ethology of the white shark *Carcharodon carcharias*. *South. Calif. Acad. Sci., Mem.* **9**: 81–91.

Tricas, T. C., and McCosker, J. E. (1984). Predatory behavior of the white shark (*Carcharodon carcharias*), with notes on its biology. *Proc. Calif. Acad. Sci.* 43:221–238.

Trillmich, F., and Mohren, W. (1981). Effects of the lunar cycle on the Galapagos fur seal, *Arctocephalus galapagoensis*. *Oecologia* **48**:85–92.

Trivers, R. L. (1972). Parental investment and sexual selection. *In* "Sexual Selection and the Descent of Man, 1871–1971" (B. Campbell, ed.), pp. 136–179. Aldine, Chicago.

Trivers, R. L. (1985). "Social Evolution." Benjamin/Cummings, Menlo Park, California.

Tschernezky, W. (1959). Distribution of sodium in compact bone as revealed by autoradiography of neutron-activated sections. *Nature (London)* 184:1331–1332.

Uchida, S. (1983). On the morphology of the whale shark *Rhincodon typus* Smith. *Aquabiol.* 5:93–101.

Uchida, S., Yasuzumi, F., Toda, M., and Okura, N. (1987). On the observations of reproduction in *Carcharodon carcharias* and *Isurus oxyrinchus*. *Jpn. Group Elasmobranch Stud., Rep.* 24:4–6.

Uchida, S., Toda, M., and Kamei, Y. (1990). Reproduction of elasmobranchs in captivity. *NOAA Tech. Rep., NMFS* **NMFS** 90:211–237.

U.S. Bureau of the Census (1992). "Statistical Abstract of the United States," 112th ed. U.S. Department of Commerce, Bureau of the Census, Washington, D.C.

Uyeno, T. (1974). A new Miocene lamnoid shark, *Carcharodon akitaensis*, from central Japan. *Bull. Natl. Sci. Mus. (Tokyo)* 17(3):257–261.

Uyeno, T., and Matsushima, Y. (1979). Comparative study of teeth from Naganuma formation of Middle Pleistocene and Recent specimens of the great white shark, *Carcharodon carcharias*, from Japan. *Kanagawa Prefect. Mus., Bull.* 11:11–29.

Uyeno, T., and Sakamoto, O. (1984). Lamnoid shark *Carcharodon from Miocene beds of Chichibu Basin, Saitama Prefecture, Japan. Saitama Mus. Nat. Hist., Bull.* **2.47**–65.

Uyeno, T., Sakamoto, O., and Sekine, H. (1989). The description of an almost complete tooth set of *Carcharodon megalodon* from a Middle Miocene bed in Saitama Prefecture, Japan. *Saitama Mus. Nat. Hist., Bull.* **7**:73–85.

van Beneden, P. J. (1882). Description des ossements fossiles des environs d'Anvers. *Ann. Mus. R. Hist. Nat. Belg.* 7(3):1–90.

Van Denise, A. B., and Adriani, M. J. (1953). On the absence of gill rakers in specimens of the basking shark, *Cetorhinus maximus* (Gunner). *Zool. Meded.* 31:307–309.

van der Elst, R. P. (1990). Marine fish tagging in South Africa. *Am. Fish. Soc. Symp.* **7**.

van der Elst, R. P., and Bullen, E. (1992). *Oceanogr. Res. Inst., Durban, Tagging News* 8:1–10.

Van Dykhuizen, G., and Mollet, H. F. (1992). Growth, age estimation, and feeding of captive sevengill sharks, *Notorynchus cepedianus*, at the Monterey Bay Aquarium. *Aust. J. Mar. Freshwater Res.* **43**:297–318.

Vazquez, J. L., and Zubillaga, J. J. K. (1989). Geological and paleoceanographic interpretation of the La Mission and Los Indios Members of the Rosarito Beach Formation (Middle Miocene), Baja California, Mexico. *Cienc. Mar.* 15(3):21–44.

Vliet, K. A. (1989). Social displays of the American alligator (*Alligator mississippiensis*). *Am. Zool.* 29:1019–1031.

Voellmy, E. (1955). "Fünfstellige Logarithmen und Zahlentafeln." Orell Füssli Verlag Zürich, Czechoslovakia.

Voisin, J. F. (1976). On the behaviour of the killer whale, *Orcinus orca* (L.). *Norw. J. Zool.* 24:69–71.

Vokes, E. H. (1970). Cenozoic Muricidae of the western Atlantic region. Part V—Pterynotus and Poirieria. *Tulane Stud. Geol. Paleontol.* 8(1):1–50.

Vokes, E. H. (1992). Cenozoic Muricidae of the western Atlantic region. Part IX—Pterynotus, Poirieria, Aspella, Dermomurex, Calotrophon, Acantholabia, and Attiliosa; additions and corrections. *Tulane Stud. Geol. Paleontol.* 25(1–3):1–108.

von Bertalanffy, L. (1960). Principles and theory of growth. *In* "Fundamental Aspects of Normal and Malignant Growth" (W. W. Nowinski, ed.), pp. 137–259. Elsevier, New York.

Walker, P., Newton, D. M., and Mantyla, A. W. (1992). Surface water temperatures, salinities and densities at shore stations, United States west coast, 1991. *Scripps Inst. Ocean., Ref.* **92-8**.

Wallett, T. S. (1973). "Analysis of Shark Meshing Returns off the Natal Coast," unpublished M.S. thesis. University of Natal, Durban, Natal, South Africa.

Wallett, T. S. (1983). "Shark Attack in Southern African Waters and Treatment of Victims." Struik, Cape Town, South Africa.

Wallis, W. A., and Roberts, H. V. (1956). "Statistics, a New Approach." Free Press, New York.

Wallner, E. P. (1990). "TIDES," version 1.66. Wallner, Wayland, Massachusetts.

Ward, D. J. (1988). *Hypotodus verticalis* (Agassiz 1843), *Hypotodus robustus* Leriche (1921) and *Hypotodus heinzelini* (Casier 1967), Chondrichthyes, Lamniformes, junior synonyms of *Carcharias hopei* (Agassiz 1843). *Tert. Res.* **10**: 1–12.

Ward, L. W. (1992). Tertiary molluscan assemblages from the Salisbury embayment of Virginia. *Va. J. Sci.* **43**(1B): 85–100.

Ward, L. W., Blackwelder, B. W., Gohn, G. S., and Poore, R. Z. (1979). Stratigraphic revision of Eocene, Oligocene and Lower Miocene formations of South Carolina. *Geol. Notes (S.C. Geol. Surv.)* **23**(1):2–32.

Watts, P. (1993). Possible lunar influence on hauling-out behavior by the pacific harbor seal (*Phoca vitulina* richardsi). *Mar. Mammal Sci.* **9**:68–76.

Webb, P. M. (1994). "Heart Rate and Oxygen Consumption in Northern Elephant Seals Diving in the Laboratory," unpublished M.A. thesis. University of California, Santa Cruz.

Webb, P. W. (1986). Locomotion and predator–prey relationships. *In* "Predator–Prey Relationships" (M. E. Feder and G. V. Lauder, eds.), pp. 24–41. University of Chicago Press, Chicago.

Welton, B. J., and Farish, R. F. (1993). "The Collector's Guide to Fossil Sharks and Rays From the Cretaceous of Texas." Before Time, Lewisville, Texas.

Welton, B. J., and Zinsmeister, W. J. (1980). Eocene neoselachians from the La Meseta Formation, Seymour Island, Antarctica Peninsula. *Contrib. Sci.* **329**.

Wendell, F. E., Hardy, R. A., and Ames, J. A. (1986). A review of California sea otter, *Enhydra lutris*, surveys. *Calif. Dep. Fish Game, Mar. Res. Tech. Rep.* **51**.

West, J. G. (1993). The Australian Shark Attack File with notes on preliminary analysis of data from Australian waters. *In* "Shark Conservation" (J. Pepperell, P. Woon, and J. West, eds.), pp. 93–101. Taronga Zoo, Sydney.

White, E. G. (1930). The whale shark, *Rhiniodon typus*. Description of the skeletal parts and classification based on the Marathon specimen captured in 1923. *Am. Mus. Nat. Hist., Bull.* **61**:129–160.

White, E. G. (1936). A classification and phylogeny of the elasmobranch fishes. *Am. Mus. Novit.* **837**:1–16.

White, E. G. (1937). Interrelationships of the elasmobranchs, with a key to the order Galeoidea. *Am. Mus. Nat. Hist., Bull.* **74**:25–138.

Whitehead, P. J. P., Bauchot, M.-L., Hureau, J.-C., Nielsen, J., and Tortonese (eds.). (1984). "Fishes of the north-eastern Atlantic and the Mediterranean." UNESCO, Paris.

Whitley, G. P. (1950). Studies in ichthyology, No. 14. *Rec. Aust. Mus.* **23**:234–245.

Whitley, G. P. (1963). Shark attacks in Australia. *In* "Sharks and Survival" (P. W. Gilbert, ed.), pp. 329–338. Heath, Boston.

Whitmore, F. C. (1994). Neogene climatic change and the emergence of the modern whale fauna of the North Atlantic Ocean. *San Diego Soc. Nat. Hist., Proc.* **29**: 223–227.

Whittaker, R. H., and Woodwell, G. M. (1968). Dimension and production relations of trees and shrubs in the Brookhaven forest, New York. *Ecology* **56**:1–25.

Wild, P. W., and Ames, J. A. (1974). A report on the sea otter, Enhydra lutris L., in California. *Calif. Dep. Fish Game, Mar. Res. Tech. Rep.* **20**.

Wilkinson, L. (1988a). "SYSTAT: The System for Statistics." SYSTAT, Evanston, Illinois.

Wilkinson, L. (1988b). "SYGRAPH: The System for Graphics." SYSTAT, Evanston, Illinois.

Williamson, G. R. (1963). Common porpoise in the stomach of a Greenland shark. *J. Fish. Res. Board Can.* **20**:1085–1086.

Wilson, E. O. (1975). "Sociobiology." Belknap, Cambridge.

Wilson, J. F., Jr. (1968). "Fluorometric Procedures for Dye Tracing: Techniques for Water-Resources Investigations of the United States Geological Survey," Book 3, Ch. A12. U.S. Geological Survey, Washington, D.C.

Wilson, R. B. (1985). "1985 Sardine Run," unpublished report. Natal Sharks Board, Natal, South Africa.

Witzell, W. N. (1987). Selective predation on large cheloniid sea turtles by tiger sharks (*Galeocerdo cuvieri*). *Jpn. J. Herpetol.* **12**:22–29.

Wolf, N. G., Swift, P. R., and Carey, F. G. (1988). Swimming muscle helps warm the brain of lamnid sharks. *J. Comp. Physiol. B* **157**:709–715.

Wood, F. G., Caldwell, D. K., and Caldwell, M. C. (1970). Behavioral interactions between porpoises and sharks. *In* "Investigations on Cetacea" (G. Pilleri, ed.), Vol. II, pp. 265–277. Brain Anatomy Institute, University of Berne, Berne, Switzerland.

Woodhouse, C. D. (1991). Marine mammal beachings as indicators of population events. *NOAA Tech. Rep., NMFS* **NMFS 98**:111–115.

Woodward, A. S. (1889). "Catalogue of the Fossil Fishes in the British Museum (Natural History)," Part 1. British Museum (Natural History), London.

Woodward, A. S. (1911). The fossil fishes of the English chalk. *Palaeontogr. Soc.* **64**(part 6):185–224.

Ximenes, I. (1962). Notas sobre elasmobranquios. I. Quadro sistematico y sinonimico provisional de los selaceos de la costa uruguaya. *Rev. Inst. Invest. Pesq.* **1**(1):35–44.

Yabumoto, Y. (1987). A new Eocene lamnoid shark, *Carcharodon nodai*, from Omuta in northern Kyushu, Japan. *Kitakyishu Mus. Nat. Hist., Bull.* **9**:111–116.

Yabumoto, Y. (1989). A new Eocene lamnoid shark, *Carcharodon nodai*, from Omuta in northern Kyushu, Japan. *Kitakyushu Mus. Nat. Hist., Bull.* **9**:111–116.

Yarnall, J. L. (1969). Aspects of the predatory behavior of *Octopus cyanea* Gray. *Anim. Behav.* **17**:747–754.

Zar, J. H. (1974). "Biostatistical Analysis." Prentice Hall, Englewood Cliffs, New Jersey.

Zlotkin, E., and Barenholz, Y. (1983). On the membranal action of pardaxin. *AAAS Sel. Symp.* **83**:157–171.

Index

Printed and bound by CPI Group (UK) Ltd, Croydon, CR0 4YY

03/10/2024

01040324-0020